EVOLUTION OF TETRAPODS

PERIODS

Quarternary
Tertiary
Cretaceous
Jurassic
Triassic
Permian
Carboniferous
Devonian
Silurian
Ordovician
Cambrian

LISSAMPHIBIA
Eutheria
Metatheria
Trituberculata
ALLOTHERIA
PROTOTHERIA
THERIA
MAMMALIA
AMPHIBIA
TEMNOSPONDYLI
ANTHRACOSAURIA
ICHTHYOSTEGALIA
SYNAPSIDA
REPTILIA
ANAPSIDA
Ichthyosauria
Sauropterygia
DIAPSIDA
TESTUDINATA
Lepidosauria
Archosauria
ARCHAEORNITHES
ENANTIORNITHES
NEORNITHES
AVES
?
?
?
?

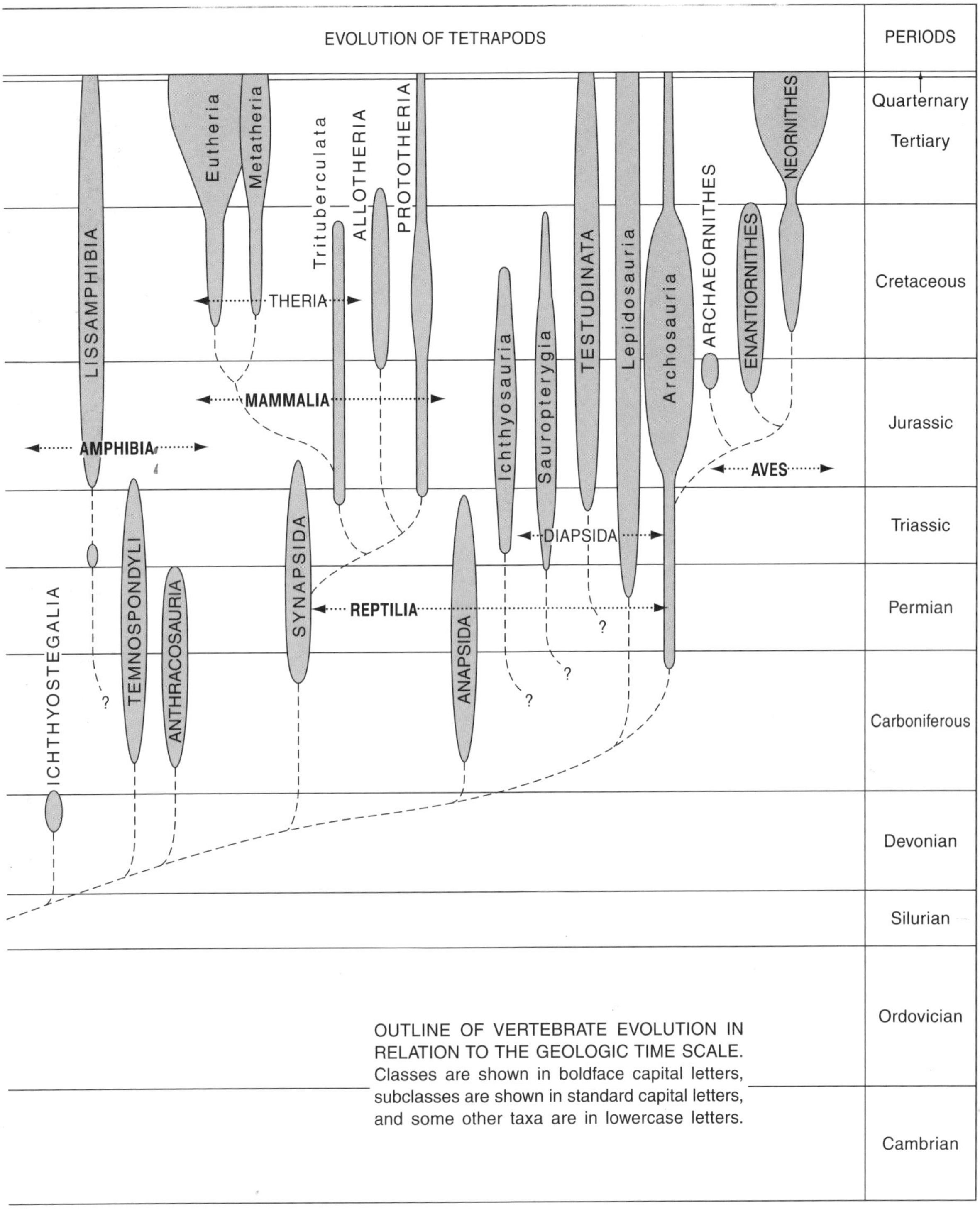

OUTLINE OF VERTEBRATE EVOLUTION IN RELATION TO THE GEOLOGIC TIME SCALE. Classes are shown in boldface capital letters, subclasses are shown in standard capital letters, and some other taxa are in lowercase letters.

ANALYSIS OF VERTEBRATE STRUCTURE

ANALYSIS OF VERTEBRATE STRUCTURE

FIFTH EDITION

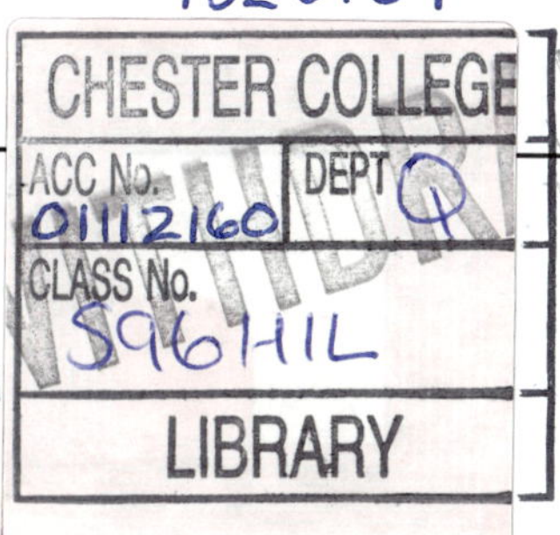

MILTON HILDEBRAND
Professor of Zoology, Emeritus
University of California, Davis

G.E. GOSLOW, JR.
Professor of Ecology and Evolutionary Biology
Brown University, Providence

Principal Illustrator
Viola Hildebrand

JOHN WILEY & SONS, INC.
New York • Chichester • Brisbane • Toronto • Singapore

Cover: Skull of a male Chinook Salmon.

ACQUISITIONS EDITOR Keri Witman

MARKETING MANAGER Clay Stone

SENIOR PRODUCTION EDITOR Patricia McFadden

SENIOR DESIGNER Kevin Murphy

PRODUCTION MANAGEMENT SERVICES Publication Services

This book was set in Times Roman by Publication Services and printed and bound by Hamilton Printing. The cover was printed by Lehigh Press, Inc.

This book is printed on acid-free paper. ∞

Library of Congress Cataloging-in-Publication Data

Hildebrand, Milton, 1918–

Analysis of vertebrate structure / Milton Hildebrand, G.E. Goslow, Jr.; principle illustrator, Viola Hildebrand.—5th

p. cm.

Includes biographical references (p.).

ISBN 0-471-29505-1 (cloth : alk. paper)

1. Vertebrates--Anatomy. 2. Vertebrates--Morphology. I. Goslow, G. E. II. Title.

QL805 .H64 2001

571.3′16--dc21

00-068604

Printed in the United States of America

10 9 8 7 6 5 4 3 2

*Dedicated to the students who use this book.
May you learn and prosper, and also be encouraged to protect
the remarkable animals with which we share the earth.*

Preface

This book is about the evolutionary and functional morphology of vertebrate animals. A first characteristic of the book is breadth and placement of emphasis. Description of structure is included because knowledge must precede interpretation. Interpretation of structure as an expression of phylogeny is stressed because organic evolution is one of the greatest stories biology has to tell, and the lineages of vertebrate animals illustrate the story with more continuity and persuasion than do the known lineages of other animals or plants. Interpretation of structure through the analysis of function is given more emphasis than usual. This is the subject of much current research, focusing on both living and extinct forms, and thus provides the opportunity to add recent advances to classical knowledge. It lends itself to analytical treatment, to which, in our experience, students respond with interest. Moderate attention is also given to the evident and engaging relationship between development and adult structure because revealing work is being done on evolutionary morphogenesis and relative growth. Interpretations of structure based on body size, age, gender, and individual variation are noted more briefly.

Second, this book presents vertebrate morphology as an active discipline. The transformations from fin to limb, from jawbone to ear ossicle, from branchial artery to carotid circulation, and many more, are classic stories that should be retold to new generations of students. However, it is recent advances and active areas of research that rightly show the discipline to be dynamic, integrative, and challenging. To this end the main text is again updated and verified, and we use insets (which we call *Comments*) to fairly evaluate points of controversy, take note of as yet inadequately understood phenomena, look deeper into basic subjects, and give historical perspective to ongoing analyses of complex systems.

Third, the book is profusely illustrated with artwork of high quality and uniform style. Of the approximately 1000 separate objects illustrated, a full two-thirds are completely original, and virtually all of the remainder are redrawn, usually with considerable modification, from other sources. For every subject we sought an appropriate compromise between illustration that is so descriptive as to introduce extraneous detail, and so simplistic as to reduce the living body to mechanical analogs. For this edition 20 chapter-opening photographs are introduced, and there are 40 new or revised figures.

The fourth characteristic of the book is its easy-to-use style. It is uniform, integrated, and thoroughly cross-referenced. Illustrations are cited in the text by figure number if in the same chapter and by page number if elsewhere. Figures are fully labeled, and legends distinguish (by capital and lowercase letters) between the point illustrated and subordinate material. Updated, annotated, and expanded reference lists conclude the respective chapters. They can give students a start on assignments and seminars, may indulge the curiosity of anyone with unanswered questions, and indicate some of our sources. The meanings of more than 150 word roots are given parenthetically where first used in the text; these are intended as aids to understanding, and hence memory, not as etymology lessons. There is a glossary of about 600 terms. We intend that the main text be sound and solid, yet not overly technical. Details that do not serve interpretations are omitted. The qualifying words "usually" and "sometimes" are used frequently for the sake of accuracy, but specific exceptions to usual structure are "usually" omitted. An effort has been made to sort concepts from illustrative material, and free use has been made of parenthetical statements to subordinate the examples and qualifications.

Part I of the book is a survey of the vertebrates. Students must be able to recognize and relate the major taxa in order to follow Parts II and III. Brief descriptions that stress typical features and recognition characteristics serve this purpose. Extinct groups are described or not, according to their relation to what will follow.

In Part II, the customary organ system approach is used to present the general structure of the classes and subclasses of vertebrates and to review the structural evidence for their evolutionary relationships. Features that do not characterize major taxa or show progressive change between successive categories are deemphasized or omitted. The treatment includes, we believe, a reasonable amount of content to be be learned in one course of study; those few students who will become professional morphologists will be motivated, we hope, to find supplementary information.

Part III presents knowledge and analysis of the major functional groups of vertebrates. Following chapters on bone-muscle mechanics, chapters consider the major locomotor and feeding adaptations and touch on energetics and scaling. Unrelated, often convergent groups of animals are taken together to see how evolution has provided for their common requirements.

Some texts work functional morphology into material otherwise organized primarily by organ system or by the phylogenetic succession of vertebrate taxa. Thus, the mechanics of swimming may be introduced with fishes or with the axial skeleton, terrestrial locomotion with early tetrapods or the appendicular skeleton, and feeding with the digestive system. We do the same to a limited degree (osmoregulation with the excretory system, thermoregulation with the circulatory system, night vision with the eye). Nevertheless, considering the uniquely thorough treatment of functional morphology in this text, most instructors surveyed agree with us that the topics here presented in Parts II and III are best kept distinct: Not all fishes use the axial skeleton in swimming, all use more than the skeleton, and most non-fishes that swim propel themselves in ways that are unusual among fishes; feeding does not directly employ the digestive system; flight crosses taxonomic and organ system boundaries. When the analysis of function is organized only by organ system or taxon, there must be omissions, compromise, and fragmentation in its presentation.

Successive editions of this book have seen modest increases in the numbers of illustrations and references, but no significant change in the length or level of the main text. We believe that textbooks can be too large and detailed. Nevertheless, this book, unlike the instructor, cannot omit entire organ systems, functions, or major taxa. In a two-term course, most of the chapters can be assigned. In a one-term course, selection is necessary.

A balance can be obtained by combining portions of Parts II and III, thus illustrating more than one of the major approaches to the analysis of structure. Unassigned chapters might become the bases for special reports.

A preparatory course in general biology or zoology is assumed; most of the requisite fundamentals and terms are reviewed here, but would come as a big dose if none were already familiar. A prior foundation in embryology, physiology, or evolution is desirable to give the student the benefit of additional familiar ground, but is not assumed. Similarly, recollection of algebra and geometry, and a course in physics would make Part III easier, but more for the security provided than for formulas remembered.

An appendix giving instructions for making anatomical preparations has been added to this edition. The rationale for its inclusion is the subject of its opening paragraphs.

In the preparation of this edition of the book, as for those that have gone before, we turned to numerous colleagues and to some students for the answers to questions and for evaluations. We are indebted to them all for willing cooperation. Viola Hildebrand prepared nearly all of the halftone illustrations and some of the line drawings. Half of the chapter-opening photographs (showing specimens formerly in the Hildebrand collection) were taken by O. Louis Mazzatenta at the Museum of Vertebrate Zoology, Berkeley, where the curator, Barbara Stein, was most helpful. Ronald E. Cole kindly provided material for the Appendix.

Milton Hildebrand

George E. Goslow, Jr.

Contents

Introduction

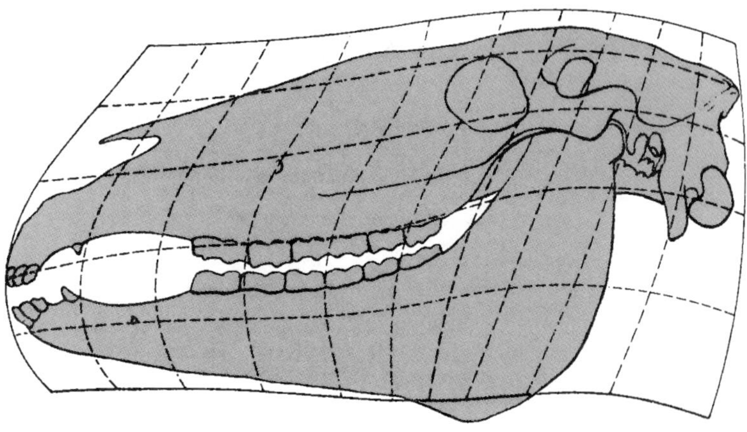

Nothing else in nature has more exquisite structure than the vertebrate body. Mankind has been fascinated by vertebrate form and function since time immemorial. Animals are incorporated the world over into art, literature, religion, and entertainment. We are amazed at the prowess of animal athletes and impressed by the diversity of their skills. A runner can dash as fast as many cars are driven on freeways. A springer can leap 14 times higher than its body length. A digger can thrust with 32 times its body weight. A climber can walk upside down on polished glass. A flyer can dart backward as well as forward. A feeder can swallow objects three times larger than its head, and another can strain from water food particles having 1/5000 the diameter of a pencil lead. There are living direction finders, strain gauges, force multipliers, flow equalizers, suction cups, bifocal lenses, locking devices, fiber-optic cables, pressure sensors, self-lubricating bearings, electric generators, gas analyzers, echolocators, depth gauges, recoil mechanisms, magnetic compasses, countercurrent exchangers, and compact computers. The continued observation and interpretation of form and function promise to increase man's pleasure and wonder over the complexity, diversity, and near perfection of the vertebrate body.

Chapter 1

The Nature of Vertebrate Morphology

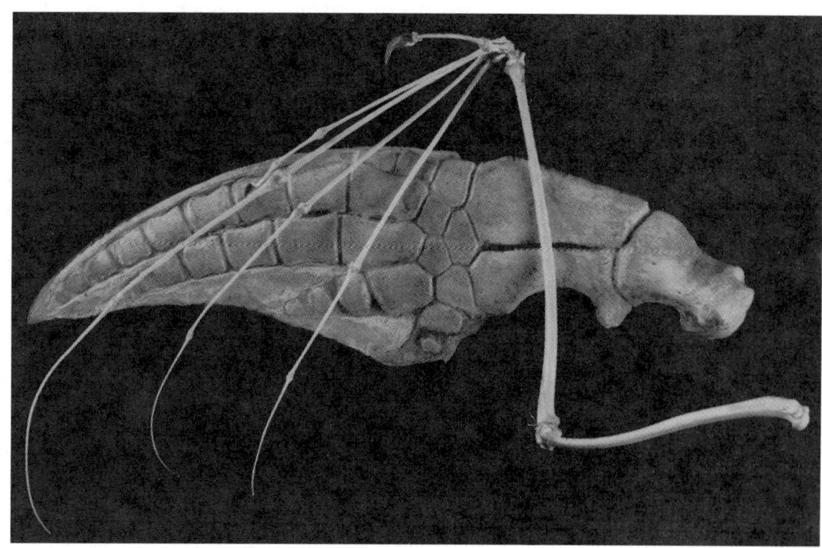

DEFINITION, SCOPE, AND RELATION TO OTHER DISCIPLINES

The science of morphology is conceptually broader than the study of structure. Morphologists are concerned not only with anatomical facts, but also with *explaining* structure and structural patterns. This requires an integration of observation and experimentation from many sources.

Modern vertebrates are modified descendants of ancestral species. As a result, the form of a particular structure or set of structural relationships may be understandable only in terms of historical change (phylogeny). Thus morphology draws much from paleontology and evolutionary biology, as well as other fields that may provide evidence of evolutionary affinity, including molecular genetics. The phylogeny of form is the principal subject of Part II of this book.

Much structure may be interpreted in terms of behavior and adjustment to the external environment. Accordingly, understanding *functional* morphology is dependent on knowledge of biomechanics, physiology, ecology, and ethology. Function is the primary subject of Part III, although related topics are included throughout the book.

Some structures and patterns of organization are best understood in terms of development and developmental mechanisms. Embryology is introduced in Chapter 5 and expanded in other chapters of Part II. Other aspects of vertebrate biology may also contribute to the understanding of structure. For example, form may be size-, age-, or gender-dependent, and individual variation may have nutritional, pathologic, or other environmental origins. Of these topics, the relation between form and body size is presented in Chapter 23, and the remainder are noted later in this chapter.

Thus vertebrate morphology relates to many other sciences. George Cuvier, the colorful but somewhat controversial "father of comparative anatomy," in his enthusiasm to bring rigor and credibility to the comparative studies of animals, encouraged students in 1800 to embrace all the established sciences of the time. Cuvier believed that from a single,

Above: Arm skeletons of a fruit bat and dolphin.

isolated tooth one can deduce all of the animal's other parts, "just as the equation of a curve determines all of its properties" (Appel, 1983)—a noble, but perhaps overly ambitious, goal. Nevertheless, our message remains the same: It is desirable for the aspiring morphologist to supplement training in the principles of biology at all levels and grounding in vertebrate structure by gaining familiarity with the concepts and methodology of several related disciplines.

In summary, this book describes the anatomy of the major structural and functional groups of vertebrate animals, and interprets their morphological differences primarily in terms of ancestry and function, employing related fields as needed to further this objective.

WHY STUDY VERTEBRATE MORPHOLOGY?

To derive full value from any course of study, the student must be convinced that the returns justify the effort. Here are some reasons that students may choose to study vertebrate morphology:

1. *To comprehend the structural basis of biology.* Knowledge of anatomy has direct application to many specializations within biology. To the physician and veterinarian, the experimental embryologist and physiologist, and the biogeographer and paleontologist, a knowledge of anatomy is valuable if not essential.

2. *To study evolution.* Vertebrate morphology provides particularly favorable evidence for the process and product of organic evolution. It contributes to the answering of questions that have long been important to people: What forces govern the stream of life? How can one gain perspective in time and space? How can one account for the complexity and competence of the animal body?

3. *To advance human health and technology.* The precision by which animals move and the elegance of the structures responsible serve as inspiration for the design and control of mechanical "mimics" of biological systems. Engineers and morphologists share their expertise to improve the design of prosthetic joints and limbs, and to fabricate limbed robots capable of terrestrial, extraterrestrial, and aquatic travel.

4. *To seek appreciation and inspiration.* The respect for biological form and the personal motivation for learning that comes with understanding (for instance) the pulley-like arrangement of the tendons and ligaments that control a claw, why a cheetah is slim and lanky, or why an elephant is not an exact scaled replica of a mouse, cannot be measured, but is highly rewarding.

SOME PRINCIPLES AND CONSIDERATIONS

Phylogenetic Homology and Analogy No other concept in vertebrate morphology is more fundamental than that of homology, yet it continues to be the subject of numerous articles and even books. Such focus alerts us that the long-accepted definition is now inadequate to encompass all advances in evolutionary, developmental, and genetic analysis. Several definitions are required according to context. The traditional concept is presently distinguished as **phylogenetic homology.** Features of two or more organisms have phylogenetic homology if they have common ancestry (see examples 1 to 4, Figure 1.1). Such homology is established if the features can be clearly linked through time by continuity in the fossil record, and is reasonably certain if they can be shown to develop similarly in the embryo from identical primordia. Homology may be difficult to establish in specific instances; nevertheless, the concept is clear, is applicable to this book, and is the kind of homology we intend where we use the word without qualification.

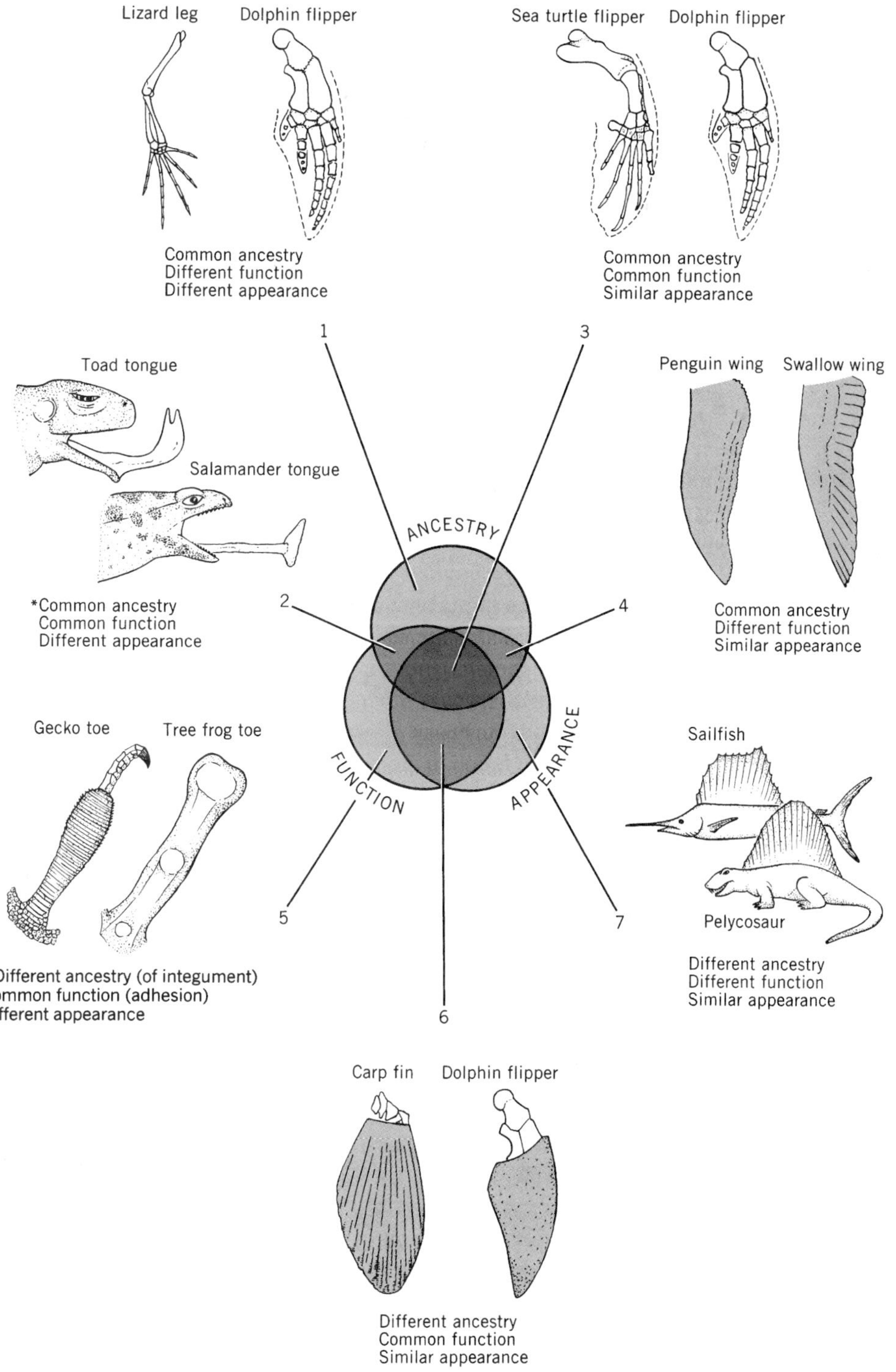

FIGURE 1.1 THE DISTINCTIONS AND RELATIONS AMONG COMMON ANCESTRY (PHYLOGENETIC HOMOLOGY), COMMON FUNCTION (ANALOGY), AND COMMON APPEARANCE. (*In 2, the tongues are homologous, but the projection mechanisms are not; in 5, the critical parts of the integument are not homologous, but the toes are.)

Before presenting other kinds of homology, we mention the concept of analogy, because if is is not distinguished from phylogenetic homology, errors can be made in the interpretation of materials.

Features of two or more organisms are **analogous** if they have common function (examples 2, 3, 5, and 6, Figure 1.1). Analogy may be inferred from structure or from comparison with other, similar features, but it is definitively established only by behavioral and biomechanical analysis. Analogous features may or may not also be homologous. Analogy can also be difficult to prove in specific instances. Acceptable levels of equivalence of function are debatable, and even the concept and its definition remain somewhat controversial. In this book the frame of reference is usually the class, subclass, or order, where the concept is exceedingly useful and seldom ambiguous.

Biological Homology and the Conservation of Process Significant advances in our understanding of early development and molecular genetics have given rise to another view of homology. **Biological homology** uses as a frame of reference the earliest developmental origins of structures: the genes. All multicellular animals share certain clusters of genes, called *homeobox genes,* which regulate the formation of major structural features such as the organization of the body into anterior and posterior ends, segmental organization, and the differentiation of limbs. The spatial location of homeobox gene clusters along their chromosomes, and the pathways of their regulatory functions, are highly conserved (i.e., little changed during evolution) from one vertebrate group to the next (e.g., fishes to mammals) and, even more remarkably, from arthropods to chordates (see Comment 5.1 on p. 73). Animals as different as a fruitfly and a bird, for example, possess clusters of virtually identical genes that dictate the early development of paired appendages. Although the wings of a fly and a bird do not arise from the same embryonic tissues (or germ layers, as explained in Chapter 5) and do not trace back to a common structure in a common ancestor (and therefore are not phylogenetic homologs), they are biological homologs because they are expressions of the same homeobox gene clusters. Both concepts of homology are useful in helping us understand the diversity of vertebrate structure. They are distinguished by context.

Other Kinds of Similarity Features of two or more organisms may also be related by similarity of *appearance* (examples 3, 4, 6, and 7, Figure 1.1). Such features are usually also analogous, and frequently are also both analogous and homologous (i.e., phylogenetically homologous). If features are related by appearance only, the resemblance is usually superficial. If two features are related by homology and appearance but not by function, one of them has usually undergone an evolutionary shift of function (see mention of *preadaptation* on p. 14).

Here are further examples of various relationships among homology, analogy, and appearance; how would you place each in Figure 1.1? (1) The quadrate bone of reptiles, which supports the jaw, and the incus of mammals, which is an ear ossicle (see the figure on p. 128). (2) The penis of mammals and the claspers on the pelvic fins of sharks, both of which convey sperm to the female (see the figure on p. 155). (3) The ear ossicles of mammals (derived from cranial bones) and the ossicles of certain fishes (derived from vertebrae), both of which transmit sound waves to the ear. (4) The large cannon bone in the lower leg of cloven-footed mammals (e.g., cows, deer, vicuna) and a similar bone in the leg of the ostrich (see the figure on p. 444); each adapts the foot for running.

Serial homology is a rather independent concept, which is mentioned here because of the similarity of terms. Structures are serially homologous if they occupy different spatial positions in a series of like structures. The separate vertebrae are serially homologous, as are the different teeth of a tooth row, the several gill arches of the series, the successive muscle segments along the back of a fish, and the many tubules in the kidney of a lower vertebrate. It is usual for the structures of such series to form gradients: Vertebrae of most mammals become larger toward the pelvis, teeth often become more complex toward the back of the mouth, gills may become smaller toward the posterior end of the series. Serial homologs have similar potential for change: Any embryonic vertebra behind the ribs will form an articulation with the pelvis if experimentally placed adjacent to it. Also, change usually affects more than one element of a series: If one tooth becomes larger or more complex, its neighbors tend to do likewise.

Structures have **sexual homology** if they develop from equivalent embryonic primordia yet are sexually dimorphic. The ovary is the sexual homolog of the testis, and the clitoris is the sexual homolog of the penis.

Adaptation Adaptation is a very important concept in evolutionary biology, yet its definition has been vigorously debated. We all know, in a general way, what is meant by "The teeth of the great white shark are adapted for shearing meat." Contemplating its serrated blades tearing large pieces of flesh from its prey, one is tempted to believe that (1) these teeth have the best design possible and (2) they were perfected through processes of evolution. Previously, it was thought that each of an animal's features is controlled by a different gene and was independently made perfect by natural selection. We now know that inheritance of traits is much more complex than that, and adaptation, however remarkable, often (or usually) is not ideal. (See Comment 29.1 on page 548 and, in References, the seminal 1979 article by S. Gould and R. Lewontin.)

Adaptation is the evolutionary process of becoming adjusted (or adapted) to a mode of life in a particular environment. Adaptedness is the state of being adapted, or the degree to which individuals survive and reproduce. More broadly, the term "adaptation" is also commonly used to indicate adaptive traits, rather than the process of acquiring them.

All individuals and species are at least adequately adapted as long as they survive. However, morphologists usually apply the concept to the structural and behavioral *features* of animals (e.g., heart conformation, feeding mechanism, mating strategy) rather than to whole organisms. Adaptive traits may be acquired rather than inherited, as are calluses on hands and feet, but morphologists are primarily concerned with the much more numerous heritable traits. Thus, it is usually stated or implied that adaptive traits are structural or behavioral features that contribute to survival of the species through natural selection.

It does not follow that all features are optimally adapted, or even that all are directly advantageous. A feature may be as it is as a compromise between competing selective advantages, or merely as a consequence of body size (see Chapter 23), or because the developmental process could not provide a better solution, or because of the random fixation of genetic factors. The same selection process may produce different features in different populations, as on adjoining islands, and there may be no selection pressure, as for a vestigial organ.

Form and Function It is clearly evident that form and function are closely correlated; appropriate structures enable an animal to perform specific tasks. However, the basis for the correlation is less evident and has been the subject of much debate. It seems to follow from Darwinian evolutionary theory that function precedes form and provides the selective advantage that guides change of form (i.e., variants that are best suited to fill an existing biological need tend to survive). However, form can precede function (see mention of *preadaptation* on p. 14), and simultaneous change in form and function must be considered. Sometimes the choice of precedence can be made, if at all, only on a theoretical basis.

Although the correlation of a specific form with a specific function can usually be made with confidence, errors have too often been made by *assuming* that a particular structure is adapted to a particular purpose because that seemed obvious, or plausible, or the only probable interpretation. Also, it is tempting to generalize from one study animal to a large group of animals without adequate attention to variation within the group. Correlations of form and function should be considered tentative until it is shown by principles of functional design, coupled if possible with experiment or field observation, that the observed form does indeed fulfill the postulated needed interaction with the environment.

Some Approaches in Morphology In seeking to understand degree of relationship among animal groups it is desirable to combine various approaches. Similarities and sequences revealed by the fossil record are of the utmost importance, but there are great gaps in the record, preservation may be poor, and soft tissues are seldom recorded. Physiology and biomechanics are important for establishing function, but have limited application to forms that are unavailable for laboratory study, including extinct forms without surviving analogs. Molecular biology has become a powerful tool for discovering the degree of affinity among surviving animals, but becomes less reliable as the isolation between contrasted groups increases. Likewise, the emerging field that explores evolution through embryology is helping to interpret certain relationships. Finally, the long-established field that interprets evolution by comparative anatomy has found new strength through rigorous new methods. When two or more of these approaches are applied to the same evolutionary puzzle, the results are sometimes in conflict (revealing their limitations and imperfections), yet in general they support, extend, and refine each other's results.

Following are some of the considerations that guide the morphologist in recognizing and distinguishing homology and analogy, and in correlating form with function.

1. It is often useful to compare forms in numerous animals that are known to be unrelated yet share the same function, such as fast swimming, climbing by adhesion, or feeding on ants. Similarly, it is frequently helpful to study form in a large group of related animals that have developed different ways of feeding, moving, reproducing, or other functioning.

2. One must not infer that analogy indicates homology. Thus, horses and cattle share large size, hoofs, and similar molar teeth because each runs well and eats grass. The common structures evolved in response to common habits, and were not retained from common ancestry. Hedgehogs, porcupines, and one of the egg-laying mammals (echidna), though unrelated, all have quills because quills provide a satisfactory defense for animals that cannot run or fight.

3. When studying the relationship between groups of vertebrates, it is important to consider all the features that they share. Correspondence of many parts bespeaks evolutionary relationship, whereas correspondence of few parts may result from other causes. For instance, numerous cartilaginous and bony fishes have electric organs. This suggests common origin, but the fishes are so different in regard to so many other characteristics that it is virtually certain that their electric organs evolved independently. Likewise, horses and cattle each have complicated enamel patterns in their cheek teeth, but their stomachs, dental formulas, and skull structures argue against close affinity.

4. The study of complicated structures, although more difficult, may be more rewarding than the study of simple ones. Ribs are just too simple to hold many secrets. The skull, on the other hand, has so many functions and so many bones, foramina, and contours that it reveals much of the habits and history of its former owner. Furthermore, corresponding structures of unrelated animals are unlikely to be similar merely by chance if they are also complicated.

5. The primitive, or ancestral, condition of a structure is more likely to be found in an animal that retains the ancestral condition of *other* parts. However, one must avoid circular reasoning. One cannot *prove* that a feature is primitive because it is found in an animal that is *assumed* to be primitive. Nor can one assume that a feature is primitive just because it is found in an animal that is *known* to be relatively primitive—even long-surviving species have some highly specialized structures. Nevertheless, an animal that is *known* to be primitive in the expression of several characters is of particular importance. For example, if one is working on phylogeny of the types of placentas found among rodents, it would be well to see what sort of placenta the mountain beaver has. It could be quite ordinary, but if it were unusual, it would be the more interesting because this animal (not really a beaver) is known by evidence of its skeleton, muscles, and fossil record to be the most primitive living rodent. Part II of this text pays special attention to several animals that retain numerous primitive features: *Polypterus* among fishes, *Sphenodon* among reptiles, the platypus among mammals, and others.

6. It is also important to note the presence of vestigial or degenerate organs having no function at all. It is safe to assume that such structures were functional in an ancestor, and this tells us something about the evolution of the descendant. The vestigial eyes of burrowing moles and cave fishes indicate that their ancestors could see, and tiny bones in pythons and whales tell us that remote ancestors of these animals had legs (Figure 1.2).

Recognition and Assessment of Variation The morphologist rarely describes an individual animal *as* an individual; single specimens are studied to learn about *kinds* of animals. Individuals of a kind vary so much among themselves that a single specimen is not adequately representative.

Most characters vary independently of one another—a specimen may be average for one yet extreme for another. However, some characters are related in such a way that variation in one can be correlated with variation in another. For instance, among certain mammals, an individual with extra-high tooth crowns is likely also to have extra-large premolars and an extra-long jaw, because these characters are functionally related.

Structures that have relatively great variation among individuals of a kind may be of relatively limited value for interpretation of form. Molar teeth and ankle bones must be

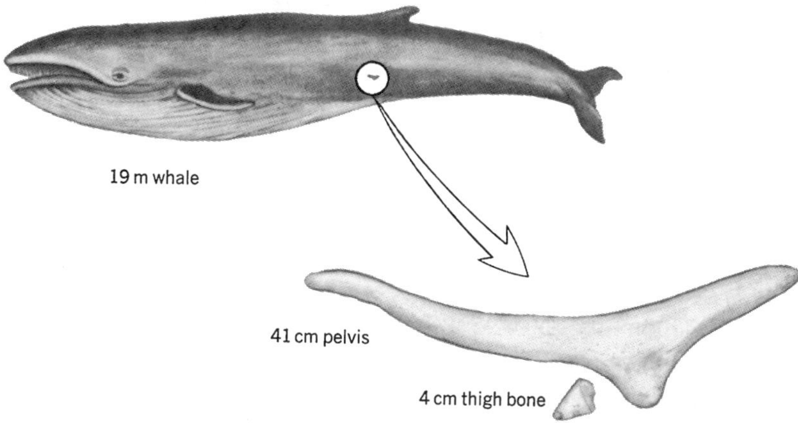

FIGURE 1.2 VESTIGIAL ORGAN illustrated by the pelvis of a finback whale.

just right to function well, so they do not vary much. Breastbones, by contrast, can be flatter, or longer, or more segmented than usual without much impairment of function. Accordingly, there is more variation in breastbones and less can be interpreted from their configurations. The cause of variation (ecological, constructional, behavioral, genetic) may be a useful subject of study.

It may be important to study **sexual dimorphism,** either so that the gender of a particular vertebrate can be established or so that sexual characteristics will not be mistaken for species characteristics or for adaptive features of another nature. Of course, sexual dimorphism relates most to the gonads, genital ducts, external genitalia, and accessory sex organs. These may be conspicuously different in the two sexes or closely similar. Sex differences of the genitalia may be much more evident at one stage of the breeding cycle than at another because of alterations in size, position, coloration, secretions, or vascular congestion.

Other examples of sexual dimorphism do not relate directly to the genital system but correlate with distinctions between the sexes in regard to sex role and behavior. These include differences in the coloration, amount, and distribution of scales, feathers, or hair; in clasping organs and pelvic architecture; and in the presence or development of crests, spurs, antlers, tusks, and scent glands.

Adult body size and rate of growth are frequently sexually dimorphic, and gender differences in body proportions and configuration are common. Thus, males commonly have coarser and relatively heavier skeletons. There may be sexual differences in the muscular and vascular systems, and in the brain and pituitary gland.

The morphologist also needs to recognize **age variation** and avoid confusing it with variation of another kind. The accretion of successive rings or layers on hard tissues often accurately reflects alternating periods of slow and rapid growth. Like tree rings, these relate to times of favorable climate or nutrition, which usually are seasonal. In favorable circumstances, the scales and otoliths (bones within the inner ear) of bony fishes, the scutes and long bones of turtles, the baleen and ear plugs of whales, and the dentine and cement of the teeth of various mammals all show marked growth rings.

Hard parts also correlate with age in other ways: The centers where developing bones first ossify appear in regular sequence, the various teeth of mammals erupt and are replaced at different but regular ages, sutures between bones close on schedule, and epiphyses (which cap the ends of many bones of juveniles) join their shafts at given times. The ages of young humans and of some other animals (for which the sequences have been worked out) can be determined within fairly narrow limits by X-ray analysis of the skull, wrists, or other joints.

The following may also be age-related, though usually with less precision: size and structure of the baculum (a bone in the penis of some mammals), structure of the cranial wall (birds), relative size of orbits, size and configuration of horns and antlers, vascularity of bones, development of crests and tuberosities on the skeleton, and relative development of rostrum or facial area.

The structure of soft tissues tends to be less precisely age-related, yet the weight of the lens of the eye (measured after careful removal and desiccation) is sometimes a reliable index of age. Changes in the skin, in the progression of molt, in the composition of muscle and in the circulatory system are also ascribed to advancing age.

If an animal is extinct, rare, or otherwise difficult to obtain, it may be necessary to work with few specimens, but the morphologist tries to obtain random samples of adequate size. We may first sort the individuals by sex, age, geographic location, or other criterion, keeping enough specimens in each subsample (about 25 is usually adequate) to assess these variables. The individuals of such subsamples still vary among themselves, however, and this **individual variation** must be analyzed if the true nature of each characteristic is to be identified. The nature of typical structure, the range of variation, and the variability, or tendency to vary, are all population characteristics; they cannot be learned from single specimens.

Statistical analysis of population characteristics should be studied by aspiring morphologists. Even several relatively simple parameters can be very helpful: The *arithmetic mean* expresses the "average" condition. The *standard deviation* shows the degree to which the separate values tend to cluster around the mean. The *standard error of the mean* enables one to judge if means derived from different populations (or samples from different populations) are significantly different. The *coefficient of variation* makes it possible to say if animals of different absolute size are comparably variable, that is, if a shrew varies as much for a shrew as an elephant does for an elephant. Morphologists commonly use these and many other, more advanced parameters, and also employ various graphical techniques.

Statistics, however important, do not substitute for attention to accuracy, reliability, and significance. Results are *accurate* if free of error, *reliable* if repeatable, and *significant* if meaningful. The grade given a student on an examination is accurate if additions and recording were done with care. It is reliable if another reader would have assigned the same grade, and if the student, taking the exam again, would be assigned a closely similar score. It is significant if the grade assigned indicates the student's progress toward desirable course objectives. Research, like student grades, is best when accurate, *and* reliable, *and* significant.

Contributions from Paleontology Paleontology, the study of prehistoric life as revealed by fossils, is an engaging and fasinating discipline that combines geology and biology. Paleontologists must be excellent morphologists who enjoy the challenge of reconstructing the biology and the lineages of long-extinct individuals, communities, or ecosystems, using the scant clues buried in rock. Some vertebrate fossils are mere trackways or body imprints made millions of years ago in mud that was later converted to stone. Rarely is an entire animal preserved: Insects trapped in ancient pitch have retained their delicate structures as the pitch gradually turned to amber, and great mammoths that stumbled into glacial crevasses have been preserved for 50,000 years in a natural deep freeze. Usually only teeth and bones are preserved, and even for these the slow process of petrification has replaced most of the original organic tissue by an exact copy in hard minerals.

To be preserved, the skeleton must be protected from the destructive forces of weathering, and this means that it must be covered up soon after it is released by decay from the tissues it supported in life. Landslides, wind-driven sand, and volcanic ash cover some skeletons, but most that become fossils are covered by silt and sand deposited in lakes or streams, or on the flood plains of wide valleys. In time, the mud is converted by pressure to shale and the sand to sandstone. Upheavals of the earth's crust, and subsequent erosion, expose the contained fossils. Sedimentary rocks are thus a gigantic filing cabinet for the fossils they contain. The oldest fossils are in the oldest strata and the recent fossils are in recent strata. Earth scientists divide earth history into *eras* and *periods* (the drawers of the filing cabinet), the boundaries of which correspond to times of relatively rapid change in the earth's crust and biota. (See the front papers of the book for the geologic time scale.)

As an interpreter of animal structure, the vertebrate paleontologist is confronted with many difficulties. He or she usually must work with broken materials and fragments; seldom finds fossils of animals that lived in the arid uplands, where skeletons disintegrate quickly in the dry air and hot sun; and, with rare exceptions, must work with a

FIGURE 1.3 CHARLES DARWIN AT AGE 40. (Print of lithograph courtesy of the Wellcome Library, London.)

single organ system. Nevertheless, paleontologists have described about three times as many extinct vertebrates as there are surviving vertebrates.

Evolutionary Theory By the time Charles Darwin (Figure 1.3) set out, in 1831, on what would become a five-year voyage around the world, he had explored, but rejected, two quite different areas of study (medicine and ministry) at two universities. He had a passion for natural history but had no idea where it would lead him; his age was 22. An unquenchable curiosity and careful observation during the voyage led him, over the next 20 years, to conclusions about the diversity of plants and animals that would galvanize biological thought.

In his *Voyage of the Beagle,* an engaging adventure story distilled from his field notes, Darwin conveys his sense of awe and enthusiasm. In 1859, with the publication of *The Origin of Species,* Darwin put forth a unifying theory for biology, but one so emotionally charged for society, that today he is perhaps *the* most contemporary of the 19th-century personalities.

The cornerstones of Darwin's theory of organic evolution are variation, competition, differential reproduction, and natural selection. He recognized that for any feature that is heritable (structure, physiological response, behavior), *variation* exists in populations. We can easily distinguish every individual we encounter when walking across campus, and a similar range of variation exists among golden retrievers, field mice, and goldfish.

Darwin predicted that within a given environment, which includes a geographical component (e.g., island versus mainland, mean temperature) as well as a biological component (presence or absence of other organisms), individuals are in *competition* for resources such as food, nesting sites, and mates. Those members of a population with favorable variants compete more successfully than those with less favorable ones, and as result, a *differential reproduction* rate occurs. Darwin concluded that the result of differential reproduction over time, particularly in an ever-changing environment, is a population wherein some characters (traits) change (evolve).

Finally, Darwin knew that in just a few generations, animal and plant breeders could artificially select for wanted variants and produce strikingly dissimilar descendents from a common ancestor (e.g., Great Dane and Chihuahua). He was fascinated by artificial selection and saw it as so integral to his thesis that he devoted the first chapter of *The Origin* to examples. He reasoned that in nature, a multitude of factors must be involved in selecting which variants will be maintained or lost through time. Darwin termed the process by which this occurs *natural selection.*

Evolutionists agree that descendant species evolved from common ancestors and that natural selection has had an important role in guiding the direction of structural change. However, they are now less unified in their interpretation of the process than they were in the 1930s and 1940s, when a "modern synthesis" of Darwinian theory emerged that seemed generally acceptable. Analysis of the molecular structure of DNA and of protein has in part supported concepts formerly invoked by geneticists but has also shown genetic variation to be far more complex than once thought. It is now recognized that both large and small genetic alterations can result in both large and small morphological changes.

The traditional belief, called **phyletic gradualism,** is that evolutionary change results within a lineage from the slow and continuous accumulation of those mutations that are favored by natural selection. This causes descendant structures, and also species, to remain well adapted to gradually changing habitats; there is no sharp demarkation between an ancestral species and its descendant.

A contrasting theory, termed **punctuated equilibrium,** holds that most species fluctuate a little in structure over time, but in sum change little for long intervals of earth history. The evolution of new species may then be relatively sudden in geological terms. Descendant species are clearly set apart from their immediate ancestors in form and function. Evolution occurs between lineages by a relatively fast branching process, rather than within lineages by slow replacement.

The debate between the proponents of punctuated equilibrium and phyletic gradualism continues, but the common ground is extensive (in sharp contrast to their shared views versus those of creationists). Most evolutionists believe that when trimmed of misconceptions and overstatement, the positions are compatible.

Evolution and Habitat Interpretation of structure is made more meaningful by familiarity with the major features of the evolutionary process, which are accepted by virtually all morphologists. A major consideration is the relation between evolutionary change and the stability of the physical and biological environment. With some exceptions, evolution results from the interplay between changing environments and adapting organisms. Each kind of animal becomes adapted to, and dependent upon, a particular kind of life (predation, seed eating, grazing) in a particular kind of habitat (marsh, stream, meadow). If the habitat is large and constant (oceans, tropical forests, coniferous forests), the animal inhabitants have time to become well adjusted. It is most advantageous for each species to remain about as it is, so natural selection tends to prevent change. Large habitats do move slowly in earth history (e.g., forests advance and recede), but most of the animals move with them, remain adapted, and change relatively little.

Although the general habitat may shift slowly in space, the restricted habitat of a population of animals may slowly change. Thus, a shoreline may gradually become more rocky, a pine-fir forest that is isolated between mountain ranges may become more alpine in character as the timberline shifts southward, or a competitor may become more abundant. The average expression of the characters of a kind of animal is then less advantageous. Natural selection therefore tends to cause the animal to become a somewhat different kind of animal. If the habitat alters as a unit, the evolutionary change is in a more or less straight line and is said to be **linear.** If the old habitat sub-divides into different units, different parts of the original animal population become isolated from one another and adapt independently, and evolution is **branching.**

When a habitat changes too fast for a resident species to adapt to its new character, then, so far as that species is concerned, the habitat does not change, but disappears. An inland sea drains away, leaving only shoreline terraces on dry hills; meadows are invaded year after year by pine seedlings and finally succumb to the forest; or an extensive area is devastated by a natural disaster. As a habitat becomes less and less satisfactory, the force of natural selection, that is, **selection pressure,** for the old way of life weakens. Finally, extreme variants in the population may be better suited for life in a new habitat than in the old, provided that one or several new habitats are physically available and not already occupied by effective competitors. Selection pressure then becomes strong and shifts away from the old lifestyle toward the new.

Several factors may contribute to success in this kind of evolution. **Specialized structures** are those that have become modified to perform restricted functions with great effectiveness, whereas **unspecialized structures** (or generalized structures) are suited to perform adequately a less restricted function or a variety of functions. Although unspecialized structures may have less pressure for change (being suitable for a wider variety of conditions), they also have more capacity for change, and hence favor survival when change is essential. Thus, the ancestral five-toed foot has been converted to a springing support, wing, paddle, grasping organ, and so on. Specialized organs are satisfactory as long as their restricted functions are needed (it is probable that there will long be ants for anteaters and krill for baleen whales), but if a new function is needed, such organs rarely can adapt.

When a species must adapt quickly to altered conditions, it is not granted time to evolve an entirely new complement of structural attributes. It must rely for a time on the intensified or altered use of attributes it already has. Natural selection may "discover" that a structure that was useful in one way before can now be useful for another purpose. Such structures are said to have **preadaptation.** For instance, it was a long and major task for evolution to convert the walking legs of proavian reptiles into the wings of birds. It was relatively quick and simple for several groups of birds to use their wings as waterfoils for swimming instead of as airfoils for flying—a shortcut of tens of millions of years in the evolution of effective paddles (example 4, Figure 1.1). Similarly, it was relatively "easy" for several groups of fishes to gulp air in air breathing because they could employ, with only slight modification, structures and behaviors that had slowly evolved for gulping food in water. It follows that one cannot always infer from the current function of a structure the remote basis for its origin. Unspecialized form, versatility, and preadaptions are good cards to hold in the game of adapt-or-become-extinct.

Just as habitats can disappear, so they can appear. A new marsh, formed close to existing marshes, will be populated from nearby, and no evolution will occur. A new inland sea, however, if extensive and isolated, is a place of evolutionary opportunity for such animals as can get a start there. Similarly, the habitat may be old, but its availability may be new. Land first became available to the vertebrates when reptiles and certain amphibians evolved. There were sufficient land plants and arthropods to provide food and shelter. There, inviting colonization, were millions of square miles of diverse new habitats having no competition from established animals. In this infrequent circumstance, evolution rapidly creates various new kinds of animals. It is said to be **radiating.**

Stasis, Change, and Extinction We have noted that the kinds of animals that are adequately adapted to relatively constant and extensive habitats may be long surviving with scant change: The reptilian genus *Sphenodon* seems to have survived for 135 million years. Other animals, changing as conditions change, form relatively complete and continuous lineages. The fossil record presents many examples. Elephants were once numerous and diversified, and camels came in all sizes and were abundant on four continents. Horses, turtles, crocodiles, the various kinds of dinosaurs, and many other groups are all assemblages of clearly related genera. Such a lineage is called a **phyletic line** and is usually represented by genera that are related in time by linear and branching evolution, and through extinction by progressive change. Ancestors disappear as descendants evolve. Systematically, a phyletic line is often a single family. Different phyletic lines evolve at different rates, each line evolves at different rates at different times, and different characters of one line evolve at different rates at the same time. With these qualifications, however, some examples can be given of the survival of lineages. The group characters of rabbits, armadillos, and turtles were each established more than 65 million years ago. Opossums have been opossums for 100 million years, and the same is true for crocodiles. Some groups of fishes (coelacanths, dipnoans, sharks) have survived for 300 million years.

Species become extinct when they cannot adapt to such shifts in their environments as climatic change, increase in competition for resources, disbalance in predator-prey relations, or alteration in host-parasite relations. Entire assemblages of animals become extinct when the scale of environmental change is extreme—periods of relatively rapid (in terms of geologic time) mountain building, major change in vegetation, or significant shift of sea level. Finally, catastrophic events such as impacts by asteroids and particularly devastating volcanic activity could cause mass extinctions. Just *how* catastrophic such an event can be is described by W. Alvarez in a gripping account of our search for an explananation of the demise of the dinosaurs 65 million years ago (see References). In 1991 a crater was recognized on the Yucatan Peninsula believed to have been made by an asteroid with a diameter of 9 km. The explosion on impact is estimated to have been the equivalent of 10,000 times the entire nuclear arsenal of the world, and the debris created in its wake was enough to block the sun's light around the world for months. At several previous (and less thoroughly documented) times in the past 500 million years there have been other episodes of mass extinction, when as many as 60 percent of all animal genera died out within a relatively short time.

Evolutionary Trends It is a striking fact that *within* phyletic lines, each adaptive change tends to progress in more or less the same direction without stopping, zigzagging, or reversing. Such gradual changes are called **evolutionary trends** or morphoclines and, though not universal, have been usual for large populations evolving at moderate rates. Evolutionary trends are oriented and prolonged by selection pressure. The characteristic continues to develop because it continues to be advantageous.

A common, though by no means universal, trend has been toward large body size. Repeatedly, and in unrelated lineages, there has been a gradual increase in size from modest ancestor to gigantic descendant. (Examples of the advantages of large size, and adaptations for supporting great weight, are presented in Chapter 23.) A common trend for the reduction in number of serial parts is exemplified by the teeth of many lineages, lateral digits of hoofed mammals (Figure 1.5), gill arches of lower fishes, and cranial bones from fishes to mammals. Conversely, teeth and muscles have increased in number in some lineages. Trends leading to change in relative size of parts of the body have been common. The gradual development of tusks, antlers, and unusual beaks are examples. Other trends involve increase in specialization of parts, such as the formation of elaborate enamel patterns in the grinding teeth of horses.

COMMENT 1.1

CONTINENTAL DRIFT AND THE DISTRIBUTION OF FOSSILS

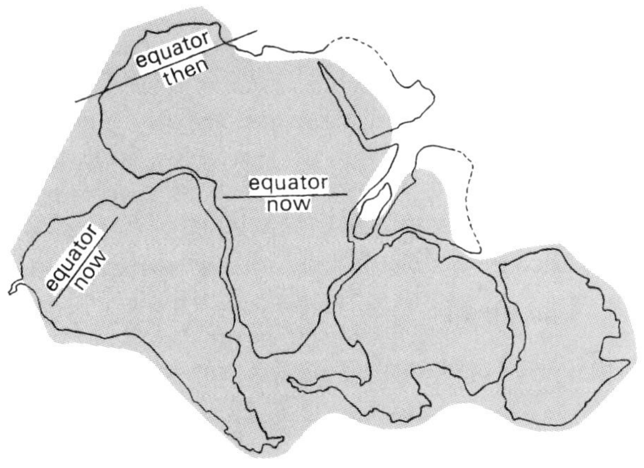

FIGURE 1.4 RELATIVE POSITIONS OF THE SOUTHERN CONTINENTS IN EARLY TRIASSIC TIMES, 245 MILLION YEARS AGO.

As recently as 50 years ago, earth scientists were deeply divided as to whether the continents have always been more or less in their current positions or were once joined. It was in 1912 that the German meteorologist Alfred Wegener placed before the (largely resistant) scientific community data in support of the idea that the continents had once been united. It remained for the development of sophisticated analytical instruments capable of measuring rates of radioactive decay in minerals, the mapping of ocean basins, and the detection of reversals of the earth's magnetic field, to bring consensus that the earth's surface is divided into a number of plates (most of them the continents) that are on the move (the concept of *plate tectonics*). There is substantial evidence that convection—the escape of deep internal heat from the earth's molten core—has driven the movements and the breakup of these plates from a single supercontinent (Pangea). The configuration of the land mass of the Southern Hemisphere (termed Gondwanaland) in the early Triassic period is shown by the stippled area in Figure 1.4. The fit of the separate plates at the margins of their respective continental shelves was much closer than that of the current shorelines, which are shown for reference. (Note that India was then far removed from other Asian land forms.)

The continental movements, and associated climatic changes (with consequent fluctuations of sea level), had a significant influence on the distribution of vertebrates. In fact, the first corroborative biological evidence in support of the geophysical evidence for continental drift came in 1968 with the discovery of a Triassic fossil in Antarctica by the eminent paleontologist E.H. Colbert (see References). The animal was a stocky, semiaquatic reptile with a distinctive skull and peculiar short tusks (genus *Lystrosaurus*) that was already known from South Africa and India. Its presence in Antarctica, coupled with recognition that the southern continents were once contiguous, explained its otherwise puzzling distribution. Further evidence is provided by various mammals. Fossils indicate that certain marsupials reached Australia from South America by way of Antarctica. A later wave of marsupials again reached Antarctica from South America but did not extend to Australia because by then Australia had drifted away from Antarctica.

Variation on a theme is inherent in our ability to recognize trends. D'Arcy Thompson demonstrated in 1917 that trends in body proportions are admirably described by progressively distorting a grid as shown in Figure 1.6. A grid of equally spaced lines, intersecting at right angles, is drawn over the presumed ancestral form and is then redrawn over the descendant forms, but this time each line touches the same anatomical landmark of the ancestral form, thus showing how features have changed.

Trends come to an end if, and when, the condition of maximum advantage is reached: The sabers of saber-toothed cats increased until the Oligocene epoch and then remained about constant for 40 million years. Furthermore, many trends have theoretical end points: Change toward longer-wearing teeth ends with ever-growing roots; change toward loss of lateral digits ends with one toe; change toward loss of vision ends with rudimentary eyes.

Still, some trends do *seem* to stop short of maximum advantage, whereas others *seem* to go too far. For instance, some extinct elephants had tusks so large they curved and

Above: Screech owl. Redrawn from J. Sparks and T. Soper, *Owls,* Gerston Point, Kingsbridge, Devon, UK.

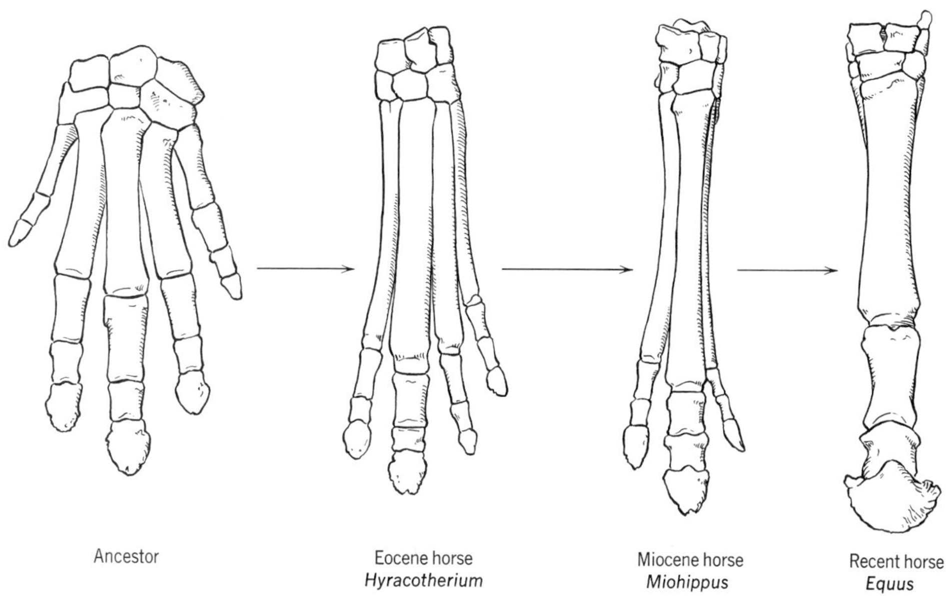

FIGURE 1.5 EVOLUTIONARY TRENDS IN PROPORTIONS AND NUMBER OF SERIAL PARTS of the left forefoot of the horse. (Not drawn to scale.) Although the trends shown are real, the exact examples may not have been direct steps. The Eocene and Miocene are divisions of the Tertiary.

crossed and could no longer effectively thrust, pry, or dig. There are several ways in which this can come about. First, natural selection makes no provision for nature's senior citizens. It improves the fitness of breeders and potential breeders, but neglects individuals that are no longer productive.

Also, if the continuance of a trend is advantageous in one respect but disadvantageous in another, then the trend will stop when the advantages and disadvantages of further change are in balance. This is common though often difficult to identify in specific instances. Increased dermal armor might give a fish added protection against predation but make it less maneuverable; increased slenderness of limb might give an antelope more speed but make it less able to avoid injury; increased curvature of the beak might make it easier for a bird to secure one kind of food but more difficult to secure another. The result is favorable compromise, not optimum structure in regard to any one function.

Parallelism and Convergence Two or more different groups of animals may be morphologically similar. This condition is called *homoplasy.* It can occur because the groups have common ancestry and, after their initial split, fail to evolve further. Usually, however, the similarity persists in spite of the fact that each of the two groups continues to evolve. Descendant lineages resemble one another closely in characters that are not present in the common ancestor. **Parallelism** is evolutionary change in two or more lineages such that corresponding features undergo equivalent alterations without becoming markedly different in degree of similarity. Descendants are about as much alike as were their ancestors, and at each stage the correspondence may be close.

The kangaroo rats of western North America and jerboas of Africa and Asia show striking parallelism. Each has long hind limbs, short forelimbs, loss of lateral toes, lax fur of tan color, long tail with white tip, large eyes, inflated bony ear capsules, and

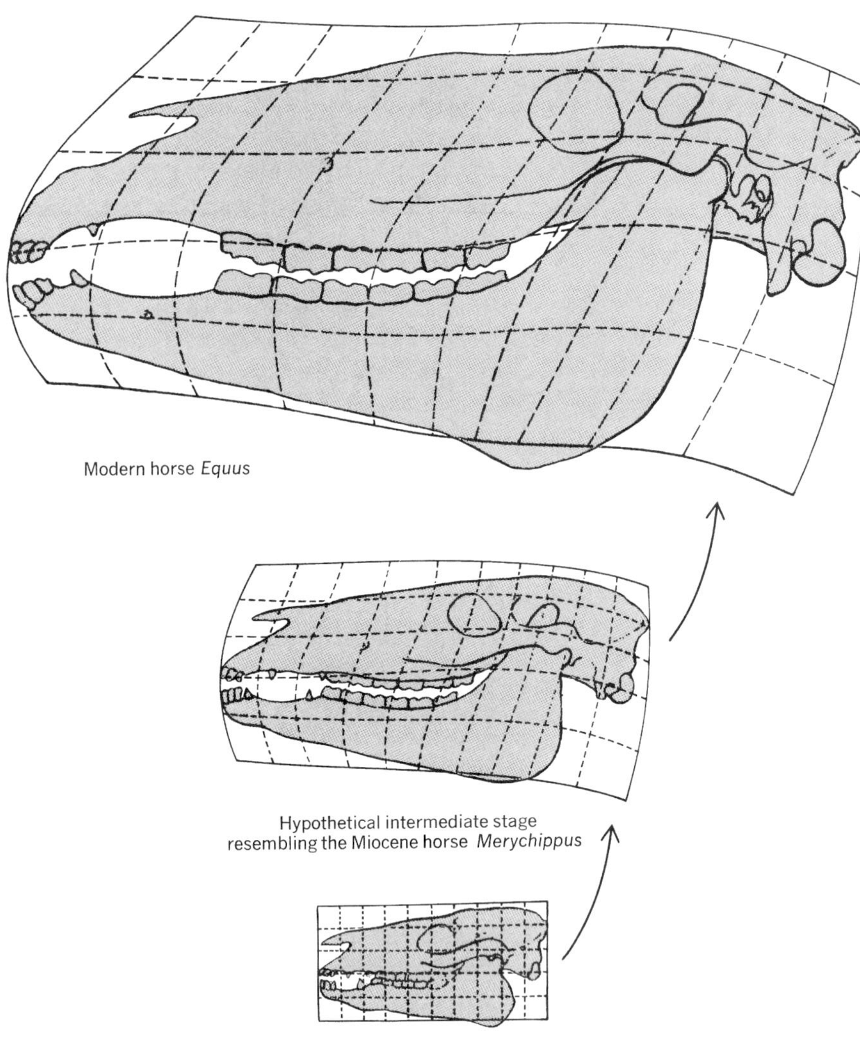

FIGURE 1.6
EVOLUTIONARY TRENDS IN PROPORTIONS AND SIZE of the horse skull shown by the progressive distortion of a grid. (Redrawn from D'Arcy Thompson.)

compacted neck vertebrae (Figure 1.7). It is unlikely that the common ancestor had any of these characters; on the basis of other features, these rodents are placed in different suborders. Similarly, the golden moles of south Africa and the marsupial moles of Australia belong to different orders yet have in common short robust forelimbs with strong claws, rudimentary eyes, loss of external ears, tough nose pads, and other characters. Parallel lineages each remain adapted to nearly identical ways and places of living—kangaroo rats and jerboas now to hot, sandy deserts; golden moles and marsupial moles to life in the soil. It is the effectiveness of natural selection that causes basically similar animals having nearly identical functional requirements to evolve similar structural adaptations.

Convergence is evolutionary change in two or more lineages such that corresponding features that were formerly dissimilar become similar. Descendants are more alike than were their ancestors, though the similarity seldom extends to correspondence of detail. Closely similar functional requirements are met in somewhat different ways because the common ancestor is often (but not always) more remote than for parallelism.

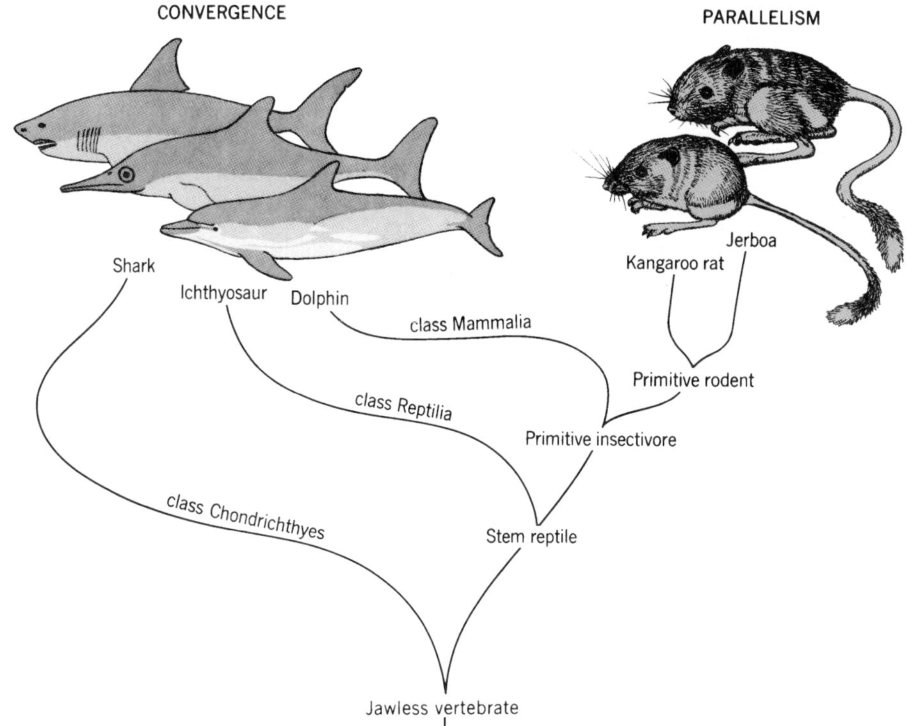

FIGURE 1.7
PARALLELISM AND CONVERGENCE both establish correspondence of structure in response to similar habits. They differ in the degree to which the common general plan extends to similarity of detailed structure.

The remarkable similarity of the shark, ichthyosaur (an extinct reptile), and dolphin is a classic example (Figure 1.7). The common ancestor was a primitive armored fish unlike any of them. One has no terrestrial ancestor; the others have dissimilar terrestrial ancestors. Nevertheless, the tail of each is a waterfoil (though one turns up, one turns down, and one is straight with lateral flukes). Each has numerous simple teeth (but they are rootless for one, rooted in grooves in another, and rooted in sockets in the third). The same sort of general resemblance with variation of detail holds for the spine, eyes, and paired and median appendages. Another example is the similarity of birds and flying reptiles: In spite of remote common origin, each has large eyes, proximal nostrils, long rostrum, long neck, short back, large breastbone firmly attached to the pectoral girdle, many vertebrae articulating with the pelvic girdle, pneumatic bones, and other common features.

The concepts of parallelism and convergence merge into one another. Each process causes taxa to become more related by function and appearance than by phylogeny. Regardless of emphasis (function or phylogeny), the morphologist will sometimes misinterpret evidence if parallelism and convergence are not recognized and interpreted.

GENERAL REFERENCES FOR THE INTRODUCTION AND PARTS I AND II

Alvarez, W. 1998. T. rex *and the crater of doom.* Vintage Press, New York. 185p. An engaging account of the scientific inquiry of a major extinction event; also serves as an introduction to the general principles of paleontology.

Appel, T.A. 1983. *The Cuvier-Geoffroy debate: French biology in the decades before Darwin.* Oxford Univ. Press, Oxford. 305p. Recounts the intellectual progress of Georges Cuvier within an 18th-century context and his convictions that all animal diversity could be understood within "a comparative context."

Bock, G.R., and G. Cardew (eds.). 1999. *Homology.* Novartis Foundation Symposium No. 222. Wiley, New York.

256p. An overview of the concepts of pylogenetic homology and biological homology.

Brown, J.H., and A.C. Gibson. 1983. *Biogeography.* Mosby, St. Louis. 643p.

Carroll, R.L. 2000. Towards a new evolutionary synthesis. *Trends Ecol. Evol. Biol.* 15:27–32. A convincing case for an integration of molecular developmental biology and changing physical factors in the study of earth history and evolutionary patterns.

Charlesworth, B., R. Lande, and M. Slatkin. 1982. A neo-Darwinian commentary on macroevolution. *Evolution* 36:474–498. Review of classic studies from paleontology, genetics, and development that support the role of natural selection in the evolution of morphology.

Colbert, E.H. 1973. *Wandering lands and animals: the story of continental drift and animal populations.* Dover, New York. 323p.

Dellman, H.D. 1992. *Textbook of veterinary histology.* 4th ed. Williams & Wilkins, Baltimore. 420p.

Desmond, A., and J. Moore. 1991. *Darwin.* Norton, New York. 808p.

Futuyma, D.J. 1997. *Evolutionary biology.* 3rd ed. Sinauer, Sunderland, MA. 763p.

Gans, C. 1988. Adaptation and the form-function relation. *Am. Zool.* 28:681–697. An eloquent case for the study of morphological adaptation.

Gould, S.J. 1977. *Ever since Darwin.* Norton, New York. 285p. Stimulating essays about Darwin, natural history, and evolution.

Gould, S.J., and N. Eldredge. 1977. Punctuated equilibria: the tempo and mode of evolution reconsidered. *Paleobiology* 3:115–151.

Gould, S.J., and R.C. Lewontin. 1979. The spandrels of San Marco and the Panglossian paradigm: a critique of the adaptationist programme. *Proc. R. Soc. Lond. [Biol.]* 205:581–598. This paper stirred considerable controversy and forced a generation of evolutionary morphologists to rethink the concept of "adaptation."

Hall, B.K. (ed.). 1994. *Homology: the hierarchial basis of comparative biology.* Academic Press, New York. 483p.

Jaeger, E.C. 1977. *A source-book of biological names and terms.* 3rd ed. Thomas, Springfield, IL. 360p.

Krstić, R.V. 1985. *General histology of the mammal: an atlas for students of medicine and biology.* Springer, New York. 415p. Three-dimensional reconstructions of tissues.

Liem, K.F. 1991. Toward a new morphology: pluralism in research and education. *Am. Zool.* 31:759–767.

Liem, K.F., and D.B. Wake. 1985. Morphology: current approaches and concepts, pp. 366–377. *In* M. Hildebrand et al. (eds.), *Functional vertebrate morphology.* Harvard Univ. Press, Cambridge, MA.

Mayr, E. 1963. *Animal species and evolution.* Harvard Univ. Press, Cambridge, MA. 797p. A classic work.

Pigliucci, M., and J. Kaplan. 2000. The fall and rise of Dr. Pangloss: adaptationism and the *Spandrels* paper 20 years later. *Trends Ecol. Evol. Biol.* 15:41–81.

Proctor, N.S., and P.J. Lynch. 1993. *Manual of ornithology: avian structure and function.* Yale Univ. Press, New Haven, CT. 340p. Superbly illustrated.

Raup, D.M., and S.M. Stanley. 1978. *Principles of paleontology.* 2nd ed. Freeman, San Francisco. 481p.

Reeve, H.K., and P.W. Sherman. 1993. Adaptation and the goals of evolutionary research. *Quart. Rev. Biol.* 68:1–32.

Rixon, A.E. 1976. *Fossil animal remains: their preparation and conservation.* Athlone Press, London. 304p. Also a useful guide to the repair and mounting of skeletons.

Rose, M.R., and G.V. Lauder (eds.). 1996. *Adaptation.* Academic Press, San Diego. 511p.

Schmidt-Nielsen, K. 1997. *Animal physiology: adaptation and environment.* 5th ed. Cambridge Univ. Press, New York. 607p.

Stebbins, G.L., and F.J. Ayala. 1985. The evolution of Darwinism. *Sci. Am.* 253(1):72–82. An evaluation of the tenets of the synthetic theory, molecular biology, and punctuated equilibrium.

Thompson, D'Arcy W. 1992. *On growth and form.* Cambridge Univ. Press, New York. 368p. A classic; first published in 1917.

Wainwright, S.A. 1988. Form and function in organisms. *Am. Zool.* 28:671–680. A thoughtful essay that urges students of functional morphology to include their knowledge of the physical sciences in the study of anatomical design.

Wake, D. 1994. Comparative terminology. *Science* 265:268–269.

Wake, D.B., and G. Roth (eds.). 1989. *Complex organismal functions: integration and evolution in vertebrates.* Wiley, New York. 449p.

Wheater, P.R., H.G. Burkitt, and V.G. Daniels. 1987. *Functional histology.* 2nd ed. Churchill Livingstone, New York. 348p. Beautifully illustrated with color micrographs, electronmicrographs, and correlative line drawings.

Zimmer, C. 1998. *At the water's edge.* Free Press, New York. 290p.

Part One

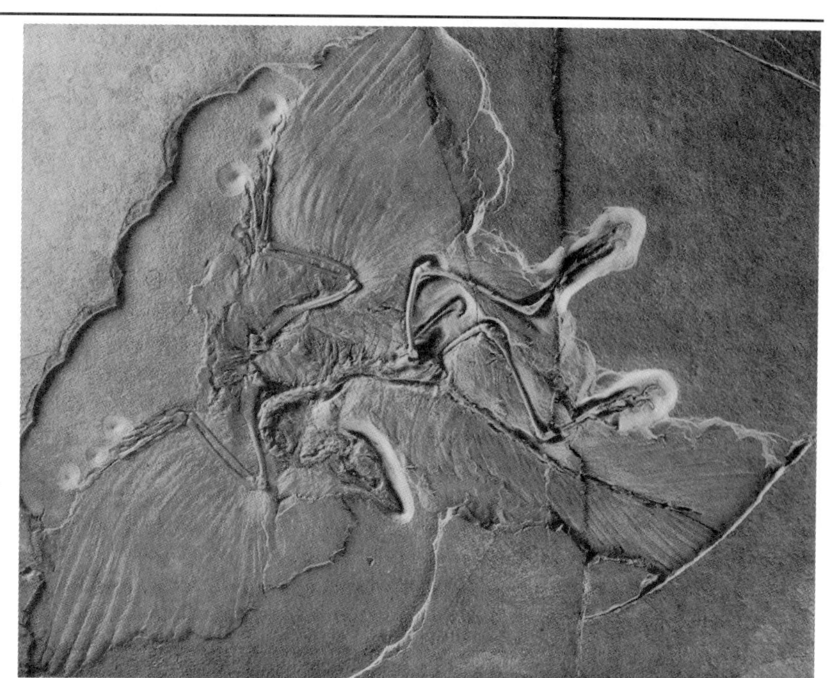

Survey of Vertebrate Animals:

The Principal Structural Patterns

Chapter 2

Nature, Origin, and Classification of Vertebrates

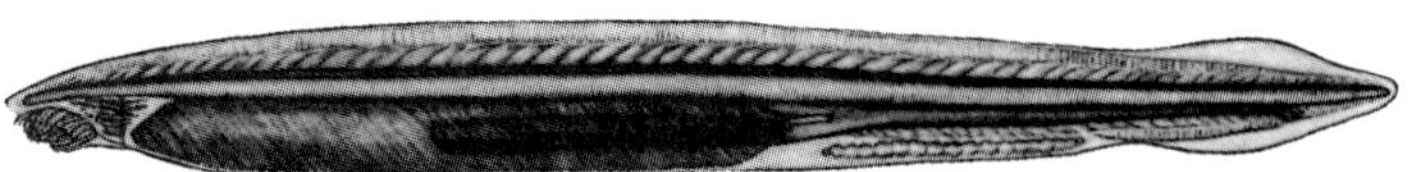

RELATION OF VERTEBRATES TO NONCHORDATES

What is a vertebrate, and how did vertebrates originate? Some animal groups can be distinguished by one or two diagnostic characters: Any animal having a wishbone and flight feathers is a bird; any animal with mammary glands is a mammal. Similarly, any animal with a cranium (skeletal brain box) is a vertebrate. However, since our objective is to understand and interpret the structure of the entire body, it is more helpful to use many characters in combination to describe vertebrates, selecting not only features that are unique to the group but also those that place vertebrates among related groups.

Remembering that VERTEBRATA is a subphylum of the phylum CHORDATA, let us start by describing vertebrates in general terms that relate their phylum to others. Vertebrates are multicellular animals derived from embryos having three tissue (or germ) layers: ectoderm outside, mesoderm, and endoderm lining the gut tube. (The few embryonic terms used in this chapter are defined further in Chapter 5 and in the Glossary.) The body has bilateral symmetry (right and left sides, anterior and posterior ends, dorsal and ventral surfaces). A body cavity, or coelom, is present and is lined by mesoderm. The gut is complete, which means that there are separate openings for mouth and anus. The anus is derived from an opening in the surface of the early embryo called the blastopore (or a point near the blastopore). There is an internal skeleton derived from mesoderm, and the mesoderm is formed at least in part from tissue derived from the embryonic gut.

These characters go far to describe vertebrates, yet they are shared by other chordates and by the phyla ECHINODERMATA (starfishes, sea urchins, sea cucumbers, etc.), HEMICHORDATA (wormlike, burrowing animals), and two lesser phyla (POGONOPHORA and CHAETOGNATHA). There are some departures: One group has no internal skeleton, another has no digestive tract, another lacks a coelom, and only the larvae of echinoderms have bilateral symmetry. In general, however, these features in combination set these animals apart from all others.

Page 21: Fossil of the Jurassic bird *Archaeopteryx*.

It is usually concluded that chordates are more closely related to echinoderms, hemichordates, and members of similar phyla than to animals belonging to other groups. Because these animals do not form the mouth from the blastopore, they are collectively called DEUTEROSTOMIA (= *second* + *mouth*).

RELATION OF VERTEBRATES TO OTHER CHORDATES

There are three subphyla in the phylum Chordata: UROCHORDATA (called tunicates and sea squirts), CEPHALOCHORDATA (called lancelets or, individually, amphioxus), and VERTEBRATA. (Many systematists now prefer CRANIATA as the name for the last subphylum, reserving the name Vertebrata for all animals in the subphylum *except* hagfishes. The reason for this is explained later in the chapter. Until we come to that section, we will retain the more familiar name Vertebrata for the entire subphylum.)

All chordates have various characteristics in common. Most distinctive is the **notochord** (= *back* + *cord*). This is a longitudinal rod of supportive tissue generally derived from the dorsal wall of the embryonic gut. Its turgid, vacuolated cells are unique. The cord is surrounded by sheaths of connective tissue. All chordates (as the name implies) have a notochord during early development. Cephalochordates and many vertebrates (but not urochordates) retain the notochord, or remnants thereof, as adults.

A second chordate character is the **dorsal hollow nerve cord** derived from ectoderm by a folding process called neurulation.

A third feature of importance is the presence, at least in early developmental stages, and usually also in adults, of a **pharynx** (expanded anterior portion of the gut), which is perforated by numerous slits that permit water taken into the mouth to be passed out of the body. Among nonchordates, pharyngeal slits are found only in hemichordates.

Another chordate character is a circulatory system having a **ventral heart** (or ventral pulsating vessel in cephalochordates) that drives blood up through the bars of the pharynx in vessels called aortic arches (these are secondarily modified in higher vertebrates) and thence caudad in a dorsal vessel. (Urochordates are exceptions in that the heart drives the blood in one direction for a time and then reverses to pump in the other direction.) The blood-vascular system of chordates is closed; that is, blood remains in vessels and does not enter the tissue spaces.

Chordates tend to have their principal sense organs concentrated in a head, a condition spoken of as **cephalization.** This characteristic goes with the combination of bilateral symmetry and motility; urochordates, which are sessile, have no head. Two other chordate features that adult urochordates fail to exhibit are a **tail** extending posterior to the anus and **metamerism,** or segmentation of some features of the body.

The distinctive features of each subphylum should also be considered, because differences as well as similarities are important to the establishment of relationships. UROCHORDATA inhabit coastal areas of all oceans. The subphylum is large and diverse. Adults are saclike or stalked, sessile, and often colonial. Colonies may be extensive, but individuals are small. The body is enveloped by a tough tunic. Between the tunic and the pharynx is a space called the atrium. Water enters the pharynx by one opening, or siphon, seeps through the complicated pores of the pharynx to reach the atrium, and leaves the body by another siphon. Larvae do not feed, but adults are filter feeders. Food particles are trapped in sticky mucus. Only the free-swimming larva has a coelom, hollow nerve cord, and notochord. Thus, it is the larva that relates urochordates to chordates (Figure 2.1). Note that the notochord is prominent, but only in the tail (Urochordata = *tail* + *cord*), where it is surrounded by muscle cells.

The subphylum CEPHALOCHORDATA has only two living genera but is known by good fossils that are 530 million years old. The surviving forms are distributed over the world, especially in coastal areas having warm shallow water. They spend most of their

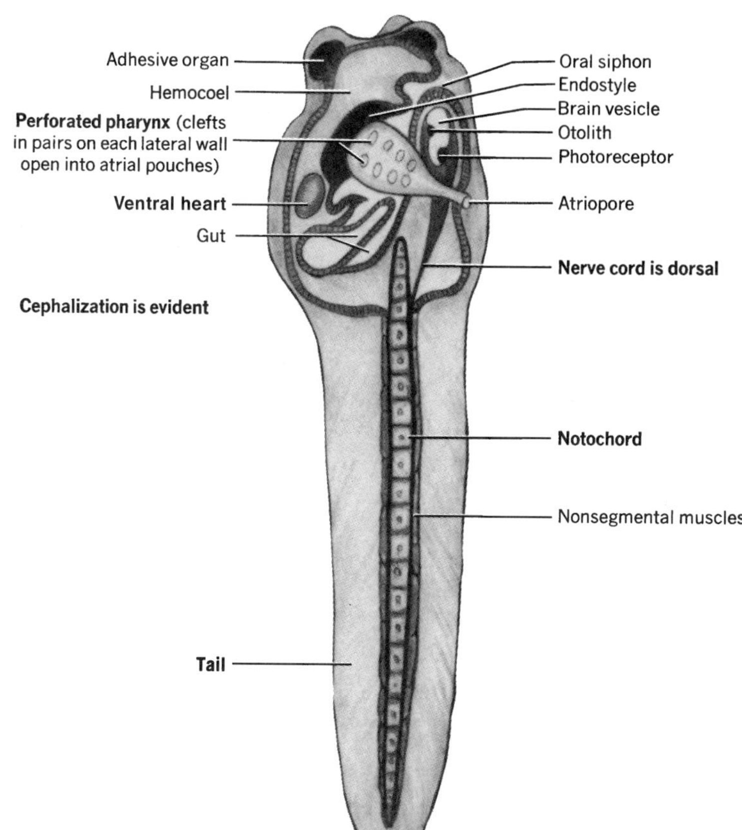

FIGURE 2.1 Stylized LARVAL UROCHORDATE drawn from a 1 mm stained specimen to illustrate internal structure. Characters of the phylum Chordata are shown by boldface labels; characters of the subphylum are shown by standard labels.

time buried nearly vertically in coarse sand with only the anterior end protruding. However, they can swim forward and backward with great agility. The lance-shaped body measures 50 to 75 mm in length. There is a low continuous dorsal and tail fin, and paired lateral body folds (Figure 2.2 and also the figure on p. 277). The persistent notochord extends anteriorly beyond the nerve cord. Other skeletal elements support the fins, pharynx, and oral structures. There are light-sensitive cells on the floor of the nerve cord. Some incipient brain structure has been identified by genetic and microanatomical research, but there is no visible brain, and compared with vertebrates amphioxus is virtually headless except for its feeding apparatus. Like urochordates, these animals are filter feeders with a complicated pharynx surrounded by an atrial cavity, although details of their structure and development are unlike those of urochordates. Water enters the pharynx from the mouth and passes through about 150 slitlike oblique clefts to reach the atrium. The ciliated bars of the pharynx move a mucus strand into the straight digestive tract. Unlike a "true" liver, the hepatic diverticulum is hollow. The circulatory system lacks a heart but is otherwise very similar to the vertebrate plan. Cephalochordates are unique and quite peculiar in the asymmetry of various organs, which results from unusual developmental processes: The segmental muscles, pharyngeal slits, gonads, and spinal nerves alternate on the two sides of the body instead of lying in successive pairs. The somewhat enigmatic excretory organs of amphioxus are mentioned on page 277. The different typefaces used in Figure 2.2 will help you sort out which features are generalized chordate characteristics and which are specialized features, unique to amphioxus and indicative of a long independent evolutionary history.

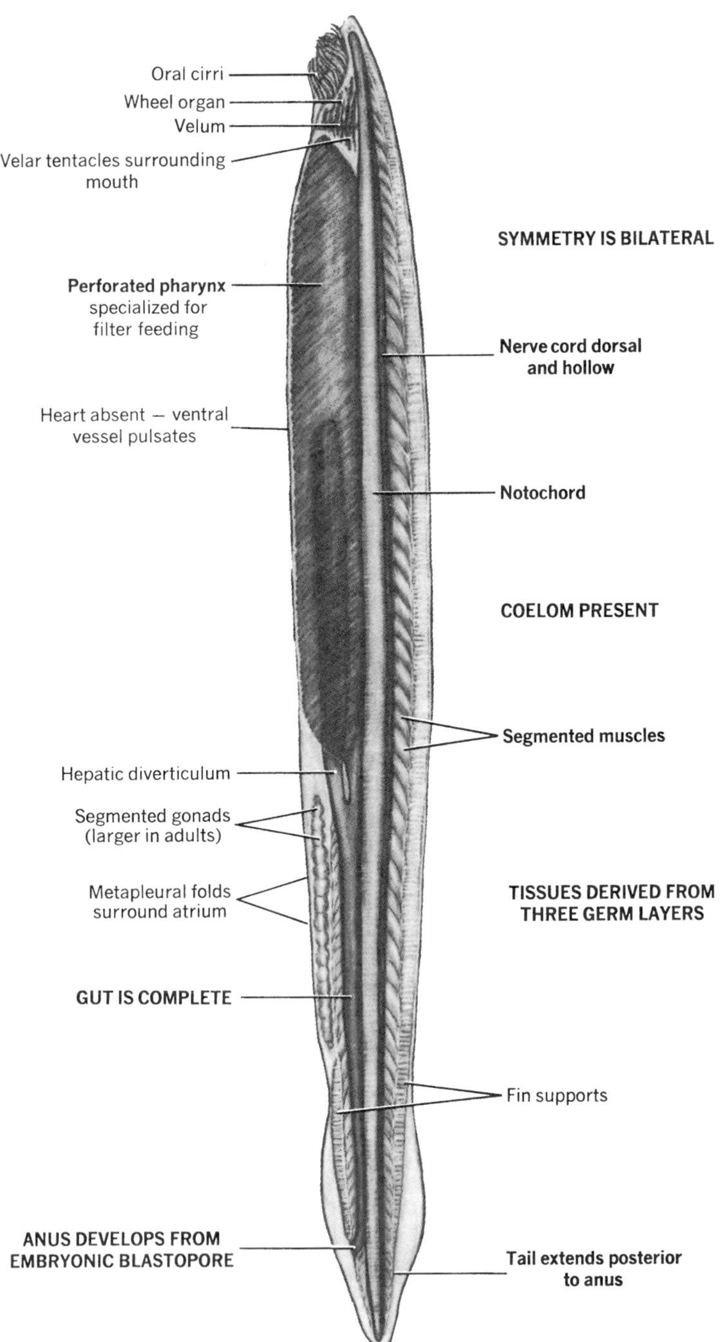

FIGURE 2.2 AMPHIOXUS, drawn from a cleared 37 mm specimen to illustrate internal structure. Characters of the Deuterostomia are shown by boldface capital labels; characters of the phylum Chordata are shown by boldface lowercase labels; characters of the subphylum Cephalochordata are shown by standard labels.

What structures of VERTEBRATA distinguish these animals from all others? Vertebrates are unique in having a true head, the development of which is associated with the evolution of embryonic neural crests (described in later chapters). Only vertebrates have a true brain, divided into several vesicles, that serves to control and coordinate the nervous responses of the body. Furthermore, only vertebrates have a skeletal structure, the cranium, that supports and protects the brain. Eyes, ears, and olfactory organs are present. Many other animals have light receptors and chemoreceptors, but the principal sense organs of vertebrates are not derived from these and rarely resemble them closely in structure. The concentration of brain, cranium, and sense organs in the head gives vertebrates a degree of cephalization that is approached only by some arthropods.

One would expect all vertebrates to have vertebrae. Most do, but some have only a few incompletely formed vertebrae, and there is reason to doubt that the first vertebrates had them at all. The term *Vertebrata* became established before this was known. Most animals have accessory digestive glands, and various of these are called "livers," but the solid liver of vertebrates is not homologous and is only partly analogous to others. The related hepatic portal system and gall bladder are equally unique to vertebrates, as are the pancreas and spleen. A heart is present and chambered. The gonads are not segmental, and the kidneys are of mesodermal origin.

Vertebrates resemble the other chordate subphyla and hemichordates in having a perforated pharynx, but the relatively simple structure, musculature, and predominantly respiratory function of this organ are vertebrate characters. Complexity of the muscular system and attendant locomotor activities are also characteristic. Not all vertebrates have paired appendages, and some nonvertebrates have them; nevertheless, paired appendages are typical of vertebrates. Finally, although vertebrates include minnows, hummingbirds, and shrews, as a group they are relatively large animals.

This discussion started with a short definition of vertebrates—"any animal with a cranium is a vertebrate"—and then expanded the concept for several pages. You might find it instructive to phrase a definition in one paragraph.

ORIGIN OF VERTEBRATES

The ancestry of vertebrates is by no means self-evident. Over the years zoologists have postulated their origin from insects, arthropods, annelids, and mollusks. In 1894 comparative embryology led Garstang to the currently favored belief that the ancestor of vertebrates was close to one of the deuterostome groups noted above. A theory that is notable for its complexity and detail postulates that the closest ancestors were not hemichordates or urochordates, but the Paleozoic echinoderms called calcichordates. However, most zoologists now believe that within the phylum Chordata the vertebrates are closer to the cephalochordates than to the urochordates, and that of the related phyla the hemichordates are closer than the echinoderms.

No surviving echinoderm could be near to the chordate ancestor. Radial symmetry, water vascular system, nerve ring, modified coelom, and other features rule them out. It is the less specialized and bilaterally symmetrical larva that resembles the larvae of simple chordates. Furthermore, no known hemichordate could be the chordate ancestor. Although hemichordates are probably closer to chordates than are echinoderms, the structure of their circulatory system, proboscis, and collar could hardly be converted to the vertebrate body plan.

Similarly, adult urochordates are far too specialized to include the sought-for ancestor. One need only recall the loss of notochord, nerve cord, and coelom, and the presence of such nonchordate features as tunic, atrium, and siphons. Amphioxus is at first glance a possible ancestor, but what of its asymmetry, atrium, and unusual pattern of nerves?

Echinoderms, hemichordates, and chordates must have diverged from a common lineage no later than early Cambrian times, some 600 million years ago. Since those remote days, each group has gone its own way, retaining some of the common features over the ages, but altering or deleting others and adding new structures in response to differing selection pressures. Likewise, the subphyla of chordates must have distinguished themselves soon thereafter. These tentative relationships are summarized in Figure 2.3.

The earliest vertebrates probably were no longer filter feeders, like their ancestors (judging by the feeding apparatus of known fossils), but instead mobile predators with relatively complex heads. Ancient fossils from the lower Cambrian of China seem to represent animals from that early stage in vertebrate history.

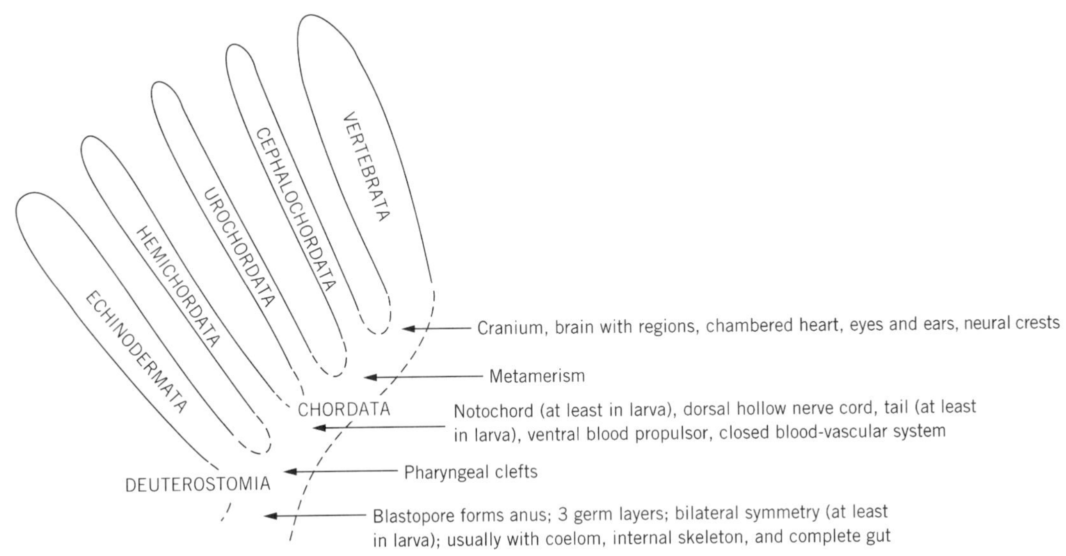

FIGURE 2.3 TENTATIVE RELATIONSHIPS of the subphyla of Chordata and of their closest nonchordate relatives.

Embryonic and larval structure must be stressed in any discussion of vertebrate ancestry. Cleavage pattern, fate of blastopore, and origin of mesoderm and coelom relate the deuterostomes to one another. Hemichordates are linked to echinoderms primarily by the similarity of their larvae, which are called tornaria. Only larval urochordates resemble other chordates, and cephalochordates are to be compared not with adult vertebrates but with embryos and with ammocoetes, which is the larva of the eel-like lamprey.

It seems that some groups evolved from the larva, not the adult, of the preceding group. Sexual development is accelerated and the development of other organ systems is arrested, so the nonreproductive larval stage of the ancestor becomes the reproductive adult stage of the descendant. An example is the amphibian *Necturus,* which is dissected in some student laboratories. This condition is discussed further in Chapter 5, where *Necturus* is illustrated.

CLASSIFICATION OF THE VERTEBRATES

The Objective About 45,000 kinds of living vertebrate animals and perhaps twice that many extinct vertebrates have been identified so far. In order to study and talk about them, it is essential to group them into like kinds and to give each kind a name. Accordingly, systematists establish hierarchies of categories ranging from a few large, inclusive groups to progessively more numerous, smaller, and more specific ones. For example, all the animals included in this book are in the category *vertebrates,* only some of which, having mammary glands and hair, fall in the more restricted category *mammals*. The subgroup of mammals adapted for eating flesh is named *carnivores,* and of these, the doglike members are *canids.* Finally, *wolf* describes a particular canid.

The preceding example, which includes only familiar and well-established categories, is straightforward. Nevertheless, there is often controversy among professional biologists over the criteria to use in defining categories and the names to assign to them. The science of classification has undergone an exciting and fruitful revolution in the last several decades. Procedural change has been profound. Feelings about how best to classify animals can run high, reflecting lively debate and active scholarship. We begin with the older procedures of classification to provide perspective and to indicate their shortcomings, but also to note advantages not yet attained by the new approaches.

The First 200 Years The scientific classification of plants and animals goes back to the Swedish naturalist Linnaeus, who, between 1735 and 1768, introduced the binomial system, giving a **genus** name and a **species** name to every recognized kind of living thing. (The species name is not used without the generic name, and each is printed in italics.) Thus, the wolf of the preceding example is *Canis lupus.* Linnaeus also created the categories **phylum, class, order,** and **family** (followed by **genus** and **species**), for successively less inclusive groups of similar organisms. (Such a group is called a **taxon**—plural, **taxa.**) His objective was to benefit recognition and create order, not to show relationship by descent; at that time the concept of organic evolution was still in the future. For Linnaeus, species were fixed, not mutable.

Following the publication of Darwin's *The Origin of Species* in 1859, and as awareness of the principles of organic evolution came to the fore, the purpose of classification expanded to reflect the branching pattern of evolutionary history. Species were seen as mutable over time, but living species were distinct enough, and extinct taxa were separated by convenient gaps in the fossil record.

Traditional Systematics The principles and procedures of classification that dominated much of the 20th century were the product of this heritage and are now referred to as *traditional* or *evolutionary* systematics. Systematists no longer consider the traditional approach to be adequate in itself; however, as we shall see, it has advantages, particularly for teaching and for communication among nonspecialists, so we switch to the present tense to characterize it further.

The binomial system and primary ranks of Linnaeus are retained. However, as more fossils have been discovered, and more groupings of related animals recognized, the primary ranks are now insufficient for large, diverse assemblages, such as bony fishes and mammals. Accordingly, new ranks are created, usually by adding the prefix *sub-* (below or under), *infra-* (below or inferior), or *super-* (over) to the usual categories. Of all levels of the taxonomic hierachy, only the species has an objective definition. In brief, a species is a group of actually or potentially interbreeding natural populations that is reproductively isolated from other such groups. A genus comprises one or more lineages of species believed to be closely related. Similarly, each higher taxon is an assemblage of related lower taxa. Animals within a species, genus, or family tend to share a common adaptation, such as climbing ability, fish eating, or mode of swimming; animals within a taxon above the family level tend to share basic structural patterns, such as four legs, wings, gill structure, or gnawing teeth.

The traditional systematist must first identify characteristics of the animals under study that might indicate ancestral relationships. Traits may be taken from the comparative anatomy or developmental stages of surviving animals, but the procedure is firmly based on paleontology and the use of fossils. If the fossil record were complete, with all the vertebrates that have ever lived being available for analysis, then, as a consequence of genetic inheritance, they could all be arranged into lineages that converge smoothly as they go back in time. Instead, the systematist has only a fragmentary record, with small gaps here and large ones there. In order to postulate a phylogeny, or evolutionary lineage, it is necessary to fill in the gaps by inference using as clues structural similarities and differences among available materials. As noted in Chapter 1, it is of the utmost importance (and sometimes difficulty) to distinguish similarities resulting from homology (shared ancestry) from those resulting from analogy (common function).

Consider this example: Sabertoothed cats were carnivores with enormously enlarged upper canines (see the figure on p. 555). It once seemed obvious that the various genera must be closely related. Then, on the basis of other similarities and differences, and of the places and times that they lived, it was concluded that sabers evolved independently at several times. Most specialists now believe that the "cats" are best classified in two families (the familiar felids and the extinct nimravids), each of which includes animals with and without sabers. All cats can be sorted into the two evolutionary categories on the basis of, for instance, a feature of the bones housing the middle ear. There have even been saber-toothed predators in two additional orders (marsupials and the all-extinct creodonts). These are striking examples of convergence (see p. 18).

To prepare a traditional classification one must sort through various traits and decide which do, and which do not, indicate an evolutionary relationship, and at what level of the hierarchy. There are no formal or rigid rules to follow; the judgment of the specialist is critical. Original studies presenting traditional classifications include, in narrative form, the rationale and explanation for the choices made. Nevertheless, one expert can dispute, but not test, the conclusions of another.

Once devised, a traditional classification is presented as a diagram called an *evolutionary tree,* or *family tree.* Originally, these often looked like thick bushes or sunbursts (Figure 2.4A). Nowadays (under the influence of the newer method described below) they tend to be trees pruned to two, or at most a few branches at each node (part B in the figure). The tree is usually displayed against the geologic time scale. Traditional classifications are also presented in tabular form (Figure 2.5, left, and back papers).

Traditional classification has its advantages. It is relatively stable. It uses relatively few and familiar terms for taxa and for their ranks. It is easily understood by laypersons and students because it extends their knowledge within a familiar paradigm.

Cladistic, or Phylogenetic Systematics Dissatisfaction with traditional systematics stems from its lack of rigor and the difficulty of postulating relationships when a fossil record is lacking or when numerous descendant groups seem to converge, somewhat vaguely, into a single ancestor. Cladistic systematics seeks to overcome these problems. The method was proposed in 1950 by the German entomologist Willi Hennig but started to become influential only when translated into English in 1966. Now, as elaborated over time, it is virtually the only method used by specialists and their students. The binomial system for naming species is retained, and groups of differing inclusiveness are again ranked to form hierarchies according to characteristics identified by specialists. Criteria and procedures, however, are different. In order to explain a first major difference, two contrasting terms must be introduced.

A lineage consisting of an ancestor and *all* of its descendants is said to be **monophyletic.** Thus, cartilaginous fishes (sharks and their relatives) have a common ancestor and no descendants that are not cartilaginous fishes. A group consisting of an ancestor and some *but not all* of its descendants is said to be **paraphyletic.** Thus, one (of several) ancestral groups of reptiles gave rise to various later reptiles (including crocodilians) and *also* to birds. In Figure 2.4B, groups IJ, HIJ, and GHIJK are each monophyletic. Group GHI is instead paraphyletic because descendants J and K are not included. Traditional systematists prefer monophyletic assemblages but accept some paraphyletic lineages; birds are considered to be a "natural" group, and not reptiles. In contrast, it is a major tenet of cladistic systematics that only monophyletic lineages (also called **clades**) are acceptable. Indeed, paraphyletic groups are not considered to be lineages at all and should not be named as taxa. Birds are not a class of vertebrates that evolved from certain reptiles; birds *are* reptiles of lesser rank than class.

Another tenet of cladistics relates to the sort of morphological attributes that are favored in the recognition of lineages. Here we must introduce some additional concepts and terminology. The presence or absence of a particular structural feature is termed a *character,* and

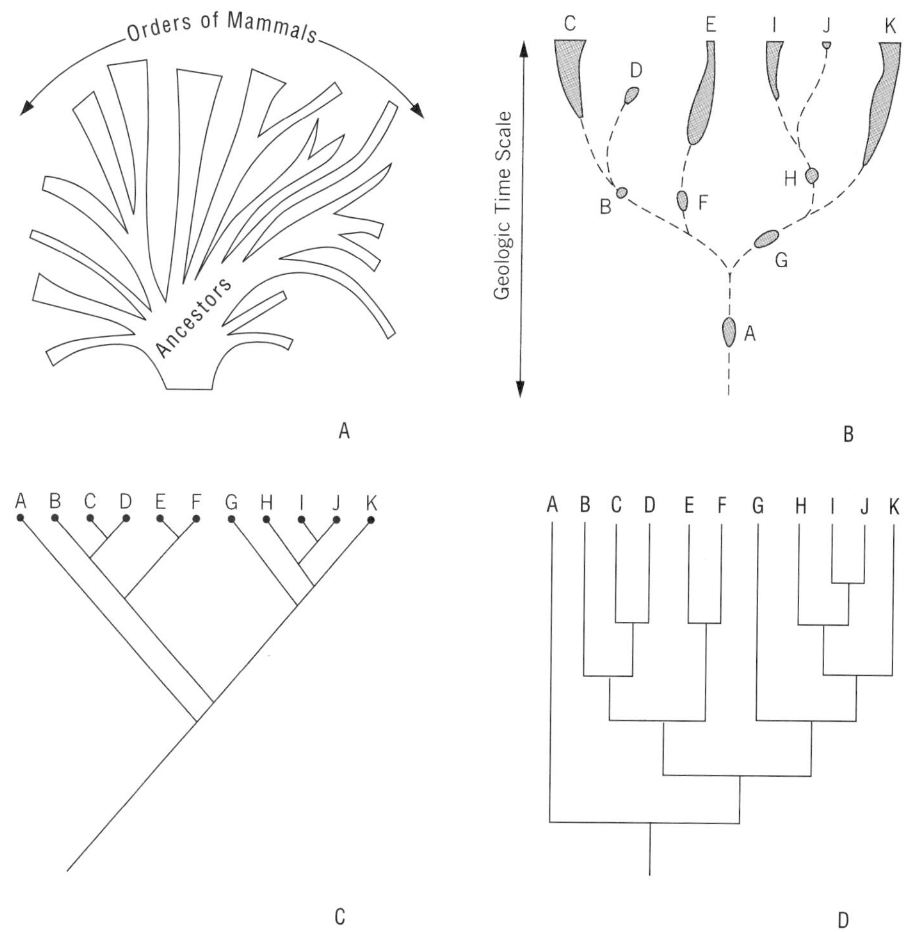

FIGURE 2.4
CONTRASTING WAYS OF DISPLAYING PHYLOGENIES. A, FAMILY TREE for the orders of mammals (names deleted) from *The Vertebrate Body*, by the eminent paleontologist A.S. Romer, the leading text in its field during the 1950s and 1960s. B, EVOLUTIONARY TREE typical of traditional classification. C and D, alternative conventional CLADOGRAMS for the same taxa shown in B.

the nature of a character is a *character-state.* Thus, the presence of canine teeth is a character of mammals not shared by fishes and reptiles. The enlargement of upper canines to form sabers is a character-state of certain mammals. Within a given lineage characters and character-states that were present early are said to be general, *primitive,* or (in the language of cladistics) **plesiomorphic** (= *close* + *form*). Character-states that evolved later in that lineage are advanced, *derived,* or **apomorphic** (= *separate* + *form*). Note that the terms must relate to a specific lineage: Among mammals the presence of canine teeth is primitive and their loss (rabbits, rodents) is derived; among tetrapods canines are derived. The condition of sharing primitive character-states by two or more lineages is called **symplesiomorphy** (= *with* + *close* + *form*). The condition of sharing derived character-states is **synapomorphy** (= *with* + *separate* + *form*). Traditional classification tends to emphasize shared primitive features, whereas cladistics stresses shared derived features. Consequently, the latter is more sensitive to change, process, and the relationships among descendant organisms.

In order to learn if an observed similarity within a particular group is primitive or derived, common or unique, several approaches are considered. For surviving taxa DNA analysis may be used, and comparative embryology may be revealing (following principles outlined on pp. 69–71). Otherwise, comparative morphology is used. The character-states of extinct members of a study group are carefully studied if there is a fossil record. The logic of cladistics does not require fossils, but they are used as available, and cladistic classification must be consistent with any lineage that is clearly (not ambiguously) supported by a fossil sequence. Finally, and importantly, cladistics make an *outgroup comparison* to see if a feature that is observed within a lineage under study extends to another group that is close, but not part of that group. This procedure helps the systematist decide which features of the lineage, or clade, under study can contribute to the definition, or identification of that clade.

Phylogenetic systematists believe that every clade evolved by dichotomous splitting from a single **sister group.** Sister groups share a common ancestor; together they constitute a monophyletic higher taxon. (One is often used as the outgroup for the other.) The objective is to identify a succession of nesting sister groups as one goes to more inclusive levels of the evolutionary hierachy. The pattern of branching is called a *cladogram.* Each cladogram is an evolutionary hypothesis. Usually several cladograms would fit the available data, in which case the most parsimonious alternative (i.e., the most direct as to evolutionary steps or consistent with character traits taken in combination) is accepted as the preferable solution. (*All* classifications are provisional pending the discovery of new fossils or other evidence.)

The two conventional ways of depicting a cladogram are shown in Figure 2.4C and D. They can be, but usually are not, shown against the geologic time scale. Hence, they usually do not show time of origin or survival. Importantly, keyed to each branch point of a cladogram is a list of at least some of the characters upon which it is based (Figures 2.3, 3.13, 4.10, 9.21, and 30.28). It follows that cladistics not merely displays phylogenies, but is the most powerful method of establishing phylogenies. Phylogenetic classifications are also presented in tabular form (Figure 2.5, right). Successive dichotomous branchings of

- Subphylum Vertebrata (vertebrates)
  - Class Agnatha (jawless vertebrates)
    - Subclass Myxinoidea (hagfishes)
    - Subclass Cephalaspidomorpha (cephalaspids)
      - Order Petromyzontia (lampreys)
  - Class Chondrichthyes (cartilaginous fishes)
    - Subclass Elasmobranchii (sharks and rays)
    - Subclass Holocephali (chimaeras)
  - Class Osteichthyes (bony fishes)
    - Subclass Actinopterygii (ray-finned fishes)
      - Infraclass Chondrostei (sturgeons and paddlefishes)
      - Infraclass Neopterygii (gars, bowfins, and teleosts)
    - Subclass Sarcopterygii (lobe-finned fishes)
      - Infraclass Actinistia (coelacanths)
      - Infraclass Dipnoi (lungfishes)
  - Class Amphibia (amphibians)
    - Subclass Lissamphibia
      - Order Apoda (limbless amphibians)
      - Order Anura (frogs and toads)
      - Order Urodela (salamanders)
  - Class Reptilia (reptiles)
    - Subclass Testudinata (turtles)
    - Subclass Diapsida (diapsids)
      - Infraclass Lepidosauria (tuatara, lizards, and snakes)
      - Infraclass Archosauria (archosaurs)
        - Order Crocodylia (crocodilians)
  - Class Aves (birds)
  - Class Mammalia (mammals)

- Craniata (craniates)
  - Hyperotreti (hagfishes)
  - Vertebrata (vertebrates)
    - Hyperorartia (lampreys)
    - Gnathostomata (jawed vertebrates)
      - Chondrichthyes (cartilaginous fishes)
        - Elasmobranchii (sharks and rays)
        - Holocephali (chimaeras)
      - Osteichthyes (bony fishes and tetrapods)
        - Actinopterygii (ray-finned fishes)
          - Chondrostei (sturgeons and paddlefishes)
          - Neopterygii (gars, bowfins, and teleosts)
        - Sarcopterygii (lobe-finned fishes and tetrapods)
          - Actinistia (coelacanths)
          - Unnamed taxon
            - Dipnoi (lungfishes)
            - Tetrapoda (four-legged vertebrates)
              - Amphibia (amphibians)
                - Apoda (limbless amphibians)
                - Batrachia (salamanders and frogs)
              - Amniota (amniotes)
                - Mammalia (mammals)
                - Reptilia (reptiles)
                  - Testudinata (turtles)
                  - Diapsida (diapsids)
                    - Lepidosauria (tuatara, lizards, and snakes)
                    - Archosauria (archosaurs)
                      - Crocodylia (crocodilians)
                      - Aves (birds)

FIGURE 2.5 CLASSIFICATION OF LIVING VERTEBRATES ACCORDING TO TRADITIONAL (left) AND PHYLOGENETIC (right) PROCEDURES.

COMMENT 2.1

THE PROBLEM OF NAMING SISTER GROUPS

The traditional tabular classification uses successive indentations for the successive levels, or ranks, of a hierachy—four to classify living vertebrates, as shown in Figure 2.5, left. Phylogenetic classification requires an indentation for each level of dichotomous branching—11 in Figure 2.5, right. If the array is enlarged to include extinct lineages, the number of indentations is increased moderately for traditional classifications and significantly for phylogenetic classifications.

Every taxon needs a name in each kind of classification. In traditional classification, each rank (each level of indentation) also has a name (class, subclass, etc.). What about phylogenetic classification, where there may be so many levels? Usually rank names are simply omitted as inconsistent with the principles followed. This omission, coupled with the multiple levels, can complicate interpretation for the general student. Scanning a traditional classification, one quickly finds taxa of equal rank (e.g., bony fishes, amphibians, mammals) and infers that they have somewhat equal "weight" or importance. From a scan of a phylogentic classification, mammals *seem* to be subordinate not only to bony fishes and amphibians, but also to numerous lesser groups. If one wishes to teach or study only the major groups, it can be difficult to pick them out.

Cladistics still needs names, lots of names, for its lineages. Note that at many branchings, one name is needed for a smaller sister group (perhaps a genus or family) and, to preserve monophyly, another name for a sister group that includes everything else not already shown in the classification. A few "everything else" names were already established: **Gnathostomata** (jawed vertebrates), **Choanata** (air breathers), **Tetrapoda,** and **Amniota.** Some phylogenetic classifications make frequent use of the words "unnamed taxon" (position holders being more needed than actual names). Often, long-established names are given new meanings, and this can be confusing. Thus, in Figure 2.5, right, bony fishes (**Osteichthyes**) means bony fishes *plus* everything to follow. Familiar names can also become less, rather than more, inclusive. For instance, surviving amphibians plus a major clade of extinct amphibians (**temnospondyls**—see Chapter 3) may be termed "amphibians," whereas another major extinct clade (**anthracosaurs**) is excluded. Since the latter are not included among reptiles (their descendants), they are not assignable to any class. Phylogenetic classifications often include genera and families that provisionally do not fall within identified higher taxonomic levels.

These are nonproblems to specialists and advanced students; the study of evolution is well served. Others may find that phylogenetic classifications are relatively difficult to learn and talk about.

---

the corresponding cladogram are shown by successive indentations. Procedures for selecting and assessing character-states are precisely defined. Numerous books present the methods (see References), and computer programs are available. We note some disadvantages of cladistics in Comment 2.1.

The classification followed in this book (see back papers, and Chapters 3 and 4) is in part a compromise, but is greatly influenced by the principles of cladistics: We use dichotomous branching in our "tree" (front papers) and avoid various groupings once widely used but now believed to be paraphyletic. Nevertheless, we retain the familiar names and their meaning for the classes (see Comment 2.1), and include a classification in the traditional style (back papers).

REFERENCES

Alverez, W., and F. Asaro. 1990. An extraterrestrial impact. *Sci. Am.* 263(4):78–84.

Carroll, R.L. 1997. *Patterns and processes of vertebrate evolution.* Cambridge Univ. Press, New York. 448p. See chapter 7 on the influence of systems of classification on concepts of evolutionary patterns.

Courtillot, V.E. 1990. A volcanic eruption. *Sci. Am.* 263(4):85–92. Volcanism in relation to mass extinctions.

Eldredge, N., and J. Cracraft. 1980. *Phylogenetic patterns and the evolutionary process.* Columbia Univ. Press, New York. 349p.

Endler, J.A. 1986. *Natural selection in the wild.* Princeton Univ. Press, Princeton, NJ. 136p.

Kitching, I.J., et al. 1998. *Cladistics: the theory and practice of parsimony analysis.* 2nd ed. Oxford Univ. Press, New York. 228p.

Lauder, G.M., et al. 1995. Systematics and the study of organismal form and function. *Bioscience* 45:696–704. Emphasizes the need for the incorporation of phylogenetic principles into studies of comparative morphology and physiology.

McKenna, M.C., and S.K. Bell. 1997. *Classification of mammals above the species level.* Columbia Univ. Press, New York. 631p. Part I is an exhaustive review of traditional and phylogenetic approaches to classification of mammals.

Novacek, M.J. 1992. Mammalian phylogeny: shaking the tree. *Nature* 356:121–125. Review article on molecular and morphological evidence.

Ridley, M. 1986. *Evolution and classification: the reformation of cladism.* Longman, New York. 201p.

Schoch, R.M. 1986. *Phylogenetic reconstruction in paleontology.* Van Nostrand Reinhold, New York. 351p.

Chapter 3

Fishes

In presenting the patterns of structure that characterize vertebrate animals, it will be necessary to refer often to the major groups of vertebrates. The classifications in the front papers and back papers present group names; now it is desirable to relate those names to animals. A brief introduction to the more obvious characters serves this initial purpose. More will be learned about the various taxa as Parts II and III of the text unfold, but it is desirable now that the student be able to visualize a representative of each group, recall several distinguishing group characteristics, and know the place of each group among the other vertebrates. If extinct and unusual animals seem to be stressed in this and the following chapter, it is not to give them special emphasis, but because one needs little introduction to animals that are already familiar.

CLASS AGNATHA

All animals that have the general characters of the subphylum Vertebrata (as reviewed in Chapter 2), but do *not* have jaws, belong to the class Agnatha. These animals are little known to most people because all of them except the lampreys and hagfishes became extinct more than 200 million years ago. Even lampreys and hagfishes are rarely seen by the public; they sometimes destroy fishes of commercial value, but otherwise are obscure. Agnathans are, nevertheless, fascinating creatures. The morphologist should become acquainted with the animals in this class because they were the first known vertebrates to evolve, and because the most primitive body plan of a living vertebrate is that of ammocoetes, the larva of the lamprey.

The jawless vertebrates have advanced over their progenitors among the protochordates in the possession of several important characters. With some equivocation for the earliest representatives (see below), all have heads with a cranium, a brain, and paired organs of sight (vestigial in hagfishes). There is a large persistent notochord. Vertebrae are at least represented in most. The internal skeletons of jawless vertebrates are largely cartilaginous, but in tiny toothlike structures of one extinct group of Agnatha, and in the heavy scales and armor of most others, some combination of dentine, enamel, and bone is

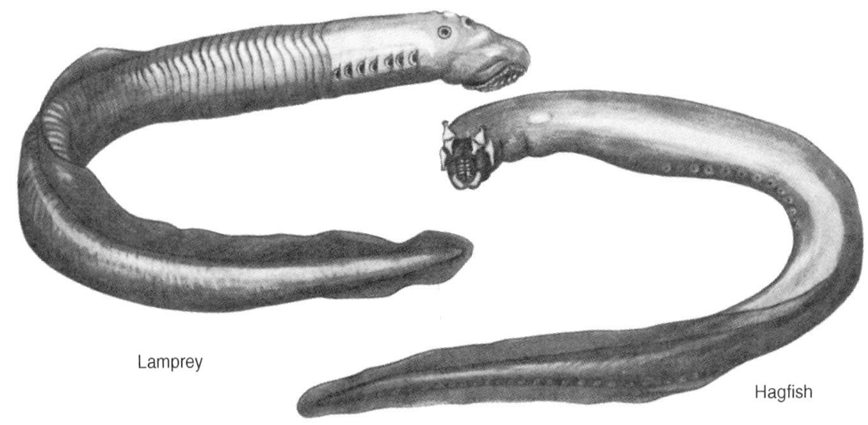

FIGURE 3.1 REPRESENTATIVE CYCLOSTOMATA showing attenuated form, rasping mouth parts, and absence of jaws, scales, and paired appendages.

present for the first time in earth history. Extinct agnathans having bony scales are often called **ostracoderms,** which means "shell skins."

The Agnatha are also noteworthy for the lack of certain characters that became typical of vertebrates higher on the evolutionary ladder: They do not possess jaws, true teeth, girdles, or typical appendages. Pectoral spikes, folds, or lobes are often present; pelvic fins are never found. The gills of surviving agnathans are located in pouches.

The extinct agnathans were all small, fishlike animals. Most lived in freshwater streams, but they are first known from fossils in marine deposits. They were abundant and diverse. Agnathans are probably not monophyletic; the term is a convenience pending a better understanding of their origins and relationships. More than a dozen orders are recognized by some specialists; the treatment here is conservative.

Later chapters will refer to jawless vertebrates in relation to the evolution of lungs, tail shapes, some special senses, scales, vertebrae, and fins.

Cyclostomata and the Subclass Myxinoidea The only surviving agnathans are the LAMPREYS and HAGFISHES (Figure 3.1). These eel-like animals have naked, slippery skin. They share such primitive characters as a persistent notochord and the absence of complete vertebrae, paired appendages, and jaws. Lampreys are semiparasitic upon bony fishes; hagfishes are primarily scavengers. Their mouths and "tongues" are adapted for holding onto their hosts and for rasping or cutting into flesh. Their skeletons are cartilaginous. Sense organs, body form, and digestive structures are also modified in response to this way of life. Because hagfishes and lampreys share these traits, and are the only survivors of their class, it is convenient to refer to them together as Cyclostomata (= *round* + *mouth*). Nevertheless, the two groups differ so markedly in the nature of their skeletons, pouched gills, sense organs, and some other characters that their relationship seems remote. In phylogenetic classifications hagfishes are designated the sister group of all animals to follow, which are then called Vertebrata, the subphylum (Vertebrata plus hagfishes) becoming Craniata. Without questioning the relationships implied, we tentatively retain the more familiar terminology: Vertebrata as the subphylum, Agnatha as a class, and Myxinoidea ("myx" = *slime*) for hagfishes as the first subclass.

Hagfishes are exclusively marine, living in burrows in soft sediments. Their fossil record is scant; it is probable that they evolved in the Lower Cambrian, at least 550 million years ago.

Subclass Petromyzontia Lampreys have no close relatives. Some experts have placed them near anaspids (see below) in the same subclass (Cephalaspidomorpha). We follow Donoghue, Forey, and Aldridge (2000) in finding them to be more primitive than any other agnathans except

COMMENT 3.1

CONODONTS AND OTHER ANCIENT VERTEBRATES

Conodonts are toothlike structures, usually several millimeters in length, that are abundant as fossils (more than 200 genera are named!) for 300 million years after appearing in the late Cambrian. Some are simple cones ("conodont" = *cone + tooth*) and others complex. Their changing configurations over time are so well known that they enable geologists to age the rocks in which they occur. Until 1982, however, nothing was known of the animals (also called conodonts) that once owned them. From faint impressions on rocks, we now know that the animals were soft-bodied, worm-shaped, laterally compressed, and often about 7 cm long, and that the toothlike structures were part of a feeding apparatus located in the head and seemingly adapted (according to species and position) for grasping, shearing, or crushing macroscopic prey. The animals had segmented muscles on the body, large eyes with extrinsic muscles, and a caudal fin with radial fin supports. It is probable that there was a brain, gill apparatus, and a notochord, but these have not been clearly identified. The complex histology of the toothlike structures is disputed. They probably contain dentine, or dentine plus enamel, but not bone. These vertebrates are now considered by most specialists to be more advanced than hagfishes and lampreys (see the article by Donoghue, Forey, and Aldridge in References).

In 1999 several genera of other ancient chordates were described from the Lower Cambrian of China, at least two of which are vertebrates. The fossils are mere imprints in fine sediments (there are no hard tissues), making interpretation of some features uncertain. One or both of the small, fusiform vertebrates appear to have had head and trunk regions, a cartilaginous cranium, a notochord, five or more gills, segmented muscles, a heart, a dorsal fin, and (probably) paired ventrolateral fins. A brain and eyes are neither identified nor ruled out.

These records of early vertebrates establish the antiquity of the subphylum but do not provide enough undisputed information to indicate *their* origins or to "connect the dots" of the ancestral tree. With patience, and some luck, more fossils will be found that allow the picture to emerge.

hagfishes. Their general features are noted above under the Cyclostomata. The lamprey is commonly dissected in the comparative anatomy laboratory. We are fortunate to have this animal to show us something of the soft tissues of the class, even though we cannot tell how representative they may be. This book will cover lampreys for what we believe to be the primitive nature of their axial musculature, circulatory system, release of sex cells, eye structure, and other parts.

The larva of the lamprey, ammocoetes, has more ancestral characters than the adult (compare Figures 3.1 and 3.2). This is particularly true in regard to mouth parts, pharynx, gonads, and some digestive organs. Ammocoetes becomes adult by metamorphosis.

Subclass Pteraspidomorpha This diverse group of agnathans appears in the fossil record more than 500 million years ago, and 60 million years earlier than any other group of vertebrates except conodonts. Its principal representative, the PTERASPIDS (order Heterostraci) have a heavy shield of armor covering the head and anterior part of the body (Figure 3.3). Most have a rostrum projecting over the mouth, and often there are bizarre spines on the shield ("pteraspid" = *wing + shield*). There are no paired fins, the eyes are lateral, and, unlike most other agnathans, there are two nostrils and a common exit from all of the gill pouches. The head is not markedly flattened dorsoventrally. One infers that pteraspids were slow swimmers but not bottom feeders. Most were marine.

Subclass Cephalaspidomorpha CEPHALASPIDS (order Osteostraci) are relatively well known for animals so long extinct; thanks to ossification of the endoskeleton the patterns of nerves, blood vessels, and sensory structures of the head are preserved in considerable detail for some genera. They had depressed heads and dorsal eyes—character-states indicating that these odd creatures moved slowly along stream bottoms to take food found in mud or sand (Figure 3.4). Cephalaspids are unique among the Agnatha for their lobelike pectoral projections. Paired lateral, and median dorsal sculptured areas on the head shield are probably sense organs. The mouth is small and ventral. The upper lobe of the tail is larger than the lower lobe.

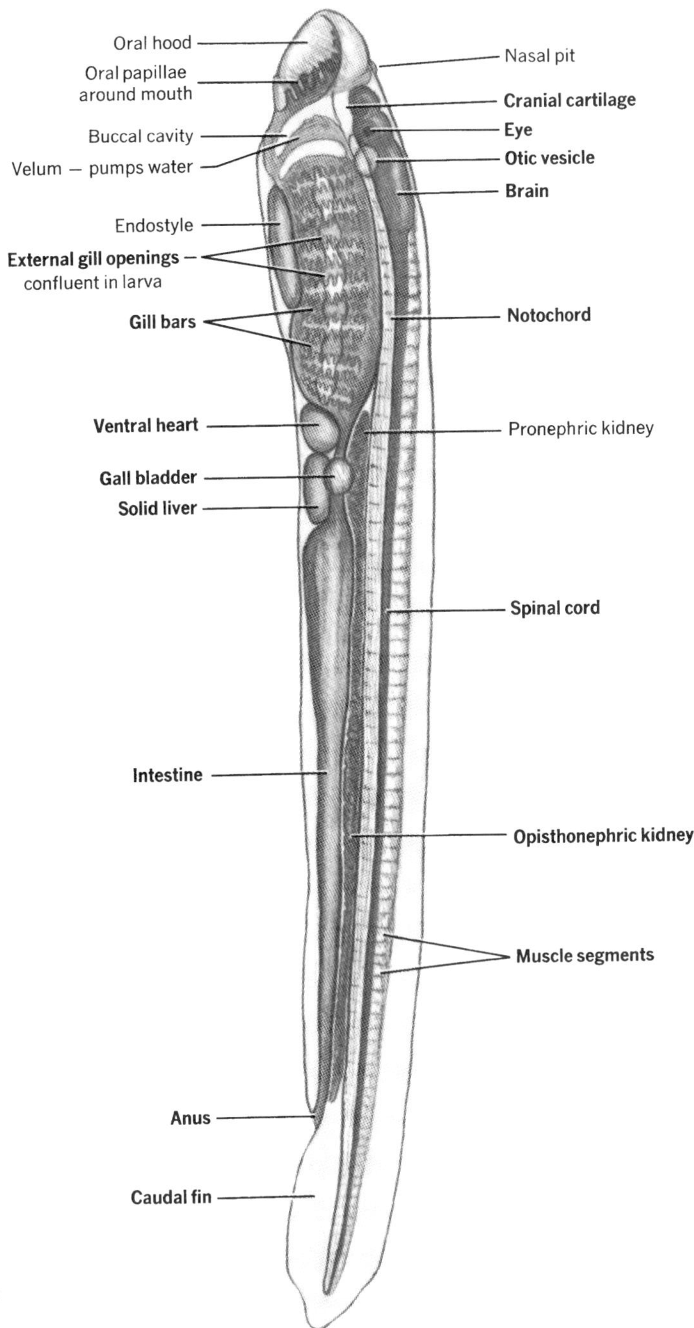

FIGURE 3.2
AMMOCOETES, larva of the lamprey, drawn from a cleared, 12 mm specimen to illustrate internal structure. Primitive and unspecialized vertebrate characters are shown by boldface labels; other characteristics are shown by standard labels.

ANASPIDS (order Anaspida), also long extinct, are the only jawless vertebrates having streamlined form. They have small, platelike scales and no head armor ("anaspid" = *without* + *shield*). Their eyes are lateral. These features indicate that they were relatively active swimmers. They have pectoral spikes, behind which are long, thin, paired fins that may have undulated to provide accurate maneuvering. Like cephalaspids and cyclostomes they have many separate gill openings, but unlike cephalaspids the openings are lateral instead of ventral.

Some Other Agnathans Five other groups of jawless vertebrates are commonly given equal rank to those described above, yet we pass over them with this brief mention because we do not return to them in chapters to follow. The Arandaspida and Astraspida

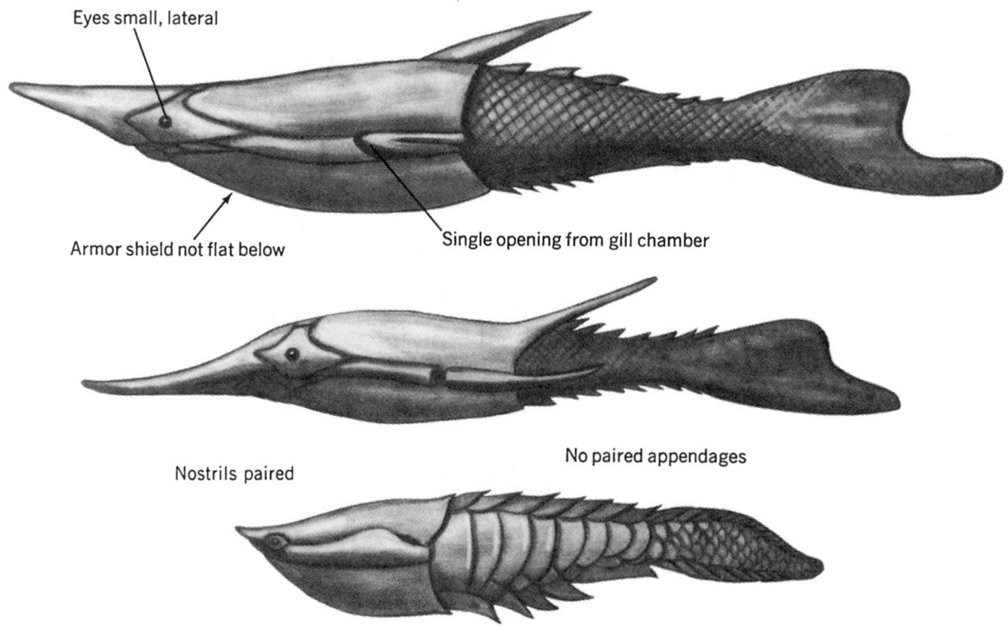

FIGURE 3.3 Restorations of REPRESENTATIVE PTERASPIDOMORPHA of the order Heterostraci. Actual size of these examples is 6–24 cm.

are of note because, other than the enigmatic conodonts, they are the only known vertebrates to flourish in the Ordovician. They had large head shields and no fins except the caudal fin. The Galeaspida had massive, one-piece shields and, like Osteostraci, acellular bone and some ossification of the internal skeleton. Pituriaspida is a poorly known group with a long rostrum on the head armor. Finally, the Thelodonti had minute, sharklike scales over the entire body. Their phylogenetic position is speculative.

JAW-BEARING FISHES

The most important evolutionary advance common to all the remaining fishes was the enlargement and adaptation of the first gill arch to function as jaws instead of gill supports. Chordates that lacked jaws could only filter microorganisms from water, grovel in mud, or rasp and cut algae or soft flesh. The evolution of jaws permitted fishes and their descendants to utilize larger and harder food, and thus enabled them to become adapted to many new and diverse ways of living. This advance was of sufficient importance that fishes and tetrapods are together called *gnathostomes* (= *jaw* + *mouth*) to set them apart from the Agnatha (= *without* + *jaw*).

A second important advance common to all jaw-bearing fishes is the presence of paired appendages. True, some agnathans had fin folds (anaspids) or lobes (cephalaspids), but it remained for jaw-bearing fishes to experiment with various types of paired appendages and perfect several models.

Most of the world's vertebrates—past and present—are jaw-bearing fishes. The diversity of their habits, form, and structure is enormous. They are important to the morphologist for the advancements and specializations found in nearly every organ system. Four classes are recognized here.

CLASS PLACODERMI

Most placoderms have either bony scales or armor plates, particularly on the forward part of the body—the word *placoderm* means "plate skin." The head is joined to the body by a hinge in the armor. These fishes, like jawless vertebrates, have persistent notochords. The

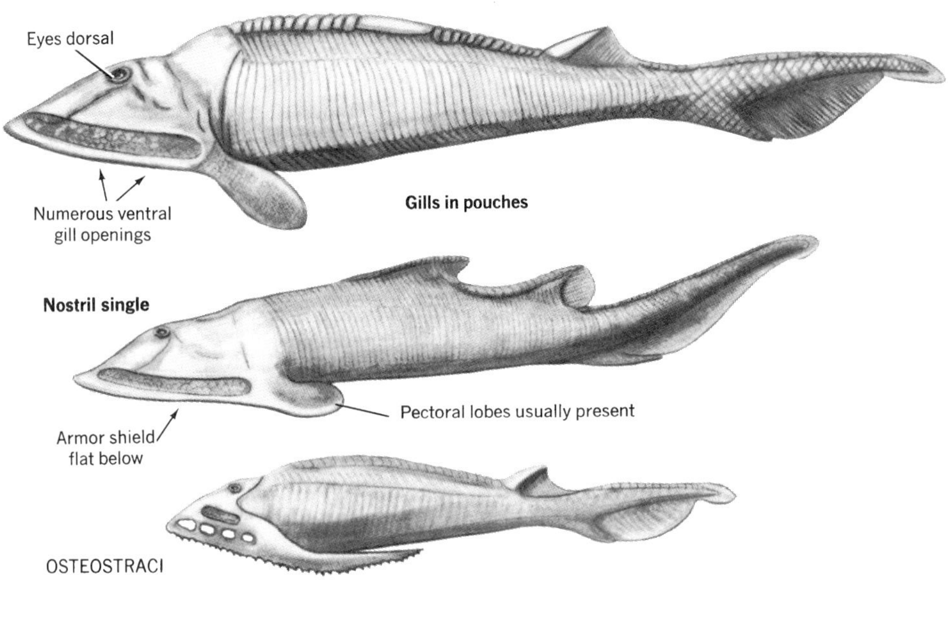

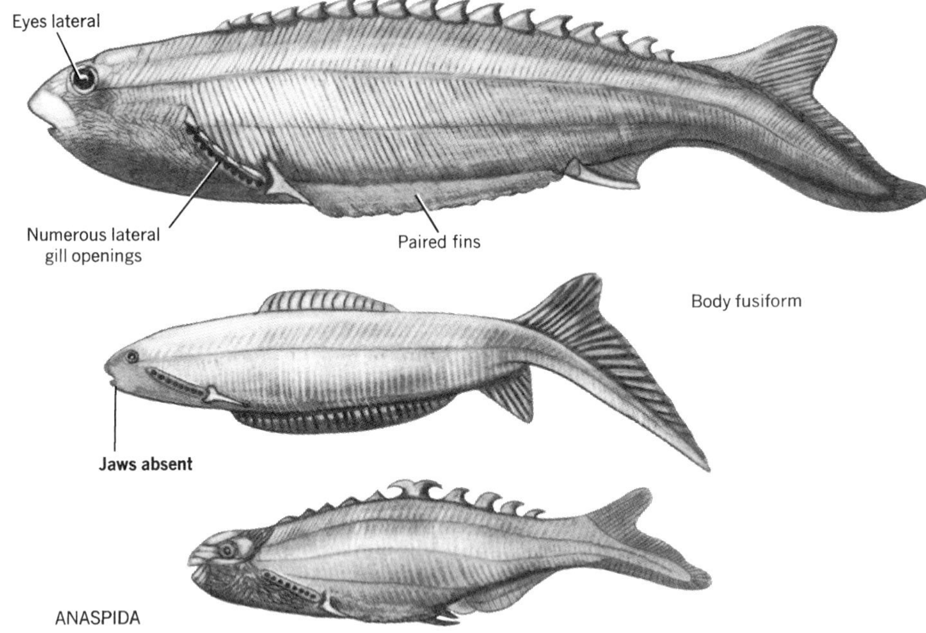

FIGURE 3.4 Restorations of REPRESENTATIVE AGNATHA OF THE SUBCLASS CEPHALASPIDOMORPHA illustrating some characteristics of the subclass (boldface labels) and of the orders Osteostraci and Anaspida (standard labels). Actual size of these examples is 10–24 cm.

internal skeleton contains some bone, and the gills are below, rather than behind, the braincase. Placoderms swam first in rivers and later in the oceans of the world for 60 million years. They reached maximum abundance about the time amphibians evolved, and became extinct some 350 million years ago, before reptiles, birds, and mammals had appeared.

These remote creatures are not well known to the general biologist, and few people have heard of them. Nevertheless, the comparative anatomist should not entirely neglect the placoderms: They were among the first fishes to evolve two of the most successful of all vertebrate structures—jaws and paired appendages—and they pioneered gas bladders that were ultimately to become lungs. Features of their scales and gill structure are noted in later chapters.

Placoderms are diverse in structure; the evolutionary relationships within the class, and between this class and others, have not been worked out. Five to nine orders are recognized. Two of the best-known groups of placoderms are the arthrodires and antiarchs. Each is worldwide in distribution. The class, as known, is not regarded as being ancestral to other classes. It seems more closely related to cartilaginous fishes than to bony fishes.

Typical ARTHRODIRA are the most spectacular of placoderms because most have large, predaceous jaws with serrated margins, and some, being 6–9 m long, were the largest vertebrates that had yet evolved when they roamed the rivers and oceans of the world (Figure 3.5). Most arthrodires have blunt heads and lateral eyes; they have heavy cephalic

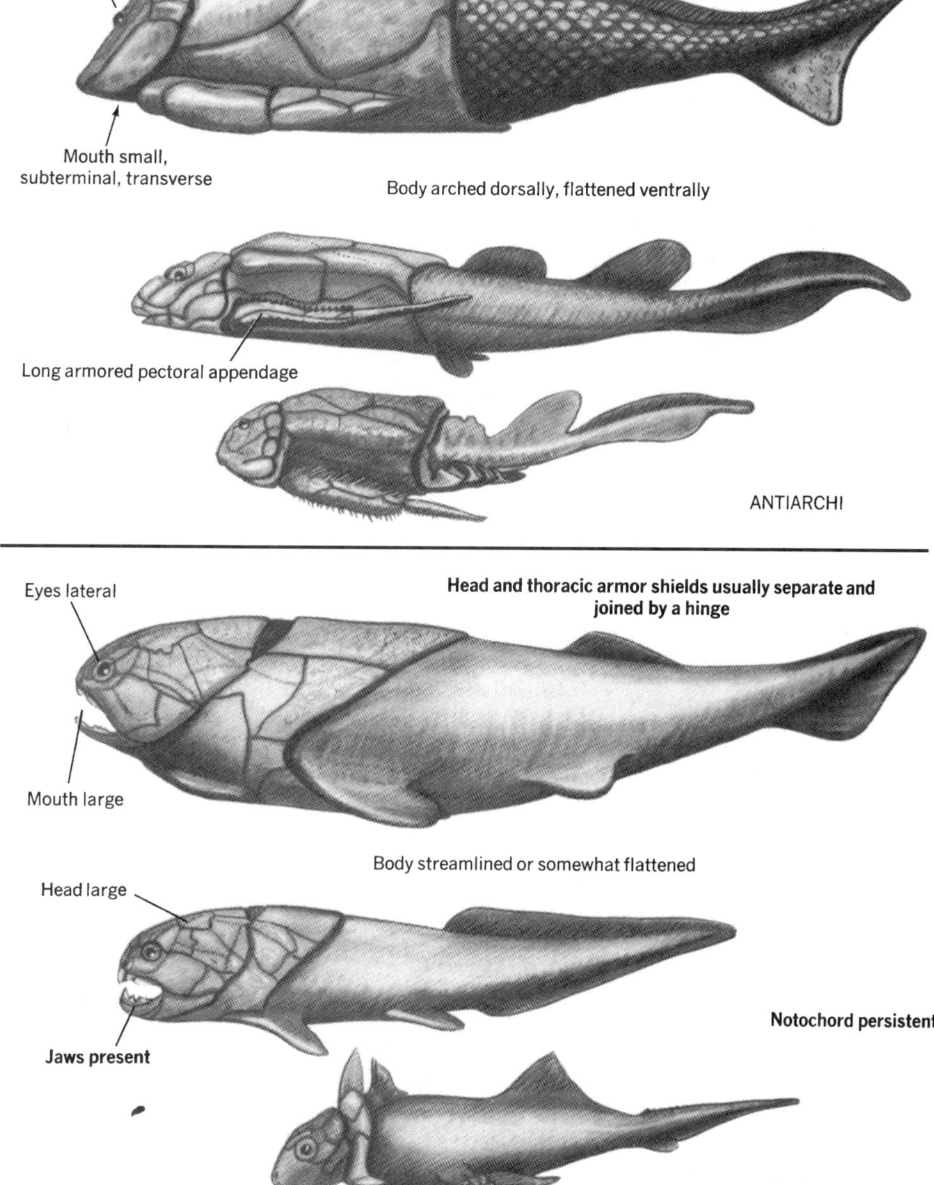

FIGURE 3.5 Restorations of REPRESENTATIVE PLACODERMI illustrating some characteristics of the class (boldface labels) and of the two principal orders (standard labels).

and thoracic shields that are hinged together ("arthrodire" = *joint* + *neck*), and the gills are usually hidden by the cephalic armor.

ANTIARCHI are the most bizarre of placoderms because of their peculiar pectoral appendages ("antiarch" = *opposite* + *arm*). These long structures are hinged to the body and covered with bony plates. It is thought that they functioned as holdfasts against the currents of streams, or were used for creeping along stream bottoms. A bottom-living habit is also suggested by the dorsal position of eyes and olfactory organs, and by the arched dorsal contour and flat ventral surface of the body. The small head and bulky thorax are encased in units of heavy armor. The jaws of antiarchs are small and weak. Overall body size is 15 to 40 cm.

CLASS CHONDRICHTHYES

This class includes the familiar sharks and rays, the less familiar but striking chimaeras, and extinct relatives of these fishes. Like the arthrodires before them, the Chondrichthyes are predominantly marine and are of medium to large size. They differ from placoderms and also from most other fishes in having calcified cartilage, but little or no bone internally, and rarely any in their scales ("Chondrichthyes" = *cartilage* + *fishes*). Cartilaginous fishes are also distinctive for their solid braincase, fin structure, branching pattern of blood vessels associated with the gills, and small toothlike scales (or none at all). Their teeth, unlike those of other fishes, are anchored to the integument and occur only at the margins of the jaws. Most of these fishes have a series of external gill openings and lack a gas bladder.

Cartilaginous fishes appear in the record some 30 million years later than bony fishes and have not been ancestral to any other class. Nevertheless, cartilaginous fishes do retain from their ancestors some features that are more primitive or unspecialized than the corresponding features of bony fishes. Examples are the structure of the heart and brain, and musculature of the pharynx.

Subclass Cladoselachii Of the several major extinct groups of cartilaginous fishes, we mention only the best known and most primitive, the cladoselachians. They were abundant in the Upper Devonian, when the first amphibians may have ventured onto land. These large, predaceous marine fishes look much like sharks except that the mouth is nearly terminal and the large tail is nearly symmetrical (Figure 3.6). This subclass is probably the sister group of other cartilaginous fishes.

Subclass Elasmobranchii Elasmobranchii (like cladoselachians) have slitlike external gill openings ("elasmobranchs" = *plate* + *gills*). We include only one of several extinct groups.

PLEURACANTHODII (also called Xenacanthodii) are freshwater fishes, nearly a meter in length, that became extinct about the time mammals evolved. A large spine borne on the back of the head is a handy recognition feature. The persistent notochord in this order and the next to be discussed is mounted by somewhat calcified vertebral arches. Pleuracanths will be mentioned in later chapters for two other characteristic features that are unusual among fishes: Their paired fins have a jointed central axis, and the axis of the tail is straight all the way to the tip.

The sharks and rays comprise the order SELACHII. All surviving cartilaginous fishes having a series of external gill slits and small abrasive scales are selachians. The first ancestral gill slit is reduced to a roundish opening called a **spiracle.** These are mostly marine fishes of medium to large size. Sharks and rays (commonly placed in about four suborders) differ markedly from one another in their habits. Many sharks are active, predaceous fishes with fusiform bodies and large, strong tails. Their paired fins have restricted bases and their gill slits are lateral in position. Most rays, in contrast, spend much time resting on the bottom or swimming sluggishly along in search of shellfishes or other relatively inactive food. Their bodies are flattened dorsoventrally and the pectoral

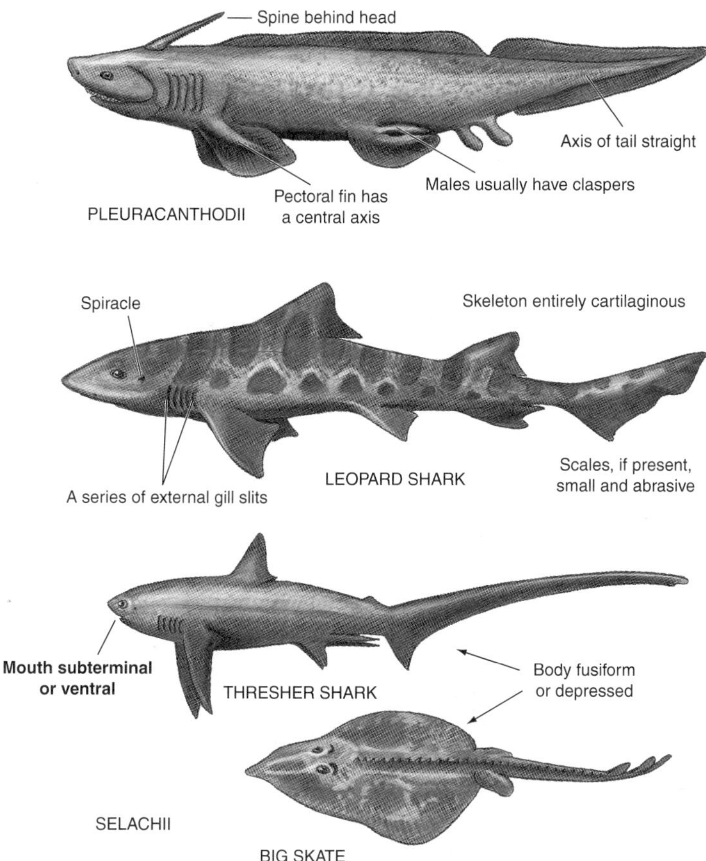

FIGURE 3.6 REPRESENTATIVES OF THE SUBCLASSES CLADOSELACHII AND ELASMOBRANCHII illustrating some characteristics of the subclasses (boldface labels) and orders (standard labels).

fins merge into the head and body. The mouth, being ventral, cannot take in water when the fish is at rest on sand or mud, so large spiracles perform this function instead.

Subclass Holocephali Surviving members of this subclass are called chimaeras. Found only at sea, they rarely come to public attention and will be mentioned infrequently in the chapters that follow. These are the only cartilaginous fishes with a fleshy **operculum** covering the gills (Figure 3.7). There are few or no scales. The notochord is persistent, and there is no spiracle. Males have a unique club-shaped clasping organ on the top of the head. The large head and eyes, huge pectoral fins, whiplash tail, and striking colors give chimaeras a bizarre appearance. We will pay attention to the structure of their solid jaws, crushing teeth, and certain sense organs.

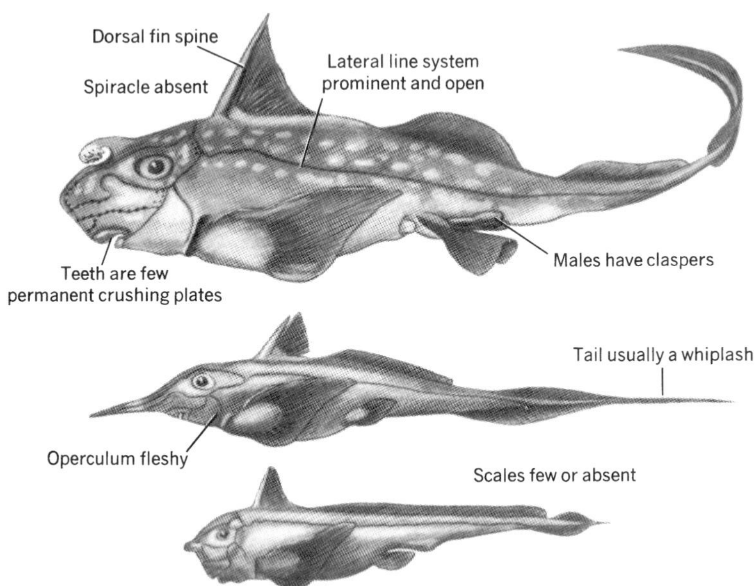

FIGURE 3.7 REPRESENTATIVE HOLOCEPHALI illustrating some characteristics of chimaeras.

CLASS ACANTHODII

Acanthodians evolved before known placoderms, cartilaginous fishes, or bony fishes. Over the years they have been included in each of those three classes, but now, although usually considered closest to bony fishes, they are elevated to a class of their own (Figure 3.8).

Acanthodians have streamlined bodies, large lateral eyes, and wide mouths. Their heads are bony and their small, distinctive scales thick and hard, but they do not have the armor of most of their jawless and placoderm contemporaries. We infer from these clues that these small, mostly freshwater fishes were active swimmers. The numerous fins of acanthodians are unique in that each has a thin membrane supported at its leading edge by a long, stout spine. This feature gives the subclass its name: "acanthodian" means "spine form."

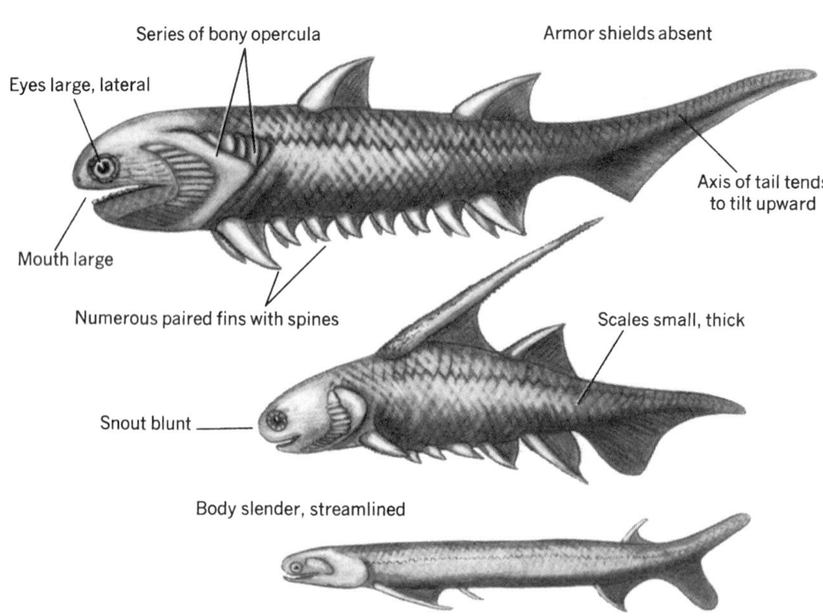

FIGURE 3.8 Restorations of REPRESENTATIVE ACANTHODII illustrating some characteristics of the class. Actual size of these examples is 8–12 cm.

CLASS OSTEICHTHYES

Bony fishes evolved from obscure ancestors a little more than 400 million years ago. For some 150 million years thereafter they were outnumbered, first by placoderms and then by cartilaginous fishes, but since the close of the Permian period, bony fishes have dominated the waters of the world. Since the end of the Mesozoic era, they have been the most abundant of all vertebrates. Their habits and structure are seemingly as diverse as adaptation to an aquatic life will permit.

Most fishes of this class have bone in their skulls, vertebrae, girdles, fin supports, and scales. Some have secondarily substituted cartilage for much of the ancestral bone, yet even such fishes retain more bone in their internal skeletons than is present in other classes of fishes. Bony fishes are the only vertebrates having the gills of each side of the body in a common chamber covered by a movable, bony operculum. They have various sorts of fins, scales, and vertebrae, yet these structures nearly always differ from those of other classes. The pectoral girdle is joined to the skull by a chain of bones. A lung or gas bladder is usually present.

Subclass Actinopterygii Most bony fishes belong to the subclass Actinopterygii, or ray-finned fishes. The membranes of the paired fins are supported by bony rays that radiate from the fin base. Consequently, the fins do not have fleshy stalks, as do the fins of fishes in the next subclass. The pattern of cranial bones and the nature of the venous system and reproductive ducts are also distinctive and clearly show that these fishes do not include the ancestors of land vertebrates.

The classification of ray-finned fishes has undergone much change in recent years and is still debated. Most specialists now recognize two principal groups, here designated infraclasses. The groups differ from one another in regard to degree of ossification of the skeleton; presence or absence, mobility, or pairing of certain bones of the skull and pectoral girdle; shape of dorsal and tail fins; and presence or absence of a spiracle.

The infraclass CHONDROSTEI includes the long-extinct but significant ancestral order of palaeoniscoid fishes, which was ancestral to other ray-finned fishes, and the surviving sturgeons and paddlefishes, which will be mentioned in regard to the axial skeleton and feeding (Figure 3.9 and figure on p. 557). Also included are the African bichirs, of which the genus *Polypterus* is the best known. Its dorsal fin is distinctive ("Polypterus" = *many* + *fins*). Bichirs are of special interest because of the primitive nature of their scales and gas bladder.

The infraclass NEOPTERYGII includes the gars and bowfins (formerly classed as holosteans) (Figure 3.10) and the huge assemblage of teleost fishes (Figure 3.11). About 30,000 species and subspecies are arranged in 400 to 600 families and are grouped

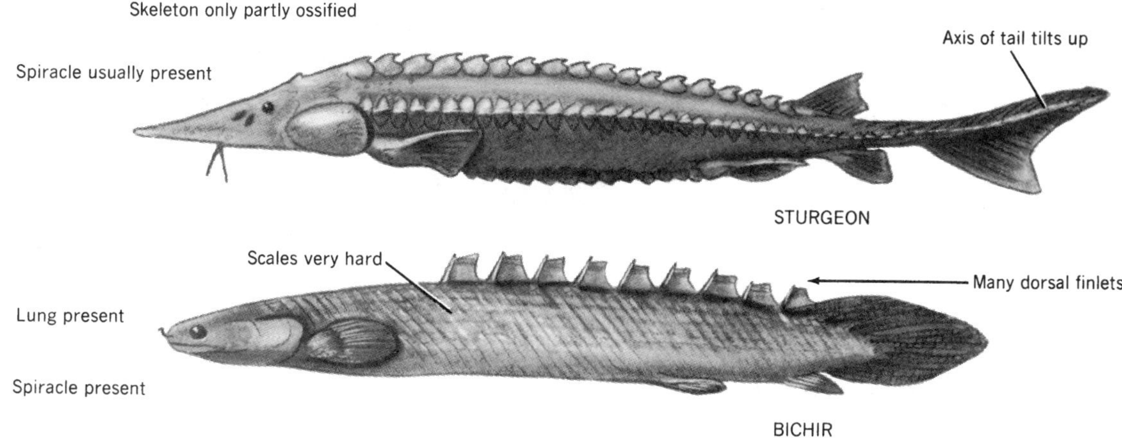

FIGURE 3.9 REPRESENTATIVE CHONDROSTEI illustrating some characteristics of two survivors.

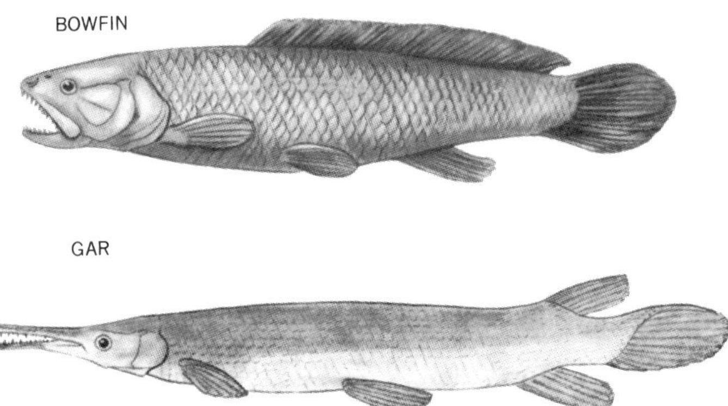

FIGURE 3.10 PRIMITIVE SURVIVING NEOPTERYGII.

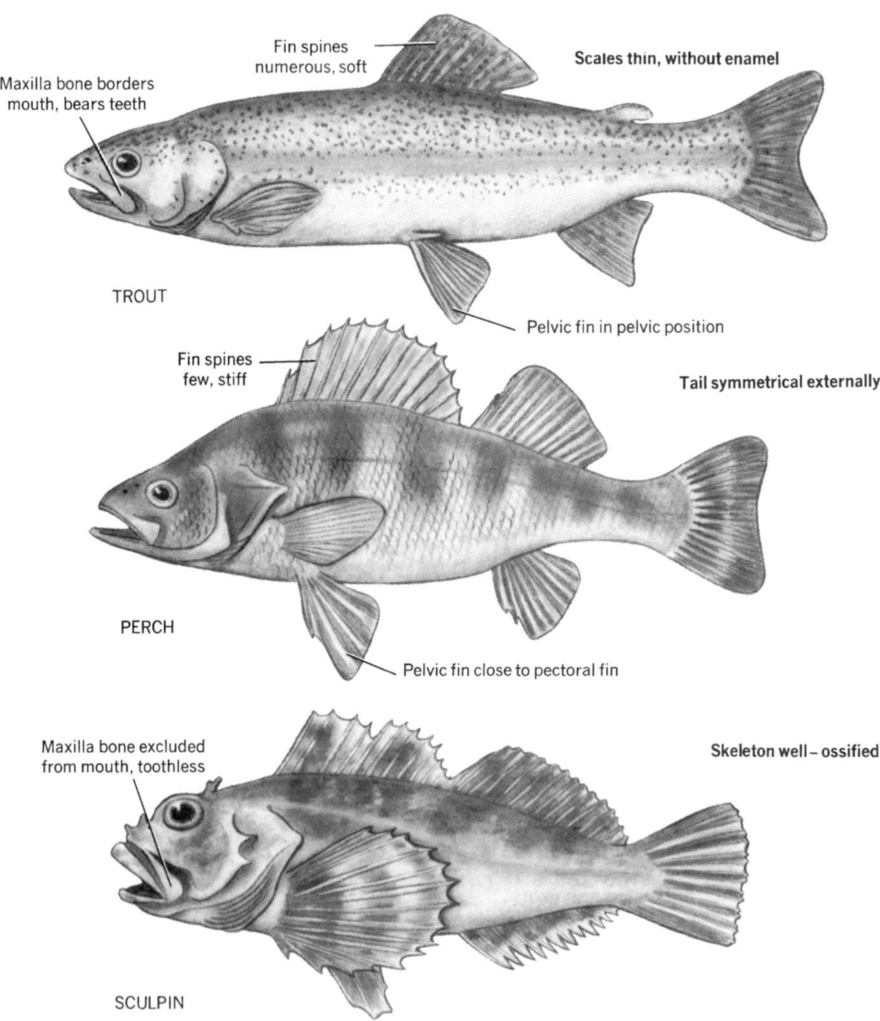

FIGURE 3.11 REPRESENTATIVE TELEOSTEI illustrating some characteristics of the group (boldface labels) and differences between more primitive (above) and more advanced species (center and below).

variously in 6 (conservative) to 30 or 40 (usual) or more orders. Teleosts are recognized as being either primitive or advanced on the basis of mouth structure, position of pelvic fins, number and stiffness of dorsal fin rays, and other characters.

Subclass Sarcopterygii Lobe-finned fishes have paired fins with fleshy stalks (the term means *flesh* + *fin*). Their jaw muscles are relatively strong. Relationships among the early sarcopterigians remain poorly understood; we simplify our classification by omitting lesser-known groups not mentioned later in the book.

The infraclass ACTINISTIA includes the surviving coelacanth (Figure 3.12) and its long-extinct yet similar relatives. Unlike other lobe-fins, these fishes lack **internal nares** (nostrils opening into the mouth). They give birth to live young and are distinctive for numerous cranial features.

Lungfishes comprise the infraclass DIPNOI. These are mostly freshwater fishes of moderate size and either "normal" or elongate form. They have internal nares, functional lungs, and (associated with their lungs) advanced circulatory systems. The paired fins of typical (extinct) dipnoans have a fleshy stalk with a jointed skeletal axis. The few teeth are distinctive,

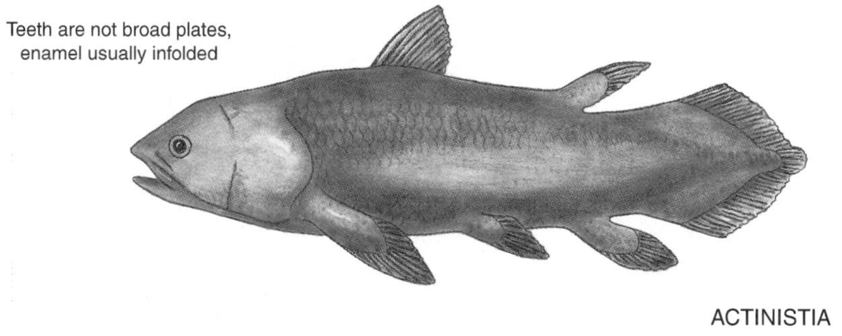

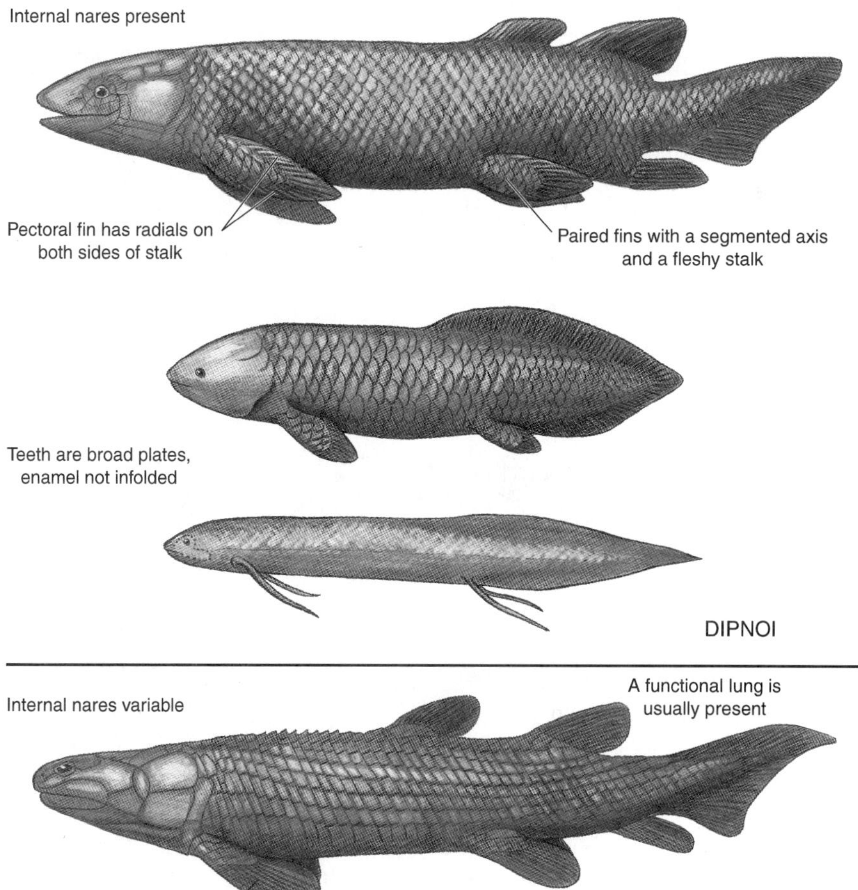

FIGURE 3.12 REPRESENTATIVE SARCOPTERYGII illustrating some characteristics of the subclass (boldface labels) and of the three infraclasses (standard labels). The most slender dipnoan is the African lungfish and the largest is the surviving genus *Latimeria*. The other examples are extinct.

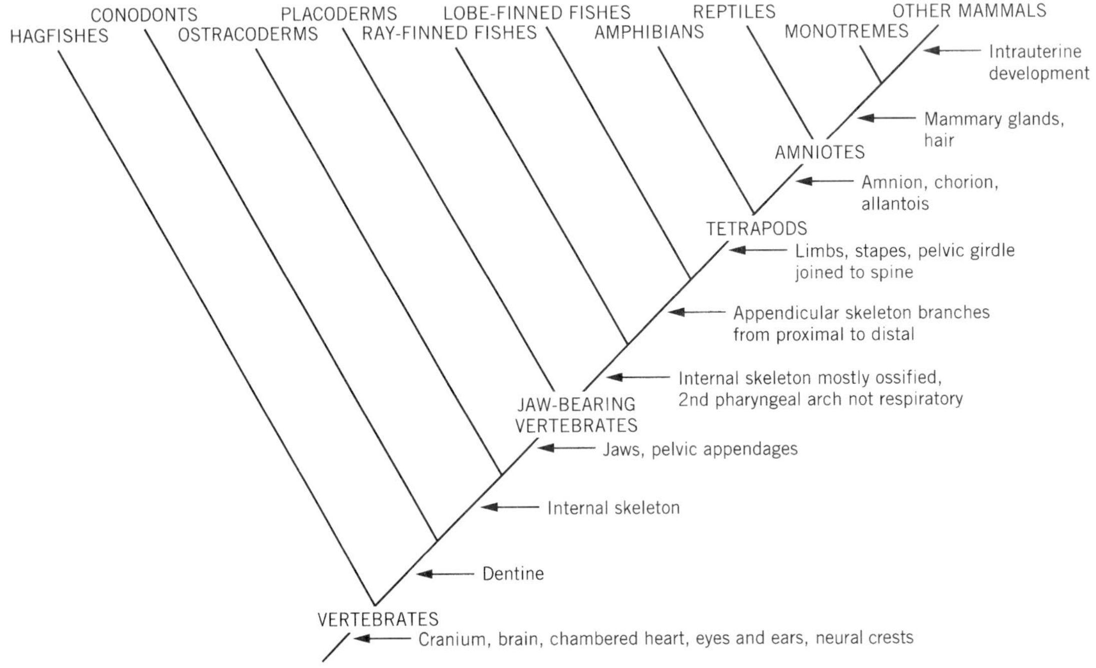

FIGURE 3.13 CLADOGRAM SHOWING RELATIONSHIPS AMONG SELECTED GROUPS OF VERTEBRATES and the successive acquisition of characters.

fan-shaped plates adapted for crushing. Braincase and vertebral column are poorly ossified. The scales of surviving lungfishes are simplified in structure and sometimes degenerate.

The classification of lobe-finned fishes being in flux, there is no clearly recognized name for the sister group OSTEOLEPIFORMES plus PANDERICHTHYIDA. The latter is virtually an amphibian with fins. Each group resembles primitive amphibians in having internal nares and teeth with infolded enamel. Panderichthyids appear to be particularly close to their tetrapod descendants in the anatomy of their skulls, pectoral fins, and ribs, and in the absence of dorsal fins.

Two additional terms are common in classifications of Sarcopterygii. *Crossopterygian* refers to all lobe-fins except Dipnoi. The grouping is now considered paraphyletic but remains useful. *Rhipidistian* refers to certain extinct lobe-fins but has been used in several ways.

REFERENCES

Bemis, W.E., W.W. Burggren, and N.E. Kemp (eds.). 1987. *The biology and evolution of lungfishes.* Liss, New York. 383p.

Benton, M.J. 1990. *Vertebrate paleontology.* Unwin Hyman, London. 377p.

Carroll, R.L. 1995. *Vertebrate paleontology and evolution.* Freeman, New York. 698p.

Donoghue, P.C.J., P.L. Forey, and R.J. Aldridge. 2000. Conodont affinity and chordate phylogeny. *Biol. Rev.* 75(2):191–251.

Evans, D.H. (ed). 1998. *The physiology of fishes.* 2nd ed. CRC Press, Boca Raton, FL. 544p.

Hardisty, M.W. 1979. *Biology of cyclostomes.* Chapman & Hall, London. 428p.

Janvier, P. 1995. Conodonts join the club. *Nature* 374(27 Apr.): 761, 762.

Janvier, P. 1996. *Early vertebrates.* Oxford monographs on geology and geophysics, no. 33. Oxford Univ. Press, New York. 393p. Authoritative, detailed, evolutionary.

Lauder, G.V., and K.F. Liem. 1983. Patterns of diversity and evolution in ray-finned fishes, vol. 1, pp. 1–24. *In* R.G. Northcutt and R.E. Davis (eds.), *Fish neurobiology.* Univ. Michigan Press, Ann Arbor.

Long, J.A. 1995. *The rise of fishes: 500 million years of evolution.* Johns Hopkins Univ. Press, Baltimore. 223p. Accurate but not technical; lavishly illustrated.

Marshall, N.B. 1972. *The life of fishes.* Universe Books, New York. 402p.

Moyle, P.B., and J.J. Cech, Jr. 1995. *Fishes: an introduction to ichthyology.* 3rd ed. Prentice Hall, Englewood Cliffs, NJ. 590p.

Nelson, J.S. 1994. *Fishes of the world.* 3rd ed. Wiley, New York. 600p. An authoritative classification and characterization of fishes.

Thomson, K.S. 1971. The adaptation and evolution of early fishes. *Quart. Rev. Biol.* 46:139–166.

Shu, D.G., et al. 1999. Lower Cambrian vertebrates from South China. *Nature* 402(4 Nov.):42–46.

Chapter 4

Tetrapods

TRANSITION TO TERRESTRIAL LIFE

Tetrapods are simply vertebrates having four legs (or at least in their ancestry). The first vertebrates to have legs instead of fins is a group of Upper Devonian amphibians that was probably as aquatic as their ancestors among fishes, which they closely resembled. They probably used their limbs to walk on the bottom where water was shallow, to paddle where it was deeper, and to crawl near the shoreline. Nevertheless, most tetrapods are terrestrial (or had land-dwelling ancestors), so it is useful to consider the changes that enable descendants of fishes to live terrestrial lives.

Out of water, the body usually no longer benefits from being streamlined. A neck becomes advantageous because the head now can turn to facilitate feeding and vision without affecting the mechanics of locomotion. Median fins are no longer useful. Deprived of the buoyancy of water, the body must be supported by the limbs, and this necessitates having appendages that are strong, girdles that are more firmly related to the axial skeleton, and a vertebral spine that can better resist bending by gravity.

Lungs and a pulmonary circulation, pioneered by air-breathing fishes, are usually retained to replace gills, which would be damaged by exposure to dry air. Gill covers can therefore be dispensed with. The superficial layer of the skin becomes sufficiently cornified to resist abrasion and drying. The eye, ear, and nose also must be modified to function in air instead of water. Oral glands are needed to moisten food that is now dry. Eggs and delicate larvae formerly were supported by water and could pass the waste products of their metabolism directly into the environment. Before becoming able to reproduce completely away from either water or moist environments, tetrapods had to accomplish the seeming miracle of evolving eggs with shells and fetal membranes to protect their embryos from desiccation and mechanical harm and to receive metabolic wastes. Some other structural changes were also necessary to make the shift to terrestrial life, and physiological and behavioral changes were also needed. It is not surprising, therefore, that the first tetrapods, the amphibians, did not fully accomplish all of these changes.

CLASS AMPHIBIA

Amphibians evolved from lobe-finned ancestors some 50 million years after bony fishes evolved. The class reached its greatest expansion after another 75 million years, in the Upper Carboniferous period, but continued to abound until the end of the Triassic period. Relatively few kinds of amphibians have survived to the present, yet they are distributed in tropical and temperate areas of all the world, and their habits and habitats are diverse.

The skin of modern amphibians is unable to withstand long exposure to dry air, and since fetal membranes are lacking, eggs must be laid either in water or in damp places. In this sense the class is not completely terrestrial ("amphibian" = *both kinds* + *life*), though many species do not utilize open water, and several live in remarkably arid places.

Adult amphibians have large mouths and a fleshy tongue that in survivors is attached near the front of the lower jaw. One of the bones that supported the jaws of the piscine ancestor has been converted to an ear ossicle, which usually contributes to hearing in air. Lungs are usually present (they are secondarily lost by one large group of salamanders), and some respiration occurs also through the skin and lining of the mouth and throat. Eyelids and glands to moisten the eyes have evolved.

Subclass Ichthyostegalia The first three subclasses to be discussed are extinct. Ichthyostegalians are the first tetrapods ever. They evolved in the water but could walk on the land. They are robust, up to 1 m long, have tail fins, large heads and mouths, and more than five digits on their feet (Figure 4.1).

Subclasses Anthracosauria and Temnospondyli These groups have teeth with infolded enamel (see figure on p. 108) and for that reason are grouped together as "labyrinthodonts." This is a useful term referring to the extinct subclasses; however, they are independent lineages. Some are small and aquatic, but most in each subclass are large (up to 2 m), stocky, and terrestrial. Anthracosaurs have deep skulls; they are in the ancestry of reptiles. Temnospondyls have large, relatively flat heads and appear to be in the ancestry of living amphibians (Figure 4.1).

Subclass Lissamphibia All surviving amphibians are lissamphibia (Figure 4.2). Most are less than 30 cm in length. Their moist skin has abundant mucous glands and only rarely

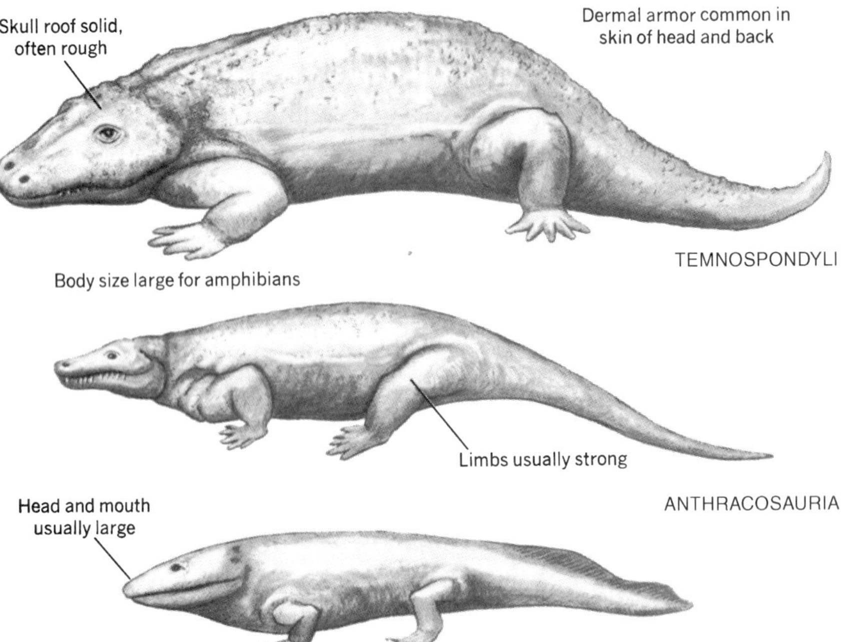

FIGURE 4.1 Restorations of REPRESENTATIVE PALEOZOIC AMPHIBIANS illustrating some characteristics of the three subclasses.

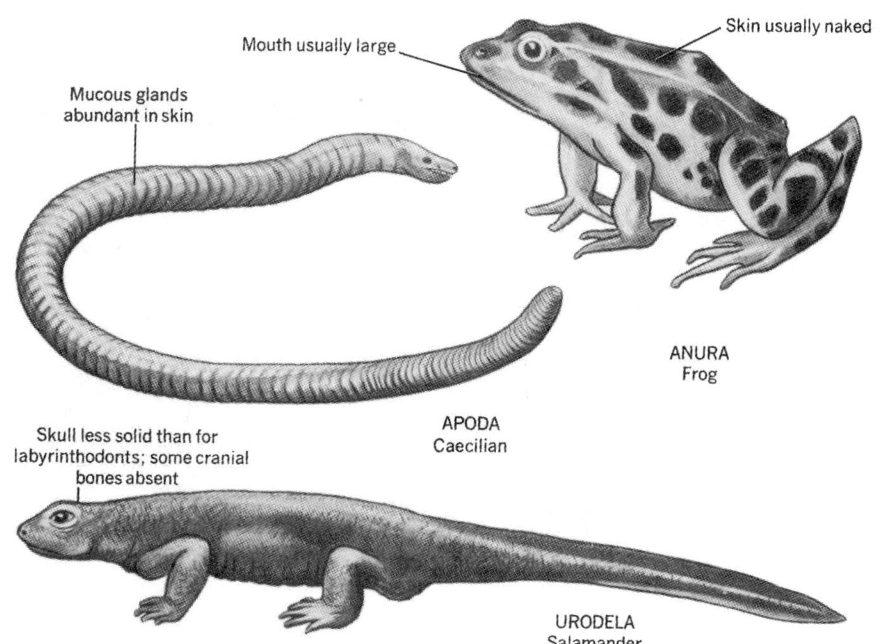

FIGURE 4.2
REPRESENTATIVES OF THE THREE ORDERS OF LISSAMPHIBIA showing some characteristics of the subclass.

supports scales ("liss" = *smooth*). The outer cornified layer of the skin is shed periodically. Parts of the skeleton, particularly of the feet, are commonly cartilaginous, and several ancestral bones have been lost from the braincase. There are only four toes in the hand. Teeth are never complicated as in labyrinthodonts and are absent from some groups.

There are three orders of lissamphibia. ANURA (= *without* + *tail*) includes frogs and toads. Salamanders belong to the URODELA (= *having a tail*). Finally, the APODA, commonly called caecilians, are, as the name tells us, legless. They are obscure animals that will merit mention in Part II of this book because of their primitive excretory organs and their scales (unusual in surviving amphibians), and in Part III because of their burrowing habits.

CLASS REPTILIA

Reptiles evolved from anthracosaurian amphibians some 60 million years after amphibians evolved. From the Permian through the Cretaceous periods they were the most abundant of vertebrates. This was the first class of tetrapods to have all the structures noted at the beginning of this chapter as requisite to fully terrestrial life, including fetal membranes and an integument that is resistant to drying. During the "age of reptiles," the different genera ranged from small to gigantic, from herbivorous to carnivorous, and from sluggish to swift. Several groups independently reverted to aquatic habitats, becoming highly skillful swimmers, and one group even invaded the air. No class of vertebrates had theretofore been so diverse in habits, and only the mammals have matched them since. Today, reptiles remain an important part of the faunas in tropical and temperate regions but are less numerous than bony fishes, birds, or mammals.

Reptiles are covered with horny scales. Excepting such specialized forms as snakes, most of them have claws, ribs that are used in drawing air into the lungs (amphibians instead use the mouth and throat as a force pump), and a vertebral column that is more differentiated into regions and more firmly attached to the pelvic girdle than in their amphibian ancestors. None of these characteristics is unique to the class, but there are features of the heart and related blood vessels that are found only in reptiles.

We list first, as a subclass, the synapsids. This extinct group is ancestral to mammals and hence, to preserve monophyly, is not considered reptilian in phylogenetic classification; mammals are instead included among synapsids. The remaining reptiles are commonly

FIGURE 4.3 Restorations of REPRESENTATIVE SYNAPSIDA of the two orders. Principal characteristics of the subclass are slender form, strong limbs, and features of the cranium noted in Chapter 8.

arranged in three to five subclasses which, in turn, fall into two major lineages. The first includes anapsids, and the other, named diapsids, includes everything else.

Subclass Synapsida The term *synapsid* refers to the distinctive position of an opening in the bones that roof over the temporal region of the skull (see figure on p. 132). Synapsids have been somewhat loosely called mammal-like reptiles. Some groups offer particularly well-documented examples of morphological transformations through time. The order PELYCOSAURIA appears in the Upper Carboniferous and becomes diverse in the Permian (Figure 4.3). They include both carnivores and herbivores. The largest reached 3 m in length. One group had sail-like structures along the back (apparently for temperature regulation) (see figure on p. 5). Pelycosaurs gave rise to the order THERAPSIDA, which also included flesh- and plant-eaters and had large representatives. One group, the CYNODONTIA of the Triassic and Jurassic, clearly was ancestral to mammals. Muscles and teeth show that they chewed their food. Their legs were relatively long and their stance upright, suggesting that they were active and may have been "warm-blooded."

Subclass Anapsida The term *anapsida* tells us that these reptiles have no openings in bones of the temporal region. They are noteworthy as the most primitive reptiles known. Accordingly, they have been called **stem reptiles** (Figure 4.4).

Subclass Diapsida This is a large and diverse subclass that is unified by (probably) common ancestry from stem reptiles and either by having two temporal openings in the skull, the diapsid condition, or by being derived from that condition through the loss of one of the ancestral openings. Three infraclasses are recognized here.

The infraclass LEPIDOSAURIA has one or two temporal openings among the bones of the palate, and usually teeth on the roof of the mouth as well as at the margins. Three or more orders are recognized, all but one of which are extinct and not useful to this course of study.

Surviving lepidosaurs (Figure 4.5 and the figure on p. 476) are placed in the order SQUAMATA. Most of the 3000 species of lizards have legs and a tail. Amphisbaenians have small forelimbs only or (usually) no limbs at all. They live underground. Snakes probably evolved from burrowing, lizardlike ancestors. They have lost almost all traces of limbs and generally have modified the skull to allow them to swallow

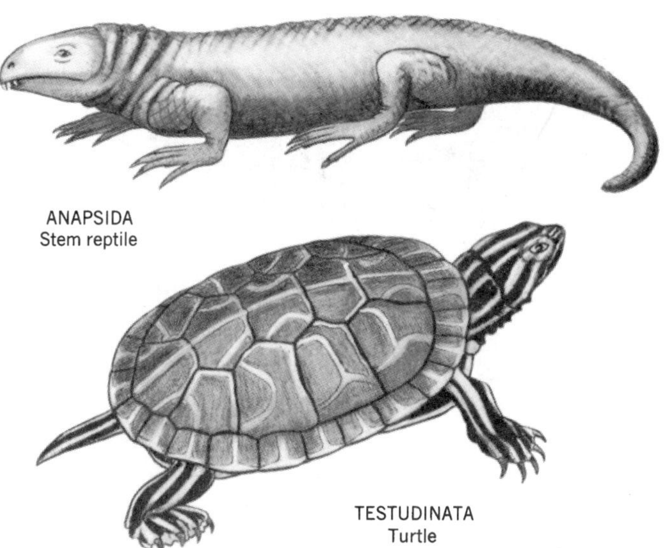

FIGURE 4.4 REPRESENTATIVES OF THE SUBCLASSES ANAPSIDA AND TESTUDINATA.

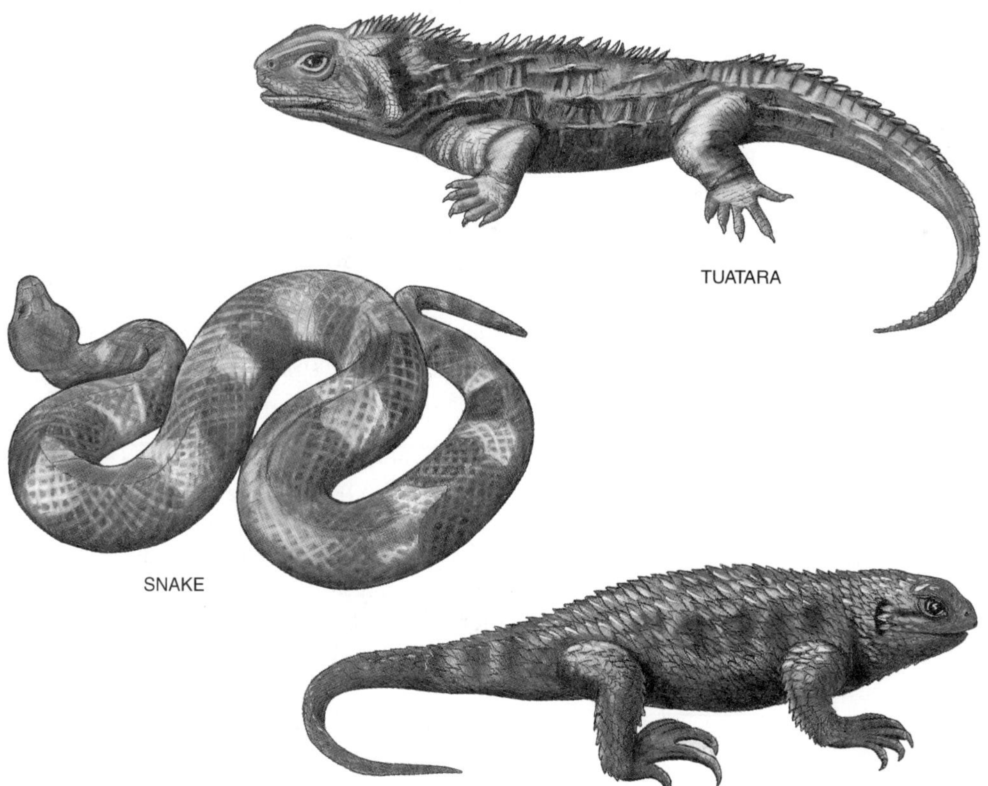

FIGURE 4.5 REPRESENTATIVES OF THE ORDER SQUAMATA OF THE INFRACLASS LEPIDOSAURIA. The principal characteristics of the infraclass are cranial features noted in Chapter 8.

FIGURE 4.6 The alligator is a REPRESENTATIVE OF THE CROCODILIA, the only surviving order of the infraclass Archosauria.

prey of diameter equal to or greater than that of their bodies. The tuatara looks like a large robust lizard. It is confined to islands off the coast of New Zealand, and is rigidly protected. The single genus, *Sphenodon,* may have survived longer (135 million years) than any other among vertebrates. It is of particular interest for the primitive nature of its skeletal and circulatory systems. (Some classifications assign this animal to its own order, Rynchocephalia.)

The large infraclass ARCHOSAURIA is characterized by retaining each of the two temporal openings and by having openings in the bony palate. Often there are also openings in the skull in front of the orbit and in the side of the lower jaw. All teeth are marginal. There are five orders. The only survivors of the class, crocodiles and their relatives, are in the order CROCODILIA (Figure 4.6). Two orders, SAURISCHIA and ORNITHISCHIA, comprise the beasts popularly known as dinosaurs (Figure 4.7). Some were the size of a turkey, but most, in each order, were large to gigantic, giving the infraclass its name: "archosauria" (= *ruler* + *lizard*). Each order had bipedal representatives. They are distinguished by features of the pelvic girdle. Flying reptiles are in the order PTEROSAURIA (= *wing* + *lizard*) (see figure on p. 523). This leaves only the order THECODONTIA, a diverse assemblage that was widespread in the Triassic. Thecodonts were apparently ancestral to other archosaurs, but the interrelationships are not clear. They ranged from small, lightly built, and probably bipedal to crocodile-like in form and size. Birds clearly evolved from archosaurs, but consensus has not been reached as to whether the ancestor was a saurischian dinosaur (of the group called theropods) or a thecodont (see Comment 4.1 on p. 61).

Subclass Testudinata As science advances, it often challenges conventional wisdom. With each revision of this book over the years it has been necessary to correct material that had long been considered secure. We come here to an example. In the first three editions, turtles and tortoises were confidently presented as an order in the subclass Anapsida. By the fourth edition they composed an independent subclass derived from Anapsida. Now many paleontologists cautiously believe that turtles may instead be allied to Diapsida, though whether the putative ancestor was in the lepidosaurian or archosaurian branch of diapsids is disputed. Perhaps the sixth edition will report a new consensus.

The subclass has one order, the CHELONIA. A turtle is often dissected in the laboratory as a representative of its class because of its availability and suitable size. Students should realize that the turtle's "shell," ribs, spine, toothless mouth, and pectoral girdle are highly specialized, and therefore not typical of the class. The skull, limbs, and (so far as we can judge) soft parts remain primitive.

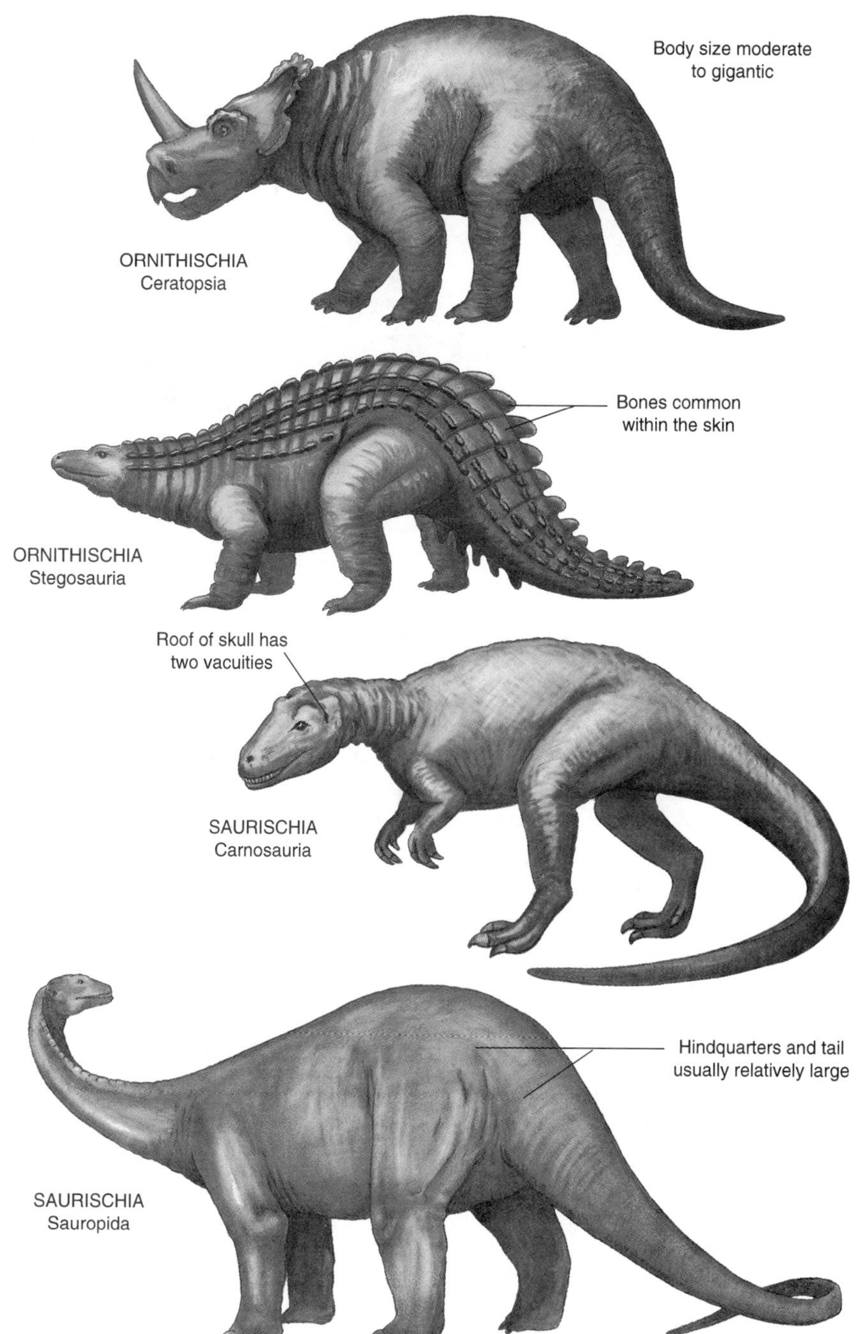

FIGURE 4.7 Restorations of REPRESENTATIVE ARCHOSAURIA of the orders Saurischia and Ornithischia and of the suborders indicated. Some characteristics of the terrestrial members of the infraclass are shown.

Two major groups of reptiles remain. Each includes highly specialized marine swimmers (mentioned in Chapter 27), and each has a single temporal opening, which is probably derived from the diapsid condition. Otherwise, the groups are very different. One, the PLESIOSAURIA, is usually placed in the infraclass SAUROPTERYGIA. Its aquatic members have broad, bulky bodies; tapering tails without lobes; paddlelike limbs; small, blunt heads; and necks that were sometimes very long (Figure 4.8). The other group comprises the ICHTHYOSAURIA. Their origin is still speculative; they may be assigned their own infraclass or even subclass. They have dolphinlike body contours, fishlike tails, large eyes, and a long rostrum.

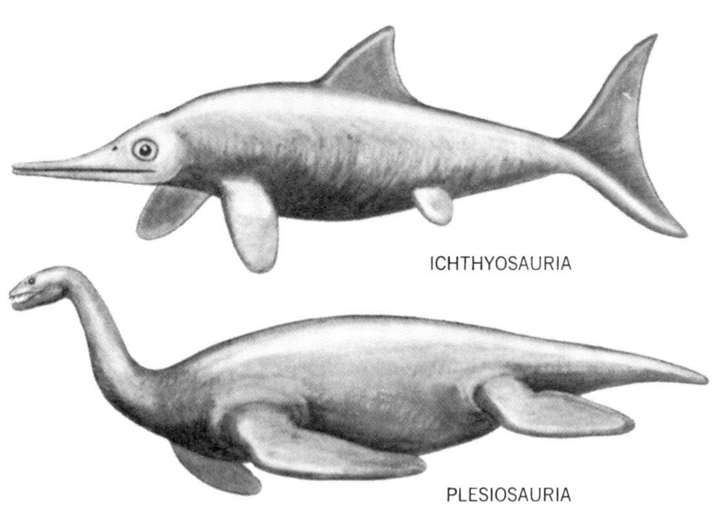

FIGURE 4.8 Restorations of REPRESENTATIVES OF TWO DISSIMILAR GROUPS OF AQUATIC REPTILES.

CLASS AVES No other locomotor adaptation requires so much structural specialization as that of flight, and all birds fly or are descendants of flyers. In striking contrast to reptiles (or, as cladists would say, to other reptiles—see page 30), birds are therefore the most homogeneous and distinctive of all tetrapod classes. However, for all their unique characteristics among the living fauna, birds are not very different from the reptiles from which they evolved. Which reptiles those were is debated (see Comment 4.1), but in any event they were probably small, slender carnivores or insectivores with long arms and legs, long necks and tails, and features of the skull and pelvis tending toward the avian condition.

Feathers are of particular importance, for birds are the most expert of flyers and the only vertebrates ever to achieve the highly successful combination of flight with bipedalism. Such flight is mechanically dependent on feathers. Furthermore, sustained flight probably requires the high metabolic rate made possible by an elevated body temperature. Contrary to former belief, many reptiles do have considerable control of body temperature; feathers may have first evolved for thermoregulation. We must not be overly presumptuous that the flying reptiles could not stay aloft for long periods, but it is unlikely that they could match the birds that replaced them during the Cretaceous period.

Over the last two decades there has been an exciting explosion in the number and kinds of known avian fossils; this is a remarkable time in the study of the origin and evolution of birds.

Subclass Archaeornithes Among the most significant, famous, and widely studied of all fossils is *Archaeopteryx*, the first known bird. This truly “missing link” is represented by seven specimens from Upper Jurassic slate deposits in Germany. *Archaeopteryx* is so distinctive that it is assigned to a subclass of its own (see the opener for Part I on p. 21, and Figures 4.9 and 4.10). Paleontologists agree that the dove-sized bird had at least limited capacity for flight; they do not agree as to whether it was an arboreal climber or a ground-living runner. It was fully feathered but, unlike present-day birds, had a long, lizardlike tail and seperate, clawed fingers (Figure 4.9). Various volant adaptations of the wing skeleton, spine, and breastbone are absent or incomplete by Cenozoic standards. The skull has the large orbits and beaklike rostrum of a bird but is reptilian in other respects, including having teeth in its jaws.

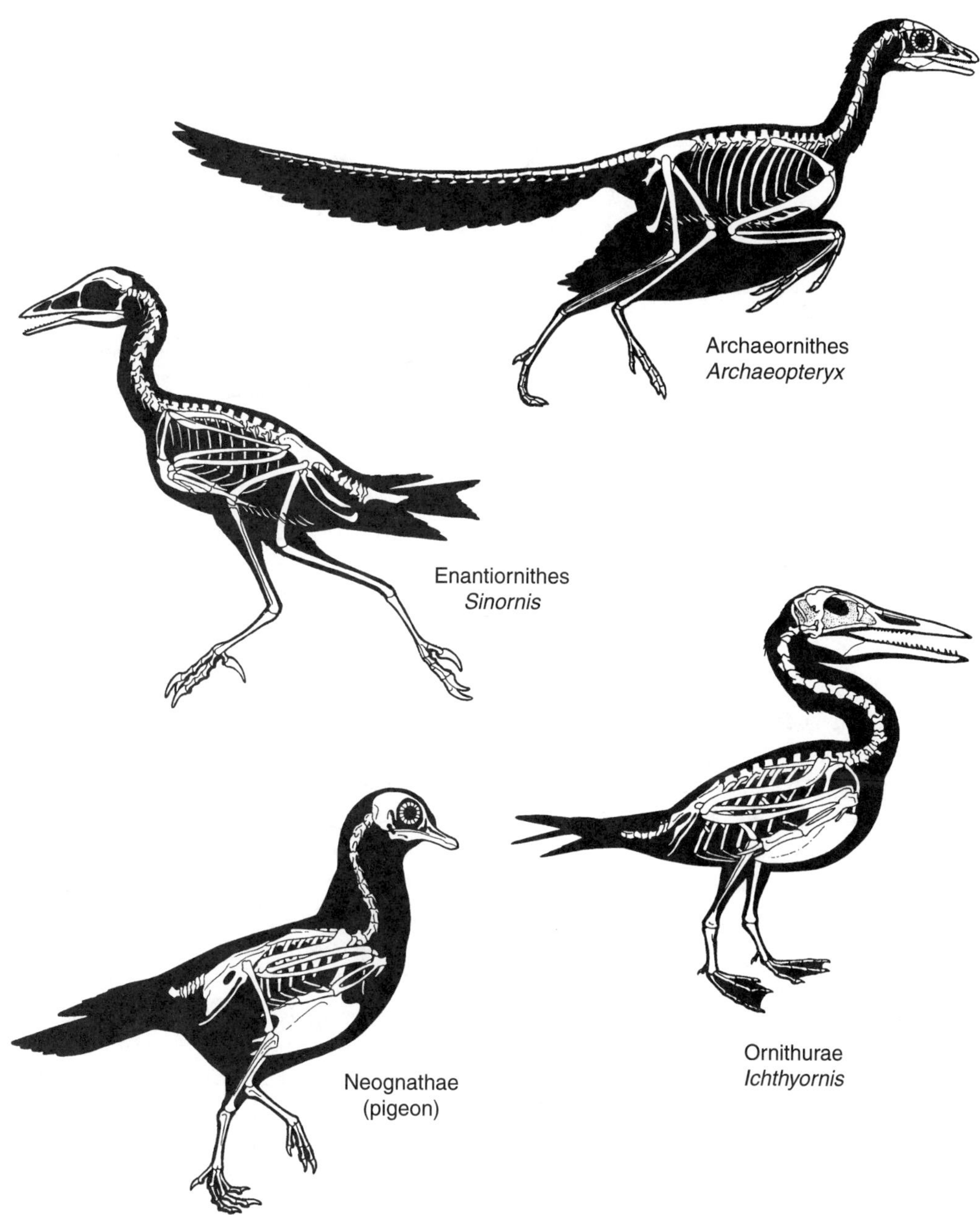

FIGURE 4.9 RESTORATIONS OF REPRESENTATIVE ANCESTRAL BIRDS. (From S. Chatterjee, 1997, *The rise of birds: 225 million years of evolution,* pp. 12, 92, 103, 110, Johns Hopkins Univ. Press.)

(It has been claimed that an even more ancient bird, *Protoavis,* lived in the Upper Triassic. The skeletal material is fragmentary and dissociated. Most paleontologists believe the alleged bird could be a composite and are reserving judgment.)

Subclass Enantiornithes This large assemblage of Cretaceous birds is known (but only since 1981) from widely distributed fossils. They range from sparrow size to turkey size.

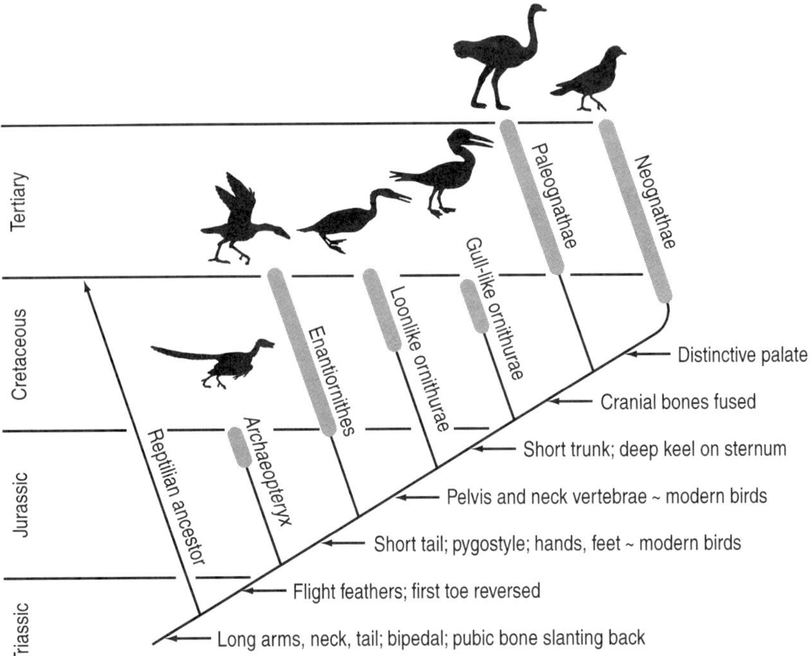

FIGURE 4.10 CLADOGRAM OF AVIAN EVOLUTION. Pygostyle = blade-like bone forming end of tail; sternum = breastbone. (Icons after S. Chatterjee, 1997, *The rise of birds: 225 million years of evolution*, pp. 110, 215, 265, Johns Hopkins Univ. Press.)

They are advanced over *Archaeopteryx* in numerous features, including fusions in the hand skeleton, a bladelike bone (the pygostyle) at the end of a short tail, and a keeled breastbone. Some retained teeth; others did not. Their perching feet indicate arboreal habits. The common name of the subclass is "opposite birds"; their differences from modern birds include a feature of the shoulder and a sequence of certain bone fusions in the lower leg.

Ornithurae and the Subclass Neornithes All remaining birds fall into four groups. Nearly all surviving birds belong to the infraclass NEOGNATHAE. The ostrich, emu, cassowary, and their large flightless relatives (all commonly called *ratites*) compose the infraclass PALEOGNATHAE. These infraclasses are distinguised by features of the pelvis and palate. The remaining two groups are both toothed, fish-eating birds of the upper Cretaceous. One group (Hesperornithiformes) includes large, foot-propelled divers that resemble modern loons but are flightless; members of the other group (Ichthyornithiformes) resemble gulls. In cladistics the four groups together compose the ORNITHURAE, to which no rank is assigned. In traditional classification all belong to the long-established subclass NEORNITHES. The surviving birds are in the infraclasses Neognathae and Paleognathae, and the Cretaceous birds are in a third infraclass designated Odontognathae. Unfortunately, the extinct groups seem not to be closely related, so the scheme is paraphyletic.

CLASS MAMMALIA

Mammals, like birds, are familiar and distinctive. Children learn that only mammals have hair and mammary glands, which give the class its name. Succeeding chapters will present unique features of the cranium, jaw, teeth, ear, pectoral girdle, pelvis, muscles, brain, and other structures. Mammals are also numerous. Some 3000 genera are known (of which 2000 are extinct).

Unlike the transition from reptiles to birds, the transition from reptiles to mammals is well recorded—so well, in fact, that the conventional boundary based on the structure of the jaw, ear, and cranium is somewhat arbitrary.

COMMENT 4.1

FEATHERS FLY OVER THE ORIGIN OF BIRDS

The origin of birds is currently among the most contentious (and fascinating) areas of vertebrate scholarship. There are two theories, each staunchly defended by its supporters. The most widely accepted theory (many authors recognize no alternative) derives birds from the reptilian order Saurischia, suborder Theropoda, and family Maniraptora. These are agile, carnivorous bipeds with long necks and tails, and relatively large brains and eyes. (The fast-running, sickle-clawed dinosaurs of the film *Jurassic Park* are maniraptors.) Cladistic analysis of dozens of character-states supports the conclusion that theropods are the most parsimonious choice as the avian ancestor. There is also spirited debate over the origin of flight. Most supporters of theropod ancestry favor the "ground up" theory of the origin of flight: Bipedal ground dwellers developed feathers for insulation, display, balance while jumping to catch flying insects, or to make the arms better able to envelop prey. Subsequently, as arm feathers enlarged, the animals could make longer jumps and finally fly. Some specialists in the theropod camp, however, prefer the alternative, "trees down" theory: Nimble climbers first jumped between branches and then, as their feathers enlarged, became able to glide and finally achieve powered flight. Supporters of theropod ancestry of birds tend to be paleontologists and cladists. For an introduction to their position see, in References, the article by Padian and Chiappe, and the book by Chatterjee.

The opposing camp derives birds from the early reptilian order Thecodontia, but has no near-avian fossils as candidate ancestors. The genera cited as possibly being close to the avian lineage are small, arboreal, climbing quadrupeds with fairly long arms, tail, and rostrum. Wrist bones, and pectoral and pelvic girdles are appropriate for the bottom of an avian lineage. Proponents of this theory support the "trees down" origin of flight. They believe that the time line is in their favor: Theropod dinosaurs, they say, occurred too late (mostly Cretaceous) to be ancestral to *Archaeopteryx* (Jurassic). They tend to be developmental biologists or comparative morphologists. See the book by Feduccia in the References.

In cladistic analysis, similarity of one character-state in two lineages does not establish shared ancestry any more than one instance of circumstantial evidence establishes guilt in a murder trial. The resemblance could be the result of a chance convergence. Therefore, the cladist uses as many character-states as possible. Much circumstantial evidence raises the chance of a correct interpretation from possible to probable. Even then, direct evidence (a videotape of the crime in progress) could take precedence. Is there any such evidence in the matter of avian ancestry? Each side thinks so.

In 1998 two genera of theropod dinosaurs were described that have clear, unambiguous feathers on arms, tail, and body. The fossils are from China and date from close to the Jurassic-Cretaceous boundary. The arms are too short for a flyer. Accordingly, feathers evolved for another purpose and are not defining for birds. Case closed in favor of theropod ancestry? Not quite. Dissenters surmise that the fossils are of flightless birds, not dinosaurs.

Maniraptor dinosaurs have three digits on the hand, and everyone agrees that they are digits 1, 2, and 3. Birds also have three digits, but comparative embryology seems clearly to show, by numerous studies over many years, that they are digits 2, 3, and 4. (In References, see the papers by Burke and Feduccia, and Hinchcliffe and Hecht.) Furthermore, in 2000 an Asian archosaur was described from the Triassic that appears to have unique, long feathers on the back. Case closed in favor of thecodont ancestry? Hardly. The dissenting majority cites what it considers to be conflicting evidence in asserting that the embryologists do not have their embryology straight, and that those long things on the strange fossil are not really feathers.

In time consensus will emerge. (Not many people still think the world is flat.)

Subclass Prototheria Several orders of Mesozoic prototheres are recognized. The only living members of the subclass make up the order MONOTREMATA, which includes the aquatic platypus and the insect-eating echidnas (Figure 4.11). These odd creatures are rare both in zoos and in their native Australia and New Guinea. If known only by the pectoral girdle, they surely would be classified as reptiles. Furthermore, they are oviparous (egg-laying), which is equally unique among mammals. The young are nourished by milk, however, and the presence of hair and a single bone in each half of the lower jaw qualify monotremes as mammals.

Subclass Allotheria These diverse animals lived at the time of the dinosaurs. They were small creatures of herbivorous or omnivorous habits. The one order, MULTITUBERCULATA, is named for the distinctive teeth. The group is not ancestral to others.

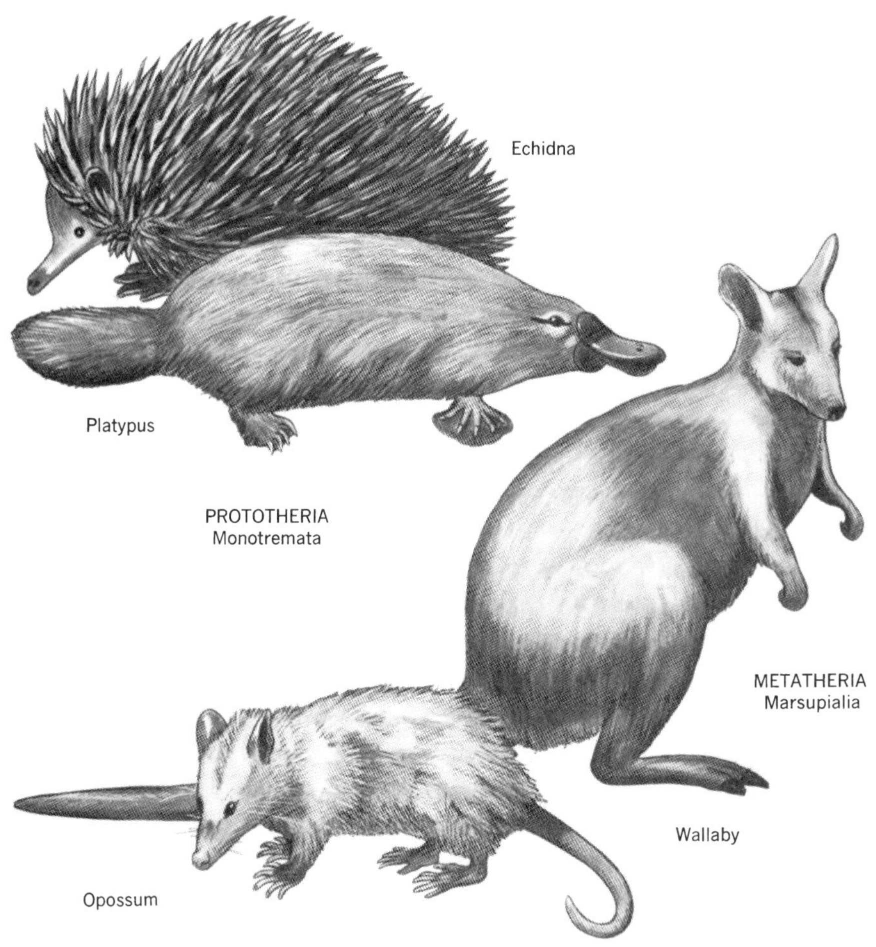

FIGURE 4.11
REPRESENTATIVES OF THE SUBCLASS PROTOTHERIA AND THE INFRACLASS METATHERIA. These taxa comprise, respectively, the orders Monotremata and Marsupialia.

Subclass Theria All familiar mammals belong to the subclass Theria. They are viviparous (give birth to live young), which sets them apart from Prototheria. The extinct infraclass TRITUBERCULATA is as ancient as the egg-laying mammals and is probably close to the ancestry of surviving mammals. Its members were small and are best known by their characteristic teeth. Two infraclasses survive to the present. METATHERIA includes the single and long-enduring order MARSUPIALIA. Opossums, bandicoots, phalangers, wombats, and kangaroos are marsupials. They give birth to tiny embryonic young, which are nourished in the pouch ("marsupium" = *pouch*) of the mother until they are able to walk about.

The other surviving infraclass, EUTHERIA (= *true* + *beasts*), comprises the animals commonly known as "placental mammals." The term is misleading. A placenta is an organ that accomplishes physiological exchange between mother and fetus. Some reptiles and even several fishes and amphibians have placentas, and so do all marsupials. The placenta of marsupials is always vascularized on the fetal side by a membrane called the yolk sac; that of eutherian mammals is usually vascularized by the allantoic membrane. However, some marsupials have both allantoic and yolk sac circulations, and some eutherian mammals have no allantoic circulation. Numerous features of the skeletal, reproductive, and nervous systems provide more technical but more exact ways to distinguish the infraclass Eutheria.

There are 17 or more surviving and about a dozen extinct orders of eutherian mammals (according to authority). Diagnosis of ordinal characters must be left for textbooks of

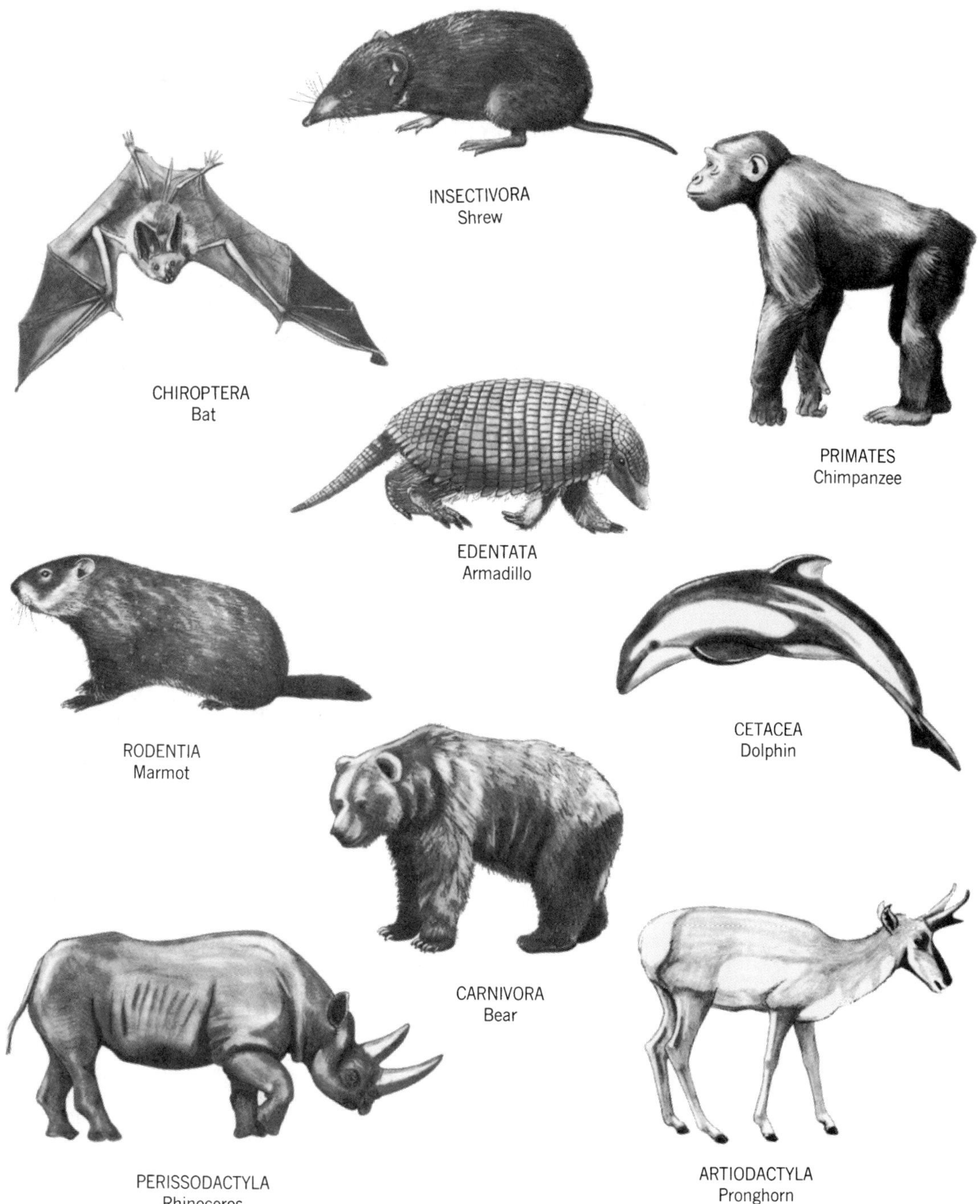

FIGURE 4.12 REPRESENTATIVES OF NINE ORDERS OF THE INFRACLASS EUTHERIA.

mammalogy. However, in Part II, and particularly in Part III, of this book, reference will be made to representatives of all surviving orders. It will be useful, therefore, to identify them here (Figure 4.12).

INSECTIVORA includes shrews, moles, and hedgehogs. All are small animals with numerous sharp teeth. This is the oldest and most primitive order of the infraclass. DERMOPTERA is represented only by the gliding colugo (see figure on p. 521). The only mammals capable of sustained flight are the bats, order CHIROPTERA. PRIMATES include lemurs, monkeys, apes, and man. As the name indicates, EDENTATA have simple teeth or none at all. They are the anteaters, sloths, and armadillos. The scaly pangolins also eat insects but comprise the PHOLIDOTA (see figure on p. 458). Rabbits and their smaller relatives, the pikas, are in the order LAGOMORPHA. The largest order is the RODENTIA, which includes squirrels, beavers, rats, mice, porcupines, and a host of other small mammals with gnawing incisors.

Whales and dolphins of the order CETACEA are among the most modified of mammals. Bears, dogs, weasels, raccoons, civets, cats, hyenas, and most other flesh-eating mammals are in the order CARNIVORA. Seals, sea lions, and walruses form the order PINNIPEDIA. Only the aardvark is in the TUBULIDENTATA. Elephants, order PROBOSCIDEA, were once more prevalent both in number and in kinds. The stocky hyraxes of Africa and Asia are in the order HYRACOIDEA. The large aquatic grazers called dugongs and manatees are SIRENIA (see figure on p. 496). Horses, tapirs, and rhinoceroses make up the order PERISSODACTYLA. The term means "odd-toed" and distinguishes them from the even-toed ARTIODACTYLA, such as pigs, camels, deer, antelopes, and cattle. Perissodactyls and artiodactyls include most hoofed mammals and are collectively known as *ungulates* ("unguis" = *hoof*), a useful term of no systematic rank.

REFERENCES

Ahlberg, P.E., and A.R. Milner. 1994. The origin and early diversification of tetrapods. *Nature* 368(7 Apr):507–514.

Anderson, S., and J.K. Jones, Jr. (eds). 1984. *Orders and families of recent mammals of the world.* 2nd ed. Wiley, New York. 686p.

Bellairs, A. d'A., and C.B. Cox. 1976. *Morphology and biology of the reptiles.* Academic Press, New York. 290p.

Benton, M.J. 1990. Phylogeny of the major tetrapod groups: morphological data and divergence dates. *J. Molec. Evol.* 3(5):409–424.

Benton, M.J. 1990. *Vertebrate paleontology.* Chapman & Hall, New York. 336p.

Burke, A.C., and A. Feduccia. 1997. Developmental patterns and the identification of homologies in the avian hand. *Science* 278:666–668.

Carroll, R.L. 1988. *Vertebrate paleontology and evolution.* Freeman, New York. 698p.

Chatterjee, S. 1997. *The rise of birds: 225 million years of evolution.* Johns Hopkins Univ. Press, Baltimore. 312p. A comprehensive reference; fine illustrations.

Chiappe, L.M. 1995. The first 85 million years of avian evolution. *Nature* 378(23 Nov):349–355. A review article.

Chiappe, L.M. 1995. A diversity of early birds. *Nat. Hist.* 104:52–55.

Cloudsley-Thompson, J.L. 1999. *The diversity of amphibians and reptiles.* Springer, New York. 254p. Includes chapters on locomotion, feeding, antipredation devices, reproduction, temperature regulation, and water balance.

Duellman, W.E., and L. Trueb. 1994. *Biology of amphibians.* Johns Hopkins Univ. Press, Baltimore. 694p. A fine reference; excellent illustrations.

Feduccia, A. 1996. *The origin and evolution of birds.* 2nd ed. Yale Univ. Press, New Haven, CT. 480p. The case that birds evolved from early archosaurs as made by an ardent champion of the thesis.

Ferguson, M.M.J. (ed.). 1984. Structure, development, and evolution of reptiles. *Symp. Zool. Soc. Lond.* 52:1–697.

Fraser, N. 1991. The true turtles' story.... *Nature* 349(24 Jan): 278, 279. The basis for according subclass rank to turtles.

Gans, C.A., et al. (eds.). 1969–. *Biology of the reptilia.* Academic Press, New York. Many volumes published; others projected.

Hinchliffe, J.R., and M.K. Hecht. 1984. Homology of the bird wing skeleton: embryological versus paleontological evidence. *Evol. Biol.* 18:21–39.

King, A.S., and J. McLelland (eds.). 1979–1989. *Form and function in birds,* vols. 1–4. Academic Press, New York.

Norman, D. 1985. *The illustrated encyclopedia of dinosaurs.* Crescent Books, New York. 208p. Skeletons, restorations, history, evolution, classification.

Nowak, R.M. 1999. *Walker's mammals of the world.* 6th ed. Johns Hopkins Univ. Press, Baltimore. 2 vols., 2015p.

Padian, K., and L.M. Chiappe. 1998. The origin and early evolution of birds. *Biol. Rev.* 73:1–42.

Panchen, A.L., and T.R Smithson. 1987. Character diagnosis, fossils, and the origin of tetrapods. *Biol. Rev.* 62:341–438.

Pough, F.H., et al. 1998. *Herpetology.* Prentice Hall, Upper Saddle River, NJ. 577p. Classification, distribution, physiology, locomotion, behavior, communication, reproduction.

Ruben, J.A., et al. 1997. Lung structure and ventilation in theropod dinosaurs and early birds. *Science* 278:1267–1270.

Shipman, P. 1998. *Taking wing:* Archaeopteryx *and the evolution of bird flight.* Simon & Schuster, New York. 336p.

Sumida, S.S., and K.L.M. Martin (eds.). 1997. *Amniote origins.* Academic Press, New York. 510p.

Van Tyne, J., and A.J. Berger. 1976. *Fundamentals of ornithology.* 2nd ed. Wiley, New York. 808p. Includes a chapter on anatomy and an account of each family.

Vaughan, T., J. Ryan, and N. Czaplewski. 2000. *Mammalogy.* 4th ed. Harcourt, Fort Worth, TX. 565p.

Welty, J.C., and L. Baptista. 1988. *The life of birds.* 4th ed. Saunders, Philadelphia. 581p.

Zug, G.R. 1993. *Herpetology: an introductory biology of amphibians and reptiles.* Academic Press, New York. 527p.

Part Two

The Phylogeny and Ontogeny of Structure:

Evolution in Relation to Time and Major Taxa

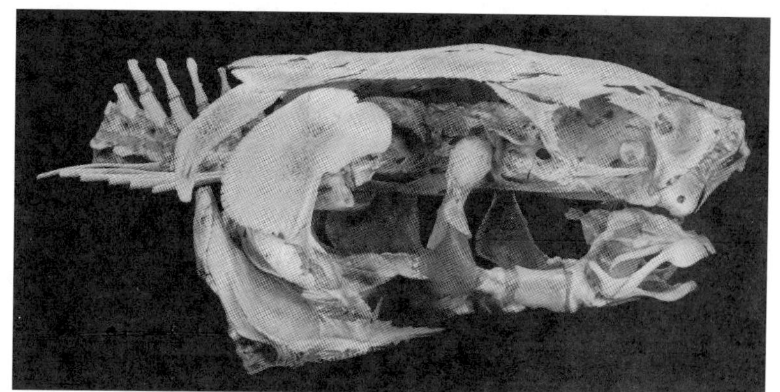

Chapter 5

Early Development

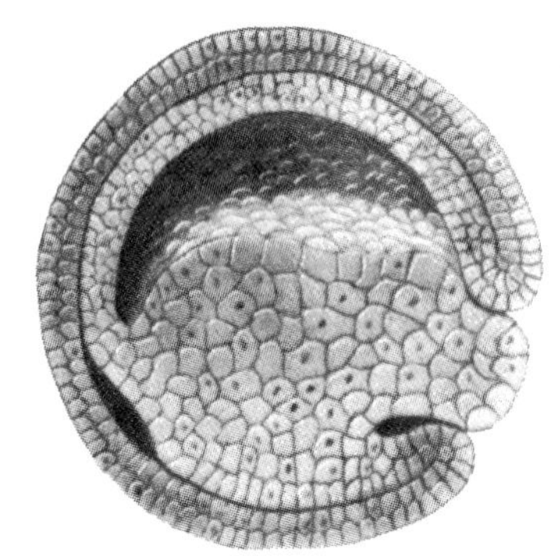

Embryology is often studied prior to, or concurrently with, morphology. This short chapter provides an introduction to relevant aspects of early development for students who have not yet studied embryology, and a refresher for those who have. Later development is included in subsequent chapters. Before a description of development, however, it is desirable to review the basis for the long, important, and close relationship between embryology and morphology.

EMBRYOLOGY AND MORPHOLOGY

Development and Ancestry It is evident to the discerning student that the development of the individual from egg to adult (*ontogeny*) and the ancestry of the species (*phylogeny*) are closely related. Following the publication in 1859 of Darwin's *The Origin of Species,* the nature of the relationship became the subject of observation and speculation. Haeckel wrote in 1866 that ontogeny recapitulates, or repeats, phylogenetic change. Study of embryology, he said, reveals ancestry. It is fascinating, for example, that the pharynx, heart, and associated arteries of mammalian embryos resemble those of fishes more closely than those of adult mammals.

In a small book published in 1930 de Beer recognized that recapitulation is not the only relationship between embryos and ancestors. For instance: (1) The embryo may have a structure that is not present in the adult of either the ancestor or the descendant. The shellbreaker used by the chick at hatching is an example (Figure 5.1). (2) A structure that was present in both embryo and adult of the ancestor may become vestigial (see figure on p. 9) or even be lost. (3) Structures that were formerly present only in the larva or embryo may come to be retained in the adult of the descendant. The gills of certain salamanders are examples (Figure 5.2). (4) The embryo of the descendant may repeat early developmental stages of the ancestor, but not late developmental stages. Thus, the gill apparatus of the embryo mammal resembles that of the larval ancestral fish, yet never recapitulates the functional gills of an adult fish. (5) The developmental sequence of the ancestor may be altered by the descendant. For example, the muscular and digestive systems of larval amphibians become functional

Page 67: Sturgeon skull. Hand-cleaned and freeze-dried.

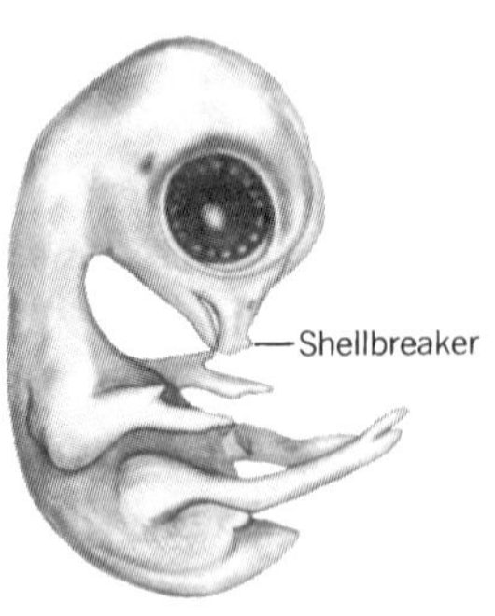

FIGURE 5.1
THE AVIAN SHELL-BREAKER illustrates an embryonic structure not present in the adult or the ancestor.

before the appendages form, whereas in birds and mammals the appendages develop first. Although not controversial, the influence of these observations was muted because they came at the onset of several decades during which genetics replaced embryology as the most exciting companion science to evolutionary morphology.

In recent years a new yet vigorous field has emerged, which has been called **evolutionary developmental biology.** Emphasis has shifted from product (the concern of de Beer) to process. Reciprocity is sought between the genetics of development and the mechanisms of morphogenesis. In some instances evolutionary theory is tested by direct experiment.

Heterochrony One important mechanism of evolutionary change is **heterochrony,** which means that the timing or rate of a developmental process changes between ancestor and descendant. There are several possibilities: (1) If the development of the somatic features (but not of the reproductive organs) of the descendant is accelerated, the result is **recapitulation** in the classical sense; that is, adult characters of the ancestor come to be juvenile characters of the descendant. Thus, the symmetrical tails of advanced fishes pass through an asymmetrical developmental stage that resembles the adult tail of the ancestor. (2) If development is accelerated only for the reproductive system (a process called **progenesis**) or is retarded for one or more somatic features (a process called **neoteny**), the resulting condition is **paedomorphosis** (= *child* + *form*), or the retention of ancestral juvenile characters by adult descendants. Paedomorphosis by progenesis involves most of the body and is cited in theories on the ancestry of vertebrates (see p. 27). Paedomorphosis by neoteny usually involves only part of the body and has been common. The example of larval gills in adult salamanders falls in this category. Neotenous features of humans include naked skin, large brain, short jaw, absence of brow ridges, and late sexual maturity. Neotenous features of post-Paleozoic lungfishes include number and shape of fins and nature of scales. The skeleton of salamanders provides further examples. (3) Finally, if the maturation of the reproductive organs is retarded, then ontogeny is extended beyond its ancestral limits, usually resulting in an animal having increased body size and complexity of certain parts. The enormous growth and branching of the antlers of a large extinct elk is a classic example. The process is called **hypermorphosis.**

Developmental Constraints on Form The developmental process puts constraints on adult form. Evolutionary change results only from alterations in development, and develop-

FIGURE 5.2
THE GILLS OF THE AMPHIBIAN *Necturus* illustrate an embryonic structure retained by the adult. Compare with the salamander shown on p. 53.

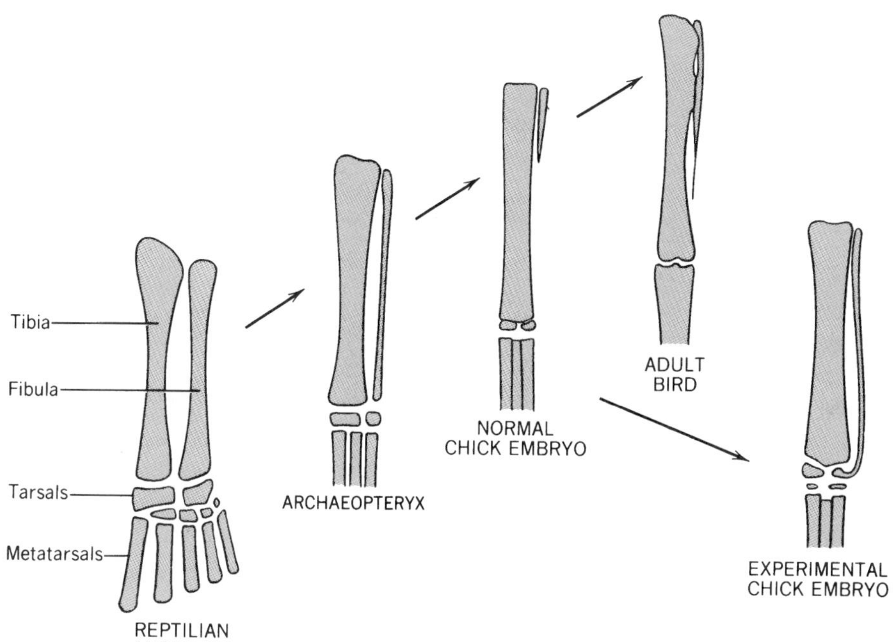

FIGURE 5.3
RESTORATION OF ANCESTRAL FORM by experimental alteration of normal development in the chick.

mental mechanisms are both stable over time and channelized. Only a fraction of the adult forms that can be imagined have ever evolved. Formerly it was believed that this is solely because selection pressure eliminates trends toward forms that do not occur. Now it is recognized that many forms are not possible because they cannot be produced by the long-persistent developmental mechanisms. True, even a small change in an established process, particularly if it occurs early in ontogeny, can lead to a marked change of adult form. Nevertheless, the kinds of changes are limited. They may be the consequence of, for instance, alteration in the concentrations of enzymes or modifications of motility of formative cells. Usually they follow from changes in the timing and sequence of developmental events.

The fact that developmental processes are more stable than their products has been demonstrated experimentally. Thus, by altering the influence of adjacent tissues, the cells covering the jaws of chick embryos can be induced to form enamel, as in ancestral toothed birds. A striking example relates to the avian leg: The repitilian ancestor of birds had a fully developed fibula bone in the lower leg, five foot bones (metatarsals), and several ankle bones (tarsals). The Jurassic bird *Archaeopteryx* retained the full fibula, three metatarsals, and two free tarsals (Figure 5.3). The three metatarsals of modern birds are free in the embryo but fused in the adult (an example of recapitulation); two tarsals can be identified in the embryo but join the tibia during development, and, importantly, the shortened fibula does not reach the ankle. By increasing, in any of several ways, the relative influence of tissue that forms the fibula, the French experimenter Hampé caused a chick to develop that had a fibula of the full ancestral length. This bone, in turn, influenced the metatarsals and several tarsals to be free. Thus a leg was created that was more ancient in pattern than that of *Archaeopteryx*. Evidently, modern birds retain, like a dormant memory, the potential to form the ancestral leg. The process has been modified over time, but not abandoned in favor of a new process.

GAMETES AND FERTILIZATION

The mature sex cells, or **gametes,** are the male **sperm** and female **ovum** or egg. Recall that each carries a haploid, or half set, of chromosomes. The sperm cells of vertebrates are highly varied in appearance (Figure 5.4), but always have a **head,** which contains the nucleus, a **middle piece** containing the mitochondria needed to provide energy, and a

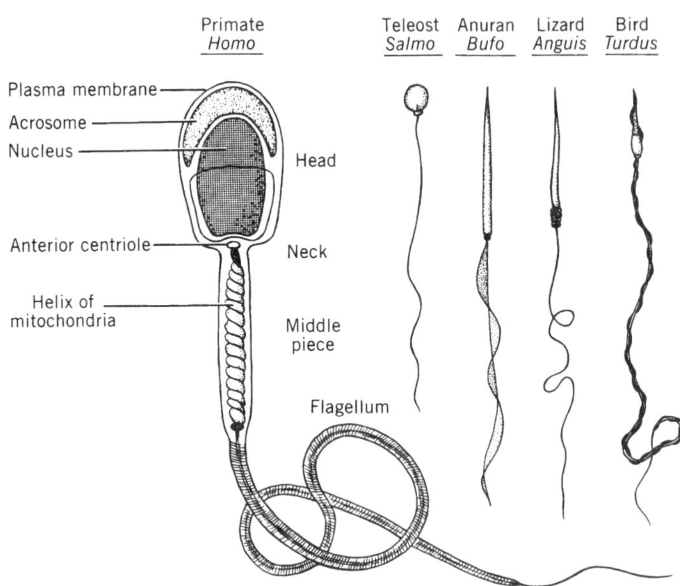

FIGURE 5.4
REPRESENTATIVE VERTEBRATE SPERM CELLS.

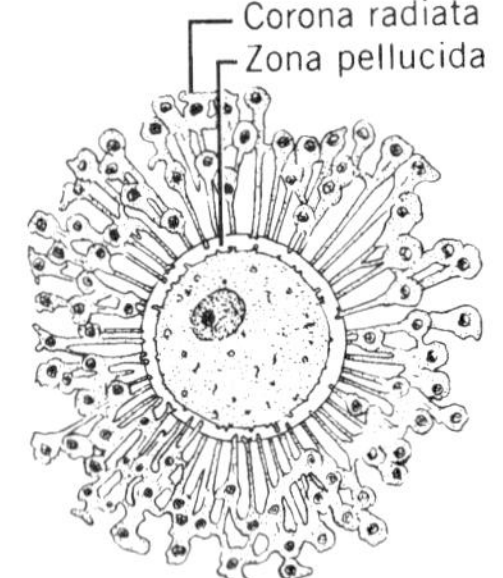

FIGURE 5.5
Section of MAMMALIAN OVUM after ovulation and before fertilization.

flagellum, which propels the cell. The head may be spherical, spatulate, hooked, lance-shaped, or spiraled. It is capped by the **acrosome.**

The small eggs of amphioxus contain little yolk and hence are said to be **microlecithal** (= *small* + *yolk*). This may be the ancestral chordate condition, but most vertebrate eggs either have moderate amounts of yolk (**mesolecithal**), as for lampreys, chondrostean fishes, and amphibians, or instead are laden with large amounts of yolk (**macrolecithal**), as for most fishes, reptiles, birds, and monotremes. Eutherian mammals, having a placenta to nourish the embryo, no longer need yolk, and their eggs have returned to the microlecithal condition (Figure 5.5). Yolk is a complex material that includes proteins, phospholipids, and neutral fats. Such yolk as is present is concentrated toward one side of the egg, the **vegetal pole,** leaving a region of clearer cytoplasm and the nucleus at the other side, or **animal pole.** Eggs with such asymmetrical distribution of yolk are **telolecithal** (= *end* + *yolk*). Metabolic activity is highest at the animal pole, and there may be a gradient in the distribution of pigment. The animal-vegetal axis imparts radial symmetry to the egg.

The egg is surrounded by a delicate **vitelline membrane.** Eggs of therian mammals are also enclosed in a thicker **zona pellucida** and a large **corona radiata,** or layer of adherent cells from the ovarian follicle. The eggs of other vertebrates may be enveloped, after ovulation, in jelly layers (amphibians), in albumen (birds), or in horny, membranous, or calcareous capsules or shells (many fishes, reptiles, birds).

Penetration of egg membranes and egg by a sperm cell is a complicated process involving both enzymatic and physical interactions between sperm acrosome and egg cortex, each of which undergoes striking changes. Entry of the sperm into the egg restores the diploid number of chromosomes, and activates the egg both to become refractory to the entry of additional sperm and to initiate development of the embryo.

CLEAVAGE

The **zygote,** or fertilized egg, is transformed by cell division, called **cleavage,** into a multicellular embryo called a **blastula.** During cleavage the individual daughter cells are termed **blastomeres.** The process of cleavage and the structure of the blastula are both closely related to the amount of yolk present (Figure 5.6).

The microlecithal eggs of amphioxus undergo **total** (or holoblastic) cleavage, which means that the cleavage furrows penetrate the entire yolk. Furthermore, cleavage is **equal,**

COMMENT 5.1

HOX GENES AND DEVELOPMENT: WINDOW INTO EVOLUTION

Because metazoan animals are believed to have evolved only once, their diverse body plans must have arisen by transformation of a single ancestral plan. Research indicates that the diversity of body architecture is the consequence of variation of a shared genetic control mechanism. The development of a multicellular organism requires that gene expression be regulated *in each cell* in the correct order and at the correct time. Important genes, called homeotic or ***Hox* genes,** control fundamental features such as polarity along an anterior-posterior axis and, in those forms possessing them, the segmental location and nature of appendages. Surprisingly, there is convincing evidence that ever since the first homeotic gene evolved, hundreds of millions of years ago, its fundamental regulatory functions have been retained in all descendent groups, including vertebrates. This provides a window through which to view not only developmental mechanisms, but patterns of evolution as well.

When a *Hox* gene is activated by "upstream" genes at different stages during development, it produces proteins, which in turn selectively activate numerous structural "downstream" genes responsible for executing the precise morphology of specific body segments. Early in the 20th century geneticists observed that induced errors in gene transcription during the early development of *Drosophila* resulted in a bizarre transformation of one body segment to another. For example, a mutant fly could grow a pair of legs on the head segment where its antennae should be, the embryonic cells expressing their *Hox* genes behaving as if confused as to their location along the anterior-posterior axis of the embryo. In *Drosophila,* rather than just one *Hox* gene (the proposed early metazoan condition), there are eight, located in tandem along a single chromosome. Together, the eight *Hox* genes form a cluster. This in-tandem arrangment on a single chromosome is an organization found in most nonvertebrates and is thought to be the result of duplication events of the single ancestral *Hox* gene. Evidence is mounting that the *Hox* genes of vertebrates (most studied in mammals, including mouse and human) possess a structural organization and functional expression that make them molecular homologs of those in *Drosophila* (and the earliest metazoans). The vertebrate condition, in which *Hox* genes are located in four clusters on four separate chromosomes (haploid state), is thought to have arisen by two duplication events and by fission. During this evolution, errors in gene duplication and transposition resulted in an alteration of function of some of the genes as well as an increase in their number. Within each *Hox* gene, there exists a sequence of base pairs (a *homeobox*) that is highly conserved from one gene to the next along the in-tandem series and across chromosomes. In any given species, it is the interactions of the multiple *Hox* genes that give rise to specific morphologies; the precise mechanism is a focal point for intense research.

In *Drosophila,* each homeotic gene in the cluster is named to reflect its primary effect (e.g., *labial* and *sex comb reduced*) and is given a corresponding abbreviation (*lab, Scr*). In vertebrates, instead of the name referring to its regulatory effect, a gene is simply noted as *Hox*-A, *Hox*-B, *Hox*-C, or *Hox*-D to designate its location on one of the four *Hox* chromosomes, and given a number from 1 to 13 to reflect its anterior-posterior position within the cluster. For example, *Hox*-A1 refers to the most anterior position on the first chromosome. In addition the conservation of an anterior-posterior order of the homeobox genes in *Drosophila* and mammals during their long evolution, it is significant (but not understood) that the relative expression of the genes follows the same pattern. For example, in a *Drosophila* of about 10 hours and a mouse of about 12 days, the molecular homologs *lab* (fly) and (by duplication) *Hox*-A1, *Hox*-B1, and *Hox*-D1 (mouse) become active, followed (in part) by the respective equivalents *Scr* and *Hox*-A5, *Hox*-B5, and *Hox*-C5.

Evolutionary morphologists are studying the effects of *Hox* genes (and of other regulatory genes not found in these clusters). The techniques available (*immunolocalization, in situ hybridization*) allow for detection of both when and where these genes influence development. Long-standing questions about the origins and evolution of paired fins, branchial arches, vertebral differentiation, and brain development (see Figure 8.2, p. 116) are being reexamined. To what extent are the regulatory genes of different major taxa true molecular homologs? What are the consequences of mutations upstream from a regulatory gene? Downstream? Do *Hox* genes evolve? If your interest is piqued, see the References.

---

because all blastomeres are of about the same size at any given time. The resultant blastula is a hollow ball of cells with a cavity called the **blastocoel.**

Cleavage of mesolecithal eggs is also total, but the greater amount of yolk in the vegetal hemisphere is an impediment that retards cell division. Consequently, development is slower there, blastomeres near the vegetal pole are larger than those near the animal pole, the blastocoel is displaced into the animal hemisphere, and cleavage is said to be **unequal.**

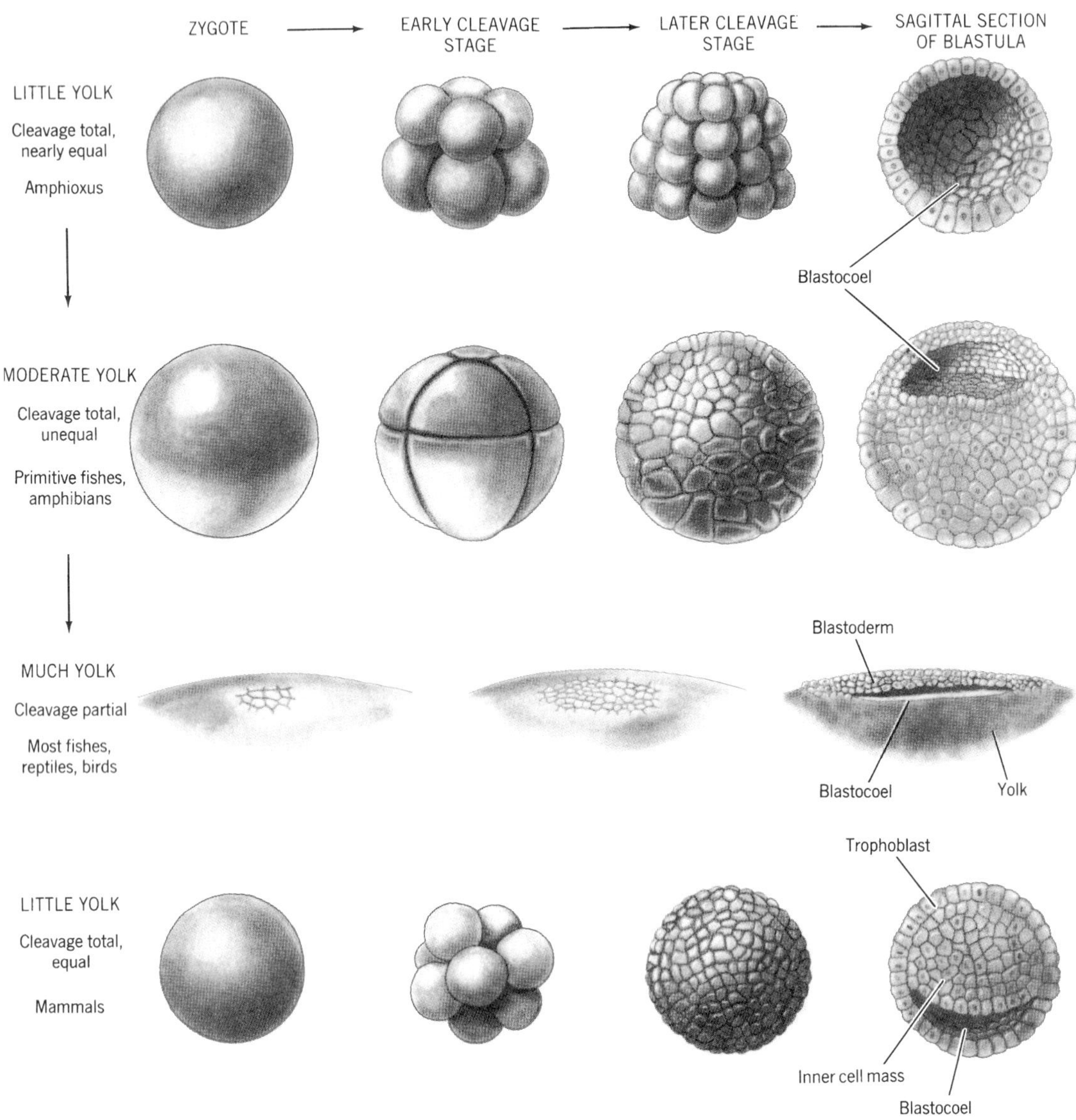

FIGURE 5.6 PRINCIPAL TYPES OF CHORDATE CLEAVAGE. Membranes, shells, and polar bodies omitted. Relative sizes only approximate.

The yolk mass of macrolecithal eggs is simply too great to be penetrated by cleavage furrows. Cleavage is therefore **partial** (meroblastic) and is limited to the relatively small yolk-free region at the animal pole. A cellular **blastoderm** comes to be separated from the uncleaved yolk by a narrow cavity.

Cleavage is again total and equal for the microlecithal eggs of mammals, though the orientation of cleavage furrows is less regular than for amphioxus. The blastula is distinctive in having a superficial layer of cells, the **trophoblast,** which surrounds an **inner cell mass** (Figure 5.6). The blastocoel is displaced toward the vegetal pole. It is significant that, regardless of size and shape, the vertebrate blastula consists of a single tissue layer made up of several hundred cells with polarity that relates to the axes of the future body. The factors that establish bilateral symmetry are not well known; according to taxon they may include gravity or the point of sperm entry.

GASTRULATION AND MESODERM FORMATION

The blastula is converted to an embryo called a **gastrula** by various processes collectively called **gastrulation.** The gastrula has at first two, and then three, tissue or **germ layers.**

Again, amphioxus exhibits the simplest and most primitive form of gastrulation. By the process of **invagination** the vegetal hemisphere of the blastula folds inward and extends to underlie the tissue layer of the animal hemisphere (Figure 5.7). In doing so, it obliterates the blastocoel and forms a double-walled cup. The lips of the cup then approach one another, as though a purse string were being drawn, leaving only a small opening, the **blastopore.** The embryo now has a new cavity, the **gastrocoel.** These tissue movements cause the embryo to rotate, in response to gravity, so that the animal-vegetal axis is horizontal rather than vertical, and the dorsal surface of the embryo is uppermost. The outer germ layer is now called **ectoderm.** Because the adult gut tube will form from much of the inner tissue layer of the embryo, it is called the **archenteron,** or primitive gut. However, as shown in Figure 5.8, the notochord forms from the dorsal wall of the archenteron, and the **mesoderm,** or middle germ layer, forms from a series of dorsolateral outpocketings of the archenteron. The cavities of the pockets become **coelom.** This kind of mesoderm and coelom formation is called

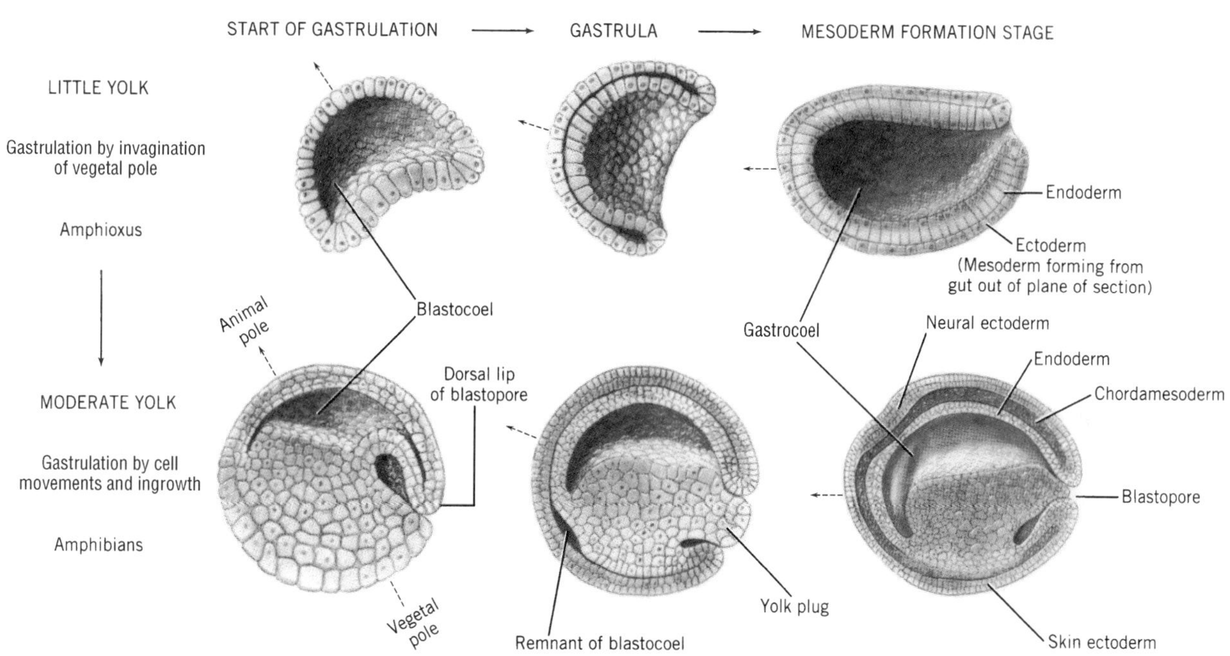

FIGURE 5.7 SOME TYPES OF CHORDATE GASTRULATION that are uncomplicated by a large yolk mass. Sagittal sections.

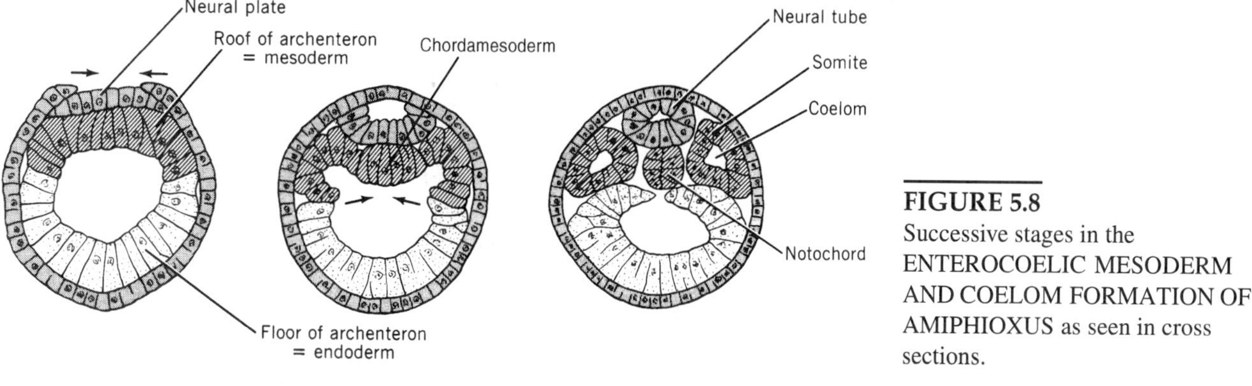

FIGURE 5.8
Successive stages in the ENTEROCOELIC MESODERM AND COELOM FORMATION OF AMIPHIOXUS as seen in cross sections.

enterocoely (= *gut* + *hollow*). It is not found in vertebrates, yet seems to link chordates and echinoderms. The remaining part of the archenteron is **endoderm.**

Blastulas derived by total but unequal cleavage can scarcely gastrulate by invagination because the yolk-laden blastomeres of the vegetal hemisphere are in the way. Instead, cells roll inward (a process called **involution**) at the site of the future blastopore (Figure 5.7). They extend into the blastocoel as a second tissue layer. Surface cells migrate to replenish the supply at the blastopore. Ingrowth is most rapid at what will be the dorsal lip of the blastopore, but ultimately cells stream inward on all sides. As in amphioxus, the blastocoel is gradually obliterated as a gastrocoel is formed. The embryo then has two tissue layers, though the ventral part of the archenteron is swollen with yolk. The roof of the archenteron is called **chordamesoderm** because it forms the notochord in the midline, and a series of paired right and left blocks of mesoderm, or **somites,** dorsolaterally (Figure 5.9). There is no outpocketing from the gut tube. Instead, coelom forms by cavitation or **schizocoely** (= *split* + *hollow*) in the somites. The remaining part of the innermost tissue layer is endoderm. It reconstitutes itself dorsally after the formation of notochord and mesoderm.

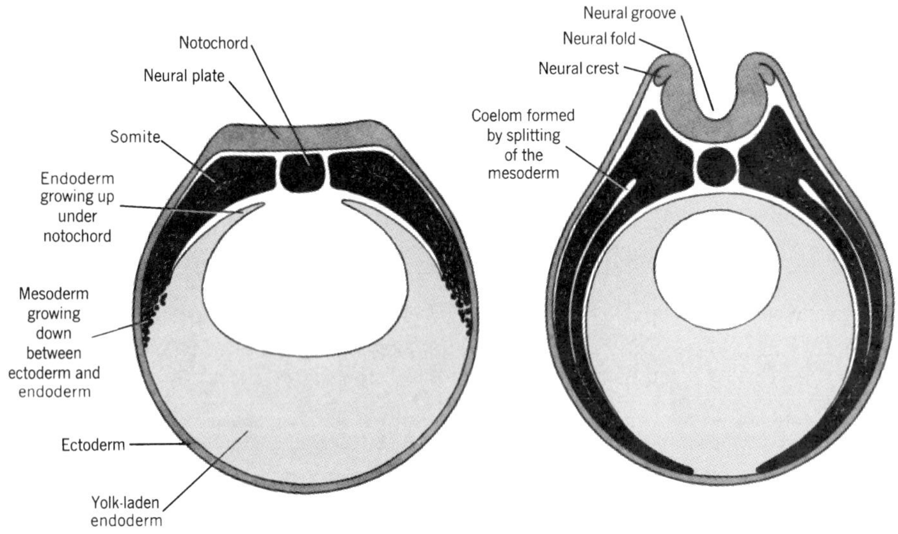

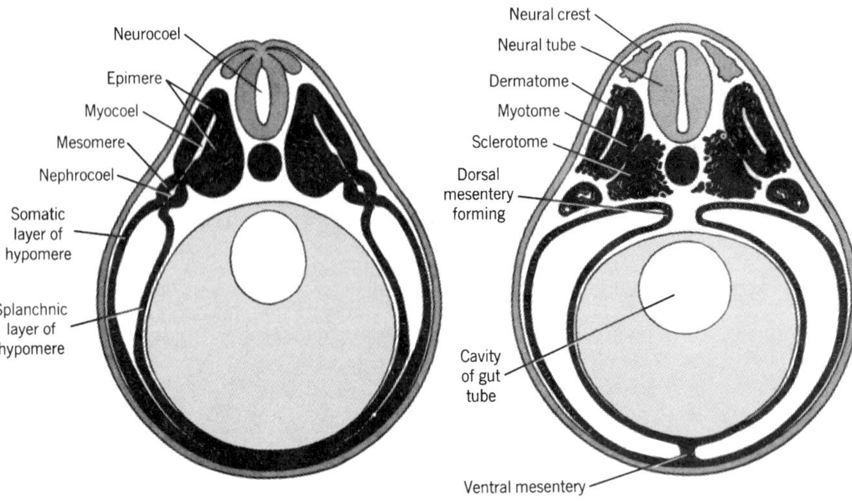

FIGURE 5.9
DIAGRAMS OF NEURULATION AND EARLY DIFFERENTIATION OF THE MESODERM as seen in cross sections of any embryo having a moderate amount of yolk.

If cleavage has been partial, gastrulation is further complicated by the mass of yolk. (The following account is generally accurate for reptiles and birds; fishes may differ.) The second tissue layer, or endoderm, forms under the blastoderm by the process of **delamination:** Some cells detach themselves from the original layer of cells and then aggregate into a second sheet. Some mesoderm may also form by delamination, but most is produced by another process: A lengthwise thickening called the **primitive streak** forms on the posterior part of the germinal disc (Figure 5.10). Surface cells stream toward the streak, where they involute and spread out between the original two tissue layers as mesoderm. The primitive streak is posterior to the **embryonic area,** where the embryo is starting to develop, and may represent a modified blastopore. Cells that involute and then move directly forward from the primitive streak will form the notochord, and alongside it some mesoderm also swings forward into the embryonic area. Mesoderm that is lateral and posterior to the streak will contribute to the fetal membranes. Coelom formation is by schizocoely.

Thus far we have noted that mesoderm may form by separation from the gut tube, by delamination, and by primitive streak. Another method of mesoderm formation is general and important: Cells may detach themselves from existing tissue layers and migrate individually as **mesenchyme.** These branched cells ultimately aggregate in the head, limbs, and other places to form mesodermal tissues.

When the mammalian embryo reaches the blastula stage, maternal fluid indirectly enters the blastocoel, causing it to enlarge greatly. (Figure 5.6 depicts an early stage before the blastocoel expands.) The embryo is now called a **blastocyst.** Endoderm forms by delamination from the inner cell mass and spreads to line the trophoblast. The flattened inner cell mass is now a blastoderm, and it forms a primitive streak to produce the notochord and most of the mesoderm for the fetal membranes and trunk of the embryo. Much mesoderm is also produced as mesenchyme. Coelom forms by schizocoely.

NEURULATION, NEURAL CRESTS, AND ECTODERMAL PLACODES

The gastrula is converted to a **neurula** by processes called **neurulation.** These events, which overlap with the formation of the germ layers as described above and with changes in body form and early differentiation of endoderm and mesoderm as described below, establish the central nervous system. We have seen that chordamesoderm extends forward in the midline from the dorsal lip of the blastopore or primitive streak. Chordamesoderm has the important function of causing, or **inducing,** the overlying ectoderm to thicken into

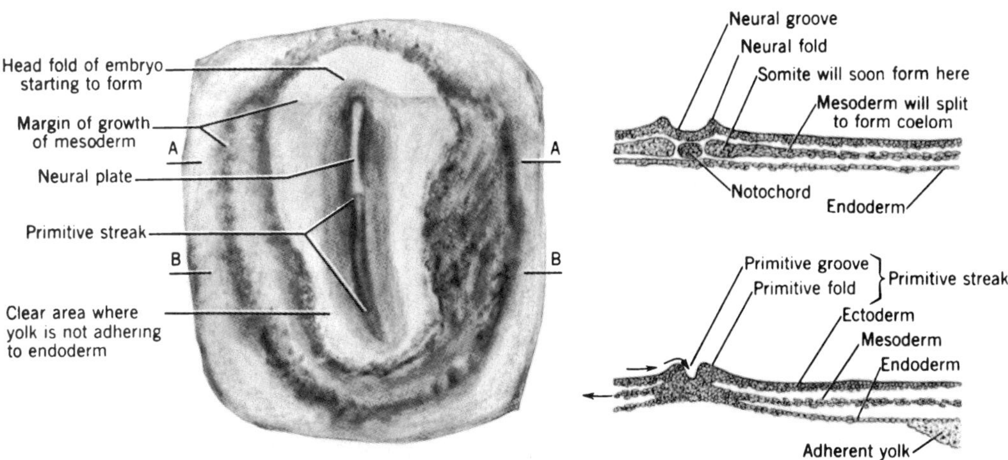

FIGURE 5.10 MESODERM FORMATION AND START OF NEURULATION IN THE CHICK. Left, surface view after about 17 hours of incubation. Right above, cross section at level A-A; right below, cross section at level B-B.

a **neural plate** (Figure 5.9) (although there is experimental evidence that notochord and neural plate may each have been determined at an earlier stage of precursor cells). Longitudinal **neural folds** form along the margins of the plate and arch inward to fuse in the midline. The resulting **neural tube** encloses the **neurocoel.** This establishes the dorsal hollow nerve cord that characterizes all chordates. (The process of neurulation may differ in fishes.)

During neurulation some cells delaminate from the neural plate and, as the neural folds lift up, aggregate loosely in the angles between the sinking neural tube and overlying ectoderm. This collection of cells constitutes the **neural crests** (Figure 5.9). Crest cells will migrate individually over much of the body, inducing other tissues with which they become associated. Their derivatives (listed in Table 5.1) are so diverse that they will be mentioned in various chapters to follow, and so extensive that neural crests have been called a fourth germ layer.

Ectodermal placodes are a series of local thickenings in the dorsolateral ectoderm of the head, paired left and right, that will form the lens of the eye and numerous sensory structures (Table 5.1), as noted further in appropriate chapters.

Neural crests and these ectodermal placodes are alike in their origin from ectoderm, their approximate time of development, their uniqueness to vertebrates, and their conservation among the classes. Note that crests and placodes both contribute to the formation of the sensory ganglia of four cranial nerves and to the sensory epithelium of the lateral line system. Placode derivatives induce the formation of nasal and optic capsules, which are crest derivatives. Most crest cells and some placode cells are alike in being highly migratory.

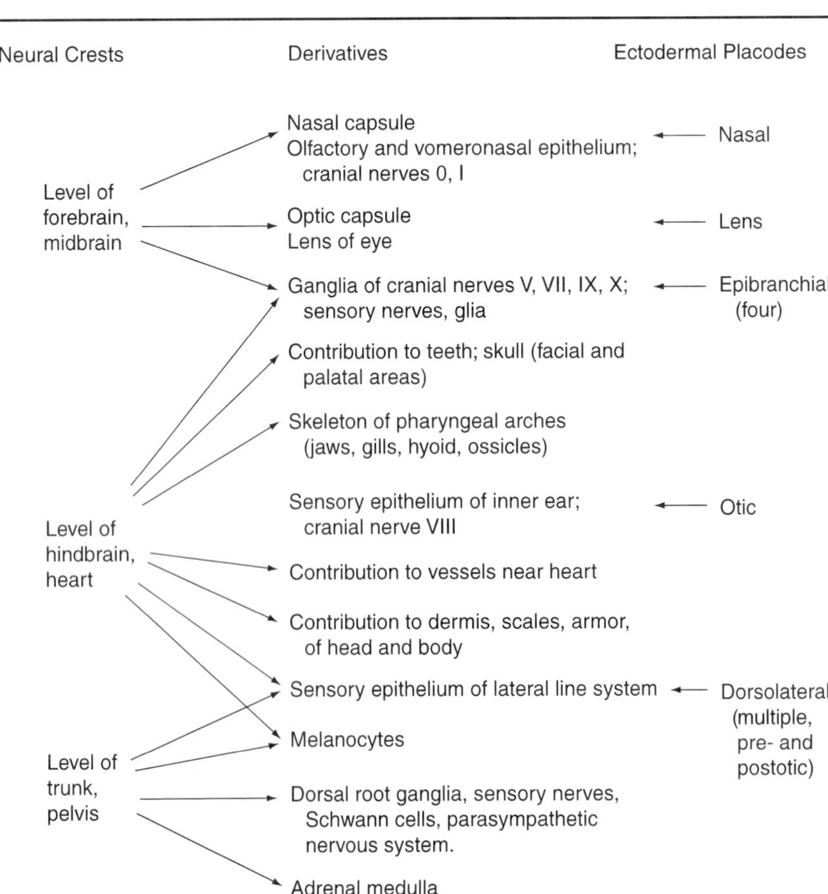

TABLE 5.1
DERIVATIVES OF NEURAL CRESTS AND ECTODERMAL PLACODES.

ESTABLISHMENT OF THE BODY PLAN AND FATE OF THE GERM LAYERS

As neurulation progresses, embryos derived by total cleavage (amphioxus, amphibians) lengthen so that a head and a tail are established. Yolk-laden cells may distend the belly area, but are enclosed within the body contour. Embryos derived by partial cleavage may be thought of as consisting for a time of three germ layers spread "face down" on the uncleaved yolk the way a bearskin rug is spread on a floor. Soon a **head fold** lifts up, like a mounted head on the bearskin rug. A **tail fold** follows, and then lateral body folds. The embryo is then broadly joined to the yolk by a **body stalk,** but is otherwise free (see figure on p. 202).

As the body lengthens, the archenteron is drawn out into a gut tube. The blastopore (or an equivalent locus) becomes the anus, a mark of all Deuterostomia (see p. 24). Later a mouth opening ruptures anteriorly. The endoderm will form the lining of the gut and of its derivative organs (lungs, liver, pancreas, some endocrine glands, yolk sac, allantois, urinary bladder).

The outer, or somatic, ectoderm will form the superficial and germinative layers of the skin and their derivatives (glands, hair, feathers, claws), the lens of the eye, olfactory organ, inner ear, part of the pituitary gland, and lining of the mouth, and will contribute to the formation of teeth and certain scales. The sensory ectoderm, or neural tube, will form the spinal cord and brain, motor nerves, retina of the eye, and part of the pituitary.

With minor exceptions, the mesoderm forms the muscular, skeletal, circulatory, and urogenital systems. Chordamesoderm contributes the notochord. Mesoderm derived mostly from the archenteron or primitive streak flanks the notochord and extends anterior to the notochord in the head, where it forms structures called **somitomeres.** These will be discussed along with the skull and muscles. In the trunk the dorsolateral mesoderm divides, on each side of the body, into three regions. Dorsally there are the discretely segmented **somites** or **epimeres.** Below these are the small, partly segmented **mesomeres** and a sheetlike, unsegmented **hypomere** (or lateral plate), which wedges its way down between ectoderm and archenteron (Figure 5.9). Coelom may be present in each of these units, but expands only in the hypomere, where it separates a lateral **somatic layer** from a medial **splanchnic layer.** The somatic layer will line the body cavity and bud off mesenchyme that contributes to the bone and muscle in the limbs. The splanchnic layer forms heart, mesenteries, and muscles of the gut (see p. 196). Where ectoderm and somatic mesoderm join, as in the body wall, they are called **somatopleure;** where endoderm and splanchnic mesoderm join, as in the wall of the gut, they are called **splanchnopleure.**

The mesomere forms only urogenital organs. The somite differentiates further into a lateral **dermatome,** a **myotome,** and a medial **sclerotome** (Figure 5.9). The dermatome becomes mesenchyme that spreads out to form the dermis of the skin and such hard tissues as are derived therefrom (e.g., scales, bone). The myotome forms muscles of the spine and throat, and much of the muscles and skeleton of the appendages. The sclerotome forms a greater or lesser portion of the vertebrae, depending on taxon.

The early differentiation of the mesoderm is outlined in Table 5.2. All of these developmental steps are carried further in chapters to follow.

FETAL MEMBRANES AND PLACENTATION

If such yolk as is present in the egg cleaves and is directly incorporated into cells of the body (amphibians), no fetal membranes are needed. The yolk supply being limited, hatching is early and the larva starts to feed.

If the yolk does not cleave, and eggs are laid in water (most fishes), then an extraembryonic circulation is needed to absorb the yolk and carry it into the body. To accomplish this, a membrane that includes all three germ layers extends over the surface of the yolk from the body stalk. This **yolk sac** quickly becomes vascularized and functional. Since respiration and excretion are by direct contact with the environment, no other fetal membrane is needed.

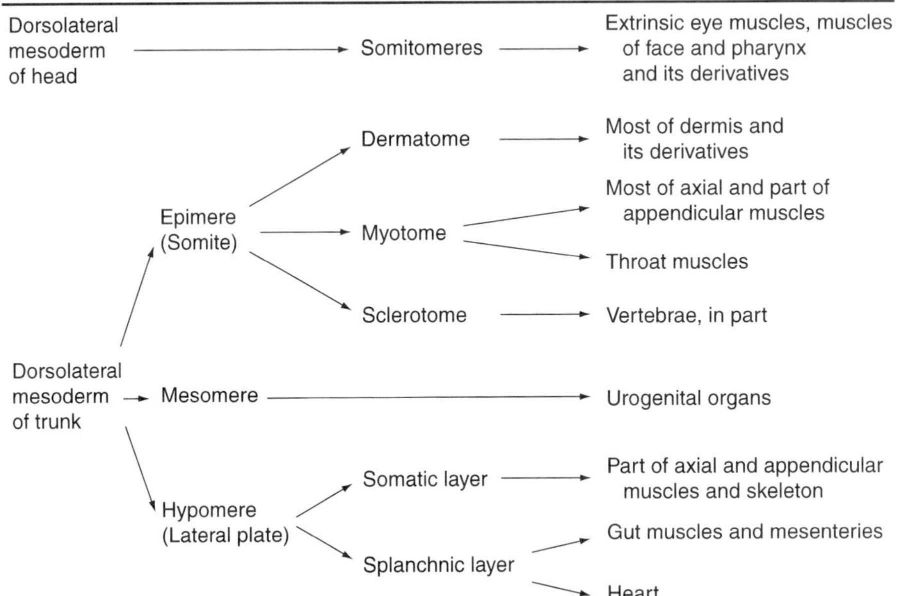

TABLE 5.2
DERIVATIVES OF EMBRYONIC MESODERM. See also p. 178.

Requirements of the embryo are much more stringent when development is on land within a shell (reptiles, birds), and hatching is delayed. The extraembryonic mesoderm derived from the primitive streak soon splits, thus establishing an extensive extraembryonic coelom (Figure 5.11). The resulting splanchnopleure is adjacent to the yolk and forms a yolk sac. The overlying somatopleure lifts into a **head fold of the amnion,** and later also into lateral and tail folds, all of which converge and fuse over the embryo. This creates an outer **chorion** and an inner **amnion.** The latter contains **amniotic fluid** that bathes the embryo and provides a sheltered space in which it can grow. Somewhat later, as the embryo becomes larger, a vesicle grows out of its hindgut and extends into the extraembryonic coelom. This splanchnopleuric vesicle forms the **allantois,** which grows out under the egg shell (from which it is separated only by the chorion and egg shell membranes), becomes vascularized, and serves the embryo for respiration. It also is a receptacle for excretory wastes, absorbs albumen, and takes some minerals from the shell.

Eutherian mammals nourish their young in the uterus by physiological exchange between fetal and maternal bloodstreams. An organ that performs such a function is a

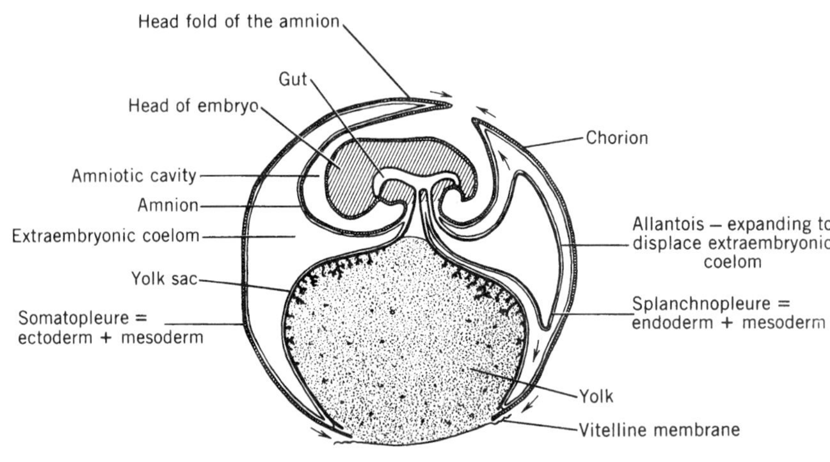

FIGURE 5.11
Stylized FETAL MEMBRANES OF THE BIRD early in development. Compare with figure on p. 202.

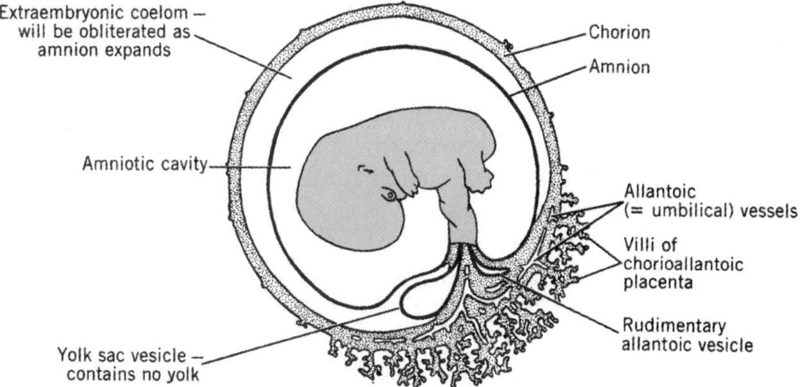

FIGURE 5.12
FETAL MEMBRANES OF THE PRIMATE TYPE illustrated by the human at about 6 weeks gestation.

placenta. It makes yolk superfluous and, hence, eutherian eggs are secondarily microlecithal. (See pp. 295 and 296 for other ways that the maternal body of other vertebrates may nourish the fetus.) The fetal membranes derived from somatopleure (chorion, amnion) do not become vascular, and therefore cannot support a placenta. Both yolk sac and allantois may become vascular, and either, or both, may contribute. The formation and nature of the fetal membranes and placenta vary greatly among mammals. A short summary serves our purpose here, although the evolution of these structures is a fascinating story in itself (see Luckett in References).

The yolk sac is the most variable of the membranes. It is the principal fetal contribution to the placenta in marsupials and some rodents. It is large and persists to birth in carnivores, is first large and then lost in ungulates, and is first small and then rudimentary in primates (Figure 5.12). The allantois makes the fetal contribution to the placenta of most mammals. The allantoic circulation of reptile and bird is homologous with the umbilical circulation of these mammals (see figure on p. 251). The vesicle of the mammalian allantois ranges from large to vestigial. Since the chorion lies between allantois and uterus, it is incorporated into the placenta (which is, therefore, described as **chorioallantoic**), although most of its tissue erodes away in some taxa. The amnion, which varies less than the other membranes, becomes a thin but tough sac. The amniotic fluid bathes the fetus, gives it freedom to move, and protects it from pressures on the maternal abdomen. Because of their common possession of an amnion, reptiles, birds, and mammals are collectively called **amniotes.** The **umbilical cord** of mammals is equivalent to an attenuated avian body stalk.

REFERENCES

Alberch, P. 1980. Ontogenesis and morphological diversification. *Am. Zool.* 20:653–667.

Alberch, P., et al. 1979. Size and shape in ontogeny and phylogeny. *Paleobiology* 5:296–317.

Carlson, B.M. 1981. *Patten's foundation of embryology.* 4th ed. McGraw-Hill, New York. 672p. Excellent general reference with good treatment of organogenesis.

de Beer, G.R. 1958. *Embryos and ancestors.* 3rd ed. Oxford Univ. Press, London. 197p.

Fink, W. 1982. The conceptual relationship between ontogeny and phylogeny. *Paleobiology* 8:254–264.

Gans, C. 1989. Stages in the origin of vertebrates: analysis by means of scenarios. *Biol. Rev.* 64:221–268.

Gilbert, S.F., and A.M. Raunio (eds.). 1997. *Embryology: constructing the organisim.* Sinauet, Sunderland, MA. 537p. One of few comparative texts. Invertebrate embryology is followed by chapters on cephalochordates, fishes, amphibians, reptiles and birds, and mammals.

Gould, S.J. 1977. *Ontogeny and phylogeny.* Harvard Univ. Press, Cambridge, MA. 501p. A significant and scholarly reexamination of a classical but ever-important subject.

Hall, B.K. 1998. *Evolutionary developmental biology.* Chapman & Hall, London. 275p. Developmental process and constraints in relation to evolutionary process and diversity, homology, timing, and principles. Historical background.

Hall, B.K. 1998. Germ layers and the germ-layer theory revisited: primary and secondary germ layers, neural crest as a fourth germ layer, homology, and demise of the germ-layer theory. *Evol. Biol.* 30:121–186.

Horder, T.J. 1989. Syllabus for an embryological synthesis, pp. 315–348. *In* D.B. Wake and G. Roth (eds.), *Complex organismal functions: integration and evolution in vertebrates.* Wiley, New York.

Lauder, G.V. 1982. Historical biology and the problem of design. *J. Theor. Biol.* 97:57–67.

Le Douarin, N.M., and C. Kalcheim. 1999. *The neural crest.* 2nd ed. Cambridge Univ. Press, New York. 445p. Documents, in some detail, the molecular underpinnings of the directed migration and differentiation of neural crest.

Luckett, W.P. 1977. Ontogeny of amniote fetal membranes and their application to phylogeny, pp. 439–515. *In* M.K. Hecht, P.C. Goody, and B.M. Hecht (eds.), *Major patterns in vertebrate evolution.* Plenum, New York.

Maderson, P.F.A. (ed.). 1987. *Developmental and evolutionary aspects of the neural crest.* Wiley, New York. 394p.

Martindale, M.Q., and B.J. Swalla. 1998. The evolution of development: patterns and process. Symposium volume, *Am. Zool.* 38(4):591–696.

Müller, G.B. 1991. Experimental strategies in evolutionary embryology. *Am. Zool.* 31:605–615.

Müller, G.B., and G.P. Wagner. 1996. Homology, *Hox* genes, and developmental integration. *Am. Zool.* 36(1): 4–13.

Northcutt, R.G., and C. Gans. 1983. The genesis of neural crest and epidermal placodes: a reinterpretation of vertebrate origins. *Quart. Rev. Biol.* 58:1–28.

Olsson, L., and B.K. Hall. 1999. Developmental and evolutionary perspectives on major transformations in body organization. Symposium volume, *Am. Zool.* 39(3): 612–694.

Schwartz, J.H. 1999. Homeobox genes, fossils, and the origin of species. *Anat. Rec.* (*New Anat.*) 257:15–31.

Smith, M.M., and B.K. Hall. 1993. A developmental model for evolution of the vertebrate exoskeleton and teeth: the role of cranial and trunk-neural crest. *Evol. Biol.* 27:387–448.

Webb, J.F., and D.M. Noden. 1993. Ectodermal placodes: contributions to the development of the vertebrate head. *Am. Zool.* 33:434–447.

Chapter 6

Integument and Its Derivatives

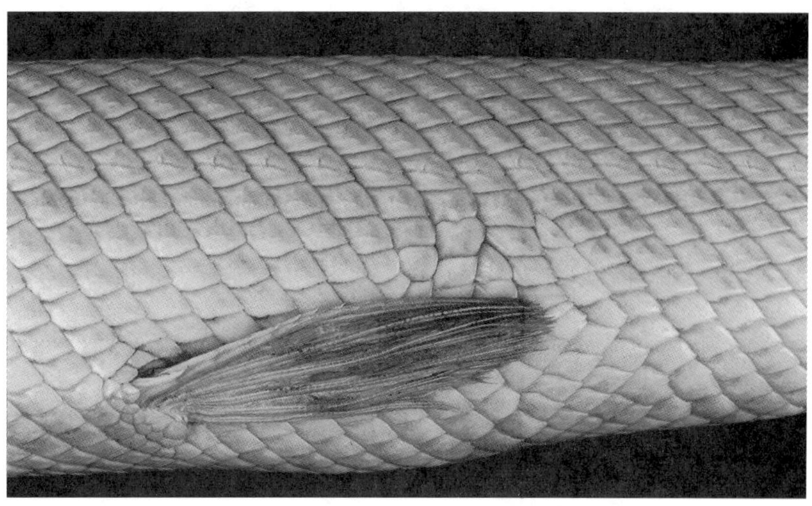

FUNCTIONS OF THE SYSTEM

The integument and its derivatives form an exceedingly varied and adaptable organ system. By weight, the system is often the largest of the body, and no other performs as many functions. It provides physical protection to delicate tissues below, thus guarding against the entry of most injurious organisms and materials and cushioning impact with the environment, as under the feet. It contributes importantly to water balance. Thus, amphibians can absorb water through the skin, even taking moisture from damp air or soil, whereas the skin of desert reptiles is resistant to water loss. Dissipation of heat to the environment may be increased by the dilation of superficial vessels and by the evaporation of sweat, whereas conservation of heat is increased by fat deposits and the erection of insulating hair or feathers. The integument provides coloration essential to identification, aggressive and sexual behavior, and camouflage. It serves locomotion through friction pads, the interlocking of scales or claws with the substrate, the provision of airfoils, and in other ways. Respiratory exchange occurs through the damp skin of amphibians, even supplanting lungs in some salamanders. Secretions of skin glands may contribute to attraction or repulsion, nutrition of the young, excretion of salts and urea, and thermoregulation. The integument houses many sense organs, contributes to the contours of the body, screens out injurious energy waves, and it may also store fat and glycogen, provide significant support and defense to the body, and synthesize vitamin D.

Because the integuments of the various vertebrates are so diverse, the system tells the morphologist much about the habits and environments of animals, and enables the systematist to identify most vertebrates. Some mammals can be identified by single hairs, and some birds from several feathers. However, because the integument is so responsive to habit and environment, it has told morphologists less than some other organ systems about phylogeny. Nevertheless, progress is being made.

Above: Thick, hard ganoid scales of a gar.

DEVELOPMENT AND GENERAL STRUCTURE OF SKIN

The skin of all vertebrates has two principal layers, a superficial **epidermis** and a deeper **dermis** (Figure 6.1). The epidermis is derived from the ectoderm on the surface of the embryo. The dermis is derived from the dermatome supplemented by contributions from the lateral and ventral somatic mesoderm. Cells from these sources migrate as mesenchyme to distribute themselves evenly under the ectoderm. Some neural crest cells also invade the developing dermis.

The epidermis is stratified into two or more layers. The deepest layer rests on the dermis and consists of closely packed, discrete cells. It is called the **stratum germinativum** because its daughter cells are pushed outward and, as they mature, are transformed to become the more superficial cells of the skin. The layer or layers of the epidermis that are superficial to the stratum germinativum are exceedingly varied according to taxon. Most are (or will become, or have been) secretory in nature and fall into two general categories, mucous cells and proteinaceous cells. The former produce various types of mucus, some kinds of poisonous secretions and, in some fishes, **photophores** (light-producing cells). The proteinaceous line of epidermal cells may produce slime, poisons, substances eliciting alarm reactions, enamel, and possibly some photophores. The principal genetic products of this cell line, however, are a complex of water-insoluble, intracellular proteins termed **keratins.** Keratins are responsible for the horny consistency of feathers, hair, claws, reptilian scales, and also the dead outermost layer, or **stratum corneum,** of the dry skin of tetrapods ("keratos"/"cornu" = *horn*). Two broadly defined molecular types, designated α-and β-keratins, are recognized, and each is the product of many genes. During development an individual epidermal cell (keratinocyte) might produce α-keratins, β-keratins, or a combination of the two.

The dermis is usually thicker than the epidermis. It has fewer kinds of cells and is characterized by a meshwork of fibers. Most abundant are **collagenous fibers,** which are constructed in strands like a rope: Groups of three polypeptide chains twist into left-handed helices, and groups of these coil into right-handed helices, thus forming microfibrils. Many of these are gathered into bundles, which, in turn, are gathered into tough, straight fibers. **Elastic fibers** are fewer in number, wavy, nonfibrillar, and branched. Relaxed elastin molecules are crumpled. They straighten out into an oriented lattice when pulled, and recoil when released.

The fibers of the dermis are arranged in specific patterns as weaves, or helices that coil around the body. This provides added toughness (as in rhinoceros hide), stiffness that resists torsion (as in sharks), or elastic recoil (whales and many fishes).

The dermis commonly has an outer, vascular **stratum spongiosum** and a deeper, thicker **stratum compactum.** These merge with one another and also bridge across to secure the skin to the connective tissue covering the muscles of the body wall. Smooth muscle fibers may be present in the dermis. Fat commonly invades the dermis or is deposited between the skin and the body. The glands of the skin are derived from the epidermis but usually penetrate into the dermis, which then adds supportive tissue to them.

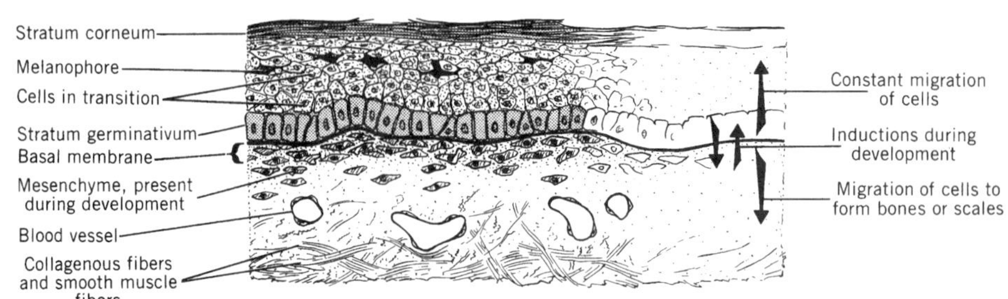

FIGURE 6.1
SECTION OF GENERALIZED VERTEBRATE SKIN.

Pigment cells are called **chromatophores.** They are derived from neural crest cells and occur, according to taxon, in any level of the skin, but tend to concentrate near the epidermal-dermal boundary. Chromatophores of the epidermis are particularly characteristic of **endotherms** (animals maintaining high resting metabolic rate and relatively constant temperature) and are of one type called **melanophores.** They have numerous migratory organelles, termed melanosomes, which contain the pigment melanin. Melanin is black, brown, or red. Color imparted by these cells may be constant or may be responsible for **morphological color change,** which is seasonal, age-related, or otherwise a relatively slow kind of change.

Chromatophores of the dermis are found almost exclusively in **ectotherms** (animals having low resting metabolic rate and usually variable body temperature). They may maintain constant color, cause morphological color change, or cause **physiological color change,** which is relatively rapid, as when a fish or lizard adapts its color to that of an altered substrate. There are three types of dermal chromatophores: Melanophores are similar to those of the epidermis. **Iridophores** have organelles called reflecting platelets that are oriented in stacks and contain crystalline deposits (chiefly of guanine) that scatter or reflect light. These cells are commonly large. **Xanthophores,** which are yellow, and **erythrophores,** which are red, have their pigments (pteridines and carotenoids) in organelles called pterinosomes.

These various kinds of chromatophores can be structurally and physiologically interrelated in the achievement of certain color effects. Their complex control may include influence by hormones of the pituitary, thyroid, gonads, and adrenals, and in some ectotherms by the nervous system as well. Color is "used" by the various vertebrates for concealment, for making themselves conspicuous (e.g., as a warning, social releaser, or sexual attractant), for control of heat absorption and conservation, for protection of the nervous system or gonads from light, and for control of the synthesis of vitamin D.

GENERAL DEVELOPMENT OF SKIN DERIVATIVES

The epidermis and dermis are separated by a thin basal membrane (Figure 6.1). During development of the skin and of its derivatives, inductions occur across the membrane between the germinal epithelium and the mesenchyme. Even if an integumentary derivative incorporates only epidermal tissue (e.g., horny scales, feathers, hair) or only dermal tissue (e.g., certain bones), both layers are essential to the formation of (probably) all derivatives, and some of them (e.g., teeth, fish scales) incorporate tissue from each layer. In the absence of underlying dermis, the embryonic epidermis degenerates, and experiments show that the kind of epidermal derivative formed (whether scale, feather, hair, or other) is controlled by the nature of the associated dermis. The influence of the epidermis on the dermis may be less universal, yet epidermis apparently triggers the dermal contributions to teeth and fish scales.

INTEGUMENT OF FISHES: EMPHASIS ON DERMAL DERIVATIVES

Soft Structures The soft part of the integument of extinct agnathans and placoderms is, of course, unknown. The epidermis of CYCLOSTOMES is thin and has several kinds of glands, all of which are unicellular. Most numerous are **mucous glands** of two types. **Club glands** produce slime of fibrous protein. **Granular glands** probably discharge at the surface of the body, but their function is not yet known. Keratin is absent. A thin noncellular **cuticle** covers the epidermis. The dermis, which may be thinner than the epidermis, consists largely of a fibrous layer that contains collagenous fibers but no elastic fibers. There is no trace of scales.

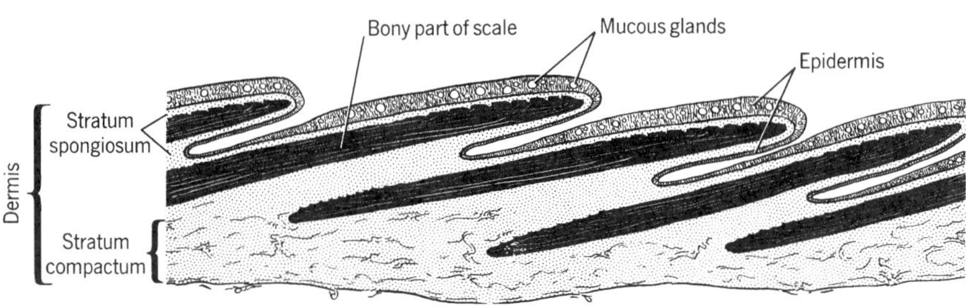

FIGURE 6.2 SECTION OF THE SKIN OF A TELEOST.

The skin of jawed FISHES is usually thin and glandular. It fits tightly over the body (Figure 6.2). With some exceptions, keratin is entirely absent. The replacement of worn epidermis is constant. Mucous glands of one or another type are nearly always abundant. They are usually unicellular but may also be multicellular. The slimy mucus they secrete cleans the body and produces a cuticle that prevents the entry of foreign material, assists in osmoregulation, and reduces resistance as the fish swims. Granular and club glands are also common. Some fishes have **poison glands** associated with fin spines; others have multicellular light organs that may even be provided with tiny lenses and reflectors. The dermis, though still thin, is divided into a stratum compactum and a stratum spongiosum (except when it covers the fins, where it is reduced to a basal membrane). When wounded, the skin of fishes heals rapidly.

Development and Structure of Hard Tissues The scales and nonglandular integumentary appendages of fishes are largely of dermal origin, whereas those of tetrapods are largely of epidermal origin. The most complex derivatives of the integument of fishes are hard scales and denticles of various kinds. Before we study their nature and phylogeny, it is desirable to know about the tissues of which they are constructed.

In historical perspective, the description and classification of the scales and hard integumentary appendages of fishes have been complicated by various factors: (1) Fish scales—particularly those of fossil fishes—include many kinds of hard tissues. (2) The various hard tissues grade into one another and combine in many ways. (3) Some hard tissues are found only in fossil fishes, thus restricting developmental analysis. (4) Certain virtually identical tissues seem to develop from different germ layers. (5) Some of the terms applied to scales describe gross shape (e.g., cycloid, rhomboid), others identify one of the tissues present (ganoid, cosmoid), and still others indicate a particular combination of tissues (palaeoniscoid, lepidosteoid). This has resulted in a confusing lack of parallel terms. However, difficulties are being overcome as molecular and biochemical analysis, physiological studies, and electron microscopy reveal details in the mechanisms of deposition and resorption of hard tissues.

Significance is attached to the embryonic origin and interactions of the precursors of hard tissues. Some neural crest cells join mesenchyme from the dermatome and contribute to the dermis of the skin and the tissues of the gums. Specifically, they form papillae that induce enamel organs from the overlying ectoderm. These organs, in turn, induce the papillae to form dentine, if this substance is to develop at all. Bone may also be induced in this way. Finally, dentine induces the enamel organ to produce enamel (these tissues are defined below). If dentine is not deposited, then the enamel organ (or its equivalent—the term is not apt in this instance) may form horny scales or any of the other derivatives of the ectoderm described earlier in this chapter. There is, therefore, a basic similarity in the mechanism of development of teeth and all skin derivatives regardless of ultimate hardness and principal germ layer of origin.

The principal kinds of hard tissue are enamel, dentinous tissue, and bone. Each of these has subtypes, and another tissue—enameloid—fits between enamel and dentinous tissue. **Enamel** is the hardest tissue of the body. It is shiny, translucent in thin sections, and composed of elongate crystals of hydroxyapatite $[3(Ca_3PO_4)_2 \cdot Ca(OH)_2]$. In therian mammals it is prismatic. Internal cells and tubules are absent. Only about 3 percent of the tissue is organic. Enamel occurs only on teeth and superficial denticles, scales, or armor plates and is usually external to any other hard tissues present. It is produced only by ectoderm—even at the back of the mouth, where the ectoderm has migrated into position. Growth is by accretion on the inner surface of an enamel organ. Hence, enamel cannot be altered or replaced once it has been deposited.

Enameloid is as hard as enamel but forms from ectomesenchyme (mesoderm derived from neural crest cells). It occurs on the surface of the scales of pteraspids and some primitive bony fishes, as well as on scales and teeth of cartilaginous fishes. **Ganoine** is an enameloid characterized by thick deposition in successive waves of growth that create a laminar structure.

Dentinous tissue is harder than bone and usually softer than enamel. The chemical composition of its inorganic salts is the same as that of enamel, but the content of organic fibers is typically about 25 percent. The generative cells usually, but not invariably, remain external to the hard tissue; their processes then penetrate the matrix via dentinal tubules. Dentinous tissue occurs only in teeth, denticles, scales, and external armor. It is present unless secondarily lost, and lies internal to enamel and usually external to bone where those tissues are also present. It is produced only by the outer surface of a mesodermal papilla, which in turn occurs adjacent to the boundary of mesoderm with ectoderm. Specialists recognize various types and subtypes of dentinous tissue according to inclusion or exclusion of cells from the hard matrix, number of tubules per cell, branching pattern of tubules, and layering of matrix. Most types of dentinous tissue (mesodentine, semidentine) are limited to fossil fishes. That of tetrapods and some other fishes is called simply **dentine** (but also metadentine, orthodentine, and osteodentine). This is the ivory of an elephant or walrus tusk. Its cells are external to the matrix, each having one tubule. It is not possible to say which of the many types of dentinous tissue is the most primitive; some, however, are probably more primitive than bone. **Cosmine** (not a parallel term) is dentine (sometimes with bone and enameloid) having characteristic tufts of tiny pore canals that radiate upward and outward from a succession of small vascular centers distributed more or less in a plane that is parallel to the surface of the scale. Cosmine is found only in certain extinct fishes. The pore canals were probably sensory.

Bone has about the same organic content as dentine (though the range of variation is greater), usually occurs internal to dentine if each is present, and develops in a deeper and less restricted part of the dermis. It usually has internal bone cells (*osteocytes*) located in small vacuities (*lacunae*) that intercommunicate by small canals (*canaliculi*) (Figure 6.3). Acellular bone (called aspidin) is common, however, in anaspids, pteraspids and teleosts. The ancestral relationship between cellular and acellular bone remains in dispute. Janvier (see References) believes that a preponderance of the evidence indicates that acellular bone is the older. Bone (like osteodentine) may be characterized by osteons (here also called Haversian systems), but when it is adjacent to internal and external surfaces it is usually deposited in laminar sheets (like orthodentine). Bone may be compact or vascular and spongy. It may have few intrinsic collagenous fibers or many that are more or less layered and make the bone soft and flexible. Other collagenous fibers (Sharpey's fibers) may penetrate bone or dentine from adjoining connective tissue to bind them together. The lamellae and osteons of bone can be resorbed and replaced at any internal or external surface, even if osteocytes are absent, and remodeling is common. Growth includes reorganization as well as accretion.

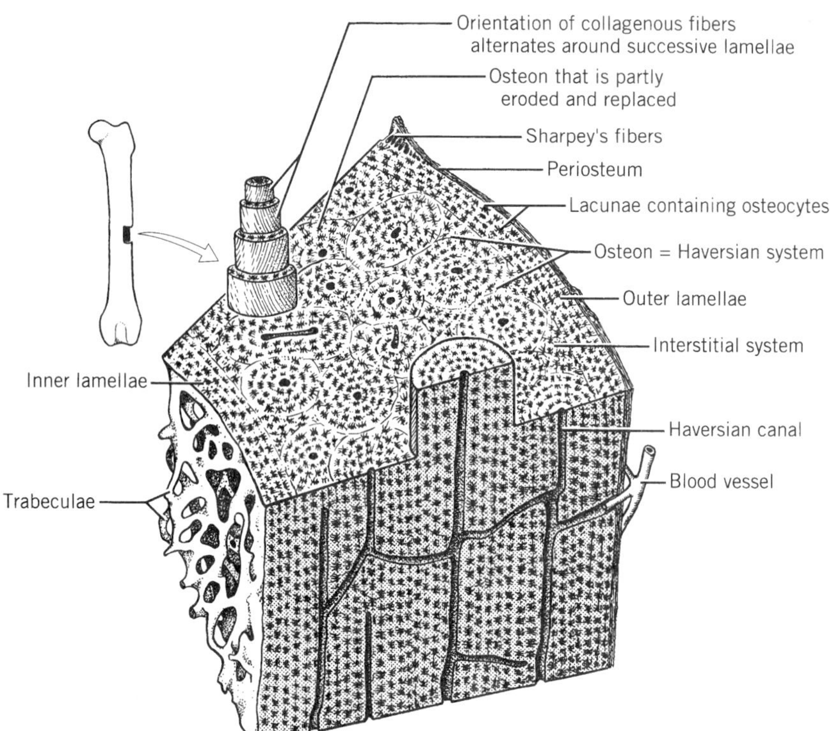

FIGURE 6.3 STRUCTURE OF MATURE COMPACT BONE. (Cross sections of 4–5 osteons span 1 mm; this is bone of a small vertebrate.)

Thus bone is a highly variable tissue; classifications are based on the presence or absence of internal cells, layering, orientation of contained fibers, vascularity, and developmental remodeling of its initial structure. (Bone is characterized further in Chapters 8, 9, and 21.)

Phylogeny of Bony Scales and Their Derivatives Hard tissues appear to have been primitive for vertebrates. The earliest known fishes and jawless vertebrates are nearly always *more* heavily armored than the descendants in their respective lineages. Enamel, dentine, and bone were all present in fragments of armor from the Ordovician period. Primitive armor may have served as a calcium reserve, for protection, for osmotic control, or perhaps to make the body heavy.

The early presence of hard tissues leaves one puzzled as to the origin of the heavy and complicated armor of ostracoderms from the (apparently) naked integument of protovertebrates. Surely there must have been an intermediate step. The theory of the aggregation of small scales and the theory that extensive dermal papillae underlying extensive enamel organs could have formed large armor plates (see p. 91) are attempts to reconstruct that step.

Armor shields of ostracoderms and placoderms (cephalaspids, pteraspids, arthrodires, antiarchs) differ only in size from the coarse scales found elsewhere on their bodies. In section, armor shows three principal layers. The surface is composed of dentine (often reduced in placoderms), which may be capped with enamel completing surface projections called **denticles** or odontodes. A middle layer is composed of bone that is riddled with anastomosing channels for small blood vessels and sensory pits. The basal layer is lamellar bone with fewer vascular channels (Figure 6.4).

Anaspids have no armor shields, and their scales have regressed in that only the basal layer is retained. It is probable that the naked skins of lampreys and hagfishes resulted from further degeneration. (There is histochemical evidence that the lamprey skeleton can mineralize under temperature-dependent experimental conditions.)

COMMENT 6.1
MORE ABOUT BONE TISSUE

Compact bone is constructed so as to optimize the relation between the strength it provides and the burden of its weight. Minimum risk is balanced against minimum mass. Optimum attributes vary with usual stress, maximum loads, the need to repair damage, and other factors. Accordingly, the fine structure of compact bone is diverse and dynamic.

The bone that is first to form during active growth is said to be **primary bone.** It may be deposited on membrane or bone, or may replace cartilage; it does not replace existing bone. Its lamellae, like the layers of an onion, are more or less parallel to its surfaces. Its few osteons are scattered where they surround small vessels. Primary bone is relatively dense and strong. **Secondary bone** replaces primary bone by a process of resorption followed by deposition (see Comment 21.2 on p. 396). Its osteons are closely packed. It is less strong but is superior at arresting and healing microcracks, so it is at least as durable as primary bone.

Bone is further classified according to the orientation of its collagenous fibers and mineral crystals. These are randomly arranged in **woven bone,** which is relatively weak but fast to form. (It is first to grow at the site of a fracture.) The fibers of **lamellar bone** display various regular patterns of organization. They may spiral at a shallow angle to the long axis of their osteon, those of successive lamellae tilting in alternate directions (see Figure 6.3). Such bony tissue has maximum resistance to tension parallel to the axis of the osteon. Alternatively, the fibers may be circumferential, or nearly transverse to the axis of the osteon. Such bone has maximum resistance to compression. Other osteons have the fibers of successive lamellae alternating between longitudinal and transverse, and additional patterns have been described.

Cosmoid scales differ in no fundamental respect from the ancient armor just described—the same basic layers are present. The term has gained wide usage, however, and is helpful for describing somewhat more advanced scales that are typically smaller, thinner, and characterized by having dentine of the cosmoid type. The surface of the scale is usually sculptured by the enamel of the denticles ("cosmoid" = *ornamented*). The scales may be **cycloid** (roundish in outline) or **rhomboid** (a parallelogram in outline). Cycloid scales are usually imbricated (overlapping). Rhomboid scales often overlap at their inner margins but are fitted edge to edge at their outer surfaces. Cosmoid scales are found on the posterior parts of the bodies of some placoderms and on crossopterygians and early dipnoans. (The modern crossopterygian, *Latimeria,* has large cycloid scales with reduced surface layers; modern dipnoans have thin scales without cosmine.)

Ganoid scales are thick rhomboid structures that evolved from cosmoid scales. There are two principal types. The more primitive is called a **palaeoniscoid scale.** The surface is thickened during successive periods of growth by laminations of the enamel called ganoine. Cosmoid dentine is retained under the ganoine. The base of the scale is lamellar bone pierced by vascular canals. This type of scale is found on primitive extinct actinopterygians and on the surviving *Polypterus.* The second type of ganoid scale was derived from the first and is called a **lepidosteoid scale.** The ganoine is the same. The cosmine is deleted. The bony base is acellular, and the canals, though present, are no longer vascular. This type of scale is found on fishes of the extinct class Acanthodii and on some primitive neopterygians, including gars (see the chapter opener). (Modern chondrosteans have degenerate scales.)

Elasmoid scales were derived from ganoid scales of the lepidosteoid type. They are restricted to teleosts. The basal layer, which now forms the bulk of the scale, remains acellular but is laced by collagenous fibers coursing in various directions. The resulting bone (called isopedine) is somewhat flexible and soft. The ancestral ganoine is absent, and in its place is a thin surface glaze derived from the enamel organ. Elasmoid scales are thin, imbricated, and cycloid or **ctenoid** (having comblike projections on the exposed margins).

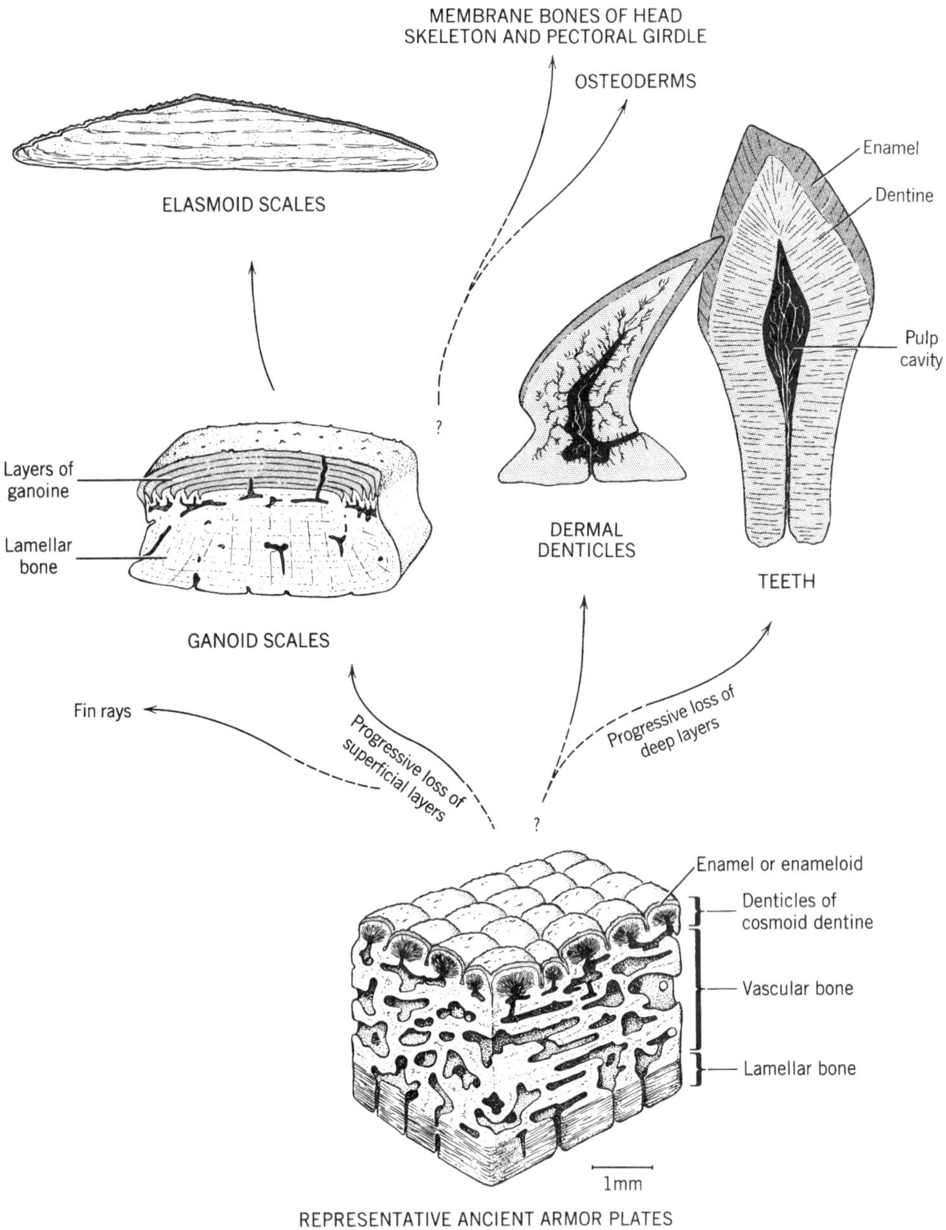

FIGURE 6.4
STRUCTURE AND RELATIONSHIPS OF DERMAL SCALES AND DERIVATIVES.

Isolated **dermal denticles** or placoid "scales" are confined to elasmobranchs and some aberrant placoderms. They evolved from cosmoid scales or possibly from the integument of a placoderm stock that sidestepped armor plates and heavy scales entirely. They lack the bony basal layer of scales and are always small and usually isolated. A central pulp cavity is surrounded by dentine, and this is capped by a tissue that is usually considered to be ectodermal enamel.

Various hard structures of tetrapods are derived from the bony scales of fishes. Caecilians have bony ossicles buried in the skin. **Osteoderms** are plates of bone that are located under the horny scutes of crocodilians, some lizards, and some extinct amphibians and reptiles (Figure 6.6). They are probably derived from dermal scales (although the identifi-

COMMENT 6.2

THE ORIGIN OF COMPLEX SCALES AND ARMOR IS A COMPLEX PROBLEM

The hard tissues of scales and body armor are so varied and complex that their classification has long been debated (see main text). The origin and evolutionary history of these structures has been equally disputed. Did large armor plates form as units as a consequence of extensive developmental interactions between neural crest derivatives and mesoderm (a position favored some years ago by the Americans M.L. Moss, A.S. Romer, and B. Schaeffer)? Or did large plates form by the accretion of small units? (The oldest fossils already have both large and small elements.)

The basic unit of the now favored small-to-large theory is the **odontode.** An odontode is a unit of dentinous tissue with vascular canals. It may be capped by enamel or enameloid and may be anchored to bone. Adjacent odontodes may coalesce in various ways to form larger, more complex units. There are two principal theories about how this could happen.

The scheme of the Scandinavian paleontologists E. Stensiö and T. Ørvig (the *lepidomorial theory*), developed in many papers from the 1950s to the 1970s, holds that larger units form by the concrescence of successive odontodes at the developmental level of the dermal papilla (like the dental papilla in the figure on p. 104). The other theory (the *odontode regulation theory*) was presented in 1982 by W.E. Reif. He believes that, once formed, individual odontodes do not grow. Each is surrounded, at least initially, by an inhibiting field that suppresses other odontodes and ensures that all remain spaced from their neighbors, as for simple teeth and the small, abrasive scales of sharks. If the inhibitory fields are lost, larger units can form and grow by the fusion of odontodes from separate papillae, either to each other or to underlying bone. According to each theory, accretion of odontodes can be more or less linear or can be in onionlike layers, and may be accompanied by partial resorption of previous generations of odontodes (Figure 6.5).

Some contributors, emphasizing the potential of versatile neural crest derivatives to form a spectrum of hard tissues, believe that the various scenarios noted above may not be mutually exclusive. In any event, the mechanisms that produce the armor of extinct fishes are difficult to investigate.

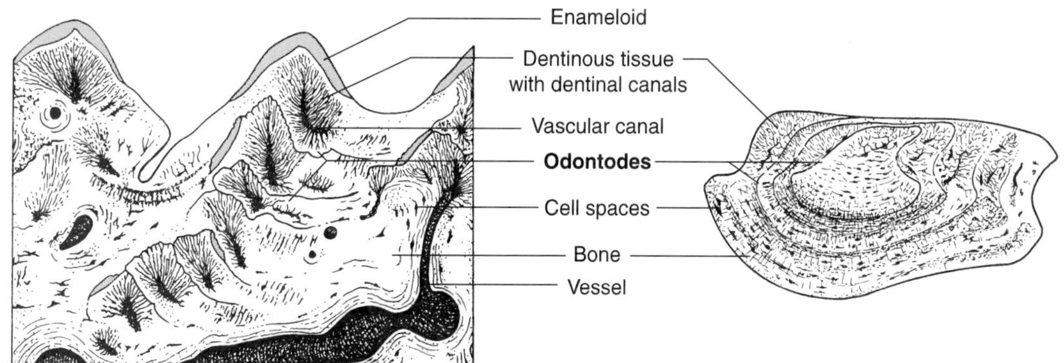

FIGURE 6.5 MUCH-ENLARGED SECTIONS OF COMPLEX SCALES HAVING ORIGIN FROM FUSED ODONTODES. On the right is a single scale of a Paleozoic acanthodian showing onionlike accretion of five generations of odontodes; on the left is a piece cut from a large scale of a Mesozoic ray-finned fish showing linear accretion of eight generations of odontodes. (These scales differ in fine structure from any shown in Figure 6.4; nevertheless, the plane and manner of sectioning of these scales is clarified by comparison, respectively, with the ganoid scale and fragment of bony armor seen in that figure.) (Redrawn from 1951 and 1967 papers by T. Ørvig.)

cation of associated cartilage during development in crocodilians raises questions about their origin). Some bones in the shells of turtles likely also originated from scales. (Other bones in the shells of turtles are flattened ribs.) Splintlike bones that lie in the muscles of the ventral abdominal wall of crocodilians (but not lizards) are also seemingly derived from dermal scales. They are termed **gastralia.** (The bones in the shells of armadillos are of similar nature, but they evolved secondarily, long after their ancestors had lost all ossifications in the skin.)

Other hard structures have also evolved from scales. Teeth certainly evolved about as dermal denticles evolved. Various bones of the roof of the skull and pectoral girdle represent armor plates that lost their enamel and dentine and sank below the skin to join bones originating in the internal skeleton. The fin rays of bony fishes are also regarded as derivatives of scales. These structures will be described in later chapters.

INTEGUMENT OF TETRAPODS: EMPHASIS ON EPIDERMAL DERIVATIVES

Skin of Amphibians, Living and Extinct The epidermis of living amphibians is thin (typically five to eight cell layers), but in response to contact with the air it has a particular mucopolysaccharide that apparently helps control desiccation, and a stratum corneum with α-keratin (Figure 6.7). Only the outermost cell layer is dead, however, and this is lost every few days, sometimes in large patches. The sloughing is under hormonal control.

Amphibians have two kinds of multicellular, alveolar (flask-shaped) glands that originate from the epidermis and grow down into the dermis. Their products reach the surface by ducts. Abundant mucous glands secrete continuously and spontaneously to clean and lubricate the skin and to keep it moist so that cutaneous respiration will be possible. Granular glands are under nervous or hormonal control. They secrete an acrid milky fluid that is distasteful, and in some instances very toxic to predators. Granular glands are grouped together in the "warts" of toads. The amphibian dermis is two-layered and may be provided with lymph spaces and muscle fibers.

Terrestrial amniotes are able to withstand abrasion and desiccation largely because of keratinized derivatives of the epidermis. Having only a thin layer of dead keratinized cells, modern amphibians must instead seek moist habitats or use behavioral adaptations to avoid drying out. It is probable that the skin of terrestrial paleozoic amphibians was thicker, drier, and more like that of some modern reptiles. Many had bony ossicles in the skin, and these are usually associated with a heavily keratinized epidermis.

FIGURE 6.6 OSTEODERM FROM THE NECK SKIN OF A LARGE CROCODILE. Actual size $7\frac{1}{2} \times 10\frac{1}{2}$ cm.

Skin of Reptiles: Horny Scales Reptiles use keratins (and lipids) of the epidermis to "airproof" their skin. The adaptation involves the fundamental patterning of the skin in that the distinctive epidermis forms a complete body covering of horny scales. Joints between scales are merely regions where the horny material is thin and folded (Figure 6.8). The epidermis of lepidosaurs is of particular complexity and interest. In these animals an entire "generation" of the epidermis is sloughed as a single unit. This occurs at

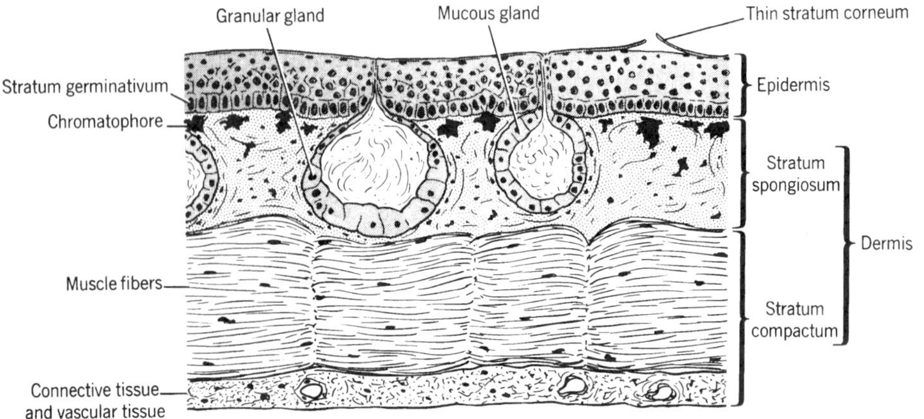

FIGURE 6.7 SECTION OF THE SKIN OF AN AMPHIBIAN.

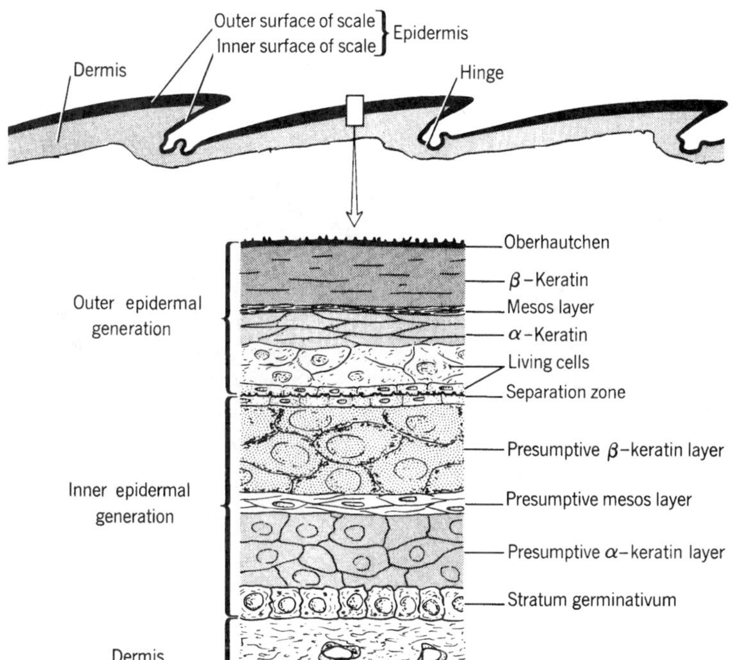

FIGURE 6.8 SECTION OF THE SKIN AND EPIDERMIS OF A SQUAMATE REPTILE shortly before a molt.

least several times a year. It is seemingly under hormonal control and may be influenced by humidity. Let us enter the cycle just after such a molt in what is termed the resting stage. The epidermis now consists of the stratum germinativum and an **outer epidermal generation** that characteristically has five layers. From the outside inward there is first a thick, dead, acellular layer heavily keratinized by β-keratins. The surface of this layer is called the *oberhautchen* and has microscopic spicules. Under this β layer is a thin mesos layer of unknown significance and then a moderately thick layer of loose, dead, anucleate material composed primarily of α-keratins. Below this are two layers of living cells: an outer layer, which will later be taken into the α layer, and an inner layer, which will later become clear and create the separation leading to sloughing.

At the end of the resting stage, the germinal epithelium rapidly proliferates the various layers of an **inner epidermal generation.** As these mature, they separate from the innermost layer of the outer epidermal generation, and sloughing follows.

The keratinous plate on the outer surface of a large, flat scale is called a **scute.** The scutes of crocodilians and chelonians are not shed. Growth adds keratinized material over the entire inner surface of a scute, compensating for wear. Each wave of growth extends beyond the previous margin of a scute to form the familiar concentric rings of the turtle shell (Figure 6.9).

The reptilian dermis is thin. Mucous glands are absent, as they are also from the skins of the other truly terrestrial tetrapods. Scent glands of various types (generation glands, preanal glands, femoral pores, etc.) occur variously on the tail (some lizards), cloacal area (most squamates), thighs (lizards), and under the jaws (crocodilians). Their secretions influence social behavior. Some bones found in the skin of reptiles were mentioned on page 90.

Integument of Birds: Thin Skin with Feathers Following a different strategy, birds have over most of the body a thin, weakly keratinized skin that is loosely joined to the underlying tissues. It is appendages of the skin—the feathers—that are heavily keratinized.

FIGURE 6.9 CARAPACE OR SHELL OF A DESERT TORTOISE, *Gopherus,* SHOWING SCUTES WITH GROWTH LINES. Dorsal view; anterior to left.

The lower leg and toes, however, are covered by horny scales, or scutes, similar to those of archosaurs. These are not shed. The **beak** is also heavily keratinized. The **shellbreaker** of birds and some reptiles is an elevation on the beak or rostrum that helps the hatchling to break out of its shell (see the figure on p. 70). (The egg tooth of snakes and lizards serves the same purpose but is a real tooth.) The **spurs** of gamecocks are horny spines covering bony cores.

With rare exceptions, glandular derivatives of the avian skin are restricted to a large, branched, alveolar **uropygial gland** above the tail ("uropygium" = *tail* + *rump*) that secretes an oil used by the bird to preen its feathers. It is most developed in water birds.

Although intermediate stages are still largely speculative, there is some evidence that feathers evolved from the epidermal scales of reptilian ancestors. However, other theories are also supported; see Comment 6.3.

There are several principal kinds of feathers and various intergrades. **Contour feathers** provide the external covering of the body, giving it contour, color, and protection from sun, rain, and abrasion. Wing feathers (remiges) and tail feathers (rectrices) make up the **flight feathers.** They resemble contour feathers but are larger and stiffer, and form airfoils for flight. (Primary remiges are supported by the hand skeleton, secondary remiges by the arm.)

These feathers have similar basic structure. The axis has a hollow, proximal (toward the body) **quill** that continues into a solid, distal (away from the body) **shaft** or rachis (Figure 6.10). The **vane** is made up of **barbs,** which branch from opposite sides of the shaft, and smaller **barbules,** which branch from the barbs. The barbules on the distal side of each barb have on their edges **hooklets** that engage the proximal barbules of the adjacent barb, much like the filaments of Velcro. The resulting web is strong, light, and flexible. If disturbed, the elements of the vane may separate, but they do not break. The integrity of the feather is restored by preening with the modified edges of the bill, which reengages the hooklets.

A variation of the contour feather lacks hooklets and therefore has no firm vane but is instead fluffy. Many birds, including the more primitive orders, have double contour feathers—a principal feather joined at its base by a shorter, softer feather called the **aftershaft.**

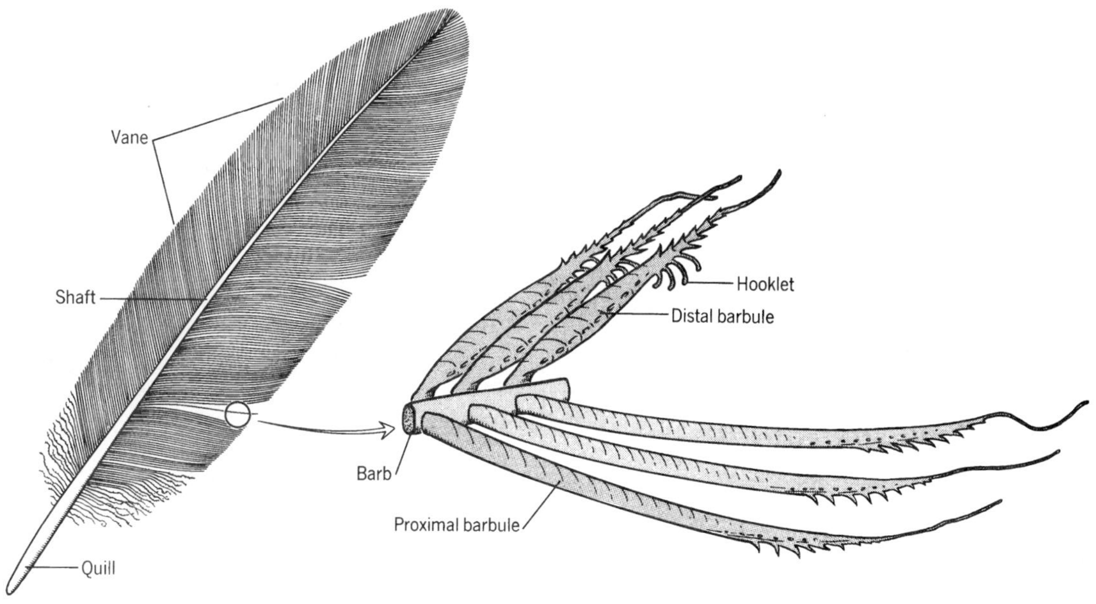

FIGURE 6.10 STRUCTURE OF A CONTOUR FEATHER.

Contour feathers are evenly distributed over the bodies of several kinds of birds (probably the primitive condition), but usually are restricted to feather tracts called **pterylae.** The feathers spread from the pterylae to cover the intervening areas. The conformation of pterylae is of use to systematists.

Down feathers have little or no shaft. Long barbs branch from the base of the feather, and there are no hooklets. The resulting feather is small and soft (Figure 6.11). Down feathers, hidden by the contour feathers, are widely distributed and not restricted to pterylae. Their function is insulation.

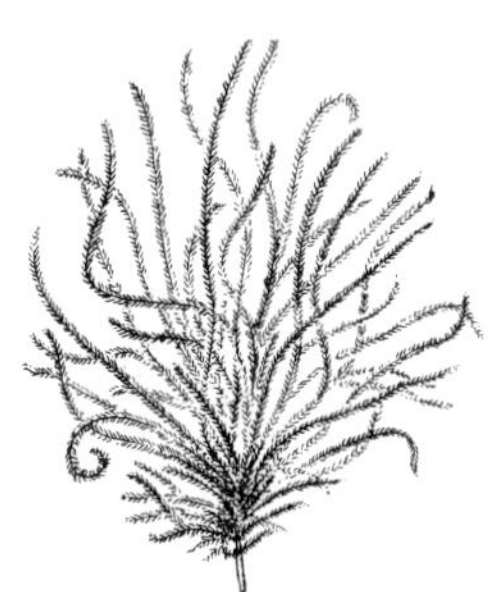

FIGURE 6.11 A DOWN FEATHER.

Bristles are short, stiff feathers that may screen foreign objects from the nostrils (hawks, blackbirds), increase the effective gape of the mouth (flycatchers), or form eyelashes (ostrich).

The colors of feathers come from two sources. Yellow, orange, red, brown, and black are the result of specific pigments introduced into the feather during its development. White results entirely from the microstructure of the feather. Blue, green, and iridescent hues result from a combination of black, yellow, or other pigment with microstructure that reflects only part of the light.

Feathers are molted and replaced once or (less commonly) twice a year. Most species drop the feathers one at a time so that function is not impaired, but ducks and some other birds lose most of the flight feathers at one time.

The development of a feather starts with a hummock of mesoderm, the **dermal papilla,** which is covered by ectoderm. This structure sinks into the skin, thus forming a narrow, double-walled depression, or **feather follicle,** all around its base (Figure 6.12). The feather is formed only by ectoderm, but the ectoderm not only must be nourished by the vascular mesoderm, but must also be activated by it. Experiments show that in the absence of mesoderm, no feather can form, and that in the presence of a papilla, ectoderm that does not normally form a feather may do so.

A superficial keratinized **feather sheath** surrounds erupting feathers and subsequently sloughs away. At the base of the follicle the germinative layer forms a **collar.**

COMMENT 6.3

THE ORIGIN AND EVOLUTION OF FEATHERS: EMERGING CONCEPTS AND DIVERGENT OPINIONS

Feathers are varied and complex in both structure and function, making them a challenge to the evolutionary morphologist. Are they developmental novelties? What was the nature of the ancestral protofeather, and what was its function? Did the reptilian ancestors of birds already have feathers? What is the evolutionary relationship of the different types of feathers? These have proven to be difficult questions to answer. Their investigation relates to emerging principles of broad application, including *Hox* genes, biological versus phylogenetic homology (see p. 4), developmental constraints (see p. 70, and the Wake and Roth book in References), and the interpretation of character-states in cladistic analysis (see p. 30).

We note here the differing opinions of several current researchers (included in the References). Brush is convinced that the feather follicle is unique and not homologous with any known ancestral structure, that the particular β-keratins of feathers are also unique and defining, that the most primitive type of feather is the down feather, and that the other types of feathers evolved from the versatile down feather follicle, independently and in no particular order. Prum concurs with Brush on most of these points, but proposes that down, contour, and flight feathers evolved in that sequence.

Maderson and Alibardi adopt the contrasting view that the feather follicle is related to the follicles of other integumentary derivatives, and is not an evolutionary novelty. They accept the cross-reactivity studies of Sawyer and colleagues that show the molecular structure of β-keratins of the scales and claws in both birds and archosaurs to be similar, suggesting homology. Maderson believes that the first feather was derived from flat, elongate, reptilian scales, and that the ancestral type of feather resembled a contour feather. Such a feather subsequently both simplified to become a down feather and increased in complexity to become a flight feather.

These contrasting views have bearing on the selective advantage postulated for protofeathers: If they were down-like, they provided insulation; if they resembled contour feathers, they protected the epidermis from abrasion and (according to work by Dyck) minimized evaporative water loss from the skin. (Other postulated functions, including protection from solar radiation, display, and camouflage, might have been provided by either type of protofeather.) The debate over the characteristics of the protofeather is also of consequence for the interpretation of featherlike structures seen on a Chinese fossil of the late Cretaceous theropod reptile *Sinosauras* (see Comment 4.1 on p. 61).

So, if *you* are attracted to the evolutionary analysis of one of the most exquisite structures in nature, plenty of intriguing questions await your attention.

The barbs of a down feather grow straight upward from the collar within the sheath. An early step in the formation of a contour feather is the development of a shaft as an outgrowth from one point on the collar. Barbs form first as branches of the shaft, and then as outgrowths of the collar itself that migrate onto the base of the shaft as it lengthens. When the sheath breaks away from the maturing feather, the barbs unfold to the right and left of the shaft to change the cylindrical, embryonic feather into the flat, mature feather.

Skin, Scales, Claws, and Integumentary Glands of Mammals The skin of mammals is relatively thick—particularly the dermis, from which leather is made. However, the thickness varies greatly according to species and location on the body (and sometimes also according to season). The epidermis thickens where the hair is sparse and also in areas subject to pressure and abrasion, such as footpads, the kneepads of camels and warthogs, and pads on prehensile tails. Between the stratum germinativum and stratum corneum there may be one or more transitional layers, the most common of which is a thin **stratum granulosum** (Figure 6.13). Bundles of smooth muscle in the dermis are related to hair follicles.

The stratum corneum may form horny scales, as on the tails of opossums and beavers. **Claws** are strong keratinized structures that wrap around the tapering terminal bones of the digits. The tip and upper and lateral parts comprise the **unguis** and are harder than the

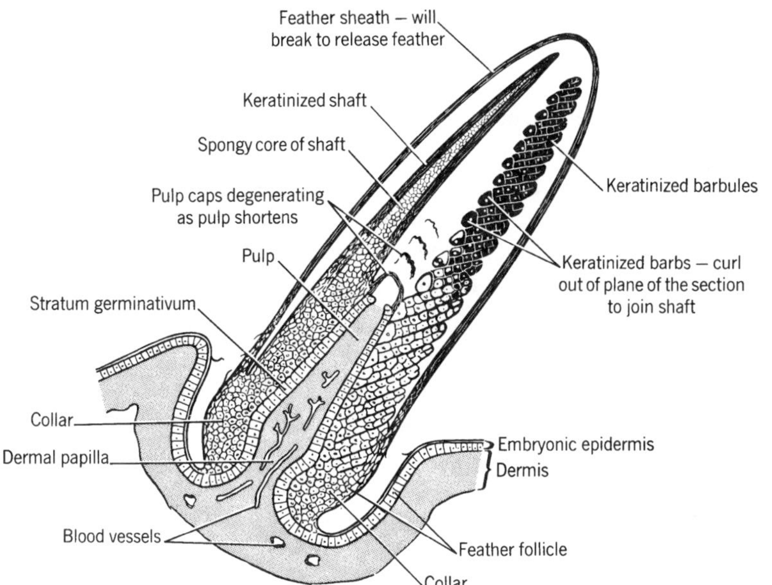

FIGURE 6.12 SECTION OF A DEVELOPING CONTOUR FEATHER.

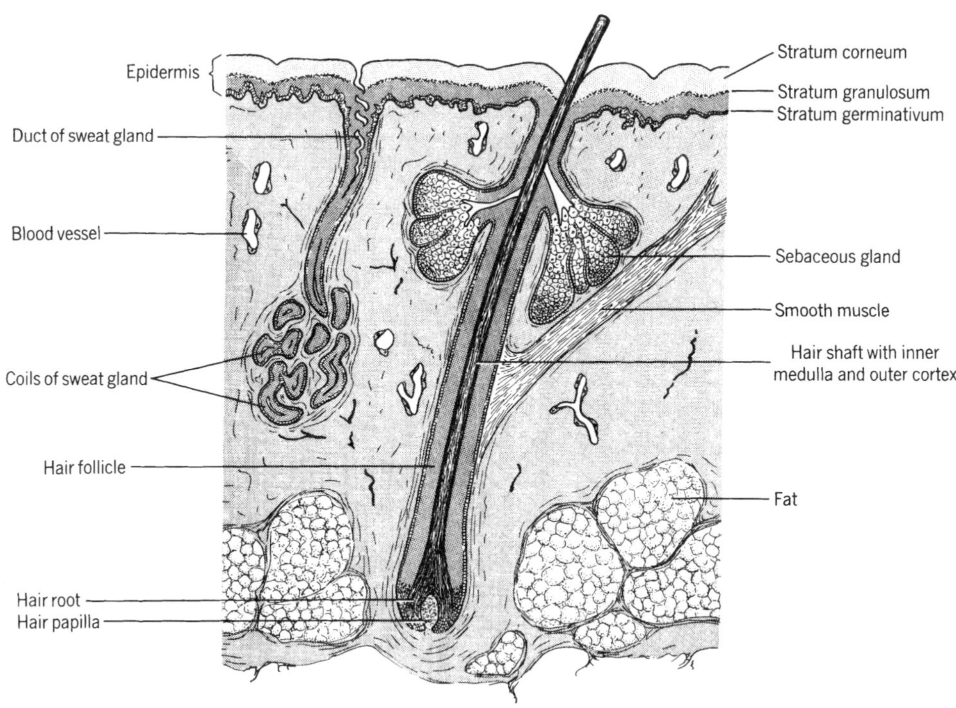

FIGURE 6.13 SECTION OF THE SKIN OF A MAMMAL.

underside (subunguis). Hooves are derived from claws. The unguis of the horse's hoof is built up of compacted horny tubules. The entire hoof spreads somewhat under the impact of a footfall. The shell of the armadillo has a heavily keratinized epidermis as well as bony dermal ossicles. The unique pangolin (mammalian order Pholidota) has markedly overlapping scales on its dorsal surface (see the figure on p. 458). These scales, which may be more than an inch long, are shed one at a time and replaced in larger sizes as the animal grows. The **baleen plates** of whales are lathlike outgrowths of the buccal epithelium that serve as strainers during feeding (see the figure on p. 557).

Sweat glands (also called sudoriferous glands) are unique to mammals. Many species have a million or more of these small glands distributed over the entire body. Others have fewer and restrict them to the muzzle or soles of the feet. Still others, including whales and manatees, which have no use for them, have none. Sweat glands are tubular, simple (not lobulated), and coiled at their inner ends. There are two kinds, which differ somewhat in structure and nature of secretion. They develop in the embryo from cords of ectoderm that sink into the dermis. The evaporation of sweat from the surface of the skin helps to prevent overheating of the body and opposes slipping of footpads over the substrate. Salt, urea, and some other wastes are excreted in the sweat. Glands in the eyelids (Moll's glands), which open near the eyelashes, and the wax glands of the external ear are considered to be enlarged and modified sweat glands.

Sebaceous glands are also limited to mammals. One or more of these branched alveolar glands drains into each hair follicle. They also occur without relation to hair on the nipples, lips, and genitalia. Their oily secretion dresses the hair and prevents excessive drying of thin skin. Lanolin, used as a base for cosmetics, is refined from the sebaceous secretion of sheep. Modified sebaceous glands (meibomian glands) occur in the eyelids, where their secretion films over the eyeball and normally prevents overflow of the tears.

Many mammals have **scent glands.** There is wide variation in their nature and distribution. They may serve for defense, recognition, or sexual attraction. The glands may be located in the anal region (weasel family), on the face (bats, antelopes), on the back (kangaroo rat), on the feet (some artiodactyls), or indeed, on any other part of the body. Some scent glands are said to be derived from sebaceous glands and others from sweat glands.

Only mammals have **mammary glands,** which secrete the milk to suckle the young. The first indication of the development of the glands is the appearance in the embryo of a pair of epidermal ridges, the **milk lines,** which extend lengthwise from the chest to the inguinal region. At intervals along the lines where the adult mammae will ultimately form, ectoderm sinks into the dermis and branches into solid cords. In females these cords enlarge at maturity, pushing under the skin and becoming compound (lobulated) and alveolar. Much of the human breast is fat. The gland becomes active at parturition under the influence of ovarian and pituitary hormones.

The number of mammae correlates with the number of young per litter and varies from one pair to about a dozen pairs. The mammae may be on the chest (primates, elephants, bats, manatees), in the inguinal area (ungulates), or at intervals in between (rodents, carnivores). Mammary glands were long considered to be phylogenetic derivatives of sweat glands. However, Blackburn believes that mammary glands are new structures sharing some characteristics with sebaceous glands.

Each milk gland sends numerous ducts to the surface. In monotremes these merge in patches called areolas whence young suck up the milk. Usually the site of emergence of the ducts is elevated into a **nipple,** which the young can hold in the mouth and suck. In ungulates it is the skin circling the point of emergence that is elevated, forming a hollow **teat** (Figure 6.14).

Hair The phylogenetic origin of hair is obscure. There is no indication that hair evolved from reptilian scales. Where hair and scales occur together (as on the tails of rats, the shells of armadillos, and the backs of pangolins), the hair grows between the scales, with the pattern of the scales imposing a pattern on the distribution of the hair. A similar pattern of hairs often occurs where scales are lacking. Evidence is presented by Maderson to support the hypothesis that hairs arose from reptilian sensory appendages of the mechanoreceptor type that were

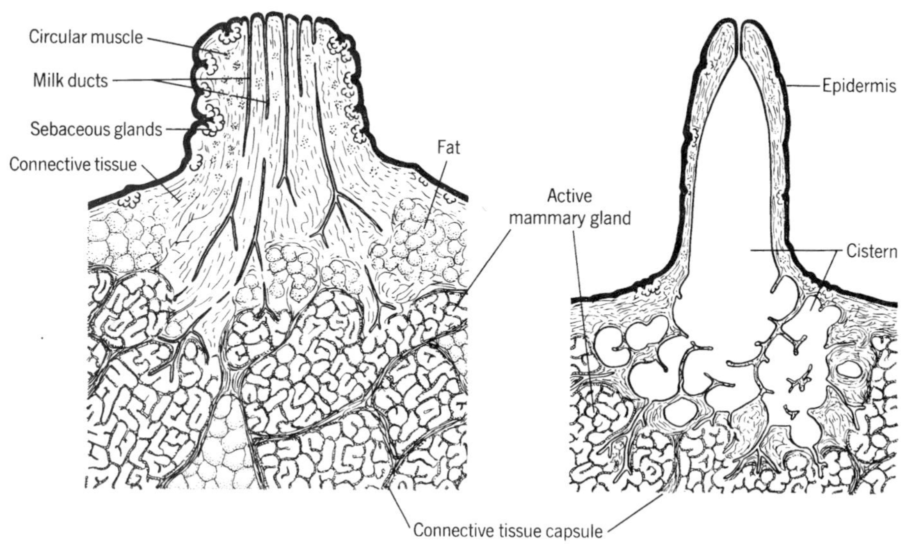

FIGURE 6.14 SECTION OF THE NIPPLE AND ASSOCIATED TISSUES OF A PRIMATE (left) AND THE TEAT OF AN ARTIODACTYL (right). Not drawn to the same scale.

located between scales and contributed to thermoregulatory behavior. It is postulated that such structures multiplied sufficiently to become useful as an insulatory body covering. Some mammal-like reptiles may well have reached such a stage of evolution.

A typical hair has an expanded root and a shaft that is hidden below the skin in an epidermal sheath or **hair follicle** (see Figure 6.13). One or more sebaceous glands usually drain into the cleft between the hair shaft and adjacent tissues. The follicle slants at an angle to the skin surface, and a tiny smooth muscle runs downward from the outer part of the dermis to insert on it at such an angle that contraction elevates the follicle and its hair. This merely gives a human "goose flesh," but for most mammals it deepens the fur coat, thus increasing the effectiveness of the fur in displays and as an insulator.

In section, hairs are seen to be made up of two or three layers. Of greatest structural importance is the **cortex,** which is relatively dense and contains the pigment of the hair. Around the outside are microscopic scales that form the **cuticle.** The size, shape, and overlapping pattern of these scales vary from species to species. Mammalogists sometimes take advantage of this individuality when they wish to identify hairs from owl pellets or droppings of carnivores. Coarse hairs also have a central pith or **medulla** consisting of shrunken dead cells and air spaces.

Guard hairs are the relatively long, straight hairs that give a pelt its apparent color and texture. They may also be specialized for other functions: Water is shed by the pelts of seals and beavers. Pronghorns have thick guard hairs containing air cells to insulate against both summer heat and winter cold. Hairs of polar bears serve as fiber-optic cables to conduct solar radiation to the skin with remarkable efficiency. Guard hairs are often grouped by twos and threes.

Parting the contour hairs of most mammals (particularly of "fur bearers") reveals shorter hairs that are very fine and numerous. These are wool hairs or **underfur.** They are usually somewhat flattened in cross section, and this makes them wavy. They often occur in groups of a dozen or more about the base of each guard hair. Underfur traps innumerable air pockets that provide insulation and may prevent water from penetrating to the skin.

Extra long and coarse hairs comprise eyelashes and manes and are found on the tails of ungulates. Whiskers, or **vibrissae,** are even coarser hairs that are specialized as tactile organs. The stiff shaft of a whisker serves as a lever that pivots at the surface of the skin to translate the slightest movement to the root. The bulb at the lower end of the root is surrounded by erectile tissue rich in nerve endings. The heaviest "hairs" of all are **quills.** Quills are hollow but can be very stiff. The cuticle at the tip of a porcupine quill is modified to produce tiny, effective barbs.

Most mammals molt once or twice a year, the winter and summer coats often differing in density, quality, and color. The new pelage usually comes in first at one or several locations and spreads over the body in a pattern characteristic of the species.

Mesoderm plays a less prominent role in the formation of a hair than in the formation of a scale or feather. A solid cord of ectoderm sinks into the dermis. The walls of the cord become the double-layered **root sheath.** A small dermal papilla forms at the enlarged base of the cord. The ectodermal cells over this papilla proliferate to form the hair itself, which pushes outward through the sheath cells to emerge from the skin.

Horns and Antlers The various types of tetrapod horns and antlers serve for recognition, display, ritual fighting, and defense. They may be carried by males only or by both sexes, but sexual dimorphism is nearly universal, reflecting the hormonal control of their development.

Rhinoceros horn is composed of keratinized fibers about 0.5 mm in diameter that are compacted into a tough, solid structure. Growth is from underlying epidermis (Figure 6.15). The horn is ever growing and is not shed. Rhino horns are medial and, depending on species, may be single or double.

The "horns" of the giraffe and okapi are properly called **ossicones.** They are permanent, skin-covered cones of bone. The bony cores are produced not as outgrowths from the skull, but as separate ossifications from dermal tissue, which then fuse with the skull as they mature. (Any permanent, bony horn-core is an *os cornu.*) The principal ossicones are paired, but some giraffes also have a smaller median ossicone.

The **antlers** of the deer family are also bony derivatives of the overlying dermis. They are shed and replaced each year. The hard, compact bone contains a little more organic material than other bone, giving it a bit more flexibility. Antlers are covered by skin ("velvet") only during growth (Figure 6.15). When full size is attained, the circulation to the velvet is cut off, causing its death and ultimate sloughing. At the end of the breeding season the bone at the base of an antler, just below a rough expansion called the burr, is weakened, and the antler is shed. Antlers are of varied shapes, often large, and usually branched in mature animals.

Pronghorns are limited to the American artiodactyls of that name. Again there are, in each sex, bony projections from the skull that are covered by skin. However, instead of producing hair, the skin forms horn. The bony core is permanent, and the horny cap is shed and regrown each year.

The **horns** of cattle, sheep, goats, and antelope (and of some dinosaurs and chameleons, among reptiles) have bony cores that are vascular and may contain extensions of the frontal sinuses. Over these cores are horny sheaths of epidermal origin. The horny substance is not filamentous like that of the rhinoceros horn and horse's hoof. The growth is by internal deposit so that the core keeps slipping outward, and growth rings may be seen around the base. The horn is "permanent," but there is some exfoliation of old horn. With some exceptions, there is only one pair of horns.

When both living and extinct mammals are considered, the number of horned taxa is much increased, as is the range of horn structure. Also, various intergrades are found between the various types of horn. It is evident that horns evolved numerous times.

Rhinoceros horn and the sheaths of true horns are alike in that each grows by the accumulation at the basal end of hard material that cannot subsequently change its shape. In this they resemble claws, tusks, and mollusk shells. When such structures grow at equal

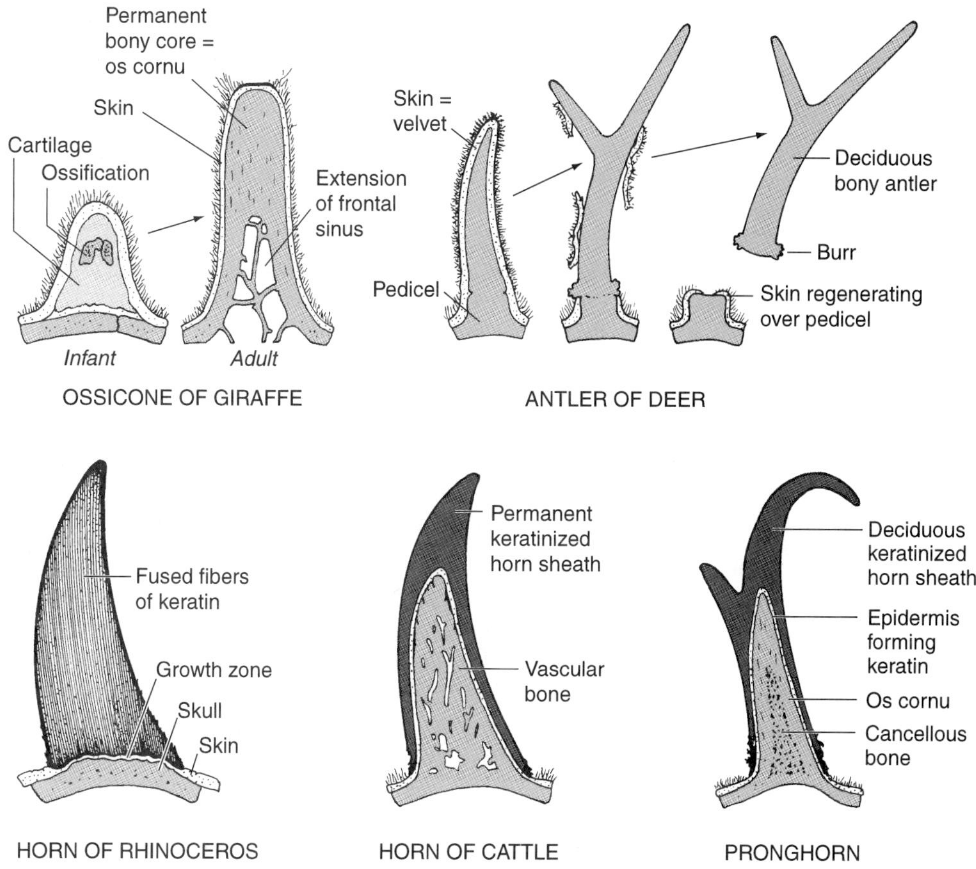

FIGURE 6.15 STRUCTURE AND DEVELOPMENT OF HORNS AND ANTLERS.

rates on all sides, they grow straight. More often the rate of growth is unequal around the base. If the point of minimum growth lies opposite the point of maximum growth, then the structure always forms a logarithmic (equiangular) spiral. The rhinoceros horn is an example. If the point of minimum growth does not lie opposite the point of maximum growth, then a helical (corkscrew) spiral in space is superimposed on the flat logarithmic spiral. The ram's horn is an example.

PHYLOGENY?

As indicated earlier in this chapter, a reasonably satisfactory phylogeny can now be prepared for the various kinds of bony scales and related structures. The evolutionary trend has been for their reduction, both in bulk and complexity. Attempts have also been made to construct a phylogeny of other integumentary structures, but the task is made difficult by multiple origins, evolutionary plasticity, parallelism, and convergence. Considerations of paleontology, development, innervation, and function have been of little help. Some morphologists believe that the various mucous glands of the aquatic vertebrates probably originated at various times, evolved along separate lines, and are not homologs of any glands of reptiles, birds, and mammals. Similarly, it is not known to what extent the granular glands of cyclostomes, fishes, and amphibians are related. The general correspondence of the keratinized skin derivatives of terrestrial vertebrates seems more evident, but truly homologous integumentary structures may not exist above the class level in modern vertebrates.

REFERENCES

Bagnara, J.T. 1998. Comparative anatomy and physiology of pigment cells in nonmammalian tissues, pp. 9–40. *In* J. Nordlund et al. (eds.), *The pigmentary system: physiology and pathophysiology.* Oxford Univ. Press, New York.

Bertram, J.E., and J.M. Gosline. 1986. Fracture toughness design in horse hoof keratin. *J. Exp. Biol.* 125:29–47.

Blackburn, D.G. 1991. Evolutionary origins of the mammary gland. *Mamm. Rev.* 21:81–96.

Brush, A.H. 2000. Evolving a protofeather and feather diversity. *Am. Zool.* 40(4):631–639.

Bubenik, G.A., and A.B. Bubenik (eds.). 1990. *Horns, pronghorns, and antlers: evolution, morphology, physiology, and social significance.* Springer, New York. 562p.

Dyck, J. 1985. The evolution of feathers. *Zool. Scripta* 14:137–154. Beautiful scanning electromicrographs.

Freedberg, I., et al. (eds.). 1999. *Fitzpatrick's dermatology in general medicine.* 5th ed. McGraw-Hill, New York. 1786p. Chapter 7 provides an excellent account of the form-function relationships of mammalian skin.

Halstead, L.B. 1974. *Vertebrate hard tissues.* Wykeham Science Series, London. 192p.

Homberger, D.G., and K.N. De Silva. 2000. Functional microanatomy of the featherbearing integument: implications for the evolution of birds and avian flight. *Am. Zool.* 40(4):553–574.

Janvier, P. 1996. *Early vertebrates.* Oxford monographs on geology and geophysics, no. 33. Oxford Univ. Press, New York. 393p.

Koller, E.J. 1972. The development of the integument: spatial, temporal, and phylogenetic factors. *Am. Zool.* 12:125–136.

Lucas, A.M., and P.R. Stettenheim. 1972. *Avian anatomy: integument.* Agriculture handbook 362, vols. I and II. U.S. Dept. Agriculture, Washington, DC. 750p. A thorough and scholarly account. A must for any student of the avian integument.

Maderson, P.F.A. 1972. When? Why? and How? Some speculations on the evolution of the vertebrate integument. *Am. Zool.* 12:159–171. Still speaks to us about good problems to pursue.

Maderson, P.F.A. 1984. The squamate epidermis: new light has been shed. *Symp. Zool. Soc. Lon.* 52:111–126.

Maderson, P.F.A., and L. Alibardi. 2000. The development of the Sauropsid integument: a contribution to the problem of the origin and evolution of feathers. *Am. Zool.* 40(4):513–529. The origin of feathers is analyzed as the emergence of complex form via modulations of embryological processes.

Maderson, P.F.A., and D.G. Homberger (organizers). 2000. Evolutionary origin of feathers. Symposium volume, *Am. Zool.* 40(4). Stimulating collection of papers.

Meinke, D.K. 1984. A review of cosmine: its structure, development, and relationship to other forms of the dermal skeleton in osteichthyans. *J. Vert. Paleontol.* 4:457–470.

Müller, G.B., and G.P. Wagner. 1991. Novelty in evolution: restructuring the concept. *Ann. Rev. Ecol. Syst.* 22:229–256. Thoughtful background and contextual discussion.

Ørvig, T. 1967. Phylogeny of tooth tissues: evolution of some calcified tissues in early vertebrates, vol. 1, pp. 45–110. *In* A.E.W. Miles (ed.), *Structural and chemical organization of teeth.* Academic Press, New York. A later paper by a major contributor to the histology and evolution of hard tissues.

Parakkal, P.F., and N.J. Alexander. 1972. *Keratinization: a survey of vertebrate epithelia.* Academic Press, New York. 59p. Excellent illustrations.

Prum, R.O. 1999. Development and evolutionary origin of feathers. *J. Exp. Zool.* 285:291–306. Provides a model for the evolutionary diversification of feathers based on development.

Quay, W.B. 1972. Integument and the environment: glandular composition, function, and evolution. *Am. Zool.* 12: 95–108.

Reif, W.E. 1982. Evolution of dermal skeleton and dentition in vertebrates: the odontode regulation theory, vol. 15, pp. 287–368. *In* M.K. Hecht, B. Wallace, and G.T. Prance (eds.), *Evolutionary Biology.* Plenum, New York. A review article.

Sawyer, R.H., et al. 2000. The expression of beta (β) keratins in the epidermal appendages of reptiles and birds. *Am. Zool.* 40(4):530–539.

Shipman, P. 1998. *Taking wing:* Archaeopteryx *and the evolution of bird flight.* Simon & Shuster, New York. 336p. See Chapter 6 for a stimulating account of suspected feather forgery and alternative theories for the functional evolution of feathers.

Sokolov, V.E. 1982. *Mammal skin.* Univ. California Press, Berkeley. 695p.

Spearman, R.I.C., and P.A. Riley (eds.). 1980. *The skin of vertebrates.* Published for the Linnean Society of London. Academic Press, London. 321p.

Wake, D., and G. Roth (eds.). 1989. *Complex organismal functions: integration and evolution in vertebrates.* Wiley, New York. 451p. Individual papers and consensus reports that discuss the concepts of evolutionary novelties and development constraints.

Chapter 7

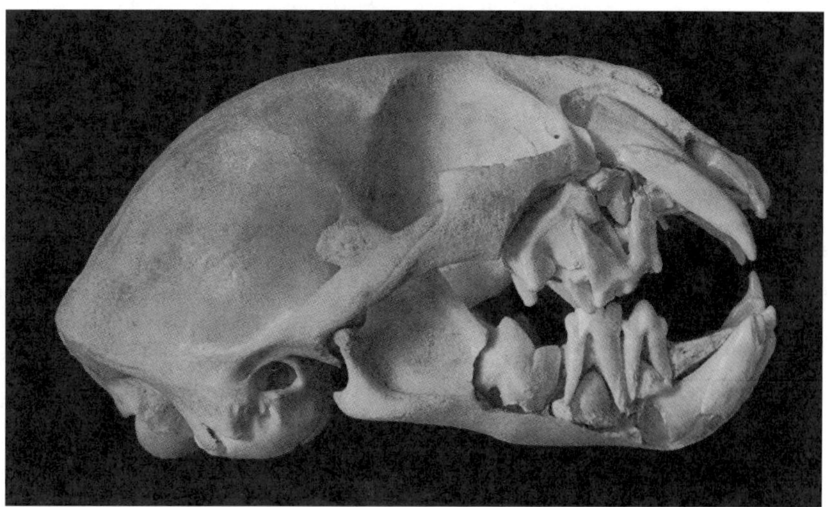

Teeth

Teeth have an importance for vertebrate morphology that is out of proportion to their contribution to the bulk of the body. This is true for several reasons. First, their durability has made them a significant part of the fossil record. Second, they are so adaptive that the diet of most animals can be approximated from their teeth. Third, great variation of structural detail among kinds of vertebrates combined with relative stability of structure within each kind makes teeth invaluable in systematics; experts can identify most species of mammals by a single cheek tooth. Finally, in spite of their adaptation to diet, teeth can often be used to trace the general course of evolution within, and among, genera, families, and orders. For these various reasons teeth have been the subject of much study.

ORIGIN AND STRUCTURE

The origin of ancient integumentary armor was mentioned in Chapter 6 and illustrated on p. 90. Such armor had surface denticles of enamel or enameloid, and dentine that merged into bone below. Teeth appear to have evolved from denticles released from armor near the margins of the mouth as ossification in the integument was gradually reduced.

The part of a mature tooth that is above the root and ultimately subject to wear is the **crown.** The **root** is hidden below the gum and is usually anchored to a jawbone. The **pulp cavity** within contains blood vessels and nerves (Figure 7.1). The bulk of a tooth usually consists of orthodentine. The bases of the teeth of some fishes contain a vascular osteodentine that may merge into the jawbone. Unworn crowns are covered by enamel, which rarely exceeds 2 mm in thickness even in large animals. The nature of these hard tissues was described in Chapter 6. Roots of teeth that are held in sockets are covered by a thin layer of **cement.** (Some teeth that are specialized for grinding also have cement on the crowns. See the figure on p. 574.) Cement is a nonvascular bone that has no osteons and is usually acellular. It is rich in collagenous fibers and is softer than dentine.

Above: Skull of a juvenile leopard with replacement teeth exposed.

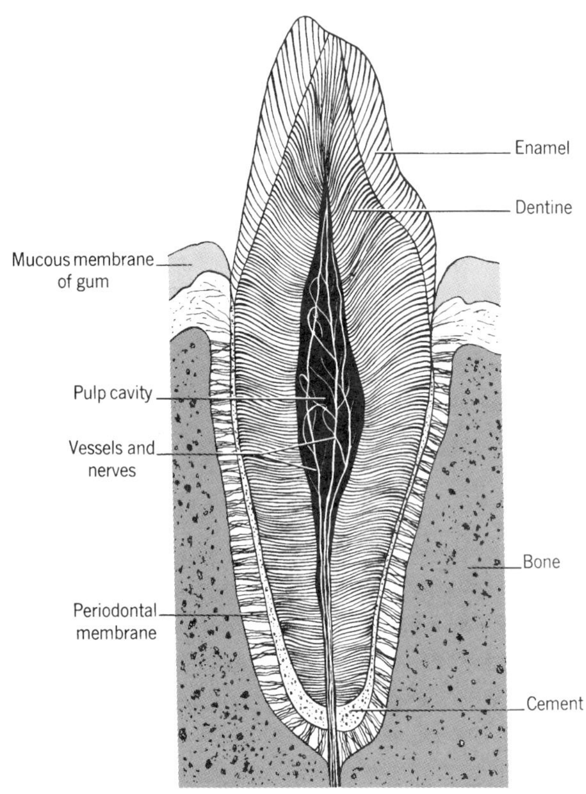

FIGURE 7.1 LONGITUDINAL SECTION OF A MAMMALIAN INCISOR TOOTH.

DEVELOPMENT

Knowledge of the process of tooth development facilitates an understanding of the mechanism of replacement and of the origin of complicated teeth from simple ones. The first step in the formation of teeth occurs in the embryo when a fold of ectoderm forms along the margins of the mouth and penetrates into the underlying mesoderm of the gums as a double-layered wall called the **dental lamina** (Figure 7.2). At intervals along the dental lamina, hummocks of tissue push into its inner edge, causing it (as seen in section across the jaw) to look like an inverted goblet. This tissue is usually regarded as mesoderm, but is derived from neural crest cells that have migrated from their site of origin (see p. 78), so it is also termed **mesectoderm** or (preferably) **ectomesenchyme.** The double-walled bowl of the globlet now constitutes the **enamel organ.** The cells of its inner layer (ameloblasts) will form enamel. Each mesodermal hummock, or **dental papilla,** develops on its surface the cells (odontoblasts) that will form dentine. The entire unit is called a **tooth bud.**

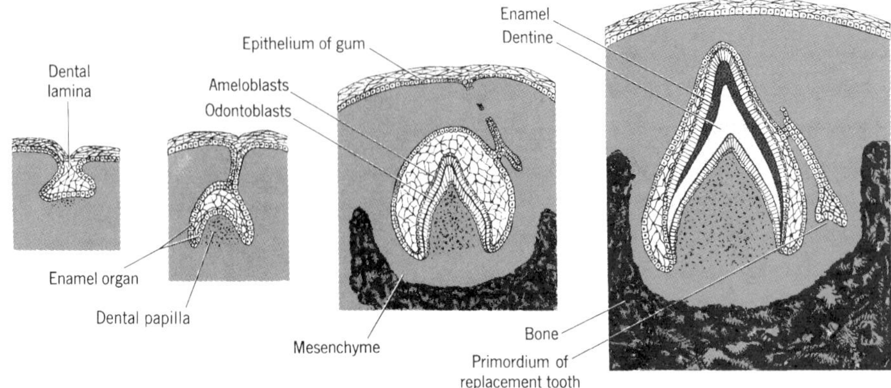

FIGURE 7.2
Diagrammatic sections showing DEVELOPMENT OF A THECODONT TOOTH.

There is a reciprocal induction (as noted also for the integument on p. 85) between the enamel organ and its papilla: each is necessary for the proper function of the other. The shape of the crown of the future tooth is determined by the conformation of the interface between enamel organ and papilla at the time the hard tissues are deposited. Until that time, the interface is gradually sculptured by differential pressures and growth rates of the various parts of the tooth bud. If there is to be more than one cusp, deposition of enamel and dentine starts at what will be the tip of the principal cusp. As maturation proceeds, the developing tooth slowly climbs up toward the surface of the gum. The enamel organ regresses ahead of the emerging crown, and no further enamel can then be deposited. The formation of dentine continues after the tooth becomes functional. Cement forms only in the presence of dentine. The papilla becomes the pulp.

ATTACHMENT AND REPLACEMENT

The teeth of cartilaginous fishes are anchored to the skin by collagenous fibers (Sharpey's fibers) that run into the dentine from the dermis. The teeth of most vertebrates, however, are more or less fixed to the bones of the jaws. Often the outer margin of each jawbone forms a thin wall having on its inner (lingual) side a series of hollows to accommodate the teeth. Each tooth touches the bone only with the outer (buccal) surface of its root. It may be joined to the jaw by collagenous fibers or by cement. This mode of attachment, which may be the primitive one, is called **pleurodont** (= *side* + *tooth*) (Figure 7.3).

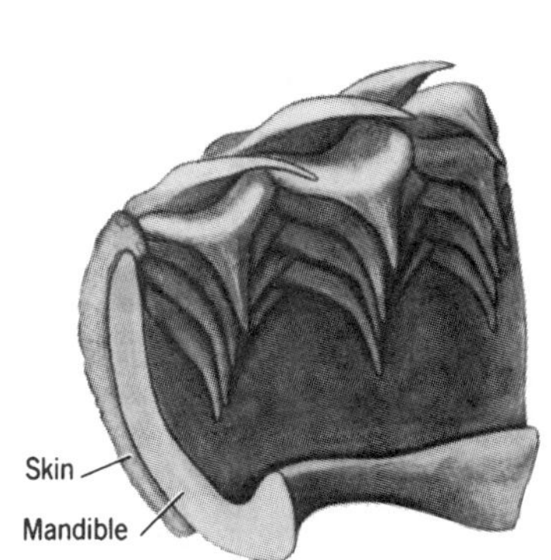

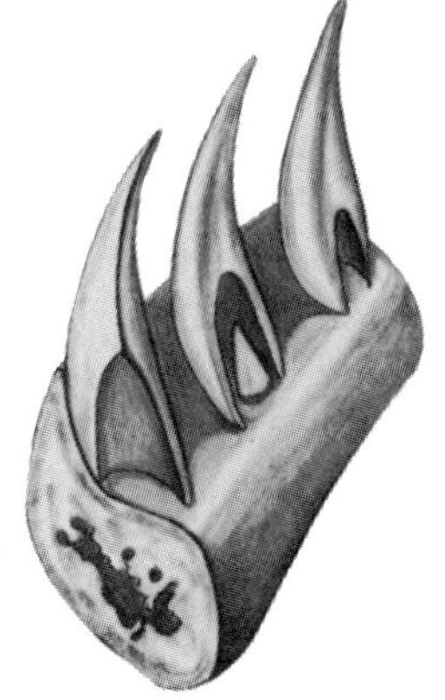

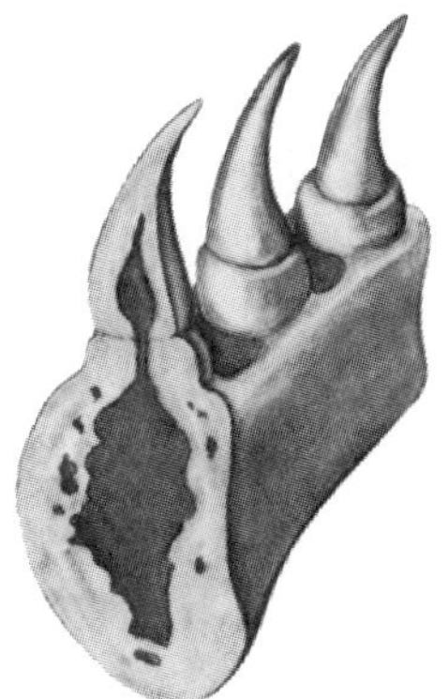

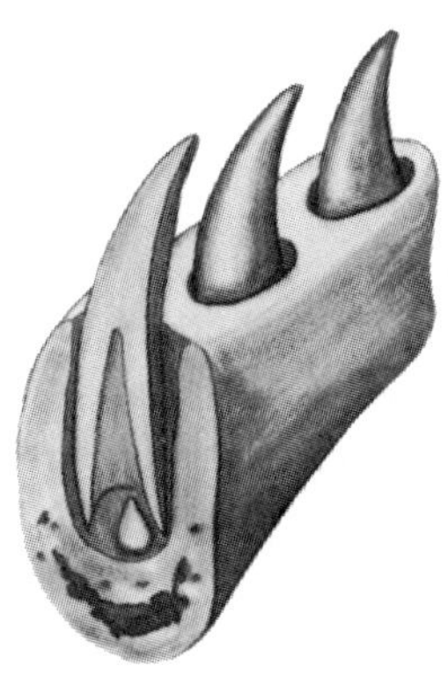

FIGURE 7.3 Sections of jaws showing TYPES OF TOOTH ATTACHMENT.

COMMENT 7.1

THE TEETH OF PREDACEOUS SHARKS

The teeth of flesh-eating sharks vary widely in form, and there are intergrades, but most are adapted either for piercing and holding or for cutting and slicing. Either way they can be formidable (Figure 7.4 and Figure 7.5, upper right).

Piercing teeth are long, slender, sharp, and smooth-edged. They usually curve inward to reduce the chance that struggling prey can escape (Figure 7.3, upper left). However, the tip of the tooth may be recurved, becoming more nearly upright; it can penetrate with least force if the bite is at right angles to the prey. (Various large snakes, having similar requirements, have similar teeth.) Slicing teeth are flatter, broader, and more bladelike. The edges are serrated. Serrations may be fine, particularly near the tip of the teeth, or coarse. These teeth, like a saw, must be drawn sideways in order to cut. By vigorously shaking its food from side to side, the shark moves its teeth against the inertia of the prey.

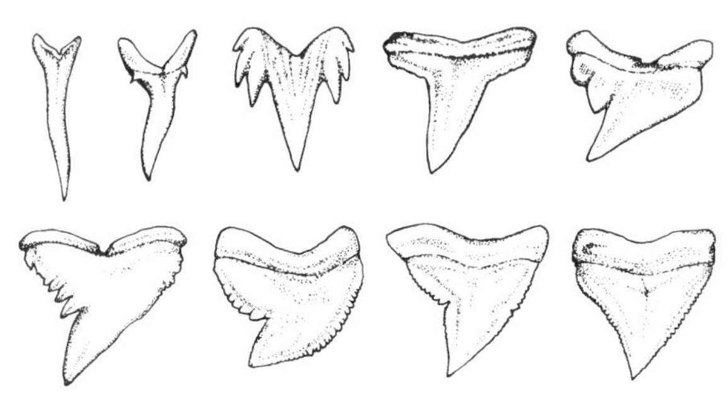

FIGURE 7.4 UPPER TEETH OF NINE KINDS OF SHARK. (Redrawn from P.R. Last and J.D. Stevens, 1994, *Sharks and rays of Australia,* CSIRO, Victoria, Aus.)

The teeth of upper and lower jaws are usually similar, but in some instances the teeth are more bladelike on one jaw than on the other. Many shark teeth have a notch near the middle of one cutting edge. It is thought that this configuration may help to prevent the teeth from lodging or binding in the prey.

For an analysis of shark teeth see Frazzetta in References.

Some other teeth scarcely have any roots and abut the rim of the jawbones to which they are joined by a continuum of hard tissue. This form of attachment, which has evolved independently several times, is called **acrodont** (= *summit* + *tooth*). Still other teeth have their roots held in sockets (alveoli) in the jawbones. This is the **thecodont** (= *sheath* + *tooth*) condition. There are intergrades among these kinds of attachment.

Replacement of teeth is necessary to provide for growth and to compensate for wear and accidental loss. Even before a first tooth is fully functional, a new tooth bud forms to initiate the development of its replacement. As the second tooth matures, the root of the first is resorbed, thus causing it to loosen and fall away. Replacements for pleurodont teeth form either just lingual or just anterior to the roots of the old teeth and move into position after the old teeth are shed. Replacements for thecodont teeth form directly under the roots of the old teeth.

Most vertebrates replace their teeth continuously, generation following generation, for as long as they live. Such animals are said to be **polyphyodont** (= *many* + *to grow* + *tooth*). Most mammals (and some mammal-like reptiles) have only two generations of teeth and are said to be **diphyodont.** Some acrodont teeth, and some that fuse to form large tooth plates, are not replaced. However, many of the animals that seem to have only one generation of teeth do have one or more other generations that are shed before birth or resorbed in the embryo.

Polyphyodont teeth are not replaced at random, seemingly because that could result in temporary impairment of function. Two conditions usually are maintained. First, the teeth are in two sets, the even-numbered teeth constituting one set and the odd-numbered teeth another. Replacement in one set is out of phase with replacement in the other. Accordingly, since adjacent teeth are at different stages of the growth cycle, vacant positions are flanked by functional teeth. Second, neighboring teeth in a set (alternate teeth in the mouth) are usually at slightly different developmental stages. Every tooth tends to be a little more mature than the next anterior tooth of its set.

Edmund and, more recently, Osborn have proposed developmental models to explain these observations. Osborn suggests that the migration of inductive ectomesenchyme into the jaws from behind initiates the first wave of tooth development from posterior to anterior. If each developing tooth somehow inhibits the development of adjacent teeth, that would account for the odd and even sets. Large teeth develop more slowly than small ones, and this influences the schedule of individual tooth replacement for animals having teeth of different sizes.

EVOLUTION OF TEETH

From Denticulate Armor to Heterodonty CYCLOSTOMES have conical, horny teeth on the oral funnel and "tongue." Their developmental origin corresponds to that of more typical teeth. CONODONTS had conical teeth with enamel and dentine (see p. 37). Some OTHER EXTINCT AGNATHA had small to medium-sized bony plates in the mouth that were roughened by surface denticles. Among PLACODERMS, the predaceous arthrodires had on the margins of their jaws bony plates with jagged edges of various configurations that sometimes contained dentine (see figure on p. 41).

The teeth of OTHER FISHES are highly varied, and evolutionary trends are scarcely found. Typically their teeth are numerous, conical or bladelike (Figure 7.5) and **homodont** (= *same* + *tooth*), or all of about the same size and shape. They are carried on the margins of the jaws and, in ray-finned fishes, may also be on the roof of the mouth, fifth gill arch, and tongue. Most ray-finned fishes are acrodont, their teeth being either fused to the jawbone or joined to it by connective tissue. (The connective tissue may form a hinge allowing the tooth to tilt.) Acrodont teeth often are not shed. The teeth of sharks are anchored to the skin and are continuously replaced as new teeth migrate up over the margins of the jaws from the inside (Figure 7.3). Departures from these general conditions (which relate to diet—see Chapter 30) include reduction or loss of teeth, fusion to form permanent crushing plates (chimaeras, dipnoans, many rays), and development of multiple cusps (cladoselachians, pleuracanths) or whorls (acanthodians).

The teeth of most CROSSOPTERYGIANS (but not the surviving coelacanth) resemble those of ray-finned fishes in most respects, but have one distinctive and important characteristic: The enamel and dentine are folded so as to form complicated patterns as seen in cross section (Figure 7.5). This structure, termed **labyrinthodont,** strengthens the tooth and makes it resistant to wear. Labyrinthodont teeth survived for 100 million years, first among early amphibians and then STEM REPTILES.

MODERN AMPHIBIANS tend to have fewer teeth than their ancestors (none in toads) and they are small, simple, pleurodont, and no longer labyrinthodont. They are supported by pedicels of dental origin, to which they are attached by zones of soft tissue (Figure 7.6). Slight inward bending of the tooth may be permitted, which is interpreted as an aid to the transport and swallowing of food.

Many REPTILES, and also the extinct TOOTHED BIRDS, are homodont. All forms of tooth attachment are observed: Some snakes and many archosaurs are (and were) thecodont, various reptiles are acrodont, and many lizards are at least partly pleurodont. Most reptiles are polyphyodont, though acrodont teeth often are not replaced. Turtles are **edentate,** or toothless.

Various reptiles, in several orders, but particularly among MAMMAL-LIKE REPTILES, have teeth of different form and function at different parts of the tooth row. Such dentition is **heterodont** (= *different* + *tooth*). This significant advance was passed on to MAMMALS and is related to chewing.

Some Consequences of Chewing Some fishes use teeth to process food. For example, white sharks tear and slice large prey into hunks that can be swallowed, stingrays break mollusk shells between flat tooth plates, and parrot fishes crush coral fragments with

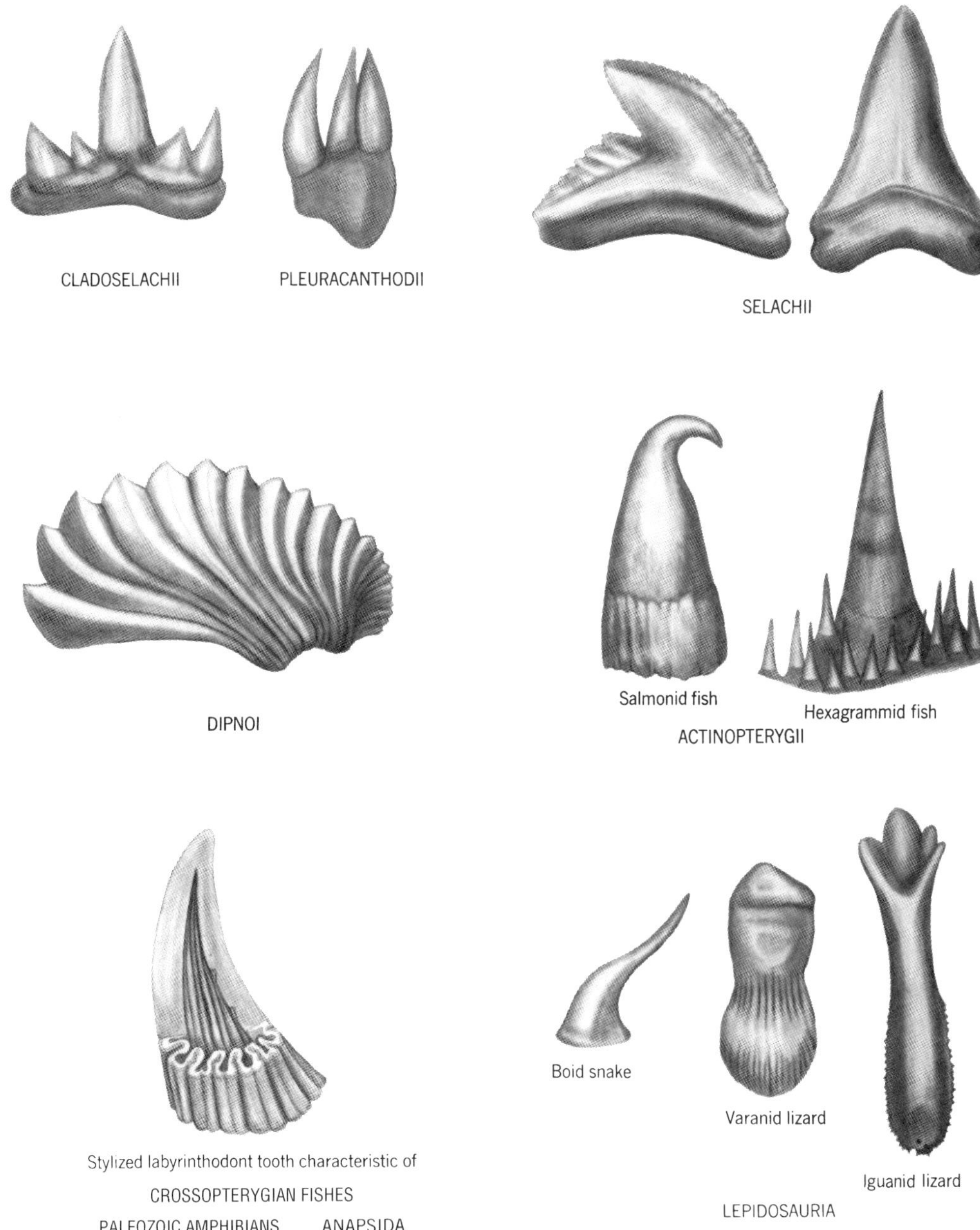

FIGURE 7.5 CHARACTERISTIC TEETH OF VARIOUS TAXA.

powerful pharyngeal teeth. Nevertheless, most nonmammals use teeth primarily for securing and holding food, which is then quickly swallowed. MAMMALS also secure and hold food, but in addition they usually shear, crush, or grind their food. These new functions are of the utmost importance because they greatly increase the surface area of ingested food and, in some instances, remove undigestible hulls, thus significantly speeding digestion and increasing the diversity of foods—particularly plant foods—that can be eaten. In order to be sheared, crushed, or ground, it is desirable for food to be retained in the mouth and chewed. This necessitates (1) modification of the ancestral jaw articulation and palate (see discussion on p. 129), (2) cheeks to hold food in the mouth, (3) a tongue capable of positioning the food, and (4) several kinds of teeth. Mammals are heterodont, unless they have secondarily reverted to homodonty in response to a diet of fish or insects (e.g., toothed whales, armadillos). Other distinctive features also relate to heterodonty: Mammalian teeth are thecodont because such teeth can best withstand shearing forces without being loosened. They are carried only at the margins of the jaws, seemingly to enable most mammals to use their marginal tooth rows somewhat independently. Specialization of the teeth leads to increased size, complexity, strength, and wearing ability. These, in turn, correlate with reduction in number of teeth that are functional at one time and reduction in number of sets of replacement teeth.

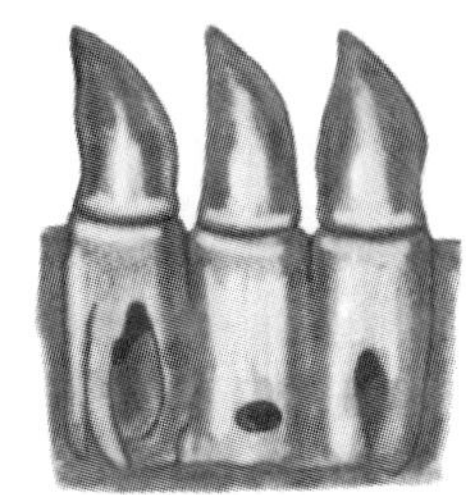

FIGURE 7.6 DIVISION OF TEETH OF LISSAMPHIBIA INTO CROWNS AND PEDICELS shown by the caecilian, *Gymnopis.*

Most mammals (and some mammal-like reptiles) are diphyodont, but this statement needs qualification. Typical mammals have a first set of teeth that consists of the temporary milk teeth (i.e., incisors, canines, and premolars) plus the permanent molars. The fact that all these teeth belong to one generation is obscured by the circumstances that they erupt in sequence and may or may not be deciduous. The second generation of teeth includes all permanent teeth except the molars. (Some marsupials replace only one tooth in each jaw, and moles replace none.)

Numbers and Kinds of Teeth Typical mammalian dentitions have three or four kinds of teeth. **Incisors** are adapted for securing food and sometimes for grooming (Figure 7.7). They may be conical spikes for holding insects or flesh, or simple blades for cutting plant stems. They are relatively small teeth and are single-rooted. Their function requires that they be forward in the mouth; the upper incisors are rooted in the more anterior of the two bones of the jaw (the premaxilla). Numbers of teeth are expressed in terms of one side of the mouth only and are commonly written as a fraction, the count of upper teeth forming the numerator and the count of lower teeth forming the denominator. Incisors of eutherian mammals are always 3/3 unless secondarily reduced. This count is exceeded by some marsupials.

The **canines** are next posterior in the mouth. They are simple spikelike teeth with single roots. If not secondarily modified, they are long and strong and serve for holding and piercing in relation to both feeding and fighting. Canines always number 1/1 if not reduced. The alveolus for the upper tooth lies in, or just posterior to, the suture between the two bones of the upper jaw.

All replacement teeth behind the canines are called **premolars,** and all nondeciduous teeth of the first generation are called **molars.** Molars tend to be larger than premolars and

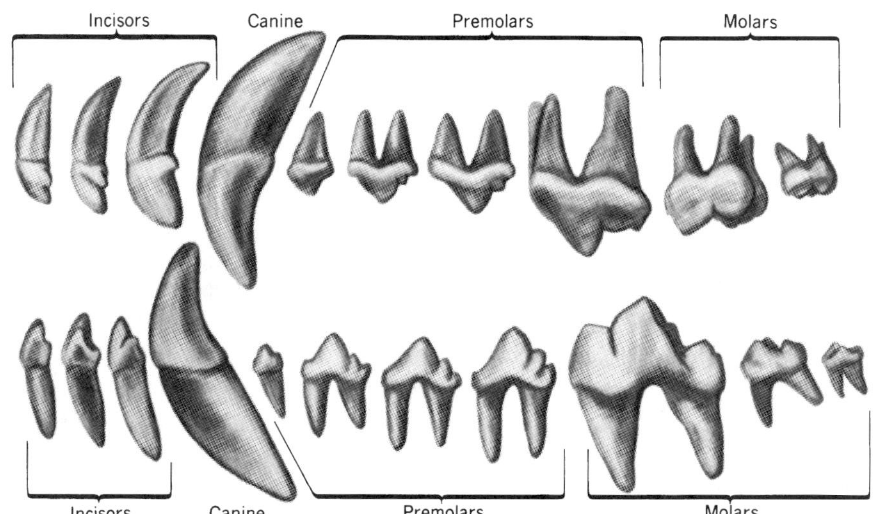

FIGURE 7.7 HETERODONT MAMMALIAN DENTITION illustrated by the teeth of the dog.

to have more cusps and roots. The numbers of premolars and molars among ancestral therian mammals were probably 4/4–5 and 7–8/7–8, respectively. The primitive counts for eutherian mammals are considered to be 4/4 and 3/3.

The distinction between premolars and molars is sometimes unsatisfactory. As a practical matter, it may not be possible to tell from an adult skull which teeth have been replaced. The entire tooth row forms a series such that teeth tend to resemble their immediate neighbors. Specialization may make the kinds of teeth indistinguishable: Artiodactyls have an incisiform lower canine and horses have molariform premolars. Furthermore, the distinction between premolars and molars breaks down if the replacement generation of teeth is incomplete or absent. For these reasons it is often convenient to speak of these teeth collectively as **cheek teeth.**

The numbers and kinds of teeth are expressed by a dental formula that consists of the numerical fractions already noted written in sequence starting with the incisors. Thus, the formula for the wolf is 3/3, 1/1, 4/4, 2/3; for the deer 0/3, 0/1, 3/3, 3/3; and for humans 2/2, 1/1, 2/2, 3/3.

More about Cheek Teeth The cheek teeth of primitive mammals (Jurassic mammals, marsupials, insectivores) and of some more advanced orders, have several cusps (see the posterior teeth, figure on p. 135). The upper molars usually have three principal cusps arranged in a triangle; the lower molars commonly have five principal cusps. Accessory cusps may also be present. The cusps are intricately arranged to interlock when the mouth is closed. Such teeth are admirably adapted for diets of flesh or invertebrates.

Speculation about the origin of multicuspid teeth from unicuspid teeth has been the subject of debate. The **concrescence theory** of Kükenthal and Röse was formulated in the 1890s. It holds that multicuspid teeth originated by the fusion of adjacent unicuspid teeth. This does actually happen in some fishes, but does not appear to have been a usual event. The favored theory, called the **differentiation theory,** was presented in the 1880s by Cope and Osborn and was revised and extended some years later by Gregory. This theory, for which there is abundant paleontological and embryological evidence, states that the single cusp of the ancestral tooth gradually became supplemented by two secondary cusps that formed from the side walls of the tooth crown. Additional, smaller cusps may form from a ridge (cingulum) that encircles the base of the crown. The patterns of cusps are various and intricate. Furthermore, cusps may become joined in various ways by ridges (see figure on p. 574). All cusps and lophs are named, and the terms distinguish between corresponding cusps of upper and lower teeth. Unfortunately, some of the terms in wide use are not apt because they do not express true homologies between upper and lower teeth.

The description and functional analysis of the principal kinds of cheek teeth is a particularly interesting subject. It will be postponed until Chapter 30 because it relates to adaptation to diet rather than to the broad sweep of evolutionary change among major taxa, which is our concern in this part of the book.

REFERENCES

Briggs, D.E.G. 1992. Conodonts: a major extinct group added to the vertebrates. *Science* 256:1285, 1286.

Butler, P.M. 1956. The ontogeny of molar pattern. *Biol. Rev.* 31:30–70.

Butler, P.M., and K.A. Joysey (eds.). 1978. *Development, function and evolution of teeth.* Academic Press, New York. 523p.

Edmund, A.G. 1960. *Tooth replacement phenomena in the lower vertebrates.* Royal Ontario Museum, Toronto. Life Sciences Division, Contribution 52. 190p.

Fink, W.L. 1981. Ontogeny and phylogeny of tooth attachment modes in actinopterygian fishes. *J. Morphol.* 167:167–184.

Frazzetta, T.H. 1988. The mechanics of cutting and the form of shark teeth. *Zoomorphology* 108:93–107.

Gregory, W.K. 1934. A half century of trituberculy: the Cope-Osborn theory of dental evolution, with a revised summary of molar evolution from fish to man. *Am. Philos. Soc. Phila. Proc.* 73:169–317.

Kurten, B. 1954. Observations on allometry in mammalian dentitions: its interpretation and evolutionary significance. *Acta Zool. Fennica* 85:1–13.

Kurten, B. 1982. *Teeth: form, function, and evolution.* Columbia Univ. Press, New York. 393p.

Moss, M.L. 1970. Enamel and bone in shark teeth: with a note on fibrous enamel in fishes. *Acta Anat.* 77:161–187.

Osborn, J.W. 1974. On the control of tooth replacement in reptiles and its relationship to growth. *J. Theor. Biol.* 46:509–527.

Osborn, J.W., and A.W. Crompton. 1973. The evolution of mammalian from reptilian dentitions. *Brevoria* 399:1–18.

Peyer, B. (translated and edited by R. Zangerl). 1968. *Comparative odontology.* Univ. Chicago Press, Chicago. 347p.

Chapter 8

Head Skeleton

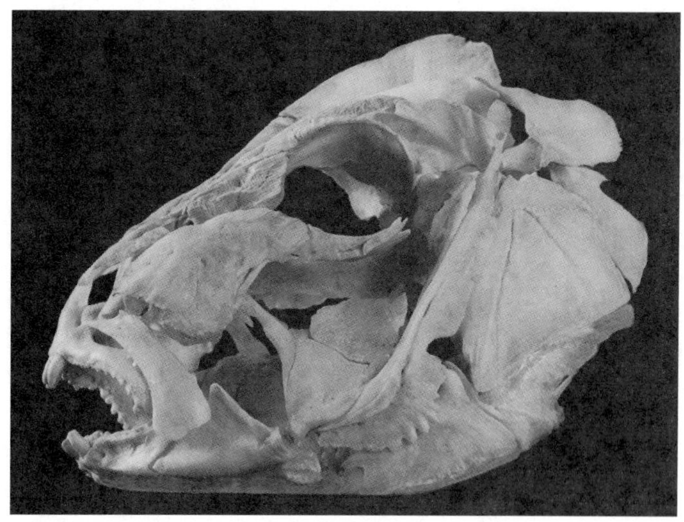

IMPORTANCE OF THE SKELETON TO MORPHOLOGY

The internal, jointed skeletal system of vertebrates is unique in the animal kingdom and is the most important of all organ systems in the study of vertebrate morphology. It is conservative enough in general pattern to show the broad outlines of vertebrate phylogeny: Homologous bones and evolutionary trends are readily demonstrated in skeletons characterizing the successive major taxa. Furthermore, the skeleton plays a central functional role. Wide variations have always been superimposed on the gradually evolving general pattern, and the skeleton has been sufficiently plastic to respond to the particular habits of the various animals. Therefore, it also provides reliable information about the specific adaptations of vertebrates: Postures and locomotor adaptations are accurately revealed, and other adaptations are sometimes indicated.

Because of its hardness and durability, the skeleton (including the teeth) becomes fossilized relatively often, and this contributes virtually (though not quite) all of our knowledge of past vertebrate life. Paleontology can be regarded as the comparative anatomy and morphology of extinct animals. No other science has contributed so much to our knowledge of vertebrate evolution, and it is, of course, limited almost entirely to the study of hard parts. Fortunately for the paleontologist, of all organ systems the skeletal system tells the trained observer the most about *other* organ systems: Most muscles take their origins and insertions on bones, and often leave tuberosities or scars to show the positions and extent of these contacts; the important cranial nerves reveal their sizes and courses by the foramina they traverse in the skull; the relative development of the different parts of the brain may be revealed by the braincase; nasal chambers, orbits, and otic cavities give some information about the sense organs they house; the nature and distribution of sensory canals on the head may be shown in detail; and even some blood vessels leave traces on the skeleton.

Moreover, of all organ systems this one is the easiest to preserve, store, and demonstrate. It is, therefore, a good one to teach and learn about.

Above: Skull of a snapper.

MORE ABOUT HARD TISSUES

Hard tissues have been described as we have come to them: horn, enamel, dentine, and bone in Chapter 6, and cement in Chapter 7. As we turn to the deeper skeleton, another hard tissue, cartilage, must be added and more should be said about the development and nature of bone.

Cartilage is a tissue that combines hardness with some flexibility. It is present in several invertebrates but is particularly characteristic of vertebrates. Its cells (chondrocytes) are dispersed in spaces (lacunae) in a matrix of proteoglycans and fibers that hold much water (about 70 percent by volume). Chondrocytes form at the surface membrane of the tissue, or perichondrium, and gradually round up, grow, and become wider spaced as the matrix they produce pushes them farther apart (Figure 8.1). Cartilage has no nerves or vessels. It is usually derived from mesoderm, but, curiously, in part of the head and in virtually all of the gill region it may also be derived from neural crests, which are initially ectodermal.

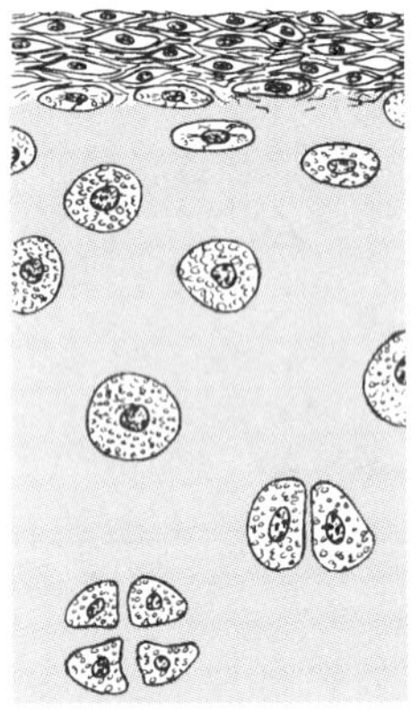

FIGURE 8.1 HYALINE CARTILAGE.

Hyaline cartilage is stiff but resilient. It has relatively much bound water and relatively small and few contained fibers. It is glassy and translucent in thin sections. Its surfaces are exceedingly smooth, as where it caps the bones at movable joints. Its important contribution to the lubrication of joints is noted on p. 400 and its role in the formation of long bones on p. 161. It is able to transform into the following three types of cartilage. **Fibrous cartilage** contains a heavy meshwork of collagenous fibers that make it cushionlike and tough, as where it forms discs that separate the vertebrae of the lower back of humans. **Elastic cartilage** is rich in elastic fibers and consequently has flexibility and resilience, as in the external ear and epiglottis. Fat cells may be incorporated. **Calcified cartilage** contains deposits of calcium salts that make it hard and firm. It is common in the skeletons of elasmobranchs. Unlike bone, calcified cartilage is not remodeled once it is formed. There are intergrades among these types of cartilage. (The compact, turgid, and vacuolated tissue of the notochord—and similar but diffuse tissue sparsely found primarily in some teleosts—has been classed as **choroid cartilage.** Similarly, various resilient tissues scattered in the heads of numerous cyclostomes and teleosts are called **chondroid,** meaning "cartilage-like.")

In terms of ontogeny, there are two kinds of bone. **Replacement** or **endochondral bone** gradually replaces cartilage that has formed earlier. Thus, in the embryonic chondrocranium the cartilaginous skeleton is eroded away just ahead of the deposition of bone. The process is described further on p. 161. Bones that are *not* preceded by cartilage are called **membrane bones,** particularly if they lie below the skin, such as the roofing bones of the skull. The membrane bones described in Chapter 6 that remain associated with the integument (armor, scales, denticles, osteoderms) are also called **dermal bones.** Once the two kinds of bone, replacement and membrane, have formed, they are identical. The distinction is of the utmost importance, however, for establishing the origins and homologies of various elements of the skeleton.

A property of bone and cartilage that conditions their other physical properties is **heterogeneity.** Disregarding minor impurities and imperfections, cast iron, ceramics, and glass are homogeneous, or uniformly the same everyplace and in all directions. Wood, by contrast, has grain that makes resistance to bending and splitting different in different planes. Similarly, bones have lamellae and osteons with specific orientations and may be compact or spongy (pp. 87 and 88). The internal cells and fibers within cartilage are unevenly arranged. Consequently, one cannot speak of *the* strength of, for instance, bone, but can report only the approximate strength of a given type of bone when loaded in a given way relative to the orientation of its components.

Cartilage and, particularly, bone are further heterogeneous in that each is a composite material consisting of two dissimilar components. In each instance, one component is an intricate meshwork of oriented collagenous fibers and sometimes of elastic fibers. The other component is a glycoprotein (for cartilage), or hydroxyapatite (for bone). Furthermore, the calcium phosphate crystals of bone are aligned differently in alternate layers of the contained fibers. As with plywood, this increases resistance to fracture. The physical properties of the composite are unlike those of either component taken alone, and are not the sum or average of the two taken together. Thus, collagenous fibers are soft, flexible, and very resistant to elongation, whereas hydroxyapatite is extremely hard, brittle, and resistant to compression. Fine grain contributes strength because threadlike strands of a material are many times stronger per unit of cross section than bulk material. Thus the composite bone is more rigid than collagenous fibers, more flexible and resistant to fracture than the mineral, and stronger for its weight and more versatile in the kinds of loads it can withstand than either component alone. Engineers make composite materials from fibers of glass in resin, boron in aluminum, tungsten in copper, and others. Also they are attempting to mimic the fine structure of biominerals using inorganic materials.

The presence of lacunae in bone (see figure on p. 88) also increases its strength. When a microfracture extends into a lacuna it tends to stop there instead of enlarging. Within skeletal elements, bone, but not cartilage, is usually spongy. The spicules of the meshwork are oriented so as to maximize strength in proportion to weight in ways explained on pp. 393 to 396. Finally, the fibers in the organic matrix of bone are woven or stratified in complex patterns that increase resistance to breakage.

ORIGIN OF THE HEAD, AND COMPONENTS OF THE HEAD SKELETON

We can speculate with confidence that as the remote protochordate ancestor of vertebrates gave up filter feeding and became a mobile predator, it also became bilaterally symmetrical, with feeding mechanism and certain sense organs concentrated at its anterior end. Presumably in parallel with these developments, the demands for more precise sensory innervation and motor control resulted in enlargement of the anterior part of the central nerve cord to form the brain. A cartilaginous head skeleton then evolved to house and protect the brain and sense organs, and to contribute to the efficiency of the feeding and respiratory mechanisms. We noted in Chapter 5 that the embryonic development of the vertebrate head is largely the product of neural crests (forming much of the head skeleton and various other structures), ectodermal placodes (forming nervous and sensory structures), and somitomeres (forming muscles). Because all of these primordia are unique to vertebrates, it follows that most of the vertebrate head (including everything anterior to the notochord) is an evolutionary neomorph (= *new* + *form*). The evolutionary history of the vertebrate skull is one of the most interesting and well-documented accounts in all of biology. It is therefore valuable to students of both comparative and human anatomy.

Since the head skeleton is complicated both as to origin and structure, it is helpful to find parts that can be described one at a time and that help unravel its evolution. The logical components to select are the chondrocranium, visceral skeleton, and dermal elements, because the structures that these contribute to the complete skull are at first rather distinct, both in their phylogeny and their ontogeny.

The **chondrocranium** supports the brain and organs of special sense. The **visceral skeleton** (or splanchnocranium) supports the gill arches and their derivatives. The **dermal elements** (or dermatocranium) complete the relatively superficial framework of the skull. The chondrocranium and visceral skeleton may remain cartilaginous or become bony, but they are always present. The dermal skeleton is always bony and is usually present, but has been secondarily lost by several major vertebrate groups. Regardless of which may have evolved first, the earliest known jawless vertebrates (except conodonts) already had all three components.

The term *skull* is sometimes used for all of the skeleton of the head. However, it is also used—and will be used here—for the single unit that forms the braincase and upper jaw and houses the nose and ear. The word is useful, but inexact.

COMMENT 8.1

THE INTRIGUING QUESTION OF HEAD SEGMENTATION

Embryology makes it abundantly clear that the trunk region of all vertebrates is fundamentally segmental. However, the nature of head segmentation, indeed its presence at all, has long been debated. Early in the 19th century two German naturalists, J.W. Goethe and L. Oken, independently postulated that the skull is derived from fused and modified vertebrae. This belief was accepted by many, including the eminent British anatomist Richard Owen. His equally eminent kinsman Thomas H. Huxley was equally sure that the skull is not vertebral. He derived the skull instead from ancestral head somites (postulated but not actually seen) numbering four anterior to the ear and five behind. In 1930 E.S. Goodrich published a scholarly book, *Studies on the Structure and Development of Vertebrates,* which included a scheme showing the derivatives of three segments anterior to the ear, two beside the ear, and three behind. His interpretation was widely accepted, at least tentatively, for nearly 60 years.

Then, in landmark research published in the early 1980s, Stephen Meier and his associates, using micro-dissection and scanning electron microscopy, convincingly demonstrated subtle segmentation of head mesenchyme in the chick. The segments appear as a series of slight hillocks, which they named **somitomeres.** These form during gastrulation and are paired, left and right, flanking the notochord and neural plate. Somites of the trunk form in sequence from anterior to posterior, and it is now recognized that all are preceded briefly by somitomeres before condensing and becoming separated from one another by clefts. In most of the head, somitomeres fail to become somites, although several somites *are* incorporated into the back of the head. Somitomeres transplanted to the trunk become somites.

Somitomeres have now been identified in the embryonic heads of representatives of all jawed vertebrates. Amniotes and teleosts have seven somitomeres, of which six are anterior to the ear and one behind. Amphibians and cartilaginous fishes have only four somitomeres. It appears that each of their segments posterior to the first is the equivalent of two adjacent segments of amniotes. Whether ancestral somitomeres split to become seven, or fused to become four, is not known.

Cranial somitomeres do not persist for long. Their metameric pattern is obscured by the spread of neural crests and head placodes and by growth and flexures of the brain. Adult derivatives of somitomeres are the extrinsic eye muscles, muscles of the face, and muscles of the pharynx and its derivatives, including the muscles of mastication. What is the relationship between cranial somitomeres and other formative tissues, both as to position and influence?

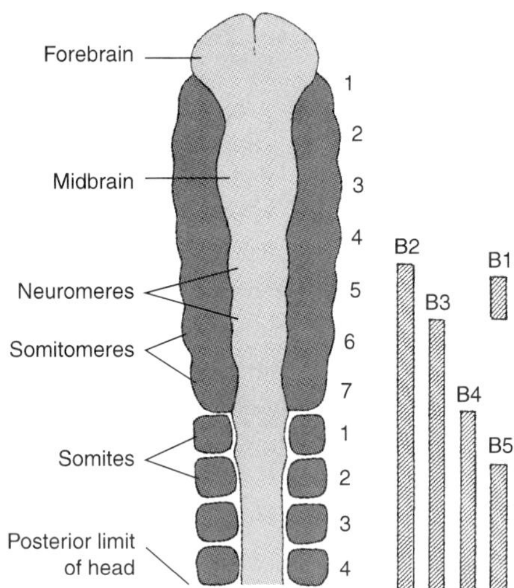

FIGURE 8.2 RELATIONSHIP OF EMBRYONIC AMNIOTE BRAIN TO SOMITOMERES, SOMITES, AND THE EXPRESSION OF SOME REGULATORY GENES. Bars on right show the domains of influence of certain *Hox* genes that control the segmentation of the head. See Comment 5.1 on p. 73.

Antone G. Jacobson has been a major contributor to the interpretation of somitomeres (see References). He believes that the segmentation of somitomeres is probably primary: they impress their segmentation onto the adjacent nervous system. Accordingly, specific somitomeres align with specific swellings along the embryonic brain called **neuromeres.** Another major contributor to the comparative embryology and biochemical control of cranial tissues is Drew M. Noden. He pioneered the experimental use of quail-chick chimeras. Quail cells have a nucleolar marker enabling the experimenter to trace cell lineages derived from transplants. (Other techniques trace cell lineages in mammals.) Noden believes the segmentation of somitomeres is probably not primary; instead, somitomeres, neural crests, and neuromeres arise more or less together. The registration of somitomeres with neuromeres is close, he says, but not exact, and may differ slightly between classes (Figure 8.2). Morphogenesis of the head is an active field of research.

Chondrocranium and Derivatives Most organs of the body can continue to function when subjected to moderate pressures by contacts with the environment or by locomotor or digestive activities. This is not true, however, of the central nervous system or organs of special sense. The spinal cord of primitive vertebrates derived some protection from the stiff notochord. The larger brain was shielded above by bony scales or plates that formed from the dermis of the skin, and was supported below and on the sides by a trough of hyaline cartilage. Capsules and rods of cartilage that protected the sense organs and stiffened the rostrum merged with this brain trough to form the chondrocranium (= *cartilage* + *cranium*). This structure has persisted in all vertebrates ever since its origin some 550 million years ago. Animals that do not have bony heads (cyclostomes and cartilaginous fishes) have relatively complete and heavy chondrocrania. Most vertebrates that do have bony heads retain a completely cartilaginous chondrocranium only in larval or fetal life. In adult life these animals replace at least part of the more delicate cartilage by bones that ossify within the cartilage.

Homologies of the major features of the chondrocranium of the different vertebrates have been established by noting their relations to such conservative landmarks as the notochord, hypophysis, and cranial nerves and blood vessels. Homologies of its numerous and complicated little rods and vacuities are more difficult to establish; seemingly, some will never be deciphered (if they are homologous at all). Embryonic recapitulation has not been very helpful. The homologies of the bones that ossify in the chondrocranium are more satisfactory.

The chondrocranium is a single unit. It is composed, however, of numerous elements that are distinct in the embryo. We will arrange these in five groups that are relatively large and constant (Figure 8.3).

The **notochord** lies within, just above, or just below the base of the developing chondrocranium. It may remain free or be obliterated, but usually its sheaths become cartilaginous and join the chondrocranium. The contribution it makes is small but important because its presence in the head may predate that of all other skeletal structures there, and because the constant position of its anterior end, just posterior to the hypophysis, serves as a reference point.

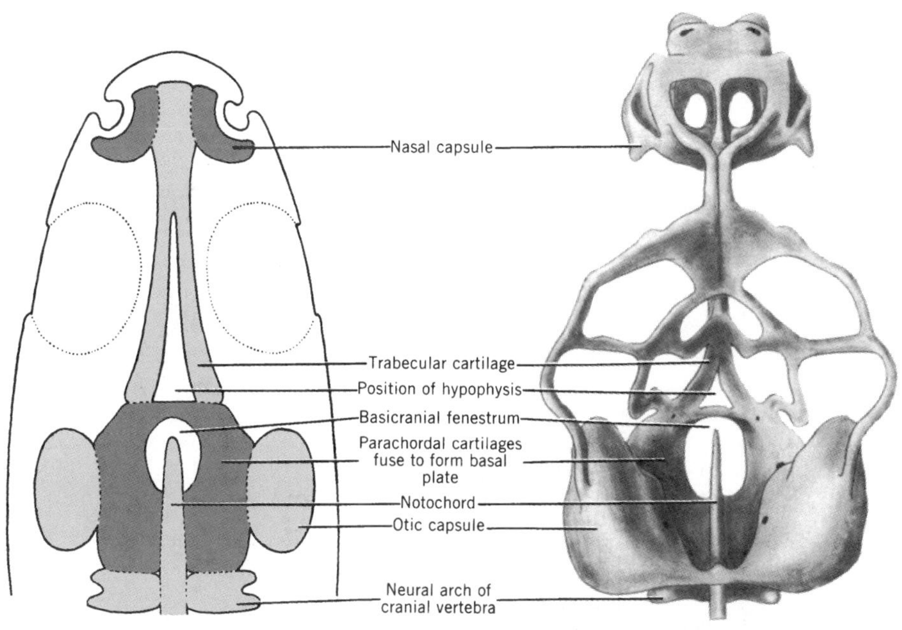

FIGURE 8.3
COMPONENTS OF THE VERTEBRATE CHONDROCRANIUM seen in dorsal view. Left, generalized early embryo; right, slightly simplified from the 25 mm stage of development of the lizard *Lacerta*.

Anterior to the notochord is a pair of bars called the **trabeculae** (= *small beams*). The hypophysis is located between their posterior ends. Sometimes, in lower vertebrates having broad heads, the two bars remain free and widely separated (the platytrabic condition); sometimes they are joined by a plate of cartilage; and usually they fuse anteriorly to form a narrower Y-shaped structure having its leg pointing forward (the tropitrabic condition). The trabecular part of the chondrocranium is related to the forebrain, nasal capsules, orbits, and rostrum. In older embryos it may be complicated by curved outgrowths that form an elaborate framework. Neural crests contribute to the formation of the trabeculae.

Behind the trabeculae, and flanking the notochord, another pair of cartilages form that are called (because of their position) **parachordal cartilages.** These soon merge above, below, or around the notochord to form the **basal plate.** The anterior part of the plate commonly encloses a large vacuity, and its lateral margins are pierced by foramina for the exit of cranial nerves. The side walls of the chondrocranium are usually incomplete. They consist of varied and sometimes complicated pillars and rods that fuse to the basal plate at its lateral margins. It is relatively clear that the basal plate is of segmental origin from the sclerotomes of embryonic head somites. It is considered to be in series with the bases of vertebrae. At the back of the basal plate are one or two projections called **occipital condyles,** which articulate with the first free vertebra of the spine. Their interesting origin from vertebral elements will be explained in Chapter 9.

A fourth contribution to the chondrocranium consists of one or more pairs of arches that rise up from the posterior angles of the basal plate to flank or encircle the spinal cord just where it enters the skull. These may be considered to be the **neural arches of cranial vertebrae.** In embryos of some fishes they even bear short ribs. The separate arches merge in late embryos and are then known collectively as the **occipital arch** ("occiput" = *back part of the skull*).

A final contribution to the chondrocranium is the cartilaginous **sense capsules** that house the nasal chambers and inner ear. The nasal capsules join the anterior ends of the trabeculae. The otic, or auditory, capsules join the margins of the basal plate just in front of the occipital arch. The eyes are also enclosed in capsules of cartilage. These remain free, however, because if they joined the chondrocranium, the eyes could not be moved independently of the head.

Bones typically develop in the chondrocrania of six of the nine classes of vertebrates (all except Agnatha, Placodermi, and Chondrichthyes). The names and positions of these bones are best learned from specimens and from Figure 8.4. No one list of such bones

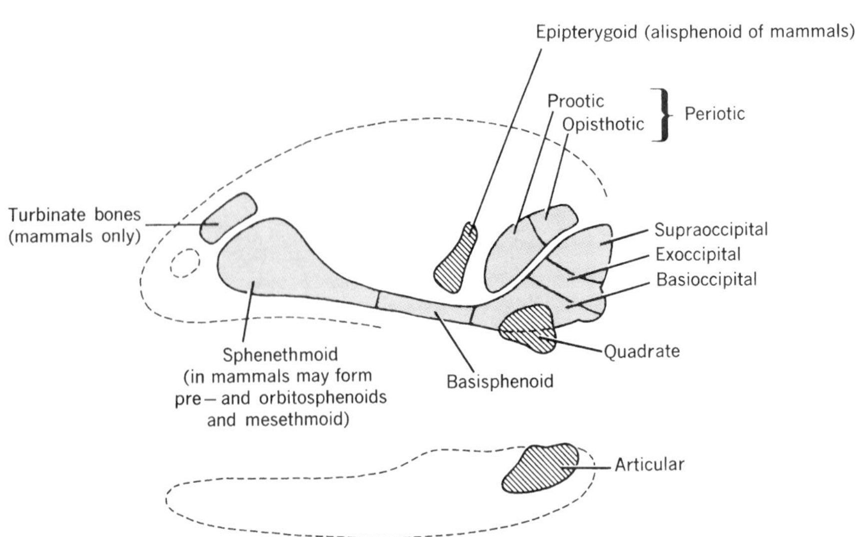

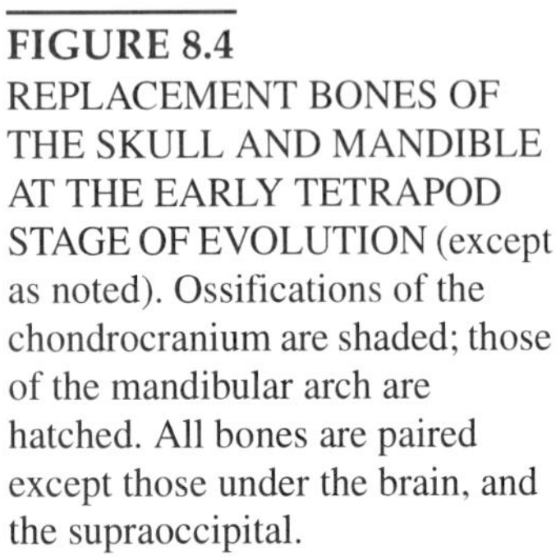
FIGURE 8.4
REPLACEMENT BONES OF THE SKULL AND MANDIBLE AT THE EARLY TETRAPOD STAGE OF EVOLUTION (except as noted). Ossifications of the chondrocranium are shaded; those of the mandibular arch are hatched. All bones are paired except those under the brain, and the supraoccipital.

serves for all vertebrates having ossified skulls; most of the bones are rather constant, but some (including the mesethmoid and orbitosphenoids) are of variable occurrence. In the higher vertebrates, the otic bones and certain of the basal bones tend to fuse together.

Visceral Skeleton and Derivatives The **pharynx** is a somewhat expanded part of the digestive tube lying between the oral cavity and the esophagus or stomach. In the earliest stages of chordate evolution the pharynx became perforated laterally by paired gill slits that served in feeding and respiration. A part of the pharynx still contributes to the feeding mechanism of all the vertebrates that have jaws, and the respiratory function of the pharynx is retained by all jawless vertebrates, fishes, and larval amphibians. When tetrapods ceased to use the pharynx for respiration, much of its musculature and skeleton became available for other uses and were adapted for the control and support of the tongue, vocal apparatus, and related structures. Thus, the pharynx has had a long and varied history.

The first protochordates may have had a dozen pairs of gills. The number was increased to as many as 100 by some protochordates (amphioxus), so that food particles could be entrapped on sticky gill bars as water was strained through the pharynx. The most ancient well-known vertebrates subsisted on larger food and did not require such a specialized pharynx. They reduced the count to 5–15 pairs of gill slits. By the time jaws evolved, the number had been further reduced and stabilized at about 6 (more in some sharks and fewer in some tetrapods).

Between each typical gill pouch, and also anterior to the first and posterior to the last, are bars of tissue that consist of skeletal elements (largely derived from neural crests), muscles, and respiratory filaments together with the nerves and vessels serving these structures. Each such bar is called a **visceral arch**—"visceral" because it forms largely from a specialized part of the gut tube. The basic number of visceral arches for jaw-bearing vertebrates is one more than the primitive number of pouches, or 7.

The relationship between the segmentation of the visceral arches and the segmentation of the head (discussed above) is of interest. The mesodermal part of the digestive tube forms in the embryo from the hypomere. Unlike the somites above it, the hypomere is not segmented. It follows that the muscles and skeleton of the visceral arches do not exhibit the primary segmentation of the body. The presence of serial gills, however, imposes on the derivatives of the pharynx a secondary segmentation that comes to be functionally integrated with the segmentation of the cranial nerves, yet need not be just the same in all vertebrates.

The presence of an ancestral preoral visceral arch is doubtful, or at best controversial. The first constant arch, therefore, is just behind the mouth. It is numbered 1, and the others follow in sequence. Each gill slit, or pouch, is given the same number as the arch that lies just anterior to it (Figure 8.5).

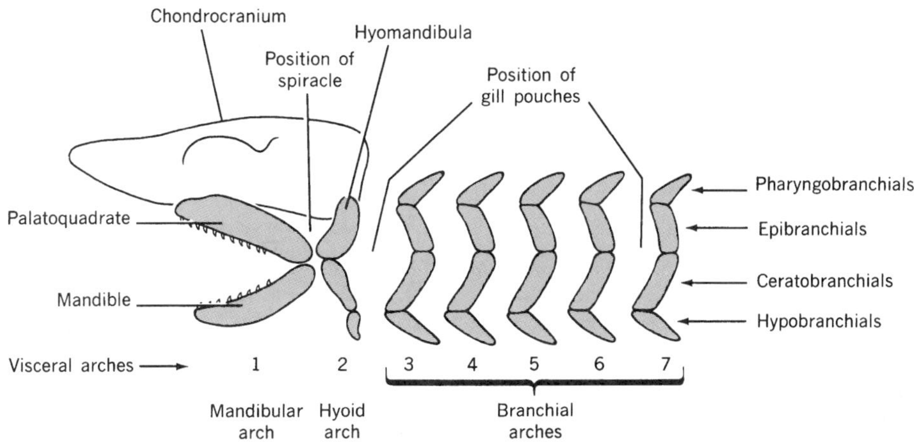

FIGURE 8.5
PRIMITIVE VISCERAL SKELETON represented by a stylized elasmobranch. Compare with Figure 8.9.

No other part of the vertebrate skeleton, except possibly the notochord itself, is as ancient in origin as the visceral skeleton. Each gill bar of jawless vertebrates is supported by a single cartilaginous rod. It may angle and bend, and usually joins its fellows above and below the gill slits, but it is not jointed. The skeleton of each visceral arch of gnathostomes is jointed. The basic number of paired segments is seemingly four per arch. Of these, the middle two, called the **epibranchial** (above) and the **ceratobranchial** (below), are relatively important for our story. There are also unpaired, midventral segments between the lower ends of the gill bars.

The first visceral arch (which probably never supported gills) enlarges to become the jaws, if jaws are present, and is then called the **mandibular arch.** Its epibranchial forms the upper jaw and takes the name **palatoquadrate.** The ceratobranchial presumably forms the lower jaw and is called the **mandibular cartilage.** The jaws become more or less firmly anchored to the chondrocranium to provide needed bracing (see below), and the second visceral arch, or **hyoid arch,** may be recruited to help in the support. The epibranchial is its key element and is called the **hyomandibula.** Succeeding arches, which serve primarily in respiration, are called **branchial arches** ("branch" = *gill*). Thus, the third visceral arch is also the first branchial arch. This terminology is clarified by Figure 8.5.

Ossifications that form in the visceral cartilages are, like those of the chondrocranium and nearly all of the postcranial skeleton, replacement bones (Figure 8.4). As noted in the following section, most of the bones that become functionally related to the jaws are not replacement bones and, therefore, are not derived from the visceral skeleton. The true first arch bones are usually small, yet no other bones of the body have a more engrossing and unexpected history. More on this later in the chapter.

Contributions from the Integument The rigid scales or heavy armor of most ostracoderms, placoderms, dipnoans, and crossopterygians was the most complete and solid over the head, where it supported the teeth and shielded the roof of the brain, the delicate gills, and other soft tissues. At first these integumentary bones were variable as to size and pattern: Anaspids had small scales; cephalaspids had huge, solid shields; arthrodires had large plates composed of individual elements joined by sutures. By the time the bony-fish stage of evolution was reached, the elements tended to stabilize as medium-sized pieces. The crossopterygians developed a general pattern of bones that was to persist throughout tetrapod history. The various classes have modified the basic pattern, however, by fusions, by deletions (particularly where the gills were lost, at the back of the head, and between head and pectoral girdle), and by sinking the bones below the skin and even underneath certain muscles.

In order to establish the homologies of these numerous bones it has been necessary for morphologists to use as many clues as possible. The paleontological sequence of various changes has indicated trends. Attention has been given to the relation of these bones to ossifications in the chondrocranium and to nerves, vessels, sense organs, and the opening often present on the top of the head that accommodates the pineal organ. The sensory canals of aquatic forms have been helpful because they tend to follow a conservative pattern and usually leave grooves, pits, or tunnels in the underlying bones. However, associations of canal to bone can alter over time as flow lines of water over the head are shifted.

The number of ossification centers, and the sequence of their appearance in the young animal, have also been useful because a particular bone usually ossifies from a given number of centers that appear in constant sequence in relation to the centers of other bones and to the regions of the brain. Indeed, the several parts of the brain induce specific ossifications adjacent to them. Since the characteristics of some centers are known to vary somewhat, however, this clue must be interpreted with caution. In these various ways it has been possible to establish with confidence the homologies of nearly all the bones derived from the integument. The names and positions of the more constant bones are shown in Figure 8.6.

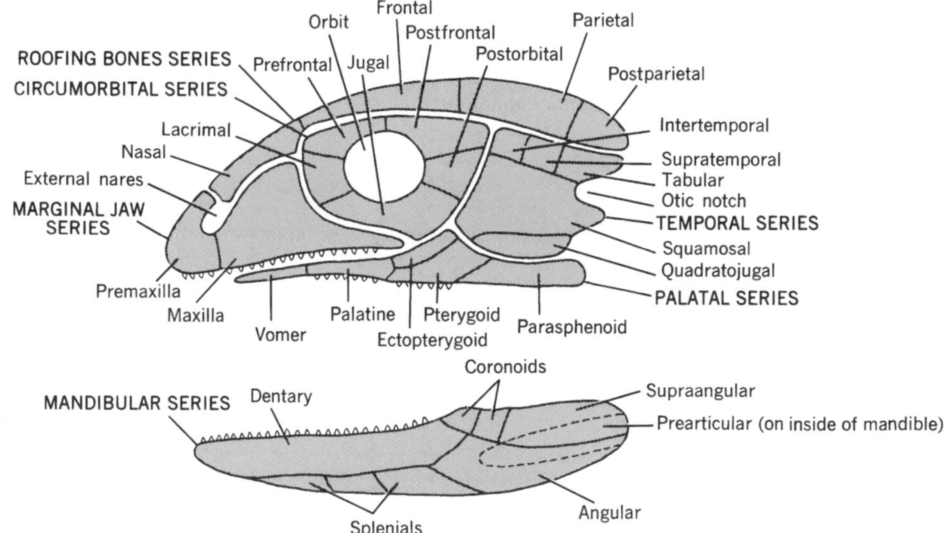

FIGURE 8.6
PRINCIPAL MEMBRANE BONES OF THE SKULL AND MANDIBLE AT THE EARLY TETRAPOD STAGE OF EVOLUTION. All bones are paired except the parasphenoid. (The sclerotic bones of the optic capsule are not shown.)

RELATIONS OF THE CRANIAL COMPONENTS

The recognition of these components of the head skeleton (chondrocranium, splanchnocranium, dermatocranium) is more than a mere convenience. As we have seen, they may retain a degree of structural independence in the adult. This is particularly true of anamniotes. The skeleton of the gills must be discrete as long as the gills are functional, and the membrane bones often remain superficial where they meet edge to edge to form a continuous covering. Furthermore, in the evolution of the various groups of animals, the bones of the skull have sometimes all become more and more strongly ossified and, in contrast, have sometimes all regressed. At other times, however, derivatives of the several components have evolved independently. Thus, it appears that the chondrocranium and visceral skeleton were emphasized by cartilaginous fishes and cyclostomes at the same time that the dermal skeleton was regressing. Conversely, the replacement bones of the skull, and these only, tended to regress in late cephalaspids, lungfishes, some primitive ray-finned fishes, and late labyrinthodonts. Evidently, replacement and membrane bones may respond differently to evolutionary forces even though they are visually identical.

In spite of original and potential independence, the replacement and membrane bones of higher vertebrates become intimately associated. Enlargement of the brain forces the chondrocranial ossifications outward, whereas enlargement of jaw musculature forces many membrane bones inward. These components meet and jointly form a single, firm unit.

When the first visceral arch was converted to jaws it became mechanically necessary to brace the arch more firmly than had been the case while it functioned only for respiration and filter feeding. This was first accomplished by attaching the palatoquadrate cartilage to the chondrocranium. Later, the hyomandibula of the second arch became a strut to further brace the jaw. Subsequently, the hyomandibula served alone, or the dermal bones covering the palatoquadrate joined other dermal bones to provide a firm anchor. The number, position, and firmness of the different points of attachment of the jaws to the remainder of the skull have been various. In fact, there are more than a dozen named patterns of jaw suspension. To further complicate the story, similar types of suspension have evolved independently in several instances. Properly interpreted, the relatively simple terminology proposed long ago by Thomas Huxley is still adequate. The terms, together with the animal groups to which they apply and the relationships among them, are shown in Figure 8.7. Some aspects of the story are amplified in the following pages.

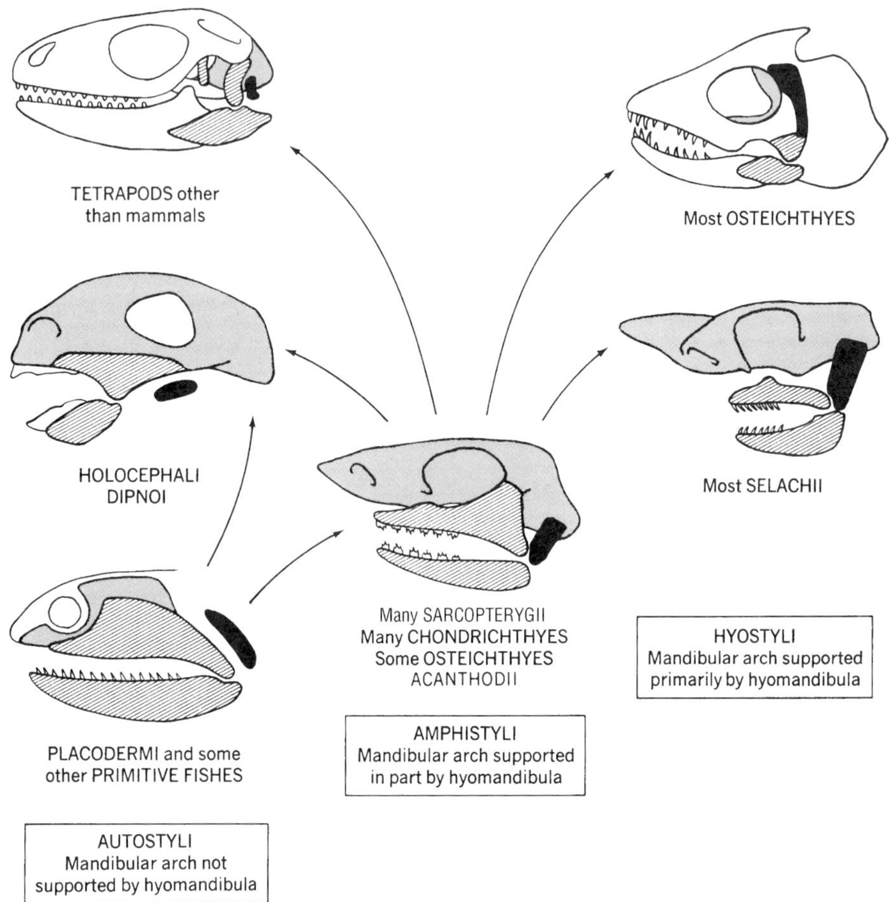

FIGURE 8.7
PRINCIPAL TYPES OF JAW SUSPENSION. Mandibular arch and derivatives are hatched, hyomandibula and derivatives are black, chondrocranium is shaded, teeth and extent of membrane bones are indicated in outline. The types of suspension may intergrade.

EVOLUTION OF THE HEAD SKELETON

Jawless Vertebrates: Innovations and Variety Conodonts are not known to have had a head skeleton. The chondrocranium of other Agnatha usually remains cartilaginous throughout life and is, therefore, poorly known for most extinct forms. In cephalaspids, however, it was often covered on all surfaces by a thin veneer of bone.

The visceral skeleton of agnathans is a continuous unit of cartilage: The individual arches tend to join above and below the gill slits, and the unit joins the chondrocranium (Figure 8.8). All arches are branchial in function. Their number is variable, but they are numerous relative to the arches of other vertebrates. (The specialized mouth parts of cyclostomes are supported by cartilages that are not derived from the visceral skeleton and are not represented in other vertebrates.)

The dermal head skeleton of agnathans ranges from armor shields (cephalaspids, pteraspids) through small scales (anaspids) to absence (conodonts, cyclostomes). The dermal elements cannot be homologized with the cranial bones of other vertebrates.

Placoderms: Enter Jaws The chondrocranium of placoderms was similar to that of agnathous vertebrates in being either entirely cartilaginous or partly ossified.

Important advances had been made in the visceral skeleton. Most significant, the first arch had become jaws. Some of the predaceous arthrodires had formidable jaws indeed. Replacement bones often formed in the large palatoquadrate. This structure was usually autostylic (= *self-supported*), as it was attached to the chondrocranium by ligaments and not by the hyomandibula. The second arch remained a typical branchial arch. As in higher vertebrates, the branchial arches of at least some placoderms were each supported by several skeletal elements.

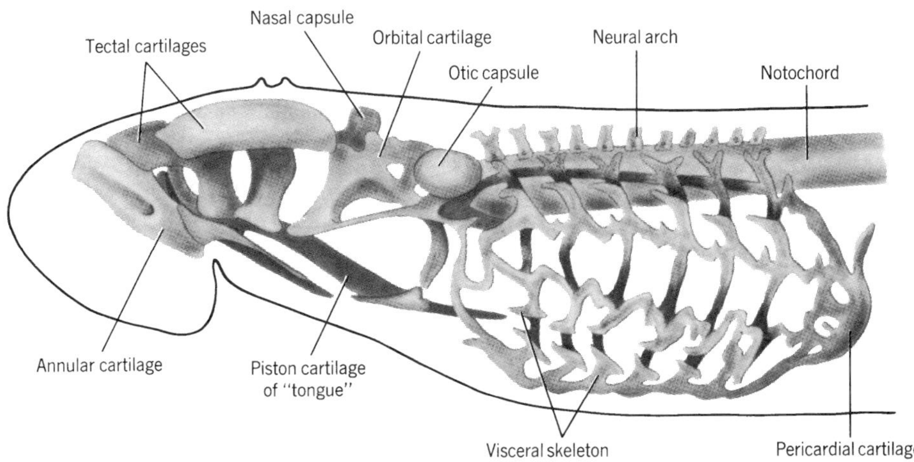

FIGURE 8.8 HEAD SKELETON OF A CYCLOSTOME shown by the lamprey, *Petromyzon,* as seen in left lateral view.

The dermal skeleton usually consisted of heavy head and thoracic shields that were joined by hinges (arthrodires and antiarchs). Individual dermal bones of placoderms cannot be homologized with those in the armor or skulls of other vertebrates.

Cartilaginous Fishes: Specialization and Regression Cartilaginous fishes have had scant bone or none throughout their long history. To provide the protection that dermal bones gave to the brains of ostracoderms and placoderms, the chondrocranium became unusually solid, with complete side walls and a roof (Figure 8.9). Although the chondrocranium never ossifies, it is sometimes so hardened with granules of calcium salts that it cannot be cut with a knife. Biology classes often study the chondrocranium of the shark, but it is not typical of that structure for vertebrates in general.

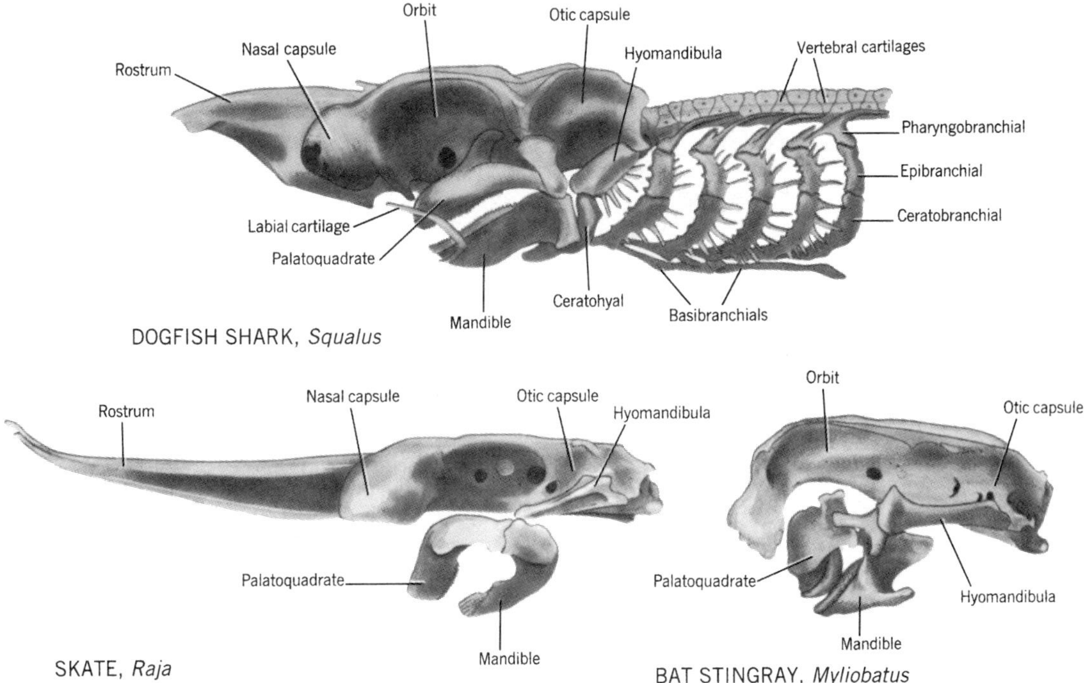

FIGURE 8.9 HEAD SKELETONS OF SELECTED CARTILAGINOUS FISHES seen in left lateral view. Above, chondrocranium, visceral skeleton, and anterior vertebrae; below, chondrocranium and jaws with their support.

Jaw suspension is usually amphistylic (pleuracanths, cladodoselachians, some sharks). It may also be hyostylic, however (other sharks), and in Holocephali, as that term implies, jaw suspension is strengthened to adapt it to a diet of shellfish by fusing firmly with the braincase—a form of autostyly. There are usually six, but as many as eight, postmandibular arches. Typical branchial arches have four paired skeletal elements, and there are also unpaired median ventral cartilages. The segmentation and configuration of the visceral skeleton of cartilaginous fishes are considered to be generally primitive for all fishes.

Bony Fishes: Diversity and Complexity In no other class is the skull so varied and complex as in the bony fishes. This is not unexpected, since the class is so large, yet other parts of the body do not exhibit as much diversity. Much of the variety results from adaptation to many diets, body shapes, and habits, which are of less concern to us here than evolutionary trends. It is important that the two subclasses have somewhat different skulls, for only one of them can include the ancestors of tetrapods, and the skull helps us to recognize which subclass this is.

The chondrocrania of the earliest known bony fishes were relatively well ossified, often, seemingly, in one unit without sutures. From this start the dipnoans had already regressed by Mesozoic times so that only the exoccipital bones ossify in an otherwise cartilaginous braincase. Crossopterygians also came to have a rather cartilaginous chondrocranium that always has a striking and unique feature: It is in two pieces (Figure 8.10). An anterior (or ethmosphenoid) unit is derived from the trabeculae and supports the orbits and rostrum. Joined to this by a movable hinge is a posterior (oticoccipital) unit that is derived from the parachordal cartilages and supports the brain. This crossopterygian trademark relates to the gape and

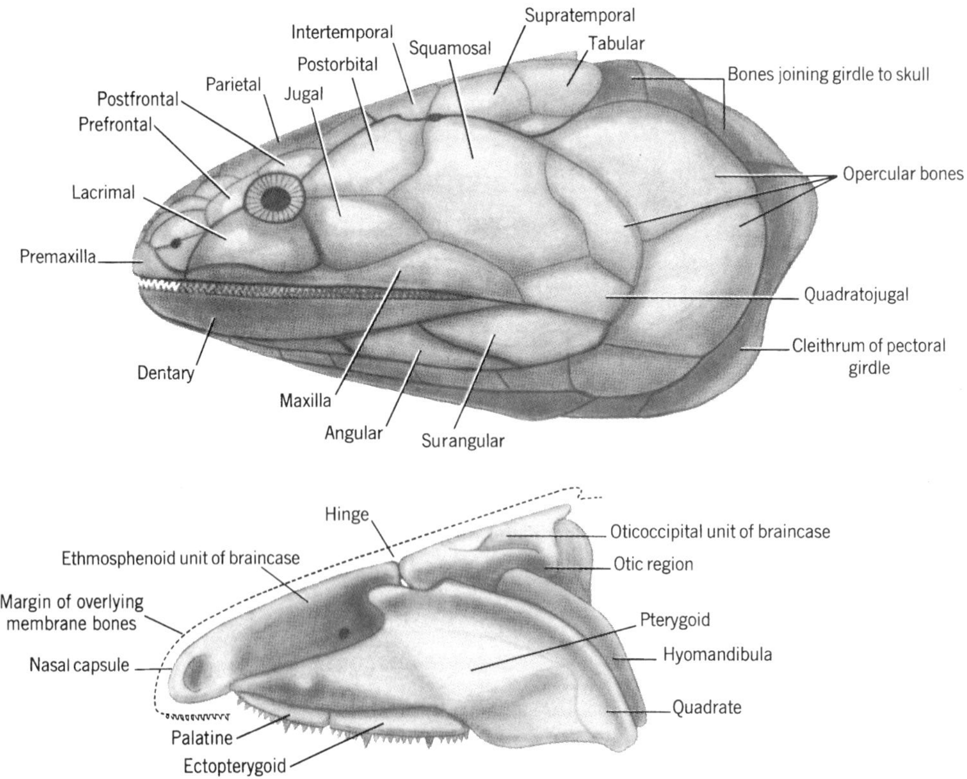

FIGURE 8.10 HEAD SKELETON OF *Eusthenopteron,* AN OSTEOLEPID. Superficial membrane bones are shown in the upper drawing; the hinged chondrocranium and some other deep parts of the skeleton are shown in the lower drawing.

power of their jaws. Typical ray-finned fishes have braincases that are more or less completely ossified with separate bones.

The visceral skeleton is more nearly the same between the two subclasses. Membrane bones support the marginal teeth; hence the mandibular arch usually regresses. Several replacement bones may form in the palatoquadrate of bony fishes. Of these, the quadrate has the most evolutionary significance. It forms the upper part of the hinge of the jaw of bony fishes, amphibians, reptiles, and birds. The only replacement bone in the lower jaw is the small articular that forms the lower part of the hinge (Figure 8.4).

The second arch is not branchial in function. Its large hyomandibula usually supports the jaw (except in dipnoans), which therefore is hyostylic (ray-fins) or amphistylic (crossopterygians). There are usually five branchial arches. The gill arches may be bony (most ray-fins) or cartilaginous (dipnoans) and do not differ importantly from those of placoderms and cartilaginous fishes.

With few exceptions, the bony fishes have complete dermal skeletons of small to medium-sized bones (Figure 8.11). A series of bones joins the pectoral girdle to the skull. A movable bony operculum covering the gills is distinctive. Each of the major taxa of bony

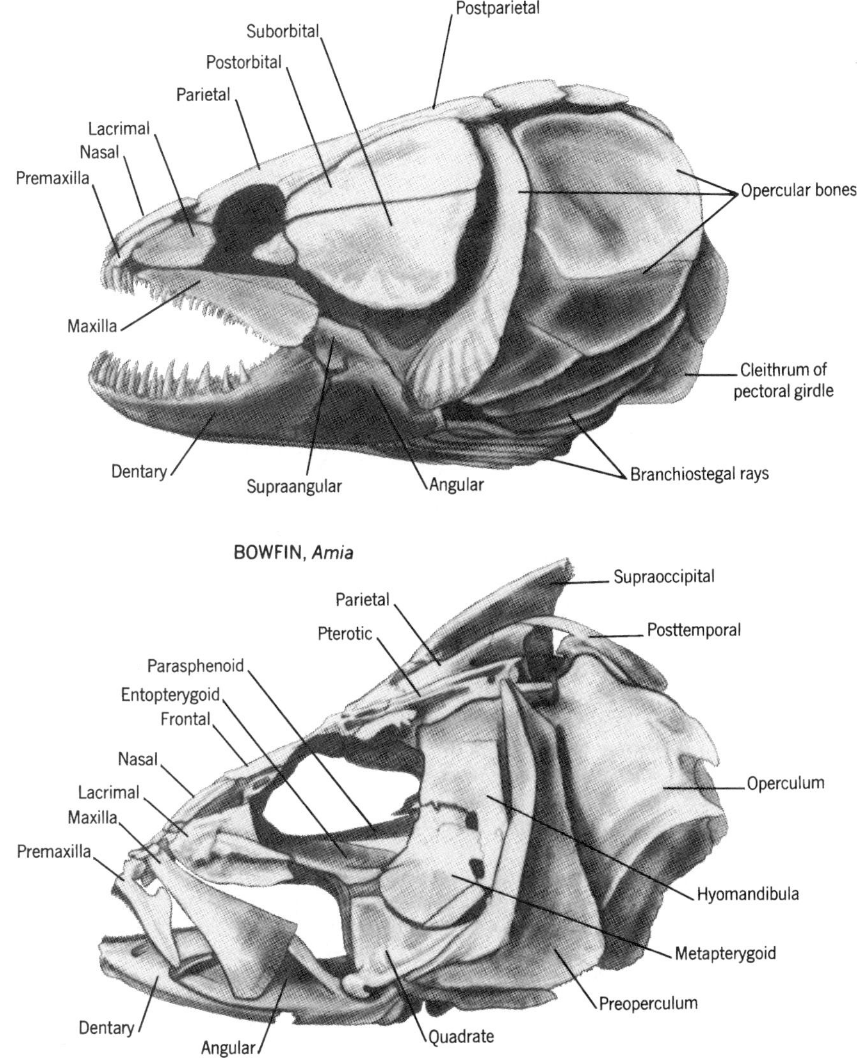

FIGURE 8.11
HEAD SKELETONS OF PRIMITIVE (above) AND DERIVED (below) RAY-FINNED FISHES of the superorder Neopterygii.

fishes has its own pattern of dermal bones. The pattern of the crossopterygians, and only this pattern, can be homologized with the pattern of amphibians. This is of primary importance and is one of several lines of evidence supporting crossopterygian ancestry for all tetrapods. The membrane bones of the other major groups of fishes are assigned names that are also applied to the bones of tetrapods, but for them the relationships are, in fact, quite uncertain.

Amphibians: Conservativism or Regression When the progenitors of amphibians crawled out of the water, they lost from their skulls such piscine characters as gill bars and operculum. Evolution of the tetrapod skull was then by no means finished—important changes were still to come to the palate, skull roof, and jaw mechanism—yet a trend toward stability of form had been initiated by crossopterygians that was to continue through all their descendants. There is less variation of skull structure in all the tetrapods together than in the bony fishes alone. Labyrinthodonts bequeathed the crossopterygian skull to the first reptiles without making striking changes. Surviving amphibians, on the other hand, have somewhat specialized skeletons. The skulls of extinct labyrinthodonts are more within the mainstream of evolution than are those of most modern amphibians (Figure 8.12).

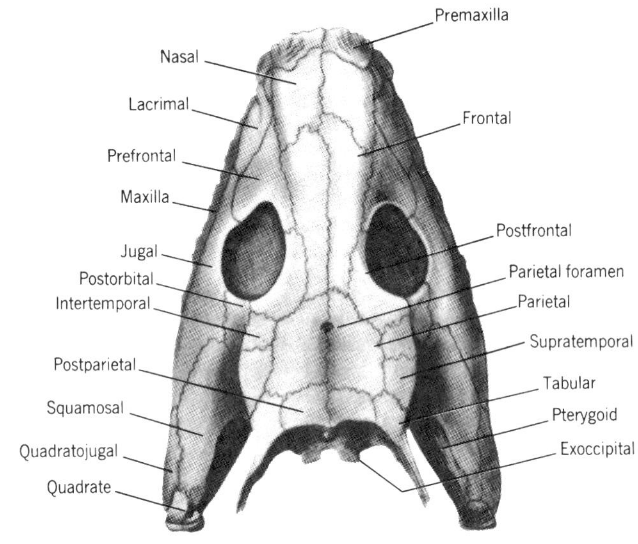

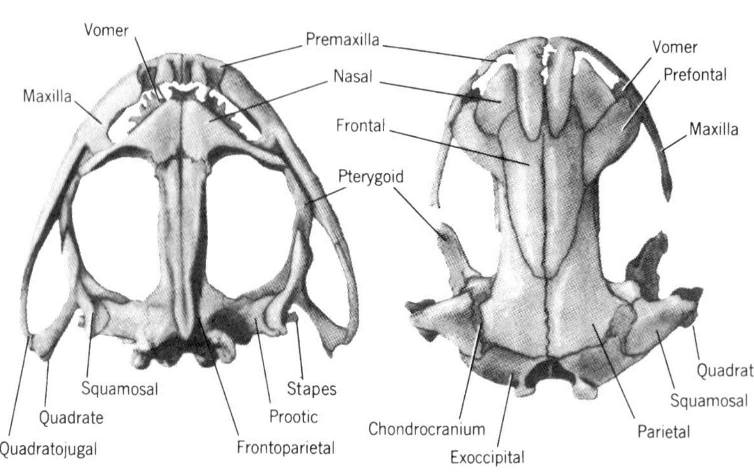

FIGURE 8.12 AMPHIBIAN SKULLS OF THE SUBCLASS LABYRINTHODONTIA (above), AND ORDERS ANURA (below left) AND URODELA (below right).

Ancestral amphibians had a nearly full complement of replacement bones derived from the chondrocranium; only the supraoccipital was missing. Lissamphibia lack even such usually prominent bones as the basioccipital and basisphenoid. The exoccipitals are retained by these animals largely to provide the paired occipital condyles. The more primitive labyrinthodonts had a single occipital condyle formed by the basioccipital.

The most significant changes that occurred in the skull during the transition from fish to amphibian concerned the visceral skeleton. The quadrate of the upper jaw now articulated with the squamosal without an "assist" from the hyomandibula. Thus, jaw suspension had become autostylic (again), as it was to remain throughout subsequent vertebrate evolution. Most extinct amphibians probably rested their large heads on the ground and conveyed vibrations in the substrate to the inner ear by bone conduction through the jaw and other structures. This mechanism was improved by incorporating a reduced, stubby hyomandibula into the line of sound transmission. The bone was then an ear ossicle and is assigned a new name, **stapes,** or columella.

Ventral elements of the hyoid arch support the tongue, as they do also in fishes, but the tongue of amphibians is larger and more muscular. The more posterior arches function as branchial arches only in larvae and in those few species that retain gills as adults. Otherwise, they are reduced in number to three and converted to help support the tongue and newly evolved larynx. The tracheal rings may have the same origin. The complex of bones that moves the tongue and suspends the larynx from the base of the skull is called the **hyoid apparatus.**

Of the membrane bones there is less to say. The operculum is lost, and the pectoral girdle is no longer joined to the skull. There has been some tendency to simplify the pattern of bones, but in general, amphibians have not introduced innovations.

Reptiles and Birds: Variations on the Basic Plan Reptiles are important to our story for modifications introduced by one or more of their lineages in relation to the structure of ear ossicles, palate, and jaw mechanism. The skulls of birds are sufficiently similar to those of their reptilian ancestors that we can here consider birds to be merely specialized reptiles.

In these classes, the replacement bones of the chondrocranium vary widely in configuration, but a full complement is typical. In birds, all of them fuse with one another and with other bones of the braincase before, or soon after, hatching, leaving no trace of sutures in the adult (Figure 8.20). Reptiles and birds have one occipital condyle.

The visceral skeleton of birds and most reptiles remains essentially the same as for amphibians: Quadrate, articular, and some cartilage form from the first arch; stapes from the second arch; hyoid apparatus chiefly from the second arch, but also from the third and sometimes fourth arch; larynx and tracheal rings probably from the sixth and seventh arches (Figure 8.13).

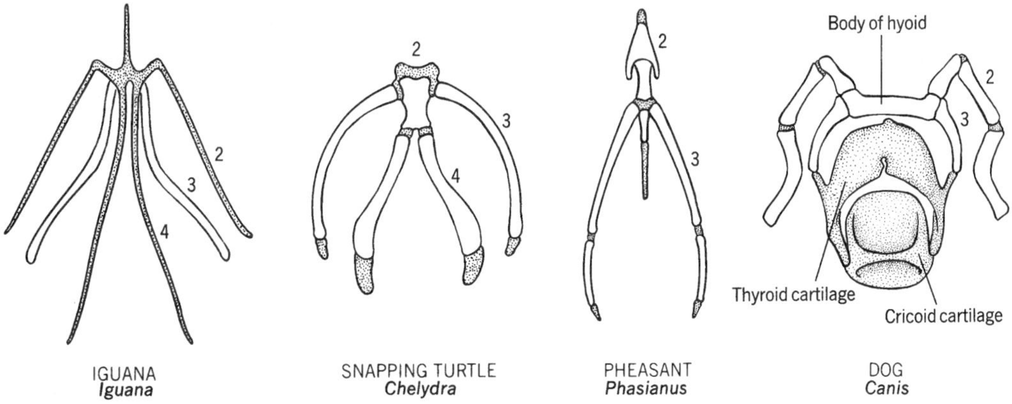

FIGURE 8.13 VISCERAL SKELETONS OF SELECTED TETRAPODS shown in ventral view. Numbers indicate visceral arch of origin. Stippled elements are cartilaginous.

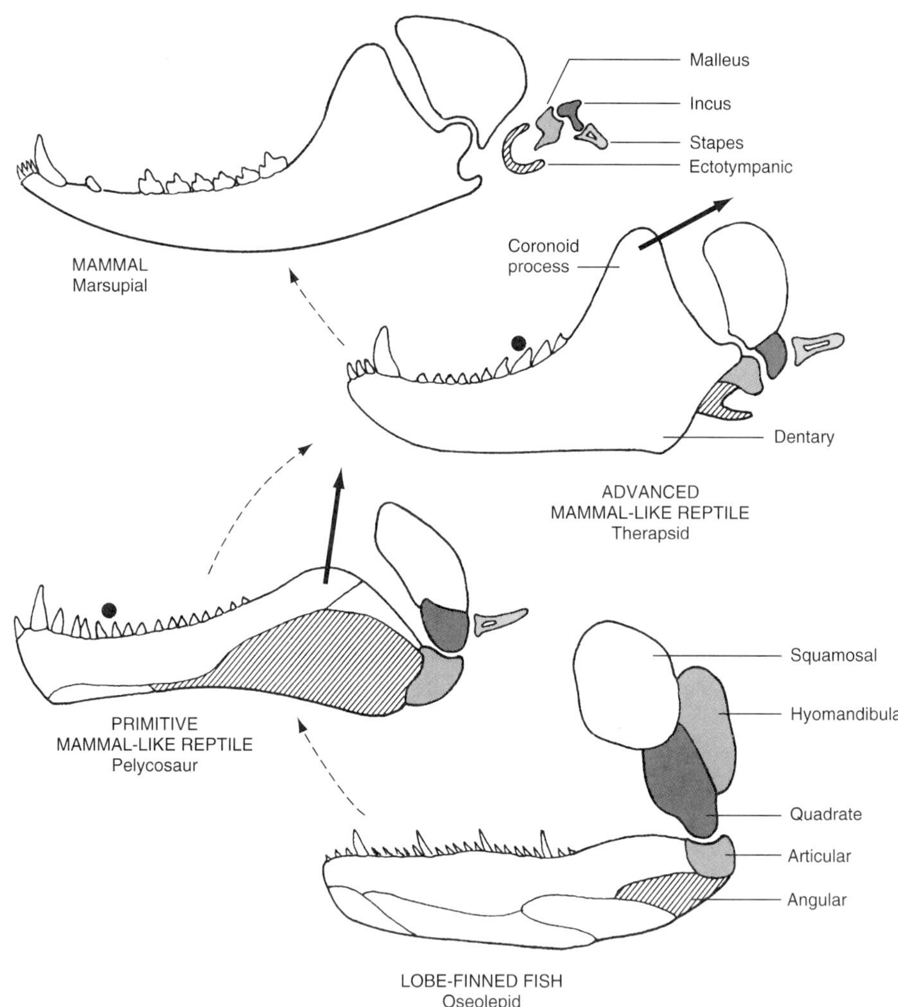

FIGURE 8.14
PHYLOGENY OF THE JAW ARTICULATION AND EAR OSSICLES. Stylized lateral views somewhat distorted for two-dimensional representation (e.g., stapes is medial, not posterior to quadrate).

Mammal-like reptiles converted certain JAWBONES to EAR OSSICLES and related structures. The transformation is of the utmost significance and is exceptionally well documented in the fossil record (see Figure 8.14). Numerous changes came together to make the shift possible (in References, see Allin and Hopson). As chewing and heterodonty evolved, the point where maximum bite force is usually applied moved backward in the mouth (black circles in the figure). The dentary bone enlarged in cynodont therapsids and evolved a dorsally projecting coronoid process. The jaw-closing muscles of pelycosaurs pulled as shown by an arrow in Figure 8.14, exerting a positive force at both the bite and the jaw joint. By the cynodont stage of evolution the jaw-closing muscles had divided. A principal part, the temporalis muscle, inserted on the coronoid process and pulled at a more backward-sloping angle. Its contraction caused the jaw to tend to pivot around the bite point (clockwise in the figure). Consequently, force was positive at the bite but minimal or even negative at the jaw joint, thus diminishing the need for bones at the back of the jaw to strengthen the joint. (This account is much simplified; in References, see the paper by Bramble.) The articular bone became smaller and was displaced until it finally lost its position as the lower element in the hinge of the jaw. Likewise, the quadrate became smaller and lost its position as the upper element of the hinge. It is probable that disproportionate evolutionary enlargement of the part of the brain called the neocortex (repeated in the ontogeny of some marsupials) contributed to the displacement of these bones away from the jaw, thus shifting its hinge to a more forward articulation between the dentary and squamosal. The articular and quadrate now became ear ossicles as the hyomandibula had done 100 million years before. The articular is given the new name **malleus** and the quadrate the new name **incus.** The angular likewise slips off of the jaw to become the **ectotympanic,** which rings the entrance to the middle ear and, in many mammals, enlarges to form the bulbous **tympanic bulla.** (Other mammals form the bulla also, or instead, form one or more new bones called entotympanics.) The stapes now articulates laterally with the incus and medially with the inner ear. These transitions are of particular interest because they relate to the recognition of the first mammals. Conveniently, but arbitrarily, mammals were long distinguished from their immediate reptilian ancestors on the basis of having an ossicle called the malleus instead of a jawbone called the articular. This distinction is no longer satisfactory for some fossils that are on the borderline.

The membrane bones of reptiles vary somewhat in number and may be heavy or delicate. Variations in configuration and relationship are more striking, however, and differences in the relation of bones on the ROOF AND SIDE WALLS OF THE SKULL to the muscles of the jaws are the basis for the naming of the subclasses of reptiles. Amphibians and fishes accommodate their jaw muscles beside the braincase and under the superficial roof bones of the skull, which may be continuous or notched. This arrangement persisted in the stem reptiles—long since extinct. Subsequent reptilian lineages developed more powerful jaws. Presumably in response to the resultant increased and altered stresses, the cranial vault strengthened in some places and weakened in others. Ultimately, one or two pairs of openings formed in the temporal region of the skull, and jaw muscles then moved out through them and took origin in part from their margins. It is evident that this process occurred independently several times, because the fenestration ("fenestra" = *window*) evolved at different places on the skull in different groups. Names for the various skull types consist of the combining form "apsid," meaning "arch," plus a prefix meaning "not," "two," "wide," etc., as appropriate.

Turtles are related to the ancestral, **anapsid** stem reptiles, and like them have no true temporal openings (Figures 8.15, 8.17, 8.18). (Sea turtles demonstrate this condition well, but other turtles have formed an emargination into the skull roof bones from behind.) Synapsida and Sauropterygia each have one pair of temporal openings, but in each it is bordered by different bones. The mammalian condition was derived from the **synapsid** arrangement of mammal-like reptiles merely by enlarging the opening. **Diapsid** skulls have the same opening as synapsid skulls and also another that is dorsal to it. This pattern is found in extinct archosaurs, surviving archosaurs (crocodilians, which have small openings, but must usually be used to illustrate the condition to students), and also in the primitive lepidosaur, *Sphenodon* (Figures 8.16, 8.17, 8.18). Other lepidosaurs, sauropterygians, and birds have modified the primitive diapsid pattern by the deletion of one or more of the arches of bone from below or between the original openings (Figures 8.15, 8.18, 8.20).

A change in the STRUCTURE OF THE PALATE that was introduced by certain reptiles and later became universal among mammals is likewise related to the feeding mechanism. Amphibians and most reptiles bolt their food. It is unimportant to them, therefore, that respired air comes through the nares into the forward part of the mouth. Also, their skulls are subjected to relatively little torsion during feeding. Other reptiles, and particularly mammal-like reptiles, tore, crushed, or even chewed their food before swallowing. To avoid interrupting their breathing, it became advantageous to deliver inspired air to the pharynx behind the chewing mechanism. This was doubly important for those that were becoming endotherms and were therefore increasing their respiratory rates. Furthermore, chewing required that their skulls become stronger. The paired vomers, which formerly had bordered the internal

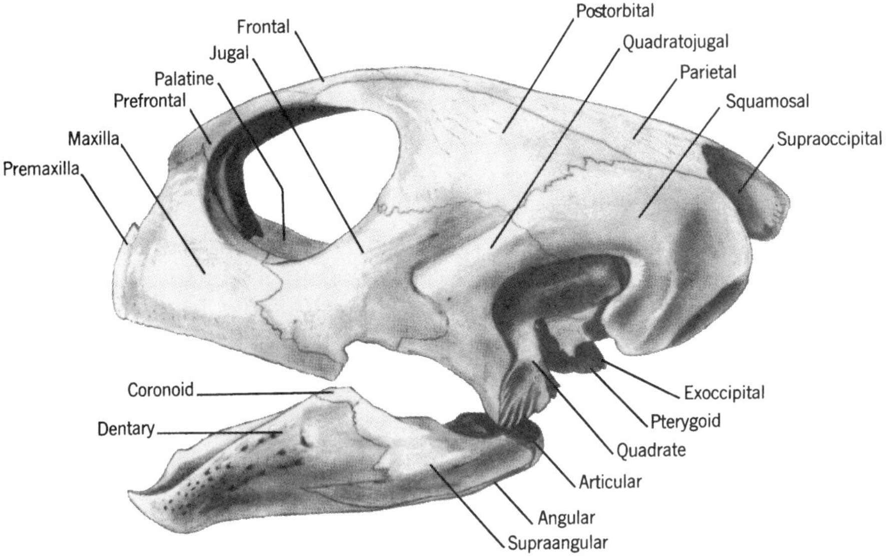

SEA TURTLE, *Chelonia*

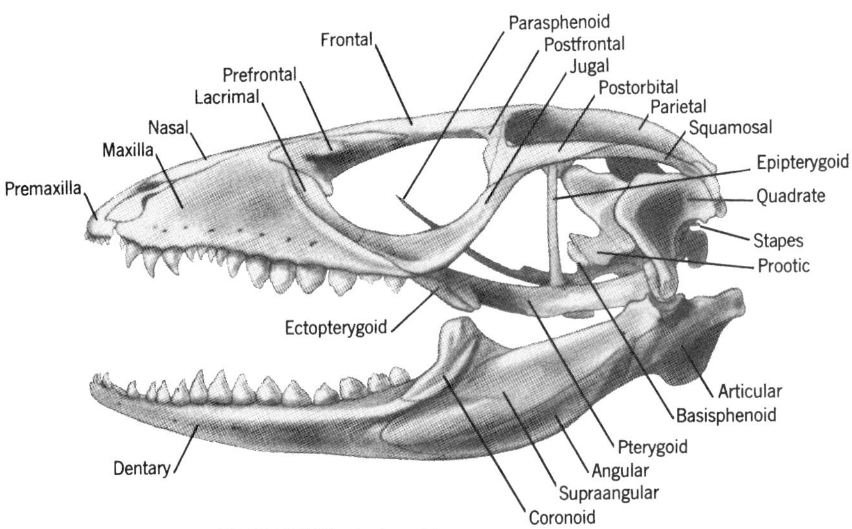

TEGU LIZARD, *Tupinambis*

FIGURE 8.15
REPTILE SKULLS AND MANDIBLES OF THE SUBCLASSES TESTUDINATA (above) AND LEPIDOSAURIA (below).

nares behind, gradually merged, migrated posteriorly, and moved dorsally above the new, more posterior internal nares (Figure 8.19). The parasphenoid was lost, thus "getting out of the way" of the relocated nares. The pterygoids shortened to the rear. Meanwhile, shelflike processes of the maxillae, and later also of the palatines, grew to the midline in front of the shifting nares to form a new, or **secondary palate.** (Actually, secondary palates evolved independently several times. Of reptiles other than mammal-like reptiles, some turtles have an incomplete secondary palate and crocodilians have a complete one.)

The palate of crossopterygians and early labyrinthodonts was complete and solid. The palates of later labyrinthodonts, surviving amphibians, most reptiles, and birds have lateral vacuities of larger or smaller proportions that lighten the skull without much loss of strength and, for some of these animals, enable protruding eyes to be withdrawn into the head on occasion to escape harm or assist in swallowing.

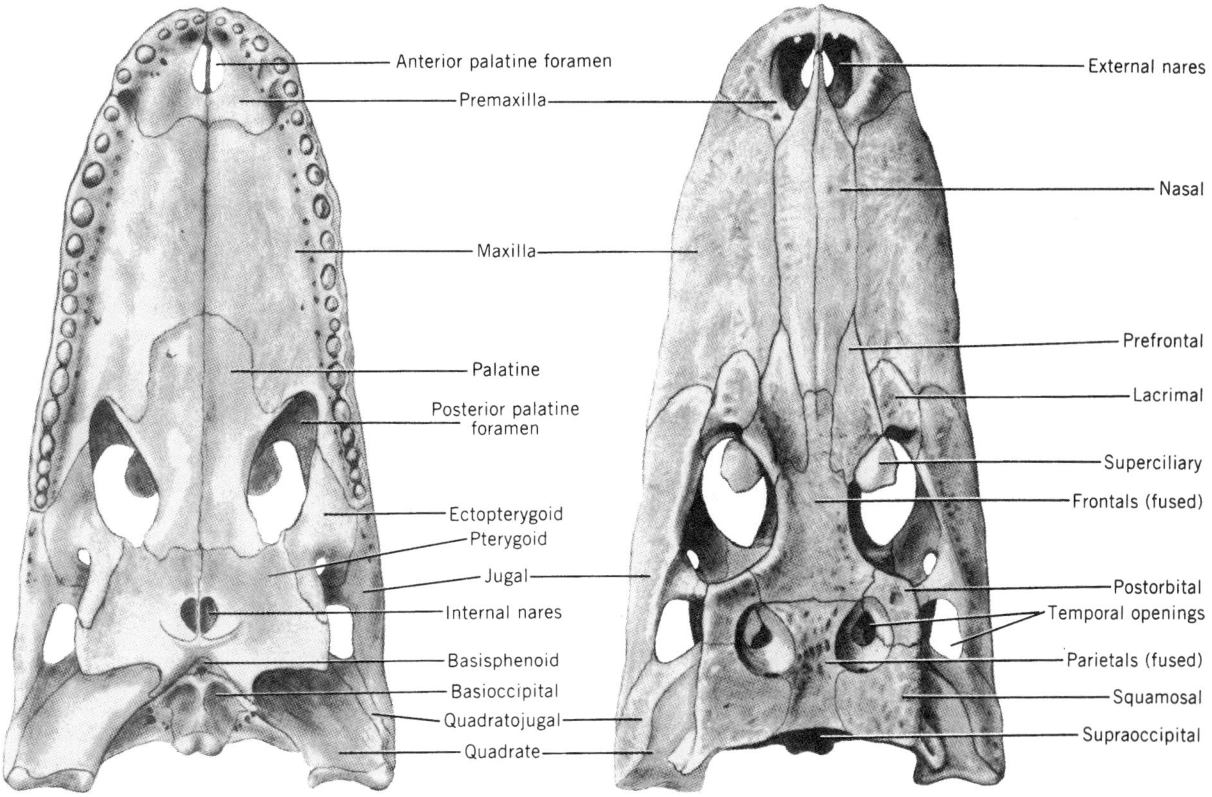

Anterior palatine foramen
Premaxilla
Maxilla
Palatine
Posterior palatine foramen
Ectopterygoid
Pterygoid
Jugal
Internal nares
Basisphenoid
Basioccipital
Quadratojugal
Quadrate
External nares
Nasal
Prefrontal
Lacrimal
Superciliary
Frontals (fused)
Postorbital
Temporal openings
Parietals (fused)
Squamosal
Supraoccipital

FIGURE 8.16 REPTILE SKULL OF THE SUBCLASS ARCHOSAURIA shown by the alligator.

Brain
Skull roof
Jaw muscles
Mandible
Zygomatic arch
ANAPSID REPTILE
DIAPSID REPTILE
MAMMAL

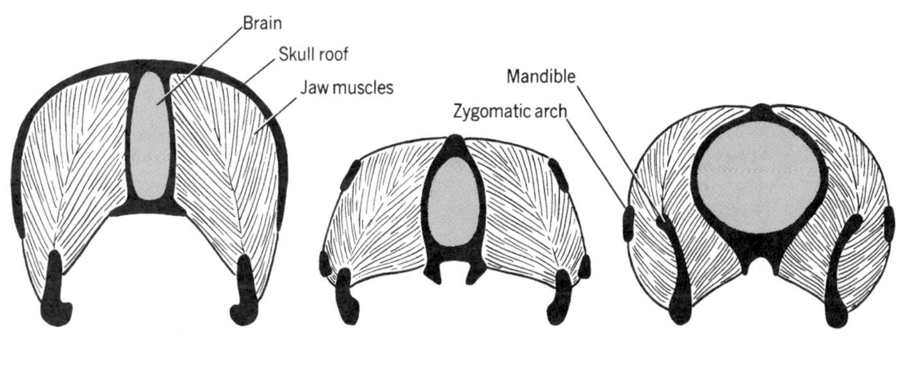

FIGURE 8.17
SOME RELATIONSHIPS BETWEEN JAW MUSCLES AND THE HEAD SKELETON shown by cross sections of the head at the epipterygoid-alisphenoid level of the skull. Diagrammatic, but based on the sea turtle, *Chelonia,* the tuatara, *Sphenodon,* and the dog, *Canis.*

Birds and some reptiles are able to move the palate forward and backward on the braincase by pivoting the quadrates to and fro and by providing a hinge in the roof of the skull in front of the orbits. This action elevates the upper jaw, which increases the gape of the mouth. Skulls that have this mechanism are said to be kinetic and are described further on pp. 561 to 563. Ornithologists divide birds into four or more groups on the basis of the relationships of the palatal bones involved in this mechanism.

Mammals: Some Further Modifications The mammalian skull is at the same time exceedingly variable in regard to adaptive features such as strength and proportions, and

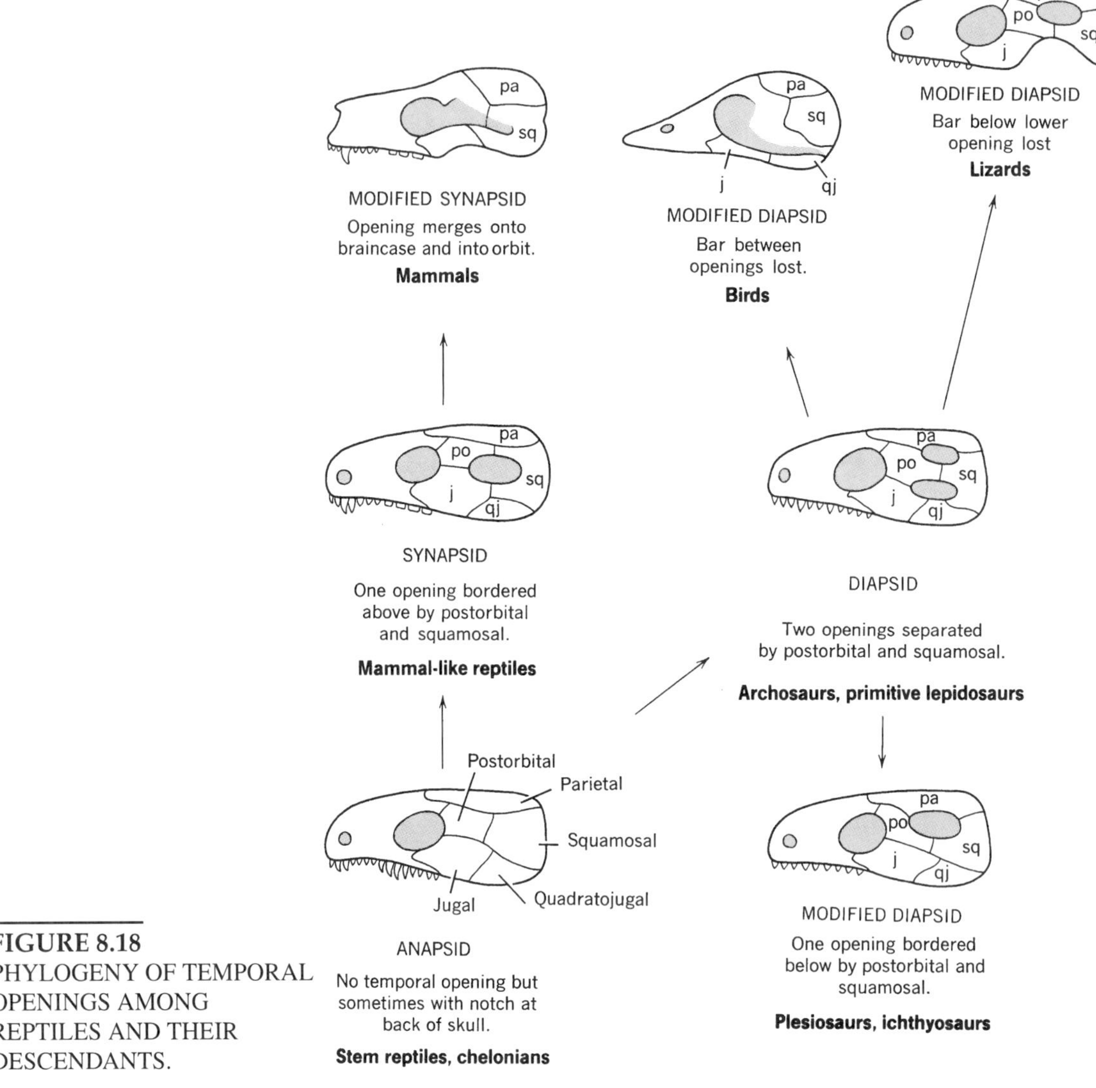

FIGURE 8.18
PHYLOGENY OF TEMPORAL OPENINGS AMONG REPTILES AND THEIR DESCENDANTS.

conservative in regard to basic plan. Temporal architecture, jaw suspension, ear ossicles, and secondary palate remain about as inherited from the most advanced synapsid reptiles (Figures 8.21–8.23). The brain is, of course, larger than for other vertebrates, a factor that contributes to complete functional integration of the replacement and membrane bones of the braincase. There are again two occipital condyles as for most amphibians. Cranial sutures are nearly always more in evidence than for birds and usually less in evidence than for reptiles. Certain combinations of bones are particularly prone to fuse early in life. Common fusions include the postparietal with the supraoccipital, basioccipital with the exoccipitals to form a single occipital, the otic bones (prootics and opisthotics) with the squamosals to form temporals, and the four sphenoid bones (basi-, pre-, ali-, orbito-) with one another to form a single sphenoid.

Mammalian nasal structure is distinctive. The anterior bony nares have merged to form a common opening. A relatively large nasal chamber is in part filled (as in some birds) by delicate scrolls of bone, the conchae, or turbinates, which are outgrowths from the inside walls of the maxillae, nasals, and ethmoids. These are covered with epithelium

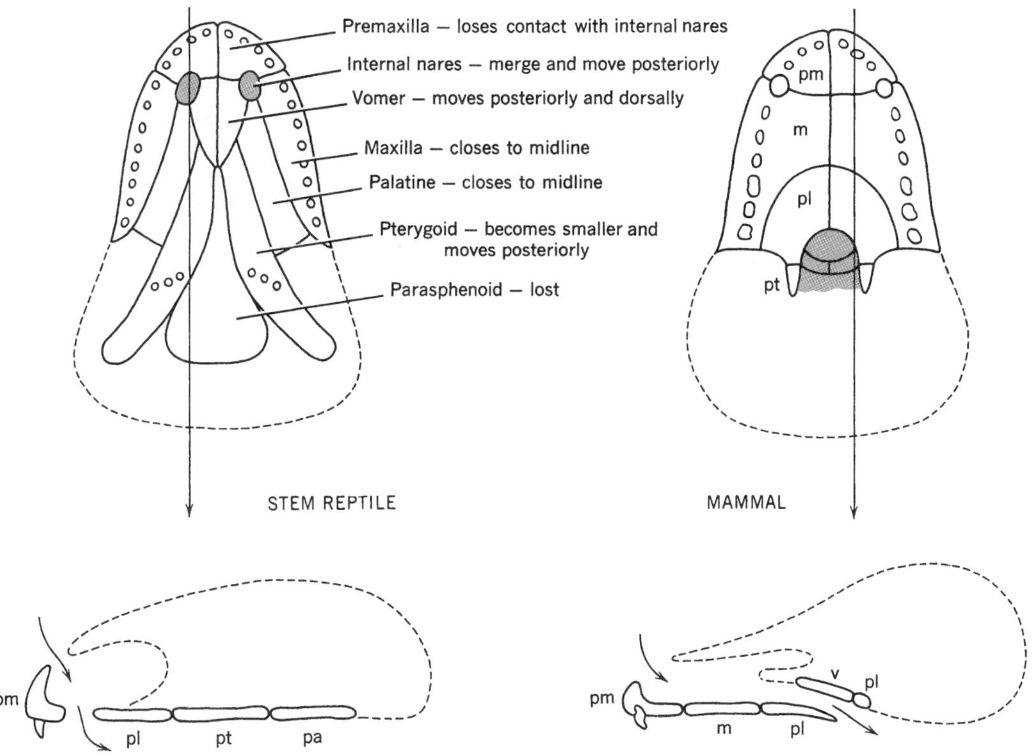

FIGURE 8.19 EVOLUTION OF THE SECONDARY PALATE. Palatal views above; parasagittal sections below. Dashed lines indicate parts of skull not involved in this evolutionary sequence. Lower arrows show path of inspired air.

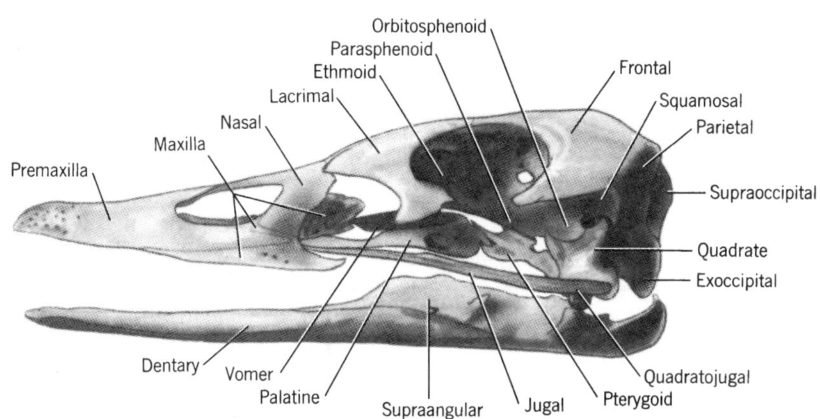

FIGURE 8.20 SKULL AND MANDIBLE OF A BIRD shown by the swan, *Cygnus*. Sutures of the braincase are obliterated in the adult.

and serve to warm and clean inspired air before it reaches the lungs. Their presence is believed to be associated with endothermy. It is, therefore, of interest that some mammal-like reptiles possessed incipient turbinates *before* the articular became a malleus.

A trend toward simplification of the skull by the loss of bones runs all through tetrapod evolution. Bones that are characteristic of amphibians and reptiles but are not represented in mammals include the prefrontals, postfrontals, postorbitals, quadratojugals, parasphenoid, and all membrane bones of the lower jaw except the angular, prearticular, and dentary (which is retained as the single jawbone). The ectopterygoid usually fuses with the pterygoid. The prearticular contributes a process to the malleus; the angular

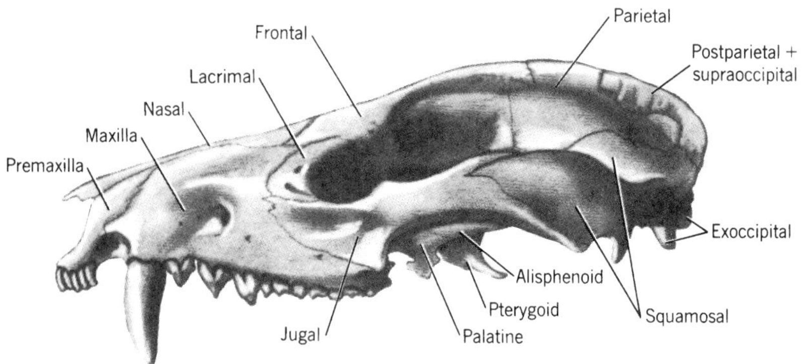

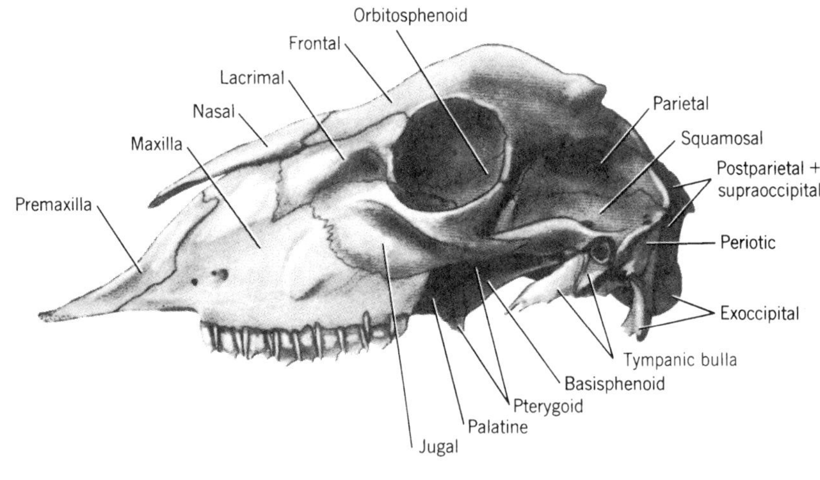

FIGURE 8.21 MAMMALIAN SKULLS OF THE PRIMITIVE ORDER MARSUPIALIA (above) AND THE ADVANCED ORDER ARTIODACTYLA (below).

moves over from the angle of the jaw to contribute to a uniquely eutherian structure, the tympanic bulla, which helps to enclose the middle ear. Details of the construction of the bulla are used in the classification of carnivores.

Reptilian bones that occur in mammals under new names may be summarized as follows:

Reptilian Bone	Mammalian Homolog
Articular + prearticular	Malleus
Quadrate	Incus
Angular	Ectotympanic
Epipterygoid	Alisphenoid
Sphenethmoid	Presphenoid + orbitosphenoid
Vomer (paired)	Vomer (unpaired)

It is doubtful that mammals have evolved any really new bones, but the entotympanic, which contributes to the tympanic bulla, seems to be a candidate.

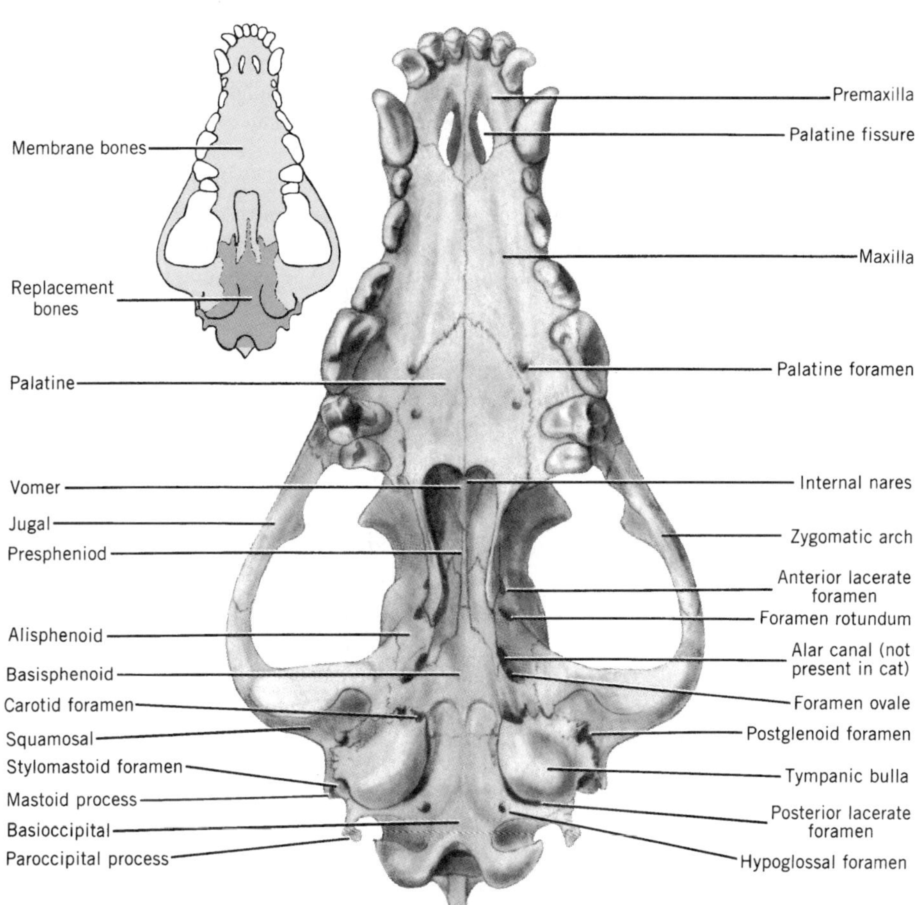

FIGURE 8.22 SOME BONES AND FEATURES OF THE SKULL OF A WOLF.

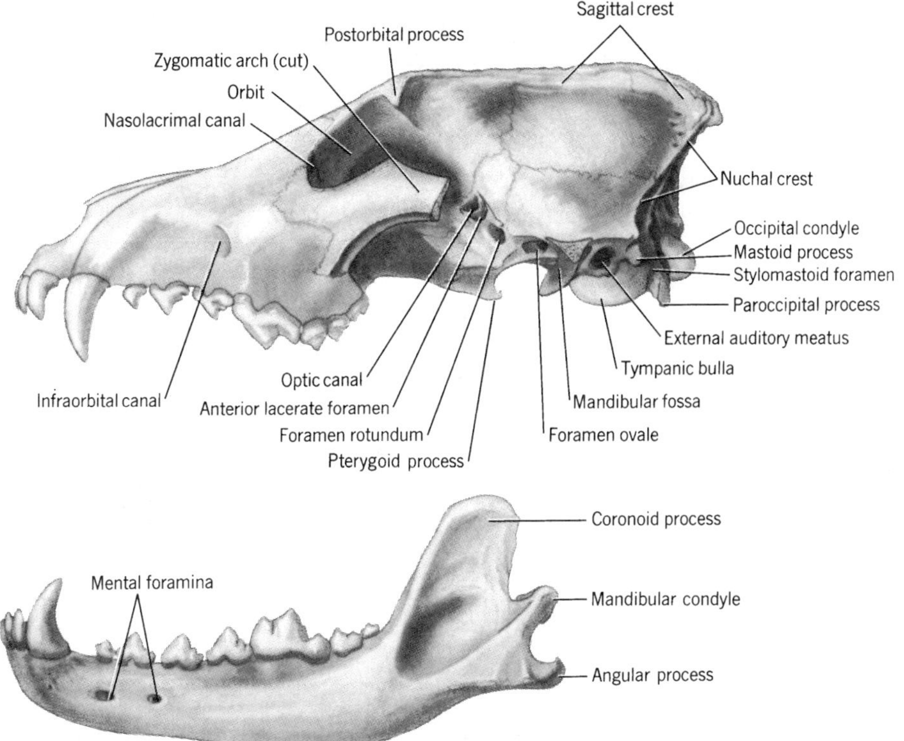

FIGURE 8.23 SOME FEATURES OF THE SKULL AND MANDIBLE OF A WOLF.

COMMENT 8.2

SKULLS IN THREE DIMENSIONS

The skull is a complex, three-dimensional (3-D) structure that is difficult to measure, visualize, and describe. Visualization in 3-D is often exceedingly useful. Stereopairs of images (images of the same specimen taken from slightly different angles) is an excellent way of 3-D depiction. An early theropod dinosaur from the Upper Triassic of Argentina serves as our example (Figure 8.24). (To see the 3-D image, hold the page 16 in (40 cm) from your eyes and stare at a point between the two photos just below the calibration bar while letting the eyes focus beyond the page. With patience, a third image, between the other two, will appear in full 3-D. Once in focus, the page may be drawn closer for inspection of detail.) The pair of deep temporal fossae are evident behind the orbit. The supratemporal fossa, the uppermost and smaller of the two, is relatively large compared with other theropods, and its anterior and medial margins are scalloped out against the temporal and parietal bones, respectively. The margins are thought to relate to the emergence and attachment of expanded jaw muscles, a character that led Sereno and Novas to conclude that this species was a carnivore.

A second way to visualize the deepest cavernous recesses of the skull is by reconstruction of serial slices. When done in the conventional way (embedding the skull and making sequential slices) the technique is labor-intensive (expensive) and destroys the specimen. With the arrival of X-ray computed tomography (CT), a nondestructive method is now available for mapping the internal structure of complex objects in three dimensions. Paleontologists are excited about using this technique for the study of fossil forms, not only because it does not destroy the specimen, but also because CT can be used (to a limited extent) for fossils still embedded in matrix. This is because the resolution of CT is dependent, in part, on the differential densities of the materials being imaged. The Digital Morphology Group at the University of Texas at Austin has established a CT facility for use by scientists around the world, and they have recently produced a large-scale project centered around a CD-ROM entitled *Alligator: Digital Atlas of the Skull.* (See Rowe in References; the CD-ROM is included in the jacket of the journal.)

FIGURE 8.24
STEREOPAIR PHOTOS OF THE SKULL OF THE TRIASSIC THEROPOD DINOSAUR *Herrerasaurus.* Scale bar equals 5 cm. (From Sereno and Novas, 1993, *J. Vert. Paleontol.* 13:454.)

Summary of Principles and Trends Many evolutionary mechanisms and principles discussed earlier in the book are illustrated by the phylogeny of the head skeleton. *Recapitulation* is seen in the ontogeny of the principal components of the skull and in the number and nature of ossification centers. *Serial homology* is exemplified by the somitomeres and head somites. *Inductions* have been demonstrated between regions of the developing brain and specific cranial ossifications. *Neoteny* is the probable explanation of gradual loss of ossification in the heads of some fish taxa. *Convergence* is manifest, for instance, in the crushing jaws of dipnoans and the unrelated holocephalians. There are classic examples of *homology* (e.g., the hyomandibula and stapes) and *analogy* (e.g., the quadrate-articular jaw joint and the squamosal-dentary jaw joint). *Preadaptation* for auditory function is seen in the bones released at the posterior end of the synapsid jaw.

Many *evolutionary trends* can be identified. Early vertebrates tended to reduce the number of visceral arches and to stabilize their number. A tendency to reduce overall ossification is observed in various taxa, and from fishes to mammals many specific bones dropped out, particularly around the orbit, at the back of the skull, on the temporal area, and on the lower jaw. This has contributed to a relative shortening of the postorbital part of the skull.

The head skeleton is of great importance to phylogeny and systematics at all levels but is particularly noteworthy for the study of fish taxa (patterns of dermal bones, jaw suspension), the subclasses of reptiles (temporal fenestration), and the reptile-to-mammal transition (origin of ear ossicles, secondary palate, nasal structure).

FEATURES OF THE SKULL

This chapter has emphasized evolutionary trends of the isolated skull. But the skull is not isolated. To be fully understood, it must also be related to other organ systems. The vertebrate skull is so complex and the range of its variation is so great that its named features are legion. Both student and professor must limit themselves to the major parts of skulls representing the major taxa, leaving the details to specialists. We have selected for illustration (Figures 8.22 and 8.23) some features of the mammalian skull and mandible that are relatively prominent and constant and that correlate with structures to be presented elsewhere in this book. Figures 8.25 and 8.26 show processes and crests in relation to muscles, and foramina in relation to cranial nerves.

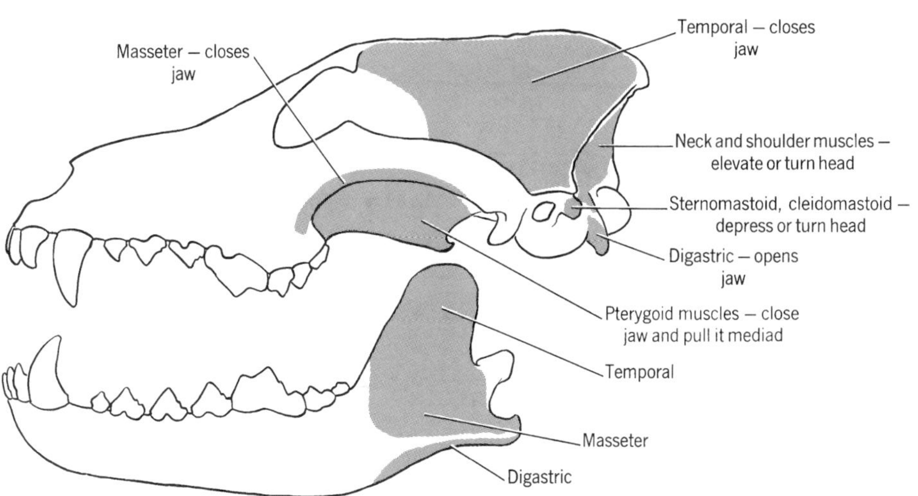

FIGURE 8.25
RELATIONS OF THE SKULL AND MANDIBLE OF A WOLF TO VARIOUS MUSCLES.

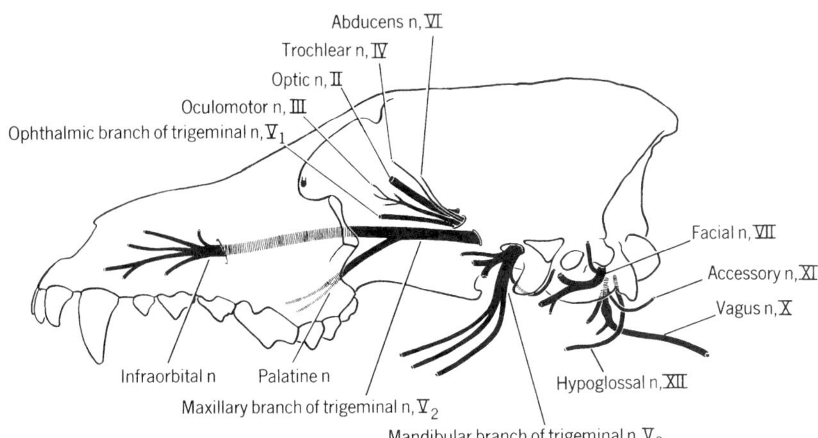

FIGURE 8.26
RELATIONS OF THE SKULL OF A WOLF TO VARIOUS NERVES.

Finally, although the elements present in a skull and the basic relationships of these parts are determined by a long evolutionary history, the relative sizes and configurations of these parts are related to specific adaptations of (chiefly) the feeding mechanism, brain, and sense organs. The importance of these factors to the architecture of the skull will be noted in other chapters.

REFERENCES

Allin, E.F., and J.A. Hopson. 1992. Evolution of the auditory system in Synapsida ("mammal-like reptiles" and primitive mammals) as seen in the fossil record, pp. 578–614. *In* D.B. Webster, R.R. Fay, and A.N. Popper (eds.), *The evolutionary biology of hearing.* Springer, New York.

Bramble, D.M. 1978. Origin of the mammalian feeding complex: models and mechanisms. *Paleobiol.* 4:271–301.

Crompton, A.W., and P. Parker. 1978. Evolution of the mammalian masticatory apparatus. *Am. Sci.* 66:192–201. Evolution of the jaw and ear, and of their functions.

De Beer, G.R. 1985. *The development of the vertebrate skull.* Univ. Chicago Press, Chicago. 853p. Originally published in 1937.

Frazzetta, T.H. 1968. Adaptive problems and possibilities in the temporal fenestration of tetrapod skulls. *J. Morphol.* 125:145–158.

Gans, C. 1993. Evolutionary origin of the vertebrate skull, vol. 2, pp. 1–35. *In* J. Hanken and B.K. Hall (eds.), *The skull.* Univ. Chicago Press, Chicago.

Gans, C., and R.G. Northcutt. 1983. Neural crest and the origin of vertebrates: a new head. *Science* 220:268–274.

Hall, B.K. 1975. Evolutionary consequences of skeletal differentiation. *Am. Zool.* 15:329–350.

Hall, B.K. 1978. *Developmental and cellular skeletal biology.* Academic Press, New York. 304p.

Halstead, L.B. 1974. *Vertebrate hard tissues.* Wykeham, London. 179p.

Hanken, J., and B.K. Hall (eds.). 1993. *The skull: patterns of structural and systematic diversity.* Univ. Chicago Press, Chicago. 587p. Best general reference; a chapter on each class; extensive reference lists.

Hunt, R.M., Jr. 1974. The auditory bulla in Carnivora: an anatomical basis for reappraisal of carnivore evolution. *J. Morphol.* 143:21–76.

Jacobson, A.G. 1993. Somitomeres: mesodermal segments of the head and trunk, vol. 1, pp. 42–76. *In* J. Hanken and B.K. Hall (eds.), *The skull.* Univ. Chicago Press, Chicago.

Jacobson, A.G. 1998. Somitomeres, vol. 11, pp. 209–228. *In Principles of medical biology: developmental biology.* JAI Press, Stamford, CT. A notably clear and concise review.

Jollie, M. 1984. The vertebrate head—segmented or a single morphogenetic structure? *J. Vert. Paleontol.* 4:320–329.

Maisey, J.G. 1980. An evolution of jaw suspension in sharks. *Am. Mus. Novitates* 2706:1–17.

Moore, W.J. 1981. *The mammalian skull.* Cambridge Univ. Press, New York. 369p.

Moss, M.L. 1968. The origin of vertebrate calcified tissues, pp. 360–371. *In* T. Ørvig (ed.), *Current problems of lower vertebrate phylogeny.* Wiley, New York.

Noden, D.M. 1987. Interactions between cephalic neural crest and mesodermal populations, pp. 89–120. *In* P.F.A. Maderson (ed.), *Developmental and evolutionary aspects of the neural crest.* Wiley, New York.

Rowe, T., et al. 1999. Introduction to *Alligator: Digital Atlas of the Skull. J. Vert. Paleontol.* 19(suppl. 2):1–8.

Sereno, P.C., and F.E. Novas. 1993. The skull and neck of the basal theropod *Herrerasaurus ischigualastensis. J. Vert. Paleontol.* 13(4):451–476.

Thompson, K.S. 1993. Segmentation, the adult skull, and the problem of homology, vol. 2, pp. 36–68. *In* J. Hanken and B.K. Hall (eds.), *The skull.* Univ. Chicago Press, Chicago.

Webb, J.F., and D.M. Noden. 1993. Ectodermal placodes: contributions to the development of the vertebrate head. *Am. Zool.* 33:434–447.

Chapter 9

Body Skeleton

The functions of the body skeleton are to protect the viscera, contribute (in amniotes) to ventilation of the lungs, serve as a store for various minerals, give rigidity to an otherwise soft body, and, importantly, to provide the series of firm, hinged segments that are essential, in conjunction with the muscles, for locomotion. An external skeleton, like that of an insect or crab, also serves in locomotion, but for large animals, particularly if terrestrial, such a skeleton would need to be prohibitively cumbersome and heavy to avoid fracture from contacts with the environment. The reception of external stimuli, thermoregulation by homeotherms, and integumentary respiration by certain aquatic forms may also be better served with an internal skeleton.

STRUCTURE AND DEVELOPMENT OF VERTEBRAE

General Structure The vertebral column is older than any other part of the postcranial skeleton except the notochord. Nevertheless, it is not so ancient as the major features of the soft organ systems, and is virtually absent in the oldest known vertebrates. Its structure and functions diversified slowly. We shall start with an account of the features of a typical vertebra to gain the terminology necessary to discuss the evolution of vertebrae.

The principal part of a vertebra of tetrapods and many fishes is the spool-like body or **centrum,** which lies just below the spinal cord, where it surrounds, restricts, or replaces the notochord (Figure 9.1). The centrum may consist of one or two elements (rarely more—in fishes). If a tetrapod centrum has two elements (many extinct amphibians, few amniotes), then the more anterior is called an **intercentrum** and the more posterior (which may be paired) is called a **pleurocentrum** (Figure 9.7). If a tetrapod has only one central element, it may be the intercentrum (some extinct amphibians) or the pleurocentrum (amniotes). As explained below, it is not yet clear if these terms are appropriate for fishes.

Extending dorsally from the centrum are pillars that flank the spinal cord, one on each side, and join above the cord to complete the **neural arch.** A **neural spine** may extend

Above: Paraffin-impregnated cartilaginous fin skeleton of a stingray.

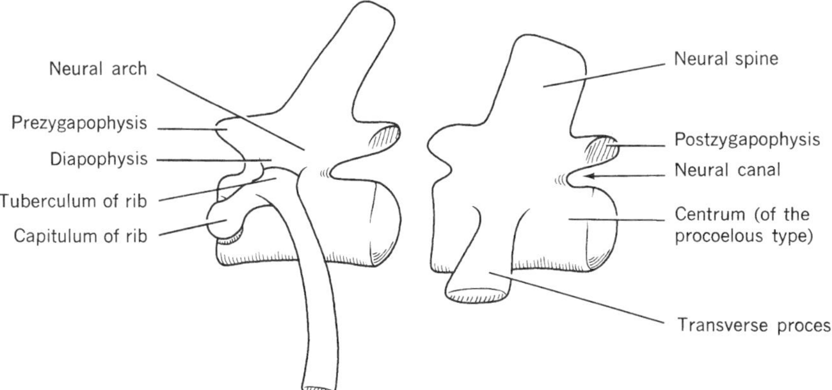

FIGURE 9.1 SOME FEATURES OF VERTEBRAE. Left side views.

from the summit of the neural arch. Frequent though less universal in occurrence is a similar **hemal arch** that extends ventrally from the centrum to surround blood vessels. In the tail it may be continued by a **hemal spine.** Neural and hemal spines are related to muscles that move the axial skeleton. Of like function are a variety of processes that project, according to species, from walls of the vertebra. Centra may accommodate the capitulum, or head of a rib, either with a process (parapophysis) or with a concavity. The part of a rib called the tuberculum (Figure 9.1) is supported by a diapophysis. Any lateral process is a **transverse process.** The most prominent of these is usually in the position of the diapophysis but only on vertebrae lacking ribs. Adjacent vertebrae articulate by their centra (if the centra are complete), and in tetrapods they also articulate by processes carried by the neural arches and called **zygapophyses** (= *joining* + *processes*). The **prezygapophyses** on the anterior aspect of one vertebra articulate with the **postzygapophyses** on the posterior aspect of its neighbor. Prezygapophyses face upward or inward, whereas postzygapophyses face downward or outward. This is useful to remember for determining which end of a single vertebra is which. (Nearly all vertebrae of snakes, and the posterior trunk vertebrae of armadillos and sloths, are said to be *xenarthrous* because they are stiffened by having a doubling of zygapophyses.)

The shapes of the articulatory surfaces at the ends of the centra are of evolutionary, functional, and systematic importance. If each surface is concave, the centrum is said to be **amphicoelous** (= *on both sides* + *hollow*). Such centra actually touch only at the periphery of the intervertebral joint (Figure 9.2). Limited motion is permitted in any direction. Within the joint is a space filled by connective tissue, cartilage, or remnants of the notochord. Adjacent spaces may be joined by a perforation through the interior of the centrum. Other kinds of centra are concave anteriorly and convex posteriorly, the bulge of one vertebra fitting into the hollow of the next. Such a centrum is **procoelous** (= *front* + *hollow*). Conversely, an **opisthocoelous** (= *behind* + *hollow*) centrum is concave posteriorly and convex anteriorly. Joints between procoelous or opisthocoelous vertebrae permit motion in any direction (except as modified by the zygapophyses) and they resist dislocation. Some centra have flat ends and are said to be **platyan** (or acoelous). These centra withstand compression and limit motion (unless the intervertebral joints are provided with thick fibrous discs). Still other centra have saddle-shaped ends. They are called **heterocoelous** (= *different* + *hollow*). They allow vertical and lateral flexion but prevent rotation around the axis of the spine. There are intergrades among these types, combinations, and occasionally doubling of the concave and convex surfaces. The functional aspects of vertebral structure are interpreted further in Part III.

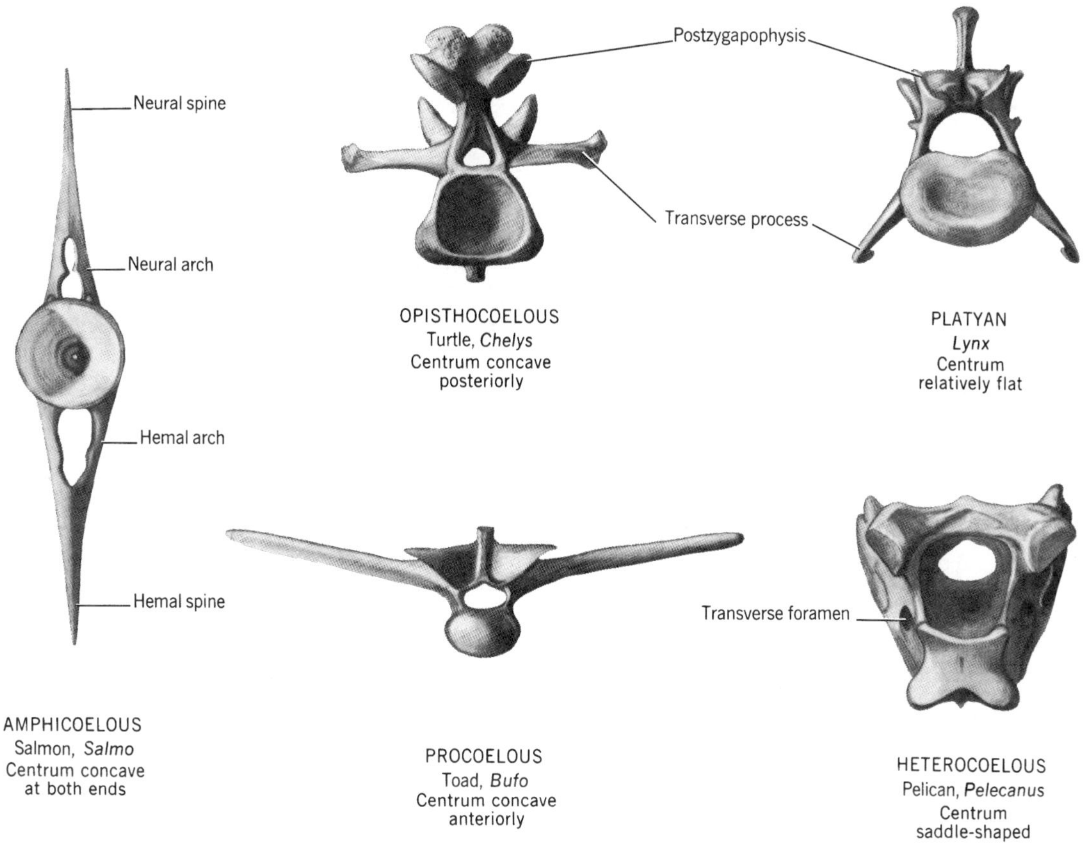

FIGURE 9.2 VERTEBRAE SHOWING VARIOUS SHAPES OF CENTRUM, and other features, as seen in posterior view.

Development and Homology Vertebrae are complex structures, and the variety of their construction is extreme. The recognition of homologous parts among vertebrate taxa is therefore desirable so that evolutionary sequences can be recognized. Contributions have come from comparative anatomy, paleontology, embryology, and molecular genetics. However, in spite of the efforts of many researchers, consensus remains elusive. We are now less confident than formerly that specific elements can be equated among fishes, amphibians, and amniotes. Nevertheless, the general correspondence of the developmental process across taxa is better understood.

The differentiation of the embryonic somite into dermatome, myotome, and sclerotome is described in Chapter 5 (see Figure 5.9 and Table 5.2). Mesenchyme from the segmentally arranged sclerotomes distributes itself around the notochord. In amniotes, at least, the cells in the anterior half sclerotomes are conspicuously less dense than those in the posterior half sclerotomes (Figure 9.3). Subsequently, the halves of one sclerotome may appear to split apart, each then joining the neighboring half segment of an adjacent sclerotome. In this manner the precursors of centra are formed that are intersegmental, and thus in position to be moved on one another by muscles derived from the segmental myotomes. (However, it is primarily the bases of the neural and hemal arches, not the centra, that must alternate with the myotomes to ensure function.)

This resegmentation of sclerotome derivatives is accepted for amniotes by most, although not all, investigators. It is denied by some authors for fishes, where the resegmentation is not

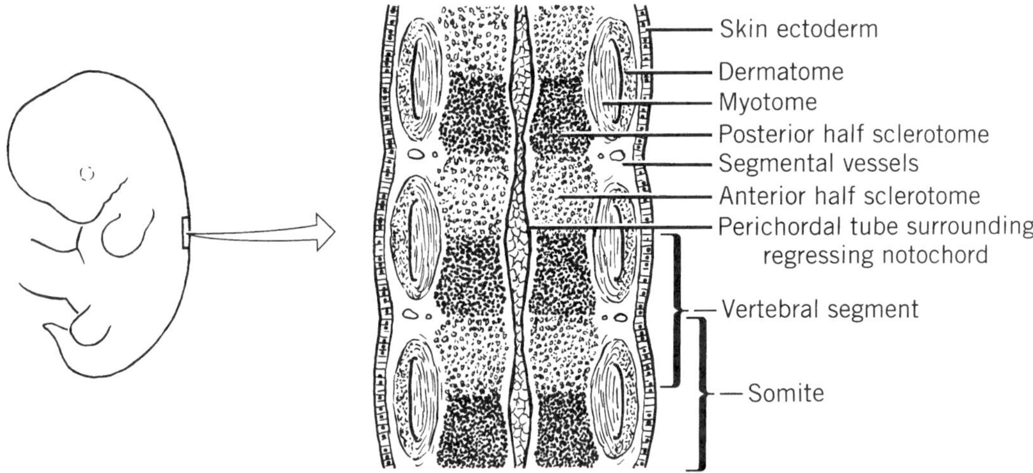

FIGURE 9.3 ONTOGENY OF AMNIOTE VERTEBRAE. Frontal sections of three body segments at the level of the notochord and developing centra.

observed. However, there are suggestions from molecular genetics that the domains of the half sclerotomes may be universal. What *is* seen in fishes is that hard tissue forms within or around the sheaths of the notochord to form centra. In amphibians there are hints of resegmentation only in caecilians. Otherwise, cells move adjacent to the notochord to form a continuous **perichordal tube.** Direct ossifications in the tube form centra, which become separated when breaks appear in cartilaginous parts of the tube.

It was long believed that specific bony elements in the centra of crossopterygian fishes and labyrinthodont amphibians could be homologized with the centra of amniotes. Unfortunately, this now seems doubtful. Indeed, centra appear to have evolved independently numerous times among anamniotes.

EVOLUTION OF THE SPINE

Beginnings: Notochord with Supplemental Cartilages The notochord is persistent in adults of jawless vertebrates, placoderms, pleuracanths, cladoselachians, chimaeras, acanthodians, lobe-finned fishes, and most members of the lower orders of ray-finned fishes. In contrast to the embryonic notochord (more often seen in the laboratory), the adult cord is large and resilient. It has an outer elastic sheath that merges with a tough, inner, fibrous sheath. The stiffness of the notochord enables muscles of the body wall to flex the body (as required for the vertebrate manner of swimming) rather than shorten it. The large notochord of the sturgeon (Figure 9.4) has sufficient internal pressure to function as a hydrostatic axial skeleton. The notochord of amphioxus is itself a contractile organ. It consists of coin-shaped cells stacked longitudinally along the cord and having contractile filaments oriented from side to side. Their shortening narrows and stiffens successive levels of the notochord, thus facilitating undulation of the body. It is possible that notochords of ancestral vertebrates had the same capacity.

All these vertebrates (except hagfishes) have neural arches; many have hemal arches, at least in the tail; and some have separate arch bases or other elements that flank, but do not restrict, the notochord (see Figure 9.4, and the figure on p. 123). It is probable that the various cartilages evolved where tendons inserted on the notochordal sheaths. Limited ossification of these elements is present in arthrodires and bony fishes. This kind of spine is primitive for vertebrates in general.

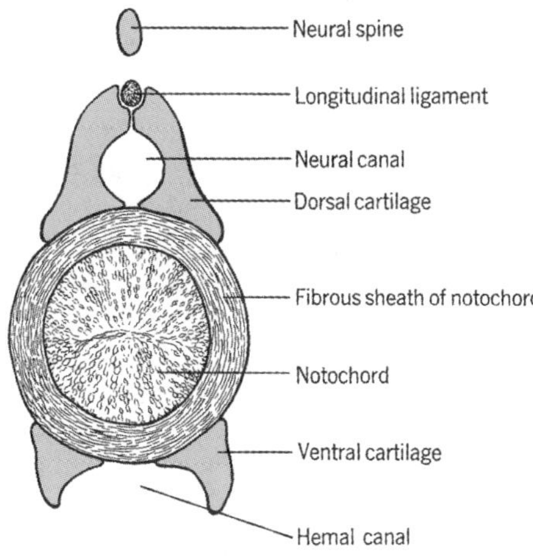

FIGURE 9.4 CROSS SECTION OF THE NOTOCHORD AND SPINAL CARTILAGES OF A CHONDROSTEAN FISH.

Advanced Fishes: The Spine Takes Over The next evolutionary steps were for the notochord to be interrupted by centra and for elaboration of arches and processes. Selachians among cartilaginous fishes, teleosts among ray-finned fishes, and some other bony fishes evolved vertebrae with firm centra that articulate with one another. Adjacent pairs of the deeply amphicoelous centra cup between them balls of tissue, derived from the notochord, around which they rotate on one another. Such a spine is stronger than the notochord alone and provides greater anchorage for muscles.

Variation of structure among the fishes is considerable. SELACHIANS are unique in having not only the centra but also the arches in continuous contact (Figure 9.5). Such an arrangement would virtually immobilize a bony spine, but the cartilaginous nature of the selachian spine provides the slight flexibility that is needed. Each vertebral centrum may have several separate pieces, and there are two principal dorsal arches (a neural plate and an intercalary plate, or interneural). The centrum chondrifies in the notochordal sheaths (a

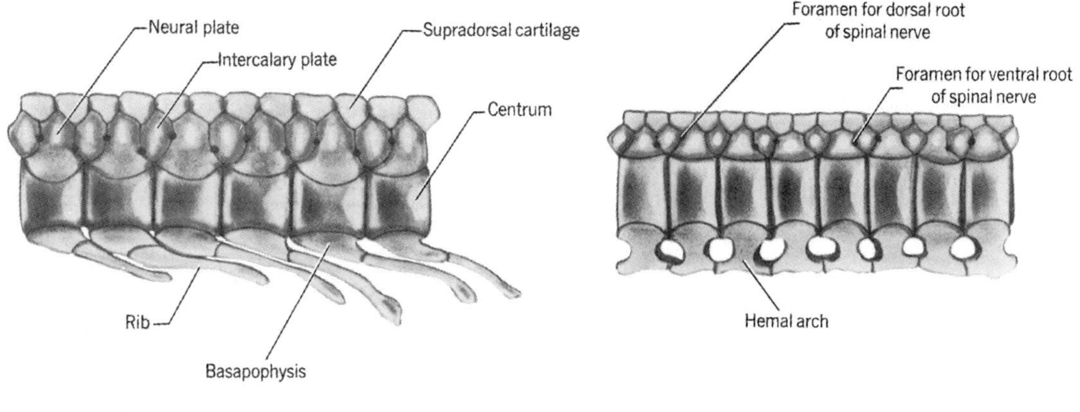

FIGURE 9.5 SECTIONS FROM THE TRUNK (left) AND CAUDAL REGIONS (right) OF THE SPINE OF AN ELASMOBRANCH in left lateral view. As the nerve foramina show, the caudal section is diplospondylous.

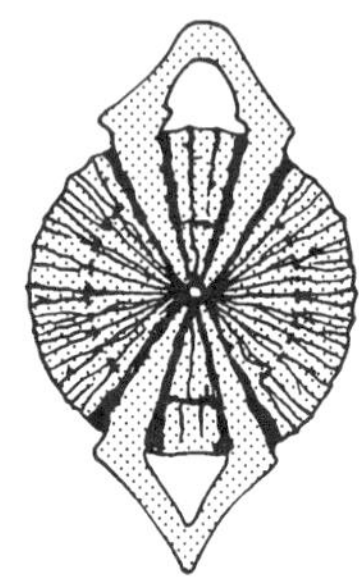
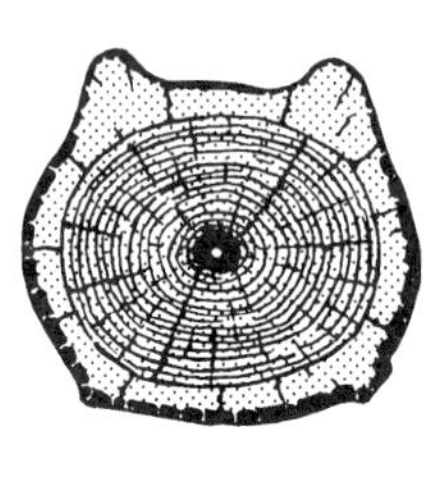

FIGURE 9.6 CROSS SECTIONS OF THE CENTRA OF TWO ELASMOBRANCHS, the white shark, *Carchorodon* (left) and angel shark, *Squatina* (right), showing patterns of calcified cartilage (black) within a matrix of uncalcified cartilage.

chordal centrum). Calcification of the centra is common, but instead of being continuous, the hard tissue is arranged in concentric cylinders around the central axis and in pillars in the positions of the primitive arch bases. Details of the pattern are various and have systematic value (Figure 9.6).

The vertebrae of BONY FISHES usually have one central element, a neural arch with spine, and, in the tail, a hemal arch with spine (Figure 9.2). The centrum usually ossifies directly from mesenchyme surrounding the notochord (a **perichordal centrum**), but may also form from cells that invade the chordal sheaths or from cartilaginous precursors.

There is little regional differentiation in the vertebal column of either cartilaginous or bony fishes. The first vertebra may be slightly modified for articulation with the skull. Behind the coelomic cavity a caudal region is distinguished by the absence of rib facets and the presence of enlarged hemal arches with spines. Some bony fishes (including the bowfin commonly seen in the laboratory) and many cartilaginous fishes have a curious doubling of vertebrae in the caudal region, giving two (sometimes even more) complete vertebrae per muscle segment and per pair of spinal nerves (Figure 9.5). The origin and significance of this condition, called **diplospondyly,** have been much discussed but remain obscure. Perhaps it increases flexibility. Rarely the centrum is single and the neural arches are doubled. Arches may be fused to the centra or free.

Amphibians: Varied Solutions to New Problems Few organs were so affected by the change from aquatic to terrestrial life as the spine. Formerly it resisted only the stresses imposed by strong axial muscles; now the axial musculature gradually became reduced but there were new stresses, imposed by gravity, that acted largely in a different plane. Formerly the paired appendages had not been related to the spine; now the slowly strengthening limbs transmitted their support to the axis of the body. Formerly it was sufficient for the spine to be nearly uniformly flexible throughout its length; now it had to resist bending in some places and provide new mobility elsewhere. The requirements were for vertebrae with firm centra, intervertebral joints that could facilitate or restrict motion according to location, processes that could increase the leverage of muscles, and a more intimate relation with the girdles. All this took many millions of years to accomplish. Indeed, it remained for the amniotes to complete the changes. Extinct amphibians seemed to experiment with different vertebral structures. Various taxa of labyrinthodonts are named and defined largely by the nature of their vertebrae.

The earliest amphibians (Ichthyostegalia) had vertebrae that were nearly identical to those of the particular lobe-finned fishes thought to be ancestral to tetrapods. The notochord was persistent. There was a neural arch, a large intercentrum encircling the notochord, and small paired pleurocentra (Figure 9.7). Other groups of labyrinthodonts no longer had a continuous notochord. Their centra were more solid, with one bony element or with varying proportions of two elements.

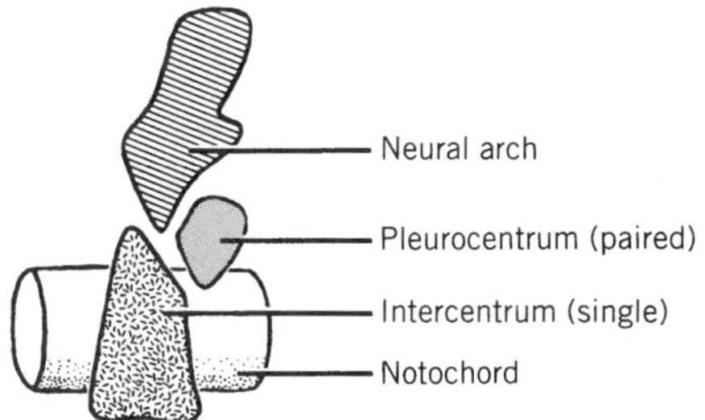

FIGURE 9.7 STYLIZED VERTEBRA OF ADVANCED CROSSOPTERYGIANS AND PRIMITIVE AMPHIBIANS as seen in left lateral view.

The vertebrae of modern amphibians remain an enigma. The pattern of development gives no useful information about the homologies of vertebral elements. Their centra (which always consist of a single unit) are probably broadly homologous to each other and to those of other tetrapods.

Amphibians, like other tetrapods, have zygapophyses to strengthen the spine and control its flexibility. Centra are amphicoelous, procoelous, or opisthocoelous. Regional differentiation of the spine is minimal for tetrapods but exceeds that of fishes. The first (and sometimes the second) vertebra is modified to give increased mobility to the head. A **cervical vertebra** is somewhat set off by the reduction or absence of ribs. **Trunk vertebrae** bear ribs (Anura excepted). A single **sacral vertebra** is enlarged to articulate (via its fused rib) with the pelvic girdle. Typical tail or **caudal vertebrae** lack zygapophyses, and have hemal arches. Anura, however, have no free caudal vertebrae but have instead a rodlike **urostyle** derived from three larval cartilages that appear (on evidence of related spinal nerves) to incorporate two or three ancestral caudal vertebrae (Figure 9.8).

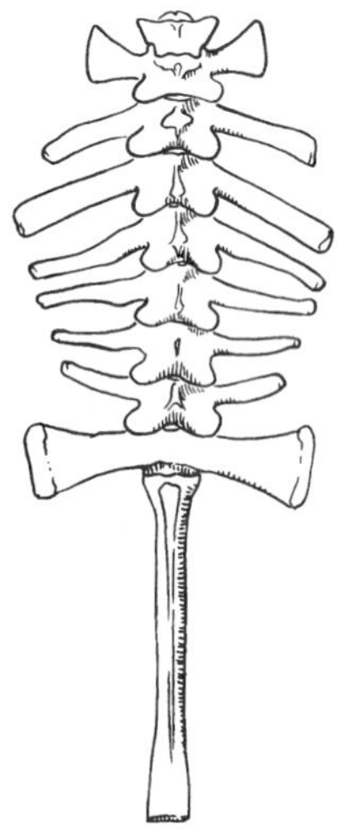

FIGURE 9.8 VERTEBRAL COLUMN OF AN ANURAN showing in dorsal view the single cervical vertebra, short trunk, absence of free ribs, single sacral vertebra, and urostyle.

Amniotes: Strength and Specialization The centrum of amniotes was long considered homologous with the pleurocentrum of labyrinthodonts, but this is now uncertain. Stem reptiles had large centra and small elements that (because of position, but not necessarily of homology) are called **intercentra.** Subsequent amniotes retain only one functional intercentrum (in the first vertebra of the neck). They have strong zygapophyses (rarely doubled for added strength) except in the caudal region. The first two cervical vertebrae are specialized to support the skull (of which more is discussed under the next subheading).

Because most REPTILES have more distinct necks than amphibians, they have more distinct cervical regions; and because their limbs are stronger, they have two (sometimes more) sacral vertebrae instead of one. Centra are usually procoelous but also take other shapes. Hemal arches are retained only in the caudal region, where they are separate bones shaped like Y's. The ancestral intercentra with which they once articulated having disappeared (or merged with them?), they articulate with the spine at the intervertebral joints. Such arches are called **chevron bones** because of their shape.

Because BIRDS have a locomotor specialization in common—flight—they have more specialized and more uniform spines than other tetrapods. The spine of Archaeornithes is transitional between those of their reptilian ancestor and Neornithes. No other class has so many cervical vertebrae—commonly 15 to 20 (10 in Archaeornithes)—and these vertebrae are distinctive for being heterocoelous. Fused to the 2 sacral vertebrae of the reptilian ancestor are 10 to 20 trunk and caudal vertebrae (5 to 6 in Archaeornithes) to form a solid unit called a **synsacrum,** which fuses with the halves of the pelvic girdle (Figure 9.9). The remaining trunk vertebrae number only 4 to 6, and there may even be fusions among some of these. The result is a short, rigid back, which, as explained in Chapter 28, is needed for flight. The free caudal vertebrae of Neornithes have been reduced to (usually) 6 or 7; Archaeornithes had a

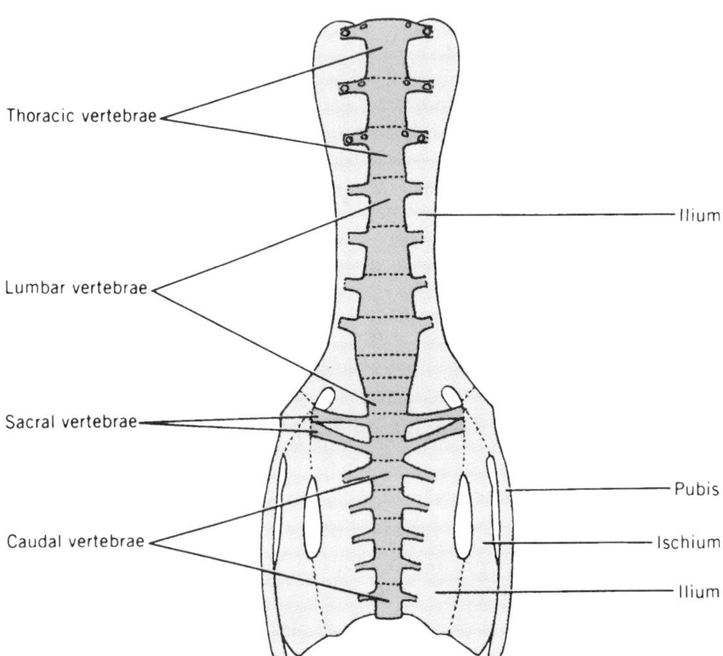

FIGURE 9.9 THE AVIAN SYNSACRUM, in ventral view, showing its origin and relationship with the pelvic girdle.

long bony axis to the tail. At the end of the tail of modern birds is a unique blade of bone called the **pygostyle,** which supports the tail feathers. It represents the fusion of 4 to 7 vertebrae. Even more embryonic somites form and disappear again in recapitulation of the long ancestral tail. Some of these features are seen in the lower figure on p. 543.

MAMMALS are unique in forming bony, platelike caps, or **epiphyses,** at the ends of their centra posterior to the first intervertebral joint. These usually fuse to the centra when the animal is mature. They are derived from the segmental thickenings of the mesenchyme (perichordal tube) that surrounds the embryonic notochord. Their function relates to the growth process. With rare exceptions (some edentates and sirenians), mammals have 7 cervical vertebrae whether the neck is short (dolphin) or long (giraffe). There are about 20 trunk vertebrae. Unlike the condition in other classes, these are sharply divided, by presence or absence of ribs, into anterior **thoracic vertebrae** and posterior **lumbar vertebrae** (see figure on p. 486). To assign isolated vertebrae to region, look for rib facets. Lumbar vertebrae tend to have larger centra, shorter and stouter neural spines, and longer transverse processes than thoracic vertebrae. (We shall see on pp. 452 and 485 that, from a functional viewpoint, there are other ways to divide the trunk vertebrae.) Mammals have 3 or more sacral vertebrae fused to form a **sacrum.** Chevron bones are usually restricted to the base of the tail.

Cranio-Vertebral Joint The occipital process of the skull of fishes usually resembles the end of a typical centrum. The head seldom has more mobility on the spine than one vertebra has on another. The cranio-vertebral joint is like an intervertebral joint.

Amphibians are the first vertebrates to have a neck, however short, and linked with this advance is a more flexible cranio-vertebral joint. The second vertebra is unchanged or only somewhat enlarged.

Reptiles and birds have a ball-and-socket joint between the small first vertebra, now called the **atlas,** and the single occipital condyle. The second vertebra, called the **axis,** is enlarged, and the joint between the atlas and axis is specialized.

The atlas of mammals has large zygapophyses, winglike transverse processes, scarcely any centrum, and no neural spine. Its articulation with the two condyles of the skull permits hingelike up-and-down motion. The axis has a bladelike neural spine and a large centrum. The joint between atlas and axis permits side-to-side motion and rotation around the axis of the spine.

In adults of some tetrapods, and in embryos of many others, small structures with no apparent function are found between the skull and the neural arch of the atlas. These form the **proatlas.**

The story of how these changes came about is of particular interest. It has been pieced together from many embryological, paleontological, and comparative anatomical studies. In Chapter 8 it was noted that several ancestral vertebrae were incorporated into the vertebrate chondrocranium. It follows that the cranio-vertebral joint falls within the primordial spine, not beyond its anterior limit. At some time (or times?) in tetrapod evolution one vertebra at the cranio-vertebral boundary became reduced in size and partly fused to adjacent units of the series. This was the proatlas (Figure 9.10). Its neural arch remains free in some labyrinthodonts and some reptiles. Otherwise, the parts of the proatlas may fuse to other vertebrae or to the skull (though there is no evidence of such fusion in modern amphibians).

The atlas of amniotes always retains its "own" neural arch and intercentrum. The axis not only retains its own neural arch and centrum, but also incorporates into its odontoid process the centrum of the atlas and even that of the proatlas if that element has not joined the skull (mammals, some reptiles). The intercentrum of the axis is lost (mammals) or also incorporated into the composite centrum of the axis.

RIBS

Ribs are intersegmental splints of cartilage or replacement bone that articulate with the vertebrae. Originally they extended the entire length of the spine to increase the direct contact of axial muscles with the skeleton. As they lengthened they came to protect underlying viscera while retaining requisite flexibility and minimum weight. Only in amniotes are they specialized to contribute to the breathing mechanism.

COMMENT 9.1

SIGNALS FOR DIFFERENTIATION

What determines the development of morphologically distinct vertebrae? How might this vertebral differentiation evolve? We noted in Comment 5.1 on p. 73 that homeotic genes associated with axial organization, for animals as diverse as insects and vertebrates, are expressed in an anterior-posterior progression. It might be anticipated that the differentiaton of the regions of the spine also is strongly linked to homeotic genes. To address this question, Burke and colleagues compared *Hox* gene expression and vertebral development of a fish, an amphibian, two birds, and a mammal (see References). Of particular interest was the transition between the cervical and thoracic vertebrae. The number of cervical vertebrae varies in the groups compared, and thus the number of somites associated with them is also variable. At the transition from cervical to thoracic vertebrae are spinal nerves that innervate the anterior paired appendages. This group of nerves, the brachial plexus (= *arm* + *braid*), invariably exits the intervertebral formina of the last three to four cervical vertebrae and the first thoracic vertebra. This enables the brachial plexus to be used as a reference for comparison across species with differing numbers of cervical somites or vertebrae. Burke's study revealed that for the species studied, the *HoxC-5* and *HoxC-6* genes maintain their anterior boundaries of expression not coincident with absolute somite number, but instead with the cervical-thoracic boundary of each species. In other words, for all vertebrates having paired appendages, regardless of neck length, the cervical-thoracic junction and the emergence of the brachial plexus are expressions of homologous *Hox* genes. The findings suggest that evolutionary change in axial formulas is due to changes "upstream" in genes that control the *Hox* genes themselves, rather than "downstream" in genes that *they* control. This concept is useful where morphological boundaries between regions have been lost, as in caecilians and snakes. However, Galis cautions that the prediction of vertebral boundaries solely on the basis of the expression of *Hox* genes is made difficult by the multiple effects of surrounding developmental processes.

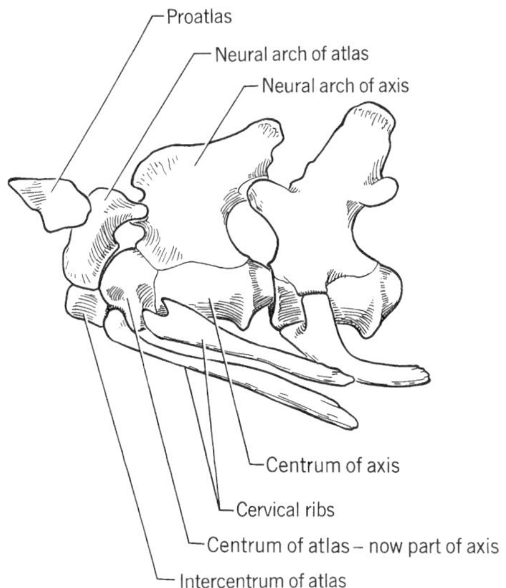

FIGURE 9.10 DERIVATION OF THE ANTERIOR CERVICAL VERTEBRAE OF AMNIOTES from ancestral elements illustrated by the crocodile, *Crocodylus,* shown in left lateral view.

There appear to be two sets of ribs among the different vertebrates. Each set evolved in the septa of connective tissue that occur between successive muscle segments of the primitive body wall. The set called **dorsal ribs** (or intermuscular ribs) evolved where these septa intersect with the partition between dorsal (epaxial) and ventral (hypaxial) muscle masses. **Ventral ribs** (or pleural ribs) form between the hypaxial muscles and the lining of the coelom. These occur only in the trunk region (Figure 9.11).

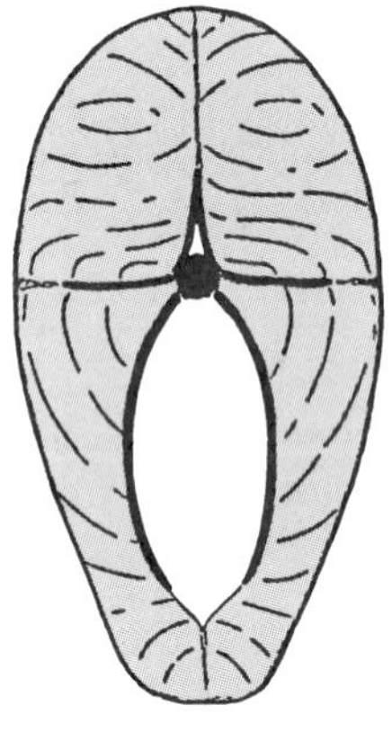

FIGURE 9.11 CROSS SECTION OF A BONY FISH SHOWING POSITION OF DORSAL AND VENTRAL RIBS in relation to muscle masses and body cavity.

The first ribs known were already complete and independent structures; paleontology gives no clue of their origin. It was formerly thought that ribs originated as outgrowths from the sclerotomes. The sclerotomes probably do contribute to the heads of ribs, but it is now thought, on the basis of their ontogeny, that they are primarily new structures not derived from the spine. The oldest known tetrapod ribs had two heads, and this condition is retained by most ribs that are strong. Weaker ribs (often more posterior on the trunk) may have a single head as the result of either fusion or loss. The more ventral head, named the **capitulum,** articulates with the intercentrum, if that piece is present, and otherwise with the centrum near the intervertebral joint. The dorsal head, or **tuberculum,** articulates with the diapophysis of the neural arch (Figure 9.1).

Jawless vertebrates and placoderms have no ribs. The functions of ribs were performed in part by the armor of some of them, and none has centra with which ribs could articulate. With few exceptions (e.g., Holocephalia), other fishes have ribs. There is disagreement as to whether the short cartilaginous ribs of pleuracanths and selachians are of dorsal or ventral origin (Figure 9.5). The long bony ribs that enclose the body cavity of ray-finned fishes are ventral ribs. Various primitive but unrelated fishes (e.g., salmon, *Polypterus*) have both ventral and smaller dorsal ribs. The ribs of tetrapods are in the position of ventral ribs; however, consensus now is that they are the ancestral dorsal set.

The short cervical ribs of tetrapods may articulate firmly with the vertebrae (labyrinthodonts, many reptiles), fuse with the vertebrae (birds, mammals), or be lost (turtles). Free caudal ribs are retained by most labyrinthodonts and some reptiles but otherwise are either lost or indistinguishably fused to the transverse processes. The same is true of the lumbar ribs of mammals. Some Anura have free ribs, but many retain only fused sacral ribs. Urodela have short ribs with unique articulations. The ribs of amniotes are in two pieces—a principal ossified segment and a shorter **sternal rib** that ossifies in birds but is otherwise usually cartilaginous. The more anterior thoracic ribs of most amniotes articulate with the sternum (breastbone) via their sternal segments (see figure on p. 437). Ribs of birds and of some reptiles have **uncinate processes** to provide anchorage for shoulder muscles (see figure on p. 543).

STERNUM

The **sternum** is a midventral skeletal element that usually articulates with the more anterior thoracic ribs (Figure 9.12). Its functions are to strengthen the body wall, help protect the thoracic viscera, accommodate muscles of the pectoral limb and, in some amniotes, aid in ventilating the lungs. Most of these functions are not relevant to fishes, and only tetrapods (and not all tetrapods) have a sternum. There is wide structural variation among species and relatively great individual variation among animals of a kind. The sternum is rarely present in fossils of amphibians and reptiles because it was cartilaginous. It contributes little to our knowledge of vertebrate evolution.

The sternum forms from either paired or midventral primordia (or both) that are now generally regarded (following considerable debate) as new structures not derived from the pectoral girdle or ribs. (The rib ends of mammalian embryos inhibit ossification of the sternum.)

It is probable that the most primitive tetrapods had no sternum. It is a simple cartilaginous plate in most living Urodela and is missing in others. In Anura it ranges from poorly developed to well ossified. The sternum is cartilaginous, and is often large in lizards and crocodilians. It is absent in snakes and turtles. Modern birds have a huge ossified sternum with a prominent keel to give origin to flight muscles (see again figure on p. 543). Mammals are unique in having the sternum divided into a linear series of about half a dozen bony segments.

ORIGIN OF APPENDAGES

Few aspects of vertebrate evolution have generated as much debate, over so many years, by so many eminent biologists as has the origin and evolution of paired appendages. New fossils, advances in methods of phylogenetic analysis, attention to developmental processes, and discovery of the regulatory effects of *Hox* genes have provided new insights, but not yet consensus. There are now two contrasting views about the origin of paired fins. Shubin, Tabin, and Kemp stress the similarities of the pectoral and pelvic appendages in all major groups, and they believe that they are serial homologs that emerged at about the

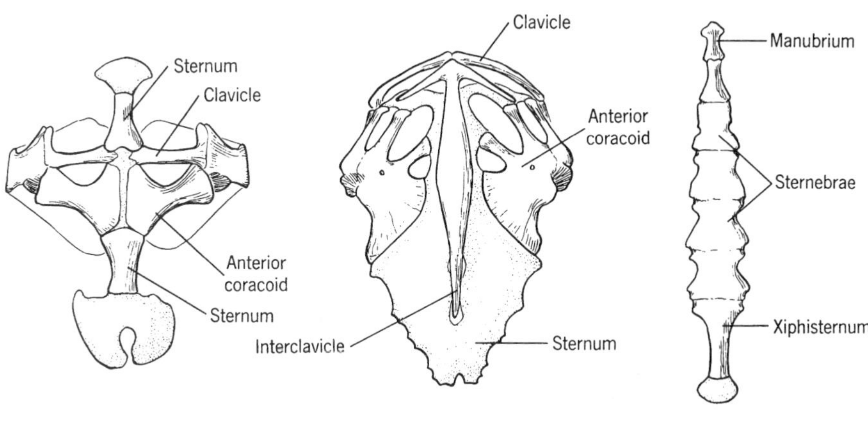

FIGURE 9.12 VARIED STERNUMS AND RELATED BONES OF THE PECTORAL GIRDLE seen in ventral view.

same time in Paleozoic fishes and that both sets of fins are under the control of a posterior series of *Hox* genes. Coates and Cohn adamantly oppose this view, contending that the patterns of fin phylogeny from the fossil record provide little evidence of simultaneous evolution of pectoral and pelvic appendages, that pectoral fins evolved first, and that gene regulation of the anterior fin is different from that of the pelvic fin. Even our most sophisticated techniques still have limitations; there is need for more fossil material and additional experimental studies.

In the early 1880s several anatomists postulated that the primitive vertebrate had continuous paired lateral fins from the gills to the cloaca, and that from there a single continuous fin ran in the midline around the tail and forward along the back to the head. It was considered that the fins of familiar fishes represent such portions of the ancestral continuous fins as remained after intervening deletions were established. Anaspids did have lateral fin folds (see figure on p. 40), and experiments on the induction of accessory limbs in amphibian larvae support the conclusion that appendages do always form where the continuous fin folds were postulated to be. In truth, we still do not know just how the appendages evolved.

MEDIAN FINS

The **dorsal fin**(s), located along the middorsal line, the **anal fin**(s), between anus and tail, and the **caudal fin** compose the median fins. They occur in nearly all jawless vertebrates and fishes. (Larval amphibians, adults of many Urodela, and some amniotes that are highly specialized for aquatic life have secondarily evolved analogous structures.)

Dorsal and Anal Fins Dorsal and anal fins function to prevent the body from yawing (turning around the vertical axis) and rolling (turning around the longitudinal axis).

The primitive condition probably was for each fin to be supported within the contour of the body by a series of rodlike **radials,** or **pterygiophores** (= *fin* + *bearer of*), arranged one per body segment (Figure 9.13). Commonly, the number is reduced (or increased) and segmental arrangement is lost. Each pterygiophore is usually divided into two or more pieces. The proximal piece is often conspicuously larger than those more distal and is then called a **basal.** Pterygiophores sometimes articulate with neural and hemal spines, but they are probably not derived from vertebrae.

The exposed membrane of the fin may originally have been supported only by dermal scales in the covering skin—those on the leading edge being larger than the rest. The fins of cephalaspids and some placoderms apparently were of this nature. The fins of more

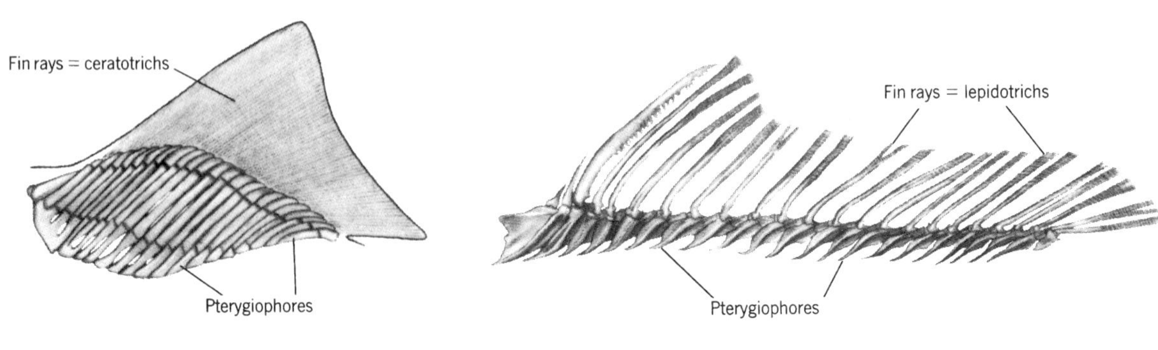

FIGURE 9.13 DORSAL FIN SKELETONS OF A CARTILAGINOUS FISH (left) AND A BONY FISH (right).

advanced fishes are supported internally by a series of slender **fin rays.** Fin rays of cartilaginous fishes are slender, unsegmented, and horny and are called **ceratotrichs** (= *horn* + *hair*). Those of bony fishes are slightly broader, segmented, paired proximally, and bony and are called **lepidotrichs** (= *scale* + *hair*). Higher teleosts have in the dorsal fin only six or fewer lepidotrichs, which have enlarged and become rigid. The leading edge of one or more median fins of many fishes (pteraspids, acanthodians, most cartilaginous fishes) is stiffened by a stout ray that serves as a cutwater and sometimes also (or instead) for defense or display. This ray may be an enlarged lepidotrich, but in cartilaginous fishes it is derived from one or more dermal denticles. Dorsal and/or anal fins may be long and continuous (some cyclostomes, pleuracanths, some teleosts), single (most ray-finned fishes), double (many selachians and sarcopterygians), multiple (*Polypterus*), or absent. A single dorsal fin is the primitive condition for ray-finned fishes.

Caudal Fin If the spine is straight to the tip of the tail, then dorsal and ventral lobes of the tail are of about equal size and the fin is said to be **diphycercal** (= *double* + *tail*). If the spine tilts upward and enters the dorsal lobe, then the dorsal lobe is longer than the ventral lobe, most of the fin membrane is ventral to the axis of the tail, and the tail is termed **heterocercal** or epicercal (Figure 9.14). If the spine enters a larger ventral lobe, the tail is **hypocercal** (see anaspids in the figure on p. 40). If all the fin membrane is posterior to the spine, then dorsal and ventral lobes are about equal and the tail is said to be **homocercal.** Several intergrades and modifications are also recognized.

The ancestral tail may have been diphycercal, but the most primitive tails of record are heterocercal (cephalaspids, placoderms, most Chondrichthyes, the more primitive Osteichthyes of each subclass). Few vertebrates have hypocercal tails (anaspids). From the heterocercal tail there evolved the (secondarily?) diphycercal tail (cyclostomes, pleuracanths, later sarcopterygians, *Polypterus*) and homocercal tail (nearly all teleosts). These tail shapes can be identified in the illustrations for Chapter 3.

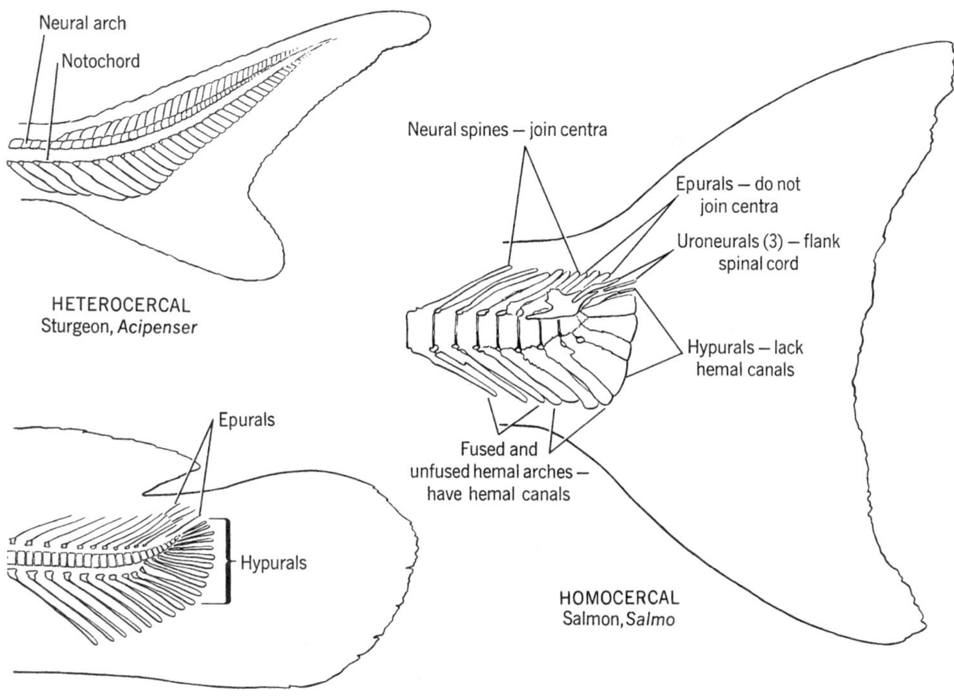

FIGURE 9.14
SHAPE AND STRUCTURE OF THE TAILS OF SOME BONY FISHES.

The caudal fin of ray-finned fishes, unlike the other median fins, is supported within its fleshy base by several modified neural arches and spines called **epurals** (= *upon* + *tail*), and more numerous modified hemal arches and spines called **hypurals** (= *below* + *tail*). These and related structures are further defined in Figure 9.14. The membrane of the fin is stiffened by fin rays corresponding in structure to those of the dorsal and anal fins of the same fish. Lepidotrichs of the caudal fin are usually branched.

STRUCTURE AND EVOLUTION OF GIRDLES

Girdles of Fishes The pectoral girdle is older, larger, and more complicated than the pelvic girdle. It includes one or more elements of cartilage or replacement bone and several dermal bones derived from ancestral scales or armor plates.

PLACODERMS illustrate the initial stages in the evolution of the pectoral girdle. A cartilaginous fin base was related to overlying plates of the dermal skeleton. The large pectoral girdle of CARTILAGINOUS FISHES is distinctive in two respects: Dermal elements are absent (presumably because of regression), and right and left halves of the girdle are fused in the midline. The result is a U-shaped girdle of one piece called the **scapulocoracoid.**

In BONY FISHES the replacement element, also termed scapulocoracoid, may ossify in one or several units. The dermal bones are identified in Figure 9.15. They join the girdle to the skull. This anchors the girdle in a manner that is not available to cartilaginous fishes (because they lack the requisite bones) and, hence, few bony fishes follow cartilaginous fishes in having the halves of the girdle joined in the midventral line. The **cleithrum** is the basic dermal element. The **clavicle** is lost in higher teleosts. The bones between the cleithrum and skull vary in number.

Placoderms had small pelvic fins at best, and a pelvic girdle has not been described. The pelvic fins of other fishes are weakly supported by a single skeletal element on each side of the body. They are bony except in cartilaginous fishes and dipnoans. The two pieces are usually separate but may overlap or articulate with one another, and in cartilaginous fishes they are joined across the midventral line by a bridge of cartilage (Figure 9.16).

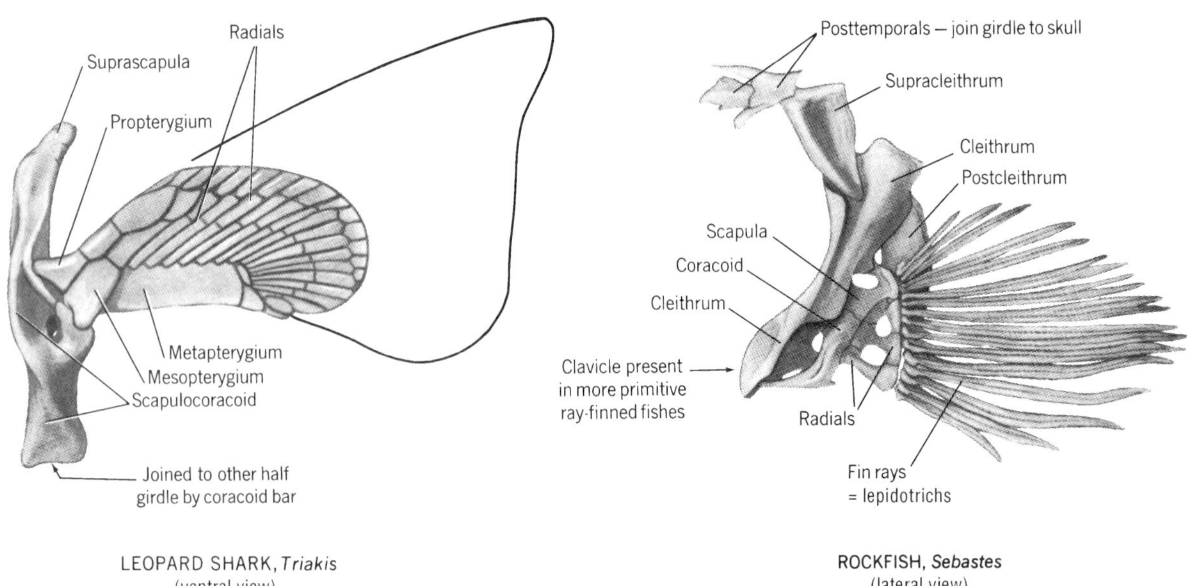

FIGURE 9.15 LEFT PECTORAL GIRDLE AND FIN SKELETON OF AN ELASMOBRANCH (left) AND A TELEOST (right).

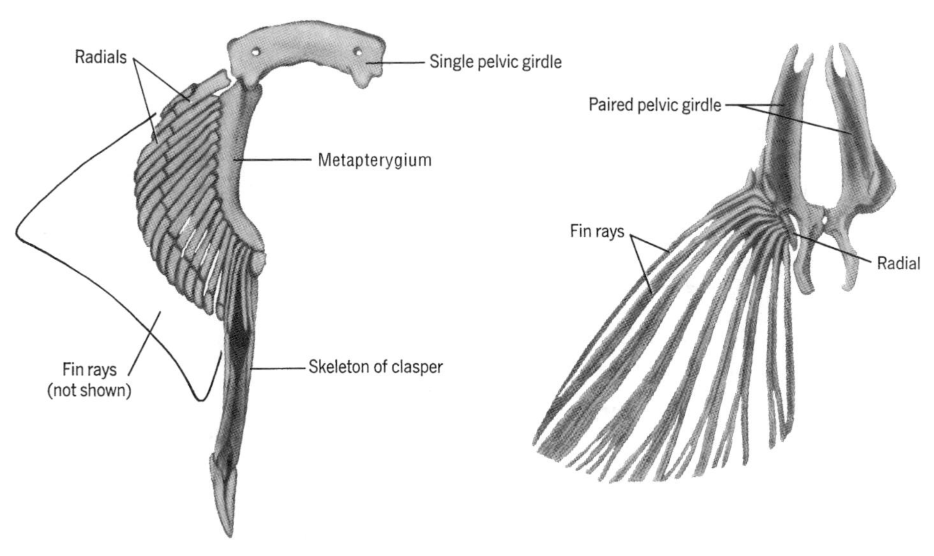

FIGURE 9.16 PELVIC GIRDLE AND LEFT FIN SKELETON OF A CARTILAGINOUS FISH (left) AND A BONY FISH (right) seen in dorsal view.

Girdles of Tetrapods The **pectoral girdle** of PRIMITIVE AMPHIBIANS differed from that of fishes in that the replacement bones were larger and the dermal bones reduced (Figure 9.17). All bones dorsal to the cleithrum were lost (except in one transitional group of primitive labyrinthodonts); thus the contact with the skull was broken and the head was freed to turn on the evolving neck. Some dipnoans and crossopterygians had a small **interclavicle** that united the two half girdles in the midventral line. This bone enlarged in labyrinthodonts, probably to compensate for loss of anchorage of the girdle to the head. There were two replacement bones: a dorsal **scapula** and a ventral bone, which for the moment we shall call the coracoid. Among MODERN AMPHIBIANS, Urodela have no membrane bones at all in this girdle, and Anura have no interclavicle and usually lack the cleithrum (Figure 9.12).

STEM REPTILES, SYNAPSIDS, and MONOTREMES are alike in having a full complement of bones. Interclavicle and clavicle are present, and the cleithrum is present (for the last time) in the more primitive reptiles. The scapula is large and there are two coracoids. Paleontological evidence indicates that the single coracoid of amphibians did not split; instead, a new bone was added behind the original coracoid. We refer, therefore, to an **anterior coracoid** (or **precoracoid**) and a **posterior coracoid** (or simply **coracoid**). OTHER REPTILES, including turtles, lepidosaurs, and archosaurs, usually lost the posterior coracoid and at least some of the membrane bones.

BIRDS have a bladelike scapula that is oriented parallel to the spine. The anterior coracoid is large and articulates firmly with the sternum. The posterior coracoid has been lost. The two clavicles fuse ventrally to form the wishbone or **furcula.** The interclavicle may be incorporated into the furcula of some birds, but is often absent.

When only one coracoid is present (labyrinthodonts, turtles, crocodilians, lizards, birds), it is usual to refer to it as *the* coracoid without qualification. We have seen, however, that with respect to tetrapod phylogeny it is the anterior coracoid.

The clavicle is the only membrane bone retained by THERIAN MAMMALS, and even this bone may be missing. It is the *anterior* coracoid that is completely lost this time. The posterior coracoid ossifies independently in the fetus and then fuses to the scapula to form the **coracoid process** of that bone. The scapula is unique in having a **spine.** The spine represents the anterior border of the ancestral bone, so in fact it is the anterior border

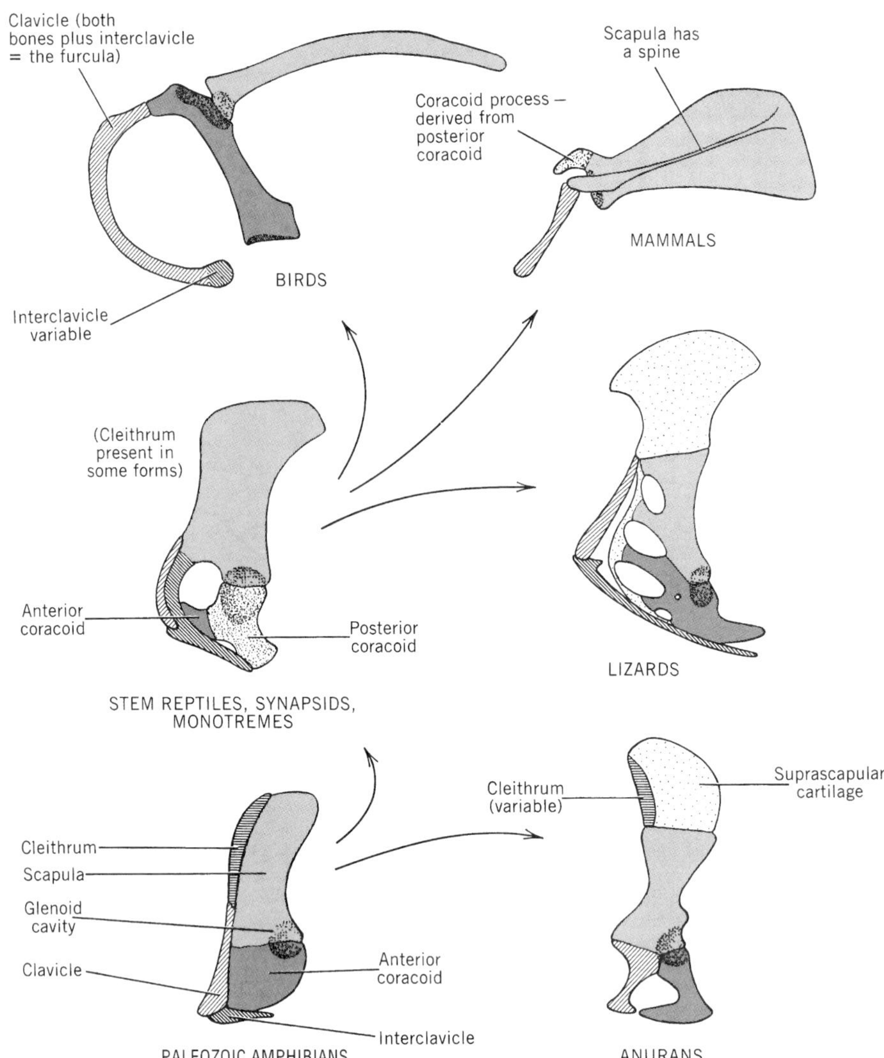

FIGURE 9.17 PHYLOGENY OF THE TETRAPOD PECTORAL GIRDLE. Left lateral views.

of the mammalian scapula that is new. The ventral end of the spine is continued as the **acromion process** to articulate with the clavicle.

The **pelvic girdle** of tetrapods is much enlarged over that of fishes and is relatively uniform in basic structure (Figure 9.18). Each half of the girdle is a single cartilaginous unit in the embryo, but three bones are constant in the adult. These are a dorsal **ilium,** which articulates with one or more sacral vertebrae, an anterior **pubis,** and a posterior **ischium.** The bones of one side commonly fuse in the adult to form the **innominate bone.** One or both of the ventral bones of the two sides usually articulate or fuse across the mid-ventral line; the contact is called the **pelvic symphysis.**

Primitive AMPHIBIANS had a solid girdle shaped like a triangle with the ilium forming the apex. The pubis can be distinguished from the ischium by having a foramen (obturator foramen) that accommodates a nerve. The atypical girdle of frogs has a long, anteriorly inclined ilium. The pubis of modern amphibians is cartilaginous.

The girdle of REPTILES takes various shapes but is like that of labyrinthodonts (Paleozoic amphibians) in basic plan. The contact with the spine is firmer. A large puboischiadic fenestrum is usually present between the two ventral bones.

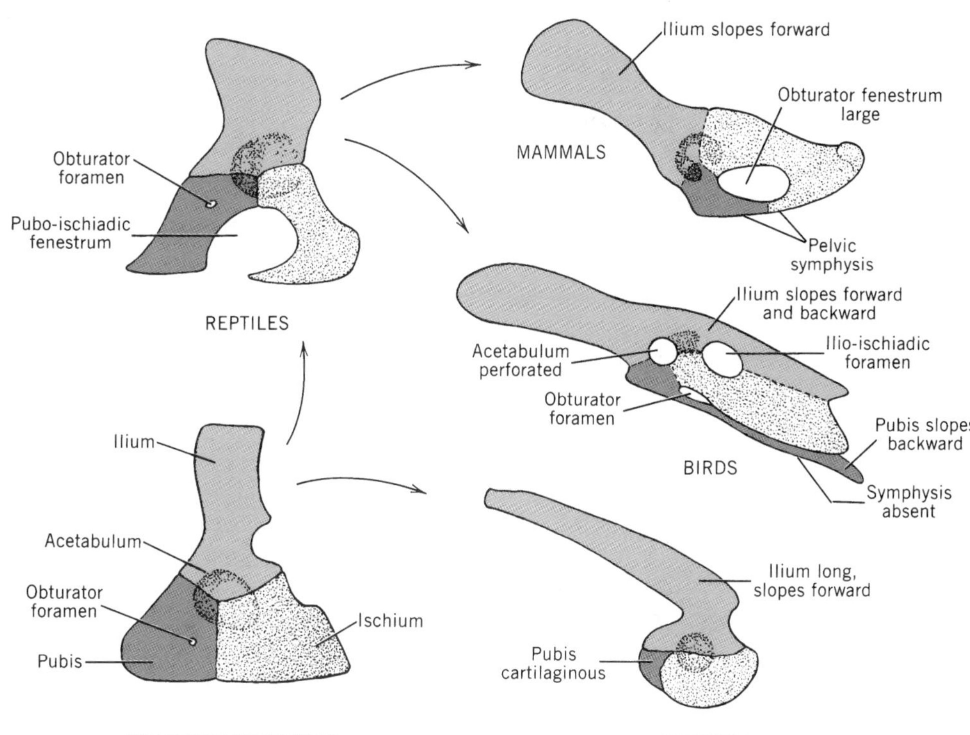

FIGURE 9.18
PHYLOGENY OF THE TETRAPOD PELVIC GIRDLE. Left lateral views.

The pelvic girdle of BIRDS is distinctive. It is large and firmly attached to the synsacrum. The long ilium extends both anterior and posterior to the socket for the femur, or **acetabulum.** The pubis is turned backward below the ischium. There is no symphysis (see also Figure 9.9).

MAMMALS have a long and expanded ilium that extends only forward from the acetabulum. The large obturator fenestrum represents both the obturator foramen and the pubo-ischiadic fenestrum of the ancestor. A symphysis is nearly always present. Monotremes and marsupials, some mammal-like reptiles and allotherian mammals, and several late Cretaceous eutherian mammals have paired **epipubic bones** (Figure 9.19) of uncertain origin that articulate with the pubic bones and extend forward in the ventral body wall. They have variously been considered to support a pouch or the weight of clinging young, or to relate to locomotion.

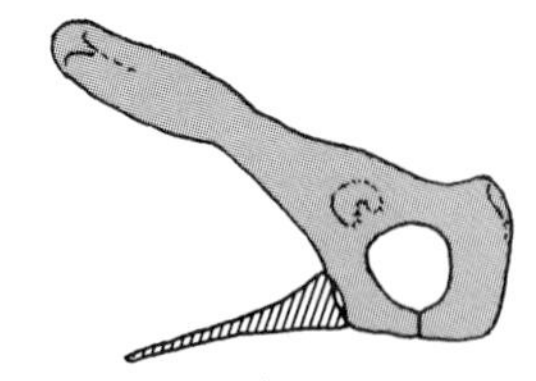

FIGURE 9.19
PELVIS OF AN OPOSSUM. Left lateral view. Epipubic bone hatched.

STRUCTURE AND EVOLUTION OF PAIRED FINS

The earliest paired fins probably had broad bases, a row of anchoring basals against the body wall, and a series of radials extending laterally into the fin. Subsequently, likely in the Ordovician, the base of the fin became constricted, increasing its mobility, and the basals, enlarged and reduced in number, were incorporated into the fin's dorsoposterior margin (Figure 9.20). The most constant basal, and usually the largest, is homologous in all fishes and is named the **metapterygium.** The metapterygial axis defines the fin axis. It always supports radials on its preaxial (= *ventral* or *anterior*) side.

The function of paired fins is usually to prevent the body from pitching (turning around the transverse axis) and rolling, and (particularly in higher fishes) to brake forward motion. The function and position of fins is discussed further in Chapter 27.

Among AGNATHOUS VERTEBRATES, cyclostomes have no trace of paired appendages, and the same was true of pteraspids. Anaspids had lateral pectoral spikes and lateral fin folds (see figure on p. 40). The spikes were of dermal origin, superficial, and not

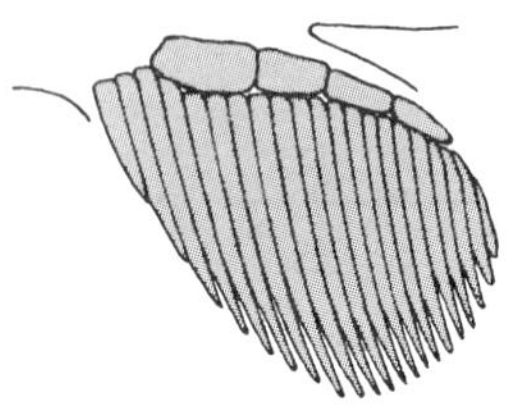

FIGURE 9.20
HYPOTHETICAL ANCESTRAL PECTORAL FIN OF JAWED VETEBRATES.

motile. The fin folds probably undulated in precision swimming. Cephalaspids had pectoral lobes behind the lateral wings of the cephalic armor. They were muscular, but seem to have had no internal skeleton.

PLACODERMS and ACANTHODIANS experimented with newly acquired paired appendages that varied from stiff fins (arthrodires) to hinged arms (antiarchs) and multiple spines (acanthodians) (see illustrations in Chapter 3). A complex dermal skeleton was sometimes supplemented by a cartilaginous internal skeleton.

At this point in the evolution of paired fins the relative importance of the dermal and internal skeleton was reversed, the latter dominating the former. Paired fins of CHONDRICHTHYES (except pleuracanths) depart little from the postulated ancestral plan, although the detailed pattern varies widely. The fin stem is supported by **basals** (usually three), from which numerous **radials** fan out (Figures 9.15 and 9.16). Ceratotrichs, like those of the dorsal fin, complete the fin skeleton.

The ACTINOPTERYGII are named according to fin structure (**actinopterygium** = *ray* + *fin*). Chondrosteans usually retain one to three basals. The radials are short and the fin skeleton is completed by lepidotrichs. Neopterygians usually have no basals and few radials.

SARCOPTERYGII are also named according to fin structure; **sarcopterygium** (= *flesh* + *fin*) refers to the fleshy nature of the bases of these fins.The ancestral fin skeleton is modified by reorienting the row of basals to project into the fin as the axis of the fleshy fin stalk. There are two principal kinds of sarcopterygium. In dipnoans and pleuracanths (the latter are not classified as Sarcopterygia but have similar fins) the radials are biserial; there is a series of radials on each side of a median axis. This fin was once thought to be ancestral to the tetrapod limb and accordingly was named an **archipterygium** (Figure 9.21). The term remains but is no longer apt. In Crossopterygii the radials are uniserial; there is a series on one side of a shorter lateral axis. This fin, called a **crossopterygium** (= *fringe* + *fin*), was unquestionably ancestral to the tetrapod limb. The archipterygium is nearly symmetrical; the crossopterygium is asymmetrical.

ORIGIN, EVOLUTION, STRUCTURE, AND DEVELOPMENT OF LIMBS

Discovery of the amphibian-like panderichthyid fishes and the fishlike ichthyostegid amphibians has narrowed the gap between fin and limb. Also, it is now known that (contrary to former belief) tetrapods evolved in and near freshwater. Their fish ancestors had already used strong fins to walk on the bottom and crawl onto damp shores to escape aquatic enemies or find terrestrial food. The transformation to limbs required the loss of fin rays, the gain of digits, and reorientation. The incipient limb extended laterally, its preaxial side forward; flexed at elbow and wrist (knee and ankle), allowing elevation of the body; and rotated medially to point the foot forward.

The paired fins of fishes always have about the same functions and vary in basic structure according to class or subclass. The pectoral fins are the stronger and more firmly related to the axial skeleton. The limbs of tetrapods, by contrast, have a variety of functions but retain a unity of basic structure. The pelvic limbs are the stronger and more firmly related to the axial skeleton. The proximal and middle elements of the tetrapod limb skeleton can clearly be identified in the fins of ancestral crossopterygian fishes (Figure 9.22). The archipterygian fin formerly appeared to be fundamentally different. However, there is now developmental evidence that the primordia of the ulna and radius fuse to become the second element of the axial series, and that the straight axis of the fin is equivalent to the arching axis of tetrapods.

The basic skeletal structure of limbs is indicated in Figure 9.22. (The terminology shown is used worldwide by comparative anatomists, but other terms are used by human antomists and mammalogists to identify the particular variations of the foot bones of

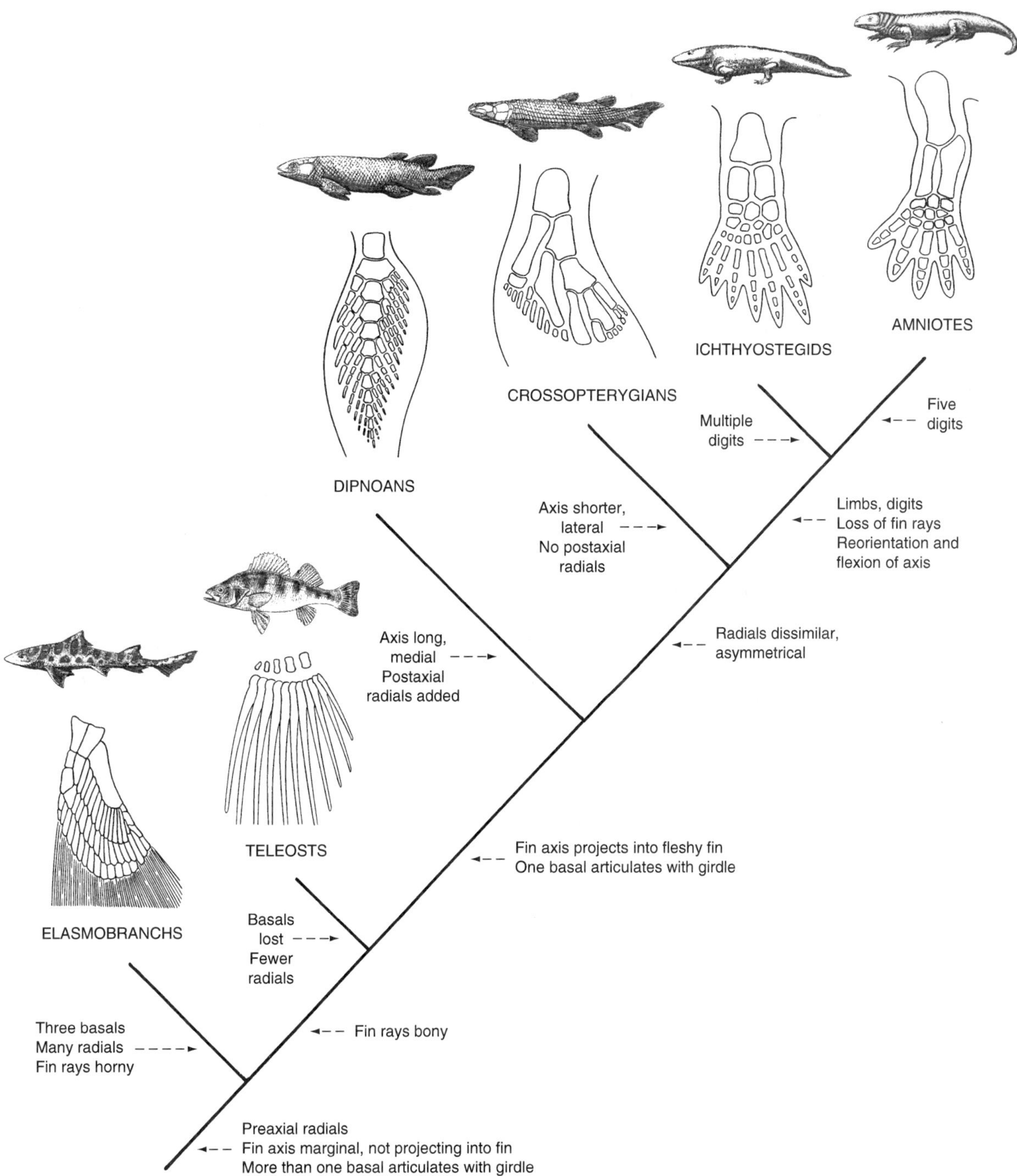

FIGURE 9.21 CLADOGRAM SHOWING EVOLUTION OF THE PECTORAL APPENDAGE. Left side in dorsal view.

their materials.) The bones of the wrist compose the **carpus;** those of the ankle compose the **tarsus.** Carpal and tarsal bones are known collectively as **podials.** Of the podials, the centralia are least constant. The forefoot is called the **manus** and the hind foot the **pes.** Metacarpal and metatarsal bones are known collectively as **metapodials.** The skeletal patterns of the various tetrapod feet are derived from the primitive pattern by deletions and fusions that can usually be verified by embryonic development.

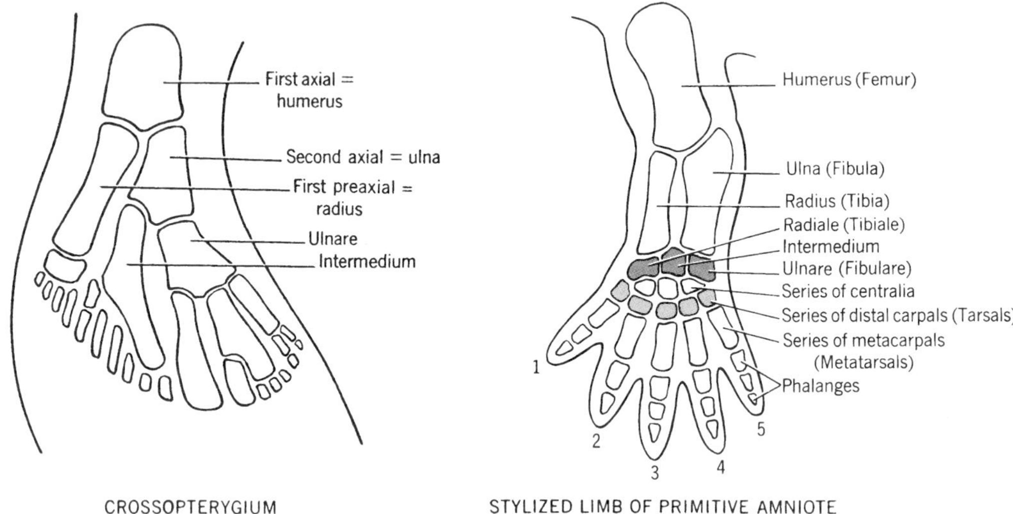

FIGURE 9.22 TERMINOLOGY AND HOMOLOGY ACROSS THE FIN-TO-LIMB TRANSITION. Left side in dorsal view. Terminology for hind leg shown in parentheses.

Shubin and Alberch (see References) used experimental and comparative embryology to clarify the development of limbs in a way that is consistent with the comparative anatomy of fossil and living lobe-finned fishes and tetrapods. The limb skeleton forms from cartilaginous elements within the developing limb bud. First to appear is a single piece derived from the ancestral metapterygium, which becomes the proximal skeletal segment (Figure 9.23). This piece then bifurcates to form the two elements of the next segment. (The second element on the axis of the dipnoan archipterygium is not paired, but it forms by the fusion of paired primordia.) The remainder of the limb skeleton develops (asymmetrically) from these paired elements, in spatial and temporal sequence from the body outward, either by segmentation (a first cartilage forming another in linear sequence with

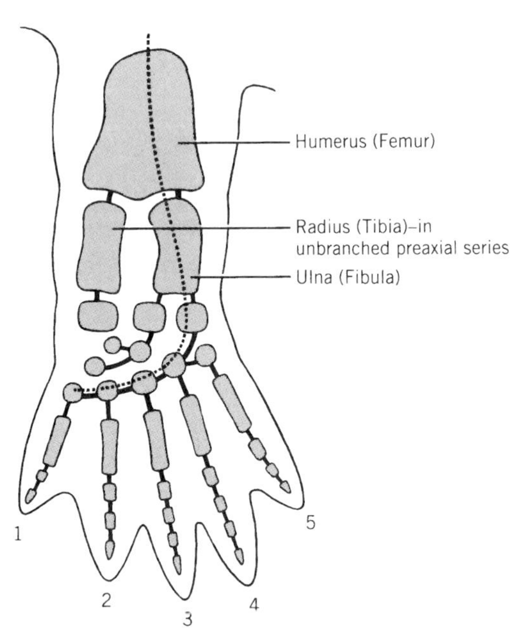

FIGURE 9.23 SCHEME OF DEVELOPMENT OF THE TETRAPOD LIMB SKELETON. Elements form in sequence, proximal to distal, from preceding elements following the paths indicated by black lines. The developmental axis is shown by a dotted line. For further terminology compare with Figure 9.22. Based on Shubin and Alberch.

itself) or by branching (a first forming two more by bifurcation). The historical and developmental stability of the appendicular skeleton decreases in the direction of its distal end. The homologies across the fin-to-limb transition shown in the figures are firm for the proximal two limb segments and probable for two bones of the foot. The many attempts that have been made to homologize additional bones across the transition now seem unwarranted. The same general plan appears to apply to all tetrapods. The limb axis runs in a curve along the wrist (not down one digit as formerly thought). The digits are not converted postaxial radials; they are instead tetrapod inventions. The first amphibians experimented with more than five. Yet a pattern of digits greater than five has never been adopted as the norm in any lineage leading to a modern species. Tabin reviews this phenomenon and provides thoughtful discussion.

STRUCTURE AND GROWTH OF LONG BONES

The larger bones of tetrapod appendages are called **long bones.** A typical long bone has a cylindrical shaft called the **diaphysis** (= *through* + *growth*), which contains the marrow cavity, and at each end an enlargement called the **epiphysis,** which articulates with adjacent bones. Epiphyses may be cartilaginous or bony. If bony, they are spongy within. Bony epiphyses usually fuse with their respective diaphyses at maturity.

The first embryonic primordium of a long bone is a condensation of mesenchyme. This then forms a one-piece model in hyaline cartilage of the future bone. Ossification of the diaphysis begins as a thin veneer of intramembranous bone deposited around the center of the model by the limiting membrane, or **periosteum** (Figure 9.24). Spicules of bone penetrate the cartilage and replace the matrix, which is gradually destroyed. Soon the incipient marrow cavity is surrounded by solid lamellae of bone arranged like the layers of orthodentine. The smallest mammals may retain this structure, but usually longitudinal canals are eroded within the developing bone tissue and the cavities reossify with the cylinder-within-cylinder construction of osteons. Only at the surface of the bone is lamellar structure retained into full maturity (see figure on p. 88). Epiphyses, by contrast, ossify (if at all) from one or more internal centers that expand outward to replace the cartilage. Their

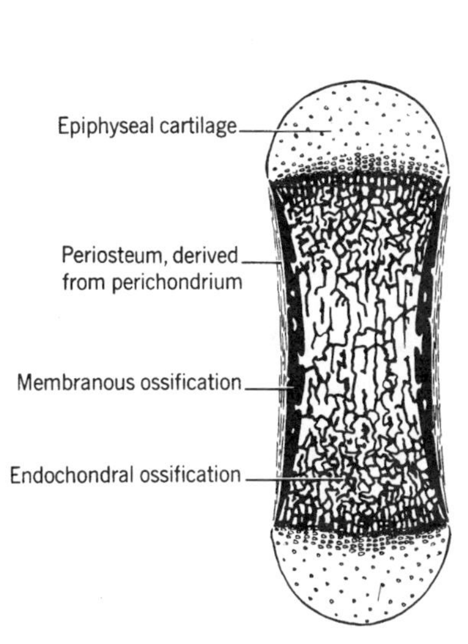

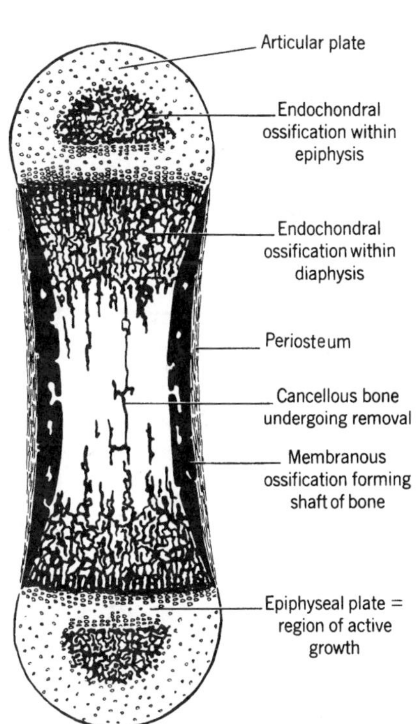

FIGURE 9.24 HISTOGENESIS OF A MAMMALIAN LONG BONE at two developmental stages as seen in longitudinal section.

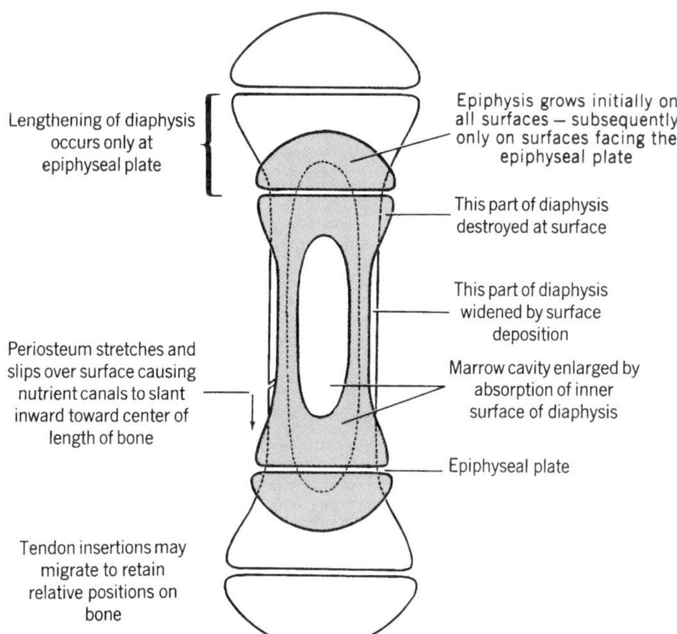
Lengthening of diaphysis occurs only at epiphyseal plate
Epiphysis grows initially on all surfaces – subsequently only on surfaces facing the epiphyseal plate
This part of diaphysis destroyed at surface
This part of diaphysis widened by surface deposition
Periosteum stretches and slips over surface causing nutrient canals to slant inward toward center of length of bone
Marrow cavity enlarged by absorption of inner surface of diaphysis
Epiphyseal plate
Tendon insertions may migrate to retain relative positions on bone

FIGURE 9.25
MORPHOGENESIS OF A MAMMALIAN LONG BONE between two development stages as seen in longitudinal section.

contained spicules are largely lamellar. The mature bone is mostly endochondral, but partly dermal in origin.

Bones grow by a complicated and wonderful process that is best known for reptiles and mammals. The diaphysis increases in length only at the cartilaginous plates that separate it from its epiphyses. If (as in mammals) these plates are obliterated at maturity, the increase in length stops. Growth in the transverse diameter of the shaft is accomplished by the deposition of bone at its outer surface and erosion of its inner surface to enlarge the marrow cavity. Some tendonous insertions must migrate to maintain their proportionate positions as a bone grows. The periosteum stretches like a sheet of rubber as the bone enlarges within; it must slip over the surface of the bone since the latter grows without stretching. Some further points are shown in Figure 9.25.

The configuration of mature bones is greatly influenced by mechanical interaction with the developing muscles. In the absence of normal muscles and muscle activity the skeleton is abnormal.

LIMB CHARACTERISTICS

AMPHIBIANS (other than the legless caecilians) usually have short limbs splayed to the sides of the body. The trunk is usually lifted from the ground when the animal walks, but only with difficulty. In urodeles undulations of the spine may be used to twist the girdles, thus helping to advance the limbs. Epiphyses are of hyaline cartilage and fit like corks into the ends of the bony shafts (most amphibians) or are calcified and fit over the ends of the shafts like match heads (Anura). The podials of modern amphibians are often cartilaginous. The principal joint of the foot is between the podials and metapodials. There are only four digits on the hand and four or five on the foot. One to three phalanges are present in each toe. The marrow cavities of the long bones of amphibians and higher vertebrates produce blood cells—a function not performed by the skeleton of fishes.

Many REPTILES still have the limbs positioned far to the sides of the body, but some dinosaurs and mammal-like reptiles placed their feet well under the body. The limbs are usually stronger than for amphibians, and the hind limbs are often disproportionately

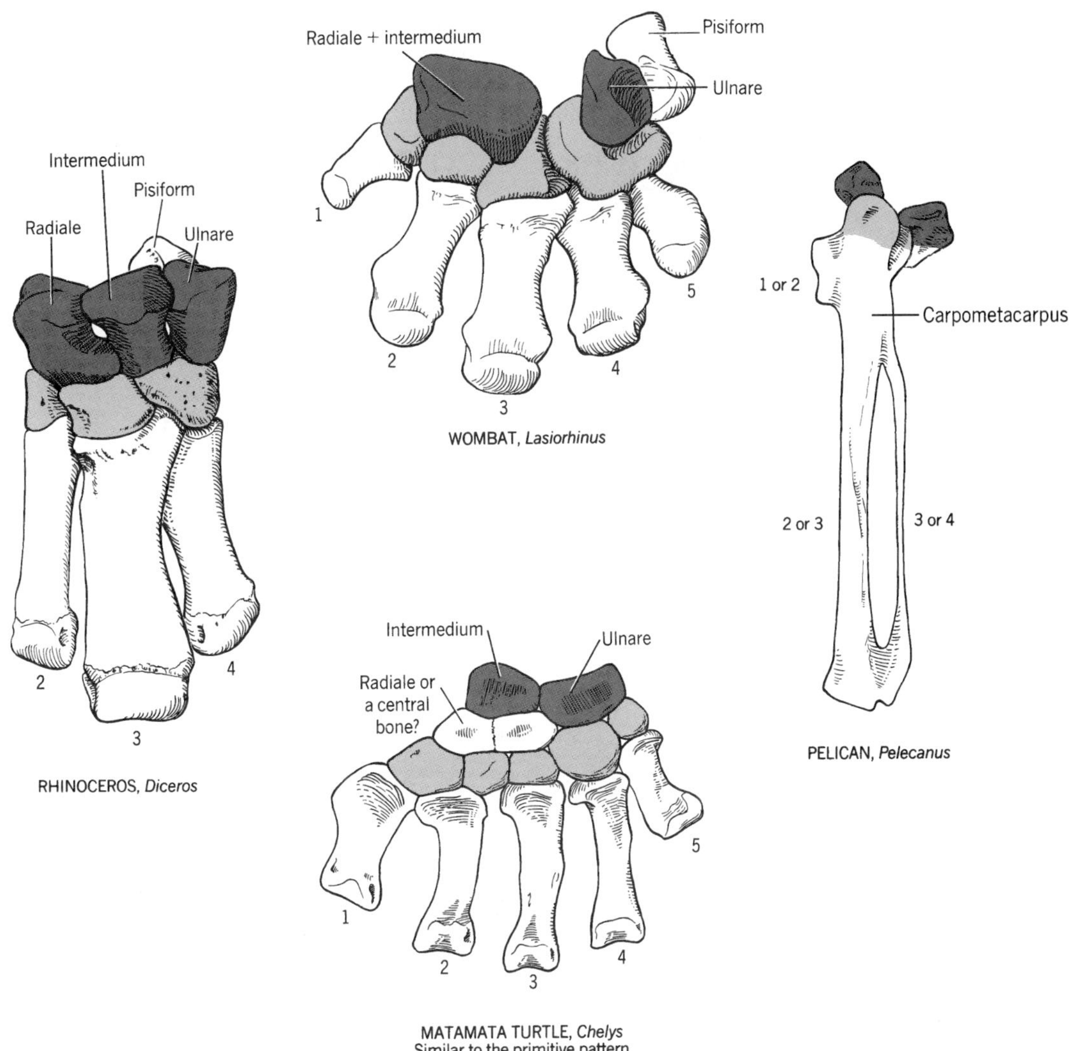

FIGURE 9.26 LEFT CARPUS AND METACARPUS in dorsal view. Distal carpals shown by light shading; proximal carpals by dark shading; and metacarpals, centralia, and pisiform by no shading. There is doubt regarding the homologies for birds.

large. Epiphyses are usually cartilaginous but may ossify in lizards. A new bone, the **pisiform** (not part of the early tetrapod pattern), may be added to the outside of the carpus, and the tibiale is no longer a free bone in the tarsus. The joint of the foot is often between podials. The **phalangeal formula** (number of phalanges per digit starting with digit 1) is 2-3-4-5-3 for the manus and 2-3-4-5-4 for the pes, if segments and digits have not been lost through specialization or degeneration.

The limb structure of BIRDS is uniform and specialized, as shown in Figures 9.26 and 9.27. Epiphyses are cartilaginous in immatures and virtually absent in adults. Note that the metacarpals retained by birds fuse, during development, to each other and to some carpal bones to form the compound *carpometacarpus.* Likewise, three metatarsals and some tarsals fuse to become the *tarsometatarsus.* (The metatarsals fuse from the bottom up in modern birds, Neornithes, but from the top down in the extinct subclass Enantiornithes—hence the common name of the latter, "opposite birds.") The avian wing has three digits.

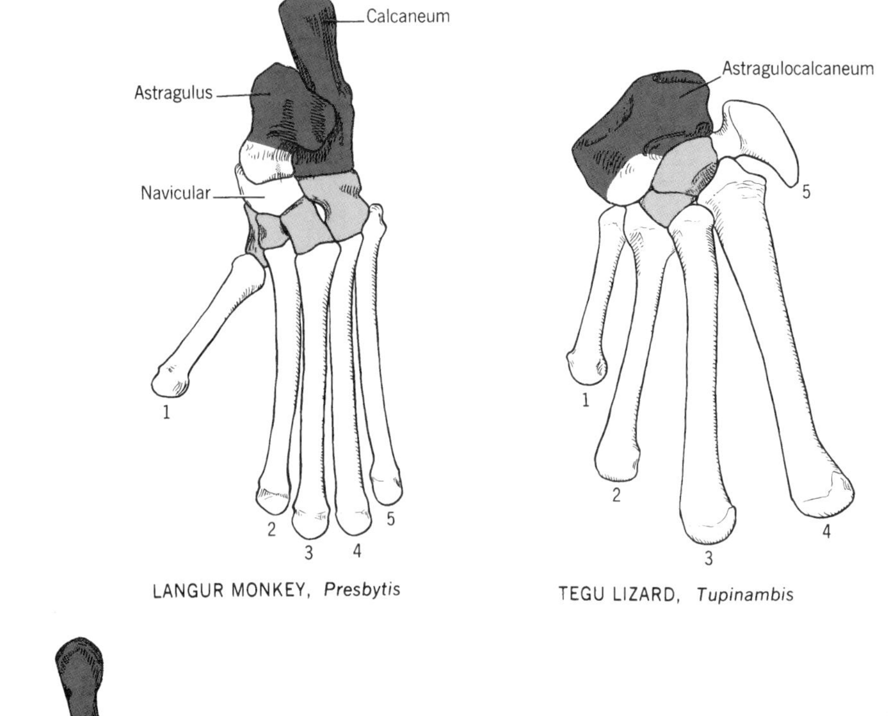

FIGURE 9.27
LEFT TARSUS AND METATARSUS in dorsal view. Distal tarsals shown by light shading, proximal tarsals by dark shading, metatarsals and centralia by no shading.

Most paleontologists identify them as numbers 1, 2, and 3, whereas embryologists have identified them as 2, 3, and 4. (See Comment 4.1 on p. 61.) The phalangeal formula of the foot is 2-3-4-5-0.

MAMMALS have bony epiphyses on each end of the long bones, on the distal ends of the metapodials, and on the proximal ends of all but the terminal phalanges. The pisiform is retained. In the tarsus the fibulare forms the heel bone, or **calcaneum** (Figure 9.27). The tibiale joins the intermedium, and the resultant large bone, called the **astragulus,** lies partly over the calcaneum.The ankle joint is between the astragulus and tibia. The ances-

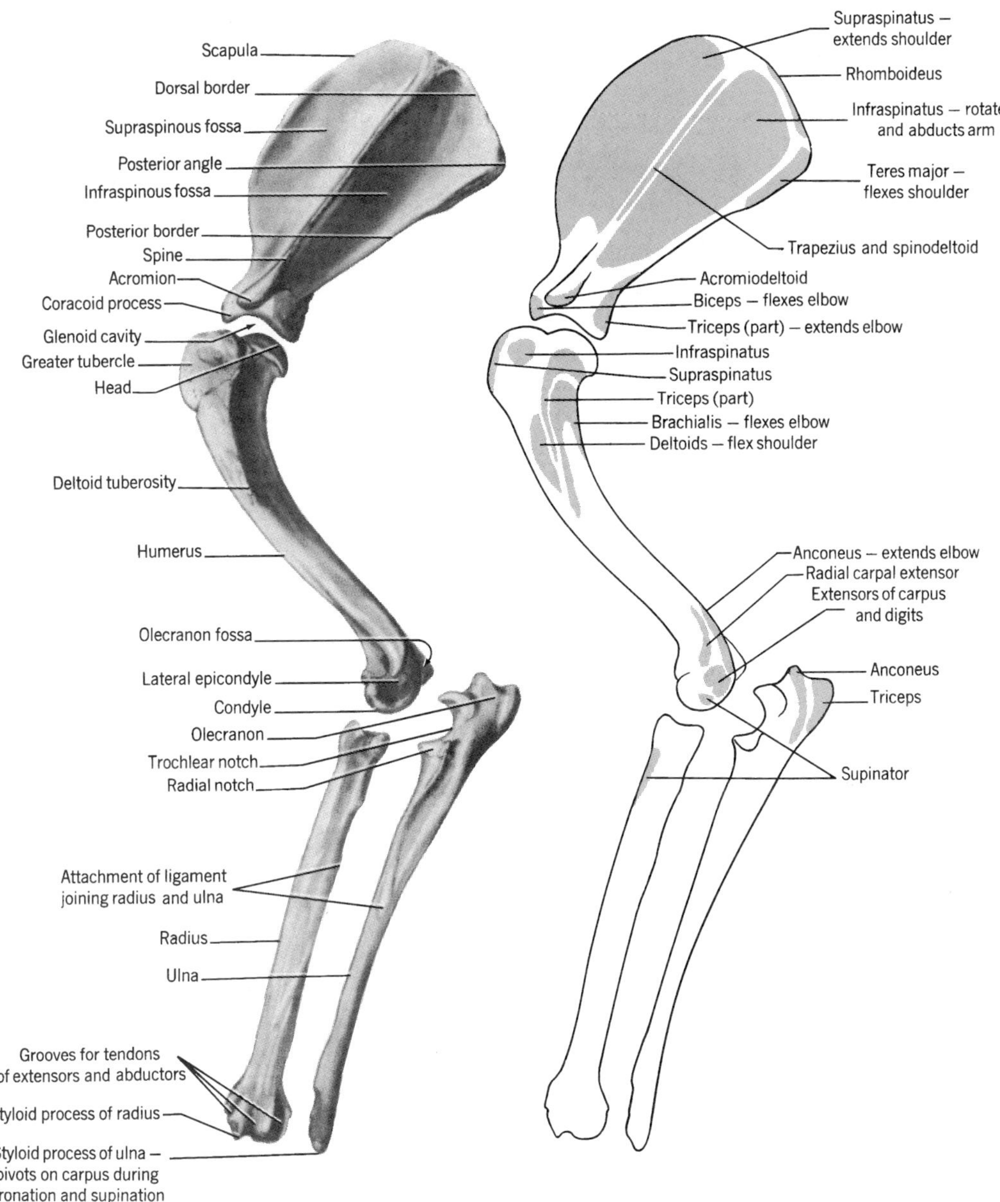

FIGURE 9.28 FEATURES AND FUNCTIONS OF THE FORELEG SKELETON OF THE DOG. Lateral view of the left leg. Not all muscle attachments are shown.

tral articulation between fibula and calcaneum is reduced or lost. Fusions among the tarsals are common. The basic phalangeal formula is 2-3-3-3-3. Features of the appendicular skeleton are identified in Figures 9.28 and 9.29, and are related to articulations and muscles. The **olecranon process** of the ulna is particularly characteristic of mammals.

SOME MISCELLANEOUS BONES

Nodules of bone tend to form where tendons play over joints. These are called **sesamoid bones** ("sesamoid" = *having the shape of a seed*). The largest sesamoid is the **patella,** or kneecap. The central tendons of some complex muscles (pinnate and unipinnate muscles, as defined on p. 170) tend to ossify—particularly in birds (they are

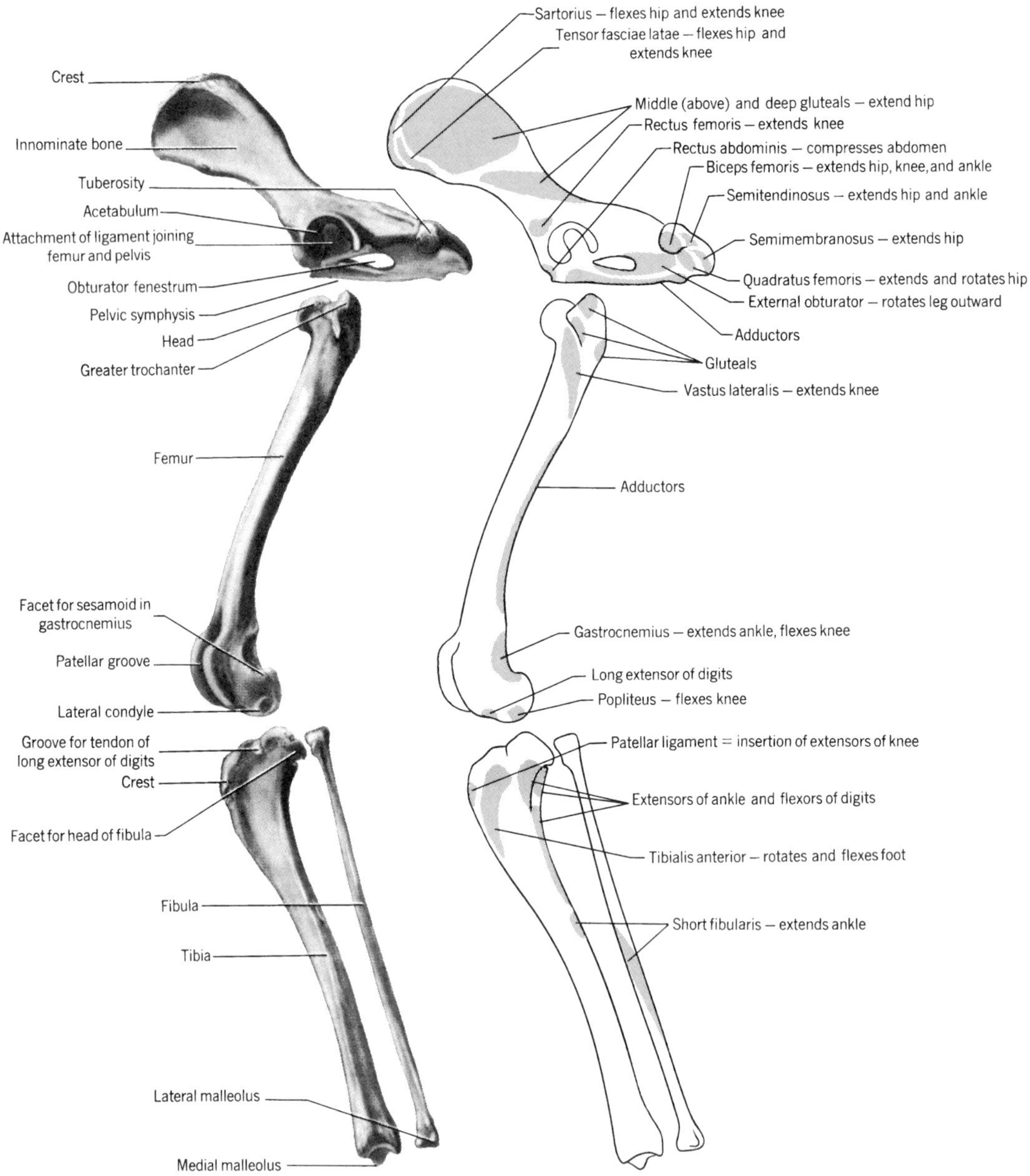

FIGURE 9.29 FEATURES AND FUNCTIONS OF THE HIND LEG SKELETON OF THE DOG. Lateral view of the left leg. Not all muscle attachments are shown.

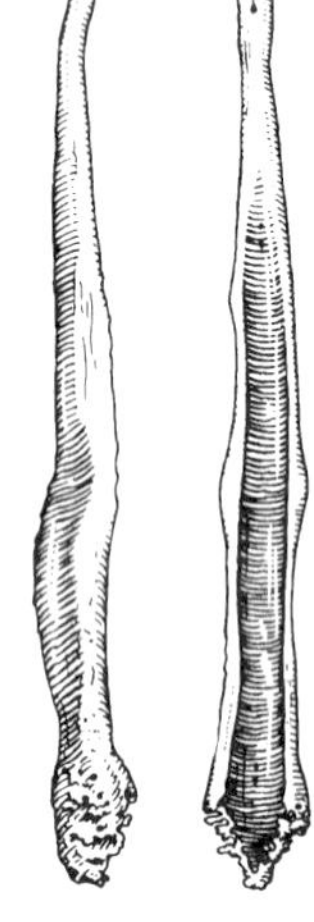

FIGURE 9.30 BACULUM of the maned wolf, *Chrysocyon,* in right lateral and ventral views.

encountered in a turkey drumstick). A bone called the **baculum** (or os penis) is present in the penis of carnivores, bats, insectivores, rodents, colugos, and some primates (Figure 9.30). Its size and shape vary widely among species and it is therefore useful in systematics. A corresponding but much smaller bone may be present in the female clitoris. Additional small bones are found here and there among tetrapods: in the eyelids of crocodilians, in the crest of a bird, in the snout of pigs, at the base of the external ear of some rodents, at the base of the aortic arch of some artiodactyls, and so on. These are of more functional than evolutionary significance.

REFERENCES

Burke, A.C., et al. 1995. *Hox* genes and the evolution of vertebrate axial morphology. *Development* 121:333–346.

Burt, W.H. 1960. Bacula of North American mammals. *Univ. of Mich. Mus. Zool. Misc. Publ.* 113:1–76.

Clack, J.A., and M.I. Coates. 1995. Romer's gap: tetrapod origins. *Bull. Mus. Natl. Hist. Nat. Sec. C Paris* 17:373–388. Recent discoveries of Devonian tetrapods reinforces the original speculations of A.S. Romer that tetrapods were originally aquatic and preceded their fully terrestrial descendents by millions of years.

Coates, M.I., and J.A. Clack. 1990. Polydactyly in the earliest known tetrapod limbs. *Nature* 347 (6 Sep.):66–69.

Coates, M.I., and M.J. Cohn. 1999. Vertebrate axial and appendicular patterning: the early development of paired appendages. *Am. Zool.* 39:676–685.

Enlow, D.H. 1962. A study of the postnatal growth and remodeling of bone. *Am. J. Anat.* 110:79–102.

Galis, F. 1999. On the homology of structures and *Hox* genes: the vertebral column, pp. 80–90. *In* G.R. Bock and G. Cardew (eds.), *Homology.* Novartis Foundation Symposium No. 222. Wiley, New York.

Gould, S.J. 1991. Eight (or fewer) little piggies. *Nat. Hist.* 91:22–29.

Haines, R.W. 1942. The evolution of epiphyses and of endochondral bone. *Biol. Rev.* 17:267–292.

Hinchliffe, J.R. 1989. Reconstructing the archetype: innovation and conservatism in the evolution and development of the pentadactyl limb, pp. 17–189. *In* D.B. Wake and G. Roth (eds.), *Complex organismal functions: integration and evolution in vertebrates.* Wiley, New York.

Jarvik, E. 1996. The Devonian tetrapod *Ichthyostega. Fossils Strata* 40:1–123. A scholarly treatment by a foremost paleontologist.

Lacroix, P. 1951. *The organization of bones.* Blakiston, Philadelphia. 235p. Excellent account of histology, development, and growth of long bones.

Laerm, J. 1982. The origin and homology of the neopterygian vertebral centrum. *J. Paleontol.* 56:191–202.

Laurin, M., M. Girondot, and A. de Ricqlès. 2000. Early tetrapod evolution. *Trends Ecol. Evol.* 15:118–123. Data from genes to fossils considered in light of newly discovered Upper Devonian/Lower Cretaceous forms.

Panchen, A.L. 1977. The origin and early evolution of tetrapod vertebrae. *Linnean Soc. Symp. Ser.* 4:289–318.

Schaeffer, B. 1967. Osteichthyan vertebrae. *Linnean Soc. Lond. Zool. J.* 47:185–195.

Schultze, H.P., and G. Arratia. 1989. The composition of the caudal skeleton of teleosts (Actinopterygii: Osteichthyes). *Zool. J. Linnean Soc.* 97:189–231.

Shubin, N. 1995. The evolution of paired fins and the origin of tetrapod limbs: phylogenetic and transformational approaches, vol. 28, pp. 39–86. *In* M.K. Hecht et al. (eds.), *Evolutionary biology.* Plenum, New York. A critical review of the morphological problems to be solved in the transition from fin to limb. Provides historical context and reviews recent contributions from phylogenetics and studies of pattern formation.

Shubin, N., and P. Alberch. 1986. A morphogenetic approach to the origin and basic organization of the tetrapod limb, vol. 20, pp. 319–387. *In* M.K. Hecht et al. (eds.), *Evolutionary biology.* Plenum, New York.

Shubin, N., C. Tabin, and N.E. Kemp. 1997. Fossils, genes and the evolution of animal limbs. *Nature* 388 (14 Aug.): 639–650.

Tabin, C. 1992. Why we have (only) five fingers per hand: *Hox* genes and the evolution of paired limbs. *Development* 116:289–296.

Verbout, A.J. 1985. The development of the vertebral column. *Advances Anat. Emb. Cell Biol.* 90:1–122.

Vorobyeva, E., and R. Hinchliffe. 1996. From fins to limbs: developmental perspectives on paleontological and morphological evidence, vol. 29, pp. 263–311. *In* M.K. Hecht et al. (eds.), *Evolutionary biology.* Plenum, New York.

Wake, D.B. 1970. Aspects of vertebral evolution in the modern Amphibia. *Forma Functio* 3:33–60.

Wake, D.B., and M.H. Wake. 1986. On the development of vertebrae in gymnophione amphibians. *Mém. Soc. Zool. Fr.* 43:67–70.

Webb, J.E. 1973. The role of the notochord in forward and reverse swimming and burrowing in the amphioxus *Branchiostoma lanceolatum. J. Zool. Lond.* 170:325–338.

Westoll, T.S. 1958. The lateral fin-fold theory and the pectoral fins of ostracoderms and early fishes, pp. 180–211. *In* T.S. Westoll (ed.), *Studies on fossil vertebrates.* Univ. London Press, London.

Williams, E.E. 1959. Gadow's arcualia and the development of tetrapod vertebrae. *Quart. Rev. Biol.* 34:1–32. Important reevaluation of a long-held theory.

Chapter 10

Muscles and Electric Organs

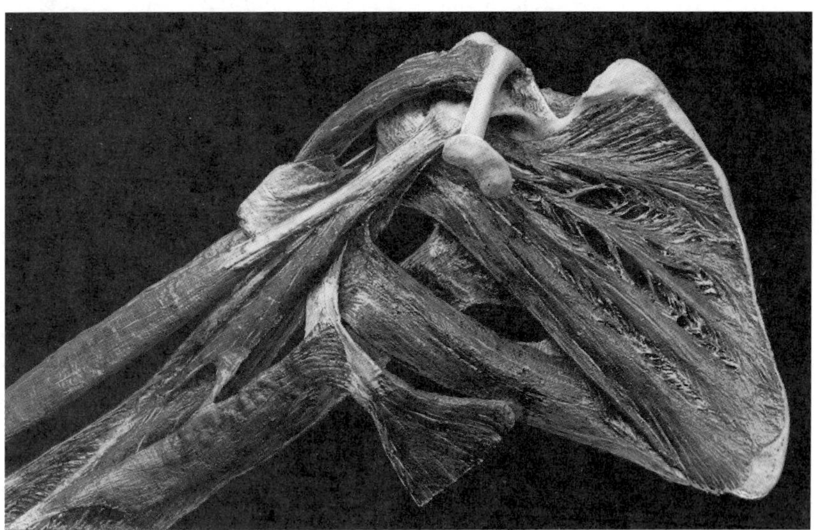

The muscular system is intuitively interesting to most of us, perhaps because our own movements, and those of other vertebrates, hold a certain fascination. This system (when studied in conjunction with the skeletal and nervous systems) is of primary importance in the functional analysis of locomotor activities, including athletic performance and recovery from injury. Homologies are nearly always evident among the muscles of related orders and lower taxa, but at the class level and higher, one must be cautious, and other organ systems must be monitored in the construction of phylogenies.

Dissection is the most important technique for the study of muscle. Knowledge of a muscle's anatomical position and its relationship to other muscles and to the skeletal system is fundamental to interpretation. Additional techniques employed to study muscle are biochemistry and histochemistry, direct stimulation, high-speed video and X-ray cinematography, and the use of a number of measuring devices such as force plates, strain and length gauges, electromyography, and computer-aided imaging.

FUNCTION AND GROSS STRUCTURE OF MUSCLES

Nearly all functions of the body are in part muscular. Without muscles, vertebrates could not move, their tissues soon would be starved or poisoned, and the products of their glands could not be distributed. Humans could not read, speak, or write to communicate their thoughts.

Muscles accomplish all this by doing only one thing—creating tension along the axis of their fibers, which tends to shorten their substance. Active muscles may shorten, thus moving a bone or constricting a space. They may also prevent motion by opposing gravity or the pull of other muscles. With little or no motion, they may function to cause a part to become more rigid. Furthermore, they may offer controlled resistance to extrinsic forces that tend to stretch them out.

The muscular system also has secondary functions: It contributes importantly to the maintenance of the body temperature of endotherms and, because of its bulk, distributes the weight of the body, influences the contours of the body, and offers protection to some of the viscera.

Above: Air-dried dissection of a chimpanzee shoulder.

There is much variation in the shapes of muscles and in the architectural organization of their fibers. Since each muscle fiber contracts only along its length, it is usually most effective for the fibers of one muscle to lie approximately parallel, thus establishing a longitudinal axis for the muscle as a whole. A *cylindrical* or *straplike* form might be taken, therefore, as a fundamental shape. Several muscles of the throat are usually of this nature. If one end of a muscle is tapered to insert on the skeleton, the shape of the muscle becomes that of a *teardrop.* Several muscles of the hip, thigh, and upper arm are commonly of this sort. Most muscles of the limbs are tapered at each end and hence are *spindle-shaped.* One end may be divided into two or more parts (or "heads"), and the fleshy part of the muscle is seldom symmetrical because it must fit against other muscles. Some muscles are spread out as *sheets* (most abdominal muscles), and others are flat at one end and gathered at the other, thus becoming *fan-shaped* (various chest and shoulder muscles). Muscles that surround orifices have curved fibers and are *washer-shaped;* muscles that surround spaces (stomach, uterus) have fibers oriented in various directions and are more or less *hollow spheres.*

Some spindle-shaped muscles do not have their fibers oriented in the long axis of the muscle but instead have fibers that slope inward to insert on a central tendon. In longitudinal sections such muscles may look like feathers and hence are called **pinnate** (see the longissimus muscle in the figure on p. 190). As explained on p. 407, for a muscle of a specific volume and length, increasing the extent of pinnation increases the cross-sectional area of the fibers and hence force production, but decreases shortening distance. Some muscles have several converging central tendons and are multipinnate (see the subscapularis muscle in the figure on p. 465) and some are unipinnate: Their fibers slant one way on to a lateral tendon or bone.

Where muscles attach directly to the skeleton, the connective tissue that surrounds and pervades them is continuous with connective tissue surrounding the bones. Where muscles do not impinge directly on the skeleton, they are joined to bones by **tendons,** which are tough cords of closely packed, parallel, collagenous fibers. (**Ligaments** join bone to bone. Their collagenous fibers are somewhat less regular and they include elastic fibers.) Some muscles (e.g., several abdominal muscles of mammals) do not attach to bones but instead distribute their forces over broad areas by means of strong flat sheets of connective tissue. These sheets are called **aponeuroses.** The loose connective tissue that binds muscle to muscle, and skin to muscle, is called **fascia.**

When a muscle contracts, it pulls equally on each end. Commonly, one attachment is relatively free to move and is then called the **insertion.** The relatively fixed attachment is the **origin.** These terms must be used with caution, however; tables of muscle origins and insertions can be misleading. Which end of a muscle is the more movable depends on posture, the activity of other muscles, and contacts with the environment. When one lifts an object from a table, the proximal end of the biceps is its origin; when one chins oneself on a bar, the distal end is the origin. During the propulsive phase of a limb's motion, its muscles move the body on the limb, not the limb on the body; contrary to usual terminology, it is then the distal ends of the muscles that are relatively fixed. Often neither end of a muscle is fixed, and sometimes neither moves.

It is necessary to have a vocabulary for describing the actions of muscles. **Flexors** are commonly defined as muscles that reduce the angle between adjacent bones; **extensors** increase the angle. These definitions usually serve but can be misleading. Considerations of position, phylogenetic origin, and innervation (as explained below) make it clear that flexors of a hind limb swing joints to the rear, whereas extensors swing them forward. At the hip, such muscles may increase or decrease the angle between adjacent bones, depending on posture. Also, since the elbow and knee bend in opposite directions, the situation is reversed for the forearm: flexion bends the elbow forward, not to the rear. Furthermore, some muscles of the back that straighten (hence extend) the spine from a hunched position also, on further contraction, arch (hence flex) the spine. If confusion threatens, it is well to avoid one-word designations of muscle function. **Adductors** move parts inward toward the sagittal plane of the body or axis of a limb; **abductors** move parts outward away from the body or axis. Spreading and closing the fingers and clapping the hands use these sets of muscles alternately. **Levators** raise and **depressors** lower such parts as the jaw or shoulders. Paired fins may alternatively be said to be depressed or adducted. **Protractors** push a part (such as the tongue or an entire limb) away from its base and **retractors** draw it back. **Sphincters** constrict openings (mouth, duct orifices) and **constrictors** compress spaces (pharynx, abdomen); they are opposed by **dilators. Rotators** turn parts about their long axes (spine, limbs). The rotators that turn the soles of the hands or feet upward are **supinators;** those that turn them downward are **pronators.** Some trunks and tongues can be stiffened by muscles that increase internal fluid pressure, thus acting as **hydrostats.**

For every action there is an opposing or restoring action. Opposing muscles are called **antagonists.** Rarely does one muscle contract alone. Muscles that supplement one another are called **synergists.** However, these terms must be used with caution: "Antagonists" may contract together to stiffen a joint or control motion; and predicted "synergists" may contract in sequence instead of synchrony, or one may fail to contract.

Muscles are given names that describe their actions (levator maxillae, flexor digitorum, adductor mandibulae), shapes (biceps, rhomboideus, trapezius), positions (temporalis, pectoralis, gluteus), or attachments, the origin being named before the insertion (geniohyoid, sternomastoid, cleidobrachialis). Most muscles were first named for the human; the terms are not always apt when applied to homologs in other animals. Nevertheless, it is easier to remember the terms when their derivations are understood.

HISTOLOGY, CONTRACTION, AND PHYSIOLOGY

Fine Structure Three types of muscle are distinguished by differences in histology and physiology (Figure 10.1). **Smooth muscle** is found in the skin, blood vessels, urogenital system, respiratory channels, and alimentary tract and the ducts of its derivatives. Its spindle-shaped, uninucleated cells lack striations. They have filaments that are oriented obliquely to their long axes and insert on their walls. **Cardiac muscle** is restricted to the heart. Its striated fibers are branched and divided into nucleated units by spaced intercalated disks. Like smooth muscle, it is relatively involuntary, although some yogis, and some persons using biofeedback apparatus, have gained limited control of such muscles.

The remaining muscles of the body are **skeletal muscles,** and it is they that concern us in this chapter. Each skeletal muscle is surrounded by a tough envelope of connective tissue, the **epimysium** (= *upon* + *muscle*), which is continuous with tendons and with such fascia as may be present (Figure 10.2). The epimysium also merges with septa, collectively

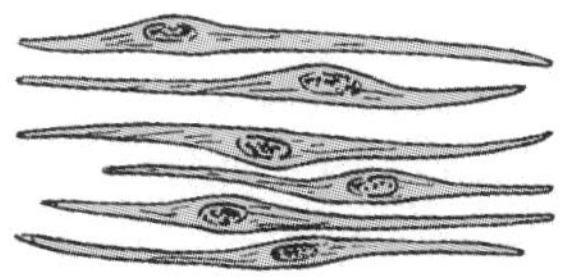

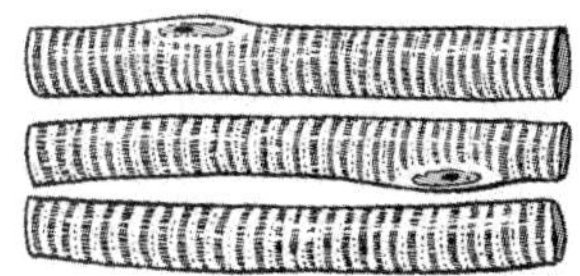

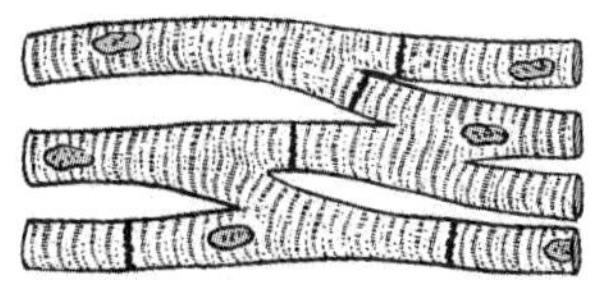

FIGURE 10.1
THE THREE TYPES OF MUSCLE TISSUE.

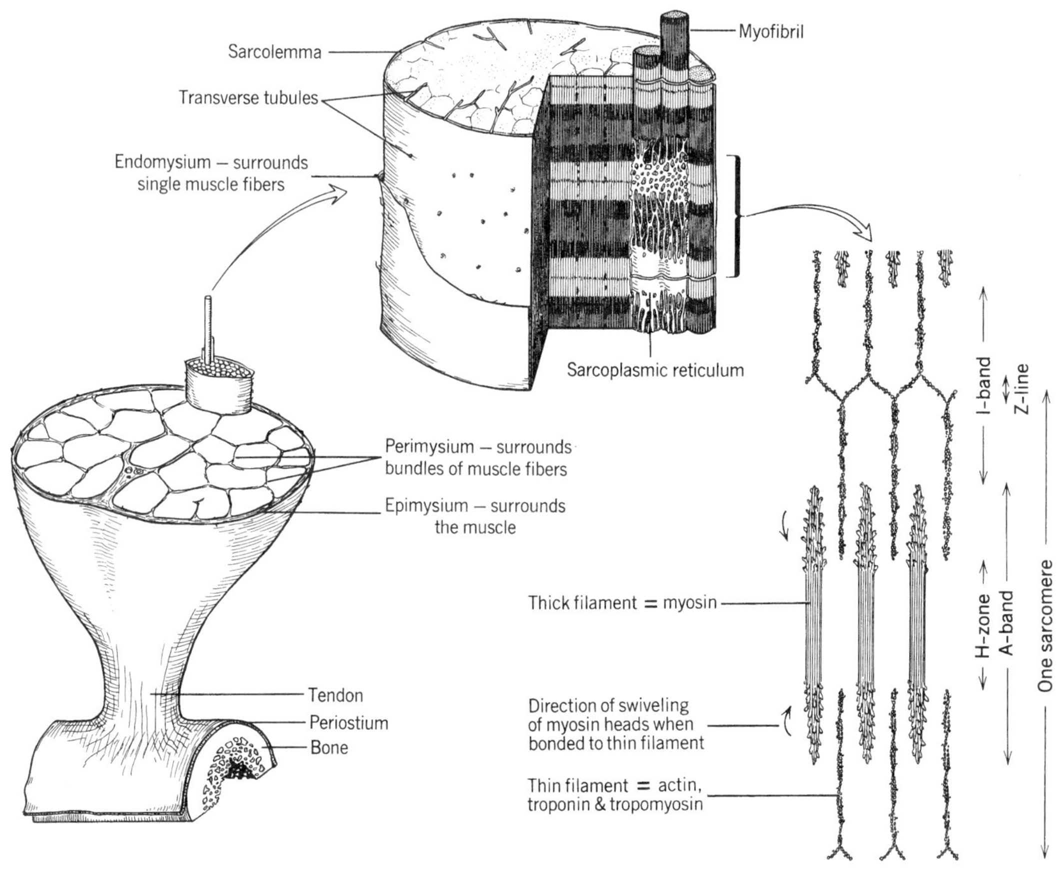

FIGURE 10.2 STRUCTURE OF SKELETAL MUSCLE. Nuclei and mitochondria omitted.

the **perimysium,** which penetrate the muscle and divide it into bundles of fibers. Such bundles may be distinct, making the muscle stringy, as for the rhomboideus of many mammals, or may be virtually absent. The perimysium, in turn, is continuous with a net of connective tissue, the **endomysium,** which surrounds the **sarcolemma** or limiting membrane of the individual muscle fibers. This continuous system of connective tissue gives muscles their shape and strength, and transmits their forces to origin and insertion.

If a fragment of muscle is macerated and pulled apart with fine needles, the hairlike **muscle fibers** can be seen. Although scarcely visible to the naked eye, they may be as long as the whole muscle, or they may be short and overlapping. This latter "in-series" organization requires several fibers in a row to span the muscle from one end to the other. In-series organization among amphibians and mammals has been documented most often, but not exclusively, in relatively long, parallel-fibered strap muscles, and they have also been documented in the pectoralis muscle of birds. They may be the rule, rather than the exception, for muscles in this class. The functional consequences of an in-series organization in locomotor muscles are questions actively being studied (see Trotter, Richmond, and Purslow in References). Regardless of length, each fiber has multiple, peripheral nuclei and is cross-banded (hence the term *striated*). The muscle fibers are penetrated by transverse tubules that open to the outside and by tiny anastomosing canals called the **sarcoplasmic reticulum.** These function in both transport and transmission. They regulate the concentration of calcium ion in the myofibril and thereby control the contraction-relaxation cycle.

When muscle fibers are stained and viewed with a light microscope it is seen that each is made up of dozens of finer strands, called **myofibrils,** each of which is about 0.001 mm in diameter. When enlarged some 400,000 diameters by the electron microscope, it is found that each myofibril is made of many filaments of two kinds.

Thick filaments are about 0.01 μm in diameter. They are composed of a protein called **myosin.** The molecules of myosin have long, slender tails arranged in cylindrical bundles, thus establishing the axes of the filaments. The short, globular, and bifid heads of the myosin molecules project outward at various angles from the filament axes. The various molecules of each filament are oriented in opposite directions. Hence, a filament has heads projecting at each end but is smooth in the middle (Figure 10.2).

Thin filaments have half the diameter of thick filaments and consist of the proteins **actin, tropomyosin,** and **troponin.** Actin molecules are small and globular, and in each filament are arranged like beads in double strands that twist about one another. Tropomyosin molecules are long and thin and lie on the surface of the actin strands. Troponin molecules are small and are spaced along the tropomyosin molecules.

Contraction occurs when myosin heads bond to actin and then swivel so as to draw the thick and thin filaments past one another. Tropomyosin and troponin control the active site on the actin molecule. Normally tropomyosin blocks this site. During activation of a fibril, the action of troponin induces a change in the orientation of tropomyosin such that the actin binding site is exposed and myosin is able to react with it, thus forming a cross bridge. The energy source is adenosine triphosphate. Bonding takes place only in the presence of calcium ion released by the sarcoplasmic reticulum following nervous stimulation. The bonds release when these ions are pumped back into the reticulum. The overlap of thick and thin filaments is responsible for the striations of skeletal muscle; the banding changes during contraction as the overlap of the filaments increases. The contractile unit of skeletal muscle is called a **sarcomere** (= *flesh* + *part of*) and is centered on a transverse band of thick filaments. The terminology used to describe the regions of a myofibril is shown in Figure 10.2.

Each muscle fiber contracts completely or not at all. From fewer than 10 to more than 1000 fibers (depending on the loads and fine control necessary for the task) are all innervated by the branching process of a single motor nerve cell and, hence, all contract together as a unit. The motor nerve cell and the muscle unit form a **motor unit,** the smallest *functional* unit of a muscle. The fibers of one unit usually occur in one region of a muscle but are interspersed with the fibers of many other units. Graded response by an entire muscle results from the recruitment of varying numbers of motor units. The tone, or **tonus** of resting muscles, is caused by the contraction of a minimal number of units.

Striated muscle fibers are classed as tonic (slow) or twitch. **Tonic fibers** are unable to propagate an action potential along their length from the point of stimulation (neuromuscular junction) by the nerve cell. Hence each fiber has multiple innervation sites to ensure that it will contract along its entire length. These fibers contract slowly (measured in seconds) and are not well understood (see Morgan and Proske in References). They occur in the appendages of all vertebrates except mammals, where they are limited to some small muscles that move the eyeball. Tonic fibers are universally highly resistant to fatigue.

A single stimulus from the neuromuscular junction is propagated along their length by the remaining **twitch fibers.** When held at a *constant length,* and given a single stimulation, these fibers twitch, or contract relatively rapidly, followed by relaxation (Figure 10.3). Of particular interest to morphologists is the contraction time, or time to peak tension. If a second stimulus follows before relaxation is complete, there is a **summation** of tension. Repeated stimuli in rapid succession cause a steady production of tension, or **tetanus.** When the muscle is not allowed to shorten, the contraction is said to be **isometric.** It is **isotonic** if tension does not increase because the muscle is shortening with constant resistance.

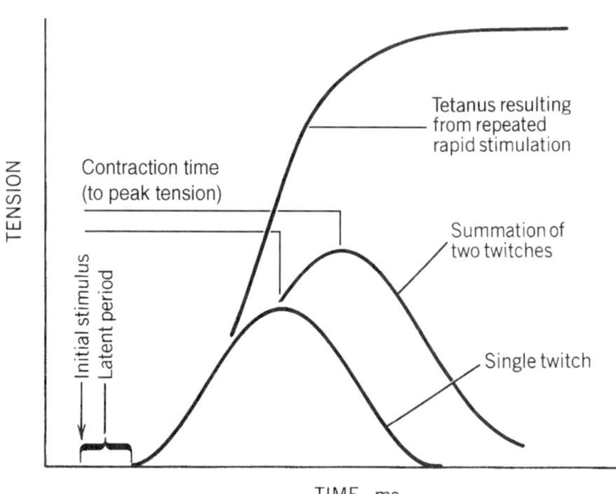

FIGURE 10.3 RESPONSE OF A MUSCLE TO STIMULATION.

Length-Tension and Force-Velocity These properties vary for muscles within a species and across species, and are of great interest and importance for studies of muscle function. In the laboratory a muscle can be stimulated **in vitro** (*in glass;* i.e., an artificial environment) or **in situ** (*in position*). Only recently are muscles being studied **in vivo** (*in the living animal*). *Length-tension* is a result of the mechanical arrangement of the overlapping thick and thin filaments; there is only a narrow range of overlap (sarcomere length) at which, upon stimulation, *isometric* tension is maximum. This length is the optimal or *resting length.* When a muscle is stimulated when at a shorter or longer length than optimal, isometric tension is less than maximum (Figure 10.4A). (On a length-tension curve, there is no time on the axes.)

When considered at the myofibril level (Figure 10.4B), the structural correlates of length tension can be illustrated more precisely. The optimum overlap of thin and thick filaments for tension (force) production when a myofibril is stimulated isometrically corresponds to a sarcomere length of 2–2.25 μm in vertebrates. At lengths shorter (due to overlap and interference of the thin filaments) and longer (due to fewer sites for attachment between the thin and thick filaments) than the optimum, force falls to a lower value.

Force-velocity of a muscle fiber, or whole muscle, refers to the relationship of the load moved (force) to the velocity of shortening during an *isotonic* contraction (Figure 10.4C). When shortening against no load, little to no force is generated and a muscle achieves maximal velocity (V_{max}). At each increment of load, velocity decreases. If the muscle can barely

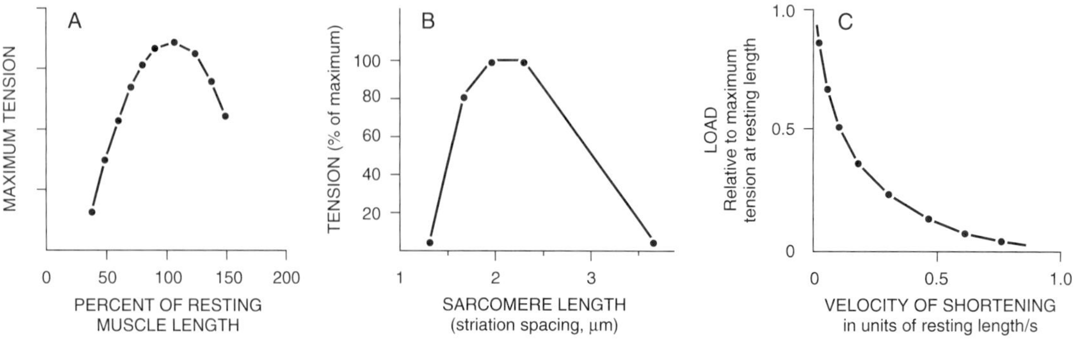

FIGURE 10.4 LENGTH-TENSION CURVES FOR WHOLE SKELETAL MUSCLE (left) AND SARCOMERES (center), AND LOAD-VELOCITY CURVE (right). (Part B modified from Gordon, Huxley, and Julian, 1966, *J. Physiol.* 184:185.)

shorten against the load, it generates much force, but velocity approaches zero. (You can illustrate this principle to yourself. Put a moderate load in your outstretched hand and contract your biceps as fast as you can to draw the load to you. Now do the same with nothing in your hand. Which is faster?) The shapes of the length-tension and force-velocity curves may vary from muscle to muscle, but the dependency on thick and thin filaments as the contractile mechanism dictates that all muscles follow these basic relationships. The challenge is to determine just where along these curves muscles operate in active animals.

Fiber Types Twitch contraction time depends on the rate at which the myosin cross bridges work. Two general physiological categories of fibers in vertebrate muscle are recognized: relatively slow-contracting fibers and relatively fast-contracting fibers. **Slow twitch fibers** (or slow oxidative fibers = SO) contract and fatigue the slowest and are of small diameter; the motor units to which they contribute generate relatively small forces. Having many mitochondria (sites of oxidative enzymes) and much myoglobin, they are reddish in color. They are economical for nearly isometric contractions that maintain posture and for slow isotonic contractions that maintain slow repetitive movements (as on the sides of slow-cruising fishes; see figure on p. 506). **Fast twitch, fatigable fibers** (or fast glycolytic fibers = FG, or white fibers) contract fast, fatigue fast, and are of large diameter; the motor units to which they contribute generate relatively large forces. Having few mitochondria, they are light in color. They are recruited for bursts of fast activity (like the breast muscles of quail and pheasants). **Fast twitch, fatigue-resistant fibers** (or fast oxidative glycolytic fibers = FOG) contract moderately fast and fatigue slowly. (FG and FOG fibers are illustrated in Figure 10.5.) They contain many mitochondria and vessels, so they resemble SO fibers in being reddish. They store more oxygen and lipids, but less glycogen, than other types of fibers and have relatively small diameters. FOG fibers are prominent in muscles capable of strong, repetitive movements, as in the flight muscles of ducks and other migratory birds. Most muscles have all types of twitch fibers in combination, and fibers intermediate between FG and FOG may be identified. All fibers in one motor unit are of the same type. The relative proportions of the various fiber types differ by animal and by muscle. In addition, an individual muscle may be compartmentalized

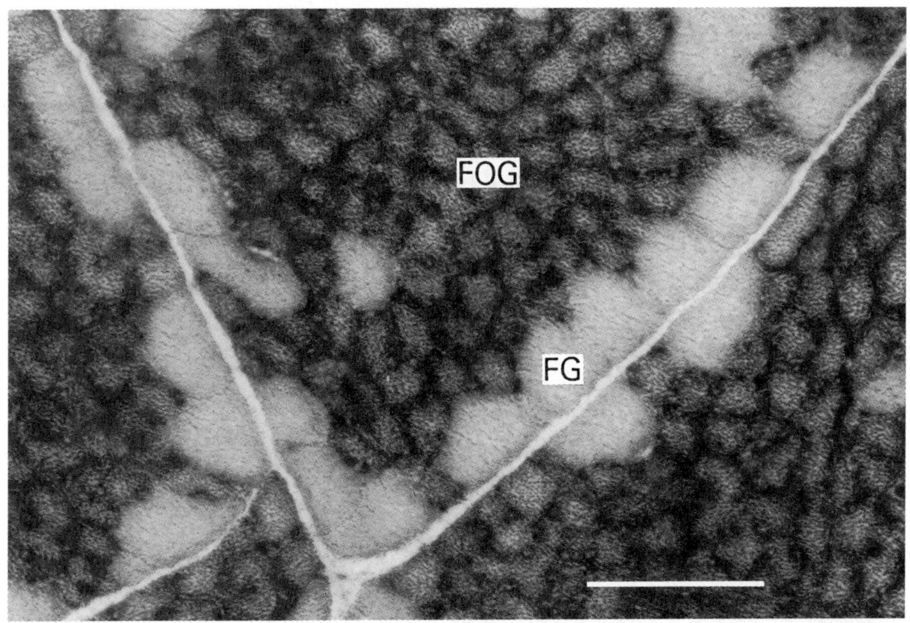

FIGURE 10.5 TYPES OF FIBERS IN THE PECTORALIS MUSCLE OF A PIGEON. FG = anaerobic, fast glycolytic fibers; FOG = aerobic, fast oxidative glycolytic fibers; scale bar = 100 μm.

into subvolumes of one type of fiber or another in relatively high density. Such an organization may reflect a functional partitioning of the muscle into individual compartments that contract independently during locomotor tasks, but this has been demonstrated only in a few cases.

Absolute twitch times vary with body size (faster in smaller animals) and locomotor adaptation. A muscle or group of muscles may be all slow twitch or all fast twitch, but the limb muscles of most terrestrial quadrupeds are a mixture. Since a twitch is measured isometrically, contraction time is not the same as V_{max}, but the two are correlated.

Various histochemical techniques are used to identify fiber types. This requires making thin sections of muscle and incubating with an appropriate substrate (Figure 10.5). Staining for adenosine triphosphatase (ATPase) activity, a glycolytic enzyme, and an oxidative enzyme (alternately in serial sections) is a quick way of establishing approximate fiber type composition.

Force (or Tension) of Muscular Contraction This is a property of great importance in the analysis of bone-muscle systems. Our intent may be to compare musculoskeletal adaptations for respiration, locomotion, or feeding, or we may have interest in constructing testable models of muscle function where estimates of force are required. How is muscle force determined? Technically it is virtually impossible to measure accurately all muscle forces in an active animal. Hence, morphologists construct best estimates. Every technique devised to date to measure muscle force has its limitations.

Some investigators measure cross-sectional area directly. The premise is that the maximum force a muscle can exert is equal to the force of contraction of one of its fibers times the number of fibers, and therefore is proportional to the cross-sectional area of the muscle. Even if this is used only as an approximation, however, errors can occur. For example, the amount of noncontractile material surrounding the fibers (mitochondria, fat, and connective tissue) varies among fiber types and should not be included in the measurement. A single cross section through a pinnate muscle does not intersect fibers at right angles and may not include all fibers. Some muscles are so complex architecturally that they defy representation even by several cross-sectional measurements. Finally, as we have just seen, tension relates to muscle length (relative to resting length) and V/V_{max} at the time of stimulation. A more realistic measurement of cross-sectional area is one that factors in as many of these variables as possible: Cross-sectional area = (muscle mass) × (pinnation angle)/ (fiber length) × (density). Muscle mass may be wet or dry, fiber length is estimated by fascicle length, and density is estimated at 1.06 g/cm^3. When used with caution, this measure of cross-sectional area can provide valuable comparative estimates of force.

Electromyography and the Measurement of Strain Muscles are not all in a state of contraction all of the time. Even individual muscles that cross a single joint (each head of the triceps, for example) may be used independently to control the same joint in different activities. **Electromyography** is a technique employed to determine when a particular muscle, or group of muscles, is electrically active (producing force) or electrically silent (relaxed). By implanting fine wire electrodes into a muscle, one can detect the small voltage changes (measured in millivolts) caused by ionic movements across the membranes of active muscle fibers. Synchronization of the movements of the animal (film, video) and the record of electrical activity provides data relevant to the timing of muscle activity in vivo. For example, the electrical activity of the primary depressor and elevator muscles of the wing of a flying pigeon are seen to be reciprocal (Figure 10.6). We also observe that the pectoralis becomes active in late upstroke several milliseconds prior to the time the wing begins its downward movements—a point significant for force production. Patterns of electrical activity during cyclic movements also provide information about the organi-

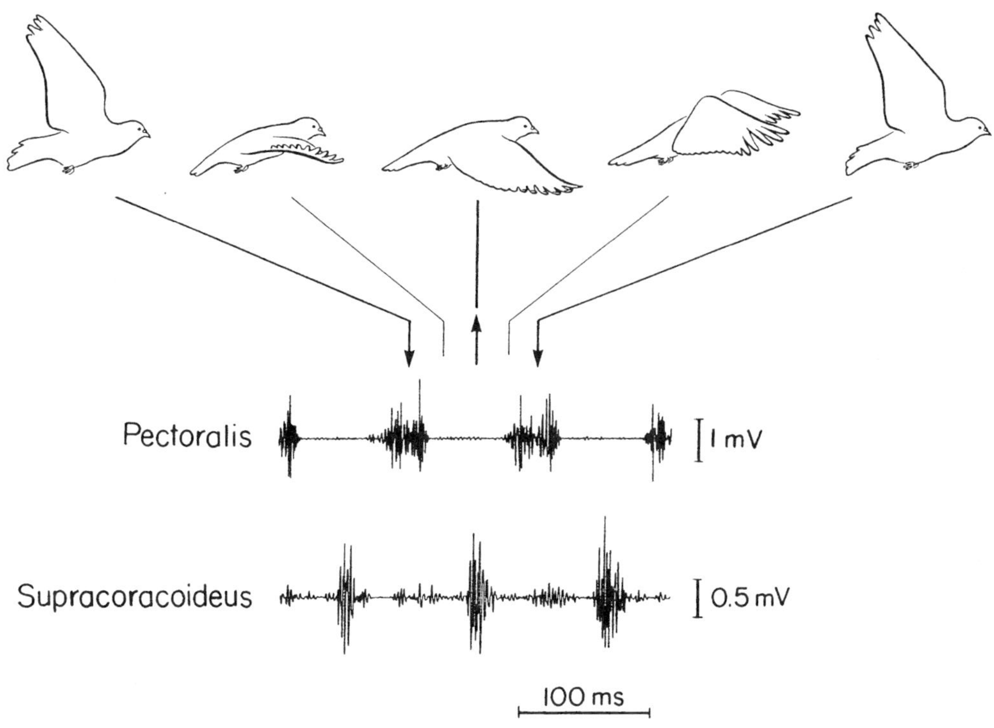

FIGURE 10.6 ELECTROMYOGRAPHIC RECORD OF MUSCLE ACTIVITY DURING FLAPPING FLIGHT OF THE PIGEON. The pectoralis muscle depresses the wing and the supracoracoideus elevates the wing. For further interpretation of this record in relation to flight see pp. 532 and 533. (From G. E. Goslow, Jr., K. P. Dial, and F. A. Jenkins, Jr., 1989, *Am. Zool.* 29:295.)

zation of the nervous system. Used in conjunction with information from dissection and fiber type, the technique is powerful. Thus, the pigeon pectoralis is composed of two types of fibers, easily distinguished by histochemical profile and size (Figure 10.5). The relatively small FOG fibers constitute 88–90% of the fiber population and have a mean diameter one-half that of the relatively large FG fibers. The hypothesis has emerged that the FG fibers are used for takeoff and landing (high forces needed) and that the FOG fibers are used for sustained flight (requiring relatively lower forces for long periods).

Electromyography provides information about the timing of force production, but little about absolute force. With the development of miniature strain gauges that can be implanted either directly on an individual tendon or on a bone adjacent to the insertion of a particular muscle, it is possible to get a good estimate of absolute force. For example, Biewener obtained a force profile of the pectoralis of a pigeon during flight by placing a strain gauge directly on a crest of the humerus and using the bone's deformation as the monitor.

SHORTENING, WORK, POWER, AND LEVERAGE

Force is the product of mass and acceleration. It is expressed in dynes (the push needed to cause 1 g to accelerate at 1 cm/s^2) or newtons (the push needed to cause 1 kg to accelerate at 1 m/s^2). However, because morphologists work with a uniform gravitational field, it is usually adequate to consider force to be a push or pull that causes motion, or must be resisted to prevent motion, and to express force in kilograms or pounds. One newton = 100,000 dynes = 100 g = 0.225 pounds weight on earth. A striated muscle can deliver to its tendon a force of about 3 kg/cm^2 (42 lb/in^2) of its cross-sectional area taken at right angles to its fibers, but the range is from 1 to 8 kg/cm^2, so remember that this is only a rough approximation.

Work is force times the distance through which it acts. When a muscle contracts isometrically, it does no work, although it does expend energy. When a muscle contracts isotonically, it does perform work. Sometimes, however, a muscle can be in a state of contraction and undergo simultaneous stretching (lengthening), in which case it also does work. The first type of work is termed positive work, the second negative work. **Mechanical power** is work per unit time. Muscles having long fibers with a parallel architecture can be expected to contract through a long distance at moderate force levels to generate high mechanical power. Pinnate muscles, in contract, which are capable of generating high forces (high number of fibers) but limited rates of shortening (few sarcomeres), might be expected to generate high forces but act over short distances.

The **lever arm** of a muscle is the perpendicular distance from its line of action to the pivot of the motion caused by its contraction. Thus, if a muscle inserts on a bony lever (e.g., the heel bone) at right angles, then the distance from the insertion to the pivot (the ankle joint in some circumstances) is the lever arm of the muscle. The turning force, or **torque** of the action, is then the force delivered to the insertion times the length of the lever arm. The leverages provided by the bone-muscle system are of the utmost importance in determining function. These and related concepts are developed further in Chapter 22.

CATEGORIES OF MUSCLES

It is both convenient and instructive to arrange the many muscles into groups for study. Several methods of grouping suggest themselves, and the student can use each with benefit. First, all muscles of one region of the body can be studied together. Thus, the muscles of the spine, forelimb, head and neck, and so on can be studied in turn. What muscles would be seen in a cross section of the thigh? What muscles would be cut in passing from the breast to the lung? What muscles have origin or insertion on the scapula? This approach is efficient at the operating table and dissection table, particularly with large animals.

A second method groups together muscles of like function. What muscles extend (protract) the forelimb? Or turn the head in a specified way? Or maintain standing posture? One discovers that a single muscle (e.g., pectoralis) can have several actions, and that some actions (e.g., swinging the thigh to the rear) can be accomplished alternatively by two or more muscles (though not with identical efficiency). This approach is practical for the functional morphologist, behaviorist, and physical therapist.

A third method is of significance for comparative studies. The various muscles can be arranged in major categories on the basis of embryonic origin. Such categories have somewhat independent phylogenies, relate to the positional and functional groups noted above, and can be distinguished by nervous innervation. Let us identify them by tracing their origins and innervations.

The mesoderm of the early embryo is differentiated into a dorsolateral, segmented epimere, a small mesomere, and a ventrolateral, unsegmented hypomere (see figure on p. 76 and table on p. 80). As has been explained in previous chapters, the epimere further differentiates into dermatome, myotome, and sclerotome. The sclerotome forms no muscles. The dermatome forms much of the skin, including any intrinsic smooth muscles that may be present there. The myotome and the hypomere are the sources of virtually all other muscles of the body.

As development proceeds, the myotomes behind the head and pharynx form much of the musculature of the body wall, or **axial muscles.** In most fishes, the axial muscles of each side of the body are clearly separated by a membranous partition, the **lateral septum,** into dorsal **epaxial muscles** and ventral **hypaxial muscles** (Figure 10.7 and figure on p. 150). On the trunk, but not on the tail, it appears (on the basis of experimental studies done chiefly on the chick) that in higher vertebrates the hypomere also contributes importantly

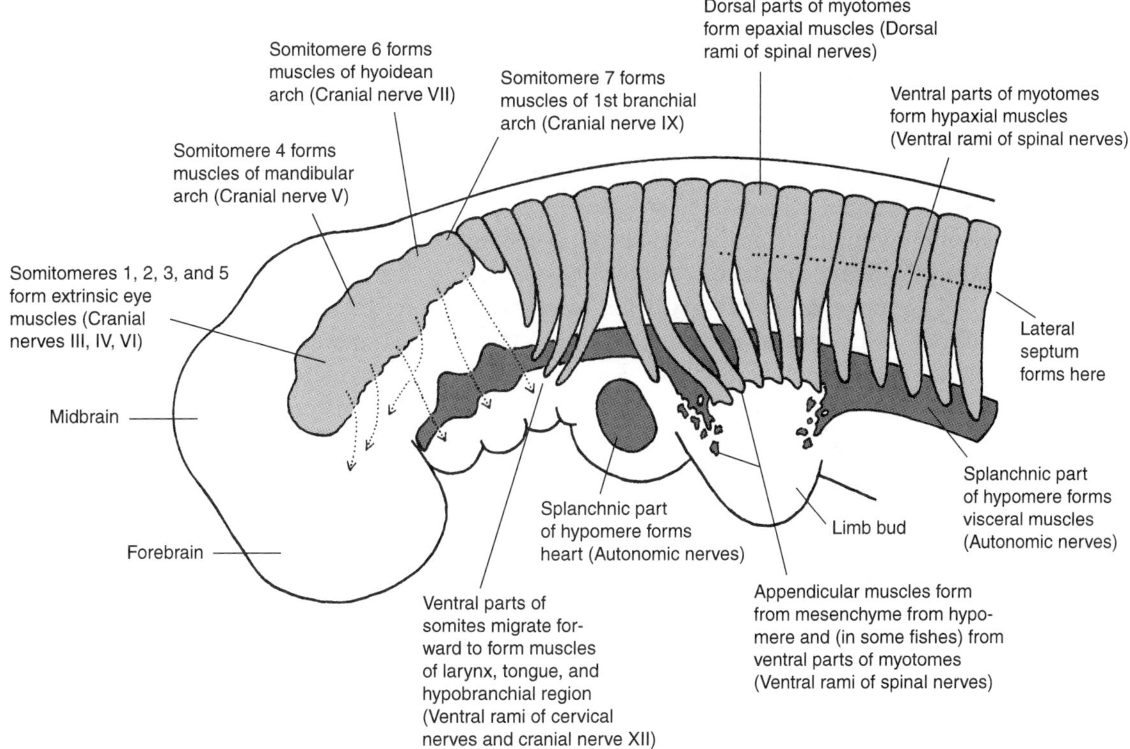
Dorsal parts of myotomes form epaxial muscles (Dorsal rami of spinal nerves)
Somitomere 6 forms muscles of hyoidean arch (Cranial nerve VII)
Somitomere 7 forms muscles of 1st branchial arch (Cranial nerve IX)
Ventral parts of myotomes form hypaxial muscles (Ventral rami of spinal nerves)
Somitomere 4 forms muscles of mandibular arch (Cranial nerve V)
Somitomeres 1, 2, 3, and 5 form extrinsic eye muscles (Cranial nerves III, IV, VI)
Lateral septum forms here
Midbrain
Splanchnic part of hypomere forms visceral muscles (Autonomic nerves)
Splanchnic part of hypomere forms heart (Autonomic nerves)
Limb bud
Forebrain
Appendicular muscles form from mesenchyme from hypomere and (in some fishes) from ventral parts of myotomes (Ventral rami of spinal nerves)
Ventral parts of somites migrate forward to form muscles of larynx, tongue, and hypobranchial region (Ventral rami of cervical nerves and cranial nerve XII)

FIGURE 10.7 STYLIZED EMBRYO SHOWING THE DIFFERENTIATION OF MUSCLE GROUPS FROM THE SOMITOMERES, MYOTOMES, AND HYPOMERE as identified on pp. 76 and 80. Ultimate innervation of the muscle groups is given in parentheses.

to the formation of hypaxial muscles. Epaxial muscles dorsoflex the spine and are innervated by dorsal rami of spinal nerves (see figure on p. 314). Hypaxial muscles ventroflex the spine and support the body wall. They are innervated by ventral rami of spinal nerves. When epaxial and hypaxial muscles of one side of the body contract together, the spine is flexed to that side.

The epaxial parts of several anterior somitomeres form the **extrinsic eye muscles.** (See again the discussion of head segmentation on p. 116.) The extrinsic eye muscles are innervated by the third, fourth, and sixth cranial nerves.

Most posterior somitomeres migrate as mesenchyme to form the **branchial muscles,** which are associated with the visceral arches and their derivatives (Figure 10.7). Anterior somites contribute muscles of the posterior branchial arches of fishes and to their minor derivatives in tetrapods. (It was formerly thought that these muscles are derived from the hypomere.) Branchial muscles are innervated by cranial nerves: The 5th nerve serves the important jaw muscles of the mandibular arch, the 7th nerve the hyoiden arch, and the 9th nerve the first branchial arch. The remaining arches are served principally by the 10th nerve, but the 11th cranial nerve and ventral rami of cervical nerves may also innervate branchial muscles.

Below the pharynx, from the pectoral girdle to the jaw, are muscles that were derived phylogenetically by the forward migration of hypaxial muscles originally located on the trunk. Because of their position they are called **hypobranchial muscles.** Their innervation by the 12th cranial nerve and by ventral rami of cervical nerves indicates their posterior origin.

There is evidence from the embryology of sharks that the muscles of the fins, or **appendicular muscles,** form as extensions from the hypaxial muscles of the body wall. This may be the primitive condition. In tetrapods, however, the appendicular muscles form in place from mesenchyme derived, at least in part, from the hypomere. Innervation is by ventral rami of spinal nerves.

The hypomere has important muscular derivatives other than the hypaxial and appendicular muscles already noted. At the level of the trunk it splits to enclose the coelomic cavity. The outer (somatic) layer forms no further muscle. The inner (splanchnic) layer forms the **heart** and the **muscles of the viscera.**

In summary, the major categories of muscles are the axial musculature (which has epaxial and hypaxial divisions), extrinsic muscles of the eye, hypobranchial muscles (which are derived from hypaxial muscles), appendicular musculature (which has dorsal and ventral divisions), muscles of the gut, and branchial muscles (which are serially related to the visceral skeleton).

EVOLUTION OF MUSCLES

Bases for Establishing Homologies In order to trace the evolution of individual muscles one must have criteria for recognizing homologous muscles in different taxa. Within orders it is usually possible, and within families it is nearly always possible, to recognize equivalent muscles on the basis of position and relationships with other muscles and with the skeleton. Thus, the supraspinatus muscle of mammals always occupies the supraspinous fossa of the scapula and inserts on the greater tubercle of the humerus. In some instances, however, the criterion of positional relationships fails: Reptiles do not have a supraspinous fossa, the attachments of a muscle may change with evolution to such an extent that the action of the muscle is materially altered, adjacent muscles of similar action may fuse; a muscle may disappear, and ancestral muscles (unlike bones of the skeleton) tend to split in the course of evolution to become several muscles.

Paleontological evidence of homology is sometimes provided by a series of fossil bones having muscle scars that evince the migration, fusion, or loss of a particular muscle. The progressive change of certain muscles in the feet of extinct horses has been learned in this way. However, the paleontologist must be careful to avoid making unjustified assumptions and is usually dependent on the muscles of surviving animals to guide analysis.

It was shown above that embryology is useful for establishing major categories of muscles. Since the pattern into which the initial muscle masses split tends to be less specialized than that of the adult, embryology is also useful for homologizing specific muscles. This approach has been applied to a variety of vertebrates and deserves further study.

The criterion of muscle homology that has received the most attention is nerve supply. In the last decade of the 19th century, a German anatomist postulated an invariable relationship between peripheral nerves and the muscles they innervated. Various authors have now studied nerve-muscle relationships in detail (notable among them are Howell, Romer, and Haines), and it is agreed that nerve supply is an important criterion of homology but that instances are known of muscles that have evolved nerve relationships differing from those of their evolutionary precursors.

The comparative myologist is well advised to use as many criteria of homology as are available.

Muscles of Primary Swimmers The muscular system of CYCLOSTOMES (particularly of lampreys) is more simple and more primitive than that of other vertebrates. A lateral septum is lacking, so the prominent axial musculature is not divided into epaxial and hypaxial divisions. The segmentation of the body is clearly evident: Each myotome con-

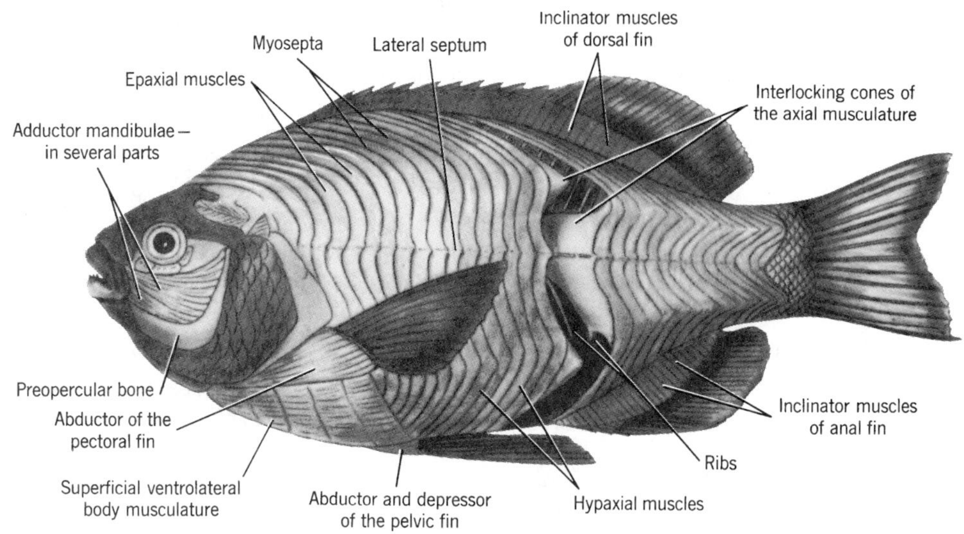

FIGURE 10.8
MUSCULATURE OF A TELEOST with two myomeres removed to show the shape of the myosepta.

tributes one muscle segment, or **myomere.** An axial skeleton other than the notochord being absent, the short fibers of the myomeres insert on partitions of connective tissue, the **myosepta,** which lie between successive myomeres. Myomeres and septa are thrown into gentle folds that are scarcely more complicated than those of amphioxus. The ventral portions of those myomeres lying close behind the pharynx turn somewhat forward, foreshadowing the hypobranchial musculature. Appendicular muscles are, of course, absent, and, since jaws are lacking and the visceral skeleton is constructed in one unit, related branchial muscles are not prominent. There is an elaborate musculature associated with the specialized mouth and tongue, but it is dissimilar in lampreys and hagfishes and cannot be homologized with muscles of higher vertebrates.

The musculature of JAWED FISHES is more advanced yet remains less complex than that of tetrapods. Strong axial muscles, which flex the spine and tail from side to side in swimming, are divided into epaxial and hypaxial portions by a lateral septum (see figure on p. 150). Dorsal ribs, if present, lie in this septum. The myomeres, although straight in the embryo, become more angled than in cyclostomes and are molded into interlocking cones (Figure 10.8 and the figure on p. 506). This arrangement directly extends the action of each myomere over several vertebrae and ensures that muscle fibers at different distances from the body axis can all shorten at about equal rates, and over nearly equal distances, in flexing the body of the fish. Tendons extending from the apices of the cones may distribute the force of contraction over additional body segments, particularly in the tails of fast-swimming fishes.

Straplike hypobranchial muscles extend from the pectoral girdle to the visceral arches and serve to open the jaws and pull the gills down and backward. The hypobranchial muscles have become distinct from the hypaxial muscles from which they evolved, but they retain the longitudinal orientation imposed by their forward migration.

The girdles of fishes lie firmly anchored within the axial musculature. Appendicular muscles have evolved with the fins and are divided into a dorsal mass of extensors (or abductors, or levators—all these terms are used) that move the fins upward or forward, and a ventral mass of flexors (adductors, depressors) that move them downward or backward.

The pharyngeal morphology of some sharks suggests that ancestral fishes, having homogeneous visceral arches, had simple and serial branchial muscles (Figure 10.9). A superficial sheet of **constrictors** was nearly continuous over the gill area and compressed

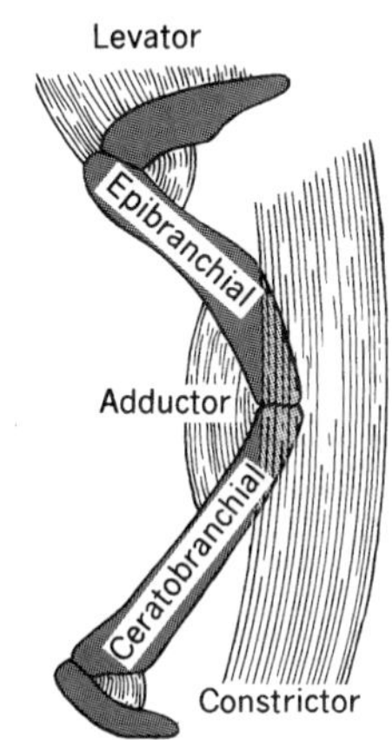

FIGURE 10.9
STYLIZED BRANCHIAL MUSCULATURE OF ONE GILL BAR OF A PRIMITIVE FISH. Compare with the figure on p. 119.

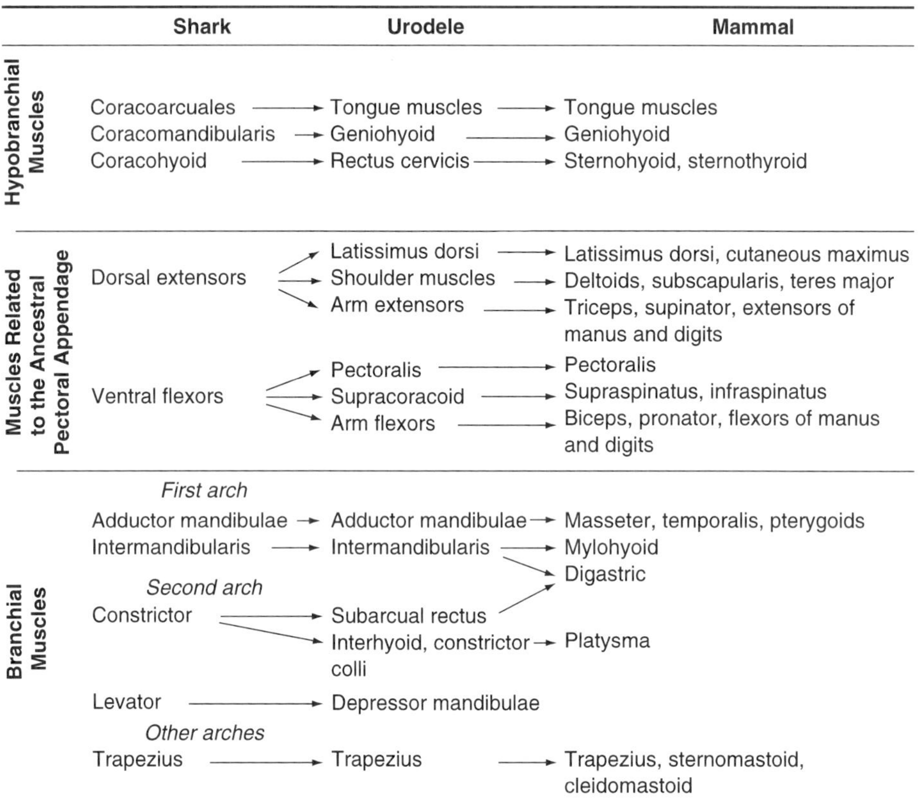

	Shark	Urodele	Mammal
Hypobranchial Muscles	Coracoarcuales →	Tongue muscles →	Tongue muscles
	Coracomandibularis →	Geniohyoid →	Geniohyoid
	Coracohyoid →	Rectus cervicis →	Sternohyoid, sternothyroid
Muscles Related to the Ancestral Pectoral Appendage	Dorsal extensors →	Latissimus dorsi →	Latissimus dorsi, cutaneous maximus
		Shoulder muscles →	Deltoids, subscapularis, teres major
		Arm extensors →	Triceps, supinator, extensors of manus and digits
	Ventral flexors →	Pectoralis →	Pectoralis
		Supracoracoid →	Supraspinatus, infraspinatus
		Arm flexors →	Biceps, pronator, flexors of manus and digits
Branchial Muscles	*First arch*		
	Adductor mandibulae →	Adductor mandibulae →	Masseter, temporalis, pterygoids
	Intermandibularis →	Intermandibularis →	Mylohyoid; Digastric
	Second arch		
	Constrictor →	Subarcual rectus →	Digastric
		Interhyoid, constrictor colli →	Platysma
	Levator →	Depressor mandibulae	
	Other arches		
	Trapezius →	Trapezius →	Trapezius, sternomastoid, cleidomastoid

TABLE 10.1 THE EVOLUTION OF SOME PRINCIPAL MUSCLES AS FOUND IN ANIMALS THAT EXEMPLIFY THREE STAGES.

the pharynx. A series of **levators** above the pharynx served to lift the gill bars. **Adductors** reduced the internal angles of each visceral arch. The regularity of this ancestral pattern is much altered among the various surviving fishes according to the type of jaw suspension, feeding mechanism, and presence or absence of spiracle and operculum. Muscles of the first two arches are the most specialized. The middle adductor of the mandibular arch is much enlarged to become the **adductor mandibulae,** which closes the jaws (Figure 10.10 and Table 10.1). The ventral constrictors of the mandibular and hyoid arches form the sheetlike **intermandibularis** muscle, which lies between the mandibles and raises the floor of the mouth. Muscles of the branchial arches are relatively unspecialized in Chondrichthyes. The levators, however, tend to mass over the more posterior gills as the cucullaris, or **trapezius** muscle, which is retained by tetrapods (Figure 10.11). Branchial muscles of the gills of Osteichthyes are usually reduced to remnants of the ventral constrictors.

Fishes have six extrinsic eye muscles (Figure 10.12). Four **rectus muscles** have their origins close together deep in the posterior part of the orbit. These rotate the eye around the longitudinal and vertical axes of the head. Two **oblique muscles** have their origins deep in the forward part of the orbit. They rotate the eye around its optical axis (the transverse axis of the head). Four of the muscles (anterior, superior, inferior recti, and inferior oblique) are derived from the premandibular somitomere and are innervated by the third cranial nerve (Figure 10.7). The superior oblique is derived from the mandibular somitomere and is innervated by the fourth cranial nerve. The posterior rectus is derived from both mandibular and hyoidean somitomeres and is innervated by the sixth cranial nerve.

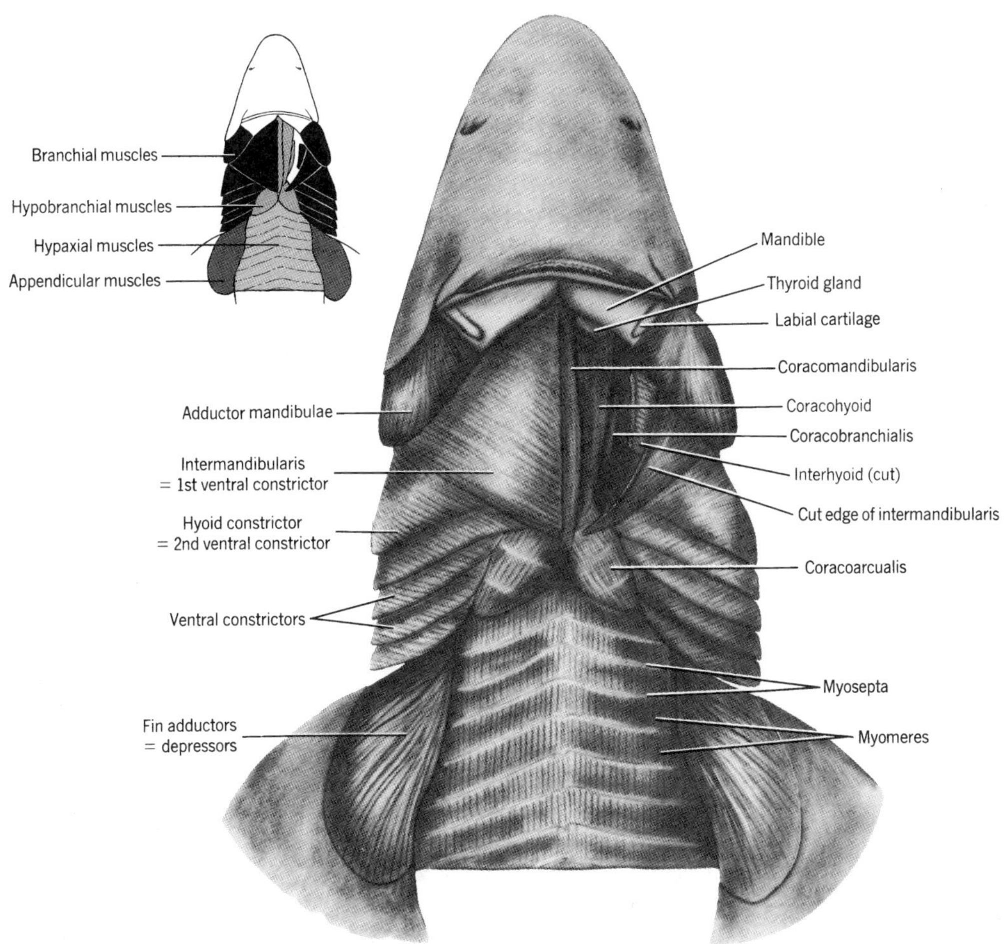

FIGURE 10.10 ANTERIOR VENTRAL MUSCULATURE OF AN ELASMOBRANCH shown by the shark, *Squalus*.

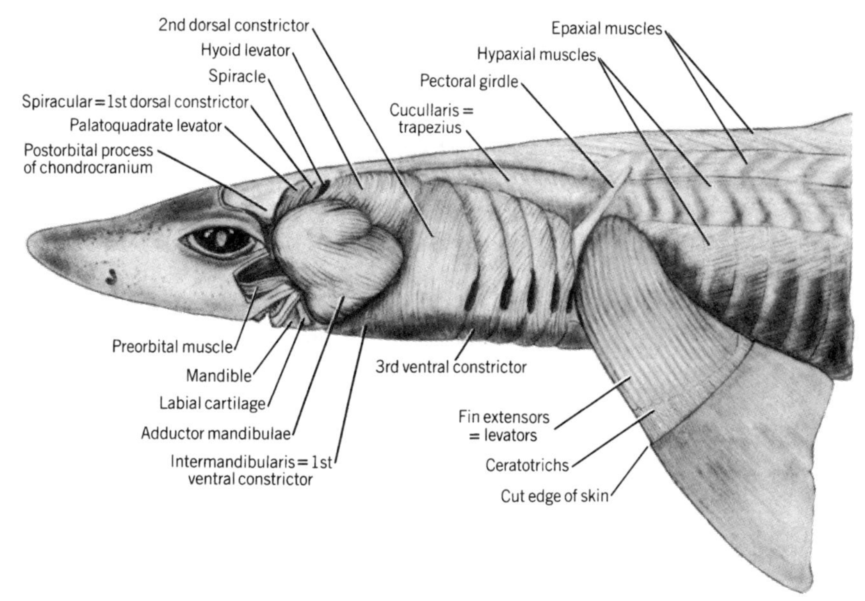

FIGURE 10.11 ANTERIOR LATERAL MUSCULATURE OF AN ELASMOBRANCH shown by the shark, *Squalus*.

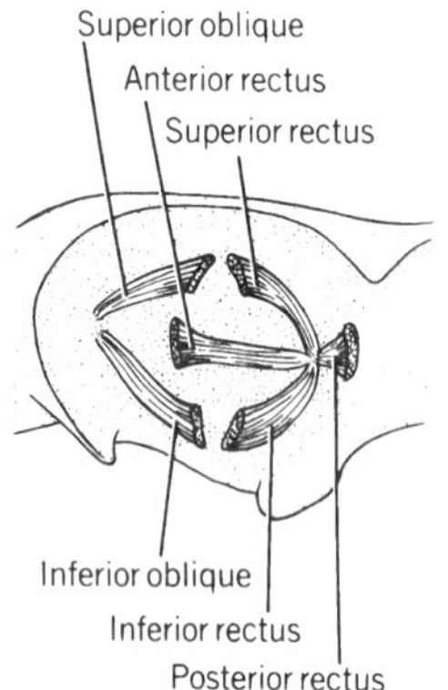

FIGURE 10.12 EXTRINSIC EYE MUSCLES OF VERTEBRATES exemplified by a shark. Left orbit.

Axial and Hypobranchial Muscles of Tetrapods Several general trends are evident in the evolution of the axial musculature of tetrapods. In fishes, these muscles, being the propulsive muscles, are the most massive of the body. As limbs take over the propulsive role, their muscles enlarge and the axial musculature diminishes. The axial skeleton of tetrapods, in contrast, becomes firmer in order to play a new supportive role and, concomitant with this trend, the remaining axial musculature becomes more intimately related to the skeleton and adds to its functions dorsoflection and ventroflection of the spine, which are rarely marked in fishes. Myosepta regress and disappear, and many muscles develop long fibers that span from two to many vertebrae. Furthermore, certain muscles tend to form sheetlike layers, and others become associated with the pectoral girdle.

The EPAXIAL MUSCLES of amphibians are conservative; myosepta are still present and are nearly vertical instead of angled, as in fishes (Figures 10.13 and 10.14). Epaxial muscles of reptiles and mammals, however, lack myosepta and have become exceedingly complex and varied in detail. Those of the cervical region tend to form layers on the now more flexible neck (Figures 10.15 and 10.16). The trunk of birds is short and relatively rigid as an adaptation to flight; consequently, axial musculature is much reduced except on the neck and short tail.

On the trunk the HYPAXIAL MUSCLES are similar in all tetrapods and are advanced over those of fishes. They are commonly classified in three groups: A subvertebral group, located below the transverse processes of the vertebrae, ventroflexes the spine. In reptiles

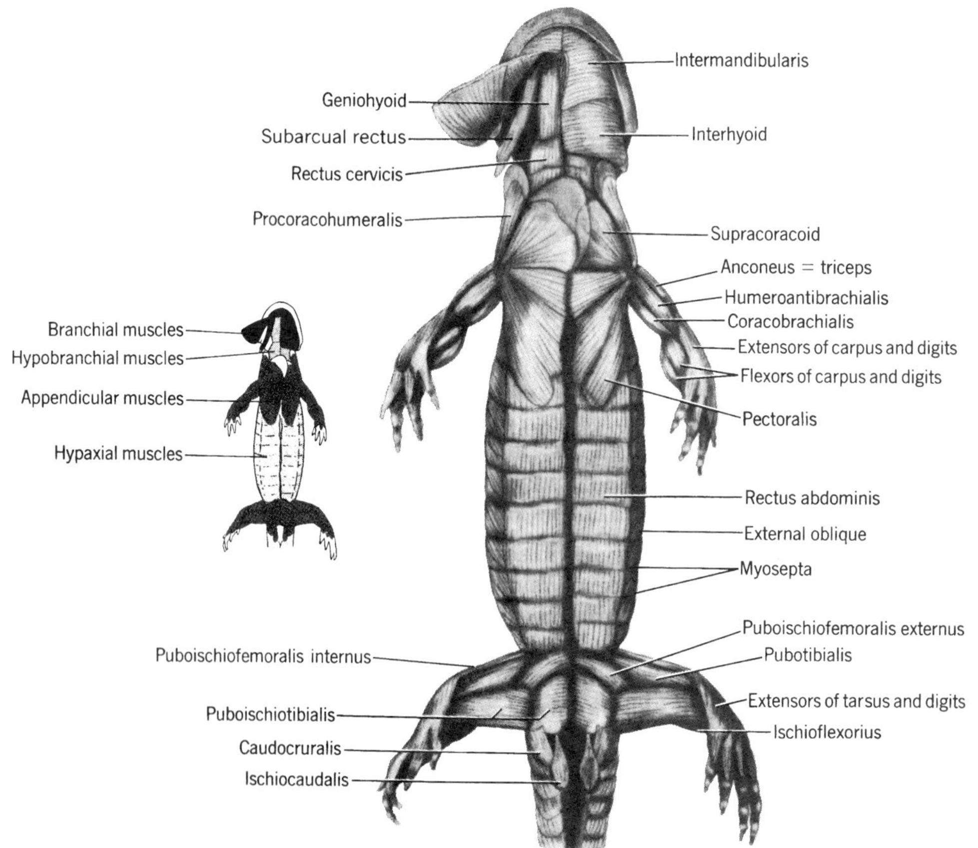

FIGURE 10.13 VENTRAL MUSCULATURE OF A URODELE shown by the tiger salamander, *Ambystoma*.

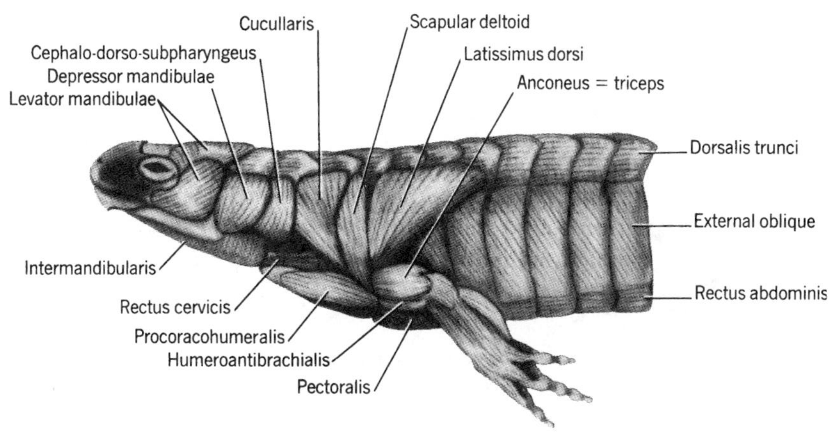

FIGURE 10.14 ANTERIOR LATERAL MUSCULATURE OF A URODELE shown by the tiger salamander, *Ambystoma*.

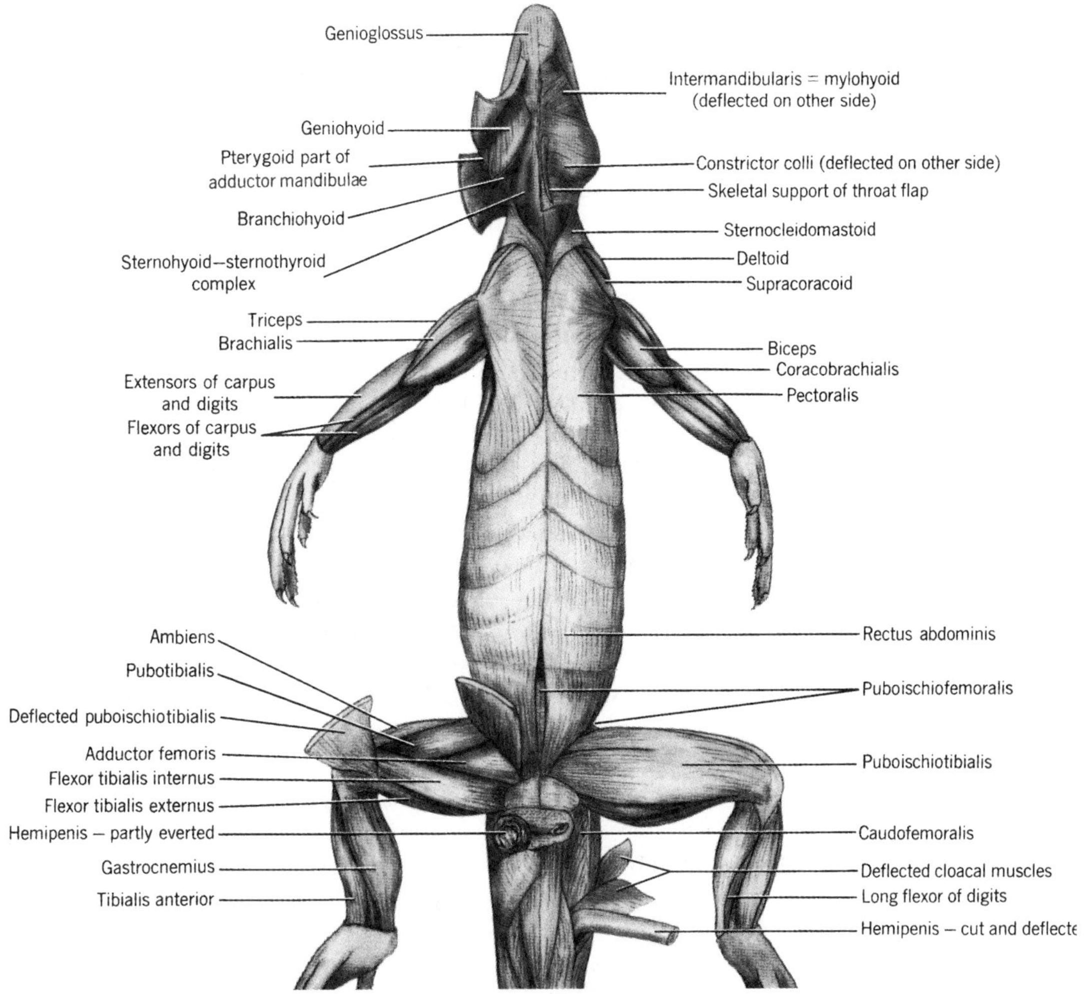

FIGURE 10.15 VENTRAL MUSCULATURE OF A SQUAMATE REPTILE shown by the iguanid, *Iguana*.

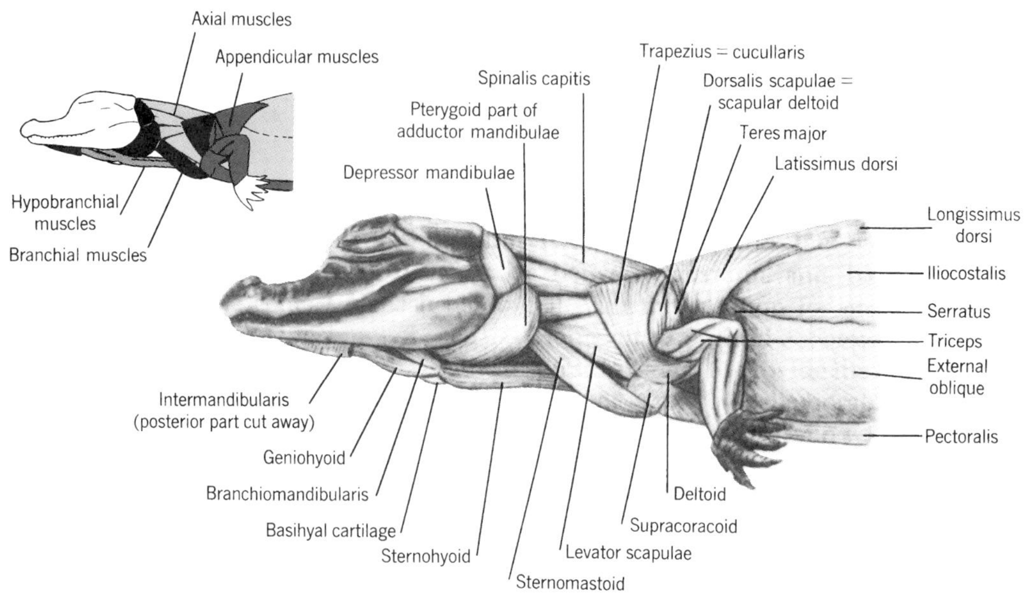

FIGURE 10.16 ANTERIOR LATERAL MUSCULATURE OF A REPTILE shown by the crocodilian, *Caiman*.

and mammals it is restricted to the lumbar area. The **rectus abdominis** muscle (or group) runs lengthwise along the ventral body wall between the two girdles. It supports the viscera and ventroflexes the body. Finally, a lateral group is located on the flanks. It breaks into (usually) three sheetlike layers, each having its fibers oriented in a different direction. Together they support and compress the body wall. The layers, in order from outside in, are the **external oblique** muscle, the **internal oblique,** and **transversus.** Anteriorly, the ribs of amniotes, enlarged over those of amphibians, penetrate the external and internal obliques, which there become the external and internal **intercostal muscles.** These contribute to the ventilation of the lungs.

The pectoral girdle of tetrapods no longer articulates with the head (as in fishes) and does not establish articulation with the spine (as does the pelvic girdle). Accordingly, several muscles evolve from the lateral group of hypaxial muscles to hold the pectoral girdle to the trunk. These include the **serratus,** which, in amniotes, runs from the ribs to the scapula to suspend the thorax, sling fashion, from the girdle; the **levator scapulae;** and the **rhomboideus.** The muscular **diaphragm,** found only in mammals, is apparently also of hypaxial origin. Its nerve, the phrenic nerve, branches from ventral rami of cervical nerves because the embryonic diaphragm originates anterior to the adult position.

The terminology and phylogeny of the principal HYPOBRANCHIAL MUSCLES identified in student laboratories are relatively straightforward. They are shown in Table 10.1.

Appendicular Muscles of Tetrapods Muscles of the PECTORAL LIMB of tetrapods have three general sources. First, one or several trapezius muscles are contributed by the branchial musculature; these are innervated by cranial or cervical nerves. Second, as noted under the previous subheading, several muscles are contributed by the axial musculature; these are innervated by ventral rami of spinal nerves that do not join the network of nerves at the base of the limb called the **brachial plexus.** Finally, most appendicular muscles of tetrapods are derived directly from appendicular muscles of fishes; these are also innervated by ventral rami of spinal nerves, but these nerves each join the plexus before entering the appendage.

When the appendicular nerves of fishes emerge from their plexuses, they tend to be arranged in a more dorsal group that runs to the dorsal mass of fin extensors and a ventral group that runs to the ventral mass of fin flexors. The appendicular muscles of adult tetrapods are numerous and complex; yet in the embryo they differentiate from dorsal and ventral masses in recapitulation of the ancestral piscine condition, and in the adult the many individual muscles usually can be identified as derivatives of the dorsal or ventral mass by their relationship to nerves emerging from the dorsal or ventral part of the respective plexus. In some instances, however, the ancestral functions of extension (for dorsal mass derivatives) and flexion (for ventral mass derivatives) become reversed.

Homologies of the muscles of the pectoral limb are well established and are outlined in simplified form in Table 10.1. It is desirable to study the table in conjunction with the illustrations in this chapter (and on p. 165), or in your laboratory manual. Note that in mammals the olecranon process of the ulna is the lever arm of the **triceps** muscle. It is useful for functional analysis to note that the **supinator** of the arm and **extensors** of manus and digits take their origin from the lateral epicondyle of the humerus, whereas the **pronator** and **flexors** arise from the medial epicondyle. The powerful **pectoralis** is the largest flight muscle of flying vertebrates. The **supracoracoid** muscle of the lower classes is importantly altered in birds and mammals. In birds it shifts to the sternum, under the pectoralis, and inserts over a bony pulley onto the upper surface of the head of the humerus, thus serving to elevate the wing (Figures 10.17 and 10.18). In mammals

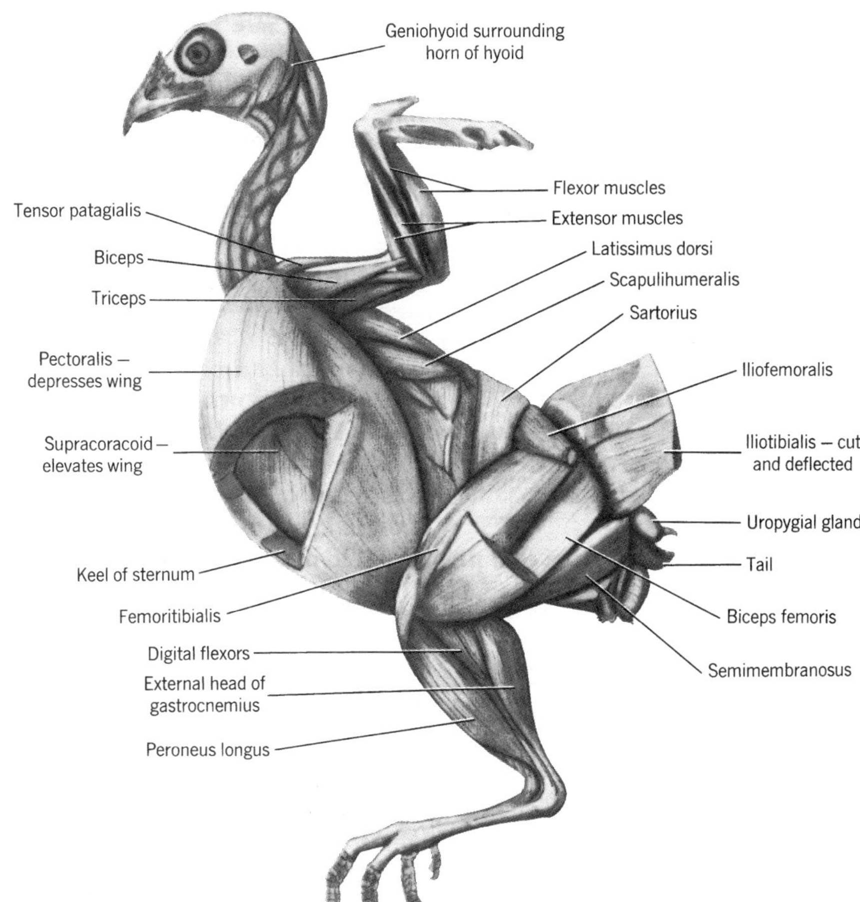

FIGURE 10.17 LATERAL MUSCULATURE OF A BIRD shown by the Japanese quail, *Coturnix.*

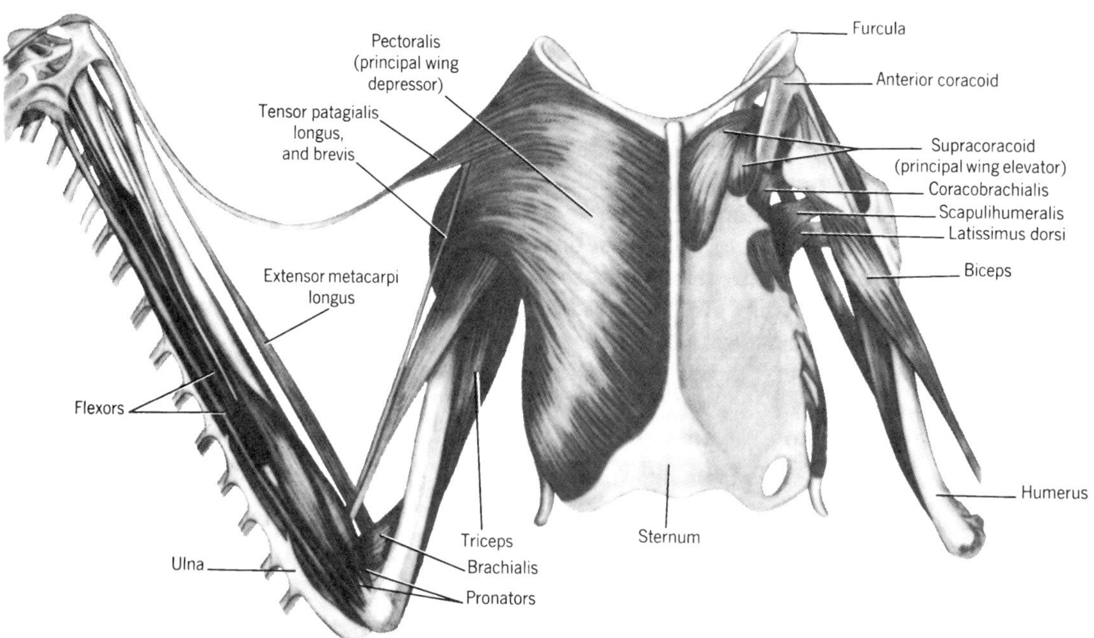

FIGURE 10.18 SOME FLIGHT MUSCLES OF A SOARING BIRD, the golden eagle, *Aquila,* seen in ventral view with the pectoralis removed on the left side. (Drawn from an air-dried dissection and hence somewhat shrunken.)

the insertion on the humerus is retained, but as the coracoid bone regresses, the origin of the muscle shifts to the scapula. The embryonic muscle grows out on each side of the spine of the scapula to become both the **supraspinatus** and the **infraspinatus.**

With minor exceptions, the muscles of the PELVIC LIMB of tetrapods are all derived from appendicular muscles of piscine ancestors. Dorsal and ventral group muscles are again recognized. Homologies between muscles of reptiles (particularly Lepidosauria) and those of mammals are relatively satisfactory; those for Lissamphibia and birds are more provisional than the commonly adopted terminologies indicate. Mammalian muscles derived from the dorsal group include the various **gluteal muscles,** the large quadratus femoris, which is made up of the **rectus femoris** and **vasti muscles,** the **sartorius, iliopsoas,** and **extensors** of the digits (Figures 10.19 and 10.20).

Mammalian derivatives of the ventral fin musculature include the **femoral adductors, semimembranosis, semitendinosus, gracilis, biceps femoris,** and **flexors** of the pes and digits. The **caudofemoralis** is an important flexor of the thigh in reptiles, but with reduction of the tail, whence this muscle has its origin, is much reduced in mammals. In mammals the strong **gastrocnemius** inserts on the newly evolved heel bone (calcaneum).

Branchial Muscles of Tetrapods The terminology and phylogeny of the most commonly identified branchial muscles are shown in Table 10.1. Note that the ancestral adductor mandibulae, already a complex muscle in lower tetrapods, becomes several muscles in mammals, the variations of which are closely related to feeding habits. The principal muscle of the second arch of all tetrapods except mammals is the **depressor mandibulae,** which supplements or replaces hypobranchial muscles as the opener of the jaw. In mammals this muscle is lost and the mouth is opened by a new muscle, the **digastric,** which is derived from the ventral constrictors of both the first and second arches. Accordingly, it is innervated by both

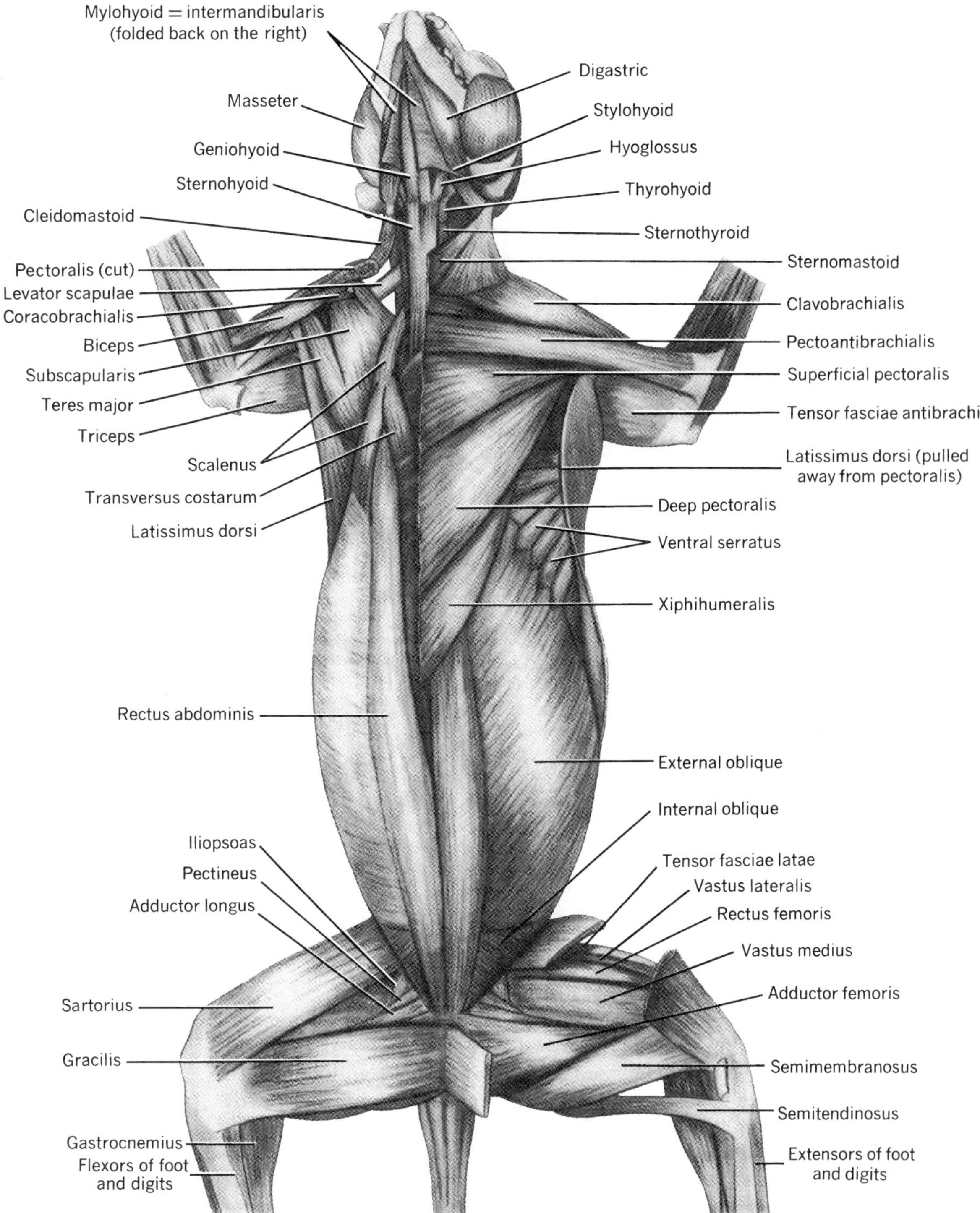

FIGURE 10.19 VENTRAL MUSCULATURE OF A MAMMAL, as seen in the cat. Sternomastoid, pectoralis complex, and tensor fasciae antibrachii removed on the right. See also pp. 165 and 166.

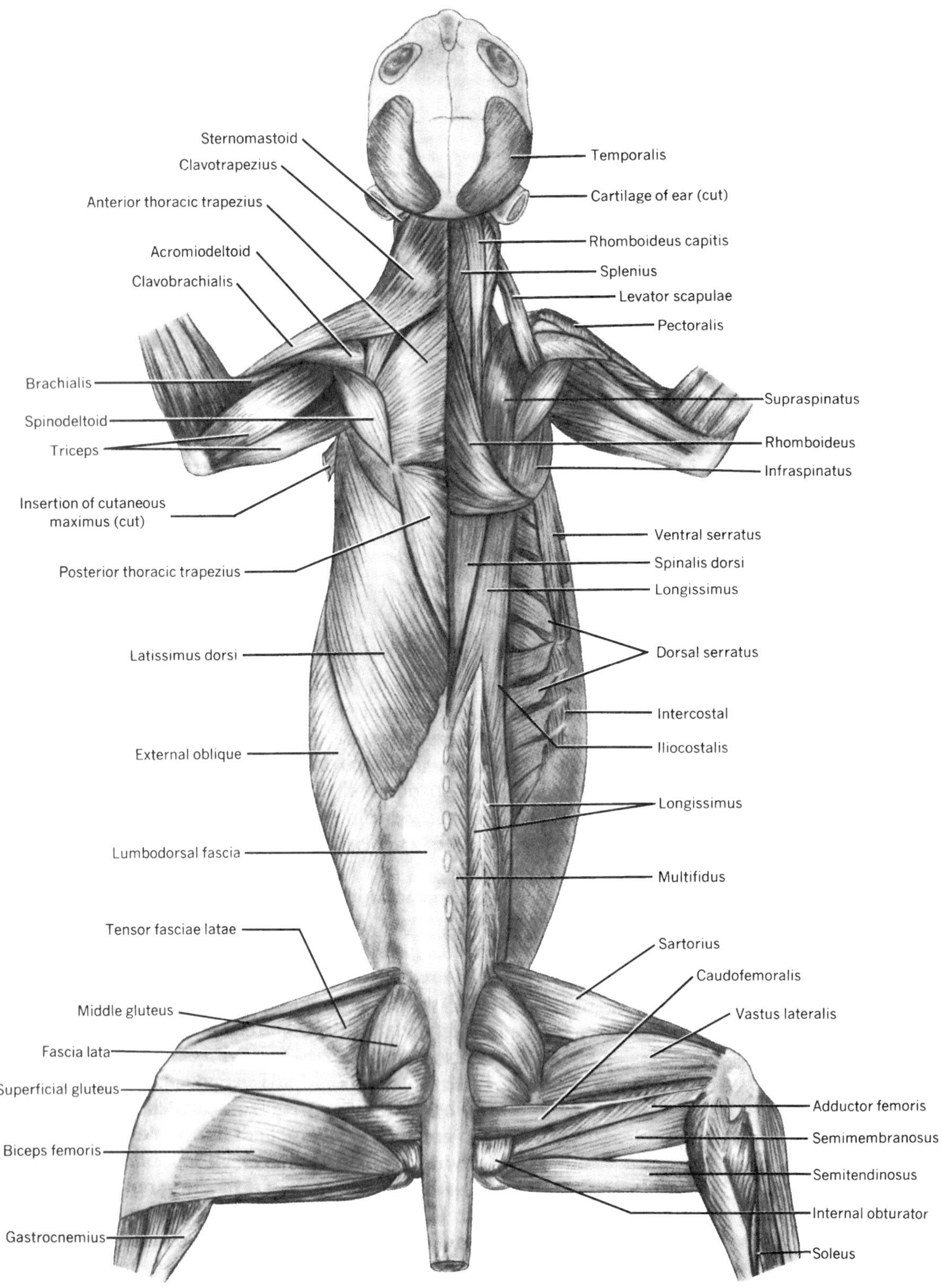

FIGURE 10.20 DORSAL MUSCULATURE OF A MAMMAL as seen in the cat. Trapezius muscles, clavobrachialis, latissimus dorsi, lumbodorsal fascia, tensor fasciae latae, and biceps femoris removed on the right. See also pp. 165 and 166.

the fifth and seventh cranial nerves. Another second arch muscle of interest is the **stapedial muscle.** This tiny muscle controls the motion of the stapes—a second arch bone. The muscles of the larynx and various constrictors of the throat are also branchial muscles.

Extrinsic Muscles of Skin and Eye in Tetrapods Muscles that run from underlying tissues to insert on and move the skin are not found in fishes or amphibians and are rare in reptiles. Snakes are a notable exception; their locomotor apparatus may include separate muscles to move each ventral scute. Birds have a muscle to tense the skin on the leading edge of the wing. Extrinsic skin muscles are most characteristic of mammals. A derivative of the second arch constrictor is the **constrictor colli,** a thin superficial muscle over the ventral and lateral parts of the neck (Figure 10.15). In mammals this muscle becomes a complex of facial muscles collectively known as the **platysma.** Facial muscles reach their highest development in human beings. The ancestral innervation by the seventh cranial nerve is retained.

A second muscle of the skin, the **cutaneous maximus,** is derived from the latissimus dorsi and pectoralis. Although vestigial in humans, this is often an extensive muscle over the trunk where it may serve to curl the body (echidna) or become subdivided for flicking insects from the skin (horse).

The six extrinsic eye muscles of fishes are retained in tetrapods with remarkably little variation (Figure 10.12). However, the eyeball usually can no longer be rotated around its optical axis, and one or more additional muscles evolve by splitting from one of the preexisting muscles. From the posterior rectus develops a **retractor bulbi,** of from one to four parts, which pulls the eyeball deeper into its socket. This action is protective and may also aid in swallowing. The muscle is marked in amphibians and some reptiles, but is lacking in many mammals.

ELECTRIC ORGANS

Electric organs are found in some 500 species of fishes belonging in seven families of Chondrichthyes and Osteichthyes. It is possible that richly innervated areas on the head shields of cephalaspids were also electric organs, though this interpretation is rejected by many paleontologists. The organs may be on the tail (electric skate, some teleosts), on the fins (electric ray), behind the eye (stargazer), or over much of the trunk (electric eel). They are usually derived from muscle cells (hence their inclusion in this chapter), but their origin from glandular and nervous tissue is not ruled out in every instance. Diversity of occurrence, location, structure, and also physiology indicate that electric organs are ancient specializations that evolved independently several times and have undergone convergence.

Many fishes are only weakly electric. The electric ray, however, can develop 50 A (the ampere is a measure of the amount of current delivered) and the electric eel can produce more than 500 V (the volt is a measure of the driving force of the current). Shocks of 2000 W (the watt, a measure of power, is the product of current flow and force) have been recorded.

Communication, orientation, and the detection of prey are the most common functions of electric organs—particularly for fishes living in murky water. One Amazonian fish can distinguish its territorial neighbors from strangers by subtle differences in the pulsed electric charges each emits. The organs of some species serve also for offense or defense; even large fishes can be electrocuted by the more powerful discharges. Electric fishes emit constant discharges (many at a high rate for their entire lives) and are highly sensitive to the disturbances that objects produce in the electric fields near their bodies. The sense organs

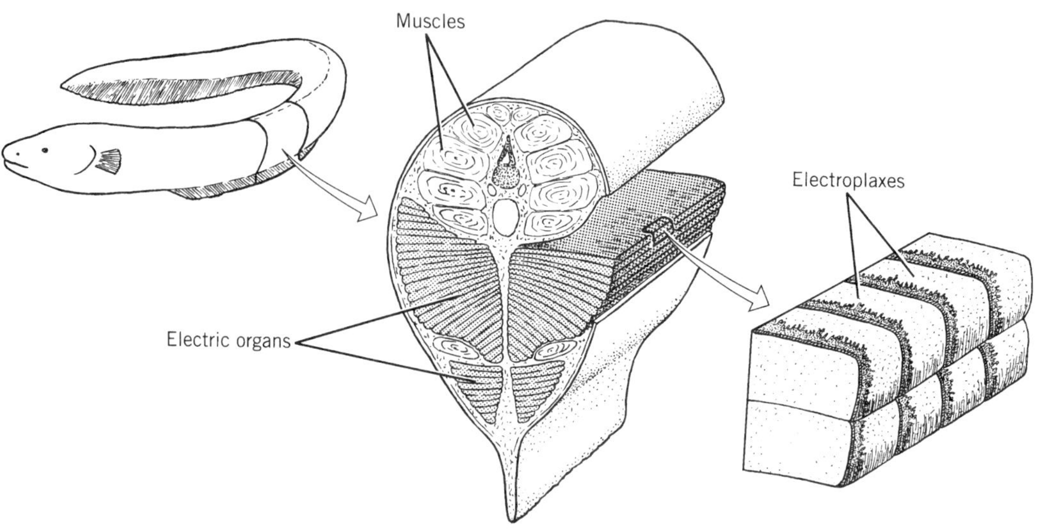

FIGURE 10.21 ELECTRIC ORGAN OF THE ELECTRIC EEL, *Electrophorus*.

that monitor the electric field are derived from the lateral line system and are located at the bases of pits in the skin (see p. 351).

The functional unit of an electric organ is the **electroplax,** a large, multinucleated cell (Figure 10.21). Usually one flat surface is minutely folded; mitochondria concentrate under this membrane. The other flat surface is richly innervated. Hundreds or thousands of electroplaxes are stacked to form a column, and many columns are commonly present in one organ. In the resting state, an electric potential develops between the inside (negative) and outside of each electroplax. When the organ is fired by its nerve, the potentials are momentarily reversed (at least in some species) so that the current exceeds the resting potentials. The organ may either be "wired" largely in series (+pole of one cell to – pole of adjacent cell, which gives maximum voltage, as is desirable for freshwater species) or largely in parallel (+ pole to + pole, which gives maximum amperage).

REFERENCES

Bass, A.H. 1986. Electric organs revisited: evolution of a vertebrate communication and orientation organ, pp. 13–70. *In* T.H. Bullock and W. Heiligenberg (eds.), *Electroreception.* Wiley, New York.

Bertram, J.E.A., and R.L. Marsh. 1998. Introduction to the symposium: muscle properties and organismal function: shifting paradigms. *Am. Zool.* 38:697–702. Entire symposium devoted to defining the next generation of muscle studies directed at improving our understanding of muscle design and function.

Biewener, A.A. 1998. Muscle function *in vivo:* a comparison of muscles used for elastic energy savings versus muscles used to generate mechanical power. *Am. Zool.* 38:703–717.

Cheng, C.-C. 1955. The development of the shoulder region of the opossum *Didelphis virginiana,* with special reference to the musculature. *J. Morphol.* 97:415–472. Illustrates use of embryology for establishing homologies of muscles.

English, A.W. 1985. Limbs vs. jaws: can they be compared? *Am. Zool.* 25:351–363. Thoughtful discussion of the properties of muscle important for comparative studies of muscle function.

Gans, C. 1982. Fiber architecture and muscle function. *Exercise Sport Sci. Rev.* 10:160–207.

Gordon, A.M., A.F. Huxley, and F.J. Julian. 1966. The variation in isometric tension with sarcomere length in vertebrate muscle fibers. *J. Physiol.* 184:170–192.

Goslow, G.E., Jr., K.P. Dial, and F.A. Jenkins, Jr. 1989. The avian shoulder: an experimental approach. *Am. Zool.* 29:287–301.

Howell, A.B. 1937. Morphogenesis of the shoulder architecture: part VI, therian mammalia. *Quart. Rev. Biol.* 12:440–463.

Jones, C.L. 1979. The morphogenesis of the thigh of the mouse with special reference to tetrapod muscle homologies. *J. Morphol.* 162:275–310.

Krstič, R.V. 1984. *General histology of the mammal: an atlas for students of medicine and biology.* Springer, New York. 404p.

Lauder, G.V. 1980. On the relationship of the myotome to the axial skeleton in vertebrate evolution. *Paleobiology* 6:51–56.

Loeb, G.E. and C. Gans. 1986. *Electromyography for experimentalists.* Univ. Chicago Press, Chicago. 373p. Detailed look at the history and theory of electromyography. Excellent practical guide for executing electromyographic studies.

McMahon, T.A. 1984. *Muscles, reflexes, and locomotion.* Princeton Univ. Press, Princeton, NJ. 331p.

Moller, P. 1995. *Electric fishes: history and behavior.* Chapman & Hall. 584p. A scholarly and fascinating account.

Morgan, D.L., and U. Proske. 1984. Vertebrate slow muscle: its structure, pattern of innervation, and mechanical properties. *Physiol. Rev.* 64:103–169. Outstanding review of tonic fibers in all vertebrate groups.

Richmond, F.J.R. 1998. Elements of style in neuromuscular architecture. *Am. Zool.* 38:729–742.

Sacks, R.D., and R.R. Roy. 1982. Architecture of the hind limb muscles of cats: functional significance. *J. Morphol.* 173:185–195.

Sokoloff, A.J., et al. 1998. Neuromuscular organization of avian flight muscle: morphology and contractile properties of motor units in the pectoralis (pars thoracicus) of pigeon *(Columba livia). J. Morphol.* 236:179–208.

Sullivan, G.E. 1962. Anatomy and embryology of the wing musculature of the domestic fowl (*Gallus*). *Australian J. Zool.* 10:458–518.

Trotter, J.A., Richmond, F.J.R., and P.P. Purslow. 1995. Functional morphology and motor control of series-fibered muscles, pp. 167–213. *In* J. Holloszy (ed.), *Exercise and sports science review.* Williams & Wilkins (Am. College of Sports Med.), Baltimore. Technical review of the literature (emphasis on mammals) regarding in-series muscle organization.

Trujillo-Cenóz, O., and J.A. Echague. 1989. Waveform generation of the electric organ discharge in *Gymnotus carapo.* I. Morphology and innervation of the electric organ. *J. Comp. Physiol. A.* 165:343–351.

Turner, R.W., L. Maler, and M. Burrows (eds. of symposium). 1999. Electroreception and electrocommunication. *J. Exp. Biol.* 202:1167–1458. Collection of papers that focus on anatomy and physiology of electroreception in a number of chordate groups.

Windhorst, U., T.M. Hamm, and D.G. Stuart. 1989. On the function of muscle and reflex partitioning. *Behav. Brain. Sci.* 12:629–681. A scholarly review of the functional implications of fiber and motor unit compartments in muscle.

Chapter 11

Coelom and Mesenteries

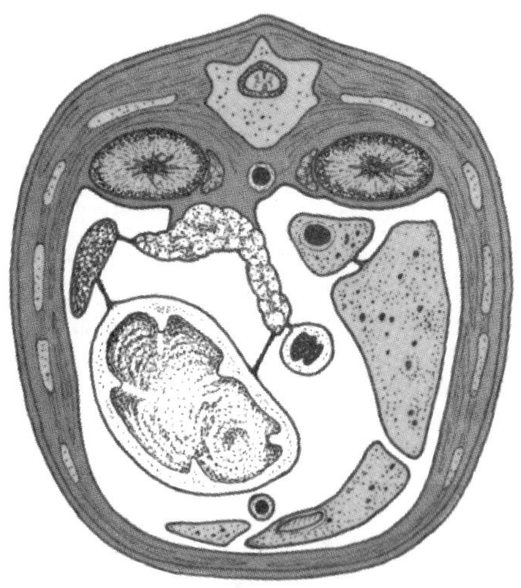

NATURE AND FUNCTION

Coelomic cavities are the spaces that surround the heart, lungs, digestive system, and certain urogenital organs. The coelom, unlike cavities of the nervous and respiratory systems, occurs within tissues of mesodermal origin. The function of coelom is to allow the internal organs to move freely and to change their relative sizes and positions as is required when the heart beats, the lungs fill and empty, the digestive tract passes food, and the pregnant uterus enlarges. Partitions of the coelom may contribute to linkages that couple respiration with locomotion (see p. 235). (For many invertebrate animals the coelom functions as a hydrostatic organ, stiffening the body to provide for locomotion.) The lining of the coelom, which covers the body walls and envelops the viscera, is a **serous membrane** composed of flat cells that secrete a fluid to lubricate the organs so they can slip easily past one another.

Mesenteries extend across the coelom from body wall to viscera. They are sheets of serous membrane strengthened by thin layers or bands of collagenous and elastic fibers. Mesenteries that join one organ to another are called **omenta** or "ligaments" (the latter is an unfortunate term, since these are not true ligaments). Mesenteries support the internal organs, without restricting function, and transmit nerves and vessels. In mammals they are commonly sites of fat storage.

The coelom has been of relatively little importance to the vertebrate morphologist because its structure is too simple and constant to contribute importantly to functional or evolutionary analysis. Mesenteries correlate with systematics at the class level, and also with marked postural differences (e.g., of dog, man, and sloth). The detailed configurations of mesenteries are often too complicated to decipher without embryological analysis.

Above: Diagrammatic cross section of a fetal pig at the level of the stomach. (Redrawn from *Anatomy and dissection of the fetal pig,* by W.F. Walker, © 1964, 1974, 1980, 1988 by W.H. Freeman and Co. Used with permission.)

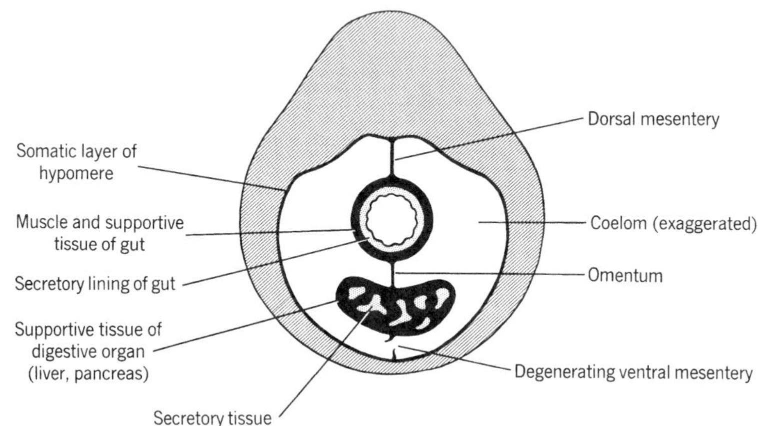

FIGURE 11.1
DERIVATIVES OF THE HYPOMERE IN RELATION TO THE GUT AND COELOM. This developmental stage follows the last shown on page 76.

DEVELOPMENT, EVOLUTION, AND RECAPITULATION

In amphioxus, echinoderms, and other invertebrate deuterostomes, the coelom forms from a series of pouches that pinch off from the dorsolateral wall of the gut tube, a process called enterocoely. This may have been the ancestral method of coelom formation in vertebrates. However, all surviving vertebrates form the coelom by the cavitation or splitting of initially solid mesoderm, the process of schizocoely (review p. 76).

Early embryos of vertebrates may have small and transitory coelomic cavities in the myotomes (**myocoels**), mesomeres (**nephrocoels**), and probably in the sclerotomes (see figure on p. 76). Nephrocoels become the adult renal capsules (see p. 273); myocoels have no known function or derivatives. The coelom of the hypomere is the **splanchnocoel,** but since it is large and persistent and gives rise to all the coelomic cavities of the adult, it is usually called simply the coelom.

This coelom splits the hypomere into an inner **splanchnic layer** and an outer **somatic layer.** As the coelom expands, right and left splanchnic layers move toward one another. They either come together in the midsagittal plane of the body or encounter the endoderm of the gut tube and its diverticula. Where they come together dorsal to the gut, they form the **dorsal mesentery;** ventral to the gut they form the **ventral mesentery,** and between the gut and its derivatives they form omenta. Parts of the ventral mesentery quickly degenerate, thus causing right and left coelomic cavities (from right and left hypomeres) to become confluent (see Figure 11.1 and figure on p. 76).

The parts of the splanchnic layer of the hypomere that encounter the gut tube and its derivatives form the smooth musculature, connective tissue, and serous membranes of the various organs. The somatic layer of the hypomere remains in contact with the body wall. It forms the serous membrane that lines the outer part of the coelom and, as noted in Chapter 10, may form mesenchyme that contributes to hypaxial and appendicular musculature.

Since coelomic cavities may occur in the ventral portions of the visceral arches, and also in the developing tail musculature of embryos, it may be inferred that the general coelom was once more extensive. However, the functional coelom of surviving vertebrates extends only from the level of the posterior part of the pharynx to the cloaca. Initially, in both phylogeny and ontogeny, the gut tube is straight and is supported by straight and continuous dorsal and ventral mesenteries. This simple structure becomes greatly complicated in adults of most vertebrates by partitioning of the coelom, and by deletions, folding, and fusions of the mesenteries as the gut tube lengthens. We will follow only the principal trends.

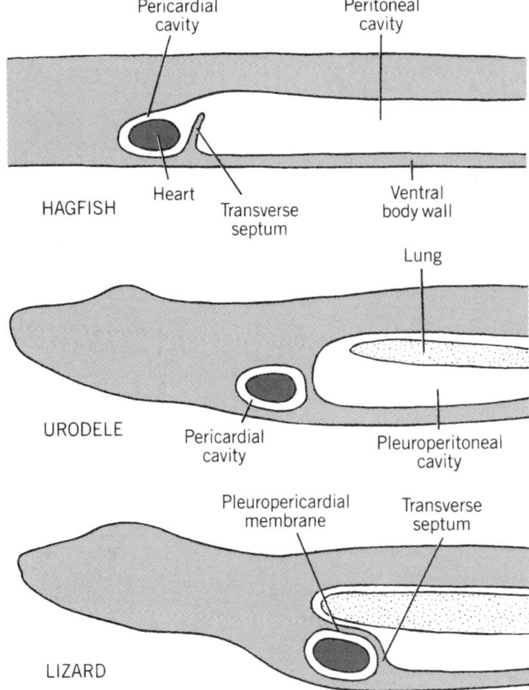

FIGURE 11.2 PARTITIONING OF THE COELOM OF REPRESENTATIVE VERTEBRATES as seen in sagittal sections. Mesenteries not shown.

The heart of fishes lies anterior to the pectoral girdle and ventral to the posterior gill chambers. In HAGFISHES a **transverse septum** extends upward from the ventral body wall posterior to the heart, partly separating an anterior **pericardial cavity** from a larger **peritoneal cavity** (Figure 11.2). In selachians the developing transverse septum temporarily separates the two cavities and then secondarily develops small orifices. Lampreys and other fishes retain a complete septum (see figure on p. 204).

These basic relationships have not been modified by URODELES. The small pericardial cavity remains far forward, where it is separated by a transverse septum from the principal coelom, which may now be called a **pleuroperitoneal cavity** because slender lungs are present. These are anchored to the lateral body wall by **lateral mesenteries.**

The heart of OTHER TETRAPODS lies at the level of the pectoral girdle or posterior to the girdle. The lungs are dorsal to the heart. (Parts of the liver intervene between heart and lungs in birds and some reptiles but not in mammals.) The heart is separated from the lungs (and liver if present) by more or less horizontal partitions that have their origin in the embryo as folds in the serous membrane of the right and left lateral body walls. These grow out to join in the midline of the body. They are called lateral mesocardia (birds) or **pleuropericardial membranes.** Posteriorly they join the transverse septum to form the adult pericardial membrane, or **pericardium.**

CROCODILIANS, SOME LIZARDS, and BIRDS partition the pleuroperitoneal coelom into additional cavities by complex outgrowths and fusions of the mesenteries. The thoracic air sacs of birds separate a ventral oblique septum from a dorsal pulmony diaphragm that is supplied with striated muscle. Their lungs grow up against the dorsolateral body walls, thus obliterating the pleural cavities in the adult.

In the partitioning of their coelom, embryonic MAMMALS resemble first early fishes (incomplete partition, posterior to heart, consisting of the transverse septum) and then reptiles (pericardium derived from transverse septum and pleuropericardial membranes).

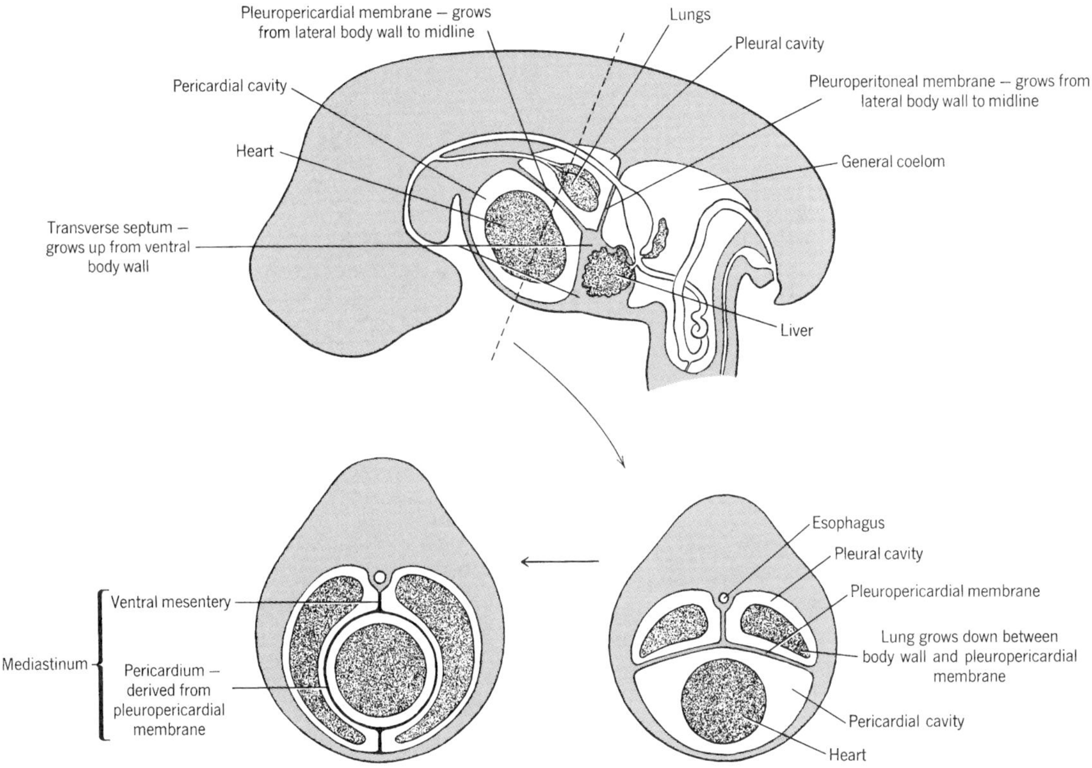

FIGURE 11.3 PARTITIONING OF THE COELOM IN A MAMMAL seen in sagittal section (above) and cross sections (below).

Mammals then separate paired **pleural cavities** from the peritoneal cavity by a **diaphragm.** The ventral portion of this organ comes from the transverse septum. The dorsal portion is derived from the dorsal mesentery and from still another pair of outgrowths from the lateral body wall, the **pleuroperitoneal membranes** (Figure 11.3). The striated muscle of the diaphragm comes from cervical myotomes (and hence is innervated by cervical nerves) because it develops at that level in the embryo and then migrates posteriorly as the neck lengthens and the lungs expand.

The pericardial cavity of anurans and reptiles is bordered dorsally by the pericardial membrane and ventrally by the body wall. In birds, the growing liver forces its way between body wall and membrane, thus wrapping the pericardial membrane nearly around the heart. In mammals, growth of the lungs does the same thing, this time obliterating the ancestral reptilian condition by removing the pericardial cavity entirely from the body wall (Figure 11.3). Right and left pleural cavities are then separated by dorsal and ventral mesenteries and by the pericardial membrane. The combined partition is called the **mediastinum** and is unique to mammals.

The mesenteries of ANAMNIOTES remain relatively straight and complete (dipnoans), are nearly absent except for portions related to stomach and liver (lampreys, selachians), or range between these extremes. Fusions and moderate folding are usual. Mesenteries of fishes are commonly pigmented—supposedly to protect light-sensitive gonads.

The embryonic liver of TETRAPODS starts to grow within the transverse septum. As it enlarges, it bulges out of the septum posteriorly and finally separates more or less completely from the developing diaphragm, trailing the **coronary ligament** behind. As the liver grows out of the septum, it grows into the ventral mesentery. The part of the ventral mesentery extending from liver to ventral body wall is the **falciform** (= *sickle-shaped*) **ligament;** the part between liver and gut tube is the **lesser omentum.** A small portion of another part of the ventral mesentery may anchor the bladder to the body wall. Between liver and bladder the ventral mesentery of tetrapods is missing. The dorsal mesentery is more complete and much complicated by folding and fusions. Between the stomach and body wall the dorsal mesentery of mammals becomes extended into a saclike **omental bursa** (= *membrane* + *purse*).

REFERENCES

Clark, R.B. 1964. *Dynamics in metazoan evolution: the origin of the coelom and segments.* Clarendon Press, Oxford. 313p. Primarily about coelom as a hydrostatic organ in invertebrates.

Feduccia, A. 1991. *Torrey's morphogenesis of vertebrates.* 5th ed. Wiley, New York. 517p.

Funayama, N., et al. 1999. Coelom formation: binary decision of the lateral plate mesoderm is controlled by the ectoderm. *Development* 126:4129–4138. A study of the pattern and molecular basis of coelom formation within chick lateral plate mesoderm. Helpful color figures.

Goodrich, E.S. 1986. *Studies on the structure and development of vertebrates.* Univ. Chicago Press, Chicago. 837p. First published in 1930. Coelom receives 44p.

Nelson, O.E. 1953. *Comparative embryology of the vertebrates.* Blakiston, New York. 982p. A chapter on coelom.

Chapter 12

Digestive System

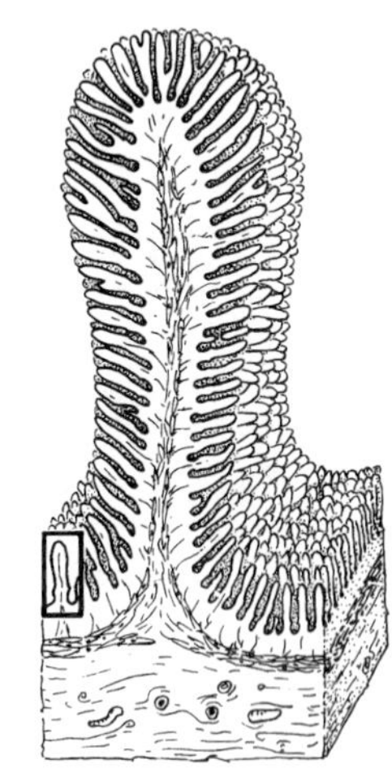

GENERAL FUNCTION AND STRUCTURE

The digestive system functions to (1) receive ingested food, (2) store it temporarily, (3) reduce it physically, (4) further reduce it chemically, (5) absorb the products of digestion, and (6) hold temporarily and then eliminate undigested wastes.

Most vertebrates are intermittent feeders; when food is available it must be taken into the body faster than it can be digested. If the food is bulky, or if feeding and drinking are infrequent and rapid, quantities of food and water must be stored temporarily. The principal storage organ is the stomach. Some birds have a storage sac off the esophagus. Many rodents, and some other mammals, have internal or external cheek pouches (opening, respectively, inside or outside the lips). If digestion is slow (as for diets of coarse vegetation), much digesting food must be retained at one time and the storage capacity of the entire tract is greatly increased.

Physical reduction of food—especially of roughage—is necessary to release nutrients from undigestable components and to increase the surface contact between food and digestive juices. Physical reduction is accomplished by the (1) chewing, rasping, or grinding of oral teeth, pharyngeal teeth (some fishes), or stomach (gizzard of many birds); (2) moistening, softening, and dissolving of food by fluids of the mouth, stomach, and intestine; (3) churning and mixing by peristalsis (anterior-to-posterior waves of contraction), reverse peristalsis, and segmentation (dividing motions) of the stomach and small intestine; and (4) emulsification of fats by secretions of the liver.

Chemical reduction of food is accomplished principally in the stomach and small intestine by enzymes produced in those organs or in the pancreas. Since the chemical nature of foodstuffs eaten by the various animals is similar, it is not surprising that the enzymes provided and the glands secreting them are also similar in the different vertebrates. Animals that employ bacterial fermentation as an aid to digestion (ungulates, some marsupials) must provide long storage in the stomach, caecum, or colon.

Absorption of the end products of digestion requires great surface contact between the digested food and the intestinal epithelium. As described further below, this is accomplished by a long intestine, folds in the lining of the gut, microscopic villi on the lining of the tract, and smaller microvilli in portions of the tract.

The digestive system reveals general dietary habits, and its structure is occasionally useful in systematics. It has been much less significant for establishing phylogenies than have most other organ systems: In part the organs are too constant in nature (small intestine), in part are too simple (gall bladder), and in part vary in ways that have limited evolutionary significance (lobulation of liver and pancreas or coiling of intestine).

DEVELOPMENT OF THE GUT

The developmental process of gastrulation provides the early embryo with an inner germ layer, the endoderm, which is nearly spherical if there is little yolk and is sheetlike if there is much yolk (see figures on pp. 75 and 77). In either event, as the embryo lengthens, the endoderm is drawn out into a tube by processes shown in Figure 12.1. This tube becomes the lining of the gut. At each end it breaks through to the ectoderm, thus establishing oral and anal openings. The midgut of embryos provided with a large yolk mass (fishes, reptiles, birds) is continuous with the **yolk sac,** which envelops and gradually absorbs the yolk (see figure on p. 80).

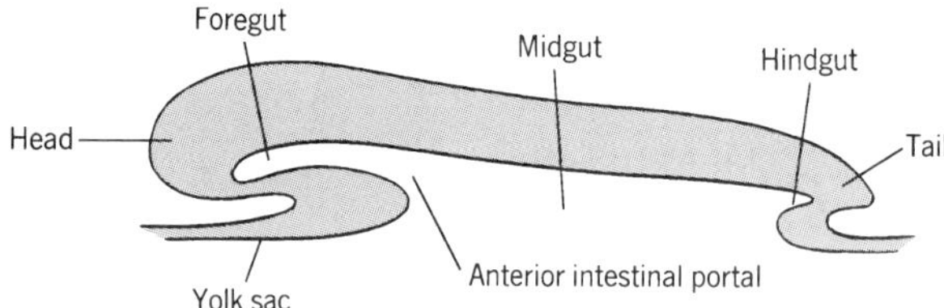

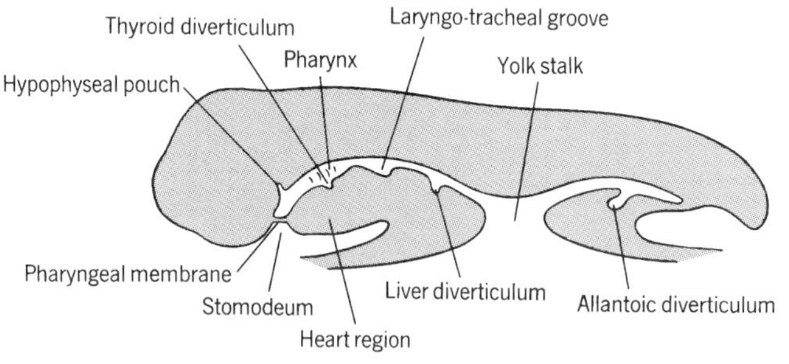

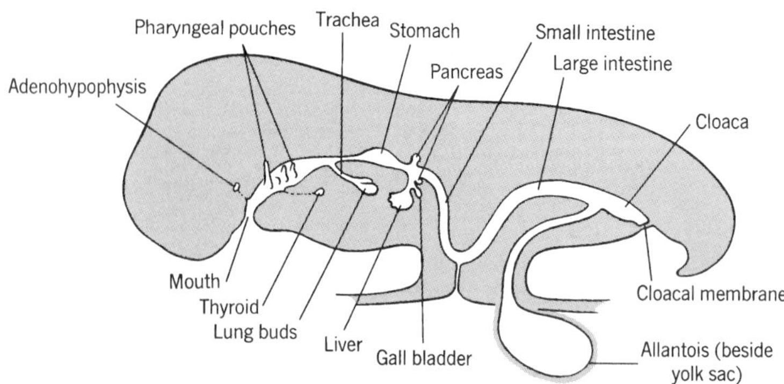

FIGURE 12.1 DEVELOPMENT OF THE GUT TUBE AND ITS DERIVATIVES IN AN AMNIOTE. Other organs and coelom are not represented. A later stage is shown on p. 198.

Initially the gut tube is straight, or nearly so, but soon it folds and coils and establishes outgrowths, or diverticula, which become the lining and secretory cells of associated organs (Figure 12.1 and the figure on p. 198). The derivatives of the complicated pouches that form in the oral cavity and pharynx are respiratory or endocrine in function and will be described in Chapters 13 and 20. Close behind the pharynx of tetrapods a ventral diverticulum foreshadows the respiratory system. Several diverticula develop posterior to the expanding stomach. These become the liver, gall bladder, pancreas, and their ducts. Near the posterior end of the gut of amniotes, a ventral diverticulum grows rapidly to become the fetal membrane called the **allantois.**

The muscular and connective tissue associated with the gut and related organs (and therefore most of their bulk) is of mesodermal origin. The differentiation of this mesoderm is related to the ontogeny of the coelom, as described in the last chapter.

STRUCTURE, ADAPTATION, AND EVOLUTION

Mouth and Oral Cavity The complicated structure of mouth, oral cavity, and pharynx cannot conveniently be discussed under the heading of any one organ system. The teeth have been described (Chapter 7), as has the evolution of the secondary palate in relation to chewing and breathing (Chapter 8). Respiratory and glandular derivatives of the pharynx will be presented in Chapters 13 and 20, and various specializations that relate to feeding will be noted in Chapter 30.

The ANCESTRAL VERTEBRATE was likely a filter feeder. Like amphioxus and the larva of the lamprey, it probably had a small mouth, virtually no oral cavity, and a large pharynx specialized to remove microscopic food particles from water by trapping them in mucus covering numerous gill bars (see figures on pp. 26 and 38). The mucus was probably moved to the intestine by cilia.

The AGNATHA have given up filter feeding, but because they have no jaws or true teeth they ingest small or soft food and have correspondingly small oral cavities. Mouth parts may be adapted for nibbling (most ostracoderms) or specialized for clinging to a host and abrading its flesh (cyclostomes). Anaspids, some cephalaspids, and cyclostomes have a rasping organ on the floor of the oral cavity that is called a tongue but is not homologous with the tongue of higher vertebrates.

The mouths and oral cavities of CARTILAGINOUS and BONY FISHES are extremely varied. Although fleshy lips are absent, the mouth parts may be highly protrusible or otherwise specialized (of which more will be covered in Chapter 30). The oral cavity and pharynx are usually distensible. Gill bars are often provided with food strainers, grinding mills, or teeth (see Figure 12.2 and the figure on p. 557). The basal elements of the visceral skeleton support a firm tongue that is little movable yet may be provided with teeth and is the partial homolog of the tetrapod tongue. Since the food of fishes is always wet, no further lubrication need be provided and oral glands are restricted to scattered mucous cells.

In general, the evolutionary trends among TETRAPODS were first to increase oral lubrication and then to add limited physical and chemical digestion. They have oral cavities of moderate or large size, depending on feeding habits. A tongue is present and is supported by derivatives of the second and third (and sometimes also the fourth) visceral arches. The tongue is usually fleshy and highly movable, but is relatively firm and fixed in many birds and some reptiles. It may function outside the mouth in securing food (see figure on p. 566) or inside the mouth in manipulating food during chewing, in swallowing, and in sound control. Mammals have lips that are moved by derivatives of the platysma muscle.

Tetrapods have multicellular oral or **salivary glands** that are compound, usually lobulated, and provided with ducts (Figure 12.3). They are named according to position as labial, lingual, palatine, nasal, maxillary, parotid (often the largest in herbivores),

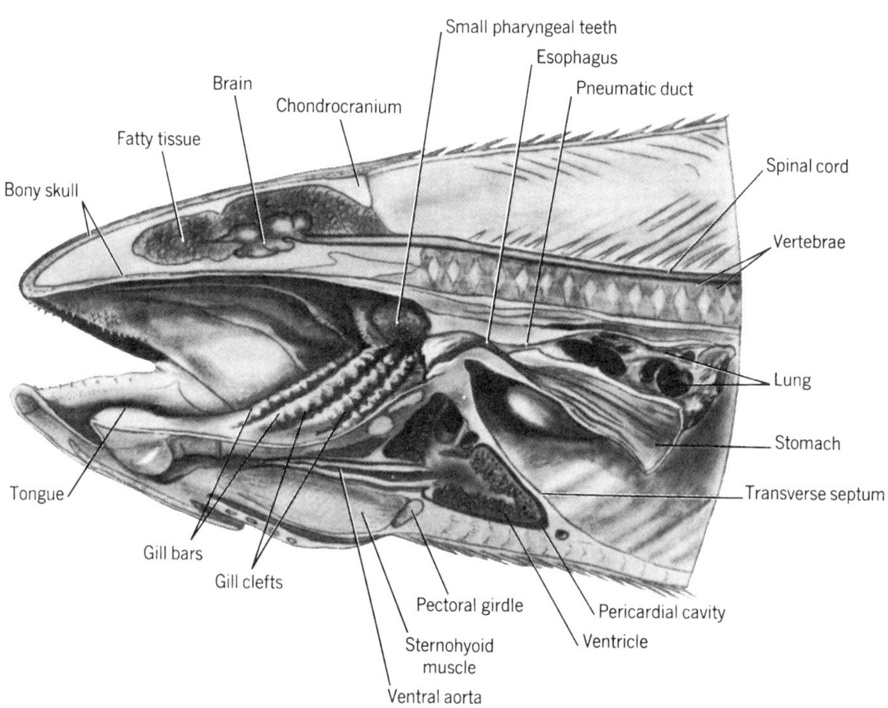

FIGURE 12.2
SAGITTAL SECTION OF THE HEAD AND ANTERIOR BODY OF A PRIMITIVE NEOPTERYGIAN FISH, the bowfin, *Amia*.

mandibular (large in carnivores), and so forth. However, the number, distribution, and detailed structure of these glands are diverse, and correspondence of name or position does not necessarily indicate equivalence. All tetrapods that are not secondarily aquatic require the mucous and serous secretions of oral glands to lubricate dry food, and this can be the only function. In some mammals (and to a lesser degree in certain other tetrapods) a starch-digesting enzyme is present, and traces of protein and fat-splitting enzymes have been identified. However, the overall importance of oral digestion is questioned. Various animals have evolved special functions for oral glands. Their sticky secretions may cause food to stick to the tongue (frogs, anteaters). Certain glands of some snakes, lizards, and shrews become poison glands; blood-feeding bats (and lampreys) secrete an anticoagulant; and the nasal glands of some marine birds and reptiles migrate to the orbits, where they function in salt excretion.

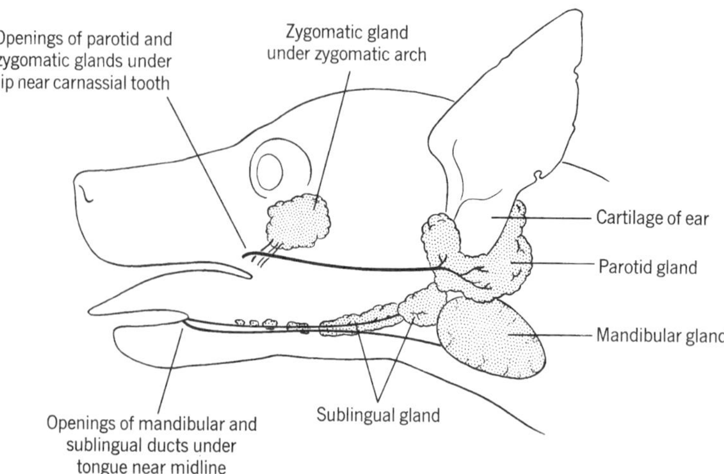

FIGURE 12.3 ORAL, OR SALIVARY, GLANDS OF THE DOG.

Fine Structure of the Gut in General The fine structure of the alimentary canal is basically similar throughout its length. The gut is constructed in layers (Figure 12.4). The innermost principal layer is the **mucosa.** It consists of a surface **epithelium,** a deeper **lamina propria,** and a **muscularis mucosae.** Cells of the epithelium may be squamous but usually are columnar with basal nuclei. Interspersed among the less specialized cells are mucus-secreting goblet cells and (according to region of the tract) unicellular or multicellular glands that secrete digestive juices. The epithelium has many folds when the tract is empty. The lamina propria is a network of loose tissue that underlies the epithelium and fills the cores of the villi. Next is the muscularis mucosae (not always present), a thin layer of smooth muscle that controls motions of the lining of the gut that are independent of the gut tube as a whole.

The second principal layer is the **submucosa,** which is a conspicuous stratum of loose connective tissue containing nerves, capillaries, lymphatic ducts and nodules, and ganglia of the parasympathetic nervous system. The larger crypts and glands of the epithelium may subtend into the submucosa.

Outside the submucosa is the third prominent layer, the **muscularis externa.** It consists of smooth muscle. An inner portion is called the circular layer because its fibers are arranged in a tight spiral around the gut tube. Their contraction lengthens and constricts

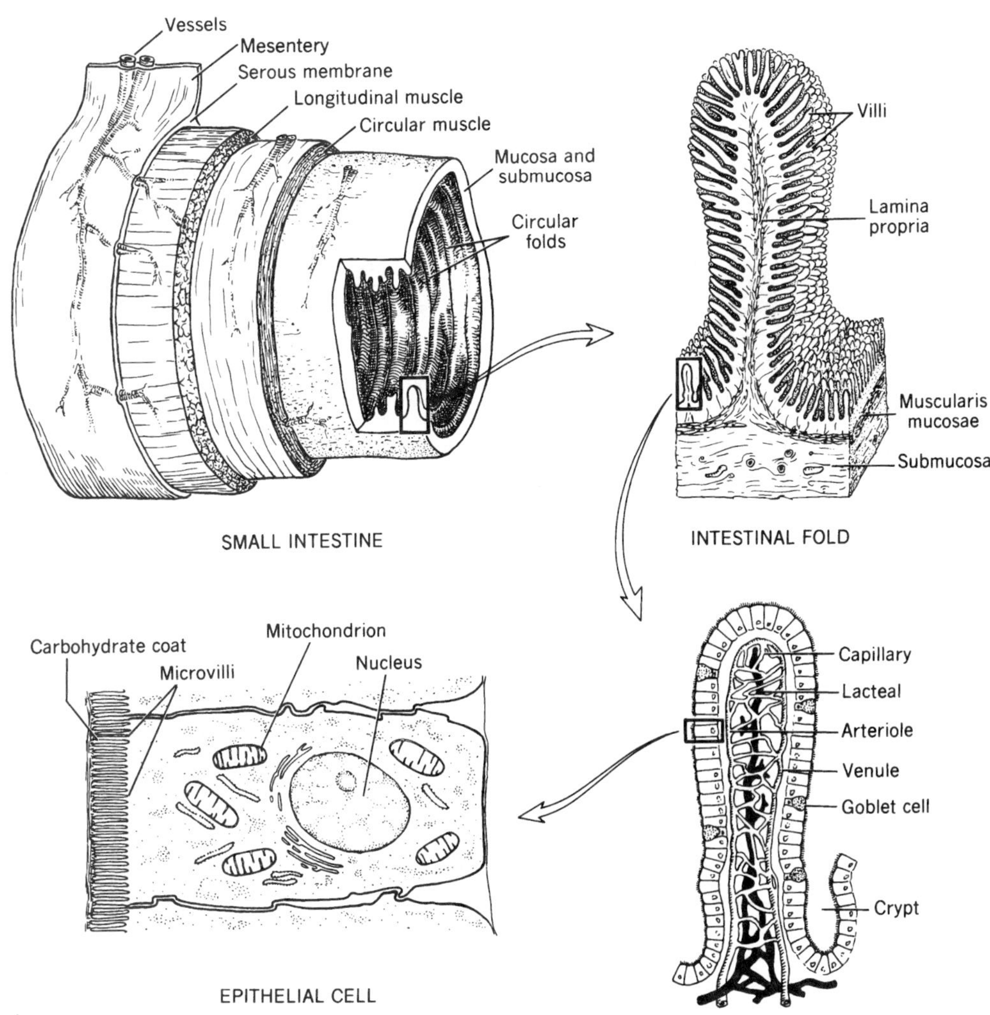

FIGURE 12.4
STRUCTURE OF THE SMALL INTESTINE.

the gut. An outer portion is called the longitudinal layer. Its fibers are nearly longitudinal and serve to shorten the tract. The coordinated action of the two layers of the muscularis accomplish peristalsis and segmentation.

The pharynx, rectum, and part of the esophagus are bound directly to adjacent structures; other parts of the tract lie in the coelom and are enveloped in serous membranes.

Esophagus and Stomach Together the esophagus and stomach constitute the *foregut*. The lining of the esophagus is usually much folded and highly distensible. Its epithelium typically consists of stratified squamous cells, which are cornified in animals that swallow coarse food. However, the epithelium is columnar and vascularized in many marine fishes and can be ciliated in cyclostomes and various other vertebrates. The muscularis externa of the esophagus has smooth muscle fiber in most classes, but may have striated fibers in both fishes and mammals.

The lining of the stomach is divided into several regions that are distinctive in fine structure and function (Figure 12.5). They may be sharply demarcated or may merge. Their relative distribution has some correlation with function, but little systematic significance. The most anterior is like the esophagus in fine structure and is called the **esophageal region,** although origin from the esophagus is doubtful. It is relatively extensive in animals that swallow coarse food. It is lined with stratified squamous epithelium, and its only secretion is mucus. A **cardiac region** is present in mammals only. It likewise secretes only mucus, but the cells of its epithelium are columnar. The digestive region of the stomach is the **fundus.** Its lining is thickened by a dense layer of straight tubular gastric glands. Their open mouths form microscopic pits. The columnar cells of the glands are of several kinds.

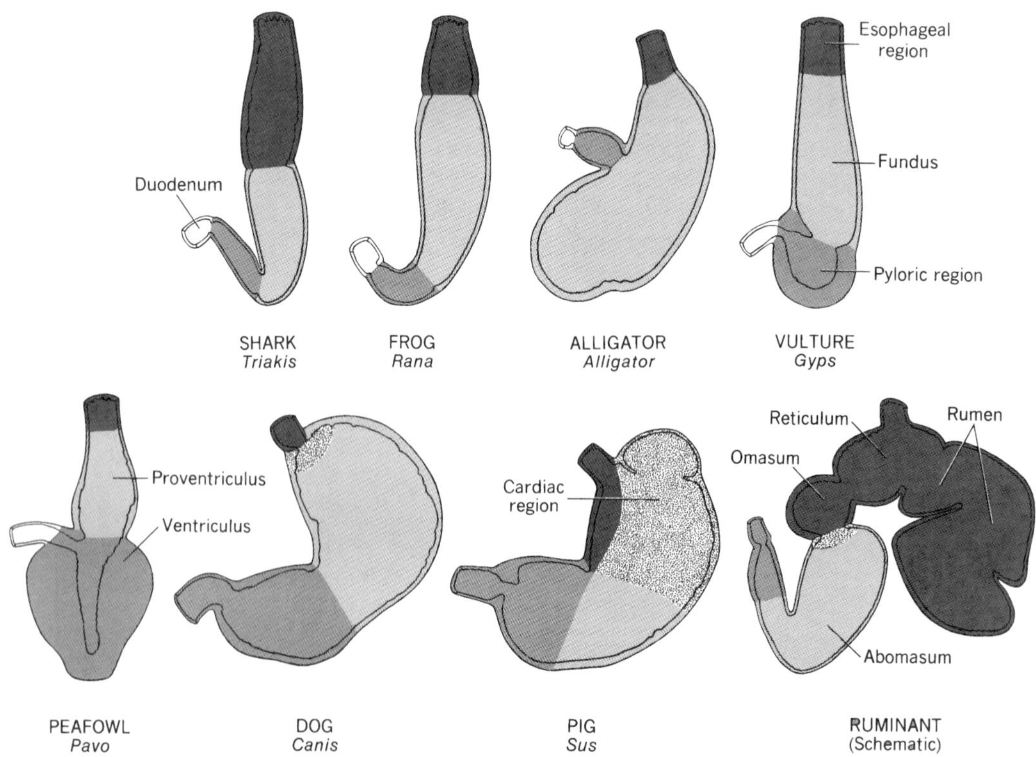

FIGURE 12.5 STOMACHS OF SELECTED VERTEBRATES showing some variations of gross form and distribution of the different types of lining. Ventral views of sectioned organs.

They secrete the enzyme pepsinogen; hydrochloric acid, which provides the acidity necessary to cleave the smaller enzyme pepsin from pepsinogen and thus initiate protein digestion; and sometimes also a fat-splitting enzyme. The necks of the glands secrete mucus. The stomachs of mammals also secrete renin, which coagulates milk. The most posterior region of the stomach is the **pyloric region.** Its coiled tubular glands secrete mucus.

The muscularis externa of the stomach is relatively thick. At the anterior end of the stomach, circular fibers form the **cardiac sphincter;** at the posterior end, the **pyloric sphincter** controls passage of chyme (digesting food) into the intestine.

No stomach can be identified in filter feeders and AGNATHA (apparently the primitive condition). The esophagus of FISHES is usually short and sometimes merges into the stomach (Figure 12.2). Commonly it has distinctive pleats or papillae directed toward the stomach. It is ciliated in most selachians. The stomach of fishes is usually either straight or bent into a J or U shape (Figure 12.6). It is particularly large in selachians and is (secondarily) lacking in some forms (chimaeras, lungfishes, etc.) that swallow only finely divided food.

The esophagus of AMPHIBIANS is short, ciliated, and well supplied with mucous glands. REPTILES tend to have a longer esophagus because of the increased length of the neck and more developed lungs. The esophagus tends to be ciliated if soft food is eaten, but is cornified in some turtles. The stomach remains simple and straight or gently curved in most amphibians and reptiles, but is rounded and very muscular in crocodilians (Figures 12.5 and 12.7).

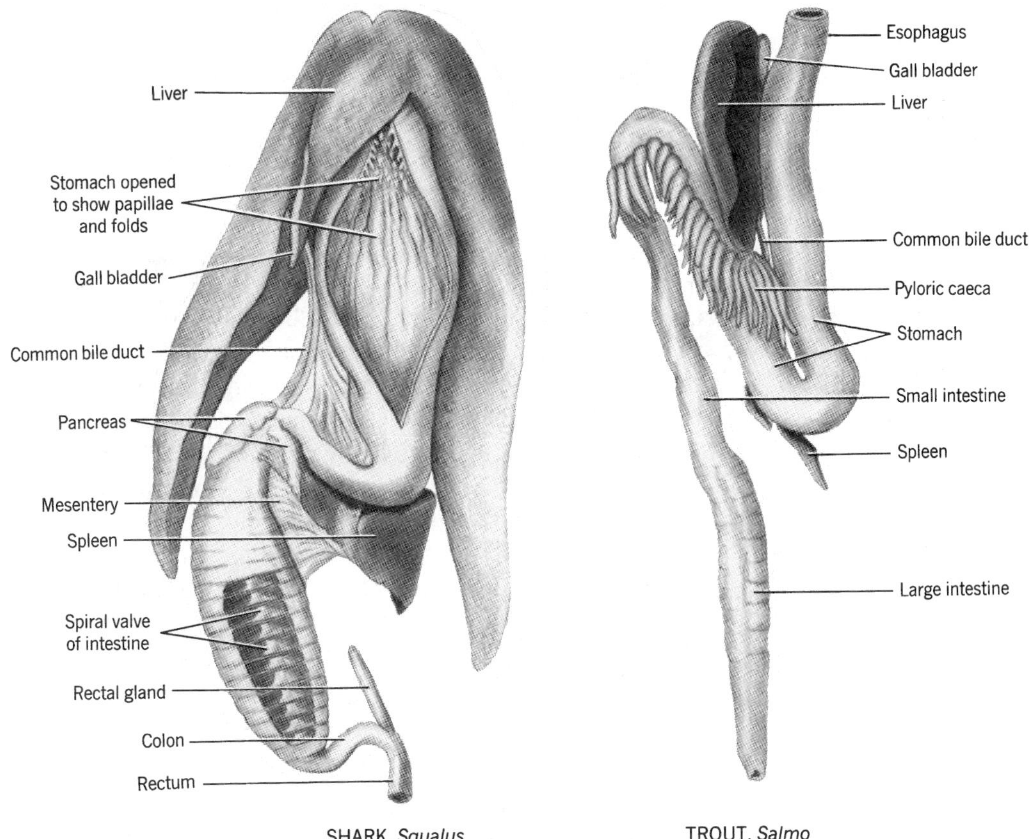

FIGURE 12.6 DIGESTIVE TRACTS OF AN ELASMOBRANCH (left) AND A TELEOST (right) that feed, respectively, largely on flesh and insects. Ventral views.

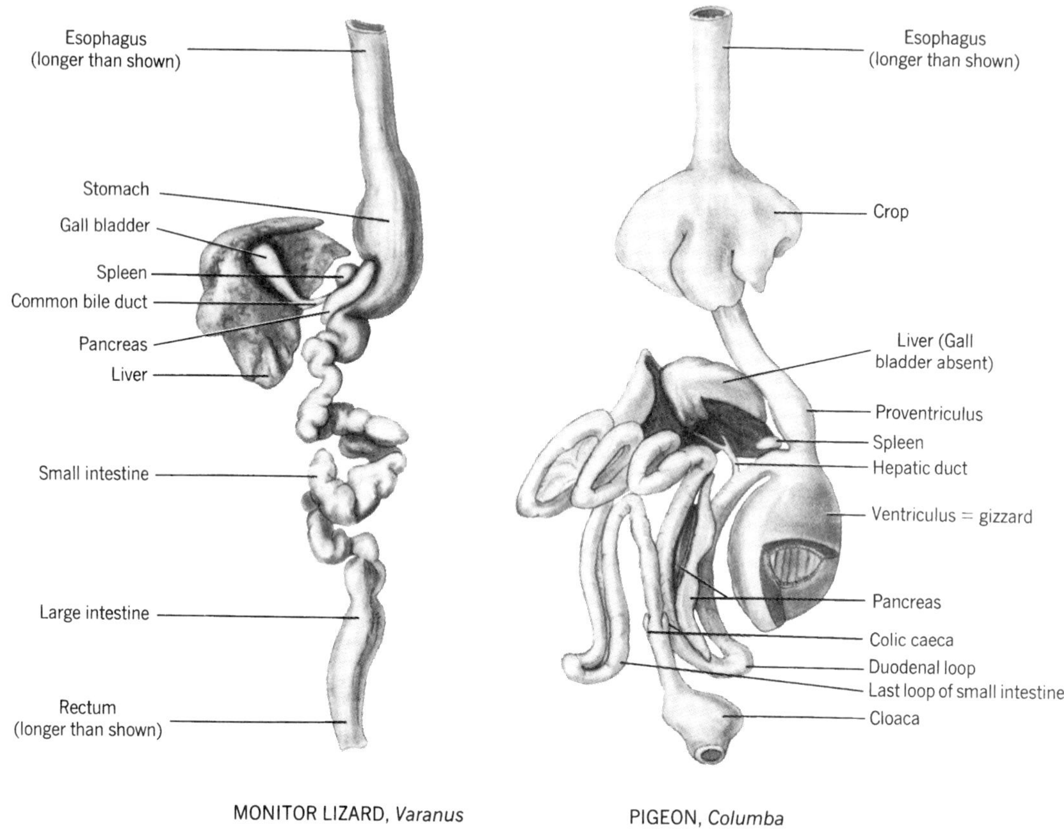

FIGURE 12.7 DIGESTIVE TRACTS OF A REPTILE (left) AND A GRANIVOROUS BIRD (right). Ventral views.

The esophagus of BIRDS is long. Its lining is usually cornified. At least some members of many families of birds have a permanent dilation of the lower part of the esophagus to serve as a storage organ called a **crop** (Figure 12.7). The dilation is usually ventral to the esophagus and sharply set off, but it may be dorsal instead and less distinct. The stomach of birds is in two parts. The anterior part, derived from the fundus but called the **proventriculus,** is very glandular and produces digestive enzymes. The posterior part corresponds to the pyloric region and is called the **ventriculus,** or **gizzard.** It may have a horny lining and be exceedingly muscular for grinding coarse food (sometimes with the aid of pebbles eaten by the bird). The proventriculus and ventriculus are least distinct in carnivorous birds and most sharply demarcated in granivorous species (Figure 12.5).

The esophagus of MAMMALS is long, devoid of cilia, and cornified in such roughage eaters as artiodactyls, perissodactyls, and rodents. The stomach may be simple and saclike (humans, many rodents, some insectivores, carnivores) or complexly compartmentalized (see below). An esophageal region is often present and is sometimes extensive. The entire stomach of monotremes is cornified, and most of the stomach of ruminants (cud chewers) is of the esophageal type. A cardiac region, found only in mammals, is characteristic but not universal.

Intestine and Caeca Digestion is completed and foodstuffs absorbed in the anterior part of the intestine. This requires an extensive surface area, which is achieved by coiling of the gut; by circular folds in the mucosal lining (absent in many small vertebrates); by fingerlike microscopic **villi,** which are packed 10 to 40/mm^2 over the lining; and finally by **microvilli** crowded 200,000/mm^2 on the exposed surface of the epithelial cells, where they, and the carbohydrate coat they support, form a **brush border** (Figure 12.4). The folds, villi, and microvilli of the digestive tube can increase its surface 600-fold. In tetrapods epithelial cells develop in the crypts between villi and migrate (in approximately two days) to the apex of a villus, where they slough into the lumen of the gut, carrying with them the enzymes that have been synthesized as the cells moved. A variety of fat-, protein-, and carbohydrate-splitting enzymes are released, but seemingly more in higher vertebrates than in fishes. Scattered goblet cells produce mucus. Hormones (gastrin, secretin, pancreozymin, cholecystokinin, enterogastrone, and others) that influence the activities of stomach, intestine, liver, and pancreas may also be secreted. Secretions of the liver and pancreas empty into a portion of the gut close behind the stomach called the **duodenum.** The combined intestinal juice is alkaline. Absorption of the products of digestion is a complex and active process that may be completed within the epithelial cytoplasm.

Inorganic electrolytes and much water are absorbed, and feces are formed, in the shorter posterior part of the intestine. Villi and microvilli are usually absent from this region. Mucous cells are abundant, and lymph nodules are often present in the submucosa. The part of the intestine that passes out of the coelom and through the pelvic girdle is called the **rectum.** The posterior-most end of the gut of many vertebrates is a **cloaca,** or common chamber for wastes of the digestive and urinary systems and for products of the gonads (see figure on p. 294).

Anterior and posterior regions of the intestine are relatively distinct in tetrapods, where they are called, respectively, the **small** and **large intestines** because of the (usually) larger diameter of the latter. They are also termed the **midgut** and **hindgut,** respectively. Furthermore, tetrapods usually have one or two pouchlike diverticula at the juncture between small and large intestines. These are called **colic caeca.** Primitively their function may have been merely to increase the surface of the gut; now they function variously for storage, fermentation, or vitamin concentration.

The intestine of CYCLOSTOMES runs straight from pharynx to cloaca, and regional differentiation is slight. These features, seen also in amphioxus, are doubtless primitive for vertebrates. In lampreys, a single marked fold of the intestinal mucosa runs lengthwise of the gut in a gentle spiral. This is the analog, if not also the homolog, of the spiral valve described below.

The nature of the digestive tract of SELACHIANS is relatively constant. The gut, now too long to run straight through the coelom, is N-shaped. One angle lies in the long stomach and one in the intestine. The three limbs run lengthwise in the body. Most of the posterior limb comprises the **spiral intestine,** or **spiral valve** (Figure 12.6). This part of the gut is large in diameter and tapers at each end. The mucosa is thrown into a single prominent fold. The fold may run lengthwise, but unlike that of the lamprey it grows out from the wall and winds on itself to form a scroll. More often the attachment of the fold to the wall of the intestine spirals as it runs down the tract. The structure of the organ resembles a spiral staircase in a circular tower. The number of turns ranges among the species from $5\frac{1}{2}$ to 50. The functional advantage is a great increase in the surface area of the epithelium. An equivalent increase could be achieved by lengthening and coiling the entire intestine, but the spiral structure of the mucosa is better adapted to the long, slender body cavity of the fish. A short rectum joins the cloaca. A dorsal appendage from the rectum is called the **rectal gland.** Its function is salt excretion (see p. 278). CHIMAERAS have no cloaca, and there is no rectal gland, although similar glandular tissue is found in the wall of the rectum.

Turning back to PLACODERMS, it is of importance that one fossil of an antiarch shows the imprint of a spiral intestine. This complicated organ is apparently primitive among jawed vertebrates.

The intestine of BONY FISHES is more variable than that of cartilaginous fishes. It is rarely straight, commonly thrown into one or two S curves, and occasionally coiled. Its length may be less than that of the body but is usually somewhat longer than in cartilaginous fishes and reaches 12 body lengths in some species. A spiral intestine is present in all modern bony fishes except teleosts. Ray-finned fishes are distinctive for another structure that increases the surface of the intestine; adjacent to the stomach the intestine develops diverticula called **pyloric caeca** (Figure 12.6). Most fishes have scores or hundreds of caeca, but some have few and several have none. The tubular caeca may open into the intestine individually or may cluster to form a compound organ. Histologically they resemble the adjacent intestine. Among bony fishes, only dipnoans and the surviving crossopterygian have a cloaca. There is no rectal gland.

Tadpoles have long, coiled intestines, but adult AMPHIBIANS have relatively short and simple digestive tracts ranging in length from $\frac{1}{2}$ to $3\frac{1}{4}$ times the length of the body. As in other tetrapods, a coiled small intestine is set off from a shorter large intestine. At the boundary between the two there may be a single small colic caecum. This structure is apparently vestigial in many surviving amphibians, yet its presence in the class must be considered primitive. A cloaca is present.

The intestine is straight in most snakes and amphisbaenians but is otherwise moderately coiled in most REPTILES. Its length usually ranges from $\frac{1}{2}$ to 2 times the body length but tends to be longer in turtles. Small and large intestines are distinct. A dorsal colic caecum of small or moderate size is present in many species but has secondarily been lost by others.

The duodenum of BIRDS always forms a long, narrow loop that lies ventral in the body cavity and is tightly joined to the pancreas. The remainder of the small intestine is relatively long and forms various complicated patterns of folds and coils that are constant within families. The large intestine is short, nearly straight, and villous. It is in a dorsal position. Two colic caeca are usual; rarely one or more than two are present. These are fingerlike in form, often of considerable length, and join the gut laterally or ventrally. A cloaca is present and has a dorsal diverticulum called the **cloacal bursa** that functions in the formation of antibodies.

The gross morphology of the intestine of MAMMALS is highly variable and correlates with systematics only in a general way (Figure 12.8). The intestine may be as short as 2 to 6 body lengths (many insectivores and carnivores) or as long as 20 to 25 body lengths (some artiodactyls and marine mammals). A duodenal loop is usually present but is less tightly bound than in birds. Furthermore, compared with birds, the pattern of folding of the small intestine is less regular and the large intestine tends to be longer and more bulky—particularly in herbivores. Several unrelated mammals have paired colic caeca. This may be the primitive condition, but a single ventral caecum is the rule. It may be small (even secondarily absent) but can be two or more times the length of the body.

Although some vertebrate taxa have distinctive features (spiral valve, pyloric caeca, gizzard, rectal gland), evolutionary trends are few in the gut tube. Functional adaptations, however, are readily found.

Adaptations of the Gut The morphology of the digestive tract correlates sufficiently well with function so that the approximate eating habits and diet of a vertebrate usually can be determined from its digestive system, though the structure of the system in some

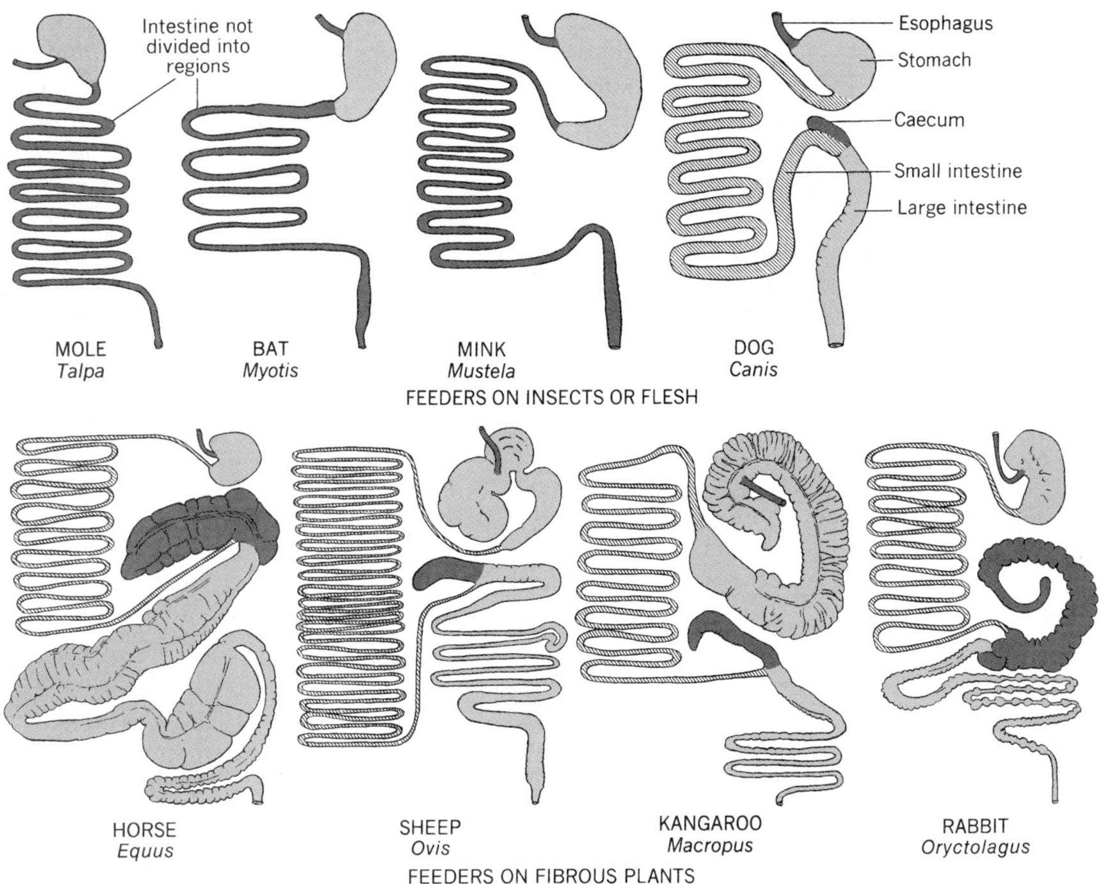

FIGURE 12.8 FUNCTIONAL VARIATION OF THE GUT IN MAMMALS. Redrawn from Stevens and Hume.

fishes and marine mammals remains puzzling. One approach is to consider its structure and function to be optimally correlated and to set up testable hypotheses (see Hume in References).

If the vertebrate feeds on nutritious food (which digests quickly), if the food is ingested as small particles, and if feeding is slow but frequent or nearly continual, then there need be no temporary storage of food, and a short gut is adequate. Many filterers and strainers, parasites that rasp the flesh of their hosts, and some feeders on algae, nectar, mollusks, and insects are in this category. They have relatively small stomachs or no stomach at all (cyclostomes, holocephalians, dipnoans), and the gut tends to be short and relatively straight (though it is unexpectedly long in baleen whales). Nectar-feeding birds have relatively nonmuscular gizzards with the entrance and exit close together so that nectar can pass directly into the intestine. The blood eaten by vampire bats passes directly from esophagus to intestine, backing up into the stomach only when the anterior intestine is full.

Carnivores, scavengers, and fish eaters also eat nutritious foods, so their guts also tend to be short, but these animals often eat large meals very quickly. Large prey may be swallowed whole, food may be available only after a successful hunt or wait, or it may be necessary to gulp a meal to prevent a fellow carnivore or scavenger from taking it away. Consequently, capacious temporary storage is essential. The esophagus of most such animals is remarkably distensible. The fishes tend to have long, straight stomachs. Carnivorous reptiles and mammals have a simple but distensible stomach. Most of the birds have crops that are pleated when empty and enormous when full. Their gizzards are relatively thin-walled. The small intestine of most carnivores and scavengers is relatively uniform and short, but in piscivorous vertebrates other than fishes it may be curiously long (penguins, pinnipeds, some toothed whales), thick-walled, and small in internal diameter. The hindgut of flesh eaters is short, and caeca are small or absent (Figure 12.8). Scales, bones, feathers, and hair are avoided by some of these animals but are swallowed by many. These may be digested (small bones), regurgitated (by raptorial birds), or passed through the gut.

Somewhat similar to the requirements of carnivores are those of feeders on swarming insects. Anteaters have a roomy stomach, which in some species is heavily cornified and muscular (one pangolin has keratinized pyloric "teeth"). Their salivary glands are very large (see figure on p. 566). Their guts are moderately short.

All feeders so far described are **faunivores:** they feed almost excusively on animal food. **Omnivores,** such as rats, pigs, and people, eat a varied diet according to taste and opportunity. Such plant foods as they select (fungi, fruits, nuts, seeds) tend to resemble animal foods in being highly nutritious. Digestion is only a little slower than for faunivores. The gut, though variable, tends to be a little longer and has more demarcation between midgut and hindgut (Figure 12.8 and Table 12.1).

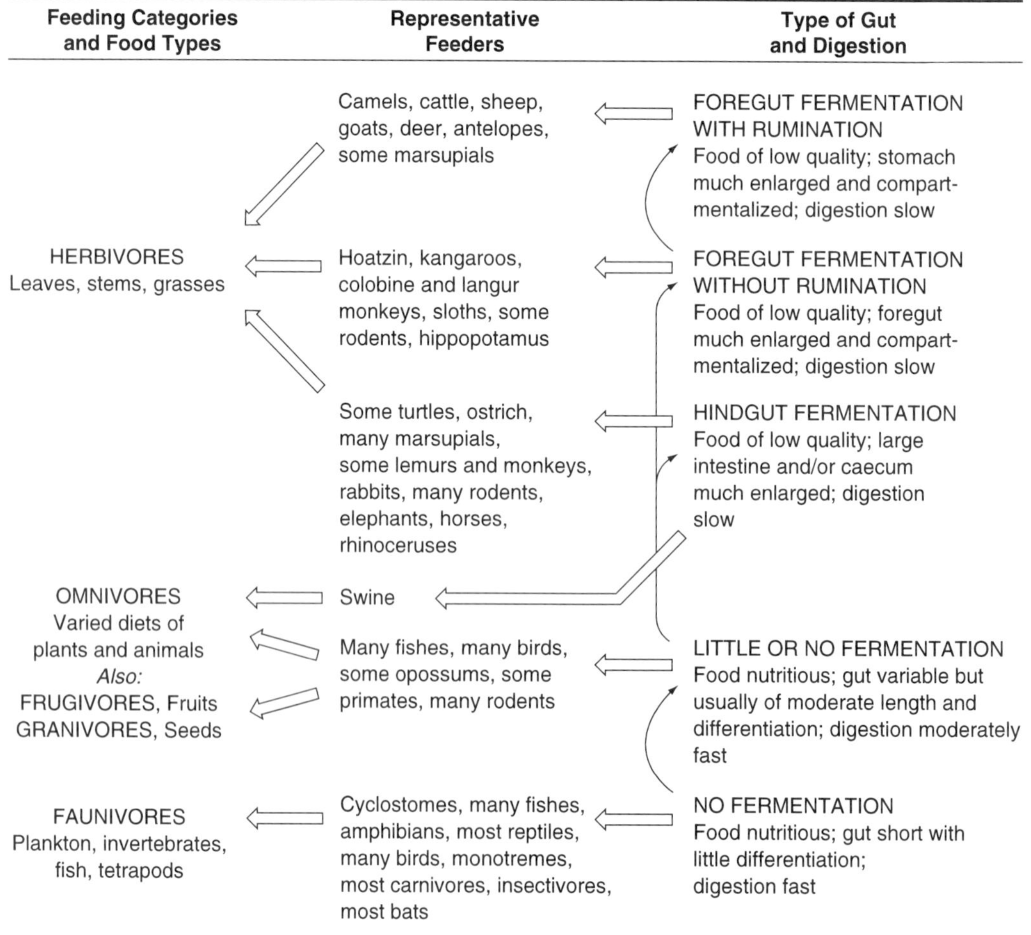

Feeding Categories and Food Types	Representative Feeders	Type of Gut and Digestion
	Camels, cattle, sheep, goats, deer, antelopes, some marsupials	FOREGUT FERMENTATION WITH RUMINATION Food of low quality; stomach much enlarged and compartmentalized; digestion slow
HERBIVORES Leaves, stems, grasses	Hoatzin, kangaroos, colobine and langur monkeys, sloths, some rodents, hippopotamus	FOREGUT FERMENTATION WITHOUT RUMINATION Food of low quality; foregut much enlarged and compartmentalized; digestion slow
	Some turtles, ostrich, many marsupials, some lemurs and monkeys, rabbits, many rodents, elephants, horses, rhinoceruses	HINDGUT FERMENTATION Food of low quality; large intestine and/or caecum much enlarged; digestion slow
OMNIVORES Varied diets of plants and animals *Also:* FRUGIVORES, Fruits GRANIVORES, Seeds	Swine	
	Many fishes, many birds, some opossums, some primates, many rodents	LITTLE OR NO FERMENTATION Food nutritious; gut variable but usually of moderate length and differentiation; digestion moderately fast
FAUNIVORES Plankton, invertebrates, fish, tetrapods	Cyclostomes, many fishes, amphibians, most reptiles, many birds, monotremes, most carnivores, insectivores, most bats	NO FERMENTATION Food nutritious; gut short with little differentiation; digestion fast

TABLE 12.1 FORM AND FUNCTION OF THE GUT IN RELATION TO TAXON AND DIET.

Herbivores eat leaves, stems, and grasses, which are relatively low in food value and hence must be eaten in large quantity. Furthermore, the cellulose wall of plant cells cannot be broken down chemically by the enzymes secreted by vertebrates. A first requirement is that there be thorough mechanical grinding in the mouth or stomach to shred the food, which facilitates its maceration and transport and increases its surface area. Second, in order that the cellulose can be utilized as food, fermentation by symbiotic bacteria and protozoa is required. The by-products are short-chain fatty acids, carbon dioxide, and methane. Nitrogenous compounds in the food are converted to ammonia, from which the microorganisms synthesize their own proteins. The host animal, in turn, may digest quantities of microorganisms, thus obtaining the protein, B-complex vitamins, and some detoxifying chemicals. Herbivores have long guts (Figure 12.8) with high motility. Contractions may be nearly constant or in rushes, and commonly move digesta in both directions. Digestion is under complex neural and hormonal control.

The digestion of low-energy plant foods by fermentation requires long retention of bulky material in either the foregut or hindgut and is favored by an elevated body temperature. The process has evolved independently many times, so the associated structures and strategies vary. They are summarized in Table 12.1 for surviving vertebrates and explained further in Comment 12.1. It has been suggested that independent evolution of foregut and hindgut fermentation occurred in mammals in response to the spread of grasslands as world climates dried and cooled in the Miocene epoch, and grasses became more fibrous.

Temporary storage is provided by cheek pouches in some rodents, primates, and bats. Having neither pouches nor crop, most herbivorous reptiles and mammals have a large stomach even if it does not serve as a fermentation chamber. Plant-eating birds have a very large crop that is positioned relatively far posterior and ventral in the body so that its mass, when full, will not interfere with the bird's stability. Furthermore, there is a muscular gizzard to substitute for teeth in grinding the food. The crop of the hoatzin is muscular itself, has a horny lining, and shreds the leaves that the bird eats. The gut of herbivores is relatively long, particularly if abundant roughage is eaten or if the animal is large. (This is a consequence of surface-to-volume ratios—see Chapter 23.) The intestine is much coiled, often in set patterns, and the long hindgut is of large diameter. The epithelial lining of the gut is subjected to considerable wear and is constantly replaced; the turnover time is 2 to 3 days.

Liver and Gall Bladder The vertebrate liver is unique to the subphylum and varies little among the classes. It is the largest organ of the body. The functions of the liver are many and diverse: It is a storage depot for carbohydrate and (particularly in cyclostomes and fishes) for fat. It converts protein to carbohydrate or fat with the release of nitrogenous waste, which is transported to the gills or kidneys for elimination. It elaborates much of the yolk that the maternal body transfers to the growing eggs. The embryonic liver (also the adult organ of fishes) produces blood cells, and the adult liver removes "old" red cells from the blood stream. Various toxicants can be removed from the blood by the liver, and substances needed for clotting are released. Several vitamins are manufactured or stored. Finally, the function that relates the liver to digestion: Bile is secreted into a duct system and delivered to the duodenum, where it emulsifies fats, thus making them digestible by pancreatic enzymes.

Liver and gall bladder develop from one or (usually) two ventral **hepatic diverticula** from the gut tube just posterior to the stomach (Figure 12.1). With few exceptions (birds), the more posterior diverticulum forms the gall bladder, and the anterior diverticulum (which may be somewhat paired in fishes) branches and expands to become the liver. However, liver tissue may also originate in adjacent tissues and migrate *toward* the gut, in which case the secretory tissue will be of mesodermal as well as endodermal origin.

COMMENT 12.1

FIBER, FERMENTATION, AND GUT FORM

Any enzymatic breakdown of energy-rich compounds is fermentation, but in the context of vertebrate digestion fermentation is taken to be the breakdown of plant fibers by symbiotic bacteria or protozoa. Such fermentation may never have occurred in jawless vertebrates. It is rare in fishes, but limited fermentation probably occurs in several. Adult amphibians do not eat fibrous plants—no fermentation there. Few surviving reptiles are herbivorous, and those that are (some turtles and lizards) must manage without efficient chewing or a constant elevated body temperature (each of which favors fermentation). What about the dinosaurs? Most were herbivores and, of course, many were huge. They must have required correspondingly huge amounts of their (probably) low-quality fodder; must have had large stomachs and bulky intestines; and, unless they were unique among vertebrates in producing the enzyme cellulase, relied on fermentation. The teeth of some were effective shredders of fibrous foods, and others had gizzardlike grinding mills in their stomachs. The large ones probably had an elevated and nearly constant body temperature (see Comment 23.1 on p. 429).

Few birds that can fly employ fermentation. It is speculated that the bulk and weight of a fermentation chamber in the posterior part of the body would handicap flight. The large flightless birds (e.g., ostrich) do have fermentation; and the hoatzin, a leaf-eating jungle bird, is unique in accomplishing fermentation in a large, modified crop.

Many mammals digest coarse vegetation by fermentation (Table 12.1). Not all are large, but all large browsers and grazers are fermenters. Fermentation may take place in the foregut or hindgut. If in the foregut, the stomach is capacious (about 200 L in a hippopotamus) and is variously divided into compartments. Some foregut fermenters (and nearly all of the really large ones) ruminate, or chew their cud; that is, they regurgitate food, one bolus at a time, from the first and largest chamber of the stomach, the **rumen,** chew it thoroughly, and swallow it again (Figure 12.9). The mash is macerated and stirred by the rumen and the second chamber, the honeycombed **reticulum,** and is recycled as cud as often as is required. Finer particles tend to settle past the reticulum into the pleated **abomasum,** which pumps the mixture on to the **omasum.** The first three chambers are derived from the esophagus and are lined with esophageal epithelium; the glandular, digestive omasum corresponds to the entire stomach of a human or dog. The stomachs of suckling ruminants have a groove with muscular lips that shunts milk directly from esophagus to omasum.

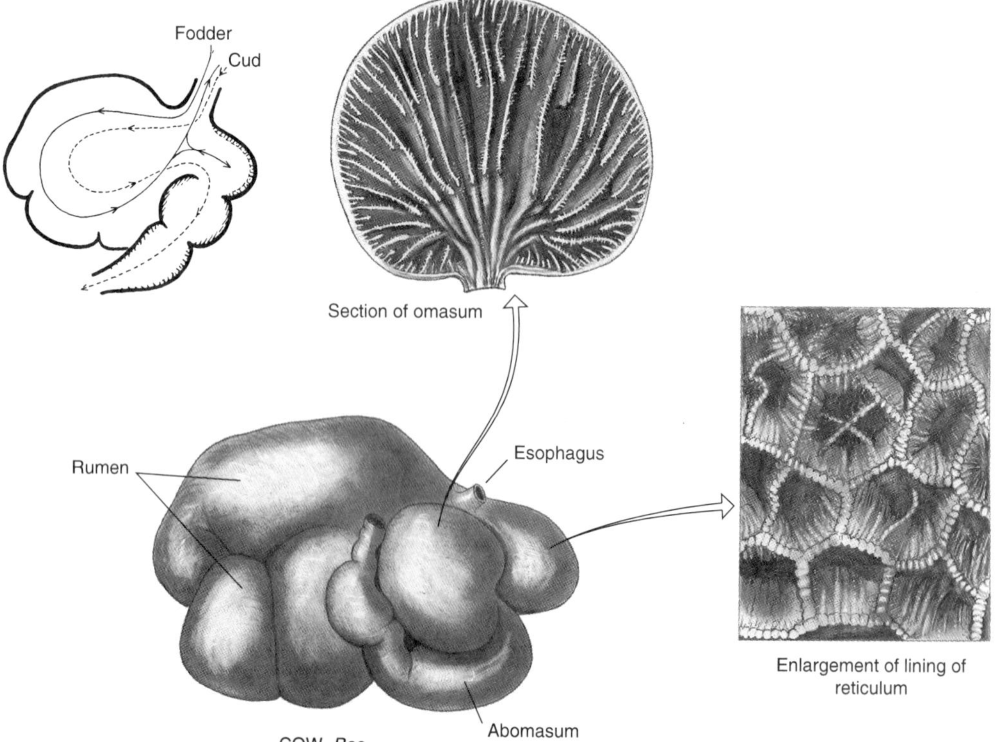

FIGURE 12.9 COMPLEX STOMACH OF A RUMINANT HERBIVORE.

Hindgut fermenters have a huge caecum, or large intestine, or both. These organs are pleated and sacculated, which slows the passsage of their contents. In only selected species (panda, and among birds the emu) does fermentation take place in an enlarged midgut.

Both foregut and hindgut fermentation have advantages and disadvantages. Foregut fermentation is relatively efficient: Products of digestion pass into the midgut, where they can be absorbed. Quantities of microorganisms are also flushed into the intestine, where they are digested, providing protein and other nutrients. Foregut fermentation seems particularly advantageous for large herbivores. However, digestion is very slow (commonly 75–100 h transit time).

Hindgut fermentation is more widespread, and apparently evolved earlier and more often. Hindgut fermenters, not needing to retain their food in their stomachs for long periods, can eat more over an extended period and process it faster (commonly 30–45 h transit time). To a degree they substitute quantity for efficiency. Hindgut fermentation is favorable for the prompt elimination of undigestable products such as silica, tannins, and resins. Soluble nutrients are absorbed in the midgut before fermentation starts, but some potential nutrients produced by fermentation cannot be absorbed at all. For this reason rabbits and various rodents eat many of their fecal pellets (the practice of *coprophagy*) in order to recover protein and vitamins.

Blood coming to the liver from the viscera in the hepatic portal system seeps through specialized capillaries called **sinusoids** before collecting in the hepatic veins to exit from the organ (Figure 12.10). The sinusoids are suspended in a labyrinth of tunnels that weave within a meshwork of interconnecting cellular walls. Sphincters control blood flow in the sinusoids. The walls between sinusoids are one cell thick in mammals and some birds; they are two cells thick in other vertebrates. Adjacent cells are held together by the tiny bile channels that follow their borders and also by projections that fit like snap fasteners into holes in adjacent cells. In higher vertebrates the vessels and ducts of the liver are arranged into polyhedral units called **lobules** (Figure 12.11).

The hepatic diverticulum of amphioxus is a single hollow organ that lies against the right side of the gut. Its fine structure and functions are unlike those of the vertebrate liver, with which it is homologous, if at all, only in a general way.

The liver tends to be lobed, and two principal lobes are frequent. However, the lobes may be side by side or one in front of the other, and the organ can have no lobes or several arranged in a variety of patterns that are without known functional or systematic significance.

Bile is usually prevented from entering the gut unless chyme (partly digested food) is leaving the stomach. At other times it backs up into the **gall bladder,** which stores it temporarily and concentrates it as much as 10 times. The **cystic duct** from the gall bladder joins

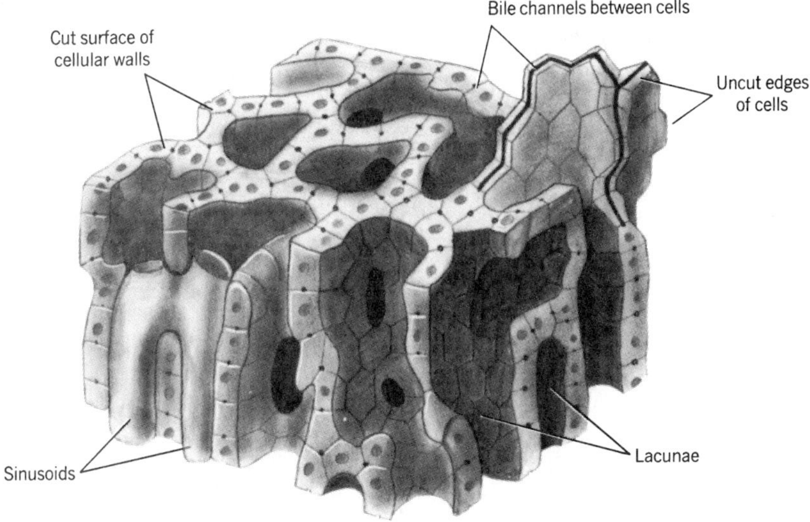

FIGURE 12.10
STEREOGRAM OF A FRAGMENT OF MAMMALIAN LIVER.

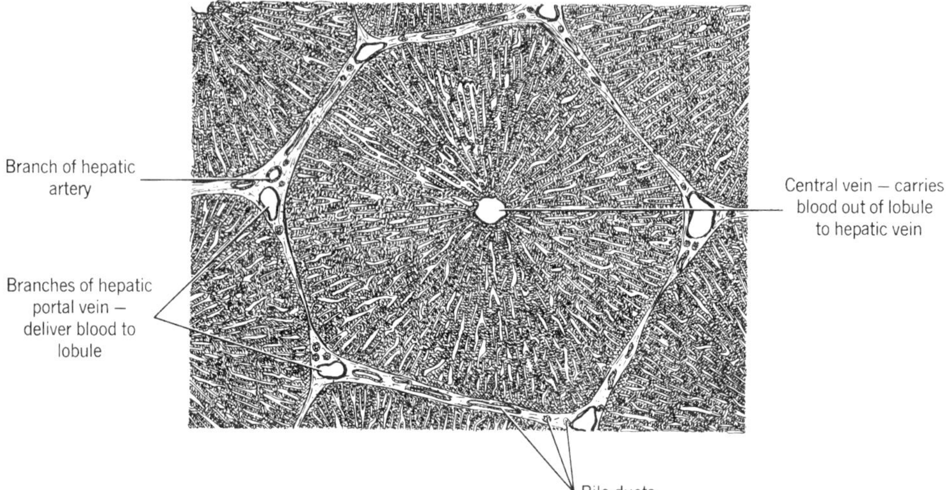

FIGURE 12.11
ONE LIVER LOBULE AND PARTS OF ADJACENT LOBULES as seen in cross section. Enlarged about $\times 40$.

the **hepatic duct** from the liver to continue to the duodenum as the **common bile duct.** The amount of bile secreted in relation to body weight varies widely among vertebrates. The presence and size of the bladder correlates with rate of bile secretion, intermittence of feeding, and fat content of the diet. The organ is always present in carnivores. It is lacking in the adult lamprey, several teleosts, and in certain herbivores distributed in five families of birds and six orders of mammals.

Pancreas The pancreas is a pale-colored organ that lies adjacent to the duodenum. It is found only in vertebrates, and all vertebrates have a pancreas. It is always a compound organ having both **exocrine** (secreting into a duct system) and **endocrine** (secreting into the blood) functions. About half a dozen enzymes are present in the pancreatic juice. These are able to digest nearly all foodstuffs. The endocrine secretions, insulin and glucagon, are essential for control of the intermediate metabolism of carbohydrates.

The pancreas has its origin in aggregations of cells in the wall of the embryonic foregut at the level of the hepatic diverticula. These form a dorsal pancreatic diverticulum and a pair of ventral diverticula (Figure 12.1). Growth brings the branching diverticula together, and they contribute in various combinations to the adult organ. In mammals, the dorsal diverticulum contributes part of the exocrine tissue and seemingly all of the endocrine tissue. It is apparent from this complicated ontogeny why the pancreas may have one, two, or three ducts with considerable individual as well as systematic variation.

The exocrine tissue of the pancreas is tubular, branched, and alveolar (Figure 12.12). Individual cells are pyramidal. Those at the ends of the alveoli contain granules that are

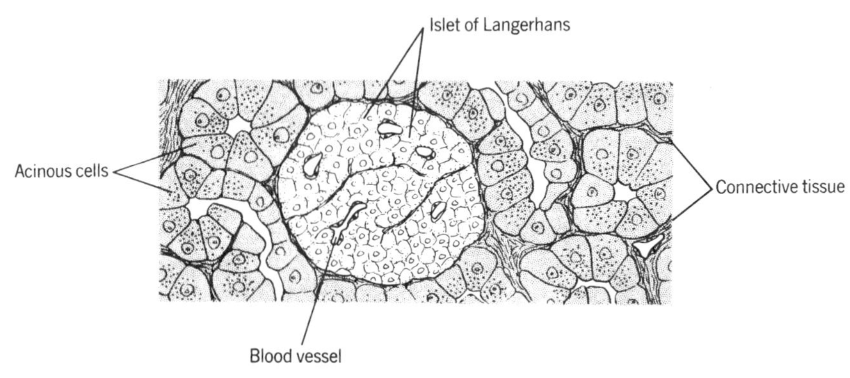

FIGURE 12.12
SECTION OF THE PANCREAS OF A MAMMAL.

converted to enzymes when released. The endocrine tissue consists of several kinds of cells that form small, scattered aggregations called **islets of Langerhans.**

Amphioxus has no pancreas, but isolated cells of the exocrine type are found in the wall of the gut around the opening of the hepatic diverticulum. In cyclostomes the pancreatic cells form aggregates under the duodenal epithelium. Dipnoans do not have a discrete pancreas but the organ is present in the intestinal wall. The organ is free and compact in selachians. It is diffuse in teleosts, being scattered as flecks of tissue in the mesenteries or along the blood vessels that penetrate liver and spleen. The pancreas of birds is somewhat three-lobed and usually has three ducts. In other tetrapods the organ is more compact and usually has two ducts or one, which may join the common bile duct.

REFERENCES

Brannon, P.M. 1990. Adaptation of the exocrine pancreas to diet. *Ann. Rev. Nutr.* 10:85–105.

Chivers, D.J., and C.M. Hladik. 1980. Morphology of the gastrointestinal tract in primates; comparisons with other mammals in relation to diet. *J. Morphol.* 166:337–386.

Chivers, D.J., and P. Langer (eds.). 1994. *The digestive system in mammals: food, form, and function.* Cambridge Univ. Press, Cambridge. 446p. Broadly comparative; relates structure and function to the physical and chemical nature of diets.

Davenport, H.W. 1982. *Physiology of the digestive tract: an introductory text.* 5th ed. Mosby Year Book, St. Louis. 245p.

Doran, G.A. 1975. Review of the evolution and phylogeny of the mammalian tongue. *Acta Anat.* 91:118–129.

Elias, H. 1955. Liver morphology. *Biol. Rev.* 30:263–310.

Gorham, F.W., and A.C. Ivy. 1938. General function of the gall bladder from the evolutionary standpoint. *Field Mus. Nat. Hist., Chicago, Zool. Ser.* 22:159–213.

Grajal, A., et al. 1989. Foregut fermentation in the hoatzin, a neotropical leaf-eating bird. *Science* 245:1236–1238.

Hill, W.C.O. 1926. A comparative study of the pancreas. *Proc. Zool. Soc. London*, 1926:581-631.

Hume, I.D. 1989. Optimal digestive strategies in mammalian herbivores. *Physiol. Zool.* 62:1145–1163.

Hume, I.D. 1998. Optimization in design of the digestive system. pp. 212–219. *In* E.R. Weibel, C.R. Taylor, and L. Bolis (eds.), *Principles of animal design.* Cambridge Univ. Press, New York.

Langer, P. 1988. *The mammalian herbivore stomach.* Gustav Fisher, New York. 557p.

McAllister, J.A. 1987. Phylogenetic distribution and morphological reassessment of the intestines of fossil and modern fishes. *Zool. Jb. Anat.* 115:281–294.

Moog, F. 1981. The lining of the small intestine. *Sci. Am.* 245(5):154–176.

Nickel, R., A. Schummer, and E. Seiferle. 1979. *The viscera of domestic mammals.* 2nd ed. Springer, New York. 401p. Fine illustrations.

Stevens, C.E., and I.D. Hume. 1995. *Comparative physiology of the vertebrate digestive system.* 2nd ed. Cambridge Univ. Press, New York. 400p.

Warner, E.D. 1958. The organogenesis and early histogenesis of the bovine stomach. *Am. J. Anat.* 102:33–63. General comments on ontogenetic and phylogenetic origins of the parts of mammalian stomachs.

Young, J.A., and E.W. van Lennep. 1978. *The morphology of salivary glands.* Academic Press, New York. 310p.

Chapter 13

Respiratory System and Gas Bladder

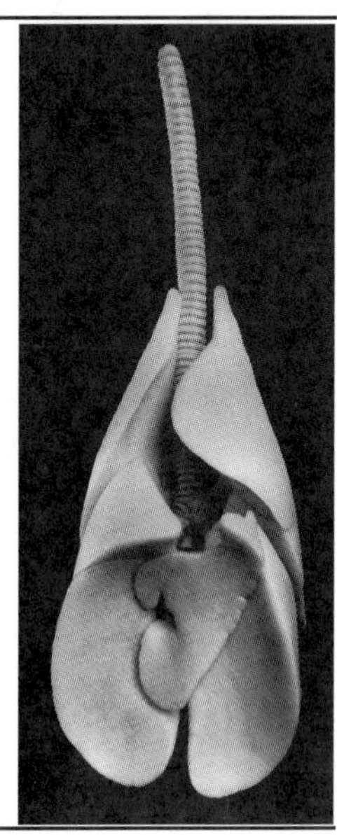

GENERAL FUNCTION AND REQUIREMENTS

All the cells of all vertebrates use oxygen and release carbon dioxide. The only way that these gases are moved in and out of tissues is by **diffusion.** Since all vertebrates are much too large for each cell to interact directly with the environment, certain organs, composing a respiratory system, are specialized to accomplish the essential gaseous exchange with the environment in behalf of all of the body. Such exchange is termed **external respiration** and may take place in certain fetal membranes, at the surface of the skin, in gills, in lungs, or occasionally in some other place. Oxygen and carbon dioxide are transported between respiratory organs and other tissues by the circulatory system. They are then exchanged with the tissues in the respective capillary beds, a process termed **internal respiration.**

Diffusion of gases in the respiratory system (i.e., the efficiency of external respiration) is increased by (1) providing a large surface of contact between the environmental medium (water or air) and the blood, (2) reducing the thickness of the barrier between medium and blood, (3) maintaining the contact of each blood cell with the medium for an adequate time, and (4) establishing for each gas a large diffusion gradient between medium and blood. It is the function of the respiratory organ to provide the first three of these requirements. The fourth requirement is met by behavioral and structural adaptations for moving both medium and blood to, from, and within the respiratory organ. The pumping of water in gills and of air in lungs is called **ventilation.** This chapter presents the structure, function, and evolution of respiratory and ventilation systems.

AQUATIC GAS EXCHANGERS

Cutaneous Respiration The chordates that were ancestral to vertebrates probably had large and complicated gills resembling those of the protochordates amphioxus and tunicates. These seem to have evolved, however, to accomplish filter feeding rather than respiration. These animals were small, probably lived in well-aerated and constantly moving water, and may have respired through their integuments.

Above: Air-dried lung of a grison (weasel family).

Cutaneous respiration is effective for small vertebrates, with low levels of activity, that live in cool, flowing water or damp air. One large group of salamanders has neither gills nor lungs as adults. Frogs and their tadpoles meet roughly half their needs for gas exchange through their highly vascular skins, and some kinds of fishes take enough oxygen in through the skin to meet the needs of the skin itself.

Nevertheless, efficient respiratory gills evolved from the pharynx very early in vertebrate history.

Development and Structure of the Pharynx Certain larval vertebrates have gills that develop from surface ectoderm and extend beyond the contour of the head. These are called external gills and are discussed in a subsequent section. Most gills lie within the contour of the head and are called internal gills. Internal gills develop in relation to the pharynx.

The pharynx is the portion of the foregut that lies between the developing oral cavity and the esophagus. It is lined by endoderm. The lateral walls of the embryonic pharynx develop six or more pairs of evaginations termed **pharyngeal pouches** (Figure 13.1). Opposite to the pouches, on the outside of the body, are shallow indentations called **visceral grooves.** Pouches and grooves are separated by thin partitions termed **closing plates.** Adjacent pouches are separated by **visceral arches,** each pair of which contains its respective pair of arteries or **aortic arches.**

Subsequent development depends on the presence or absence of jaws, gills, and related structures in the adult. In agnathans the first pharyngeal pouch becomes a typical **gill chamber.** In other fishes it is lost or is reduced in size and modified in function to become the cavity of the **spiracle.** Succeeding pouches of agnathans and fishes form gill chambers. The first pouch of tetrapods becomes the cavity of the middle ear; more posterior pouches form gill chambers in larval amphibians, but otherwise lose their pouchlike character after contributing glandular and lymphatic tissues from their epithelial linings (see Chapter 20). At the placoderm level of evolution, however, a posterior pair of pouches probably became the primordia of air bladders and lungs.

The closing plates of gill-bearing vertebrates rupture in the embryo to establish communication between the gill chambers and the outside of the body. Tetrapods retain the first closing plate as the eardrum; succeeding closing plates and visceral grooves of amniotes and of most adult amphibians lose their identity and leave no derivatives.

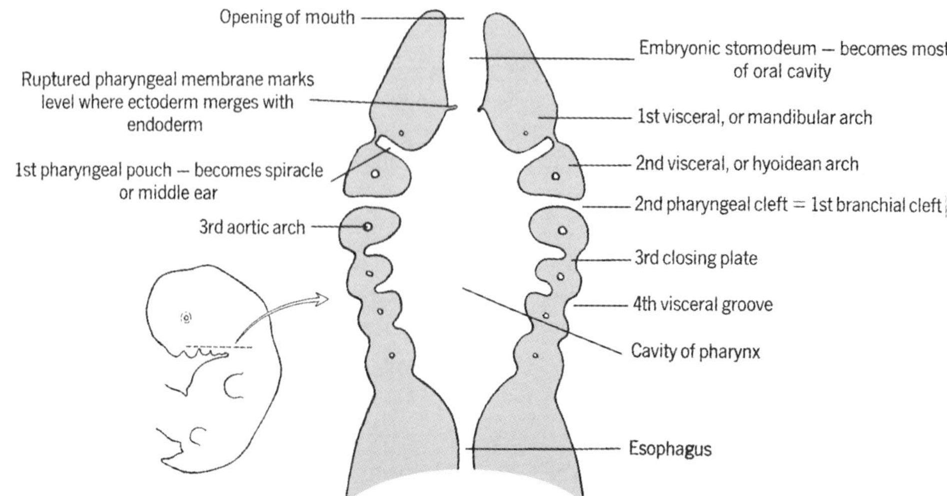

FIGURE 13.1 FRONTAL SECTION OF THE HEAD AND PHARYNX OF THE LARVA OR EMBRYO OF A JAW-BEARING VERTEBRATE at a developmental stage corresponding to that of the lower drawing on p. 202. Compare also with the figure on p. 119.

The first visceral arch (the mandibular arch) becomes the jaws of gnathostomes. The second (or hyoid) arch usually supports the jaws in fishes and contributes to the middle ear in tetrapods. As was explained in Chapter 8, the more posterior visceral arches support the **gill bars** of fishes and have various derivatives in tetrapods. The respiratory epithelium of gills develops from the margins of visceral arches near the positions of the embryonic closing plates; it is considered to be endodermal in cyclostomes but probably is more often ectodermal in gnathostomes. As explained in Chapter 10, musculature related to derivatives of visceral arches develops in part from somitomeres; that of the floor of the pharynx is hypobranchial in origin.

General Structure and Function of Gills The basic structure of internal gills is remarkably similar for all fishes. Each gill bar consists of a part of the visceral skeleton, blood vessels derived from the corresponding aortic arch, the associated cranial or cervical nerve, intrinsic branchial muscles, and the related epithelium (Figure 13.2). The bars may or may not be extended by **gill septa** of supportive tissue. These are sometimes short, but in elasmobranchs they extend to the body surface, thus separating adjacent gill slits. On their pharyngeal (inner) margins most bars carry one or more rows of **gill rakers,** the function of which is to prevent food particles from entering the gill chambers.

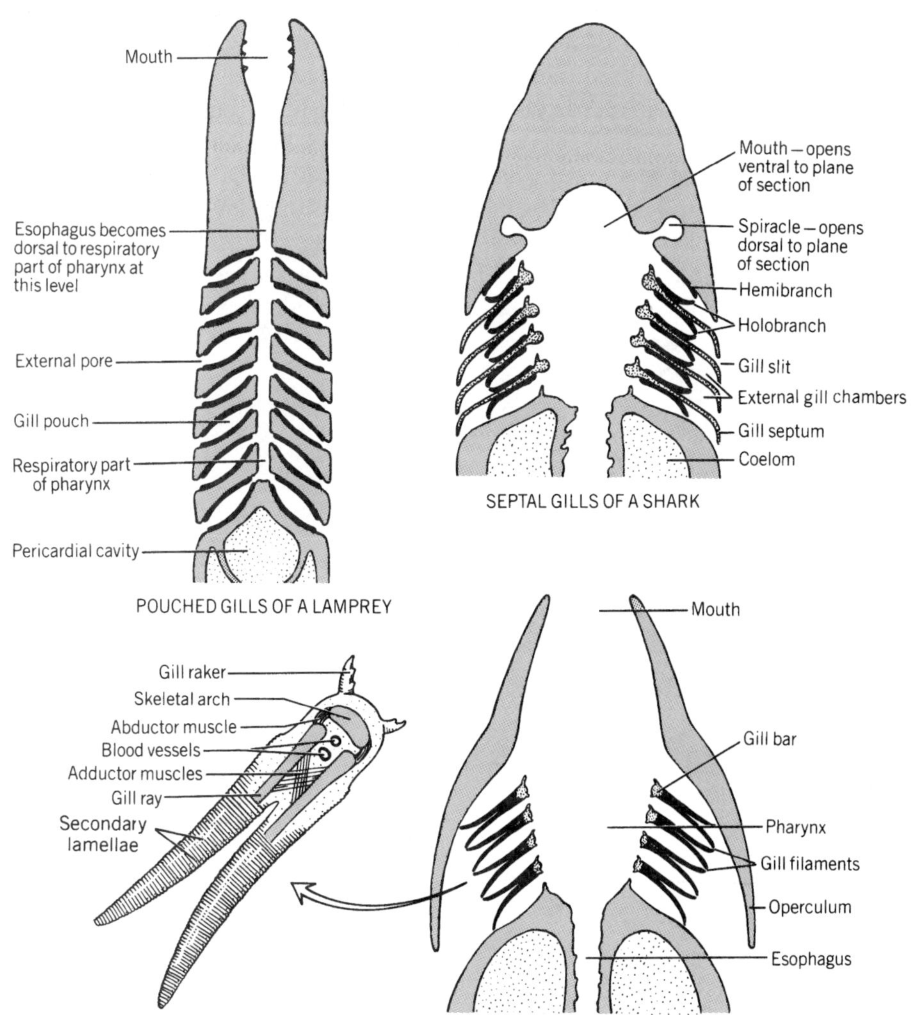

FIGURE 13.2 TYPES OF VERTEBRATE GILLS shown by frontal sections of the head and pharynx, and STRUCTURE OF A GILL BAR AND FILAMENTS.

A typical gill bar bears two rows of **gill filaments** (or primary filaments), which are on opposite sides of the septum if a septum is present. Each row may be likened to a comb, with the long delicate filaments corresponding to the teeth of the comb. Each filament is stiffened by fine skeletal gill rays and is subject to muscular control. A bar with filaments (and septum if present) forms a partition between adjacent gill chambers. It follows that the anterior row of filaments faces into one chamber, with the other row facing into a succeeding chamber.

A single gill bar with its anterior and posterior rows of respiratory filaments is called a **holobranch** (= *entire* + *gill*). If a bar bears filaments on one surface only, it is a **hemibranch** (= *half* + *gill*). The filaments on the posterior surface of the mandibular arch (facing into the first or spiracular gill chamber) are often modified to serve a nonrespiratory function and then compose a **pseudobranch** (= *false* + *gill*).

The gill filaments bear, on both their upper and lower surfaces, tiny parallel **secondary lamellae** numbering in each series about 20/mm of filament, but ranging from 10 to 40/mm of filament. It is important that the tips of the filaments from adjacent bars touch to close the gill sieve, and that lamellae of neighboring filaments be close together. In some fishes cavernous blood channels within the filaments serve as a hydrostatic mechanism to ensure that the filaments remain stiffly in position. Thus water passing through the gills must seep through the minute crevices of the mesh. The total respiratory surface per unit weight of fish varies about 10-fold according to the activity of the species. For a sea bass, which has an intermediate value, an individual weighing 20 kg has a respiratory surface of about 9.2 m^2 (60 ft^2 for a fish of 44 lb).

An **afferent branchial artery** ("afferent" = *toward* + *to bear*) enters each gill bar from below. As it extends upward it gives off **filamental vessels,** one of which loops to the apex of each filament and returns to drain into an **efferent branchial artery** ("efferent" = *away* + *to bear*), which continues upward and out of the gill. Capillary beds in the lamellae arch between the loops of the filamental vessels. Each capillary is scarcely larger in diameter than a single red blood cell, and is constituted by the enfolding arms of pillar cells (Figure 13.3). The nuclei of the flat epithelial cells lie opposite the nuclei of the pillar cells so they will not retard diffusion. The resulting water-to-blood barrier is only about 1 μm thick, and each blood corpuscle is exposed to the water on both sides of a lamella.

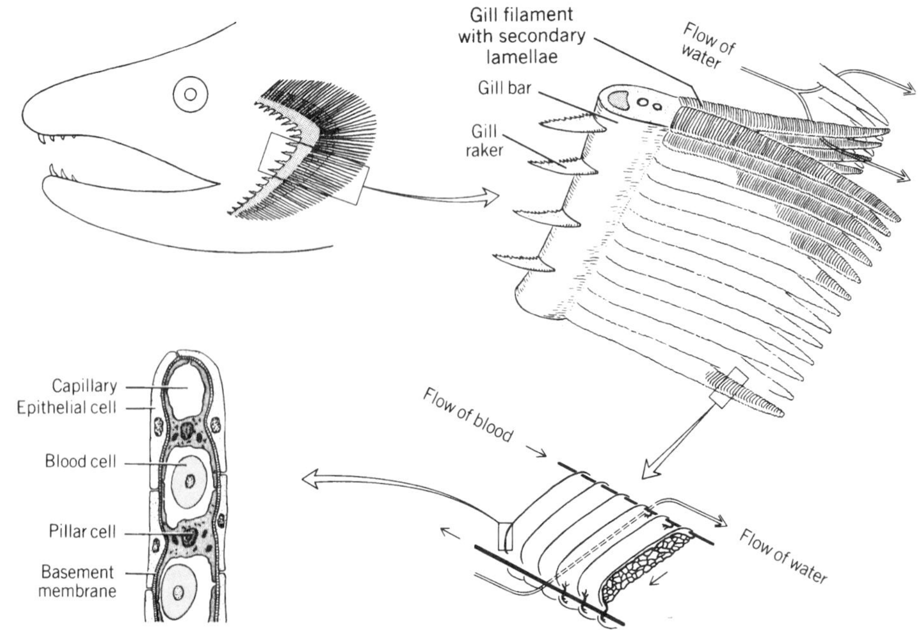

FIGURE 13.3
STRUCTURE OF A GILL BAR, GILL FILAMENTS, AND LAMELLAE showing the structural basis for large surface area, countercurrent exchange, and thin water-to-blood barrier.

Blood passes through sphincters both on entering and on leaving the lamellae, and pillar cells contain actin and myosin, suggesting that they may be contractile. Thus flow of blood is controlled, even at the lamellar level.

Lamellae are oriented parallel to the stream of water, which always passes between them from the inside of the gill apparatus to the outside. Blood always flows in the lamellae from the outside of the apparatus inward. Consequently, when blood first enters the area of respiratory exchange, it contacts water that is almost depleted of oxygen. This blood, however, has even less oxygen content, so a gradient exists and oxygen immediately diffuses into the blood. As the blood continues through the lamellae, its oxygen content increases, but so does the oxygen content of the ever "fresher" water with which it comes in contact. Thus, a diffusion gradient is maintained and gaseous exchange continues until the blood leaves the lamellae. The diffusion of carbon dioxide from blood to water is simultaneously favored by the same mechanism (Figure 13.4). The transfer of heat or diffusible materials between currents of gas or liquid passing one another in opposite directions is called **countercurrent exchange.** It was perfected several hundred million years ago in the gills and gas bladders of fishes and, as will be explained elsewhere in this book, is the basis for other important adaptations of vertebrates.

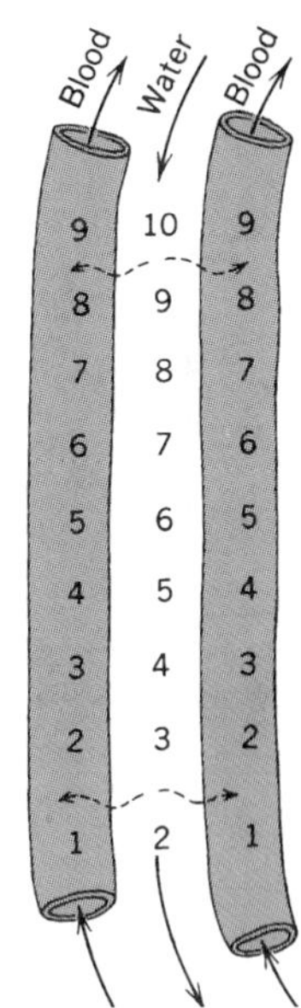

FIGURE 13.4 COUNTERCURRENT EXCHANGE between capillaries of two adjacent gill lamellae and the water flowing between them. Because blood and water flow in opposite directions, a diffusion gradient is maintained as exchange progresses.

In summary, diffusion is favored in many ways. The combined surface area of the lamellae is enormous, and the combined surface area of lamellar capillaries is correspondingly large. The barrier between blood and water is exceedingly thin. The fine mesh of the gills slows the water stream, and both water and blood are slowed by the expansion of their combined respective channels at the site of the exchange. This provides time for diffusion to approach equilibrium. Countercurrent exchange assures that the diffusion gradients for oxygen and carbon dioxide will scarcely diminish as diffusion proceeds. The hemoglobin of fish blood combines more readily with oxygen than does that of tetrapods. Finally, as explained in a following section, ventilation moves water through the gills in a nearly continuous flow.

The teleost gill is the most efficient respiratory organ in the vertebrate world. Up to 80 percent of the oxygen in water is removed. Efficiency is essential because water contains only about one-thirtieth as much oxygen as air (depending on temperature), and diffusion is many thousand times slower.

The universal respiratory function of gills is exchange of oxygen and carbon dioxide between blood and environment. Gills also function importantly in excretion and osmoregulation. Bony fishes excrete nearly all their nitrogenous waste from the gills. It is transported in the blood as urea but is mostly converted in the gills to ammonia, which is lost by diffusion. Gills of freshwater bony fishes passively admit water and actively absorb salts; gills of marine bony fishes pass little water and actively excrete salts. Cartilaginous fishes are unique in that their gills are nearly impervious to urea, which is retained in high concentration in the blood and tissues. These differences among the fishes will be correlated with kidney function and osmoregulation in Chapter 15.

Evolution of Internal Gills Gills are commonly classified in three categories according to arrangement and relation to supportive and protective tissues. Some gills, however, are intermediate in nature.

Pouched gills are characteristic of agnathans. The gill filaments are arranged over the surface of discrete spherical or lenticular pouchlike gill chambers (Figure 13.2). Each pouch may have its own external pore (never a large slit) to the surface of the body (cephalaspids, anaspids, lampreys), or the pouches on each side of the body may communicate with the outside by a common duct and pore (pteraspids, some hagfishes). Similarly, each pouch characteristically has its own internal pore to communicate with a typical pharynx (hagfishes, probably cephalaspids) or with a division of the pharynx that is separate from the path of food (lampreys).

The number of gill chambers ranges from 5 to 15, being variable even within orders. The evolutionary relationship of pouched gills to other types is not clear.

Septal gills differ from pouched gills in that the gill chambers tend to be larger, to communicate more widely with the pharynx internally, and to communicate with the outside of the body through vertical **gill slits** instead of pores. The serial slits cause the supportive tissue that joins the gill bars to the surface of the body to be in the form of platelike **gill septa** that share in the support of the gill filaments. This is the basis for the term **elasmobranch** (= *plate* + *gill*), which is applied to fishes with septal gills. These are Selachii, Cladoselachii, and Pleuracanthodii.

The first gill chamber and cleft of elasmobranchs is reduced to a spiracle. On the anterior face of the spiracular chamber is a vascular pseudobranch that receives oxygenated blood and may monitor it in some way before delivering it to the eye.

Elasmobranchs nearly always have a hemibranch on the anterior face of the first branchial chamber and then four holobranchs. There are four posthyoidean gill-bearing arches and five gill clefts. Three genera of sharks have six clefts and one has seven. Perhaps the greater number is the ancestral condition, but this is not known. In contrast to jawless vertebrates, stability, not variability, in number of clefts should be stressed for elasmobranchs.

Holocephalians have gills that are intermediate between septal gills and those of the next category. Septa are present but are much reduced so that effective serial gill slits are absent. Instead, a unique fleshy operculum covers the gills. These fishes are also distinctive for having an anterior hemibranch, three instead of four holobranchs, and then a posterior hemibranch. The fifth cleft is closed.

Opercular gills are characteristic of Osteichthyes. Septa are usually shorter than their filaments, and may be virtually absent so that only the gill bars remain to anchor the gill filaments. A bony operculum is present to protect the otherwise exposed filaments and to contribute to the pump and valve system of the gill apparatus.

A spiracle is present in the living coelacanth and is usually present in Chondrostei. These fishes have a pseudobranch on the anterior face of the spiracular chamber. Other bony fishes lose the spiracle but may retain the pseudobranch, which then fills a depression on the wall of the pharynx or moves back to the anterior face of the first branchial chamber. The structure of pseudobranchs is various and their function is not adequately known: They may have a reduced number of free filaments, may be covered by epithelium, or may assume a glandular character and sink well below the surface. They are thought to contribute to the special metabolic requirements of the eye and other organs in ways that include enzymatic control of the dissociation of carbonic acid in the blood.

A hemibranch is present on the anterior face of the first branchial chamber (posterior wall of hyoid arch) of Dipnoi and some Chondrostei, as it is also in elasmobranchs. Otherwise this hemibranch is missing in bony fishes (although, as just noted, the pseudobranch of the spiracular chamber may move to a corresponding position when its "own" chamber is lost).

The functional gills of bony fishes usually consist of four holobranchs (starting with the third visceral or first branchial arch). The complement is reduced to three, two, or even one holobranch (plus a posterior hemibranch in some instances) by some Dipnoi and some Teleostei for which the gas bladder has come to function in respiration.

Few trends can be noted in the evolution of gills—particularly when it is remembered that septal gills, as exemplified by elasmobranchs, are not in the ancestry of opercular gills. There has been a general tendency to reduce and to stabilize the number of gill bars and chambers. Within the Osteichthyes, but not the Chondrichthyes, there was an early trend toward loss of the spiracle and loss of septa (but among amphibians, some larval caecilians and several salamanders still retain a spiracle). Other variations relate to habit or are seemingly nonadaptive (for instance, the determination of which holobranchs will be retained as the series is reduced).

External Gills External gills develop from skin ectoderm of the branchial area but are not directly related to the visceral skeleton or branchial chambers. They are filamentous or featherlike, and their epithelium may be ciliated. Blood supply is indirectly from the second aortic arch if there is a single pair of gills, and from several aortic arches if there are more. Muscles wave the gills to provide ventilation in quiet water.

External gills occur in the larvae of two of the three Recent genera of dipnoans, the chondrostean *Polypterus*, one teleost, and amphibians. They occur also in adults of **perennibranchiate** (= *through the year + gill-bearing*) urodeles, which, however, are neotenous in this respect (see Figure 13.5 and the figure on p. 70). The principal function of external gills is respiratory, but they develop early in ontogeny—before the pharyngeal slits are formed—and in some instances absorb yolk or other nutrients before the fish hatches. It is clear that external gills are a larval adaptation that evolved early in the phylogeny of bony fishes; they probably occurred in crossopterygians. There is no evidence that external gills ever occurred in adult fishes.

The terms "internal" and "external" in reference to gills are not always apt. Some larval rays have extensions from their "internal" gills that stream far out of the gill slits. Young tadpoles of anurans have "external" gills that are hidden under a fleshy operculum. Some older tadpoles have gills of uncertain homology.

VENTILATION OF INTERNAL GILLS

Since water contains much less oxygen than air, and diffusion is very much slower, effective ventilation of the gills is essential, yet the greater density and viscosity of water place a greater burden on the ventilation apparatus.

Ostracoderms probably took water into the pharynx by way of the mouth, passed it through the gill pouches, and expelled it from the head through the branchial pores. Cyclostomes also expel water through the branchial pores, but are unique in that the specialized use of the mouth for feeding and prehension precludes its constant use in respiration. Instead, water may enter and leave the gill pouches from the external branchial pores. Flow is thus tidal, though water probably passes the lamellae in only one direction. The visceral skeleton is a continuous, unjointed lattice of cartilage. Water is actively expelled from the pouches by muscular contraction; it reenters the pouches as the result of the inherent resilience of the visceral skeleton.

AFRICAN LUNGFISH, *Lepidosiren*

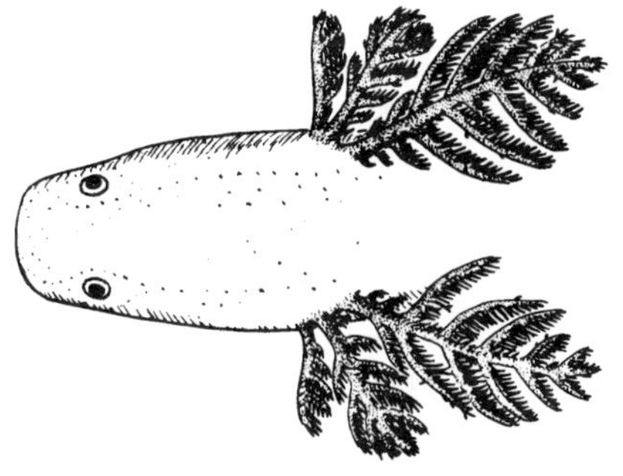

SALAMANDER, *Pseudobranchus*

FIGURE 13.5 EXTERNAL GILLS in larvae of a fish and an amphibian seen, respectively, in lateral and dorsal views.

Sharks draw water into the pharynx through mouth and spiracles. This is accomplished largely by elastic recoil of the visceral skeleton following the previous contraction of the pharynx, though hypobranchial muscles may contribute. At the same time, branchial muscles enlarge the external gill chambers (identified on Figure 13.2) between the filaments and the gill slits. This suction pump closes the valvelike slits and pulls water through the gills even though the slits are closed. Next, oral and pharyngeal chambers are compressed, converting them from a suction pump to a pressure pump and moving water through the gills to the external gill chambers. When the pressure there surpasses the outside pressure, the gill slits open, water is forced out, and the cycle is completed. The entire sequence can be reversed, forcing jets of water out of spiracles or mouth to clean the gills or reject food. When rays rest on the bottom, their ventral mouths are against the substrate. They may then suck water into the pharynx only through their large, dorsally positioned spiracles.

Ventilation of opercular gills resembles that of septal gills (Figure 13.6). Again, the passage of water through the filaments is nearly continuous, even though the operculum opens and closes. A flap of tissue on the inside of the operculum near its margin functions as a valve to seal the external gill chamber when it is acting as a suction pump. Some active teleosts (tuna, mackerel) swim continuously and water constantly enters the open mouth, whence it is forced through the gills by the swimming process.

AERIAL GAS EXCHANGERS

Origin and Development Warm or stagnant water retains much less oxygen than cold and moving water. Accordingly, the many fishes that live in swampy areas (probably including the ancestors of tetrapods) find it difficult or impossible to meet their needs for oxygen with gills alone. Carbon dioxide is more soluble in water, and diffuses more rapidly, so it is not the limiting factor. In oxygen-poor water, therefore, many fishes turn, at least in part, to aerial respiration. More than 20 genera of bony fishes are habitual air breathers, and some suffocate if prevented from breathing at the surface. The fish can make fewer trips to the surface if gill respiration is retained to rid the body of carbon dioxide, but some of these fishes have reduced gill area and, when not needed, perfusion of the gills by blood is probably reduced. In times of drought, gills are actually detrimental because they dry out easily.

Numerous fishes have "experimented" with gulping air into the pharynx, or pouches off the gill chambers, or gut where it comes in contact with specialized, highly vascularized epithelium. By far the most successful aerial gas exchanger, however, has been the lung.

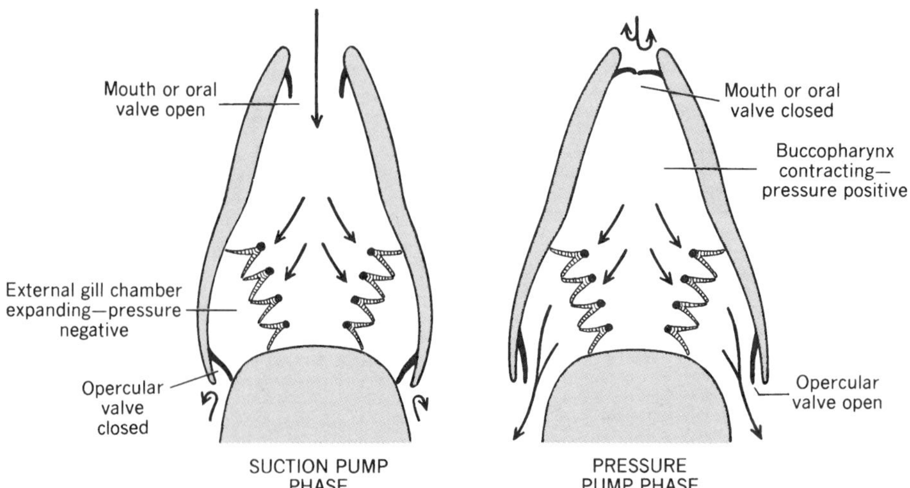

FIGURE 13.6
VENTILATION OF TELEOST GILLS. Frontal sections.

Internal organs derived from the gut tube that are filled with air and function primarily in respiration are here called lungs, whether they are the familiar paired structures of tetrapods or the unpaired structures of certain fishes. Internal organs that are filled with gas but are not respiratory are called gas bladders. These occur only in bony fishes.

The respiratory system of amniotes develops from a single, ventral evagination of the gut tube close behind the pharynx. The initial primordium quickly bifurcates to form two primary lung buds, which, in turn, divide further to produce whatever respiratory tree is to be present. Mesoderm contributes the supportive tissues. The lungs of Dipnoi and the chondrostean *Polypterus* also develop ventral to the gut and may be single, bilobed, or paired. Lungs of air-breathing Actinopterygii (except *Polypterus*) develop dorsal to the gut and are single. The same is true of gas bladders. How can these differences be explained? The following observations are useful: (1) A fossil of an antiarch shows imprints of paired, ducted bladders or lungs. (2) The tissue that evaginates to form the respiratory system of amniotes takes its origin from paired areas of endoderm located just posterior to the pharyngeal pouches. (3) In anurans and caecilians among amphibians, the lungs do in fact develop from paired lateral evaginations of the gut tube.

From these clues it appears that (1) the evolution of lungs extends back at least to the Devonian period; (2) lungs were initially paired lateral organs that developed in series with the pharyngeal pouches; (3) these ancient organs shifted ventrally to form the respiratory structures of tetrapods, dipnoans, and some primitive ray-finned fishes; and (4) in most ray-finned fishes either the ancient paired organs shifted dorsally to merge over the gut and become a single bladder, or one of the initial lungs was lost, leaving the other to shift around the gut to right or left. Embryology and the morphology of related blood vessels indicate that different groups of fishes made the shift in different ways.

Lungs and Gas Bladders of Fishes Because of its mode of origin, the embryonic lung or gas bladder is always joined to the gut by a **pneumatic duct.** If the organ is to function as a lung, the duct is retained (see figure on p. 204). Such a lung or bladder (and also a fish having this condition) is said to be **physostomous** (= *bladder* + *mouth*). The surface area of fish lungs is much increased by pockets and partitions in their walls. A gas bladder (or fish) having no pneumatic duct is **physoclistous** (= *bladder* + *closed*). Physostomes are usually fishes that are relatively primitive or unspecialized and live in freshwater. Conversely, physoclists tend to be relatively specialized and to be marine.

Gas bladders and lungs of fishes may be located high in the body cavity above the animal's center of mass. This arrangement enables the fish to remain upright without expending muscular effort. Of fishes having the gas bladder below the center of mass, most must use their fins to counter a tendency to roll, but one catfish has solved the problem another way—it habitually swims upside down.

Gas bladders constitute about 4 to 11 percent of the body by volume. They may be long or short, straight or curved, simple or partitioned into two or three more or less distinct parts. Gas is secreted into the bladder from the blood. Rarely (and primitively?), an extensive part of the epithelium of the bladder is vascularized for this purpose. Typically, the secretory area is limited to one or several anterior **gas glands** where the epithelium is columnar and tightly folded on itself (Figure 13.7). Underlying each gas gland is a **rete mirabile** (= *net* + *marvelous*) consisting of as many as tens of thousands of uniquely long afferent capillaries, all oriented in the same direction, among which pass a like number of efferent capillaries. Gas gland and rete mirabile together are called a **red body.** In the rete the surface contact between blood approaching and leaving the bladder may be more than a square meter. This countercurrent multiplier exchanges gases from the outgoing capillaries to the incoming capillaries, thus

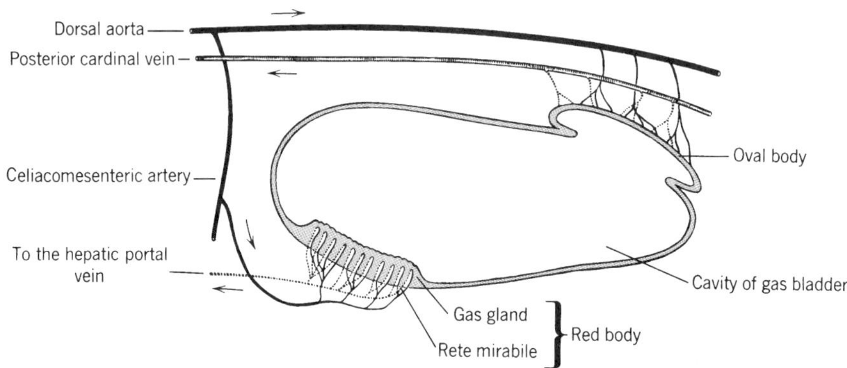

FIGURE 13.7
GENERAL STRUCTURE AND BLOOD SUPPLY OF A PHYSOCLISTOUS GAS BLADDER.

maintaining maximum gas content in the gas gland. However, the amazing physiology of the gland cannot be explained by diffusion gradients alone; in contrast to simple diffusion, which controls gas exchange across all respiratory surfaces, there are *active* metabolic processes at work. The oxygen content within the bladder of certain deep-sea fishes reaches nearly 1000 times the oxygen content of the surrounding seawater, and the partial pressure of oxygen in the bladder ranges as high as 200 atm.

Gas is resorbed from the epithelium of the posterior part of the bladder or from a portion thereof, which may be set off and then takes the name **oval body.** Arterial blood reaches the red body from the celiacomesenteric artery (which delivers to the anterior viscera) and reaches the oval body from the dorsal aorta. It leaves these structures, respectively, in the hepatic portal and posterior cardinal veins (described in the next chapter). This is in sharp contrast to the blood circulation of lungs. The red body is innervated by the vagus nerve and the oval body by the sympathetic nervous system. The lining of the gas bladder may be thin or fibrous. In some fishes the organ is more or less covered by striated muscle.

The most important function of virtually all physoclistous bladders, and also of some physostomous bladders, is **hydrostasis.** When this is true, the common term "swim bladder" is less apt; the mixture of oxygen, nitrogen, and carbon dioxide in the bladder approximates that of air when the fish is near the surface but changes as the fish descends, reaching nearly 90 percent oxygen in some instances.

By secreting or resorbing gas, the fish can alter the volume of its bladder. This changes the mass per unit volume of the fish and allows it to adjust its density to that of the environment. Adjustment is under reflex control. The change in the density (and hence buoyancy) of the fish that results from activity of its gas bladder is not rapid enough to be used as a means of moving from less dense surface water to more dense deep water (or the reverse), but once the fish has changed its level by swimming, it can maintain that level with the gas bladder. Some fishes make daily vertical excursions of as much as 1000 m. Fishes that inhabit shallow rapid streams, and marine fishes that remain at constant depth, tend to reduce or lose the gas bladder. Some deep-sea fishes have a bladder, and some do not. Bottom fishes such as flounders and halibuts have no bladder.

A second function of the gas bladder is **sound production.** The layperson's conception of fishes as being silent is not generally correct; many fishes produce buzzing, rasping, squeaking, clicking, or other sounds that function in aggressive, warning, or reproductive behavior. The bladder may serve as resonator for sounds produced by rubbing the pharyngeal teeth together or scraping certain bones together. Vibrations of up to 100 Hz may be produced by action of the intrinsic or extrinsic muscles of the bladder itself, and some physostomes produce sound by the controlled passage of air between the bladder and gut.

Another function of the bladder is **sound** and **pressure reception.** The soft tissues of the fish cannot receive sound waves because their density is nearly the same as that of water. Compact bone may serve, but the compressible gas bladder is far superior and has been adapted for that purpose by several orders and many families of fishes. Sound vibrations received by the bladder must be transmitted to the inner ear in order to be sensed. This is accomplished either by paired long extensions of the bladder into the back of the skull (cods, herrings) or by paired chains of four (sometimes three) ossicles that impinge on the bladder posteriorly and on the perilymph of the internal ear anteriorly (minnows, catfishes, carps). The ossicles are derived from processes of anterior vertebrae and are called the **Weberian apparatus.**

Evolution of Lungs from Amphibians to Mammals Typical lungs of tetrapods differ from fish lungs in being paired, in having a higher surface-to-volume ratio, in joining the ventral side of the gut tube by a duct termed the **trachea,** in receiving blood of low oxygen content in vessels that are related in development to the sixth aortic arches, and in returning oxygenated blood directly to the heart without prior mixing with blood from other organs. The principal evolutionary trend has been adaptation to increased body size or metabolic rate by increasing the compartmentalization of the lungs.

ANURA have large but short lungs (Figure 13.8). The interior of the lung is an open sac, but the walls have partitions of the first, second, and sometimes third order, providing a total respiratory surface of about 1 cm^2/g body weight, but varying inversely with the effectiveness of cutaneous respiration. The very short trachea divides into two short **bronchi,** one leading to the apex of each lung. The epithelium of these ducts is ciliated and thus is able to clean the respiratory system. Cartilage may support the walls of trachea and bronchi against collapse. The opening from trachea to pharynx is called the **glottis.** It is a longitudinal slit that is usually flanked by a dorsal pair of **arytenoid cartilages** ("arytenoid" = *cup-shaped*), which support **vocal cords,** and a ventral pair (often fused) of **cricoid cartilages** ("cricoid" = *ring-shaped*). These cartilages are regarded as derivatives of posterior visceral arches of ancestors. Together they constitute the **larynx,** a structure characteristic of tetrapods. The larynx is joined by ligaments to the hyoid apparatus.

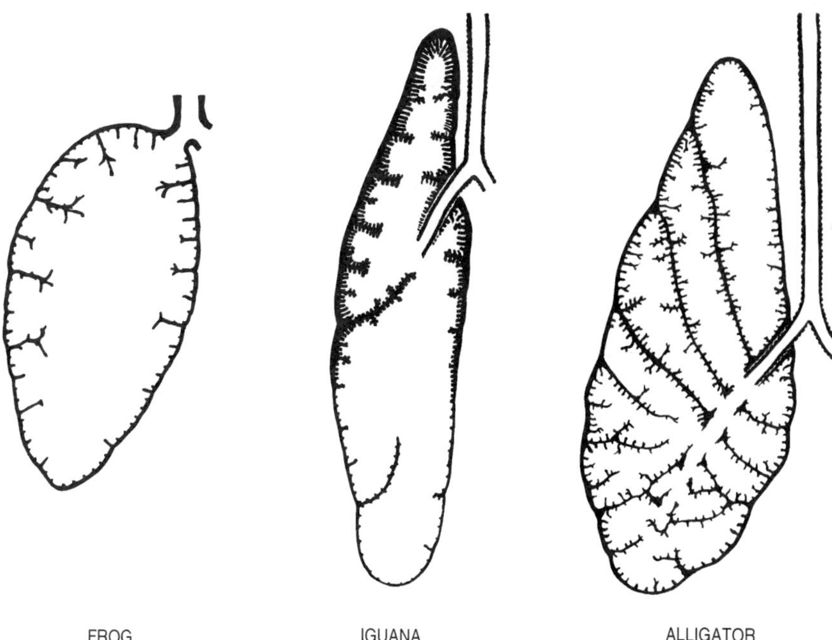

FIGURE 13.8 GROSS LUNG STRUCTURE OF SOME LOWER TETRAPODS as seen in frontal section. Somewhat stylized and two-dimensional. A finer order of branching is not visible at this scale.

APODA, having long, slender bodies, usually retain only the right lung. (Their larvae have both lungs.) Otherwise, their respiratory system resembles that of Anura. The more terrestrial LABYRINTHODONTS probably had the most advanced lungs of the class, likely resembling those of large reptiles. Lungs of URODELA, by contrast, have regressed; most species have lost their lungs entirely. Where present, they are often long, slender sacs with smooth walls. Their respiratory function is diminished (skin or gills serving instead), and they act as hydrostatic organs.

Lungs of REPTILES are large and varied. One of the pair may be reduced or rudimentary in long-bodied forms (snakes, amphisbaenians). There may be only one chamber in each lung and limited partitioning of the walls (most lizards), but, in order to maintain an adequate ratio of lung surface area to body weight, the larger reptiles (turtles, monitors, crocodilians) have lungs with many compartments and partitions along their walls (Figure 13.8). The lungs of some reptiles are of intermediate complexity (the iguana in the figure). Partitions can be sparse or dense, shallow or deep, and evenly or unevenly distributed. The anterior part of the lung is commonly partitioned into smaller, more vascularized, and stiffer chambers than the posterior part.

Trachea and bronchi are longer than for amphibians (much longer in snakes) and are supported by cartilaginous rings that may be closed or open dorsally. Each bronchus enters its lung near the middle or near the anterior end, but not at the apex. The bronchus continues within the lung without further division. The larynx again consists of cricoid and arytenoid cartilages (variously shaped and fused) that (except in snakes) are joined to the hyoid apparatus. Vocal cords are present only in some lizards. Many reptiles hiss by passing air slowly through a partly closed glottis, at the edge of which may be an erectile, sound-producing flap. (Certain snakes hiss by rubbing the scales of adjacent body regions together and amplifying this sound via the inflated body.)

Lungs of MAMMALS are much more finely and homogeneously divided. Therefore they are more efficient, as is required by the higher metabolic rate of these endotherms. Lobulation of the lungs is variable and without systematic import. Lobes may be absent (horse, whales, sea cows, some bats), but usually there are at least two lobes on the left and at least three on the right (Figure 13.9). The lobes may or may not be divided into lobules, which can be conspicuous or faint.

Of more functional importance is the nature of the respiratory tree. The trachea (like that of reptiles) is supported by rings of hyaline or fibrous cartilage, which are here incomplete dorsally. Elastic connective tissue joins ring to ring and completes the tube where cartilage is absent. The resultant structure is ideal for holding the airway open yet allowing the tube to twist as the neck is turned, to change length with swallowing, and to change diameter if there is marked alteration of internal pressure, as in coughing. The trachea is lined by ciliated epithelium. Smooth muscle and mucous glands are present in the walls.

The trachea divides into right and left primary bronchi, each of which enters its lung somewhat anterior and dorsal to the center. Mammalian lungs differ fundamentally from those of other tetrapods in that within their lungs, the primary bronchi divide into smaller bronchi, which divide again through numerous generations of branching. All internal bronchi are supported by cartilage, which, however, may be in plates rather than rings. The airways are continued by branching **bronchioles** that are membranous, not cartilaginous. These, in turn, open into nonciliated **respiratory bronchioles,** where gaseous exchange begins, and finally into **alveolar duct systems,** which are clusters of about 20 hemispherical **alveoli** all opening into a common terminal chamber (Figure 13.10).

Lung volume is nearly proportional to body size, but alveolar size (and hence respiratory surface) varies with metabolic rate. Total surface ranges from about 8 cm^2/g body weight (some primates) through 50 cm^2/g (mouse) to 100 cm^2/g (bat). In other terms, the

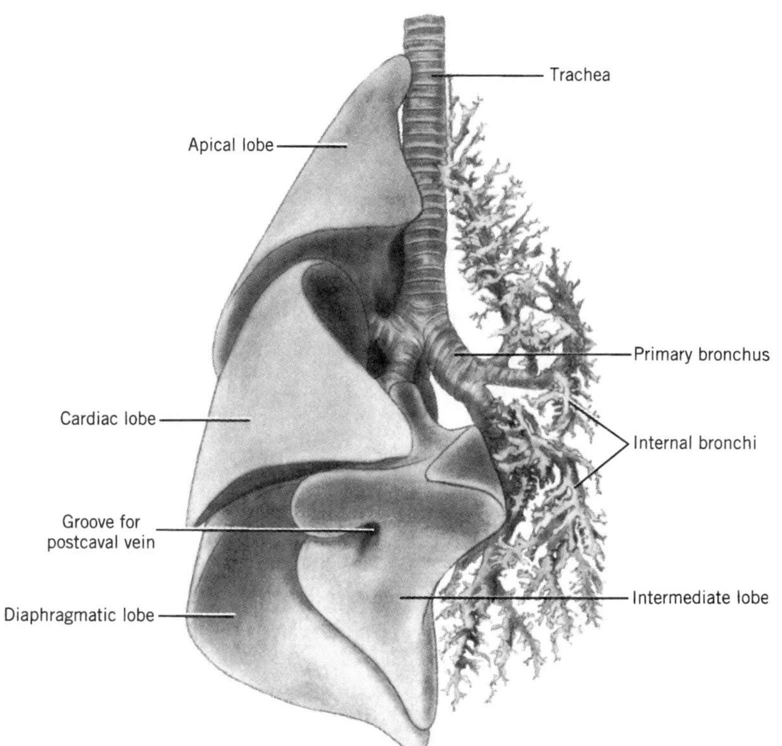

FIGURE 13.9 MAMMALIAN LUNG illustrated by the dog. Ventral view. Dissected on one side to show the bronchial tree.

respiratory surface of the human lung is about 150 m^2 or 85 times the surface area of the body, and a 20 kg dog has a respiratory surface that is 9 times greater than that of a moderately active fish of the same weight.

Surface tension of the liquid film within the small alveoli would cause the lungs of mammals and many other tetrapods to collapse were a lipoprotein not secreted that reduces the tension.

Pulmonary vessels may follow the airways, or the arteries only may follow according to species. Alveoli are richly supplied with capillaries where blood is separated from air only by the endothelium of the capillary, a basement membrane, and an exceedingly thin alveolar epithelium.

The large larynx is attached to the hyoid apparatus. Paired arytenoid cartilages help support and control the vocal cords. The cricoid cartilage is single. Two additional cartilages are present that are lacking in other vertebrates: a large ventral **thyroid cartilage** ("thyroid" = *shield-shaped*) and a cartilage in the **epiglottis.** The epiglottis is a stiff valve-like flap that guides air between posterior nares and glottis during respiration and, to keep food out of the respiratory system, closes the glottis during swallowing.

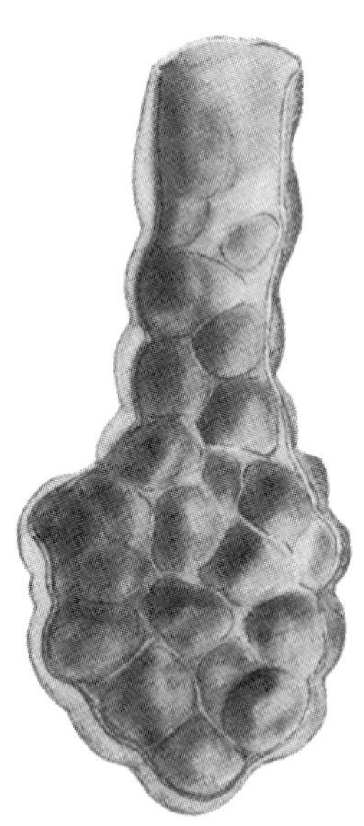

FIGURE 13.10 RESPIRATORY BRONCHIOLE AND RELATED ALVEOLAR DUCT SYSTEM OF A MAMMAL.

Avian Lungs and Air Sacs Birds match their uniquely high activity and metabolic rate with a respiratory system that is unique in structure, function, and high efficiency. Avian lungs are small and compact, and instead of moving freely in pleural cavities they adhere tightly to the dorsal ribcage. They are nearly rigid and have a virtually fixed volume. Large air sacs join the lungs and serve to ventilate them. In most of the lung air is not tidal (moving in and out of the lungs along the same pathway), but instead moves only in one direction (see below under the next heading). One would expect, therefore, that the air-blood circulation would be countercurrent in nature, but it is instead cross-current (Figure 13.11). Compared with mammals, there is more blood in the lung per

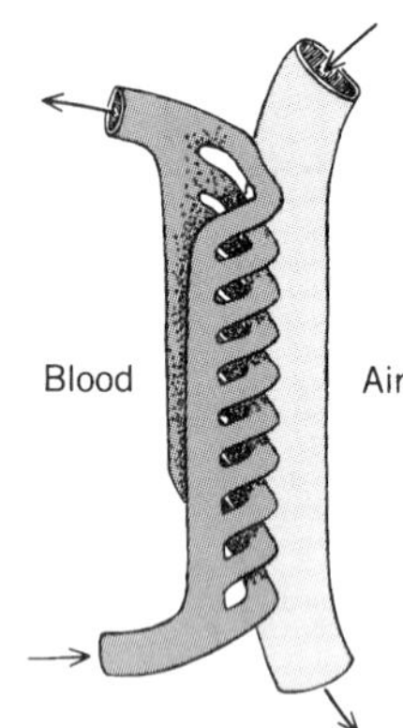

FIGURE 13.11 CROSSCURRENT CIRCULATION. Compare with countercurrent circulation illustrated on pp. 223 and 266.

unit volume, more of that blood is in capillaries, and the ratio of respiratory exchange surface to volume is about 10 times greater. Some birds can fly steadily at elevations at which mammals at rest become comatose!

The trachea, like the neck, is long. Indeed, in certain large birds it may form one or more loops near or within the sternum. Its resting diameter may vary at different levels.

Each **primary bronchus** enters its respective lung ventrally somewhat anterior to the center of the organ. Just inside the lung four secondary bronchi, or **ventrobronchi,** join the primary bronchus (Figure 13.12). These follow the ventromedial contour of the organ, branching as they go. Seven to 10 **dorsobronchi** next join the primary bronchus. They branch over the dorsolateral surface of the lung. Ventrobronchi and dorsobronchi are connected by thousands of **parabronchi,** which are about 1 mm in diameter. The walls of parabronchi are honeycombed with pockets (atria), which, in turn, have alcoves (infundibula) from which branch interlacing and cross-linking **air capillaries** (Figure 13.13). These are a mere 3 to 10 μm in diameter, which is much less than the diameter of mammalian alveoli.

Lungs of birds function in relation to air sacs that are devoid of respiratory epithelium but contribute to ventilation of the system. The primary bronchi terminate in large paired **abdominal air sacs.** Extensions from these sacs penetrate the synsacrum, femur, and thigh muscles. Paired **posterior thoracic air sacs** join the primary bronchi by way of **laterobronchi;** paired **anterior thoracic air sacs** usually join the third ventrobronchus. These sacs also have secondary connections to the duct system by small recurrent bronchi. Smaller paired **cervical air sacs** pneumatize cervical vertebrae and muscles; they join the first ventrobronchus. Finally, an unpaired **interclavicular air sac** sends diverticula into the humeri and among muscles of the axillae and shoulders. The entire system is somewhat variable.

Birds have a larynx of characteristic form. Its cartilages are the same as those of reptiles. Vocal cords are absent. Sound is produced instead by a unique **syrinx** located at or near the bifurcation of the trachea (Figure 13.14). A portion of the outer or (more often) mesial wall

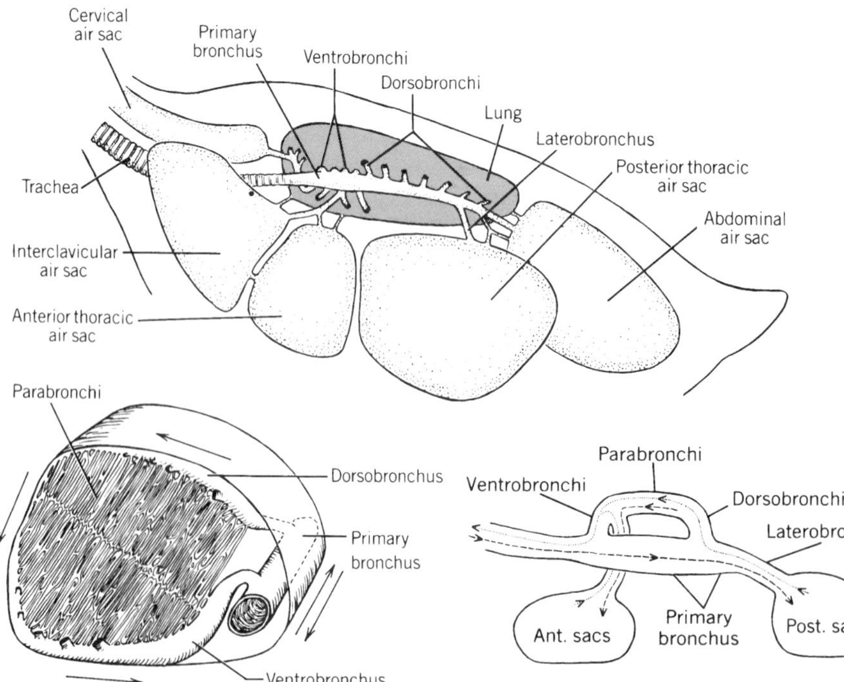

FIGURE 13.12 AVIAN RESPIRATORY SYSTEM. Stylized left lateral view of the entire system, cross-sectional view of the left lung, and diagram of airflow. Pathways of the neopulma not shown.

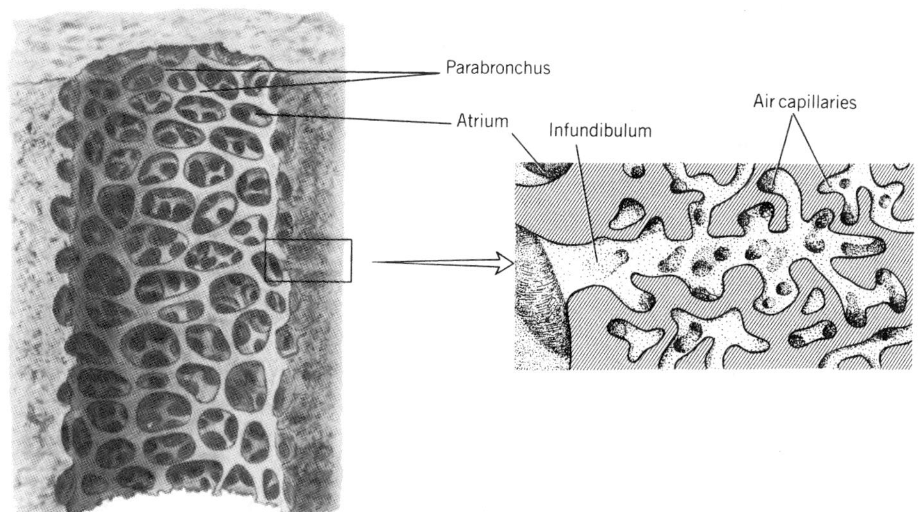

FIGURE 13.13 SECTION OF A PARABRONCHUS AND OF AIR CAPILLARIES.

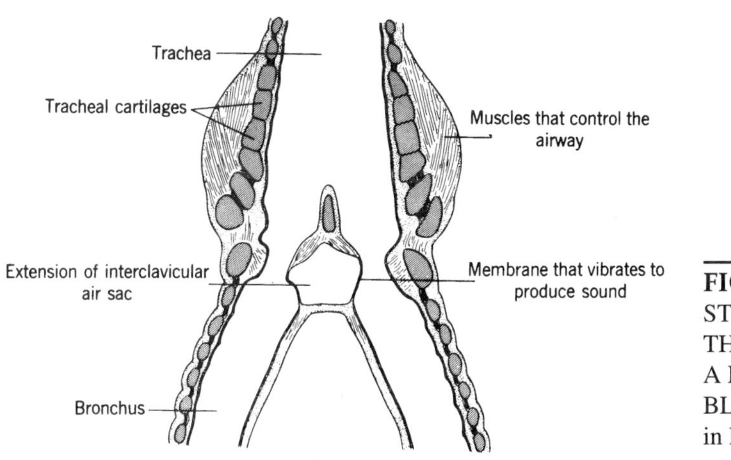

FIGURE 13.14 STRUCTURE OF THE SYRINX OF A EUROPEAN BLACKBIRD as seen in longitudinal section.

of each bronchus (or, less frequently, the wall of the trachea or of trachea and bronchi in combination) is a thin membrane supported at its margins by modified cartilaginous rings. This membrane impinges on air sacs and thus is free to vibrate in the stream of air passing through the system. The organ is sometimes asymmetrical and can be conspicuous in birds with resonant voices (Figure 13.15). Complicated intrinsic and extrinsic muscles control tension of the membranes and also the characteristics of the resonating system. The muscles are variable among species and may be used in systematics.

VENTILATION OF LUNGS: NATURE, EVOLUTION, AND LINKAGES

AIR-BREATHING ACTINOPTERYGIAN FISHES use a two-cycle (or four-stroke) sequence in ventilating their lungs. This mode of breathing minimizes the mixing of incoming and outgoing air. First, the pharynx expands, drawing expired gas from the lungs into the buccal cavity. Second, the pharynx contracts, forcing this gas to exit either through the open mouth or gill chambers. Next, fresh air enters the buccal cavity through the mouth as the pharynx again expands. Finally, with mouth closed, the pharynx forces the fresh air into the lungs. (This has been termed unmixed-air buccal pumping.)

Most SARCOPTERYGIAN FISHES AND AMPHIBIANS use a one-cycle (or two-stroke) manner of ventilation (sometimes modified by vocalizations or accessory strokes).

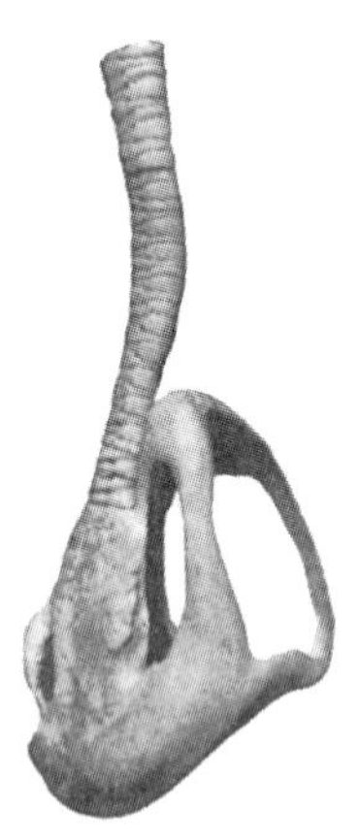

FIGURE 13.15
SKELETAL SUPPORT OF THE SYRINX OF A MERGANSER, *Mergus*, seen in ventral view.

This mode of breathing (mixed-air buccal pumping) allows for some mixing of inspired and expired air. ANURA serve to illustrate this kind of pulse pump. With glottis closed and nares open, fresh air is sucked into the large buccopharyngeal space by lowering the throat. Then, with glottis and nares both open, expired gas escapes from the body under pressure from the resilient lungs. In the process it passes through the top of the buccopharyngeal space but mixes little with the fresh air already stored in the ventral part of the space. The nares then close, and raising of the throat forces the fresh air into the lungs via the open glottis. Thus, the tidal flow of respired air causes mixing of fresh air with some residual gas in the lungs. The greater diffusivity of carbon dioxide allows much of this gas to escape through the moist skin, independent of the breathing apparatus.

Ventilation by *aspiration* is more efficient. Air is sucked into the lungs by negative pressure created in the thoracic or abdominal cavities. Exhalation is either passive or results from constriction of these cavities. Aspiration breathing frees the oral cavity and pharynx to adapt principally to feeding, is a one-cycle system and, because it is rapid enough to rid the body of carbon dioxide as well as supply it with oxygen, lends itself to a simpler circulatory pattern and allows the skin to be dry.

Aspiration breathers tend to have strong ribs and strong intercostal and abdominal muscles. LABYRINTHODONTS were probably the first such breathers. It is noteworthy that the same bilateral musculature that would have contracted synchronously, right and left, to inflate the lungs would have contracted alternately to produce the side-to-side bending that helped (as for salamanders) to advance the legs. Thus breathing was probably interrupted by bouts of locomotion.

The evolution of different patterns of lung ventilation has recently been approached through experiential studies, often constructed in a cladistic framework. It is assumed that the buccal pumps used in air ventilation must be derived from the aquatic buccal pump. Brainerd, Ditelberg, and Bramble studied lung ventilation in the urodele *Necturus* by cinevideography, measurement of pressure in buccal and pleuroperitoneal cavities, and electromyography of hypaxial musculature. Pending further comparative data about lung development and ventilatory motor patterns, they could not choose among hypotheses for the origin of mixed-air and unmixed-air mechanisms. Subsequent studies of ventilatory motor patterns of aquatic and terrestrial forms are addressing these hypotheses (see Simons, Bennett, and Brainerd in References). Nevertheless, the data support the conclusion that aspiration breathing evolved from buccal pumping in two stages: In the first, abdominal musculature was used only for expiration (as for *Necturus* and other anamniotes) and is, therefore, primitive; in the second, abdominal musculature was used for both expiration and inspiration.

Experiments show that moderate- and high-speed running interferes with the breathing of most LIZARDS, but in at least one species of *Varamus,* gular (throat) pumping of air into the lungs during running overcomes this problem. CROCODILIANS, by contrast, have overcome the conflict between respiration and locomotion in another way. They have a diaphragmatic muscle (not homologous with the mammalian diaphragm) that pulls the liver toward the pelvis. The pistonlike liver, in turn, sucks on the lungs, causing inhalation. In water there is enough pressure on the body that exhalation can be passive; on land other muscles pull the liver forward. Crocodilians also possess pubic bones that are connected to the ischia at movable joints. Farmer and Carter have shown that in the American alligator two abdominal muscles act in ventral rotation of the pubic bones to increase abdominal volume (inspiration) and two act in dorsal rotation to decrease volume (expiration). They conclude that these specializations reflect selection early in the evolution of the crocodilian lineage for high aerobic activity and the ability to run and breathe at the same time. TURTLES inhale when muscles crossing the limb apertures of the shell enlarge the internal cavity. Exhalation may be passive or active. Inhalation and most of exhalation is forced in SNAKES and results from the action of muscles inserting on the ribs and ventral skin. Any part of the long body can serve if other parts are distended by food.

Different muscles are used by BIRDS for respiration and flight, so these requirements are uncoupled (as they must have been also for pterosaurs). Ventilation is best known for birds at rest, of course, and must be increased enormously during flight. The sternum is actively rocked downward and forward for inspiration and is raised for expiration. The uncinate processes of the ribs (see figure on p. 543) are lever arms for the external intercostal muscles used (among others) during inspiration.

Two respiratory cycles are required for air to make one complete circuit. The first inspiration draws air through the trachea, primary bronchus, and laterobronchi into the expanding posterior air sacs. The first expiration pushes this volume of air out of the posterior air sacs, through the dorsobronchi, and into the parabronchi. During inspiration of the second cycle, this air moves through ventrobronchi to fill the anterior air sacs. Finally, the second expiration pushes the air out through the trachea. Thus, air is tidal in the trachea, air sacs, and much of the primary bronchi; nevertheless in most of the lung of most birds it passes in only one direction. The differential flow of air from the primary bronchi *into* dorsobronchi, but *out of* ventrobronchi, may be the consequence of the different angles those ducts make with the primary bronchi (Figure 13.12, lower left) or, as Wang and colleagues showed for geese at rest, aerodynamic valving within the pulmonary bronchus. Air reaches the air capillaries by diffusion from parabronchi. (The part of the avian lung just described, which has unidirectional airflow through parallel parabronchi, is predominant and is called the *paleopulmo.* A usually smaller, ventrocaudal part of the lungs of many birds, particularly songbirds, has anastomosing parabronchi and tidal airflow. It is called the *neopulmo.* The functional significance of the two kinds of lung structure is not clear. Crosscurrent exchange is equally effective regardless of the direction of airflow.)

Lungs of resting MAMMALS are inflated by the negative pressure that results from contraction of the dome-shaped diaphragm. They deflate largely by their inherent elasticity. The internal and external intercostal muscles also contribute, but their coordination with the respiratory cycle probably varies among species. Additional muscles of the thorax and abdomen function in forced breathing. Studies on the trotting and panting dog by Bramble and Jenkins indicate that, when the animal is not at rest, ventilation of the mammalian lung can be complex: Right and left apical lobes of the lung may ventilate out of phase with each other, and posterior lobes fill before anterior lobes. Posterior gas may even be recycled forward. Pistonlike displacements of the visceral mass may be more important than the diaphragm in driving the cycle, and the primary function of the intercostal muscles is to brace the thorax.

Mammals are like birds and crocodilians (and unlike amphibians and many reptiles) in having sufficiently independent muscular control of locomotion and respiration so that respiration can continue during movement. However, the two activities are usually not independent, and during sustained or rapid locomotion there often is a 1:1 linkage between the locomotor and respiratory cycles. This integration has been observed for numerous species, ranging from small to large, during the time that they are trotting, galloping, and hopping. Human runners, at least, may also use 2:1, 3:1, and other ratios. Coordination of the two systems appears to be facilitated by a common command center in the brain. The mechanism may (according to species) be the consequence of inertial motion of the viscera, of bending the thorax, or of flexing and extending the back. For running dogs, at least, the coupling of ventilation and locomotion benefits primarily the latter.

REFERENCES

Brainerd, E.L., J.S. Ditelberg, and D.M. Bramble. 1993. Lung ventilation in salamanders and the evolution of vertebrate air-breathing mechanisms. *Biol. J. Linnean Soc.* 49:163–183.

Brainerd, E.L., K.F. Liem, and C.T. Samper. 1989. Air ventilation by recoil aspiration in polypterid fishes. *Science* 246:1593–1595.

Bramble, D.M., and D.R. Carrier. 1983. Running and breathing in mammals. *Science* 219:251–256.

Bramble, D.M., and F.A. Jenkins, Jr. 1993. Mammalian locomotor-respiratory integration: implications for diaphragmatic and pulmonary design. *Science* 262:235–240.

Coates, M.I., and J.A. Clack. 1991. Fish-like gills and breathing in the earliest known tetrapod. *Nature* 352:234–236.

Demski, L.S., J.W. Gerald, and A.N. Popper. 1973. Central and peripheral mechanisms of teleost sound production. *Am. Zool.* 13:1141–1167.

Duncker, H.R. 1971. The lung air sac system of birds; a contribution to the functional anatomy of the respiratory apparatus. *Ergeb. Anat. Entwicklungsgesch.* 45(6). 171p. Excellent coverage and superb illustrations.

Duncker, H.R. 1989. Structural and functional integration across the reptile-bird transition: locomotor and respiratory systems, pp. 147–169. *In* D.B. Wake and G. Roth (eds.), *Complex organismal functions: integration and evolution in vertebrates.* Wiley, New York.

Farmer, C.G., and D.R. Carrier. 2000. Pelvic aspiration in the American alligator (*Alligator mississippiensis*). *J. Exp. Biol.* 203:1679–1687.

Fedde, M.R. 1976. Respiration, pp. 122–145. *In* P.D. Sturkie (ed.), *Avian physiology.* Springer, New York.

Feder, M.E., and W.W. Burggren. 1985. Cutaneous gas exchange in vertebrates: design, patterns, control and implications. *Biol. Rev.* 60:1–45.

Feder, M.E., and W.W. Burggren. 1985. Skin breathing in vertebrates. *Sci. Am.* 253(5):126–142.

Frazer Sissom, D.E., and D.A. Rice. 1991. How cats purr. *J. Zool. London* 223:67–78.

Gaunt, A.S. (ed.). 1973. Vertebrate sound production. *Am. Zool.* 13:1139–1255.

Henry, R.P., and N.J. Smatresk (symposium organizers). 1994. Current perspectives on the evolution, ecology, and comparative physiology of bimodal breathing. *Am. Zool.* 34(2):117–299. Contributions address various structural and functional adaptations relating to respiratory transitions from water to air in fishes and amphibians.

Houlihan, D.F., J.C. Rankin, and T.J. Shuttleworth. 1982. *Gills.* Cambridge Univ. Press, New York. 228p.

Hughes, G.M., and M. Morgan. 1973. The structure of fish gills in relation to their respiratory function. *Biol. Rev.* 48:419–475.

Jones, F.R.H., and N.B. Marshall. 1953. The structure and functions of the teleostean swim bladder. *Biol. Rev.* 28:16–83. Comprehensive.

King, A.S., and J. McLelland (eds.). 1989. *Form and function in birds,* vol. 4. Academic Press, New York. 591p. The entire volume is devoted to respiration.

Laurent, P. 1982. Structure of vertebrate gills, pp. 25–43. *In* D.F. Houlihan et al. (eds.), *Gills.* Cambridge Univ. Press, New York.

Liem, K.F. 1985. Ventilation, pp. 185–209. *In* M. Hildebrand et al. (eds.), *Functional vertebrate morphology.* Harvard Univ. Press, Cambridge, MA.

Liem, K.F. 1987. Functional design of the air ventilation apparatus and overland excursions by teleosts. *Fieldiana Zool.* NS 37:1–29.

Liem, K.F. 1988. Form and function of lungs: the evolution of air breathing mechanisms. *Am. Zool.* 28:739–759.

Liem, K.F. 1989. Respiratory gas bladders in teleosts: functional conservatism and morphological diversity. *Am. Zool.* 29:333–352.

Maina, J.N. 1998. *The gas exhangers: structure, function, and evolution of the respiratory processes.* Springer, New York. 420p. A comprehensive, interdisciplinary reference at the graduate student level. Extensive bibliography.

Maina, J.N. 2000. Comparative respiratory morphology: themes and principles in the design and construction of the gas exchangers. *Anat. Rec. (New Anat.)* 261:25–44. Broad review of factors contributing to the evolution of respiratory systems in animals; vertebrates given good treatment.

McLaughlin, R.F., W.S. Tyler, and R.O. Canada. 1961. A study of the subgross pulmonary anatomy in various mammals. *Am. J. Anat.* 108:149–158.

Owerkowicz, T., et al. 1999. Contribution of gular pumping to lung ventilation in monitor lizards. *Science* 284(4):1661–1663. Illustrates effective use of X-ray video to visualize the extent and timing of air intake from the mouth during locomotion and how it relates to oxygen consumption.

Perry, S.F. 1983. Reptilian lungs: functional anatomy and evolution. *Adv. Anat. Embryol. Cell Biol.* 79:1–81.

Perry, S.F. 1998. Lungs: comparative anatomy, functional morphology and evolution, pp. 1–92. *In* C. Gans and A.S. Gaunt (eds.), *Biology of the reptilia,* vol. 2. Academic Press, New York.

Randall, D.J., et al. 1981. *The evolution of air breathing in vertebrates.* Cambridge Univ. Press, Cambridge. 133p.

Schmidt-Nielsen, K. 1972. *How animals work*. Cambridge Univ. Press, New York. 114p. Concerns respiration, thermoregulation, and countercurrent exchange.

Sellers, T.J. (ed.). 1987. *Bird respiration,* vol. 1. CRC Press, Boca Raton, FL. 160p.

Simons, R.S., W.O. Bennett, and E.L. Brainerd. 2000. Mechanics of lung ventilation in a post-metamorphic salamander, *Ambystoma tigrinum*. *J. Exp. Biol*. 203:1081–1092.

Vitalis, T.Z., and G. Shelton. 1990. Breathing in *Rana pipiens:* the mechanism of ventilation. *J. Exp. Biol.* 154:537–556.

Wake, M.H. 1990. The evolution of integration of biological systems: an evolutionary perspective through studies on cells, tissues, and organs. *Am. Zool.* 30: 897–906. Discusses linkage of respiratory and locomotor systems.

Wang, N., et al. 1992. An aerodynamic valve in the avian primary bronchus. *J. Exp. Zool.* 262:441–445.

Wood, S.C., et al. 1992. *Physiological adaptations in vertebrates: respiration, circulation, and metabolism.* Marcel Dekker, New York. 419p.

Chapter 14

Circulatory System

GENERAL FUNCTION AND NATURE OF THE SYSTEM

Animals must have a system of internal transport unless their size and structure place all their cells in close proximity to an aqueous environment. Vertebrates, being large and solid, have the most highly evolved circulatory system in the animal kingdom. It functions to transport respiratory gases, nutrients, metabolic wastes, hormones, and antibodies. It serves (in conjunction with the kidneys and some other organs) in maintaining the internal environment. It removes toxic and pathogenic materials from the body, and may function (with muscles, integument, and behavior) in temperature regulation. Furthermore, it has the capacity to repair leaks, compensate for damage, and respond with amazing versatility to the varying requirements of the moment.

The vertebrate circulatory system has two components: a **blood-vascular system** and a **lymphatic system.** The former consists of the heart, blood vessels, and blood. **Arteries** distribute blood from heart to tissues, and **veins** return blood from tissues to heart. Small arteries (arterioles) are joined to small veins (venules) by **capillaries,** which form a network within the tissues (look ahead to Figure 14.10). Capillaries are usually about 1 mm long, scarcely larger in diameter than a single red blood cell, and only a fraction of a millimeter distant from one another. Physiological exchange between blood and tissues takes place through their thin walls. In several places in the body (digestive organs, kidneys, hypophysis) blood that has passed a capillary bed elsewhere enters a second capillary bed before reaching the heart. The veins between two capillary networks constitute a **portal system.**

The blood-vascular system of vertebrates (unlike that of invertebrates, except annelids) is a continuum of ducts and is therefore said to be a **closed system.** However, fluid constituents of the blood leak out of the capillaries, driven by diffusion, osmosis, the hydrostatic pressure produced by the heart, and, in some places, gravity. Fluids would accumulate in the tissues and cause swelling were they not drained away by the second

Above: Corrosion cast of the ducts and vessels of the kidneys and liver of a monkey.

component of the circulatory system, the lymphatic system. Tissue fluids enter netlike or blindly ending **lymphatic capillaries,** where they constitute **lymph.** Lymph passes slowly into larger and larger **lymphatic vessels** until it is discharged into the venous system at several points. Lymphatic capillaries in the gut are called **lacteals.** They absorb the digested long-chain fats; their lymph is whitish after a fatty meal ("lacti" = *milky*).

The heart provides little pressure to drive venous blood and none to drive lymph. Their passage is assisted by pressures on their vessels resulting from respiratory and other motions of the body and probably by their own intrinsic musculature. Most veins and lymphatic vessels of tetrapods have valves to prevent backflow. Many lower vertebrates also have lymph sinuses, some of which beat weakly as **lymph hearts.**

Ontogeny and phylogeny are more strikingly related in the circulatory system than in any other. Embryonic hearts, arteries, and veins of the higher vertebrates closely resemble the corresponding organs of remote ancestors. The pattern of circulation changes in development much as it must have changed in evolution, and it is fascinating to observe the sometimes marked changes in progress because the system functions continuously throughout.

The circulatory system also has more individual variation than any other. This is particularly true of the veins and is the combined result of a complicated ontogeny (which provides many opportunities for deviations) and the near functional equivalence of various departures from the norm.

Furthermore, the system is exceedingly adaptable. A piece of a vein grafted into an artery transforms structurally to become an artery. Nearly any vessel of the body can be tied off without serious inconvenience to the animal if the obstruction is made slowly enough for the system to compensate by enlarging alternative routes. Blood can, and does, flow in either direction in several vessels. Blood can be diverted toward or away from any given part of the body, the volume of circulating blood can be changed, and the rate of circulation can be varied about fivefold.

In spite of marked individual variation, parts of the system are useful in systematics. This is particularly true of the heart and major circuits at levels of the class and subclass, and of patterns of arteries in the limbs and basicranial area at levels of the order and family.

THE HEART

Development The circulatory system is the first of all organ systems to become functional during development. This is hardly surprising since differentiation and growth of all parts of the body are soon dependent on internal transport. The chick heart starts to pulsate at about 30 hours of incubation, when the body is not yet large enough to enclose the organ; the human heart becomes functional at 4 weeks, when the embryo is scarcely 5 mm long.

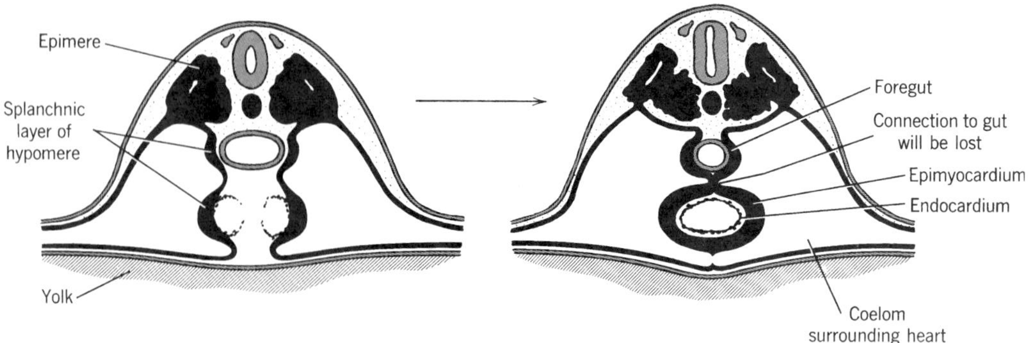

FIGURE 14.1 FORMATION OF THE HEART IN AN EMBRYO HAVING MUCH YOLK.

COMMENT 14.1

INSIGHTS INTO CIRCULATORY FORM AND FUNCTION TAKEN FROM FLUID MECHANICS

The flow of blood can be laminar or turbulent. If it is *laminar,* particles of blood at all positions in the vessel move parallel to each other and to the vessel walls. Flow is smooth and even. If flow is *turbulent,* there are eddies and mixing everywhere. The energy that produces the eddies must come from the heart, so laminar flow is more efficient. The likelihood that the flow of a liquid in a pipe will be turbulent was quantified in 1883 by the Englishman Osborne Reynolds. His index, now called the *Reynolds number,* is equal to flow velocity times diameter of pipe times density, all divided by viscosity. Thus, turbulence is most likely for fluids of low viscosity moving rapidly through large pipes. The vertebrates studied so far are for the most part below the threshold at which turbulence starts: burst of turbulence do occur in the heart and largest vessels, but flow is otherwise laminar. Such flow surely helps prevent mixing of oxygenated and unoxygenated blood in the single ventricle of dipnoans and amphibians (see below). Because it takes time for turbulence to develop, the pulsatile nature of arterial blood favors laminar flow. (We will revisit turbulence and the Reynolds number in the chapters on swimming and flight.)

The ultrathin layer of fluid that is in contact with a vessel wall does not flow at all. Successive, concentric, cylinder-shaped layers of fluid positioned farther and farther from the wall shear past one another faster and faster, reaching maximum velocity at the center of the vessel. If a parabola is positioned within a vessel, then the flow rate at any given position on the cross section of the vessel is proportional to the height of the parabola at that distance from the wall of the vessel. Red blood cells tend to be swept into the faster-flowing stream at the center of a vessel—particularly a small one. Consequently, the cells circulate a bit faster than the plasma, and the viscosity of blood increases slightly toward the center of the flow.

The variables affecting the laminar flow of fluids in rigid, unbranched pipes are related by the *Poiseuille equation* (named for the French physician who presented it in 1842 and 1846). Volume flow (F) varies with the pressure drop over a unit length of pipe $\Delta P/L$, the radius (r) of the pipe, and the viscosity (η) of the fluid. The relationship is $F = \Delta P \pi r^4 / 8L\eta$. The equation is not exact for blood in vessels (which are neither rigid nor unbranched), but nevertheless helps us to interpret the vascular system. In a closed system, volume flow must remain constant. Hence, rate of flow $\times$ cross-sectional area of all vessels at each level of the circuit (branching arteries, capillaries, converging veins) must also be constant. If r_1 is the radius of the parent vessel, and r_2 and r_3 the radii of its branches, then, to exactly maintain flow volume, $r_1^3 = r_2^3 + r_3^3$, and the body conforms well to this relation . It follows that at each level of arterial branching mean flow velocity falls off, vessels become smaller, and their total cross-sectional area increases. For veins, the sequence repeats in reverse as parent vessels merge.

The pressure in a closed cylinder equals the tension in its wall divided by its radius of curvature. This explains why large arteries need much thicker walls than small ones, even though blood pressure in the respective vessels is not very different. (It also explains why bicycle tires must be inflated to several times the pressure of car tires to achieve comparable stiffness.) When a cylindrical balloon is blown up, it develops a local bulge before inflating elsewhere. Healthy blood vessels would do the same (forming an aneurism) were it not for the fact that their walls are constructed to gain in stiffness as they inflate. Wavy elastic and (particularly) collagenous fibers are first straightened out, and then, gaining resistance, they stretch.

There are numerous books about hemodynamics. Two examples (Milnor and Vogel) are listed in the references.

The part of the splanchnic layer of the hypomere that is just posterior to the pharynx and ventral to the gut (or will soon come into that position—there are complicating changes of position in mammals) becomes markedly thicker on both sides of the body (Figure 14.1). These mesodermal folds approach one another in the midline and fuse to form a longitudinal tube. The tube is fixed to surrounding tissues at each end but otherwise becomes free as it passes through an expanded portion of the coelom. The free section establishes four chambers, which begin to contract in sequence and thus become the heart.

When the embryonic heart is first established it has two layers, an internal **endocardium** and an external **epimyocardium.** The endocardium of the mature heart has a thick layer of elastic connective tissue under an endothelial lining. The epimyocardium matures—while functioning—to have an external **epicardium** and a **myocardium.** The former becomes the enveloping serous membrane of the organ and is underlain by connective tissue. The myocardium is the muscle of the heart. Recall that cardiac muscle fibers are striated and branched and have central nuclei and intercalated disks (see figure on p. 171). It is involuntary and, as explained below, has the inherent capacity to contract rhythmically.

Primitive Heart: A Single-Circuit Pump Amphioxus has no heart. It has instead a pulsating vessel in the position where the heart evolved in vertebrates. This vessel approximates the embryonic primordium of the vertebrate heart and is apparently homologous.

The structure of the adult ancestral vertebrate heart can be inferred from the structure of the embryonic heart of descendants. It is a nearly straight tube having four chambers, which contract in sequence. Importantly, it pumps a *single stream* of *unoxygenated* blood forward in the body (Figure 14.2). A thin-walled **sinous venosus** receives blood from the great veins and empties it through a simple **sinuatrial valve** into the **atrium.** The atrium, also thin-walled but muscular, expels blood through one or more rows of **atrioventricular valves** into a large, thick-walled **ventricle.** The ventricle pumps into the **conus** (also called conus arteriosus or bulbus cordis), which looks like an enlarged artery and is lined with several rows of cup-shaped **semilunar valves.**

Hearts of CYCLOSTOMES and FISHES (except dipnoans) vary widely in detailed structure but depart little from the general ancestral plan. In fishes the heart is relatively far forward in front of the pectoral girdle and under the posterior gills (see figure on p. 204). The sinus venosus ranges from large (most sharks) to small (cyclostomes). The atrium is relatively large and usually shifts to a position dorsal to the ventricle (Figure 14.3). The inner layer of the ventricle is exceedingly spongy. In teleosts this chamber is conical, with apex pointing posteriorly. The conus varies from long and active as a pumping organ (cartilaginous fishes, *Polypterus*, bowfin) to virtually absent (cyclostomes, teleosts). When large it contains two to eight rows of semilunar valves (some of which, however, may not span the channel of the organ). The conus prevents backflow of blood as the ventricle fills. Teleosts have a **bulbus arteriosus** within the pericardial cavity in the position of the conus of other fishes. This organ has smooth, not cardiac, muscle but is highly elastic and passively evens the flow of blood into the afferent branchial arteries.

The pericardial cavity of elasmobranchs, being bordered in part by the skeleton, is semirigid. Consequently, as the ventricle contracts, blood enters the sinus venosus and atrium partly by suction. Hearts of fishes are relatively small because fishes have relatively

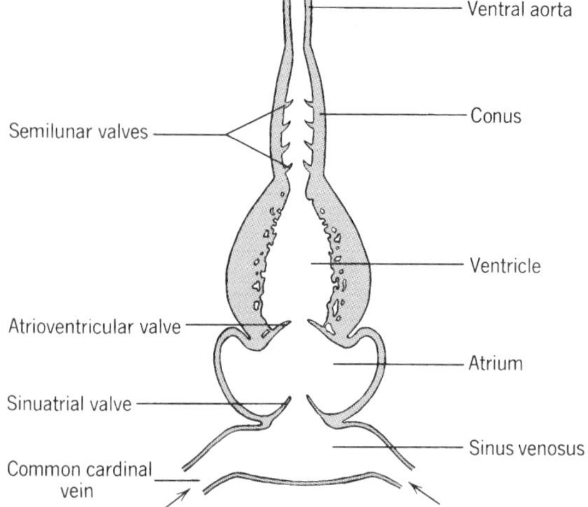

FIGURE 14.2 HYPOTHETICAL ANCESTRAL VERTEBRATE HEART as seen in frontal section. This structure is closely approximated in the ontogeny of all vertebrate hearts.

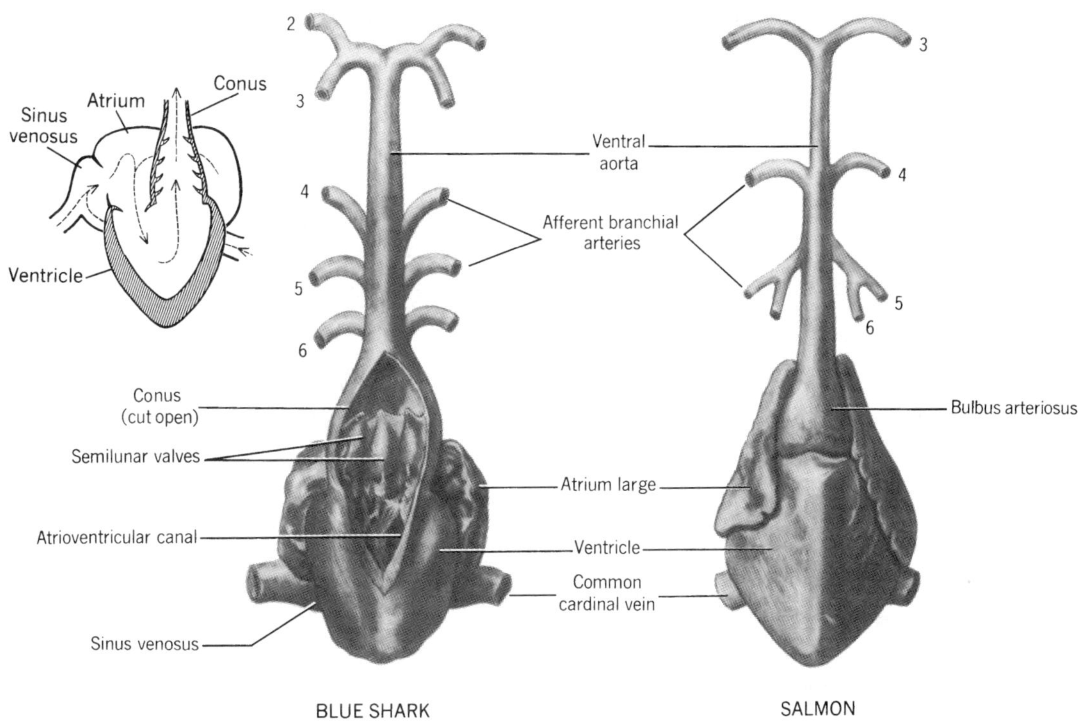

FIGURE 14.3 HEARTS AND RELATED VESSELS OF REPRESENTATIVE FISHES. Ventral views. Aortic arch derivatives are numbered.

small blood volume. As expected, active fishes have larger hearts than their sluggish relatives. Cyclostomes and many fishes have accessory hearts or pumping mechanisms elsewhere in the body. Some of these are mentioned later in this chapter. Change in the cardiac output of fishes usually results from a large change in stroke volume and only a small change in heart rate.

Lungfishes to Reptiles: Intermediate and Facultative Hearts We have just seen that the heart of most fishes pumps one stream of unoxygenated blood. The heart of birds and mammals pumps a stream of unoxygenated blood and also a completely separate stream of oxygenated blood. By contrast, the heart of lobe-finned fishes, amphibians, and reptiles is intermediate in that it usually receives both "kinds" of blood yet does not provide complete structural separation of the two streams, thus allowing mixing under some conditions.

Because the intermediate heart has probably survived for nearly 400 million years, and varies widely in both structural and functional details, it is not accurate to think of it as imperfect or merely a transitional evolutionary step toward a better mechanism. The amount and place of mixing of oxygenated and unoxygenated blood is varied by the animal according to circumstance. It is desirable to send unoxygenated blood to the lungs when they are functioning. However, it is preferable to send much of the same blood elsewhere if respiration is temporarily occurring primarily in gills (dipnoans in freshwater) or skin (submerged anurans), or simply because the lungs are not active (diving turtles, sea snakes, and crocodilians; turtles hibernating in mud). The heart of these animals is facultative: It adapts, at least in some degree, to a range of conditions not encountered by lungless fishes, birds, or mammals.

[Farmer presents another, somewhat novel interpretation of the mixing of oxygenated and unoxygenated blood in the single ventricle of air-breathing fishes, amphibians, and reptiles. In these vertebrates much of the myocardium of the ventricle is composed of a spongy matrix. Some have no coronary arteries to provide oxygen to this matrix; it then receives its oxygen supply directly from the circulating blood. (Other fishes do have small coronary arteries; see Figure 14.13 and p. 263) Farmer proposes that the *primary* purpose of the lungs in these species, particularly in times of high activity, is to deliver oxygen-rich blood to the heart tissue. Mixing of the oxygen-rich and oxygen-poor blood in the incompletely divided ventricle is necessary for adequate oxygenation of the entire organ. Farmer suggests that lungs may have evolved initially to serve the heart with oxygenated blood, rather than to provide oxygenated blood to the entire body. In birds and mammals this problem is resolved by the development of a system of coronary vessels and fully separated ventricles, and their myocardium is not spongy.]

The facultative heart has been much studied, but so far only in selected animals. The difficulty of analyzing the function of these hearts, particularly under stressful conditions, is great. Since both conformation and blood flow of the functioning heart are exceedingly complex and constantly changing, the figures are somewhat diagrammatic to provide clarity. This is particularly true of the divisions or derivatives of the conus.

The atrium of DIPNOI is partly divided by an **interatrial septum** into right and left chambers. The sinus venosus delivers unoxygenated blood to the right chamber, and the pulmonary veins supply oxygenated blood to the left (Figure 14.4). The ventricle (in genera that use the lung frequently) is partly divided by a large muscular **interventricular septum,** and studies show that mixing of the two streams of blood remains surprisingly low. The conus, which is large and no longer contractile, is also partly divided by a **spiral fold,** or flap of tissue that is anchored to the wall of the conus along a spiral path. Most unoxygenated blood is normally shunted to the fifth and sixth aortic arches, which deliver to the posterior gills and lungs. Oxygenated blood goes to the third and fourth arches, which send it to the body.

The atrium of ANURA is completely divided into right and left chambers. Again, blood in the left chamber has usually been oxygenated in the lungs, and blood in the right is relatively unoxygenated. However, blood returning from the skin joins systemic veins, so there is mixing of blood on the right side (Figure 14.4). The ventricle is undivided (probably a primitive rather than a degenerate condition), yet blood flows through it almost without eddies, and mixing of the two streams in the ventricle is minimal. The conus has a spiral fold that normally shunts right-side, unoxygenated blood to the lungs and skin, and left-side, oxygenated blood to the right systemic arch, from which branch

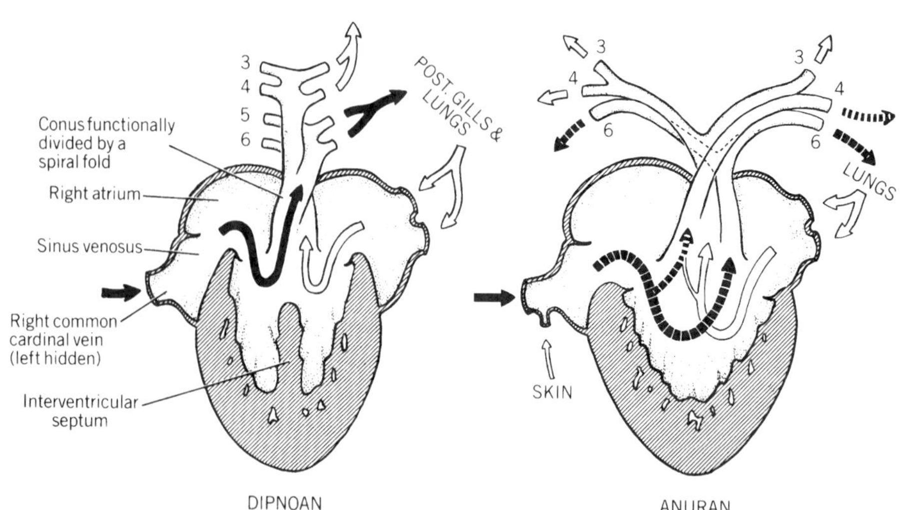

FIGURE 14.4 DIAGRAMS OF INTERMEDIATE HEARTS. Ventral views of frontal sections. Somewhat distorted by two-dimensional representation—sinus venosus and pulmonary veins enter atria more dorsally; divisions of conus spiral in space. There is variation within the taxa represented. Aortic arch derivatives are numbered. *Open arrows:* oxygenated blood. *Dark arrows:* unoxygenated blood. *Hatched arrows:* mixed blood.

the arteries to the head and forelimbs. Both types of blood may enter the left systemic arch, the mix depending on the resistance in the pulmonary circuit.

It is not surprising that URODELA, being less dependent on pulmonary respiration, have a less effective double-circuit pump. Regression has gone so far in lungless salamanders that the interatrial septum may be lost.

The conus of REPTILES is unique in that the embryonic organ divides completely into *three* channels, a pulmonary trunk and independent right and left systemic trunks. (A "trunk" is a division of the conus; an "arch" is a continuation extending to the dorsal aorta.) The sinus venosus varies from large (turtles) to small or vestigial, and again joins only the right side of a completely divided atrium. The ventricle of reptiles other than crocodilians is incompletely divided into dorsal and ventral chambers by a horizontal septum, and there is also a smaller vertical septum. Crocodilians have a complete interventricular septum, but there is a foramen between the bases of the two systemic trunks.

The flow of blood through the complex ventricle under various physiological conditions is complicated and difficult to illustrate; this paragraph, and Figure 14.5, provide only the consequences of that flow. (Several articles cited in the references give more details.) When reptiles respire in air, pulmonary vessels are dilated, making pulmonary resistance low. Hence, blood flows to the lungs, and, for reptiles other than crocodiles, the circulation is as shown in the left-hand drawing of Figure 14.5. If it becomes disadvantageous to drive much blood through the lungs, as when breathing is interrupted, then pulmonary resistance increases, flow decreases (or even stops), and the exit of blood from the heart is shifted as shown in the right-hand drawing. (It is probable that intermediate patterns occur, and variation among species is to be expected.) Crocodilians achieve a similar result in response to change in pulmonary resistance but do so in a somewhat different way. During air breathing, oxygenated blood enters the left systemic trunk from the right trunk through the foramen joining the two. During dives, little blood passes through the foramen; instead, blood enters the left trunk from the right ventricle.

Hearts of Endotherms: A Double-Circuit Pump Adult BIRDS and MAMMALS have completely double circulations: a low-pressure pulmonary circuit using the right side of the heart and a high-pressure systemic circuit using the left side (Figure 14.6). Pulmonary pressure is low to prevent edema and damage to delicate lung tissue; systemic pressure is high to drive blood through tissues that (like contracting muscle) may have their own internal pressure.

The sinus venosus is vestigial in birds and absent in adult mammals, the embryonic sinus having merged with the right atrium. The atrium is divided and relatively smaller than in many fishes: The ventricle is divided and disproportionately strong on the left side

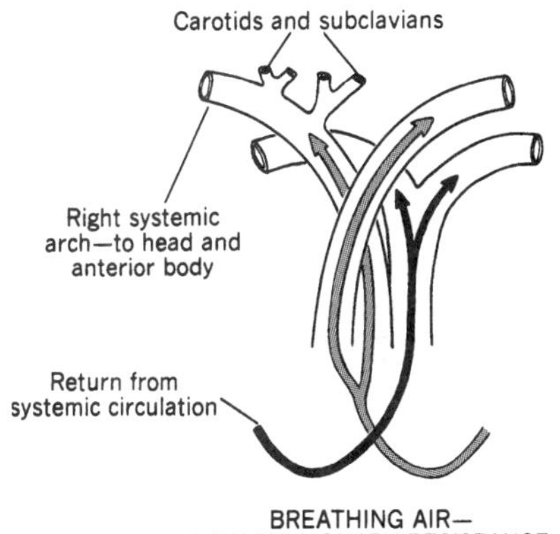

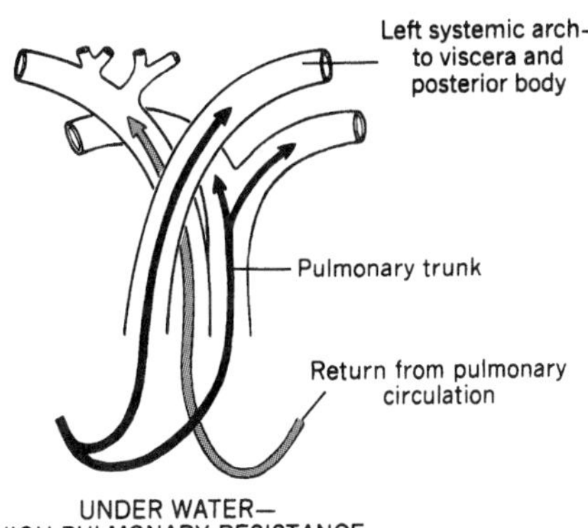

FIGURE 14.5 DIAGRAM OF THE FLOW OF BLOOD OUT OF THE FACULTATIVE HEART OF A TURTLE. Ventral views, distorted by two-dimensional representation—the trunks spiral in space. Compare with Figure 14.19.

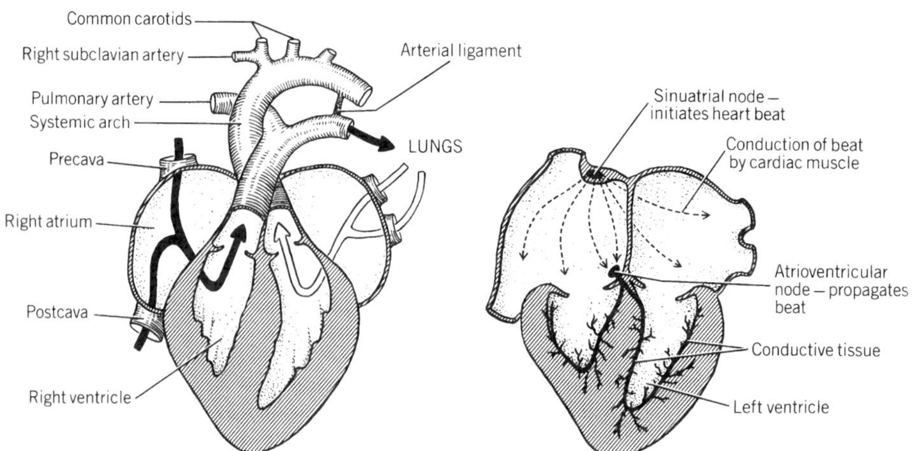

FIGURE 14.6 HEART OF A MAMMAL. Ventral views of frontal sections. Somewhat diagrammatic.

because of the greater resistance it must overcome there. The embryonic conus divides into a pulmonary trunk joining the right ventricle and a systemic trunk joining the left. Unlike the condition in lower vertebrates, the adult systemic arch is single. It loops to the right in birds and to the left in mammals.

The developing mammalian heart recapitulates evolutionary stages, having a sinus venosus that joins the right atrium, incomplete interatrial and interventricular septa, and an undivided conus (Figure 14.7).

Control of the Heart Beat The beat of the adult heart of amniotes is influenced by the autonomic nervous system, some hormones, and temperature. However, cardiac muscle has the inherent capacity to contract rhythmically. The embryonic ventricle is the first chamber to beat. When the atrium starts, it passes its own faster rhythm to the ventricle, and finally when the sinus venosus pulses, its still faster rhythm is followed by the entire heart. In lower vertebrates and in embryos, a **sinuatrial node** develops in the sinus venosus and initiates the beat, which is then transmitted over the heart by the muscle tissue itself, not by nervous tissue. The node is derived from muscle cells but is unlike either muscle or nervous tissue. Its pacemaker mechanism is not known.

When the embryonic sinus venosus of endotherms merges with the right atrium, it carries the sinuatrial node with it, and a second, **atrioventricular node** forms (Figure 14.6). This node distributes the beat over the ventricle and controls vascular contractility

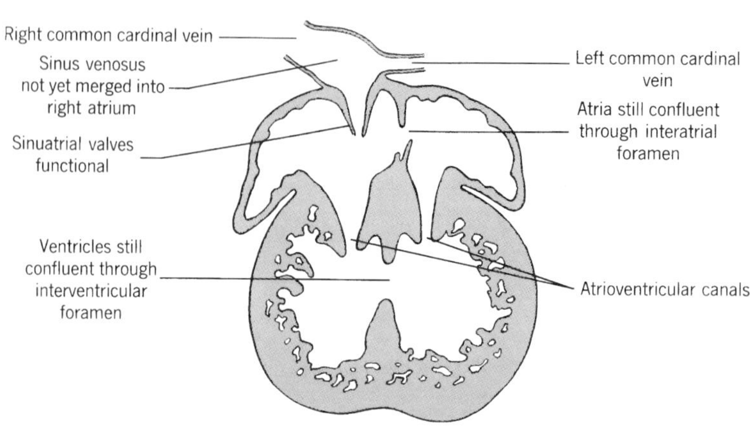

FIGURE 14.7 RECAPITULATION IN THE HEART OF A MAMMALIAN FETUS as seen in a ventral view of a nearly frontal section. Not seen in this plane of sectioning are the small pulmonary veins and the conus, which is starting to divide into pulmonary and systemic trunks.

in the systemic trunk. Unlike the sinuatrial node, it has a special conductive tissue to carry out its function.

Summary of Cardiac Evolution The ancestral vertebrate heart may have been straight, but known hearts are folded to place the atrium dorsal or anterior to the ventricle. The single-circuit heart is nearly symmetrical; others have marked asymmetry. The sinus venosus of intermediate hearts joins the right atrium, or the right side of an incompletely divided atrium. In birds and mammals it merges into the wall of the right atrium, contributing its node and valves, and possibly forming part of the coronary sinus, which drains heart muscle.

The atrium is partly divided in dipnoans and urodeles, and completely divided in other tetrapods. The incomplete interventricular septum of some dipnoans apparently is not homologous to the septum of tetrapods. The origin of the complete septum of crocodilians, birds, and mammals is debated. Holmes believes that these are broadly equivalent to each other and to the horizontal septum of turtles and lizards, although other tissue may augment the horizontal septum as it is converted.

The conus is partly divided by a spiral fold in dipnoans and anurans. It is completely divided into three trunks (internally if not also externally) in reptiles, and into two trunks in birds and mammals. This has seemed to some morphologists to be a difficulty in deriving birds and mammals from known reptilian lineages. However, the embryonic conus of all these animals is at first undivided, then divided into two, and finally, in reptiles, into three channels. The adult two-channel condition could have evolved from the three-channel condition by neoteny, or the deletion of a last developmental step.

BLOOD AND BLOOD-FORMING TISSUES

Vertebrate blood consists of blood cells of various types suspended in a fluid called **plasma.** Plasma is an aqueous solution of nutrients, metabolic wastes, salts, hormones, and proteins. Blood proteins are probably formed by the liver. They include albumens, the most abundant; fibrinogen, which contributes to the complicated clotting reaction; and several globulins, which respond to the entry of certain foreign materials into the body. Lymph is essentially a plasma with reduced protein content. Serum is the fluid remaining after blood has clotted.

Blood cells are of three principal kinds: red cells or **erythrocytes,** white cells or **leucocytes,** and **thrombocytes.** Erythrocytes occur only in blood vessels. They tend to be smaller than leucocytes, yet vary considerably in size, being relatively large in amphibians and small in mammals (Figure 14.8). They are usually flat, oval or round, and nucleated, but in nearly all mature mammals are round and enucleate. Erythrocytes are rich in the red

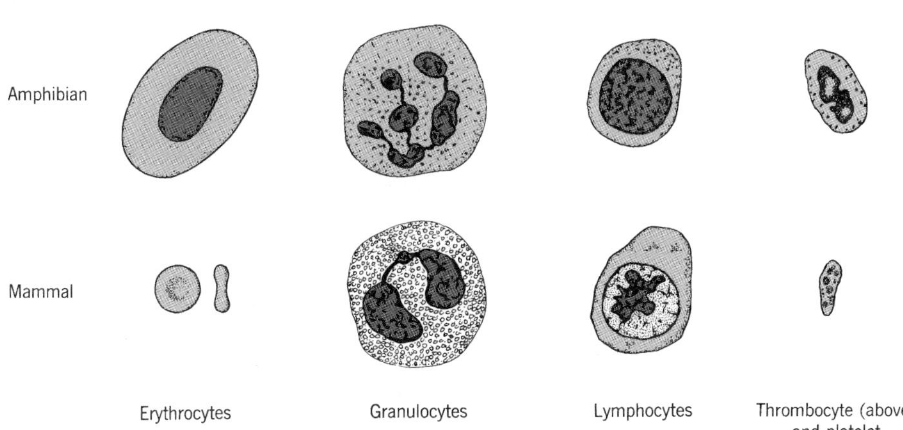

FIGURE 14.8 REPRESENTATIVE VERTEBRATE BLOOD CELLS.

protein hemoglobin, which combines readily with oxygen and is responsible for the efficiency of these cells in oxygen transport. They also transport carbon dioxide.

There are only 1 percent (mammals) to 10 percent (fishes) as many leucocytes as erythrocytes in the blood stream, but only leucocytes occur in the lymphatic system. These cells actively move through capillary walls and quickly aggregate at sites of local infection. Some leucocytes destroy foreign bodies by engulfing them—a process called **phagocytosis** (= *to eat* + *cell* + *process*). These cells are also involved in the immune response. It appears the leucocytes that enter the tissues may change function and become typical connective tissue cells; blood is usually classified as a special type of connective tissue.

Many varieties of leucocytes have been identified among the different vertebrates, and intergrades make classification difficult. Two main kinds are granulocytes and lymphoid cells. **Granulocytes** are large cells. The nucleus is subdivided into two or more lobes and the cytoplasm is highly granular. Lymphoid cells have a central, unlobed nucleus and lack cytoplasmic granules. The most abundant kind of lymphoid cell in all vertebrates is the small **lymphocyte.**

Thrombocytes are small, nucleated, spindle-shaped cells of the blood stream. They occur in all vertebrates except mammals, which have instead enucleate cell fragments called platelets. When thrombocytes or platelets escape through a cut in a blood vessel, they adhere to other tissues and disintegrate, thereby releasing a material that initiates the clotting process ("thrombocyte" = *clot* + *cell*).

Unlike other cells, blood cells are short-lived; they survive for days, weeks, or several months and then for some reason are destroyed. Consequently, they are produced and removed continuously from early embryonic life to death. For the most part, cell formation (**hemopoiesis**) and destruction occur in the same tissues. Many of these tissues also contribute to antibody production and clean the blood by filtration and the removal of many kinds of pathogens.

Solid, isolated masses of mesoderm, called **blood islands,** form on the yolk sac of the early embryo of most vertebrates. These produce the first blood. Blood also forms in conjunction with vessels of the body generally, and vessels continue to produce red cells in adult fishes. The digestive tract, thymus, kidney, liver, and parts of the nasopharynx produce blood in the embryos only of some vertebrates and in embryos and adults of others. **Lymph nodes** occur sparingly in certain water birds and are otherwise limited to mammals, where they are abundant along the lymphatic vessels in certain areas (Figure 14.9). They are whitish lumps of variable size and shape that produce quantities of lymphocytes.

Islets and cords of splenic tissue occur under the submucosa of the gut of hagfishes and within the lengthwise intestinal fold of lampreys. In other vertebrates the **spleen** is a discrete reddish organ located in the dorsal mesentery. It may be elongate or compact, and relatively small (birds) or large (mammals). In mammals it produces lymphocytes,

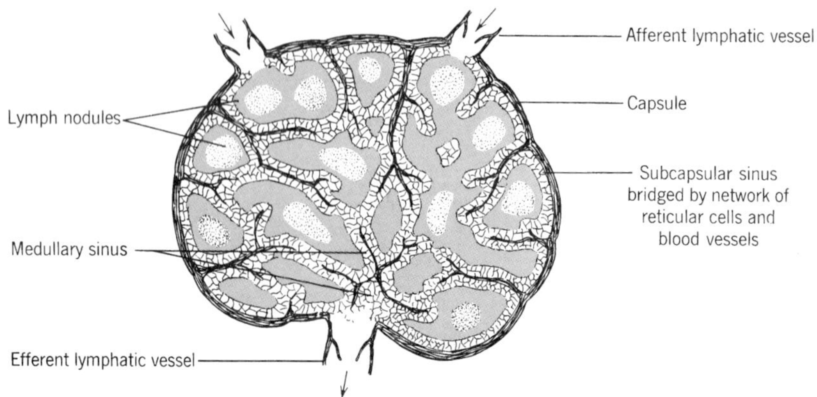

FIGURE 14.9 DIAGRAM OF A SECTIONED LYMPH NODE.

destroys erythrocytes, and also stores quantities of erythrocytes for release when the body has need of them. In all other classes the spleen produces all kinds of blood cells. Ungulates and some other mammals also have **hemal nodes,** which are small nodules of splenic tissue distributed along blood vessels of the gut, kidney, and liver.

There are two kinds of bone marrow: Yellow marrow occurs in the larger cavities of the long bones and is fatty; red marrow occurs in the ribs, vertebral centra, and epiphyses of long bones. The types are not always distinct. Red marrow is hemopoietic in tetrapods.

In general, hemopoietic tissue consists of a matrix of connective tissue intimately related to a rich blood supply. The supportive framework often includes **reticular tissue,** the stellate cells of which are phagocytic and can differentiate into various kinds of blood cells. Commonly, the blood is carried in **sinusoids** that differ from capillaries in being larger and more irregular. They anastomose freely and carry venous blood. (An **anastomosis** is a netlike intercommunication of vessels.) Similar lymphatic sinusoids carry lymph in the lymph nodes.

BLOOD VESSELS

Development and Structure The first visual indication of the formation of the circulatory system is the appearance of blood islands on the yolk sac. The peripheral cells of adjacent islands gradually become contiguous and form a network of tiny vessels. The deeper cells separate from one another and become blood cells.

Vessels soon form also from mesenchyme within tissues of the body. They are initially all of about the same size and together compose a continuous network in relation to organs of rapid growth such as the central nervous system and eyes. Gradually certain channels enlarge or merge to become the first arteries and veins. All parts of the system are initially paired and symmetrical. However, the primordia of the heart and certain arteries fuse in the midline very soon after their appearance, and asymmetries are established early in certain veins. Some major lymphatic vessels develop in relation to veins and subsequently become independent.

FIGURE 14.10 ARTERIOLE, VENULE, AND NETWORK OF CAPILLARIES.

Arteries, veins, and capillaries are at first indistinguishable histologically. Each is a tube formed from thin, flat endothelial cells that are loosely wrapped on the outside in a meshwork of connective tissue. This is the definitive structure of capillaries (Figure 14.10). Arteries and veins each retain these tissues as a **tunica interna** but add more peripheral tissues as they mature (Figure 14.11). Arteries develop a thick **tunica media** that usually consists of circularly arranged smooth muscle fibers, but in the largest arteries consists instead of yellow elastic fibers. These provide the strength and resilience needed to keep blood moving under high pressure. Arteries are completed by a thinner **tunica externa** of longitudinally oriented connective tissue. The scanning electron microscope reveals that at least some arteries of at least some mammals are lined by a meshwork of irregular projections that vastly increase their surface.

Veins are usually larger in diameter and thinner-walled than corresponding arteries. This relates to their function of containing most of the blood of the body and transporting it under relatively low pressure. Veins are also more variable in structure, however, and in some instances are much like arteries. The tunica media is typically thin and may be indistinct. The tunica externa, by contrast, is as thick or thicker than that of arteries. At death, the more muscular arteries squeeze most of the blood into the veins. When cut, the arteries then stand open but empty, whereas the veins usually are either gorged with blood or collapsed.

Lymphatic capillaries resemble blood capillaries but are larger, more irregular in shape, and more permeable. Their endothelial cells are relatively large. Lymphatic vessels are constructed somewhat like veins, although the three layers tend to be less distinct.

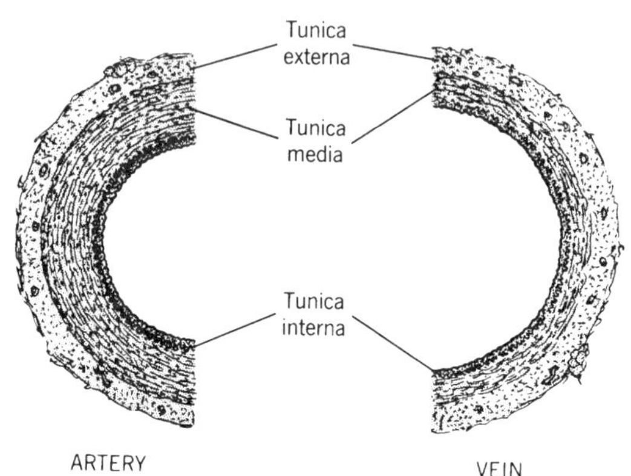

FIGURE 14.11 STRUCTURE OF BLOOD VESSELS AS SEEN IN CROSS SECTION.

The Initial Pattern of Arteries The pattern of vessels that make up the initial functioning system of the embryo is basically the same for all vertebrates. Virtually the same pattern is also found in adult amphioxus. It can, therefore, confidently be considered primitive for vertebrate phylogeny as well as ontogeny (Figure 14.12). It should be learned as our point of departure.

The heart pumps blood forward under the pharynx in the **ventral aorta** (also called truncus arteriosus), a vessel that is single where it leaves the heart but sometimes paired under the anterior branchial arches. In embryos, the ventral aorta distributes its blood to paired **aortic arches** that run upward through the visceral arches. In adults of lower vertebrates, the aortic arches have differentiated into proximal (toward the heart) **afferent branchial arteries,** gill capillaries, and distal **efferent branchial arteries.**

The principal distributing vessel of the body is the **dorsal aorta.** This vessel is paired throughout its length when it first differentiates. It may remain paired dorsal to the pharynx, but at more posterior levels immediately fuses in the midline. Blood entering the paired dorsal aortae from the anterior aortic arches (or anterior branchial efferents) runs forward and continues into the head in extensions of the aortae called **internal carotid arteries.** Blood entering the aortae from the posterior arches (or efferents) runs posteriorly into the unpaired dorsal aorta, whence it is distributed by paired dorsal and lateral branches, and by ventral branches, which are unpaired except for the **vitelline arteries** to the yolk sac and **allantoic arteries** to the allantois.

Evolution of Anterior Arteries Embryos of jawed vertebrates nearly always have six pairs of aortic arches, but the first (located in the mandibular visceral arch) is always lost or reduced to remnants in the adult.

CYCLOSTOMES have a long ventral aorta and eight or more aortic arches. The dorsal aorta is single over the gills (lampreys) or paired (hagfishes). Among JAWED FISHES, the Chondrichthyes, Dipnoi, and Chondrostei retain the second aortic arch to serve the hyoidean hemibranch. In other fishes, the second arch is modified or lost (Figure 14.13). Thus, cartilaginous fishes usually have five branchial afferents, whereas Actinopterygii have four to serve their four holobranchs (Figure 14.3). Actinopterygii also lack the branchial loops and shunts that complicate the gill circulation of cartilaginous fishes (Figure 14.13). The air-breathing dipnoans have pulmonary arteries that branch from the efferent segments of the sixth arches. Furthermore, their fifth and sixth afferents branch from the ventral aorta just anterior to the partly divided conus that diverts into them the relatively unoxygenated stream of blood (Figure 14.4). The paired dorsal aortae of fishes extend forward under the brain as internal carotid arteries. Jaws, orbits, mouth, and pseudobranchs are supplied by arteries of varied patterns, doubtful homologies, and many names.

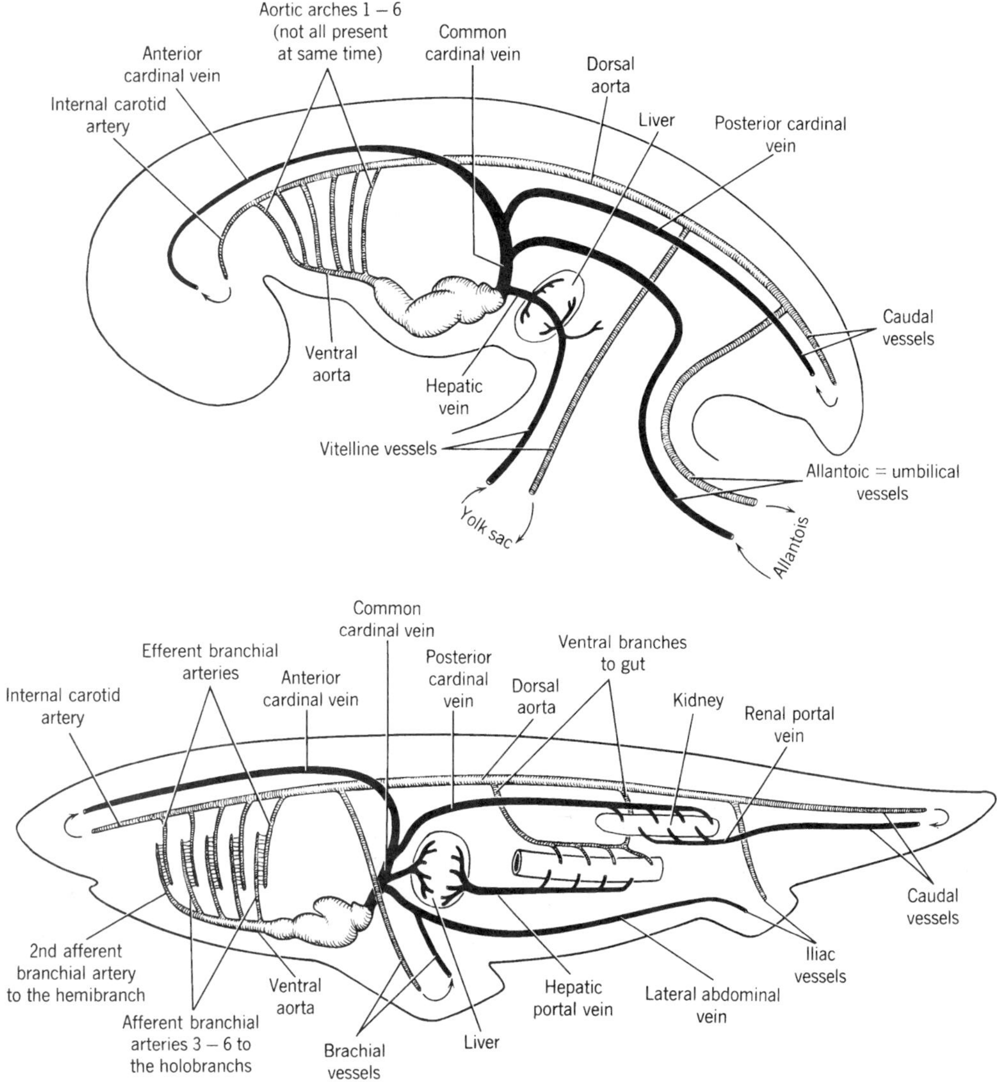

FIGURE 14.12 BASIC PATTERN OF THE VERTEBRATE CIRCULATORY SYSTEM as seen in an amniote embryo (above) and an adult fish at the shark level of evolution (below). Not shown: subcardinal and supracardinal veins, dorsal and lateral branches of dorsal aorta. All vessels are paired except dorsal and ventral aortae, caudal vein, and vessels of the gut.

Adult TETRAPODS lack both first and second aortic arches. A characteristic carotid system delivers blood to the head. It consists of (1) common carotid arteries derived from segments of the ancestral (and embryonic) paired ventral aortae, (2) external carotid arteries that supply the throat and ventral part of the head, and (3) internal carotid arteries that supply the brain and much of the head. They are derived from the third aortic arches and anterior extensions of the paired dorsal aortae.

Adult AMPHIBIANS also retain paired fourth arches, which are systemic arches delivering blood to the posterior part of the body. The short segment of the paired dorsal aortae between the dorsal roots of the third and fourth arches (called the carotid duct) is thus a useless shunt between diverging streams of blood; although retained by most urodeles, it is lost in anurans. Salamanders that keep their gills as adults retain also the fifth and sixth aortic arches (all their arches having afferent and efferent branchial arteries). Anura lose the fifth

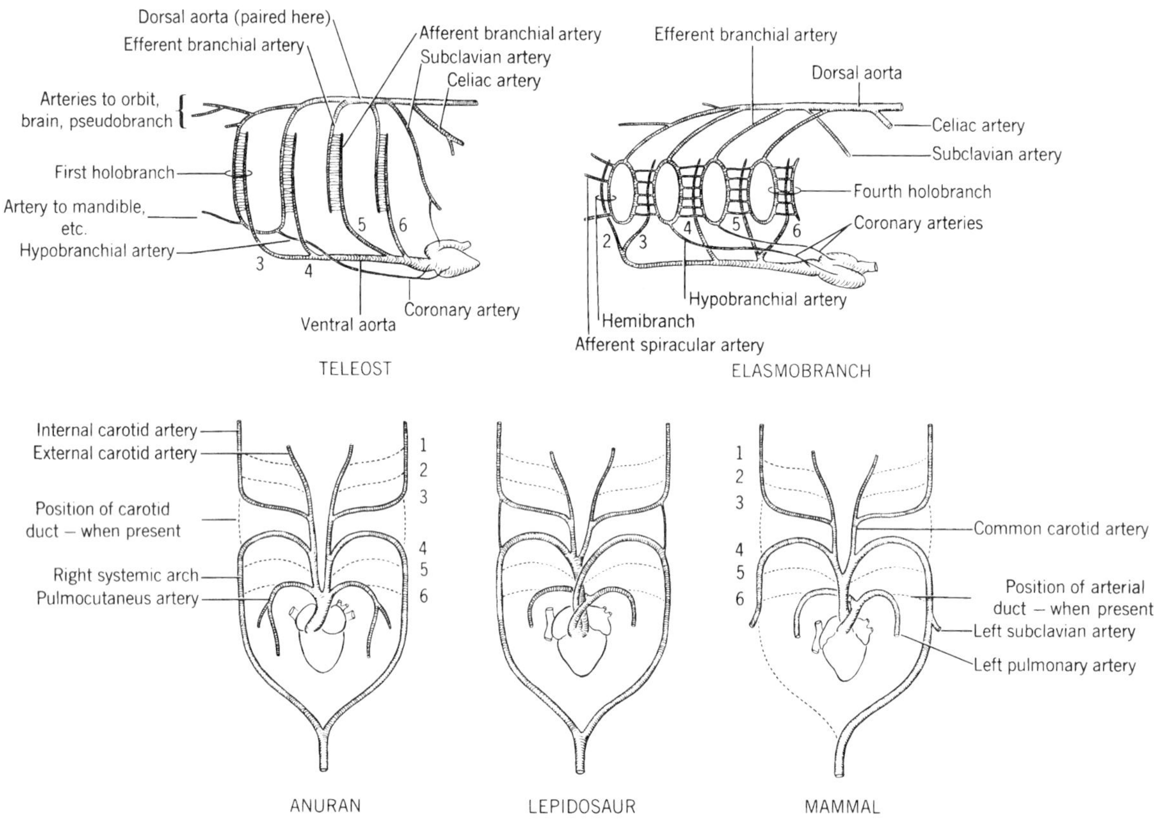

FIGURE 14.13 DERIVATIVES OF AORTIC ARCHES AND RELATED VESSELS OF REPRESENTATIVE VERTEBRATES. Semidiagrammatic. Left lateral views (above) and ventral views (below). Aortic arches are numbered.

arch. Pulmonary arteries branch from the sixth arches (which become their bases). The useless (usually detrimental) part of the sixth arch that is between the pulmonary artery and the aorta (called the arterial duct) is also lost. (*Necturus*, which is often dissected in laboratory, unfortunately is atypical in regard to its pattern of anterior arteries.)

The basic pattern of aortic arch derivatives in modern REPTILES is the same as for anurans, with the important difference that the conus has divided into three instead of two channels (Figures 14.5 and 14.19). The distribution of blood to these vessels was presented above with the heart. The carotid system (which is variously modified in the different groups) relates only to the right arch, which may have the more oxygenated blood. It is curious that the carotid duct still persists in some lepidosaurs, and the arterial duct in *Sphenodon* and several other genera.

BIRDS modify the carotid system so that internal carotids substitute for common carotids in the long neck (Figure 14.21). Right and left vessels lie side by side in the chick, and it is of interest to systematists that both or either may be retained in the different orders. The right, fourth aortic arch becomes the systemic arch. Characteristic brachiocephalic arteries, which branch from the systemic arch, give rise to carotid and subclavian arteries, the pattern of branching being various.

MAMMALS retain only the left systemic arch (Figure 14.22). The carotid system is less modified than in birds, but the proximal relations of the common carotids to the arch and the subclavian arteries are often asymmetrical.

Evolution of Posterior Arteries Posterior to the pharynx the **dorsal aorta** is the most conservative vessel of the body. It is always a large, median longitudinal artery lying ventral to the notochord or spine. It extends into the tail as the **caudal artery.** Moreover, there is little evolutionary significance to the variations of the branches of the aorta. **Ventral visceral branches** may be numerous (amphibians) but typically include only a **celiac artery** to stomach, duodenum, liver, and pancreas (i.e, the foregut) and one or several **mesenteric arteries** to the remainder of the gut (hindgut).

Lateral visceral branches serve the urogenital organs. They are numerous if these organs are long (most anamniotes) and otherwise few. They take their names from the organs served—renal, ovarian, spermatic.

Dorsal somatic branches of the dorsal aorta serve spinal cord, muscles, and skin. They are segmental in fishes, but in tetrapods tend to be modified for functional reasons noted below. The large **subclavian arteries** extend into the pectoral appendage as **brachial arteries** (do not confuse with branchial arteries); the corresponding **iliac arteries** extend into the pelvic appendages as **femoral arteries.**

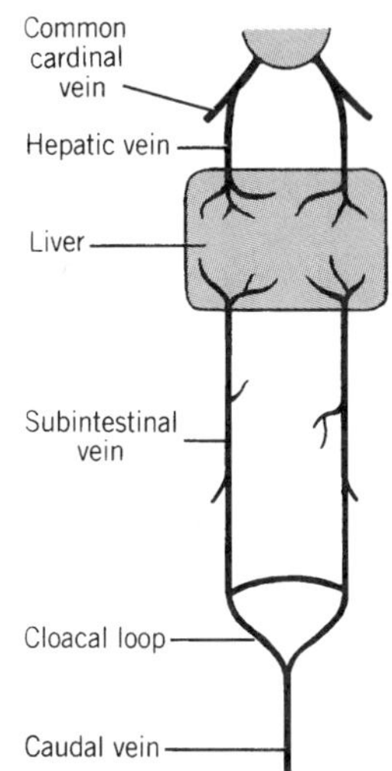

FIGURE 14.14
THE ANCESTRAL SUBINTESTINAL SYSTEM OF VEINS.
Ventral view.

The Initial Pattern of Veins The initial ontogenetic and phylogenetic pattern of veins comprises three somewhat independent systems. The **subintestinal-vitelline system** drains the tail, digestive tract, and yolk sac (Figure 14.14). A **caudal vein** runs forward to the cloacal area. **Subintestinal veins** continue the drainage forward, receive tributaries from the digestive tract and, after penetrating the liver, empty into the common cardinal veins (see below) close to the heart. This completes the subintestinal component of the system as it may have occurred in ancestral vertebrates. A vitelline component consists of large **vitelline veins,** which come from the yolk sac and enter the subintestinals (Figure 14.12). Morphologists find it convenient to think of the vitelline veins as large precocious branches of the subintestinals. Embryologists are more likely to regard the subintestinals as belated branches of the vitelline veins.

A second system of veins is the **cardinal system,** which drains the head, dorsal body wall, and kidneys (Figure 14.12). Its principal vessels are the **anterior cardinal veins,** which are lateral to the internal carotid arteries; **posterior cardinal veins,** which are located in or adjacent to the dorsal portion of the kidneys, and short **common cardinal veins,** which are formed by the confluence of the preceding vessels. After being joined by veins of other systems these return all blood to the heart. (Several additional pairs of longitudinal veins are related to the embryonic kidneys. These anastomose freely with one another and drain into the posterior cardinal veins. Two that contribute to the story to follow are the subcardinal and supracardinal veins.)

The third system of veins is the **abdominal system,** which drains the ventral body wall and paired appendages. It consists of paired **lateral abdominal veins,** which receive **iliac** and **subclavian branches** from the posterior and anterior appendages and drain into the common cardinal veins near the entrances of the subintestinal veins (Figures 14.12 and 14.15).

Evolution of Anterior Veins Vessels draining the anterior part of the body have been conservative. They represent part of the ancestral cardinal system and associated veins. Anterior cardinal veins drain the brain and part of the head in cyclostomes and fishes (Figure 14.15). In tetrapods the same vessels are called **internal jugular veins** and are somewhat modified peripherally by various combinations of the parent vessels with their embryonic tributaries (Figures 14.16, 14.17, and 14.18). The more ventral and external parts of the head are drained in fishes by veins of uncertain homologies usually called inferior jugular veins. The same regions are drained in tetrapods by **external jugular**

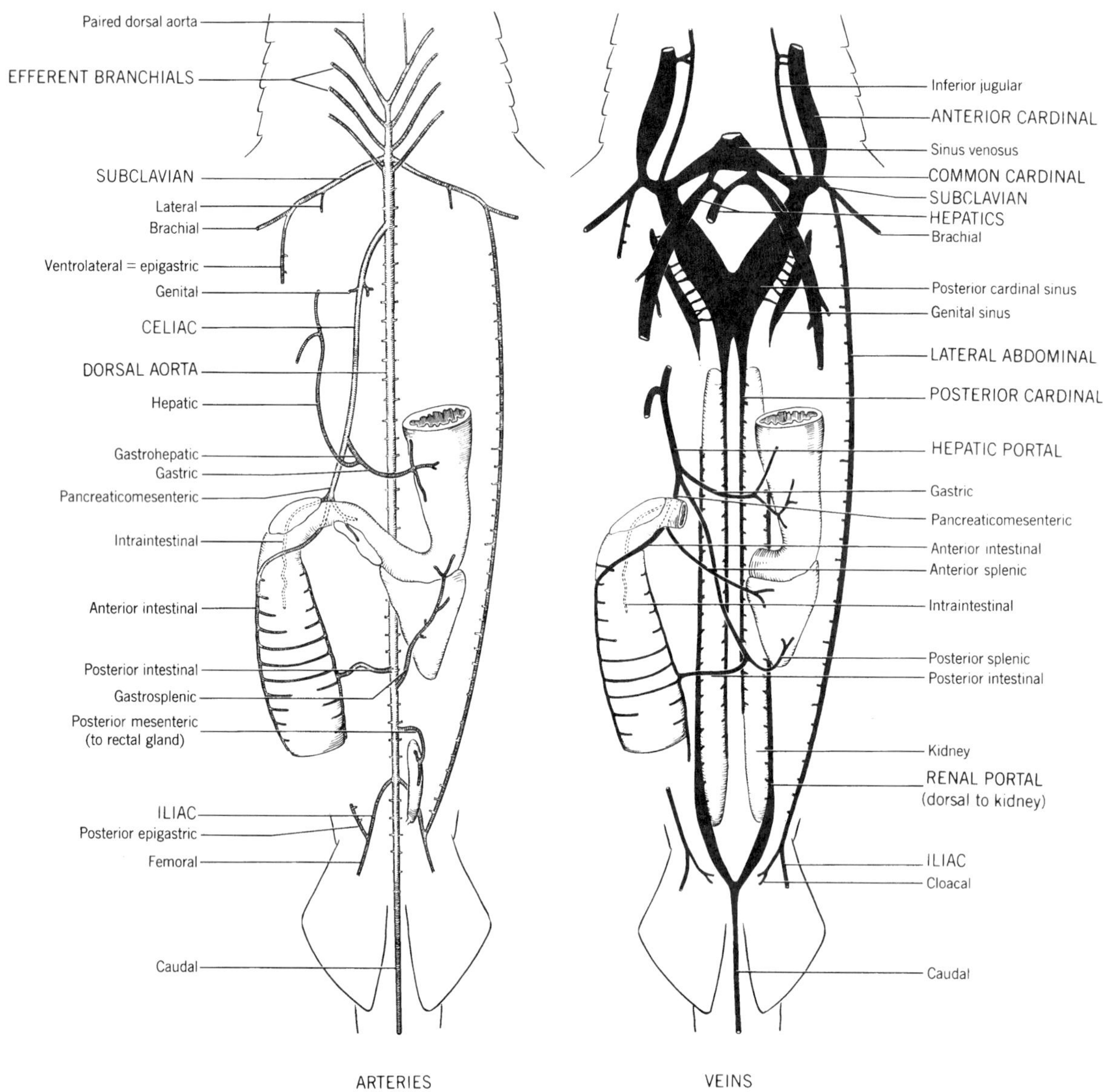

FIGURE 14.15 CIRCULATORY SYSTEM OF THE SHARK *Squalus*. Ventral views. Principal vessels identified by capital letters. For the ventral aorta and branchial vessels see Figures 14.3 and 14.13.

veins, which join the internal jugulars in the neck. The common cardinal veins are paired in fishes; one or the other is lost in adult cyclostomes. Venous sinuses are common in the cardinal and jugular drainages—particularly in cyclostomes, cartilaginous fishes, and (adjacent to the brain) in mammals.

Between the heart and the confluence of internal and external jugulars the derivatives of the cardinals of tetrapods are called **precavae** (also anterior vena cavae). These vessels are paired all the way to the heart in tetrapods other than mammals, and also in some primitive mammals (most marsupials and rodents, insectivores, and some others). In other mammals a shunt called the **brachiocephalic** (or **innominate**) vein carries blood from the left jugulars to the right precava. The left precava is then lost except for a contribution to the coronary sinus (Figure 14.23).

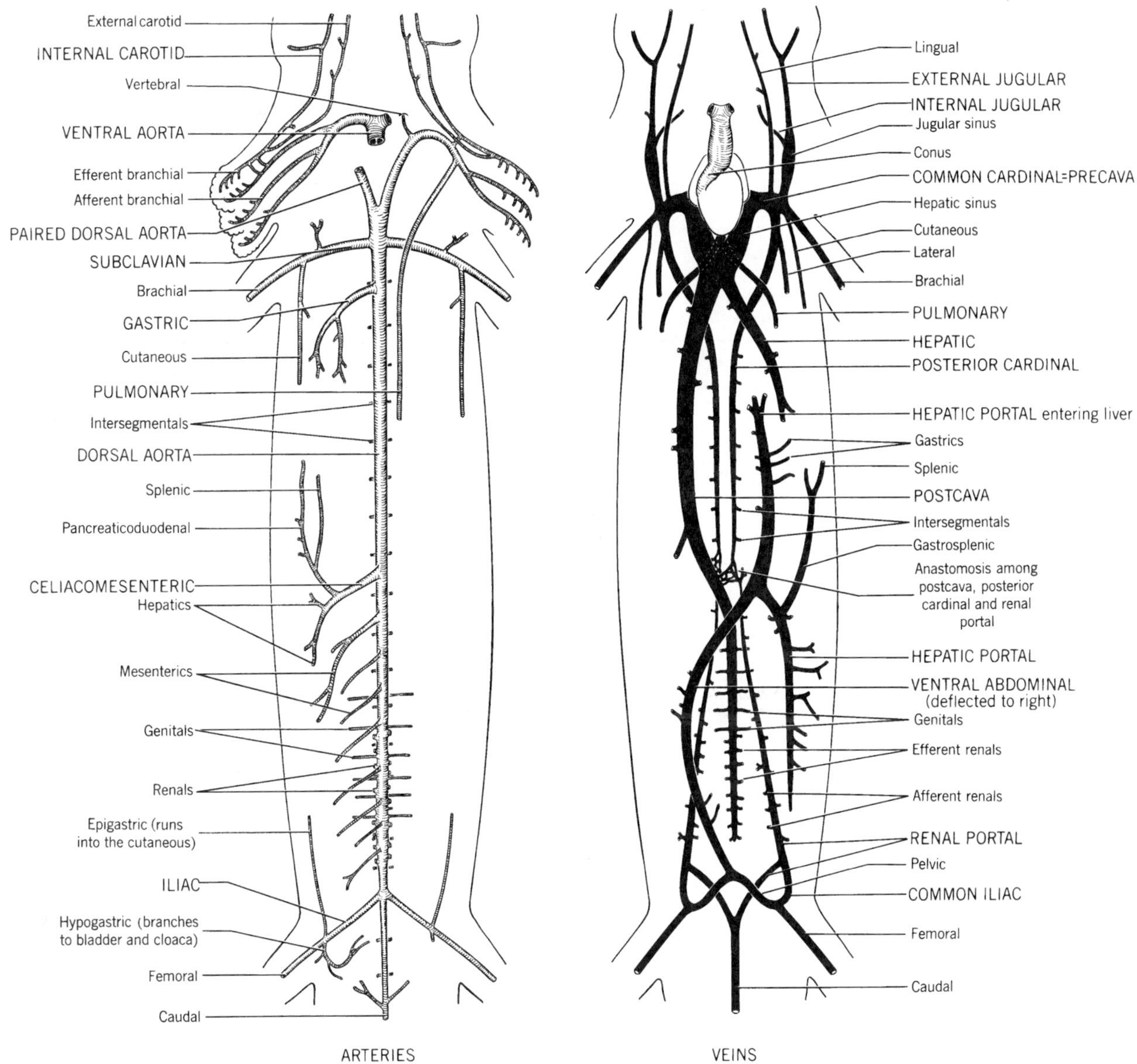

FIGURE 14.16 CIRCULATORY SYSTEM OF THE URODELE *Necturus*. Ventral view. Principal vessels identified by capital letters.

The anterior appendages are drained by subclavian veins, which enter the anterior or common cardinals or their equivalents, the precavae or brachiocephalic.

Evolution of the Hepatic Portal System The evolution of the drainage of the digestive viscera has also been conservative. It relates to the subintestinal system of the ancestral circulation (Figure 14.14). Adults of all vertebrates establish a single large **hepatic portal vein** ("hepatic" = *of the liver*) by the selective retention of parts of the left and right subintestinals and of several anastomoses that occur between them within and just posterior to the liver. Inside the liver the system breaks up into sinusoids, and anterior to the liver it continues toward the heart as one or more **hepatic veins** that drain the liver (Figure 14.23).

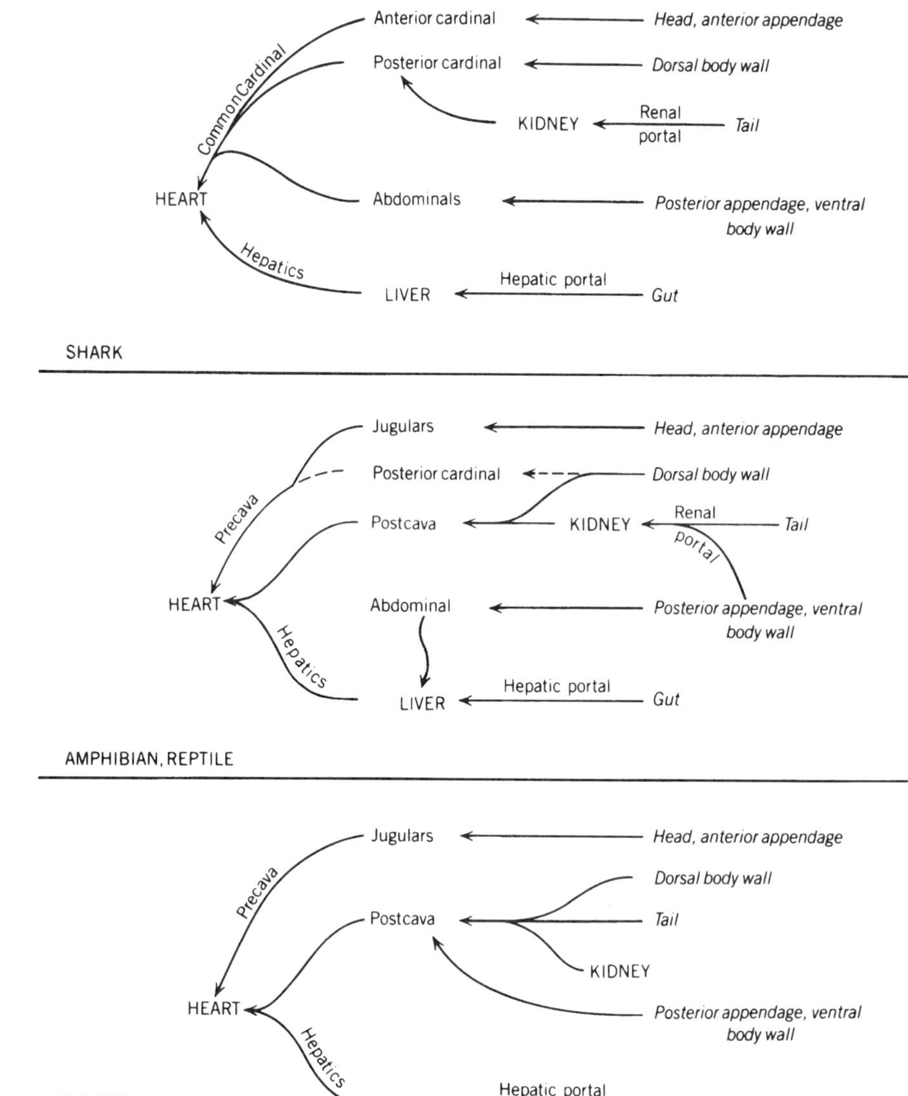

FIGURE 14.17 DIAGRAM OF THE EVOLUTION OF VENOUS BLOOD FLOW.

Primitively, the hepatic portal vein drains not only the gut but also the tail. This condition is retained by hagfishes and some teleosts. In other vertebrates the hepatic portal loses its connection to the tail. Consequently, the caudal blood must enter another system of veins (see below). (As noted later, the liver of amphibians and reptiles may also receive portal blood by way of the abdominal system of veins.)

Evolution of the Renal Portal System The renal portal system evolved from the ancestral posterior cardinal veins and associated channels. Cyclostomes and larval fishes retain the ancestral condition: The posterior cardinal runs from the anal area to the common cardinal, receiving tributaries from urogenital organs and body wall.

In FISHES an interruption develops in the posterior cardinals just anterior to the kidneys (Figure 14.18). Blood from the posterior part of the body usually flows into the *posterior* segments of the posterior cardinals—now called **renal portal veins**—passes into the tissues of the kidneys, and emerges into subcardinal veins, which transmit it to the

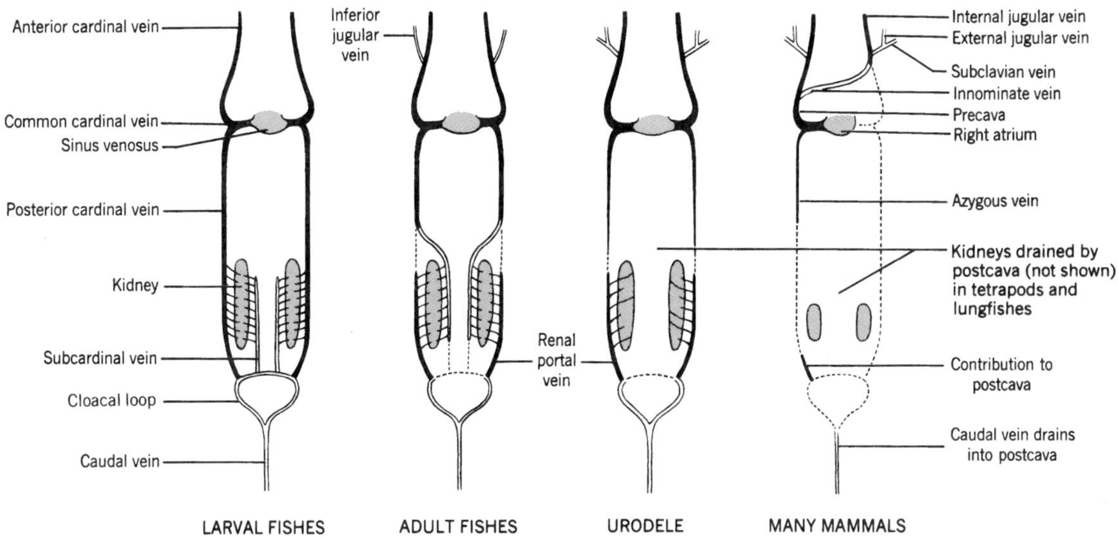

FIGURE 14.18 EVOLUTION OF THE CARDINAL SYSTEM OF VEINS (solid black) AND SOME RELATED VESSELS (open). Ventral views.

anterior segments of the posterior cardinals. The **renal portal system** is thus established (which, as we shall see later, is of great physiological importance).

DIPNOANS and URODELES retain the same pattern of vessels but add a new vessel, the postcava (see below), which receives most of the blood from the kidney. The reduced anterior segments of the posterior cardinal veins drain the body wall. ANURANS and REPTILES are further advanced in that the anterior segments of the posterior cardinals are usually represented only by variable **vertebral veins,** which drain the anterior part of the thorax. All blood in the renal portal veins now enters the kidneys, but in some species part of the blood is shunted through the organ to the postcava without entering a capillary bed.

The situation is the same in BIRDS except that nearly all blood shunts through the kidney to the postcava. It enters kidney tissues only if a valve closes and occludes the direct route. MAMMALS have no renal portal system at all. An **azygous vein** and, in some species, a hemiazygous vein, which drain part of the thorax, are the only derivatives of the anterior segments of the posterior cardinals.

Evolution of Posterior Somatic and Placental Veins The tail, posterior appendage, and body wall are drained by veins having a relatively diverse evolutionary history. We have already noted that blood from the tail drains into the *subintestinal system* in hagfishes and some teleosts, but that this ancestral connection is otherwise lost. We have also seen that posterior segments of the *posterior cardinals,* as renal portal veins, carry caudal blood to the kidneys in all vertebrates except mammals. The renal portal veins of tetrapods also receive at least part of the blood from the hind limb. Furthermore, anterior segments of the posterior cardinals (or derivatives of one or both of them) drain part of the body wall.

Another drainage, the *abdominal system,* drains the pelvic fins and ventral body wall of sharks, and this seems to be the primitive condition (Figures 14.15 and 14.17). This system is curiously absent in cyclostomes and ray-finned fishes. In adult dipnoans, amphibians, and reptiles, abdominal veins (single or paired) are present, but with the difference that they join the caudal circulation posteriorly and (except in dipnoans) enter the liver anteriorly. Thus, blood from the hind limbs of amphibians and reptiles may be diverted

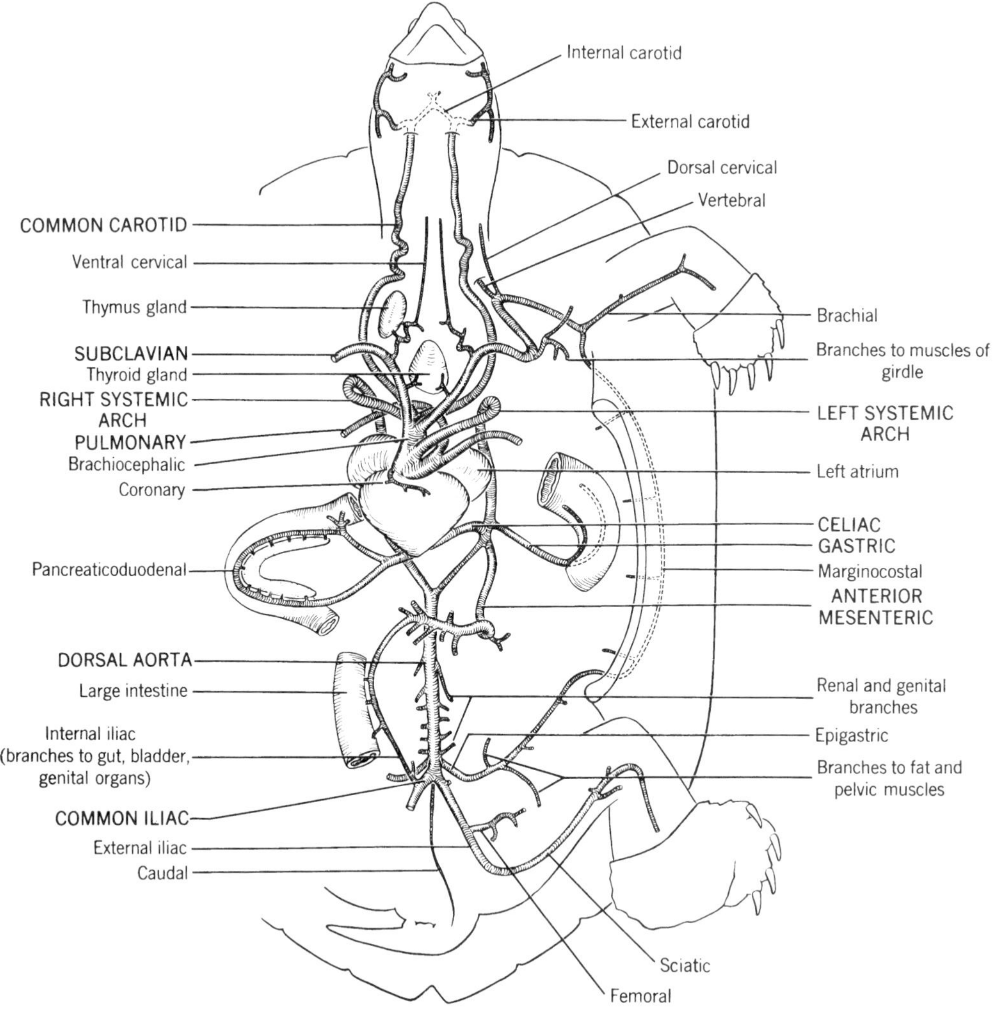

FIGURE 14.19 ARTERIES OF A TURTLE, family Testudinidae. Principal vessels identified by capital letters.

either into the renal portal veins on its way to the kidneys, *or* into the abdominal veins whence it goes to the liver (Figures 14.17 and 14.20).

Adult birds and mammals lack the abdominal system, but embryonic amniotes have paired abdominals. Fetuses retain both (reptiles) or the left only as **allantoic** or **umbilical veins.** These vessels enter the liver as their antecedents did in ancestral amphibians. However, within the liver a **venous duct** usually connects them directly with hepatic veins, so the fetal circulation is not functionally portal.

Finally, and importantly, another major posterior somatic vein, the **postcava** (or posterior vena cava), is present in dipnoans, the living coelacanth, and tetrapods. This is not a new vessel in an evolutionary sense, but is instead derived from segments of preexisting systems. In dipnoans it forms from parts of the right posterior cardinal and the adjacent subcardinal. (The postcava is foreshadowed in other fishes by asymmetries in these channels.) In mammals, right hepatic and supracardinal veins, and anastomoses among the primordia, also contribute (Figure 14.24). In amphibians, reptiles, and birds, the postcava extends only to the posterior end of the kidney (Figures 14.16

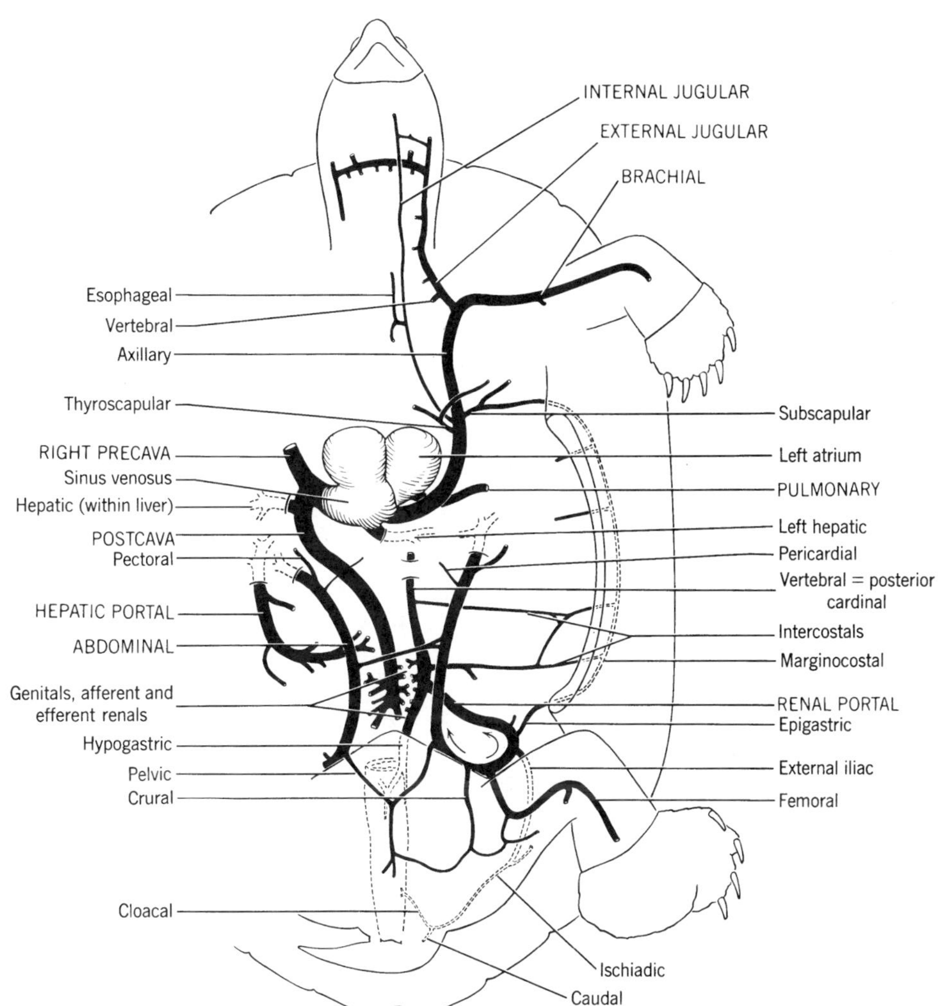

FIGURE 14.20 VEINS OF A TURTLE, family Testudinidae. Principal vessels identified by capital letters.

and 14.20). In mammals, there being no renal portal system, the postcava collects directly from the tail and hind limb (Figure 14.23). Considering its complicated ontogeny and phylogeny, it is not surprising that the postcava and its tributaries show much individual variation.

SOME FUNCTIONAL CONSIDERATIONS

It is important that the heart and blood vessels be studied not merely as a pattern of channels, but also as a functional system. One approach is to select any two organs of a vertebrate and name, in sequence, the vessels and heart chambers that would be traversed by blood in moving from one organ to the other. The exercise can be repeated for the equivalent organs of several other vertebrates.

It is also instructive to seek functional explanations for certain configurations of the system.

Response to Special Needs of Tissues and Organs As a first example, the conservative **circulation to the liver** always receives all of the blood from the digestive tract and may (in amphibians and reptiles) also receive venous blood returning from the tail and hind limbs. The supply of arterial blood, by contrast, is a trickle. Why so much blood, and why mostly

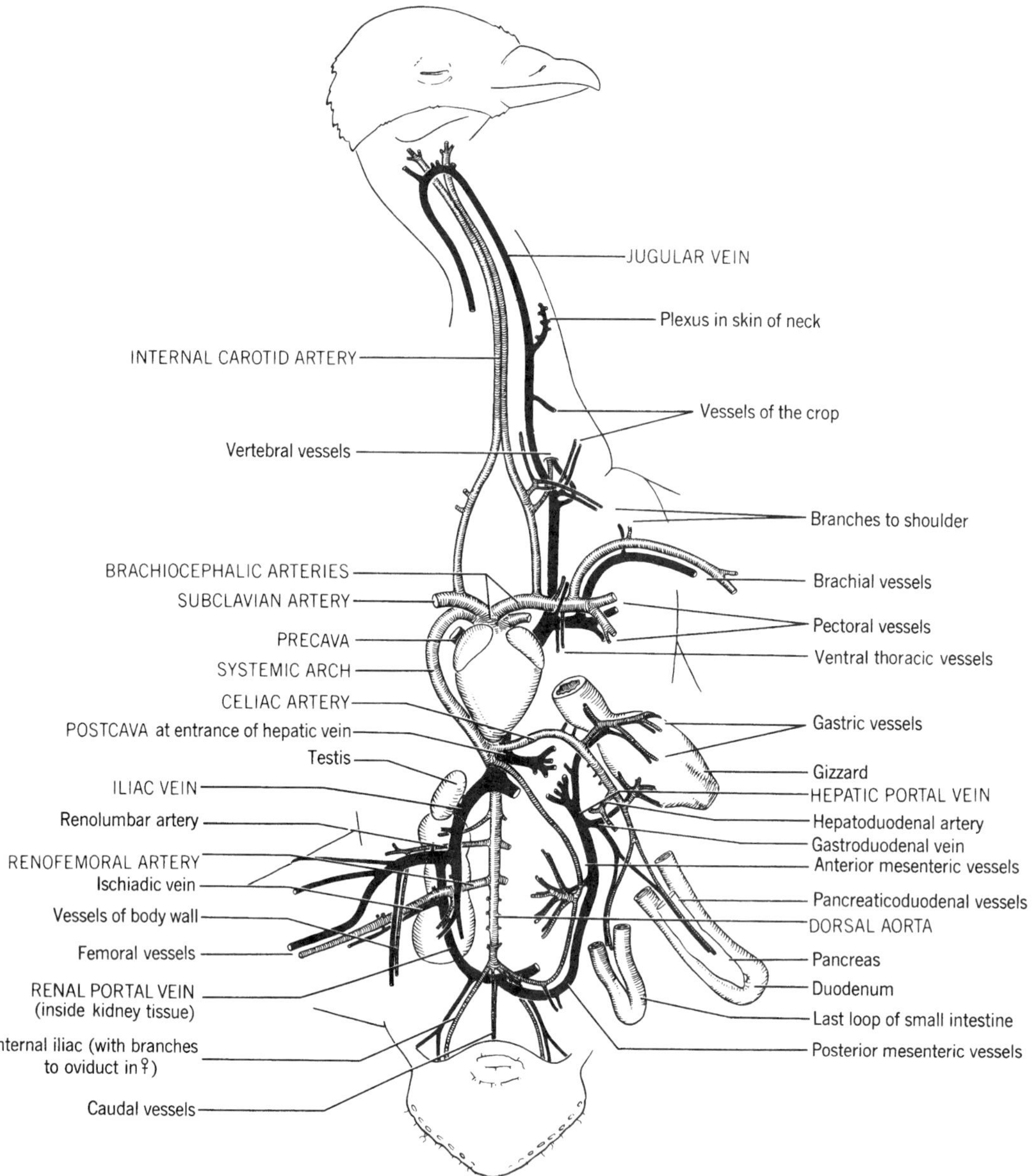

FIGURE 14.21 CIRCULATORY SYSTEM OF THE PIGEON. Principal vessels identified by capital letters.

venous? The many functions of the liver (see p. 213) require that it process quantities of blood. It receives the blood from the gut because some of the nutrients absorbed there would not be useful, or would even be detrimental in the general circulation if not first transformed by the liver. Another reason for furnishing the liver with portal blood is that since it requires little arterial blood to do its job, energy would be wasted if a large quantity of high-pressure, oxygenated blood were provided.

Turning to the **circulation of the kidney,** we note that most vertebrates have a functional renal portal system, but that mammals have none, and some others shunt part or all of their portal blood directly into the postcava. Furthermore, some vertebrates direct a

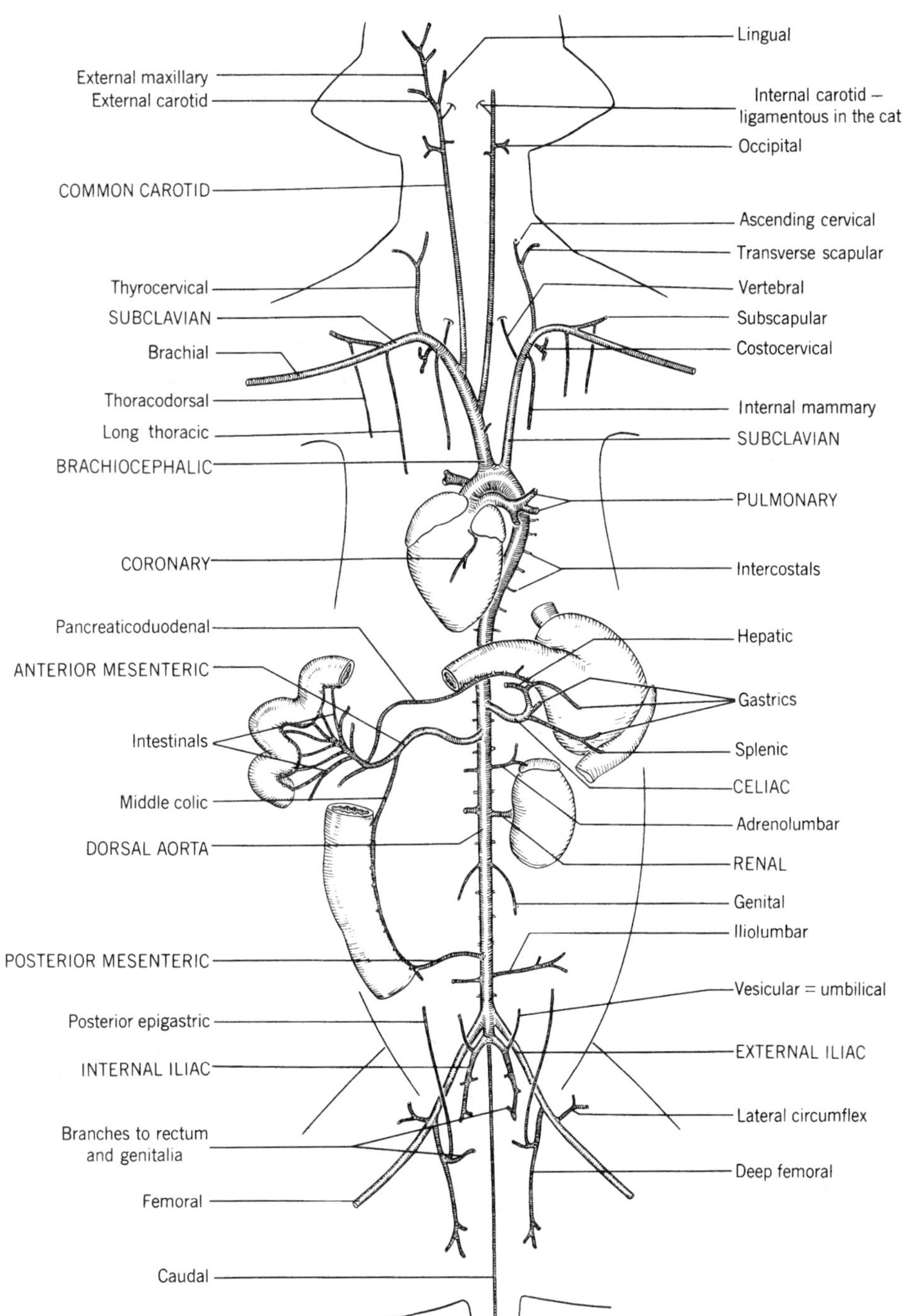

FIGURE 14.22
ARTERIES OF THE CAT. Ventral view. Principal vessels identified by capital letters.

large supply of arterial blood to the kidney, and some do not. How are these differences explained?

The vertebrate kidney usually has two capillary beds. One is organized as tiny knots called glomeruli, which lie within the renal capsules and are associated with the filtration phase of kidney function. The other is organized as nets that surround the nephric tubules and are associated with excretion and selective resorption (Figure 14.25). Animals that produce a large volume of urine (freshwater fishes, amphibians, mammals) must have a large

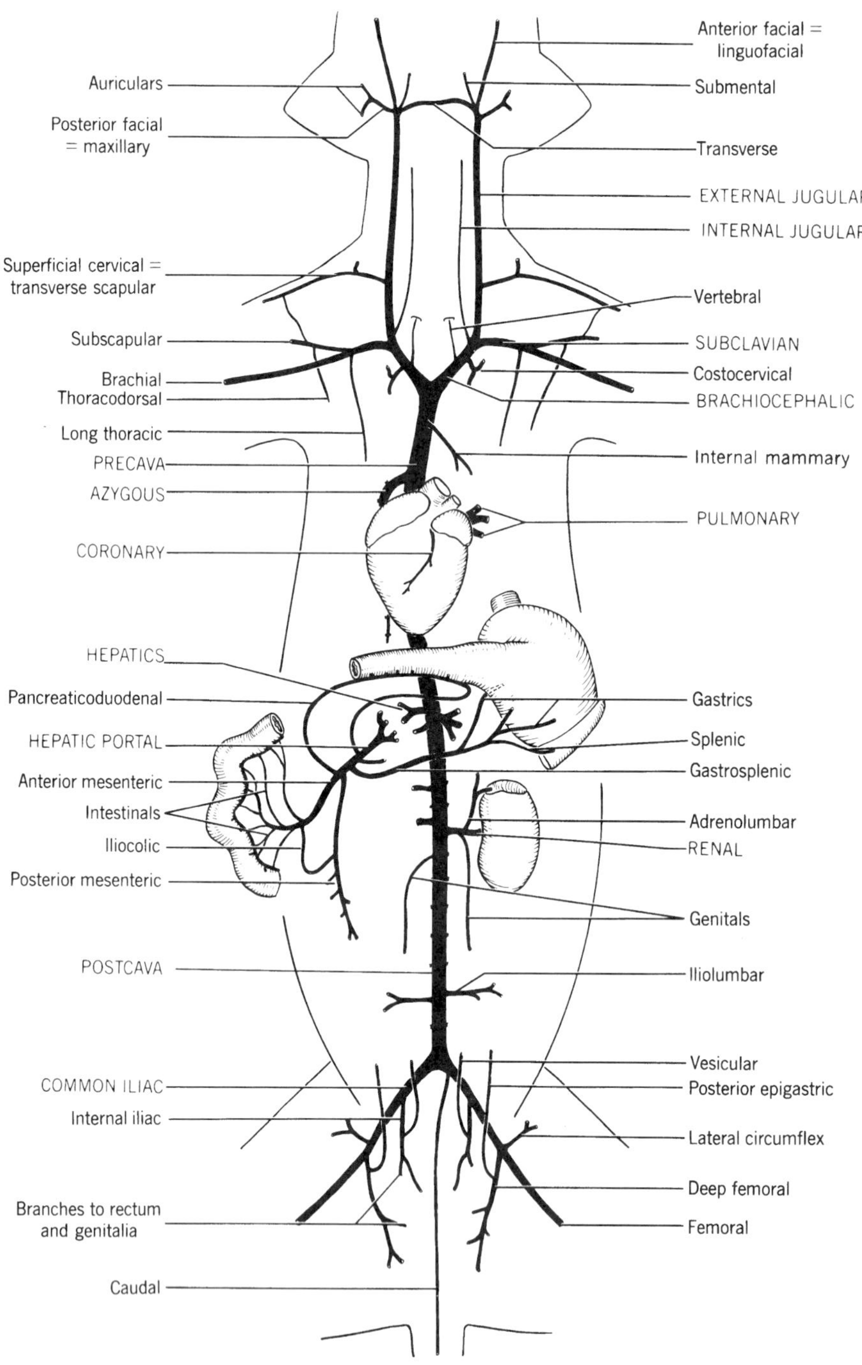

FIGURE 14.23 VEINS OF THE CAT. Ventral view. Principal vessels identified by capital letters.

high-pressure (arterial) flow of blood to the glomeruli, because it is this pressure that is the driving force of the filtration process. These animals can either reuse the same blood for the tubules (mammals) or use portal blood instead. The oxygen content of this blood is unimportant; it is low pressure that is required so reabsorption can take place. Animals that conserve water and salts (marine fishes, reptiles, birds) have only a small high-pressure flow to the glomeruli (or none in some fishes), so filtration will be limited. Therefore, in order to supply enough low-pressure blood to the tubules, a renal portal system is needed.

The function of the kidney in osmoregulation is discussed further on p. 278.

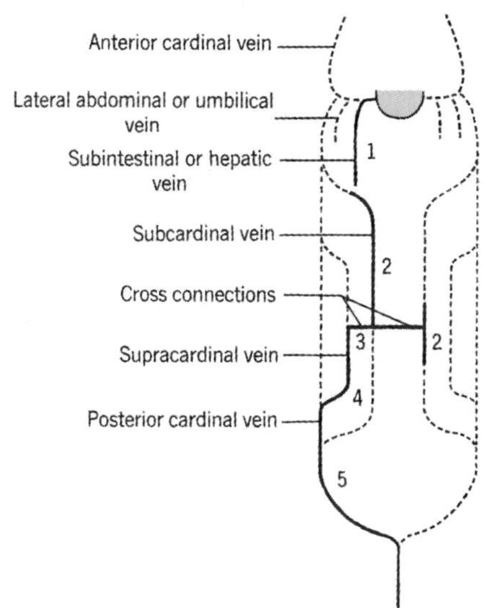

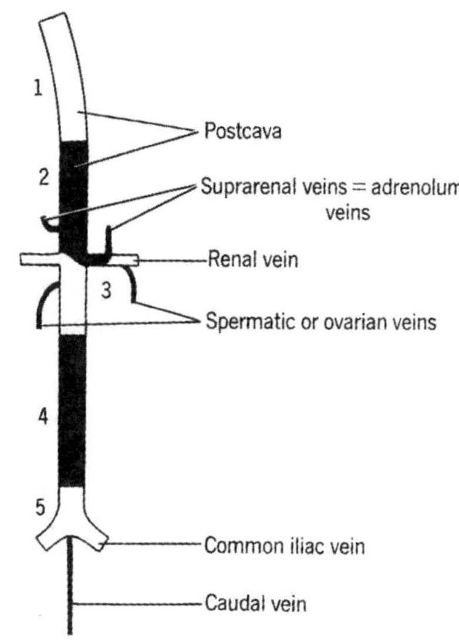

FIGURE 14.24 FORMATION OF THE POSTCAVA IN A MAMMAL. Left, embryonic and evolutionary primordia; right, adult derivatives. Ventral views. Contributions of the different primordia vary among tetrapods.

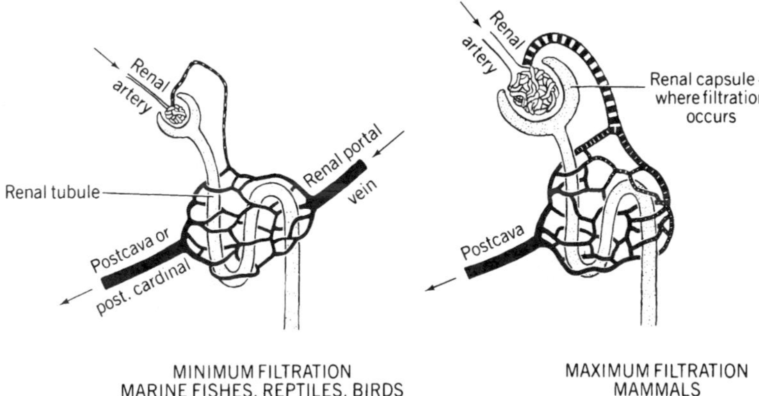

FIGURE 14.25 BLOOD CIRCULATION TO THE EXCRETORY UNITS OF THE KIDNEY IN RELATION TO FUNCTION.

As noted on p. 244, the **coronary circulation** differs markedly between vertebrates using gills and those using lungs. The heart muscle of fishes cannot be supplied by blood from adjacent major vessels because that blood is low in oxygen and high in carbon dioxide. It is supplied instead by the circulating blood or by long coronary arteries that branch from branchial efferents, subclavians, or dorsal aorta (Figure 14.13). The coronary arteries of vertebrates returning oxygenated blood to the heart from lungs can, by contrast, be short vessels that branch from the base of the systemic arch (or right systemic arch if there are two).

In order for the **respiratory circulation** to be most effective, one would think that blood delivered to the respiratory organ should have minimum oxygen content and maximum carbon dioxide content, and blood leaving the organ should reach the metabolizing tissues without prior mixing with oxygen-poor blood. These conditions often pertain, but not always, and the departures clearly serve very well for the animals in question. Blood oxygenated in the gills of fishes and branchiate amphibians does continue directly to the tissues of the body. However, a consequence is that blood pressure in the dorsal aorta can be only one-half or one-fourth of the pressure in the ventral aorta. This is adequate for fishes because their hearts need not pump against gravity and their metabolism is low. Also, the swimming muscles or ventilation muscles of various fishes assist in moving the blood.

One might expect air-breathing fishes to supply blood to their lungs from the ventral aorta or branchial afferent arteries, those vessels having oxygen-poor blood. It is surprising that none does. Instead, they supply their lungs with blood coming either from the dorsal aorta or the efferents of the fifth or sixth aortic arches. Either arrangement can be effective only if the more posterior gills have not already accomplished a significant degree of oxygenation of this blood before it is sent on to the lungs. This circumstance would pertain in stagnant water or if much blood were shunted past the gill capillaries. In any event, the taking of blood to the lungs from dorsal aorta or posterior branchial efferent arteries has what appears to be a serious weakness: These fishes mix oxygen-rich blood from the lungs with oxygen-poor blood from the tissues and then send the mixture to both lungs and tissues. (Air-breathing fishes eliminate most carbon dioxide by way of the gills, not the lung. Their need for gills is continuous; their need for lungs is intermittent.)

This general plan also serves the needs of the embryos and fetuses of amniotes. The respiratory organ is a yolk sac or allantois (or placenta, which is derived in part from the yolk sac or allantois). The vitelline arteries to yolk sac, and allantoic or umbilical arteries to allantois or placenta, are like the pulmonary arteries of some fishes in being branches of the dorsal aorta. For fetuses also, therefore, the posterior part of the body must get along with blood having the same oxygen content as that going *to* the respiratory organ. Evidently fetuses, like the fishes, get along fine nonetheless.

Fetuses alter this situation by abandoning their fetal respiratory organs at birth or hatching. Adult tetrapods do it by keeping oxygen-rich and oxygen-poor blood separate (or nearly so) in the heart and then sending the latter only to the lungs by way of the sixth aortic arch (Figure 14.13).

The evolution of an effective respiratory circulation included one further change for air-breathing vertebrates: Blood returning to the heart from the lungs was to be kept separate from systemic blood. Most fishes drain blood from the gas bladder or lung into the postcardinal or hepatic portal veins, which are far from the heart. Blood returning from the respiratory skin of amphibians also mixes with systemic blood. The pulmonary veins of dipnoans and tetrapods, at last, enter directly into the left, or "oxygenated side" of the heart.

Some features of the **flow of blood to and from the tissues** remain to be mentioned. Fishes maintain a nearly constant body configuration; vessels may be interrupted by injury but not occluded by accidents of posture. Furthermore, gravity does not affect blood circulation of fishes because of the buoyancy of the surrounding water. Tetrapods temporarily restrict or occlude vessels as posture is changed, and gravity does affect the distribution of blood. Several provisions are made to ensure that tissues will not be deprived. Arterial blood pressure tends to be higher than in fishes, and blood volume is relatively greater. Most parts of the body have a double blood supply by vessels that are parallel (e.g., ulnar, radial, and interosseus arteries of forearm) or that form **distribution loops** having inflow at each end (e.g., volar arches that deliver blood to digits, arches along curvatures of stomach and intestine, confluence of internal mammary and epigastric arteries in body wall, confluence of internal carotid and vertebral arteries in a common distribution loop ventral to the brain). Moreover, throughout the body, adjacent major arteries are joined by frequent small anastomosing vessels, and isolated major arteries have frequent small branches that form somewhat meandering and crisscrossing adjacent channels. These small vessels make up the **collateral circulation.** They are standby channels in case of occlusion or injury to the major channels. Fishes also have a collateral circulation, but it is enhanced in tetrapods.

The venous drainage of the more active tissues is similarly primitive in fishes. The veins have valves not spaced along their length (parietal valves), but only where they are joined by tributaries (ostial valves). This is apparently because the channels are relatively direct (there being no long neck or appendages) and the flow of blood is not complicated by gravity. Nevertheless, venous pressure is low in fishes, and some of them have evolved accessory venous pumping mechanisms. Caudal musculature of elasmobranchs squeezes blood from segmental veins through ostial valves into the caudal vein, which, being in the rigid hemal canal, is not constricted by the muscles. In hagfishes (and perhaps some other fishes as well) respiratory movements hasten venous return in nearby vessels.

The veins of tetrapods differ in that they, like the arteries, tend to develop rich collateral circulations. Parietal valves are usually present; the pull of gravity contributes to their necessity.

Further Response to Gravity Most tetrapods can adapt their circulations to gravity, both at rest and during postural change, by selective vasoconstriction and adjustment of heart function. Larger species that adopt vertical orientation of an elongate body, or that have long necks and limbs, need further adaptations to deliver blood far above the heart and return it from far below. Snakes that climb up tree trunks are an example. Lillywhite has shown that compared with ground snakes, and particularly sea snakes, the climbers have high blood pressure and hearts located far forward (near the head when ascending). The body is slender and the skin is tight, so blood cannot pool toward the tail. Body muscles help squeeze venous blood back to the heart. The vascular tissue of the lung is located only near the heart to avoid edema. (It runs the length of the body cavity in sea snakes.)

The giraffe illustrates similar adaptations but remains a source of lively disagreement among functional morphologists and fluid engineers. Giraffe arterial pressure near the heart is twice that of humans, and its heart mass is greater than 2 percent of body mass, compared with about 0.5 percent of body mass for most animals. These features seem necessary to move blood to the brain, located 1.5 m above (see Seymour et al. in References). Hicks and Badeer, however, argue that since the vertebrate circulatory system is a closed system, normal gravitational effects are not operating. Based in part on Poiseuille's equation (see Comment 14.1), these authors are convinced that the siphon principle operates to relieve the heart of extra work associated with pushing the blood to an elevated head, except for overcoming the resistance of a long circuit. These contrasting arguments emphasize the need for more study. For example, if gravity is a factor, does the high mean aortic pressure at the heart aggravate an already elevated pressure in the legs and feet (the feet are 2–3 m below the heart!)? In giraffes the skin and connective tissue of the legs are tight (like support stockings), presumably to prevent swelling, and the blood vessels in the legs have unusually thick walls. Lymphatic vessels are prominent in the legs, and elastic fibers are particularly dense in the long jugular veins. Again, the muscles act as pumps and, unlike snakes, giraffes have numerous valves along appendicular veins. Based on these observations, we can surmise that the huge sauropod dinosaurs had similar solutions for distributing blood (see Gunga et al. in References).

Role in Thermoregulation The circulatory system participates in most kinds of thermoregulation by selectively conserving, dissipating, and distributing heat in the body. Heat that enters or leaves the body by conduction (a snake resting on a warm rock), or convection (a whale fluke moving through cold water), or radiation (a basking lizard), or the combination provided by evaporative cooling (a panting bird or sweating mammal) alters the temperature of the skin or of the epithelium of the tongue, mouth, pharynx, nasal chamber, respiratory tree, or gill. Much heat is produced by active muscles, and many mammals have a kind of fat called brown fat that may also produce heat. The circulatory system can extend the exchange of heat to other parts of the body by the dilation of small vessels at the affected area and the transport of the warmed blood elsewhere. It can limit the heat exchange to the local area by constriction of the local vessels. Several examples will illustrate the effectiveness of the mechanisms.

When subjected to heat stress, many kinds of mammals, including cats and artiodactyls, cool the nasal epithelium by rapid shallow breathing. The surface area of the vascular epithelium may be increased by the complexity of the turbinate bones. Blood from the nasal chamber goes to a **carotid rete** (= *mesh*). There the carotid artery, which carries blood to the brain, breaks up to vessels of 200 to 300 μm in diameter. These are bathed by a pool of blood from the nasal chamber. The two streams of blood flow in opposite directions, thus forming a countercurrent heat exchanger that can be very effective: The blood in the brain of a desert-living gazelle may be 2.9°C cooler than the blood leaving the heart. Ceratopsian dinosaurs may have had a similar mechanism. Conversely, during dives to deep, cold water the swordfish and its relatives keep the brain as much as 10°C warmer than the water by using blood to transfer heat to the brain from a calcium-mediated heat-producing organ near the eye.

The fins, flippers, and flukes of whales are potential heat exchangers having enormous surface. An active whale dissipates excess heat by deflecting blood into small surface vessels, where it can be cooled by the water. A less active animal in very cold water must prevent heat loss. Arteries entering the flukes divide into small arterioles that pass through parallel venules running in the opposite direction (shown diagrammatically in Figure 14.26). In this countercurrent exchanger blood entering the fin is cooled by the returning blood and, conversely, the returning blood is warmed by the outflowing blood. The consequence is that the fluke, but not the body, is cooled by the arctic sea. The gray whale also has heat exchangers in its enormous tongue to conserve body heat during feeding in frigid water. Similarly, there are countercurrent heat exchangers in the legs of penguins and many other birds, the tails of beavers, and the limbs of some primates. In each instance the exchanger can be used or bypassed according to need.

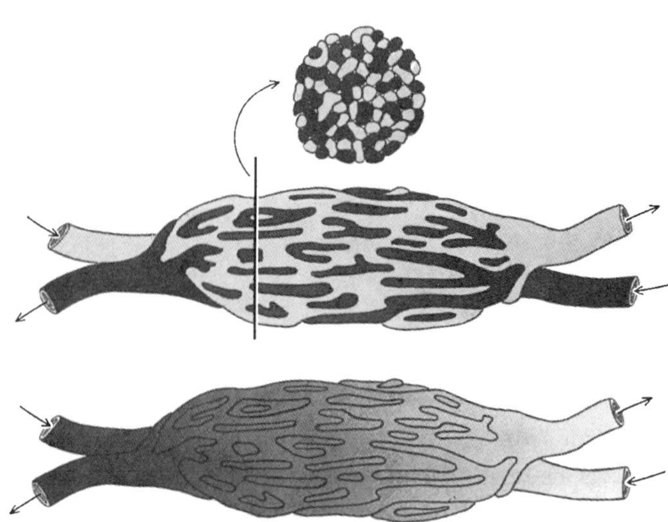

FIGURE 14.26
VASCULAR COUNTERCURRENT EXCHANGER. Above, shaded to distinguish arterioles from venules; below, shaded to indicate the transfer of heat (or diffusible substance if the barrier is sufficiently thin).

Certain fishes (e.g., bluefin tuna, mako shark) swim continuously using dark muscles deep in the body (see p. 175 regarding red muscle fibers). The metabolic heat produced by these muscles is retained by countercurrent exchangers. The central part of the body may be 10 or even 20°C warmer than the water. The elevated temperature enables the muscles to contract more rapidly and probably also more powerfully and efficiently. The fish can control the mechanism according to need.

Antarctic fishes, some pond turtles, and various frogs endure subfreezing temperatures by adding to their tissues antifreeze proteins that bind to the surface of ice crystal nuclei, thereby inhibiting their growth.

EVOLUTION OF THE LYMPHATIC SYSTEM

Amphioxus has no lymphatic system. CYCLOSTOMATA, CHONDRICHTHYES, and CHONDROSTEI have networks of fine vessels that resemble lymphatic vessels but differ in containing some red blood cells and in joining the veins and venous sinuses in many places. The term **hemolymphatic system** is appropriate for these vessels. The more superficial channels tend to be more like veins and the visceral channels more like lymphatics. **Hemolymph propulsors** are present in hagfishes but absent in these other fishes. These valved reservoirs passively propel hemolymph as extrinsic muscles impinge on them. The hemolymphatic system of these vertebrates almost certainly represents an early stage in the evolution of the true lymphatic system.

The lymphatic system is incomplete in Dipnoi but fully developed in NEOPTERYGII (Figure 14.27). Four **subcutaneous ducts** are typical: one dorsal, one ventral, and two lateral. One or usually two **subvertebral ducts** (also called cardinal ducts) extend the length of the body cavity. Cranial and visceral networks complete the system. Small pairs of **lymph propulsors** are usually present in the tail and near the pharynx. There are no valves in the vessels. Lymph usually enters the venous system by a pair of openings in the anterior cardinal veins and another pair near the iliac veins.

AMPHIBIANS and REPTILES are similar to teleosts in having well-developed lymphatic systems with subcutaneous, subvertebral, and visceral vessels. Valves are absent. Urodeles usually have numerous small pairs of segmentally arranged **lymph hearts** (with their own intrinsic musculature) and also a larger pair near the bases of each set of limbs. Anurans have fewer small hearts but retain the two larger pairs. They have large **lymph sinuses** under the skin. Reptiles have the posterior pair of hearts that drain into the renal portal veins. They are located at the ends of transverse processes of basal caudal vertebrae.

The subvertebral ducts of BIRDS are also called **thoracic ducts.** They drain into the precavae. Smaller ducts may enter the iliac veins. Valves may be present but are not numerous. Small lymph hearts appear in the fetus but are rarely present in the adult.

Most MAMMALS have a large single thoracic duct. It is derived from the subvertebral ducts, drains most of the body, and enters the left jugular or subclavian vein. A small duct on the right drains the forelimb and shoulder on that side and enters the right subclavian vein. Numerous bicuspid valves are present in the vessels. Lymph hearts are absent. **Lymph nodes** (see Figure 14.9), sparsely represented in the other tetrapod classes, are numerous in mammals—particularly among the viscera, in the neck, and at the bases of the limbs. The system has many anastomoses and parallel channels. Lymphatic vessels penetrate most tissues but are absent from the central nervous system, bone marrow, deeper parts of the liver and spleen, epithelium of the skin, cartilage, and placenta.

The **thymus gland** occurs in all vertebrates except possibly the cyclostomes (see figure on 374). It is derived from the epithelium of one or (usually) several pharyngeal pouches and comes to lie in the gill area, neck, or anterior thorax. It may be diffuse or discrete. The thymus produces lymphocytes and also establishes the immunological potential of the young animal.

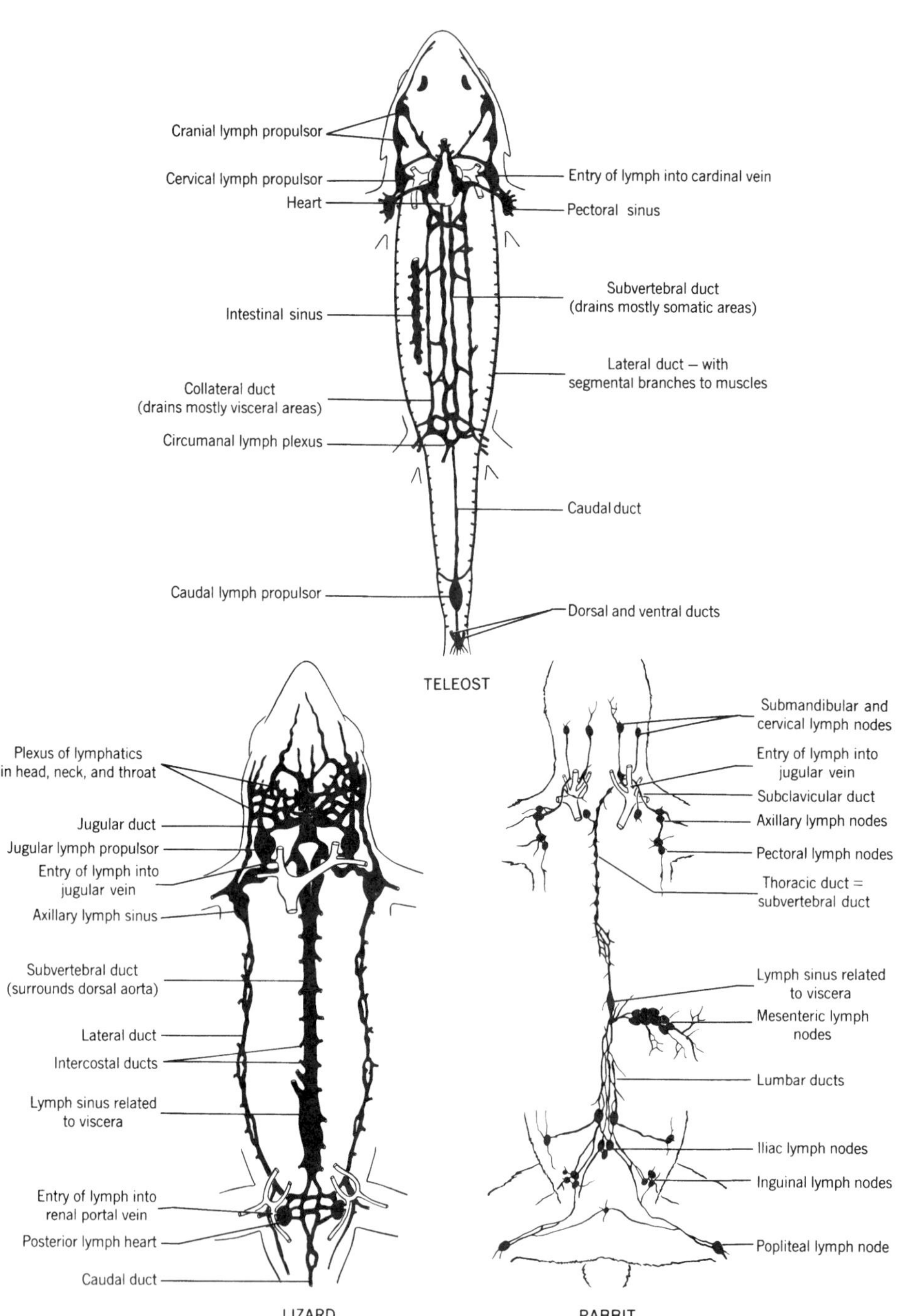

FIGURE 14.27
LYMPHATIC SYSTEM. Ventral views. The teleost and lizard are composite drawings. Modified from Kampmeier.

REFERENCES

Baker, M.A. 1979. A brain-cooling system in mammals. *Sci. Am.* 240(5):130–139. The relation between nasal veins, carotid rete, and brain temperature.

Block, B.A. 1991. Endothermy in fish: thermogenesis, ecology and evolution, pp. 269–311. *In* P.W. Hochacka and T.P. Mommsen (eds.), *Biochemistry and molecular biology of fishes,* vol. 1. Elsevier Science, New York. Fascinating review of the circulatory and musculature adaptions associated with maintaining a high brain and body temperature in tunas and sharks.

Bourne, G.H. (ed.). 1980. *Hearts and heart-like organs* (vol. 1, *Comparative anatomy and development*). Academic Press, New York. 415p.

Burggren, W.W. 1987. Form and function in reptilian circulations. *Am. Zool.* 27:5–19.

Burggren, W.W., and B.B. Keller (eds.). 1997. *Development of cardiovascular systems.* Cambridge Univ. Press, Cambridge, UK, and New York. 360p. Part II devoted to a comprehensive treatment of vertebrate hearts.

Farmer, C. 1997. Did lungs and the intracardiac shunt evolve to oxygenate the heart in vertebrates? *Paleobiology* 23(3):358–372.

Graham, J.B. 1997. *Air-breathing fishes: evolution, diversity, and adaptation.* Academic Press, New York. 299p. Comprehensive phylogenetic treatment of the comparative morphometrics and functional morphology of fish lungs and respiratory gas bladders including microcirculation (ch. 3), general circulation (ch. 4), and blood respiratory properties (ch. 7).

Gunga, H-C., et al. 1999. Body size and body volume distribution in two sauropods from the upper jurassic of tendaguru (Tanzania). *Mitt. Mus. Nat.kd. Berl., Geowiss. Reihe* 2:91–102. Following estimations of the body size of *Brachiosaurus* and *Dicraeosaurus* (stereophotography and laser scans), estimates are made of heart volume and pressures needed to overcome an 8.0 m heart-brain distance.

Hicks, J.W. 1998. Cardiac shunting in reptiles: mechanisms, regulation and physiological function, pp. 425–483. *In* C. Gans and A.S. Gaunt (eds.), *The biology of the reptilia,* vol. 19, *The visceral organs.* Soc. for the Study of Amphibians and Reptiles, Academic Press, New York.

Hicks, J.W., and H.S. Badeer. 1992. Gravity and the circulation: "open" vs. "closed" systems. *Am. J. Physiol.* 262 (*Regul. Integ. Comp. Physiol.* 31):R725–R732.

Holmes, E.B. 1975. A reconsideration of the phylogeny of the tetrapod heart. *J. Morphol.* 147:209–228.

Kampmeier, O.F. 1969. *Evolution and comparative morphology of the lymphatic system.* Thomas, Springfield, IL. 620p.

King, A.S., and J. McLelland. 1989. *Form and function in birds.,* vol. 4. Academic Press, New York. 591p.

LaBarbera, M. 1990. Principles of design of fluid transport systems in zoology. *Science* 249:992–1000.

Lillywhite, H.B. 1988. Snakes, blood circulation, and gravity. *Sci. Am.* 259(6):92–98.

Milnor, W.R. 1982. *Haemodynamics.* Williams & Wilkins, Baltimore. 390p

Pedley, T.J., B.S. Brook, and R.S. Seymour. 1996. Blood pressure and flow rate in the giraffe jugular vein. *Philos. Trans. R. Soc. Lond. Biol.* 351(1342):855–866.

Rennick, B.R., and H. Gandia. 1954. Pharmacology of smooth muscle valve in renal portal circulation of birds. *Soc. Exp. Biol. Med. Proc.* 85:234–236.

Rowlatt, U. 1990. Comparative anatomy of the heart of mammals. *Zool. J. Linnean Soc.* 98:73–110.

Satchell, G.H. 1991. *Physiology and form of fish circulation.* Cambridge Univ. Press, New York. 304p.

Sedmera, D., et al. 2000. Developmental patterning of the myocardium. *Anat. Rec.* 258:319–337. Reviews the pattern of trabecular development in a number of vertebrate species and discusses its significance within a functional context; excellent illustrations.

Seymour, R.S., A.R. Hargens, and T.J. Pedley. 1993. The heart works against gravity. *Am. J. Physiol.* 265 (*Regul. Integ. Comp. Physiol.* 34):R715–R720.

Vogel, S. 1992. *Vital circuits: on pumps, pipes, and the workings of circulatory systems.* Oxford Univ. Press, New York. 315p. A nontechnical introduction.

Wood, S.C., et al. (eds.). 1992. *Physiological adaptations in vertebrates: respiration, circulation, and metabolism.* Marcel Dekker, NewYork. 419p.

Chapter 15

Excretory System and Osmoregulation

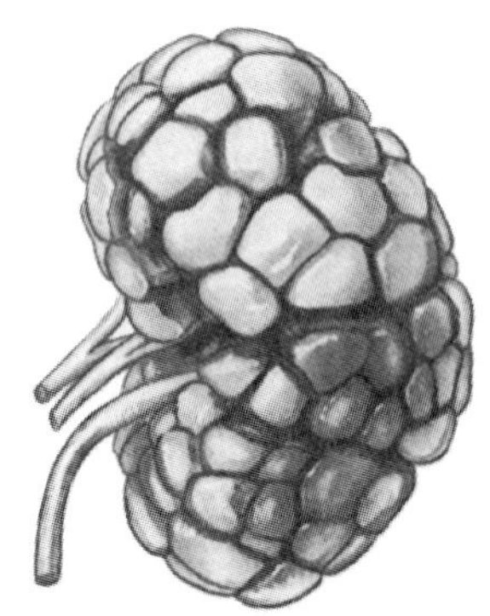

GENERAL NATURE AND DEVELOPMENT OF KIDNEYS

Function and Characteristics Kidneys are the primary adult excretory organs. Other organs that may contribute to the elimination of wastes from the body are the gills, lungs, skin, parts of the digestive system, and various salt glands. Together, these organs perform two related and essential functions: (1) They remove the nitrogenous waste products of protein metabolism and also many other harmful substances, and (2) they eliminate controlled amounts of water and salts, thus maintaining the internal environment within the narrow limits necessary for life.

Like the pharynx and circulatory system, the urinary system is outstanding for the recapitulation that takes place during its development. The system was much studied several generations ago in a mostly unfruitful effort to elucidate vertebrate origins. Furthermore, the urinary system is of interest because it provides striking examples of gradients. Embryos and ancestors each form a series of excretory structures of increasing complexity that replace one another in orderly sequence in time and space.

The kidneys of vertebrates are compact organs derived from mesoderm and consisting of numerous **nephric tubules** (also called renal tubules—"nephros" and "ren" both mean *kidney*) that open either into the general coelom or into cup-shaped spaces called **renal capsules,** which are coelom derivatives (Figure 15.1). Quantities of water and other constituents of the plasma leave the blood by filtration from small knots of capillaries called **glomeruli** (= *small balls*). If the capillary bed is surrounded by a renal capsule, it is said to be an **internal glomerulus.** Glomerulus and capsule together form a **renal corpuscle.** If the capillary bed discharges instead into the general coelom, it is then called an **external glomerulus** and the filtrate reaches the tubules indirectly. A second capillary bed meshes around each tubule, where many constituents of the filtrate are selectively reabsorbed and returned to the blood stream (see figure on p. 263). A tubule with related corpuscle is the functional unit of the vertebrate kidney and is called a **nephron.** There are thousands or even millions of nephrons in adult kidneys. Urine, the product of excretion by the kidneys, moves out of all the tubules into a long pair of **nephric ducts** that open into the cloaca or a derivative of the cloaca.

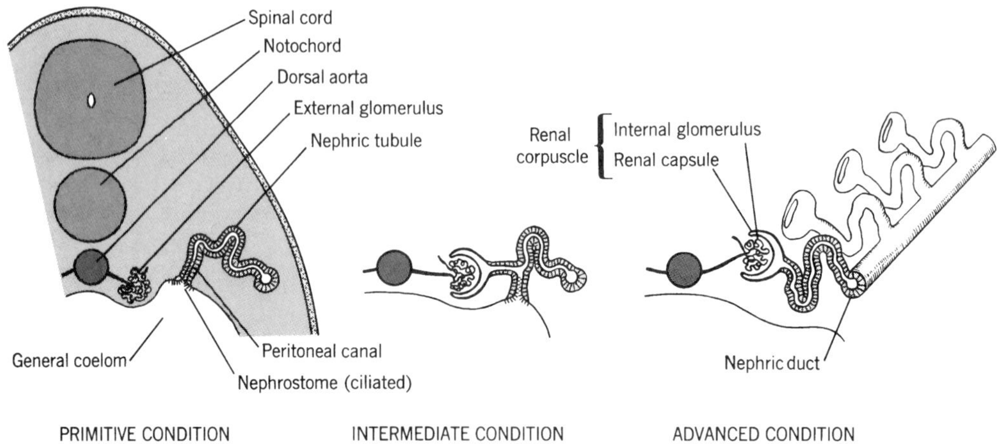

FIGURE 15.1 RELATIONS OF THE NEPHRIC TUBULE TO THE GLOMERULUS AND GENERAL COELOM. Cross sections, with a perspective view added on the right.

Development The embryonic mesoderm differentiates on each side of the body into a segmented dorsal epimere, a small mesomere in an intermediate position, and an unsegmented ventrolateral hypomere (see figure on p. 76). Earlier chapters have presented the derivatives of the epimere and hypomere; we now come to the relatively inconspicuous but important mesomere.

The mesomere is as long as the general coelom, or even a little longer. Gradually it pinches off from the overlying epimere. Anteriorly it remains relatively thin and becomes segmented into units called **nephrotomes.** Posteriorly it does not become segmented and is called the **nephrogenic cord.** Nephrotomes and nephrogenic cord merge into one another at a level that varies according to species and is unimportant to subsequent development (Figure 15.2). At mid and posterior levels the developing kidney bulges somewhat into the coelom to form a **nephric ridge** (see figure on p. 284).

The differentiation of the mesomere is subject to control by induction: It will not form normal organs in the absence of either the adjacent epimere or hypomere, and undifferentiated mesoderm from elsewhere in the body will form urogenital organs if transplanted to the position of the mesomere.

The general coelom or splanchnocoel occurs in the hypomere. It is initially continuous with small coelomic spaces or **nephrocoels** in the nephrotomes. The narrow channels that join splanchnocoel to nephrocoels are called **peritoneal canals,** and their openings

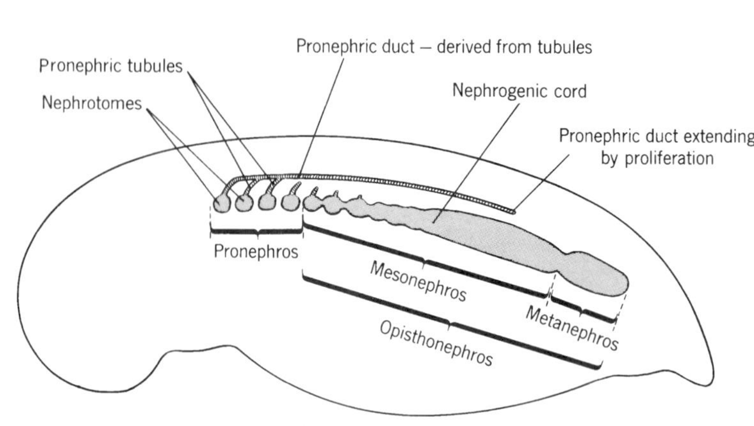

FIGURE 15.2 DEVELOPMENTAL ORIGIN OF THE ELEMENTS OF THE EXCRETORY SYSTEM. Left lateral view of a generalized vertebrate embryo. (Pronephric duct is actually lateral, not dorsal to the nephrogenic cord.)

into the splanchnocoel are **nephrostomes** (Figure 15.1). Functional peritoneal canals and nephrostomes are ciliated. Posteriorly, at levels of the nephrogenic cord, peritoneal canals may be present, but in most amniotes they fail to form. In their absence, spaces regarded as nephrocoels form within the cord by a process of cavitation. Nephrocoels are important because they become the renal capsules of the mature kidney.

Nephric tubules develop either as outgrowths of the walls of nephrocoels (probably the primitive method) or from small solid masses of mesenchyme. Gradually they lengthen and become more or less convoluted. The anterior tubules are the first to mature, and their outer ends join to form a nephric duct on each side of the body, which then extends itself caudad by terminal proliferation until it has grown all the way to the cloaca (Figure 15.2). As the more posterior tubules mature, they merely join the preexisting duct. Indeed, the duct induces their development.

(The developmental steps outlined above are typical, but not constant.)

EVOLUTION AND STRUCTURE OF VERTEBRATE KIDNEYS

Holonephros: Ancestral Kidney The general nature of the early vertebrate kidney can be approximated from embryological and comparative morphological evidence. This somewhat hypothetical organ is called a holonephros (an older name is archinephros). It was derived from the entire mesomere and hence was long. It was segmented throughout most or all of its length and had one pair of nephrocoels and one pair of tubules per pair of nephrotomes. Glomeruli were large. It is not certain if they were initially internal or external. In either event, nephrostomes were present.

A holonephros is found only in larvae of hagfishes and caecilians. Usually, the developing mesomere has a marked tendency to become subdivided into regions. Development always sweeps along the mesomere from anterior to posterior, but the wave is interrupted at two levels so that anterior, middle, and posterior kidneys form in sequence. The more anterior regions usually degenerate as posterior regions become functional. Furthermore, there is a tendency for the tissues to fail to mature between adjacent regions, thus establishing gaps. The more posterior break may be omitted, however, so that two instead of three pairs of kidneys are formed in sequence. The various resultant kidneys will now be reviewed in order.

Pronephros: Larval Kidney or Specialized Remnant The most anterior kidney, and the first to form, is the **pronephros** (Figure 15.2). Relatively few nephrotomes are incorporated—1 to 12 pairs according to species, but rarely more than 4. All participating segments form discrete units in the early larvae of some species (notably in hagfishes and caecilians), but before becoming functional the pronephros is usually modified by the degeneration of the first and last pairs or so of tubules and by the partial or complete fusion of the remaining segments to form a giant corpuscle, or **glomus,** on each side of the body. Right and left primordia even fuse in the midline to form a single glomus in many Actinopterygii. Pronephric tubules are relatively simple unless related to a glomus, in which case they may become long and coiled. Glomeruli are internal (caecilians, most bony fishes), external (amphibians except caecilians), or even intermediate (some birds and reptiles). Ciliated nephrostomes are often present and may open into that portion of the anterior splanchnocoel that will become the pericardial cavity in the adult.

The pronephros appears in at least rudimentary form in all vertebrates. It is functional in the free-living larvae of bony fishes and amphibians and possibly briefly in embryos of some reptiles. Renal corpuscles fail to form in cartilaginous fishes. A somewhat modified pronephros—usually compacted into a glomus—remains functional in adults of hagfishes and several bony fishes, where it may be called the **head kidney** because it is usually not the only functional kidney of these animals. Where not functional in the adult, pronephroi may vanish as development proceeds (cartilaginous fishes, most amphibians, birds) or may leave derivatives that are lymphatic or glandular in nature. The duct of the pronephros, which is identified as the **pronephric duct,** is the most constant part of the organ and is not lost even though the tubules degenerate.

It is seen that the pronephros is highly variable in development and structure. This is the result of ancient origin and reduced function. Few phylogenetic conclusions can be drawn, and the nonspecialist will not find it profitable to memorize individual peculiarities.

Opisthonephros: Kidney of Anamniotes It was explained above that there is an anterior break in both time and space in the development of each mesomere and that the kidney that forms anterior to the break is the pronephros. If a second break forms in the mesomere, then middle and posterior kidneys develop in sequence. The middle kidney is called the **mesonephros** and is present only in fetuses of animals that retain the posterior kidney, or **metanephros,** and none other as adults. These are the amniotes. If all or most of the mesomere posterior to the pronephros instead forms one kidney, then that kidney is called an **opisthonephros.** Such a kidney is typical of late larval and adult anamniotes (see figures on p. 285).

[In spite of the distinction made above, the terms "mesonephros" and "opisthonephros" are used somewhat interchangeably, not entirely without reason. Some vertebrates complicate the concept of multiple kidneys by having an opisthonephros that excludes most of the posterior segment of the mesomere (anurans) or that is incompletely divided into anterior and posterior portions (dipnoans, many urodeles, some elasmobranchs, some teleosts). Some other teleosts have completely separate middle and posterior kidneys and retain each as adults. These organs are not identified as mesonephros and metanephros (as would be logical), but instead as an opisthonephros divided into a trunk kidney and a tail kidney. The more posterior kidneys tend to be larger and more complex than anterior kidneys, but there are no single structural distinctions between adjacent organs. Nephrons at opposite ends of an opisthonephros may be more dissimilar than one at the anterior end of the opisthonephros and one of the pronephros. The paired organs of teleost embryos sometimes fuse to become a single median kidney. Also, the anterior end may be lymphoid and not excretory. Students should be cognizant of the concept of variable yet graded development along the length of the mesomere and not be unduly dismayed by the multiplicity of terms or difficulties in their definitions.]

The opisthonephros, then, develops later in time than the pronephros and forms from much or all of the long middle and posterior portions of the mesomere. It is initially segmental, at least anteriorly, and remains so in hagfishes. In other vertebrates, 1 to 20 generations of secondary tubules develop from the primary tubules, thus causing the organ to lose any segmental character. Opisthonephric tubules tend to be more coiled than unfused pronephric tubules. The entire organ bulges more into the coelom. Anteriorly, the glomeruli may be external, but usually all are internal and relate to renal capsules. Nephrostomes are usually present in early development. Some of them may be retained (some sharks, several primitive bony fishes, some amphibians), but usually all are lost.

Opisthonephric tubules usually tap into the preexisting pronephric duct, which then can be called the **opisthonephric** (or in amniotes the mesonephric) **duct.** In cyclostomes, some amphibians, and some bony fishes, however, the opisthonephric tubules contribute to the formation of the posterior portion of the duct, and in caecilians the duct sends buds to meet the growing tubules. Which mode of origin is the most primitive is uncertain.

The shorter but otherwise equivalent *mesonephros* is functional in fetuses of amniotes and even briefly after hatching or birth in some reptiles, monotremes, marsupials, and some ungulates. The organ varies from small (some rodents) to medium (carnivores, primates) to large (some ungulates), its development being inversely proportional to the excretory efficiency of the placenta.

Metanephros: Kidney of Amniotes The metanephros is the most posterior of the kidneys and is the last to develop in both ontogeny and phylogeny ("meta" = *later in time*). The part of the nephrogenic cord from which it develops lies opposite to only one or two body segments. The organ itself is never segmental, and hundreds of thousands of nephrons are formed by successive generations of budding. Glomeruli are always internal, and nephrostomes are absent. Metanephroi differ from other kidneys in being of dual origin. Early in development a **ureteric bud** grows out from each mesonephric duct near its entrance into the cloaca and approaches the differentiating nephrogenic cord (Figure 15.3). The lengthening neck of the bud becomes the duct of the metanephros, or **ureter,** and the end branches within the kidney to form the **collecting tubules** that become continuous with the nephric tubules derived from the nephrogenic cord. The entire organ migrates forward somewhat as it matures, particularly in mammals. It is of interest that the tail kidney of certain teleosts, and also the posterior part of the opisthonephros of some other anamniotes, are drained by independent ducts. These are called **accessory ducts** rather than ureters, but they are nonetheless probably homologous with the amniote ureter.

The metanephroi of reptiles are of varying irregular shapes and may be lobulated. The kidneys of birds fit into hollows in the synsacrum and are usually three-lobed (see figure on p. 290). Kidneys of mammals have a collecting basin called the **renal pelvis** into which urine oozes from one or more **renal papillae** that project into the pelvis. The organs are bean-shaped, simple, and smooth in primitive and small mammals and in primates and most carnivores (see figure on p. 291). In some other species they are smooth externally but have numerous lobes internally and multiple or otherwise complicated papillae. In marine mammals and some ungulates and other large species, they are compound organs resembling clusters of rat kidneys (Figure 15.4). The development of compound kidneys in large animals seems to be a means of avoiding very long collecting tubules that would require excessive pressure for driving the contained fluid.

In summary, trends in the evolution of the vertebrate kidney include the loss of nephrostomes and external glomeruli, the occurrence of more numerous and nonsegmental nephrons, and development from a progressively more posterior, and ultimately shorter part of the nephrotome.

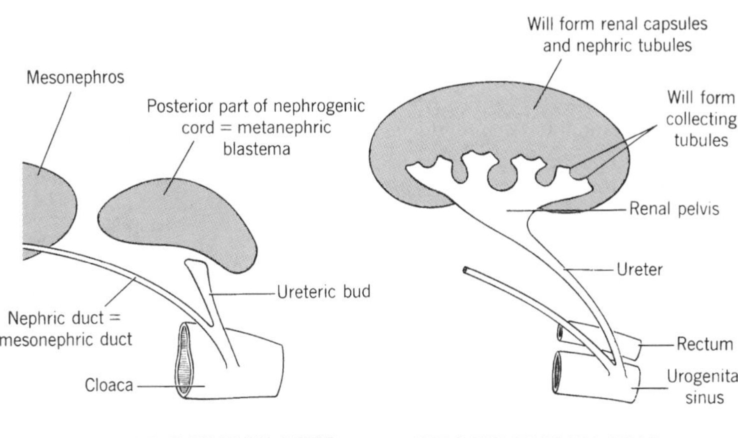

FIGURE 15.3 STAGES IN THE DEVELOPMENT OF THE METANEPHROS AND URETER.

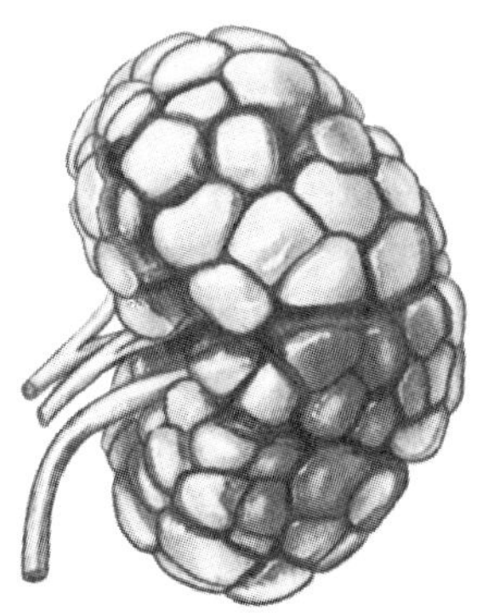

FIGURE 15.4 THE COMPOUND KIDNEY OF THE SEA OTTER, *Enhydra.*

KIDNEY STRUCTURE IN RELATION TO OSMOREGULATION

A more detailed account of kidney structure is now desirable to provide a basis for interpretation of anatomy in relation to function. The glomerulus is a tuft of capillary loops and anastomoses that hangs into the renal capsule. The capsule is thus cup-shaped, with an outer, or **parietal wall** and an inner, or **visceral wall** (Figure 15.5). The parietal wall is of squamous or cuboidal epithelium. In higher vertebrates the visceral wall enfolds the capillaries of the glomerulus. Its cells, named **podocytes,** have numerous fingerlike projections called **pedicels** ("pod"/"ped" = *foot*). Small spaces remain between the interdigitating pedicels, and there are very fine pores in the capillary endothelium. Accordingly, the only unbroken barrier between the blood in the capillaries and the cavity of the renal capsule is the glomerular basement membrane. This membrane is the ultrafilter of the kidney. It passes water, ions, sugars, amino acids, hormones, and various waste products, which together make up the *primary urine.* The filter holds back proteins, cells, and fats. The driving force of the filter is the relatively high pressure of arterial blood (review p. 260). In mammals the total filtration surface is at least half the body surface, and the daily production of filtrate is equal to many times the volume of the blood of the entire body.

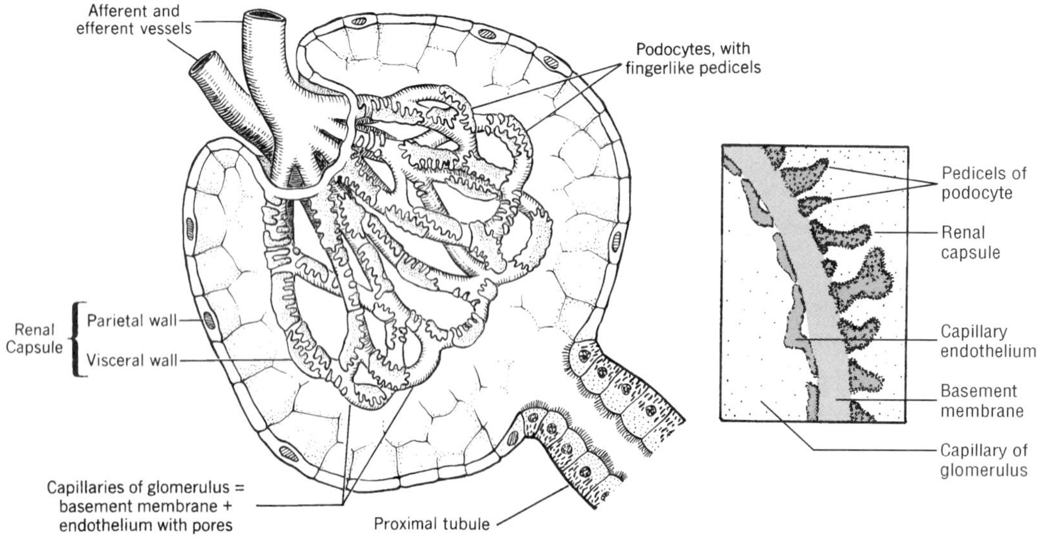

FIGURE 15.5 RENAL CORPUSCLE OF A MAMMAL (left) AND MUCH-ENLARGED CROSS SECTION OF THE BLOOD-CAPSULE BARRIER. For relationship to entire kidney, see Figure 15.8.

COMMENT 15.1

AMPHIOXUS AND THE ORIGIN OF THE VERTEBRATE NEPHRON

The ancestral origin of the vertebrate nephron was a clouded subject for half a century. The excretory organs typical of invertebrate animals having bilateral symmetry were designated **protonephridia.** These have tubules that discharge (directly or by a duct) to the outside of the body. At their inner ends the tubules terminate blindly within body tissue, or a pseudocoel, or coelom. They are capped by *terminal cells* that have flagella extending into the tubules. The beating of the flagella (not blood pressure) creates the pressure gradient needed to activate the system. Protonephridia are derived from ectoderm.

The kidneys of vertebrates are **metanephridia.** As described in the main text, these discharge into a coelom derivative (nephrocoels). There are no terminal cells and no flagella. Specialized cells, the podocytes, create an ultrafilter that is activated by blood pressure in glomeruli. Metanephridia develop from mesoderm.

How did evolution bridge these differences? For clues one naturally turns to amphioxus, the nearest known relative of vertebrates. Amphioxus has terminal cells of a type named **solenocytes.** These have flagella. The organ discharges outside the body (indirectly through an atrial cavity). The germ layer of origin is ectoderm. Hence, the excretory organs of amphioxus were long considered to be protonephridia, suggestive of those of annelids. How could amphioxus resemble vertebrates in other respects yet have such disparate excretory organs?

Electron microscopy has narrowed the difference (Figure 15.6). Solenocytes have flagella, but *also* pedicels, making them probable homologs of podocytes. They seem to accomplish filtration from arterial blood after all, and are apparently functionally related to the coelom they span. Ruppert (see References) believes that the germ layer question may also be moot: The contributions of ectoderm and mesoderm to the formation of excretory organs and ducts vary widely among animals. Position and function may be of more significance than germ layer in establishing homology. Many coelomate invertebrates appear to have filtration excretory systems with cells that are probably precursors of podocytes. Amphioxus need no longer be embarrassed about its kidneys.

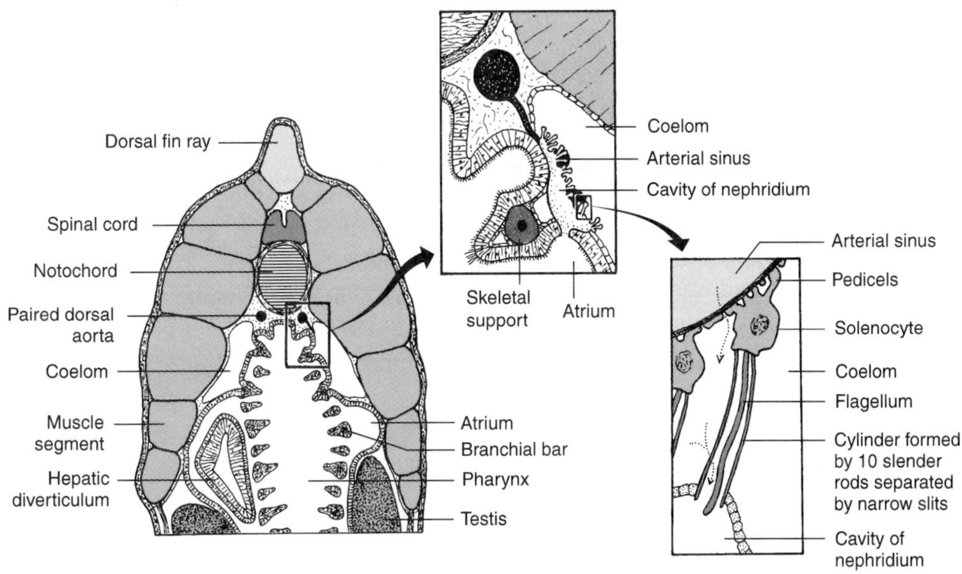

FIGURE 15.6 A NEPHRIDIUM AND RELATED STRUCTURES OF AMPHIOXUS as seen in a cross section near the center of the body. Compare with figure on p. 26.

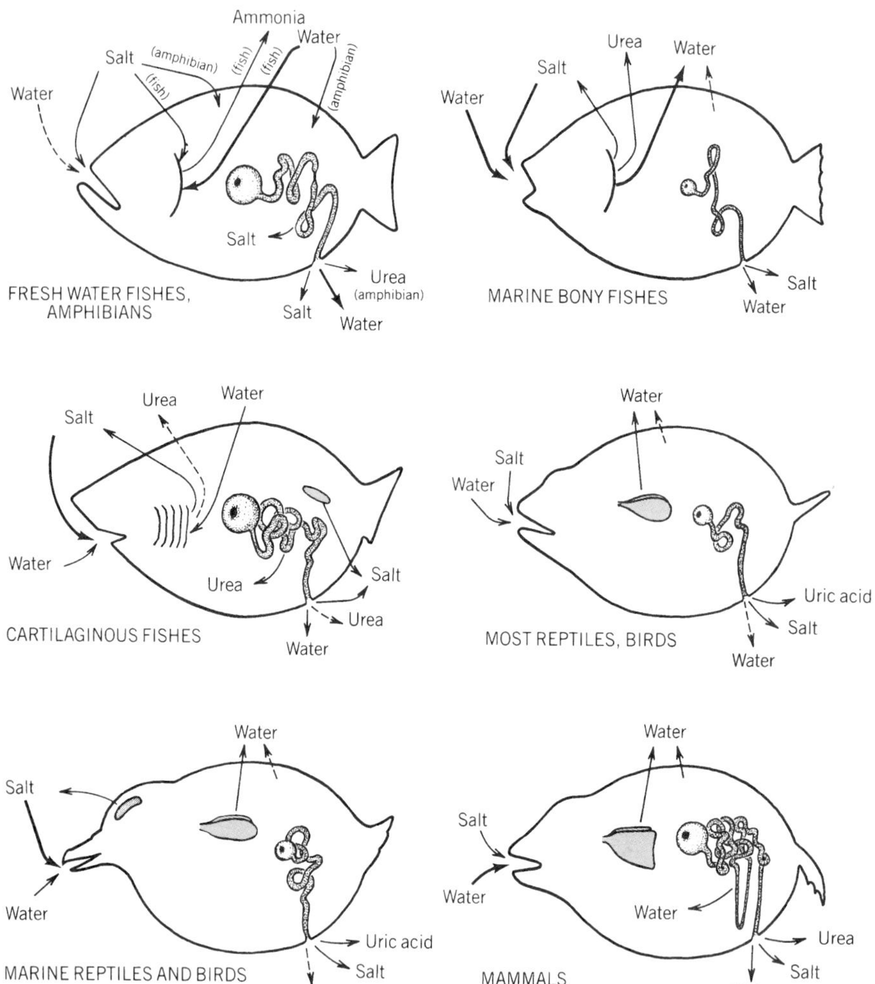

FIGURE 15.7 DIAGRAM OF THE STRUCTURE AND FUNCTION OF THE NEPHRON IN RELATION TO EXCRETION AND OSMOREGULATION. (There is variation within the groups illustrated.)

Renal capsules of endothermic vertebrates are usually separated from their convoluted nephric tubules by narrow ciliated **neck segments** (see Figure 15.8). Nephric tubules are characteristically divided into a **proximal tubule** and a **distal tubule,** which are separated by a short yet sharply demarcated **intermediate segment.** The proximal tubule has cuboidal or low columnar cells with microvilli. The intermediate segment is thin and ciliated. Cells of the distal tubules are cuboidal and have few microvilli or none. The proximal tubule returns sugars, amino acids, vitamins, and various salts to the blood stream and may secrete certain foreign materials into the filtrate. The distal tubule acidifies the filtrate and removes sodium and chloride ions. Water may be returned to the blood by both tubules. The details of tubule function are varied and complicated. Excretion is influenced by hormones in at least some classes.

Most marine invertebrates have body fluids that are isotonic (equal in osmotic concentration) to seawater. Little water enters or leaves their bodies, and their excretory organs lack filtering devices for removing water from their body fluids. Most vertebrates, regardless of habitat, have body fluids that are hypotonic (lower in osmotic concentration) to seawater but hypertonic (higher in concentration) to freshwater. Consider first the FRESHWATER FISHES and AMPHIBIANS. Water constantly enters their bodies from the environment because of the osmotic gradient (Figure 15.7). The skin may be moderately waterproof, but the gills and oral membranes admit much water. There is also water in their food. Consequently, even though these animals scarcely drink at all, they must eliminate copious quantities of urine to maintain their water balance. Accordingly, they have prominent renal corpuscles. Short proximal and distal tubules return solutes to the blood stream, yet although the distal tubules resorb some salt, the urine is more salty than the environment, and salt intake is essential. Some salt is taken with the food. Many amphibians can selectively absorb salt through the skin, and most of the fishes can extract salt from freshwater with the gills. Nitrogen is eliminated from the body by the gills as ammonia (fishes, larval amphibians) or by the kidneys as urea (adult amphibians).

MARINE BONY FISHES have the opposite problem with osmosis: Their body fluids constantly tend to leak away into the sea, particularly through the gills. In order to conserve water, they must pass little urine and therefore have relatively small and poorly vascularized renal corpuscles. Some species have only rudimentary renal corpuscles or none at all. In order to replace water, these fishes drink freely, which introduces an excess of salt. Special chloride cells in the gills excrete monovalent ions, and the proximal tubules of the kidneys excrete divalent ions. Distal tubules are usually absent. The gills also actively excrete urea.

CARTILAGINOUS FISHES are also marine but have solved the problem of water balance in a different way. Nitrogenous wastes of vertebrates are produced and excreted in several forms but in the bloodstream occur primarily as urea. Urea is eliminated by the kidneys of amphibians, many turtles, and mammals. These animals excrete urea because they do not need it, but it is scarcely toxic and may to a degree be eliminated because it diffuses readily and cannot be retained. The only vertebrates that can retain urea are the cartilaginous fishes (and marine toads and the modern crossopterygian, *Latimeria*). An unknown mechanism makes their gills impervious to the substance, and their large, distinctive nephric tubules are able to return urea to the blood from the filtrate. Consequently, the blood contains enough urea to be slightly hyperosmotic to seawater. Some water enters the gills. Little water is drunk, and urine volume is moderate. Renal corpuscles are nevertheless large. Excess urea is excreted by the kidneys and gills. Excess salt is also excreted by the kidneys which, however, may fall behind in the elimination of the monovalent ions. These are then excreted by the rectal gland, which discharges into the hindgut.

CYCLOSTOMES are largely marine, but some lampreys migrate into freshwater. Hagfishes have very large, segmentally arranged renal corpuscles. Nephric tubules consist only of short neck segments. Lampreys have longer tubules with all the usual segments. Hagfishes control water intake by having blood of uniquely high salt concentration. In freshwater, lampreys absorb some salt through the skin and excrete much urine.

AMNIOTES use water to carry out excretory wastes and to condition the surfaces of the lungs. They also lose water through the skin. Since drinking is the only source of water in a usually dry environment, it is desirable for the animal to conserve water. This is done in various ways, but most importantly by the kidneys. The kidneys of mammals do the job in one way, whereas those of reptiles and birds do it in another.

Urea is highly soluble. It flushes from the body in water and escapes from aquatic embryos into the environment. (Some sharks must provide their embryos with waterproof cases to conserve urea for their special needs.) Most BIRDS and REPTILES have insufficient water available to pass dilute urine, and their terrestrial eggs have no way to dispose of soluble urea. They excrete instead uric acid, which is insoluble and can be passed from the body in a semisolid state with very little water. Since the formation of quantities of filtrate would be detrimental, these animals have small, poorly vascularized renal corpuscles. Nephric tubules are short in reptiles and of moderate length in birds. Most of these animals can eliminate excess salt in the urine. However, various marine species of both reptiles and birds take quantities of salt with the food and have only seawater to drink (if they drink at all). Many of these animals (sea snakes, marine iguanas,

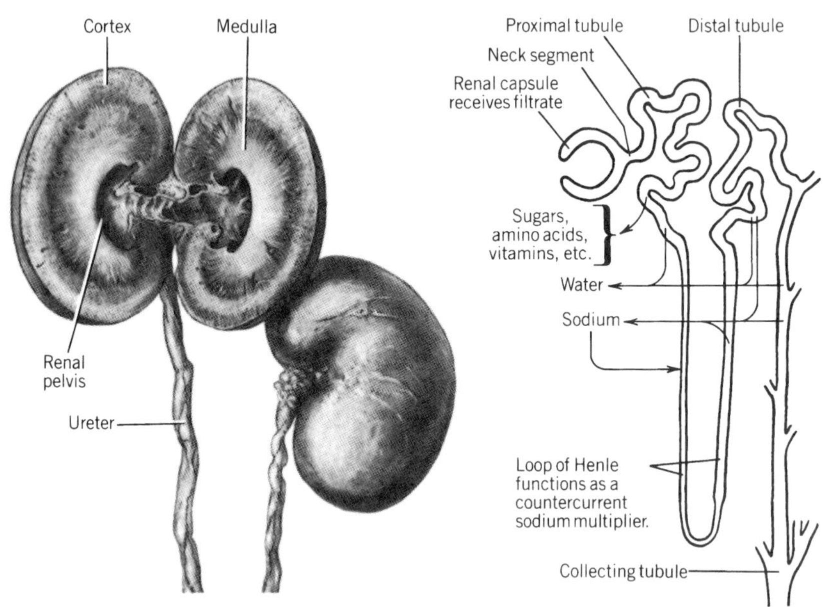

FIGURE 15.8 STRUCTURE AND FUNCTION OF THE MAMMALIAN KIDNEY AND NEPHRON illustrated by the dog. Renal capsules and convoluted tubules are in the cortex; loop of Henle and collecting duct are in the medulla. Compare with Figure 15.9.

sea turtles, cormorants, albatrosses, petrels, gulls, terns, sea ducks, etc.) have salt glands that excrete a very concentrated brine. These glands may be derived from lacrimal glands or nasal or orbital glands and are variously located near the eyes, jaws, or tongue, or in the nasal chamber. Their function is under complex nervous and hormonal control.

MAMMALS excrete urea, which is removed from the blood by prominent renal corpuscles. Quantities of dilute filtrate are thus produced. The mammalian kidney, however, is more effective than any other at returning water from the filtrate to the blood. Only about one one-hundredth of the filtrate is passed as urine. Concentration is dependent on the distinctive **loops of Henle** (Figure 15.8), which are derived from parts of the proximal and distal tubules and extend toward the pelvis of the kidney. (There is no intermediate segment.) Some loops are long and some of moderate length. The thin descending and thicker ascending limbs of a single loop are straight and immediately adjacent to one another. This is of functional significance because one of the means of modifying the urine is the cycling of sodium by countercurrent multiplication between the limbs (the countercurrent mechanism is explained on p. 223). Mammals are the only vertebrates to pass urine that is more concentrated than the blood. Birds also have loops of Henle (apparently independently evolved), but their loops are short and are borne by a minority of nephrons.

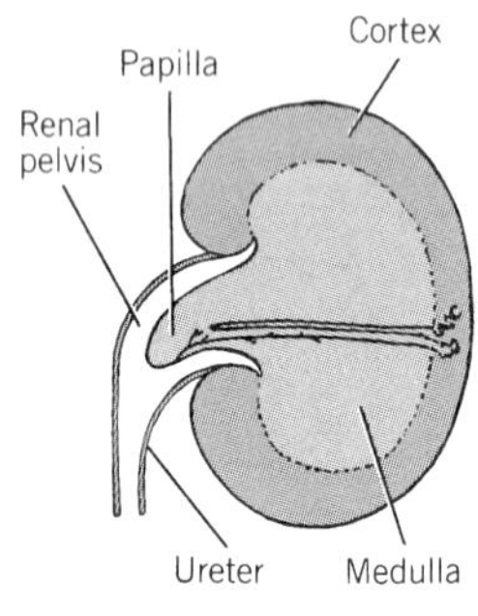

FIGURE 15.9 KIDNEY OF A SMALL MAMMAL FROM A DRY HABITAT showing the position of one nephron.

Renal corpuscles and convoluted parts of the tubule are located in the **cortex,** or outer zone of the kidney ("cortic" = *bark*). The loops of Henle are located in the **medulla,** or inner zone ("medull" = *pith*). Desert-living mammals that must excrete a particularly concentrated urine accentuate medullary tissue to the extent that it commonly extends into the renal pelvis as one or more **papillae** (Figure 15.9).

URINARY BLADDERS

The urinary bladder provides temporary storage for urine and, except in mammals, often modifies the concentration and composition of urine by selective absorption or secretion.

Most FISHES have urinary bladders, but even if much urine is passed, their bladders tend to be of small or moderate size because the aquatic environment makes long retention of urine unnecessary to accomplish sanitation. Enlargement of posterior segments of the nephric ducts provides temporary storage for dipnoans, primitive ray-finned fishes, and female elasmobranchs. The two ducts may fuse to form a single median bladder. The accessory ducts instead enlarge to form bladders in male elasmobranchs. Similar expansions of the posterior ends of the urinary ducts serve as bladders in lampreys and teleosts, but in this instance a pocket of the embryonic cloaca is considered to contribute to the adult bladder, which is, therefore, largely a median structure.

AMPHIBIANS evolved a different kind of bladder of evolutionary importance. It is a large ventral outpocketing of the cloaca. Nephric ducts enter the cloaca dorsally, so urine must cross the cloaca to enter the bladder. Almost certainly it was the enlargement and more precocious development of this bladder that produced the allantois of amniotes. Anurans that live in arid habitats (e.g., desert toads) can store quantities of water in the bladder as dilute urine and resorb most of it as needed.

Some REPTILES (turtles, some lizards, *Sphenodon*) have an identical cloacal bladder, but this time it is formed by the retention in the adult of the base of the fetal allantoic stalk. Other reptiles and BIRDS understandably have no urinary bladder because they excrete a semisolid urine containing uric acid. Ureters enter the sides of the cloaca, and urine mingles with fecal material before being discharged.

The ureters of MAMMALS join the ventrolateral surfaces of the embryonic cloaca and hence are associated with the urodeum after the cloaca is divided (see figure on p. 294). Parts of the urodeum and allantoic stalk together form the urinary bladder. The bladder is lined by a distinctive kind of epithelium that thins as it stretches. Smooth muscle in the walls of the bladder contracts to empty the organ.

Some of the ducts that serve the excretory system also serve the reproductive system. Accordingly, the urogenital ducts are discussed in the next chapter.

REFERENCES

Bentley, P.J. 1971. *Endocrines and osmoregulation: a comparative account of the regulation of water and salt in vertebrates.* Springer, New York. 300p.

Beuchat, C.A. 1996. Structure and concentrating ability of the mammalian kidney: correlations with habitat. *Am. J. Physiol.* 271 (*Regul. Integr. Comp. Physiol.* 40):R157–R179. Comparative anatomy and physiology of kidneys from shrews to whales with emphasis on how little we know.

Braun, E.J. 1993. Renal function in birds, pp. 167–188. *In* J.A. Brown, R.J. Balment, and J.C. Rankin (eds.), *New insights in vertebrate kidney function.* Seminar series (Society for Experimental Biology), no. 52. Cambridge Univ. Press, New York.

Dantzler, W.H. 1989. *Comparative physiology of the vertebrate kidney.* Springer, New York. 198p.

Fox, H. 1977. The urogenital system of reptiles, pp. 1–157. *In* C. Gans and T.S. Parsons (eds.), *Biology of the reptilia,* vol. 6. Academic Press, New York.

Hentschel, H. 1987. Renal architecture of the dogfish *Scyliorhinus caniculus* (Chondrichthyes, Elasmobranchii). *Zoomorphology* 107:115–125.

Hicks, R.M. 1975. The mammalian urinary bladder: an accommodating organ. *Biol. Rev.* 50:215–246.

Holmes, W.N., and J.G. Phillips. 1985. The avian salt gland. *Biol. Rev.* 60:213–256.

Lacy, E.R., and E. Reale. 1985. The elasmobranch kidney: gross anatomy and general distribution of the nephrons. *Anat. Embroyl.* 173:23–34.

Moffat, D.B. 1975. *The mammalian kidney.* Cambridge Univ. Press, New York. 263p.

Pang, P.K.T., R.W. Griffith, and J.W. Atz. 1977. Osmoregulation in elasmobranchs. *Am. Sci.* 17:365–377.

Ruppert, E.E. 1994. Evolutionary origin of the vertebrate nephron. *Am. Zool.* 34:542–553.

Ruppert, E.E., and P.R. Smith. 1988. The functional organization of filtration nephridia. *Biol. Rev.* 63:231–258.

Skadhuage, E. 1981. *Osmoregulation in birds.* Springer, New York. 203p.

Wake, M.H. 1986. Urogenital morphology of dipnoans, with comparisons to other fishes and to amphibians. *J. Morphol.* (suppl. 1):199–216. Indicates how little comparative data exists for this group and points the way for future studies.

Chapter 16

Reproductive System and Urogenital Ducts

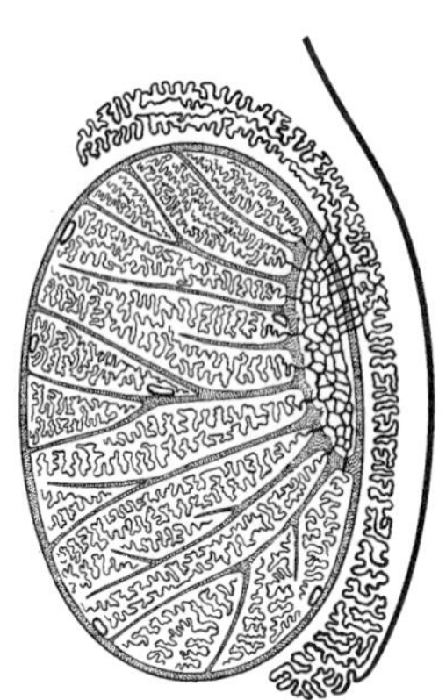

The functions of the reproductive system are to produce the sex cells, or **gametes,** bring egg and sperm cells together, provide for the nourishment of the embryo or fetus until hatching or birth, and release eggs or young from the maternal body. Structural adaptations of other parts of the body and behavioral adaptations may subsequently furnish defense, care, nourishment, or warmth to the eggs or young.

EARLY DEVELOPMENT AND ANCESTRY OF GONADS

Embryonic primordia of the sex organs, or **gonads,** come from two diverse sources. The first is the mesomere. After opisthonephroi, or mesonephroi, as the case may be, are well established, **genital ridges** appear on their mesial surfaces. These ridges are shorter than the nephric ridges that preceded them, yet are longer than the definitive gonads—particularly in amniotes. Anterior and posterior parts of the genital ridges that do not form gonads form instead fat bodies and mesenteries that support the gonads. The latter are called **mesorchia** in males and **mesovaria** in females. Where reflected over the genital ridges, the lining of the coelom thickens to become the **embryonic germinal epithelium.** At the center of the developing organ is another type of tissue called **blastema** (Figure 16.1). The germinal epithelium soon subtends sheets of tissue called **primary sex cords** that penetrate into the blastema.

The second source of gonadal tissue is unexpected and provides one of embryology's more fascinating stories. Several thousand cells near the base of the yolk sac become distinctively large. Some of these stay behind, but many migrate individually and actively into the splanchnic mesoderm of the gut, up the body stalk and mesenteries, and laterally into the genital ridges. These are the **primitive sex cells.** They divide in route, yet some are lost along the way. The remainder space themselves in the germinal epithelium and sex cords, where they have important derivatives, as noted below. They have been much studied, yet the cause of their migration and the developmental and evolutionary significance of their remote origin remain unknown.

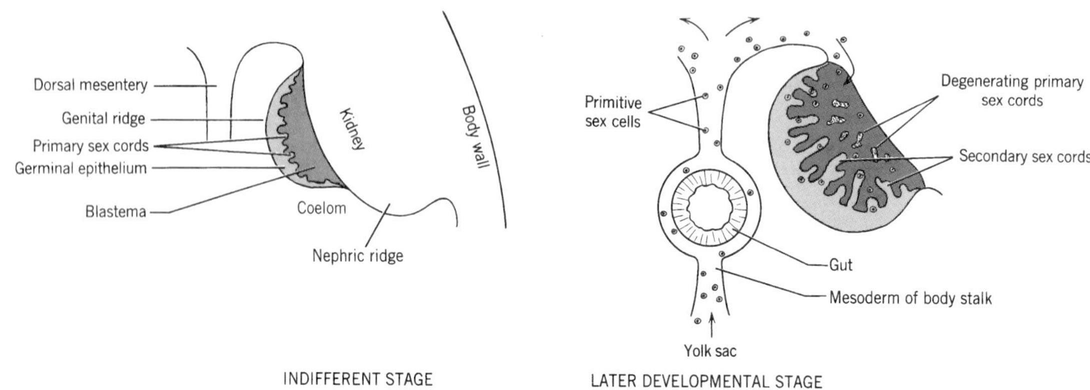

FIGURE 16.1 STAGES IN THE DEVELOPMENT OF THE AMNIOTE OVARY. Cross sections of the body.

In summary, the early gonad consists of a superficial germinal epithelium and a deep blastema. Sex cords penetrate the blastema, and primitive sex cells are distributed within the epithelium and sex cords. The outer portion of the organ is called the **cortex;** the inner portion, including blastema and sex cords, is the **medulla.** Although the sex of an individual is determined at fertilization, development is identical in the two sexes to the point described, and this phase of development is called the **indifferent stage.** The indifferent stage may be completed early in embryonic development (mammals) or delayed until adult body size is approached (hagfishes).

Gonads of the ancestral vertebrate probably extended the entire length of the coelom. They may have had gonocoels (coelomic cavities), though these are absent from the adult organs of known vertebrates. Furthermore, ancestral gonads may have been at least partly segmental, as they are in amphioxus and some embryonic urodeles and caecilians. Finally, the ancestral organs may have been bisexual. Dominance of the cortex produces a female, and dominance of the medulla produces a male. At the end of the indifferent stage, cortex and medulla are in potential physiological competition. The genetic sex usually masks the other, but under abnormal or experimental conditions the nongenetic sex may prevail.

STRUCTURE OF GONADS

The structure of gonads varies within some taxa (Teleostei, Reptilia), but in general is uniform for each class and too conservative to provide evidence for evolutionary lineages.

Ovary The germinal epithelium of the indifferent-stage ovary becomes the thin peritoneal covering of the organ and the important adult **germinal epithelium.** The latter is one cell layer in thickness and includes thousands of **oogonia,** some of which sink below the surface in each consecutive breeding cycle to enlarge and undergo maturation divisions and thus become **ova.** It is probable that the primitive sex cells within the epithelium form all ultimate germ cells. Primary sex cords degenerate and are replaced by **secondary sex cords** (Figure 16.1). These form **follicle cells,** which surround, nourish, and support the ripening eggs, and **theca,** or envelopes that enclose the follicles. Following ovulation, the follicles of mammals (and probably of some elasmobranchs and birds) are temporarily converted to **corpora lutea.** The inner theca and corpora lutea have endocrine functions (see p. 376). The blastema becomes only the **stroma,** or matrix of connective tissue and vessels within the ovary.

Ovaries vary widely in character, being long or short, compact or flat, smooth or lumpy, solid or flabby, according to species (Figures 16.2 and 16.3). Eggs differ in

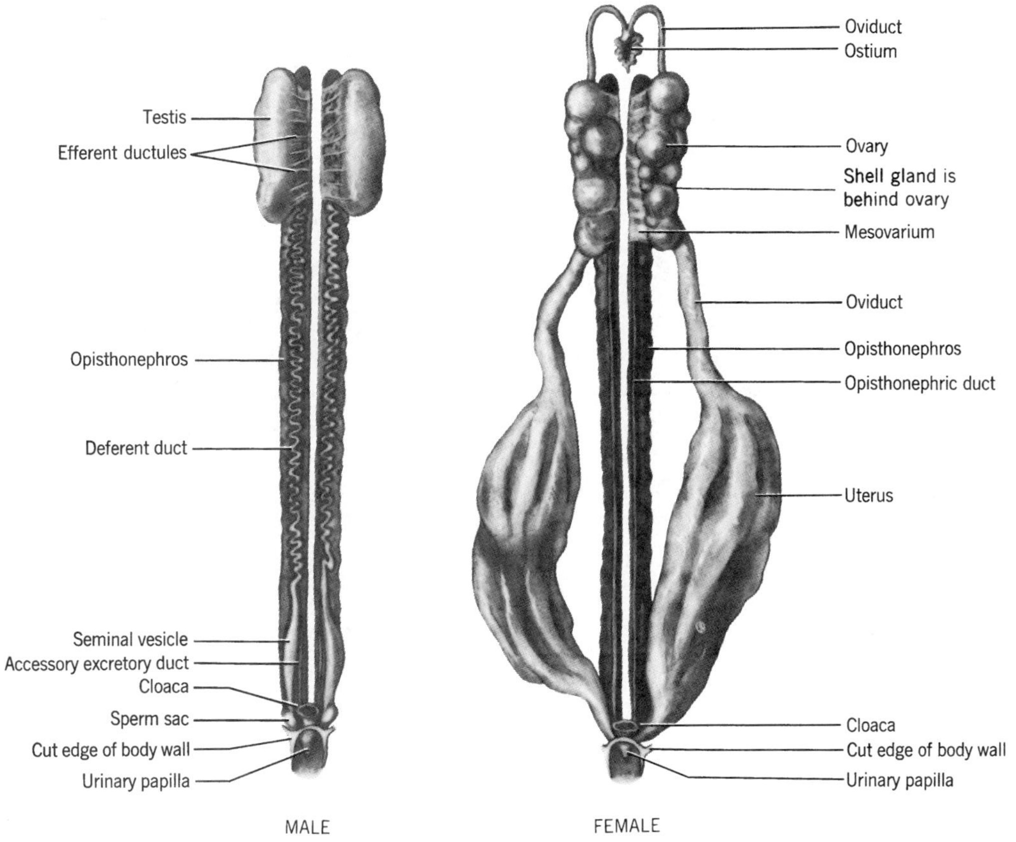

FIGURE 16.2 UROGENITAL SYSTEM OF THE ELASMOBRANCH *Squalus.* Ventral views. In the male, sperm are transferred from the cloaca to grooves in the claspers (see figure p. 155).

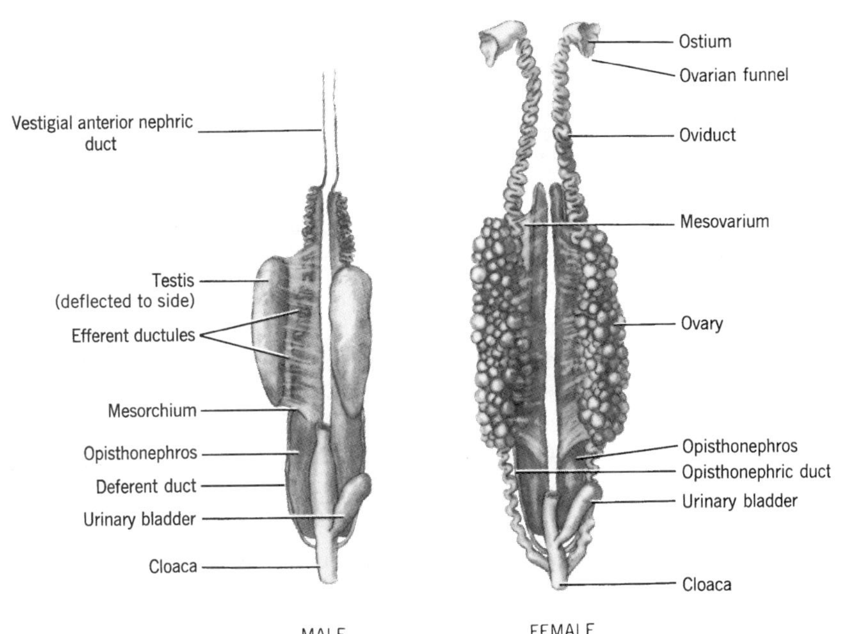

FIGURE 16.3 UROGENITAL SYSTEM OF THE AMPHIBIAN *Necturus.* Ventral views.

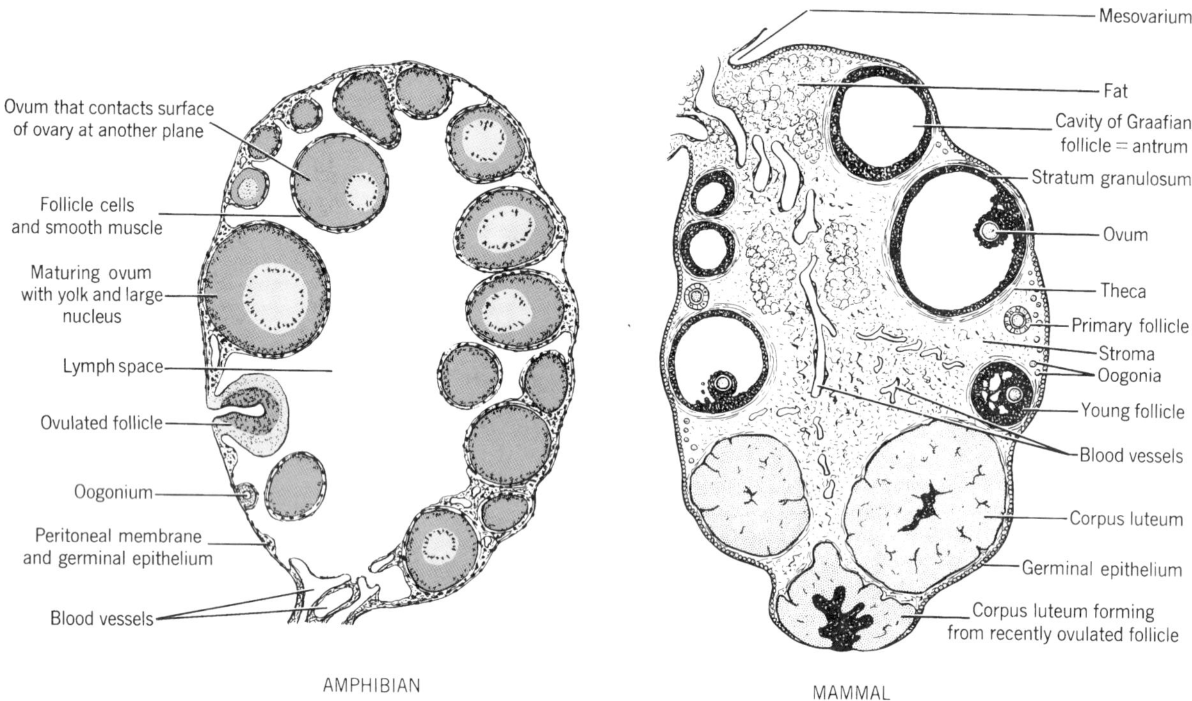

FIGURE 16.4 SECTIONS OF CONTRASTING TYPES OF OVARIES. (It is unlikely that all developmental stages shown would be present at the same time in the reproductive cycle.)

size in proportion to the amount of yolk present. Relatively few eggs ripen at one time if the eggs are large (cartilaginous fishes, reptiles, birds) or if the animal retains the eggs in the body until hatching (many fishes, some caecilians, and some reptiles) or gives birth to live young (mammals). The production of eggs is always cyclic, and the size of the ovaries of some vertebrates fluctuates markedly according to breeding condition. Long-bodied amphibians have long, slender ovaries. Nevertheless, paired ovaries of large size are not easily accommodated in a streamlined body. Consequently, the two embryonic organs of some animals fuse more or less completely in the adult (lamprey, many teleosts), whereas other animals partially or completely suppress the left (many cartilaginous fishes) or right organ (hagfishes, most birds).

Ovaries of crocodilians, turtles, birds, and mammals are solid with relatively much stroma. Those of cyclostomes, cartilaginous fishes, dipnoans, and some primitive ray-finned fishes are also solid but less compact. In amphibians, regression of the blastema leaves one (urodeles) or several (caecilians and anurans) large lymph spaces within the ovary (Figure 16.4). Stroma is then virtually absent, and the organ is soft and pleated. Ripening eggs hang into the central cavity but at ovulation rupture outward into the coelom. Smaller central lymph spaces occur also in ovaries of Lepidosauria.

Most teleosts have hollow ovaries, but the cavity is of different origin. A thin margin of the developing organ curls over and fuses either to the body wall or to the ovary itself, thus sealing off a bit of the general coelom. Ovulation is into the resultant cavity. In a restricted sense, the ovary only appears to be hollow. The coelomic space that is adapted to receive the eggs is continuous with the special oviducts of these fishes, so eggs never reach the unmodified coelom.

FIGURE 16.5
CROSS SECTION OF A SEMINIFEROUS TUBULE OF A MAMMAL.

Still another kind of cavity is present in ovaries of mammals. Other vertebrates have solid follicles, but in mammals a space appears within each maturing follicle, which is then called a **Graafian follicle.** Corpora lutea are more prominent in mammalian than in other ovaries, and in this class there is a tendency for the ovaries to migrate posteriorly as they mature.

Testis Testes tend to develop a little earlier than ovaries. Their development emphasizes derivatives of the medulla of the indifferent-stage gonad; the germinal epithelium of the embryonic organ forms only the peritoneal covering of the adult testis. Primary sex cords do not degenerate, as they do in females, and secondary sex cords are not formed. The primary cords separate from the overlying germinal epithelium and become hollow, producing slender, coiled **seminiferous tubules** (amniotes), saclike **seminiferous ampullae** (cyclostomes, urodeles), or intermediate structures.

Seminiferous tubules usually branch, and may end blindly (reptiles) or anastomose peripherally (mammals). There are dozens or even hundreds of tubules in each testis. Their walls, when in breeding condition, consist of stratified epithelium having germ cells at various stages of maturation (Figure 16.5). Undifferentiated **spermatogonia** are peripheral. As the **sperm cells** are maturing, they are held for a time by supportive cells scattered among the germ cells. Finally, the mature sperm are released into the duct system of the reproductive tract.

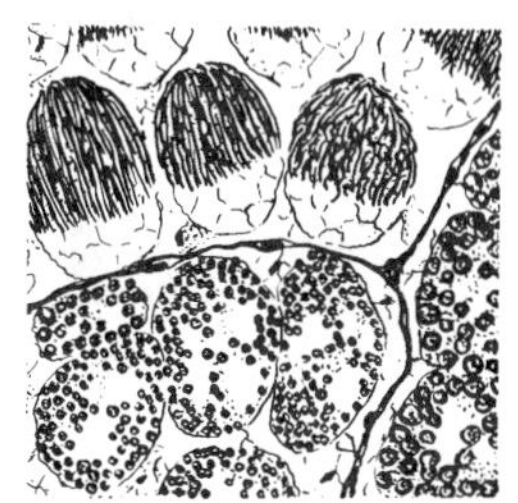

FIGURE 16.6
SECTION OF A SALAMANDER TESTIS SHOWING AMPULLAE with primary spermatocytes (small, spherical), secondary spermatocytes (large, spherical), and spermatids (linear).

Spermatogenesis within ampullae differs in that potential germ cells enter an ampulla from adjacent tissue and then divide many times to form clusters of cells all of which mature together (Figure 16.6). The result is masses of sperm that will all be evacuated at once, leaving each ampulla empty and collapsed.

As in the ovary, primitive sex cells are considered to play the critical role in the formation of germ cells. The blastema of the indifferent-stage gonad forms supportive tissue and, in tetrapods, also cords that hollow out to become **rete tubules** (Figure 16.7). These come to be continuous with the seminiferous tubules. They form an anastomosis within the testis and conduct sperm to the margin of the organ.

Testes are usually smoother, firmer, and smaller than the ovaries of the same species. Usually they are paired, but the two organs may be partially fused in elasmobranchs and are completely fused in adult cyclostomes. Testes are elongate in slender vertebrates (cyclostomes, most fishes, caecilians, urodeles) but compact and ovoid in some cartilaginous

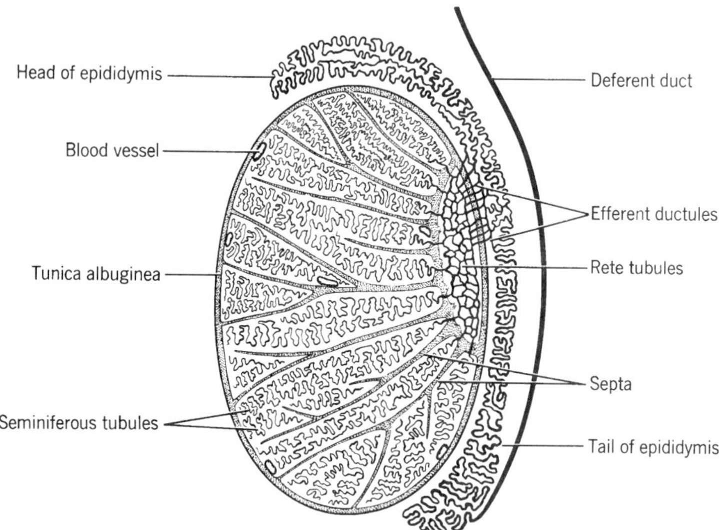

FIGURE 16.7 STRUCTURE OF THE MAMMALIAN TESTIS. Diagrammatic longitudinal section.

fishes, anurans, and amniotes. Some asymmetry in the size and position of the two members of a pair is not unusual.

Mammalian testes are distinctive for the degree to which several features are expressed: The organ is enclosed in a tough envelope of connective tissue called the **tunica albuginea** (= *tunic* + *white*). Internal lobulation is more pronounced than for most other vertebrates, and **septa** separate adjacent lobules. **Interstitial cells,** which produce male hormones, are packed in the interstices among the seminiferous tubules (see Figure 16.5). Lower vertebrates also have male hormones and doubtless also interstitial tissue. However, in them the tissue is never as prominent as it is in mammals, and for many it has not been identified at all.

Spermatogenesis ceases in testes that are warmer than about 36.5°C. Diurnal temperatures of birds usually exceed this figure by several degrees, but at night the body is cooler, and even during the day the testes are probably somewhat cooled by the abdominal air sacs. The testes of mammals are abdominal or pelvic in position if body temperature is relatively low (monotremes, edentates, whales, elephants). In many mammals (primates, most carnivores, most ungulates, and others) the testes descend at maturity out of the abdominal cavity into a cooler pouch of skin in the inguinal area called the **scrotum.** Small pockets of coelom extend into the scrotum to facilitate the descent of the testes into the organ and to give them some freedom of motion within their sac. The paired canals joining scrotal coelom to general coelom usually close, but in such seasonal breeders as rodents may remain open so the testes can descend when active and return to the abdominal cavity at other times. Some temperature control is provided by scrotal muscles, which raise or lower the testes, and by countercurrent exchange in the blood supply. The adaptive significance of the scrotum and the evolution of the descent of the testes in mammals have long been of interest. An extensive phylogenetic analysis led Werdelin and Nilsonne to conclude that it is not the presence of the scrotum in various groups of mammals that requires explaining (the focus of many studies) but instead, why the scrotum has been lost in so many groups. It is their conclusion that the scrotum evolved in concert with endothermy *before* the evolution of mammals. The external genitalia of males and females show sexual homologies, as do the gonads and ducts. In females, the **large labia,** which flank the genital area in some species, are sexual homologs of the scrotum.

UROGENITAL DUCTS AND ACCESSORY ORGANS

Origins Cyclostomes are unique in having nephric ducts but no genital ducts. Eggs and sperm are both released into the coelom whence they exit into the cloaca or urogenital sinus by way of a pair of **genital pores.** Similar openings, also called **abdominal pores,** are said to be present but not functional in various higher vertebrates: some sharks, dipnoans, and several primitive ray-finned fishes. The evolutionary origin of abdominal pores is obscure. Some anatomists believe they represent a posterior pair of ancient, segmentally arranged genital openings. If so, their presence is a primitive rather than a specialized condition, a view that is supported by their wide distribution.

All other vertebrates have two pairs of urogenital ducts that remain identical in the two sexes to the end of the indifferent stage of development and then are modified according to sex and taxon. One pair is the **nephric ducts,** which have already been described as the ducts of fetal kidneys other than metanephroi (Figure 16.8). The other pair is the **paramesonephric** ducts ("para" = *beside*). The phylogenetic origin of paramesonephric ducts is again obscure. In sharks and urodeles they form by a lengthwise splitting of the more precocious nephric ducts. In other vertebrates they develop either from long solid cords at the surface of the nephric ridges or from shorter anterior primordia that extend themselves back to the cloacal area. It is not possible to say what method of origin is primitive.

Other structures that contribute to the urogenital duct system of adults are the more anterior of the opisthonephric or mesonephric tubules, the ureter of amniotes, the accessory ducts of certain anamniotes, and, in teleosts, still other structures to be identified below.

The retention, loss, or modification of these fetal structures to produce the adult male and female patterns of ducts in the different taxa will now be summarized. To memorize the varied arrangements can be a bewildering and useless exercise, but a review of the kinds of arrangements with attention to several principles that are illustrated is worthwhile. First, one must be impressed by nature's resourcefulness in adapting these "raw materials" to her needs. Evolution is opportunistic and in this instance has selected a variety of patterns of near functional equivalence. Second, the developmental mechanism of hormonal influence is well illustrated by the maturation of the urogenital ducts: The indifferent-stage ducts are modified under the influence of sex hormones elaborated in the fetal gonads. Male and female patterns can be altered or even reversed experimentally, and variation between the two norms is common. Finally, the concept of **sexual homology** is exemplified. Organs of the two sexes that are derived from identical primordia of the indifferent stage, yet become different in structure and function, have an equivalence that differs from phylogenetic homology.

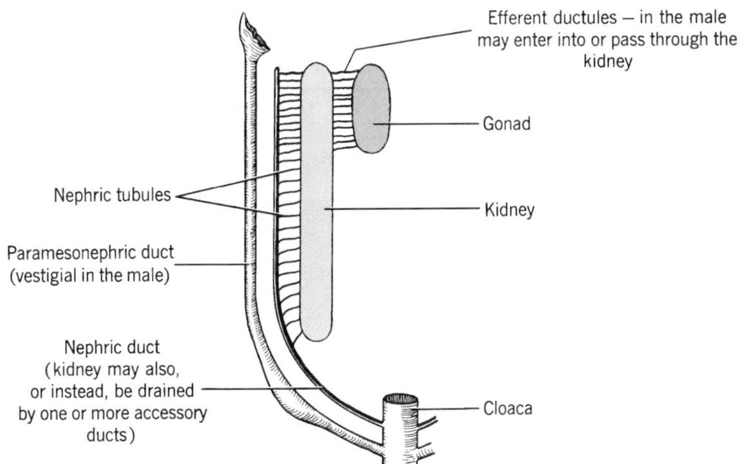

FIGURE 16.8 THE PRINCIPAL UROGENITAL DUCTS AS THEY OCCUR EARLY IN BOTH PHYLOGENY AND ONTOGENY. Ventral view of right side only.

Male Ducts The fetal paramesonephric ducts of males always regress. Vestiges are left that may be prominent (elasmobranchs and amphibians—particularly caecilians) but usually are not.

Except in cyclostomes, sperm are not released into the coelom. Instead they are conveyed in a closed system of ducts that usually is appropriated, at least in part, from the urinary system. If so, sperm cells leaving each testis enter a dozen or so **efferent ductules** derived from anterior nephric tubules. These ciliated ducts are small and cross the mesorchium to enter the nephric duct. The nephric duct takes the new name **deferent duct** when it carries only sperm or both sperm and urine. It has ciliated cuboidal or columnar epithelium and walls of smooth muscle. In addition to conveying sperm, it provides temporary storage and contracts during mating to ejaculate its contents. Various accessory organs are established by modification of parts of the deferent duct.

Typical CHONDRICHTHYES have paired deferent ducts that convey only sperm (Figure 16.2). The anterior end of each duct is convoluted to form an **epididymis** (= *upon* + *testes*) where sperm are stored and fluids augmented. The posterior end of each duct expands to become a **seminal vesicle,** which also stores sperm. The two vesicles may fuse before emptying into the cloaca. Urine is transported in several accessory ducts.

Various patterns serve the OSTEICHTHYES. Several primitive ray-finned fishes and one dipnoan pass sperm into the anterior end of the deferent duct, which conveys both sperm and urine. Other primitive ray-fins and dipnoans have short, paired sperm ducts, without apparent counterpart in other vertebrates, which run to the posterior ends of the opisthonephroi. There, efferent ductules cross into the nephric ducts, which carry both urine and sperm for part of their length only. In teleosts the same distinctive sperm ducts, formed by folds of the peritoneum, extend posterior to the kidneys and either enter the nephric ducts just before their exit from the body or else exit independently.

All AMPHIBIA have deferent ducts, but in some species of each major group they convey only sperm and in others they convey urine also (Figure 16.3). Accessory ducts are present or absent as needed. When present they drain principally the posterior parts of the kidneys. Seminal vesicles are usually present; epididymides (plural of epididymis) may be present.

Deferent ducts of AMNIOTES carry only sperm (Figures 16.9 and 16.10). Each duct forms an epididymis, though this organ is reduced in birds. Seminal vesicles are present in

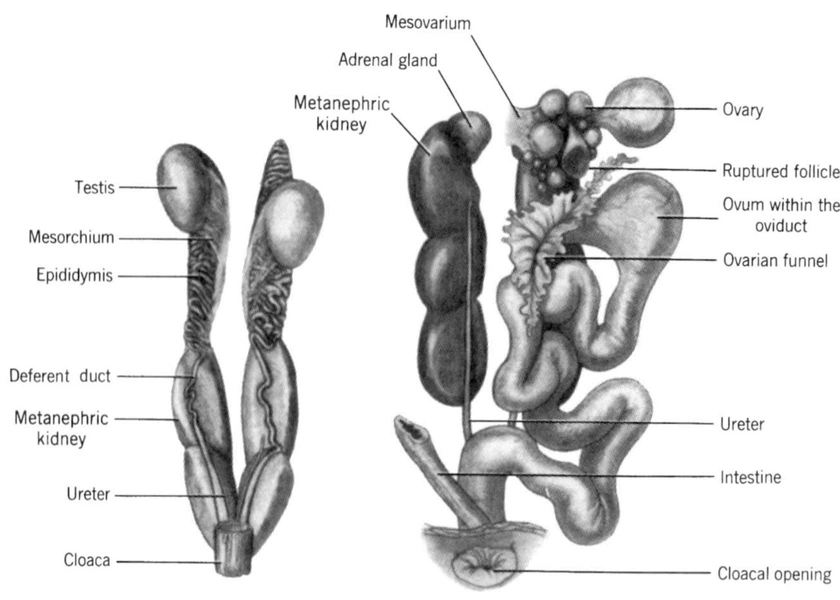

FIGURE 16.9 UROGENITAL SYSTEM OF THE LEPIDOSAUR *Crotaphytus* AND THE BIRD *Gallus*. Ventral views.

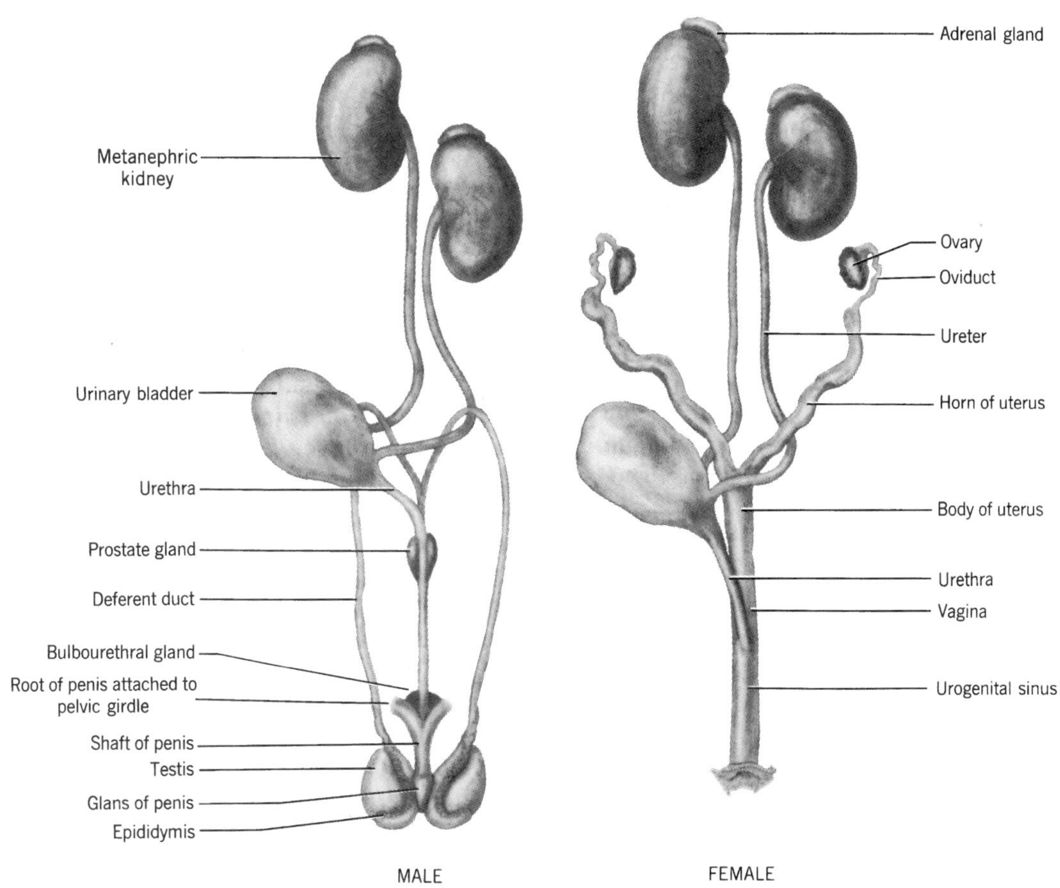

FIGURE 16.10 UROGENITAL SYSTEM OF THE CAT. Ventral views.

reptiles and birds. Mammals have glandular outgrowths from the deferent ducts that augment seminal fluids but do not store sperm. They are commonly called seminal vesicles, but are better termed **vesicular glands.** They are variable in size and form, and sometimes are lacking. In mammals the **prostate gland,** which is an outgrowth from the urethra, also augments the seminal fluid.

Female Ducts Variation is less extreme in the patterns of ducts of females, and urinary and genital systems are more independent of one another.

Opisthonephric ducts and ureters convey only urine in the female, but drainage of opisthonephroi is curiously supplemented by accessory ducts in some sharks and urodeles. The mesonephric ducts leave vestiges in adults of some amniotes.

It was noted above that the developing ovaries of TELEOSTEI fold to enclose pockets of coelom into which eggs rupture. Similar folding of the parts of the genital ridges posterior to the ovaries extends the same spaces as short oviducts, which are apparently without counterpart in other vertebrates. This arrangement correlates with the prodigious numbers of small eggs released by most of these fishes.

Eggs of other gnathostomes rupture into the coelom before entering a duct system derived from the paramesonephric ducts. The anterior ends of the passages enlarge as **ovarian funnels** that have openings called **ostia.** The funnels of at least some species are thought to represent an anterior pair of pronephric tubules, and the ostia the nephrostomes of those tubules. Much of the paramesonephric ducts, including all the anterior portions, become **oviducts.** These are lined by ciliated columnar epithelium with interspersed goblet cells. Eggs are moved along by peristaltic contractions of smooth muscle in the walls of the ducts. Posterior portions of the paramesonephric ducts usually become more muscular to expel eggs and more glandular to provide nutritive or protective coats to eggs or to nourish unborn young. The physical and physiological readiness of the female tract to perform its functions vary markedly with the breeding cycle.

The ovarian funnels of CHONDRICHTHYES lie relatively far forward in the body and often fuse in the midline (Figure 16.2). Oviducts lead into **shell glands** that envelop eggs in albumen and, in some species, a horny shell. Some sharks retain their developing eggs in expanded **ovisacs** or **uteri.** Nourishment of the embryo by yolk may be supplemented by a simple yolk-sac placenta.

DIPNOI and some PRIMITIVE ACTINOPTERYGII have paired oviducts apparently derived from paramesonephric ducts, though some doubt exists about the homologies of the ducts of most bony fishes.

AMPHIBIA have glandular and somewhat coiled oviducts. The lining of the coelom is provided with cilia that beat toward the ostia. Posterior portions of the tract apply coats of jelly to the eggs, which are usually held temporarily in ovisacs, though many caecilians bear living young that develop in the oviduct.

The genital tracts of female REPTILES, BIRDS, and MONOTREMATA vary widely in detailed structure, but since all have large eggs they agree in general form. The right side of the tract tends to be larger in reptiles, whereas the same side is vestigial in birds (Figure 16.9). Ovarian funnels are large and pleated. The glandular upper part of the tract applies albumen to the eggs as they spiral along their course. Near the cloaca the tract enlarges as a combined shell gland and ovisac.

In female THERIAN MAMMALIA the genital tract is divided into three regions. First are the oviducts. Their ciliated epithelium is thrown into folds, but externally the ducts are relatively straight and slender in correlation with the small eggs, without albumen or shells, which they convey. The second region is the **uterus** (or uteri) that houses the fetus during pregnancy and provides the maternal contribution to the placenta. It is lined by a mucous membrane called the **endometrium.** When functionally active the endometrium is thick, soft, glandular, and highly vascular. Following each pregnancy or breeding cycle it regresses by sloughing or resorption. The uterus has a thick wall, the **myometrium,** composed of smooth muscle having circular, longitudinal, and oblique fibers. Posteriorly the uterus is closed by a muscular neck called the **cervix.** The third region of the tract is the **vagina** (= *sheath*), which receives the penis during copulation and serves as the birth canal. The vagina is soft and distensible. Its stratified epithelial lining may be glandular or cornified according to breeding condition and taxon.

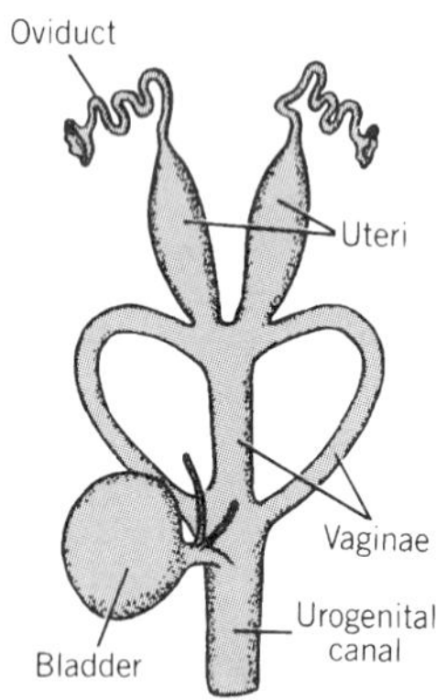

FIGURE 16.11 REPRODUCTIVE TRACT OF A FEMALE MARSUPIAL showing tripartite vagina. Ventral view.

The entire genital tract remains paired in monotremes. In marsupials the terminal part of the tract is fused to form a single urogenital canal. Anteriorly the two embryonic uteri and vaginae are retained, and in addition a third, or pseudovagina, grows down from near the cervixes toward the urogenital canal (Figure 16.11). In other mammals fusion of the embryonic primordia includes all of the vagina and usually extends to the uterus. If the uterus is completely double (and there are two cervixes), it is said to be **duplex,** the condition found in monotremes, marsupials, elephants, many rodents, and some other groups. If the uterus is Y-shaped externally but nearly divided internally (except for the cervix), it is said to be **bipartite** (most ungulates, most carnivores, and others). If fusion is more nearly complete, yet does not include the anterior end of the organ, the uterus is **bicornuate** (some members of several orders). Finally, if there is a single uterine chamber, the organ is **simplex** (most primates, some edentates). Degree of fusion does not follow clear evolutionary lines, and departures from the norm are not unusual within species.

The *estrus cycle* of placental mammals is the periodic cycle of growth, maturation, and release of the egg(s), which is under the control of hormones (Chapter 20). In humans, and a few other species of primates, the end of the ovulatory cycle is accompanied by a breakdown of the lining of the endometrium and bleeding, a process known as **menstruation** (= *monthly*). Menstruation as a process has recieved much historical attention, but its biological significance is debated. Theories to explain its origin, including pathogen control, energy conservation, and reasons of immunity, have been proposed and recently reviewed by Strassmann and Finn.

Seminal vesicles and vesicular glands have been mentioned as common male derivatives of the nephric ducts, and similarly, albumen and shell glands have been mentioned as adjuncts of the oviducts. Additional accessory glands of the reproductive system (which unfortunately have in some instances been given the same names) may be derived from the urethra, urogenital sinus, and adjacent tissues. Products of these glands augment the seminal fluids, soften and lubricate the genital organs to facilitate copulation, and provide scents that act as sexual attractants.

CLOACA AND DERIVATIVES

The embryonic hindgut forms before urogenital ducts have developed. When the pronephroi are established, the nephric ducts extend posteriorly to enter the hindgut, and later the paramesonephric ducts do likewise. The common passageway for products of the digestive, urinary, and reproductive systems is then called the **cloaca** (= *sewer*). This primitive arrangement is retained by adults of hagfishes, elasmobranchs, dipnoans, amphibians, reptiles, and birds (Figure 16.12). The posterior part of the gut of adult lampreys, chimaeras, and bony fishes, by contrast, is no longer joined by the urogenital ducts and is called a **rectum.** Their nephric and genital ducts either exit from the body independently or join each other to exit at a common papilla. In either instance, urine and gametes are discharged posterior to the anus.

A different evolutionary path has been followed by mammals. In monotremes the embryonic cloaca is partly divided by a septum that wedges between the gut and allantoic stalk. The result is a dorsal **coprodeum** (= *dung* + *divide*), a ventral **urodeum** ("ur" = *urine*) that is joined by the ureters and paramesonephric ducts, and a common posterior **proctodeum** ("proc" = *anus*). In therian mammals the embryonic septum continues to push back until a dorsal rectum is completely separated from the urogenital structures, which then exit anterior to the anus. In males, urine and sperm are discharged by a common **urethra.** In females of most mammals, urinary and genital tracts exit by a common **urogenital sinus.** In primates and some rodents, however, fetal eversion of the common passageway eliminates the urogenital sinus by separating an anterior urethral opening from a posterior vaginal opening.

COPULATORY ORGANS

CYCLOSTOMES and most BONY FISHES lay their eggs in water, where the attendant male promptly discharges his sperm over them. Successful fertilization is dependent on behavioral rather than on structural adaptations. Copulatory organs are lacking, though the two sexes of some species entwine their bodies during egg laying and rarely some kind of holding organ is present. Teleosts that retain their eggs during development, or bear live young, however, require internal fertilization. Males of most such species have a margin of the anal fin modified as a copulatory organ called a **gonopodium.** It is enlarged, rigid, and movable. When inserted into the female tract it transmits sperm through a duct or groove. Several cottid fishes have instead an enlarged genital papilla that serves as a penis.

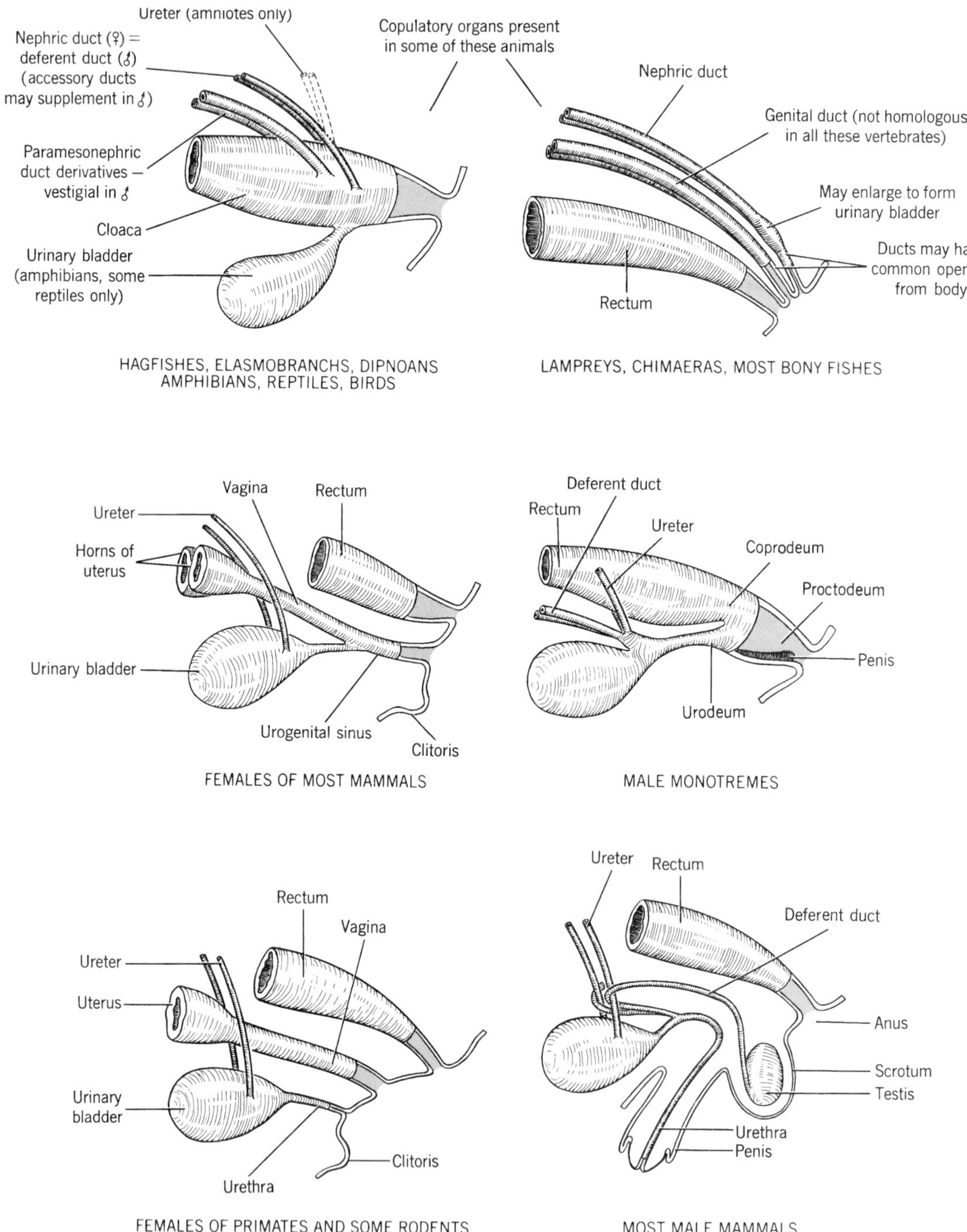

FIGURE 16.12 REPRESENTATIVE DIVISIONS OF THE CLOACA AND THEIR RELATIONS WITH UROGENITAL DUCTS AND THE URINARY BLADDER. Left lateral views.

ELASMOBRANCHS (and one group of placoderms) also have internal fertilization. This time it is the pelvic fins of males that develop copulatory organs called **claspers** (see figure on p. 155). They are supported by the fin skeleton and may be of elaborate configuration. In some rays they include erectile tissue. Again, grooves convey sperm into the female cloaca.

Many AMPHIBIANS return to water to breed. Fertilization is usually external in anurans, so copulatory organs are not needed. Fertilization is usually internal in urodeles but external genitalia are still lacking. Either the two sexes press their cloacas together to transfer sperm, or the male deposits packets of sperm, which are later picked up by the cloaca of the female. Male caecilians use an evertable extension of the cloaca for internal fertilization by means of copulation.

Internal fertilization is necessary if copulation takes place out of the water, if there is internal development of the young, or if eggs are provided with shells before deposition. With few exceptions amniotes must have internal fertilization for the first, and either the second or third of these reasons. REPTILES evolved two kinds of copulatory organs. Male lepidosaurs have paired structures called **hemipenes,** which lie concealed in long sacs opening to the outside of the body on each side of the cloacal aperture (see figure on p. 185). During copulation one hemipenis is everted (sides alternating on successive matings) and introduced into the female cloaca. Most other reptiles have evolved the **penis,** which in this class is a grooved organ located internally on the floor of the cloaca. It is largely composed of two long, spongy, vascular bodies called **corpora cavernosa.** A **glans penis** caps the end of the organ. During sexual stimulation sphincters reduce the outflow of blood from these structures, thus causing engorgement and enlargement. The groove then closes over, and the penis protrudes from the cloaca to serve as an intromittent organ. Female turtles and crocodilians have a small **clitoris,** which is the sexual homolog of the penis. The penis of MONOTREMES is like that of reptiles except that the sperm channel is permanently separated from the cloaca.

Most BIRDS copulate by pressing the cloacas together for the transfer of sperm. Avian ancestors doubtless had a penis, however, because the ostrich and its relatives and ducks and geese—all relatively primitive birds—have a small penis.

At the indifferent stage of development THERIAN MAMMALS have a genital tubercle anterior to the cloacal opening. When the cloaca becomes divided, its ventral portion contributes to the urinary bladder anteriorly and forms the female urogenital sinus or male pelvic urethra posteriorly. The genital tubercle of females becomes the clitoris and is not penetrated by the urethra. In males the pelvic part of the urethra is extended by a phallic part that grows into the tubercle. The tubercle enlarges to become the penis. Corpora cavernosa and glans are present (Figure 16.13). The organ may be hidden under the skin but is usually at least partly external even when not erect. The glans is variously shaped and is forked in monotremes and marsupials to correspond with the divided vagina of the female. The penis may be stiffened by a bone, the baculum (see figure on p. 166).

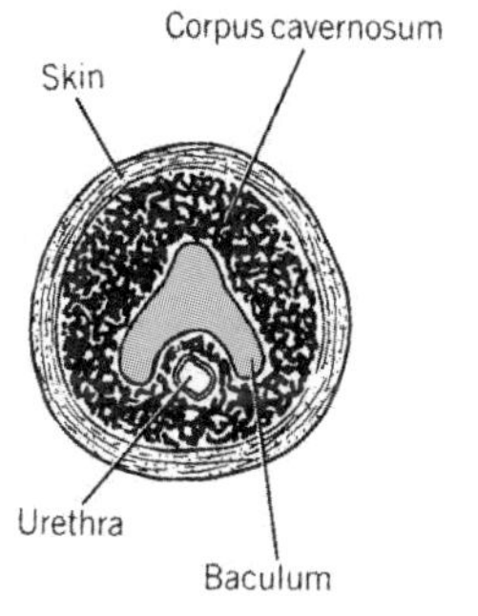

FIGURE 16.13
CROSS SECTION OF THE PENIS OF A DOG.

REPRODUCTIVE STRATEGIES

The diverse reproductive strategies of vertebrates often relate directly or indirectly to structure.

Nearly all vertebrates are **dioecious** (having separate male and female individuals), but hagfishes and some teleosts are **hermaphroditic** (have both sexes functional in the same individual). Eggs and sperm are then produced in different parts of the gonads and usually at different times, thus ensuring cross-fertilization. Several teleosts, however, are self-fertilizing. Some perches, darters, basses, and lizards are **parthenogenetic,** the eggs maturing without entry of sperm.

Most vertebrates are **oviparous;** that is, they lay eggs that mature outside the maternal body. Included are cyclostomes, most bony fishes, holocephalians, some elasmobranchs, most amphibians, most reptiles, birds, and monotremes. The number of eggs laid per season ranges from 1 to 20 (some birds), 100 (pythons), or even 28 million (several teleosts), the number relating to the hazards of development. The laying of few eggs tends to be associated with large egg size, much yolk, precocious hatchlings, and parental care.

Brooding of eggs is common. It usually provides protection from predation, and according to species may also provide aeration (some fishes), humidity control (terrestrial amphibians, some squamates), protection from mold (some amphibians), temperature control (some reptiles, birds, platypus), and turning to prevent adhesions (birds). Most birds develop a naked, highly vascular **brood patch** on the breast or under the body.

Some species retain the eggs in the female reproductive tract until development is partly completed, thus reducing exposure to the vicissitudes of the environment. **Ovoviviparous** species retain their eggs in the body until hatching; embryos are nourished by egg yolk, yet eggs are not laid. Examples are found among fishes, amphibians of each order, and squamates. Since there is a tendency for such animals to reduce or eliminate the ancestral egg shells, they are not sharply demarcated from **viviparous** species, which nourish the embryo by physiological exchange with maternal tissues, and also give birth to "live" young (a poor term because viable eggs are also alive).

Viviparity evolved several dozen times among teleosts yet characterizes only 2–3 percent of bony fishes (including rockfishes, sea perches, blennies, and cyprinodonts). Viviparity is found in 55 percent of elasmobranchs, in some amphibians (many caecilians, several marsupial frogs), many lizards and snakes, and all therian mammals. This reproductive strategy is a drain on the female, but provides maximum survival for larva or fetus. The evolution of viviparity poses interesting questions (see Packard et al. and Wake in Refences).

Viviparous teleosts either nourish the fetus within the hollow ovary or within the ovarian follicle. Other viviparous vertebrates retain their fetuses in the oviducts or uterus. Respiratory gases and water are always exchanged between fetus and mother, and nourishment is usually provided from "milk" secreted by the female reproductive tract (some fishes), juices released by the lysis of maternal epithelia or blood (early stages in mammalian development), or physiological exchange between fetal and maternal blood streams. The latter mechanism is facilitated by a **placenta,** which is an organ where there is an intimate juxtaposition of fetal and maternal tissues in such a way as to ensure a large area of contact (see figure on p. 81). A wide variety of pleats, folds, filaments, and villi have evolved for the purpose. The yolk sac is commonly involved on the fetal side, though the allantois supplements or substitutes in most mammals.

Various fishes and mammals have evolved provisions for the storage of sperm cells in the male body long after the seasonal loss of testicular function, or within the female body for as long as 10 months after copulation. Some mammals in at least five orders postpone development instead by delayed implantation of the blastocyst (corresponding to the blastula of other vertebrates). These mechanisms time birth with seasonal food, dormancy, or return to breeding grounds.

Vertebrates with short brooding periods (11 days for several birds) or gestation periods (13 days for some marsupials) tend to have small, dependent hatchlings or young. Other animals have long brooding periods (about 79 days for one albatross) or gestation periods (22 months for the elephant) and usually have larger, more active young (ducks, grouse, ungulates, cetaceans). Sexual maturity is reached by several fishes by the time of hatching, and in humans and some large vertebrates only after many years.

REFERENCES

Duellmann, W.E. 1992. Reproductive strategies of frogs. *Sci. Am.* 267(1):80–87.

Duellmann, W.E., and L. Trueb. 1994. *Biology of amphibians*. Johns Hopkins Univ. Press, Baltimore. 694p.

Finn, C.A. 1998 Menstruation: A nonadaptive consequence of uterine evolution. *Quart. Rev. Biol.* 73(2):163–173. Reviews the evolution of the female reproductive tract in vertebrates and stresses the importance of immunity in the evolution of menstruation.

Fox, H. 1977. The urogenital system of reptiles, vol. 6, pp. 1–157. *In* C. Gans (ed.), *Biology of the reptilia,* Academic Press, New York.

Hogarth, P.J. 1978. *Biology of reproduction.* Wiley, New York. 189p.

Jones, R.E. (ed.). 1978. *The vertebrate ovary: comparative biology and evolution.* Plenum, New York. 853p.

Nagahama, Y. 1983. The functional morphology of teleost gonads, vol. 11A, pp. 223–227. *In* W.S. Hoar, D.J. Randall, and E.M. Donaldson (eds.), *Fish physiology.* Academic Press, New York.

Packard, G.C., C.R. Tracy, and J.J. Roth. 1977. The physiological ecology of reptilian eggs and embryos, and the evolution of viviparity within the class Reptilia. *Bio. Rev.* 52:71–105

Packard, G.C., et al. 1989. How are reproductive systems integrated and how has viviparity evolved?, pp. 281–293. *In* D.B. Wake and G. Roth (eds.), *Complex organismal functions: integration and evolution in vertebrates.* Wiley, New York.

Potts, G.W., and R.J. Wooten (eds.). 1984. *Fish reproduction: strategies and tactics.* Academic Press, New York. 410p.

Setchell, B.P. 1978. *The mammalian testis.* Cornell Univ. Press, Ithaca, NY. 450p.

Seymour R.S. 1999. Respiration of aquatic and terrestrial amphibian embryos. *Am. Zool.* 39(2):261–270. Illustrates how the requirements for gas exchange have influenced the evolution of amphibian eggs and egg masses.

Strassmann, B.I. 1996. The evolution of endometrial cycles and menstruation. *Quart Rev. Biol.* 71:181–220.

van Tienhoven, A. 1983. *Reproductive physiology of vertebrates.* 2nd ed. Cornell Univ. Press, Ithaca, NY. 491p. Includes comparative morphology.

Wake, M. 1992 Evolutionary scenarios, homology and convergence of structural specializations for vertebrate viviparity. *Am. Zool.* 32(2):256–263. An instructive overview of how methods of moderm comparative biology are employed to address questions of evolutionary origin.

Werdelin, L. and A. Nilsonne. 1999. The evolution of the scrotum and testicular descent in mammals: a phylogenetic view. *J. Theor. Biol.* 196(1)0:61–72.

Wourms, J.P., and I.P. Callard. (eds.) 1992. A retrospect to the symposium on evolution of viviparity in vertebrates. *Am. Zool.* 32(2):251–255. Thoughtful overview of how integrative studies of vertebrate reproduction assist our understanding of anatomical and physiological diversity.

Wourms, J.P., and J. Lombardi. 1992. Reflections on the evolution of piscine viviparity. *Am. Zool.* 32:276–293.

Chapter 17

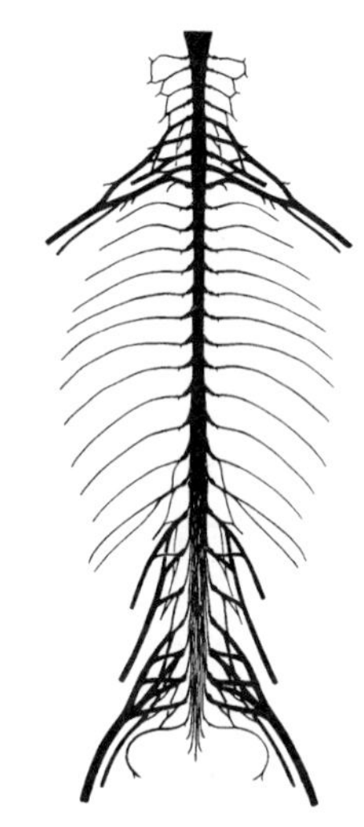

Nervous System: General, Spinal Cord, and Peripheral Nerves

Nerve cells, or **neurons,** are uniquely adapted for conducting information from one place to another, and form a truly amazing network in the vertebrate body. In general, this information is obtained from the environment, both external and internal, by specialized receptor cells. These cells transmit to the spinal cord and brain by way of sensory neurons; these in turn activate either motor neurons, which carry out the motor functions, or, more commonly, interneurons, which transmit within the nervous system. They may activate other interneurons, in an exceedingly complex pattern, before sending command signals to motor neurons, which alter some ongoing activity or initiate an action by activating effectors, which are muscle or gland cells (Figure 17.1). Communication of neurons with other neurons, or with effector cells, is accomplished by a process known as synaptic transmission. Thus, the nervous system (with help from the endocrine glands) determines responses of the body to changes in its environment. It is the body's messenger and coordination system for most activities—for all activities that are rapid or complex.

Neuroscience has emerged in the last 15–20 years as a most integrative discipline that is attracting an increasing number of researchers. The nervous system is of particular interest to morphologists because it is the most complex system of the body, yet in some aspects it is remarkably conservative. Students of comparative anatomy who specialize in animal or human health sciences are particularly struck by the conservation of structures, at the gross anatomy level, across species. Studies of respiration, feeding, and locomotion have even

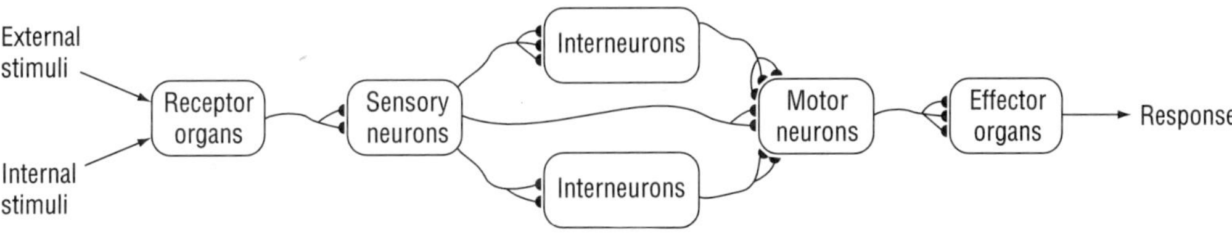

FIGURE 17.1 BASIC FLOWCHART FOR THE NERVOUS SYSTEM.

shown that in some cases (but not all), neuromotor patterns (as measured by electromyography) are conserved during major morphological and functional transformations (e.g., from ventilating and feeding underwater to performing these functions in air). Although the brain is too soft to be directly preserved in the fossil record, its size and shape, and the distribution of cranial nerves, are remarkably well recorded by fossil skulls. This record, along with comparative study of surviving vertebrates, and developmental studies, enable morphologists to construct the outlines of the phylogeny of the system. The general habits of an animal can also be determined (where information is adequate) from the nervous system.

ELEMENTS OF THE NERVOUS SYSTEM

Neurons and Neuroglia Neurons range from short to uniquely long, yet are always small in bulk. Each has a **nerve cell body** that may be oval in outline or irregularly star-shaped (Figure 17.2). Within the nerve cell body are the nucleus and many distinctive Nissl granules that contribute to the very high rate of protein synthesis characteristic of nervous tissue. Most nerve cell bodies, particularly those of the brain and spinal cord, support many filamentous processes and are termed multipolar. Neurons related to the nose, eye, ear, and lateral line usually have only two processes (i.e., are bipolar). The sensory neurons bringing information into the spinal cord via the spinal nerves also have two primary processes, which, however, because of peculiarities related to their development, exit from the nerve cell body at the same place. They are considered to be specialized bipolar cells.

Each neuron has one to many tapering processes called **dendrites** (= *treelike*). These transmit information toward the cell body and receive information from other neurons or, in the case of the sensory neurons of the spinal nerves, directly from receptors. Dendrites of motor neurons or interneurons may be numerous, with complex branching patterns, and they are relatively short. The one dendrite of the specialized sensory neuron may be quite long. Impulses are transmitted away from the nerve cell body by a single process, the **axon** (= *axis*). Axons may be short or long; some travel the distance from the spinal cord to muscles of the foot. Axons tend to branch less than dendrites, yet in some regions they give off collaterals and usually have twigs (called telodendria) at their far ends, where they communicate with other cells. Axons lack the organelles (ribosomes) that synthesize proteins. Newly synthesized macromolecules are assembled into organelles within the cell body and are moved along the axon to the terminals by a process called axoplasmic transport. Neuroanatomy was revolutionized by the relatively recent discovery that individual axons will take up various plant proteins (horseradish peroxidase is widely used), bacterial toxins, and

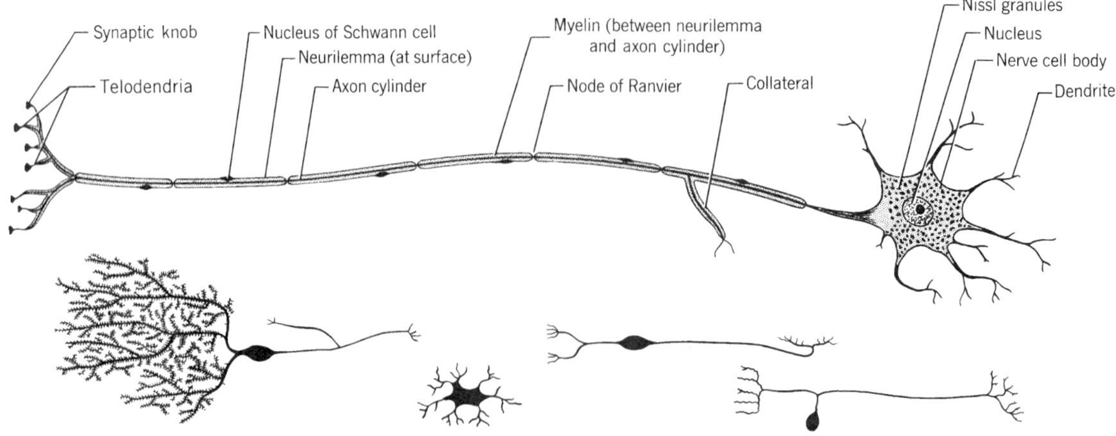

FIGURE 17.2 STRUCTURE AND SHAPES OF REPRESENTATIVE NEURONS.

fluorescent dyes at their terminal ends and transport these proximally to their cell bodies and finest dendritic processes. Prior to the discoveries of these anatomical tracers, it was painstakingly difficult to map the pathways and interconnections among neurons.

Between their short initial segment and their terminal twigs, most axons are sheathed. **Schwann cells** arrange themselves along developing axons located outside the brain and spinal cord. A fold of each cell then wraps around the axon much as a window shade wraps around its wooden core, taking as many as 70 turns. Thus, the axon is ensheathed in lamellae derived from the cellular membranes of Schwann cells (Figures 17.3 and 17.4). These enclose lipids and proteins. The entire coating is called **myelin.** Myelin is essential for rapid conduction and the maintenance of a healthy axon. It is interrupted at intervals of 50–1000 μm along the enclosed protoplasmic strand, the **axon cylinder,** to form **nodes of Ranvier.** Outside the brain and spinal cord, axons are also covered by a second sheath that is external to the myelin. This delicate transparent membrane is also formed from Schwann cells and is the **neurilemma.** (There are some long unmyelinated fibers in the peripheral nervous system—particularly in cutaneous nerves.)

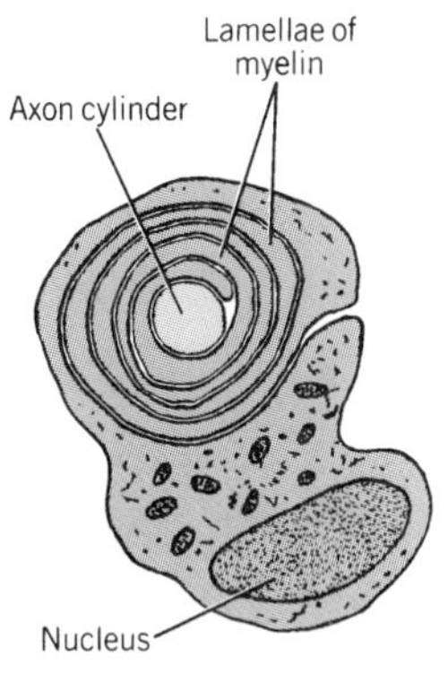

FIGURE 17.3 SCHWANN CELL ENSHEATHING AN AXON WITH MYELIN.

Within the brain and spinal cord neurons may be in contact with cells called **neuroglia** (= *nerve* + *glue*). Three kinds of neuroglia—astrocytes, oligodendroglia, and microglia—are distinguished histologically. They fill interstices among neurons, bind fiber to fiber, and contribute to the energetics of neurons in various ways, including ion transport, nutrition, excretion, regeneration, and repair. Astrocytes, which are much branched, maintain potassium balance and contribute to the metabolism of the neurotransmitter glutamate (see below). They have footplates on vessels and contribute to the blood-brain barrier. Oligodendroglia apply myelin to axons of the central nervous system. Microglia are phagocytic. About half of the bulk of the brain is neuroglia.

Nerve Impulse and Synapse A nerve impulse is an electrical phenomenon that passes as a wave along the surface membrane of a nerve fiber. The fluids bathing the inside and outside of the membrane differ chemically. The crucial difference is that there is about 30 times more potassium on the inside of the resting membrane and about 10 times more sodium on the outside. The potassium leaks out through the membrane, but in some unknown way the membrane resists the entrance of sodium. The consequence is a difference in electric potential of about 60 mV across the resting membrane, with the inside negative. A local change in the resting potential can either excite or inhibit the nerve fiber. When the fiber is locally excited, the membrane suddenly but briefly allows sodium to

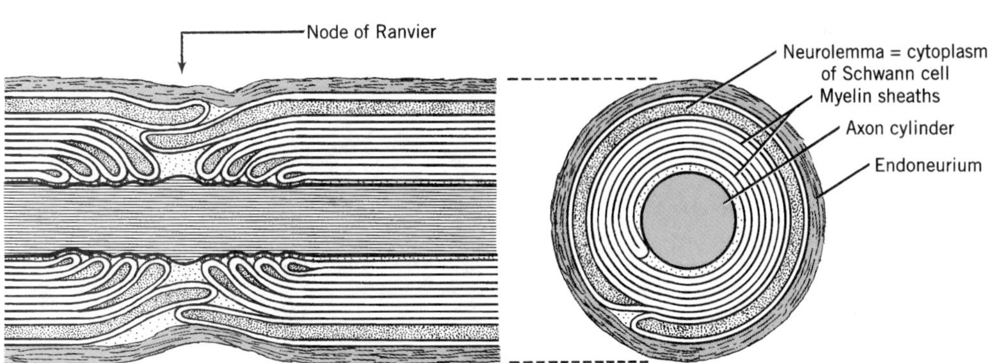

FIGURE 17.4 LONGITUDINAL AND CROSS SECTIONS OF AN AXON AND ITS SHEATHS, including a node of Ranvier.

rush in, and the inside of the membrane changes to a positive potential of about 50 mV. The excited part of the membrane now differs from adjacent parts, and a tiny eddy of current is set up between the excited portion and adjacent portions of the membrane. This, in turn, depolarizes the resting membrane and in this manner the impulse is propagated along the fiber, switching charges ahead and restoring them behind as it travels. Progression of an impulse is continuous along unmyelinated fibers. Where there are nodes of Ranvier, waves of depolarization jump from node to node. An excised fiber can transmit thousands of impulses without replenishing its stored energy. In time, however, enough sodium would pass into the fiber to destroy the mechanism. To avoid this, the membrane pumps sodium out, in a way that is not yet fully understood, and this does use energy.

Nerve fibers of vertebrates range in diameter from about 0.5 to about 22 μm. The rate of travel of an impulse increases with fiber size and temperature. In mammals the rate ranges from about 1.0 to 120 m/s, with values below 1.5 being characteristic of unmyelinated fibers.

An impulse is propagated without decrement or is not propagated at all, and all conducted impulses are alike. It is by the frequency of impulses in each fiber, number of active fibers, and connections made by the neurons that the system decodes the messages. Other electrical charges are not conducted, especially in dendrites in the brain, but instead affect excitation. These may have graded responses.

There is no cytoplasmic continuity between neurons. The functional union of an axon of one neuron with a dendrite or nerve cell body of another neuron is called a **synapse.** Some neurons have few synapses, but others have thousands. The synapse transfers impulses only in the direction away from the axon, and only if a threshold level of excitation is achieved. At its terminus, the axon enlarges to form a **synaptic knob** (Figure 17.5). The knob is separated from the adjacent dendrite or nerve cell body by a narrow cleft (10–30 nm wide). In the knob are mitochondria and **presynaptic vesicles** that contain some chemical known as a **neurotransmitter.** When the transmitter is released into the cleft by the arrival of a nerve impulse at the synaptic knob, it alters the permeability of the postsynaptic membrane to the ions bathing it. If the transmitter's effect on the postsynaptic membrane is an excitatory one, it causes propagation of the nerve impulse in the connecting neuron. If the transmitter's effect is an inhibitory one, the threshold for depolarization of the postsynaptic membrane of the connecting neuron is elevated (resisting depolarization). Numerous neurotransmitters have been identified (neurochemistry is an active area) and

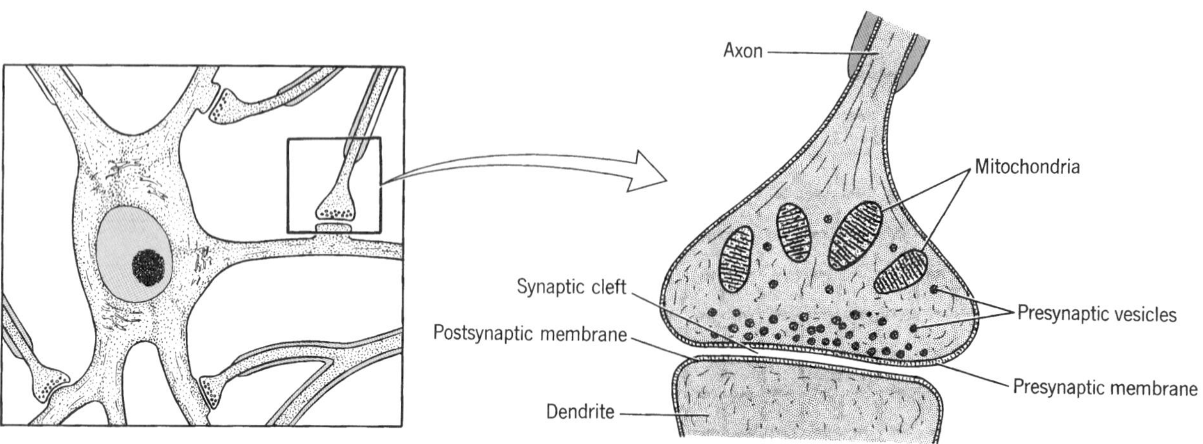

FIGURE 17.5 SYNAPTIC KNOB AND SYNAPSE.

include acetylcholine, noradrenalin, serotonin, dopamine, morphine, glutamic acid, adenosine triphosphate, nitric oxide, and many peptides. Each is associated with certain parts of the system. Some transmitters also have endocrine functions (see also p. 378).

The synapse between axon and muscle cell (neuromuscular junction) or gland cell tends to be broader than those between neurons, but apparently the two function in a similar manner.

Mammals (and perhaps other vertebrates) retain throughout their lives the ability to modify the number, nature, and level of activity of their synapses. This plasticity is particularly evident following injury, but turnover of synapses is an ongoing process.

Tracts, Nerves, and Ganglia Nerve cell bodies of functionally related neurons tend to mass together, and their fibers, if long, tend to run parallel to one another in bundles. Within the cord such bundles are called **tracts.** Myelinated tracts are whitish and together make up the **white matter,** whereas nerve cell bodies and associated unmyelinated fibers (sensory dendrites and motor axons), being of darker color, together make up the **gray matter.**

Outside the brain and cord, nerve fibers, with their myelin, are surrounded by a delicate, membranous **neurilemma.** Groups, or fascicles, of fibers are supported by a **perineurium.** Finally, bundles of fascicles make up **nerves,** which are enveloped by a tough **epineurium** (Figure 17.6). Aggregates of nerve cell bodies cause marked swellings on nerves and are termed **ganglia.** At the levels of the appendages, several spinal nerves may come together and exchange bundles of fibers, thus weaving to form a **plexus** (= *braid*) (Figure 17.11).

Some Divisions of the System The parts of the nervous system are functionally interrelated to a remarkable degree. Nevertheless, it is convenient to recognize structural and functional divisions. Brain and spinal cord compose the **central nervous system (CNS),** leaving nerves and ganglia to the **peripheral nervous system. Afferent,** or **sensory fibers** of the peripheral nervous system carry impulses from receptor organs to the central nervous system; **efferent,** or **motor fibers** carry impulses from the central nervous system to effector organs. Nerves may be entirely sensory or motor, or may be mixed, having each kind of fiber. Within the central nervous system there are also **interneurons** (also called **association neurons**) that make up local circuits and are not themselves afferent or efferent. Interneurons far outnumber sensory and motor neurons.

Somatic fibers (sensory and motor) relate to the skin and its derivatives and to voluntary muscles. **Visceral fibers** (again sensory and motor) relate to involuntary muscles and

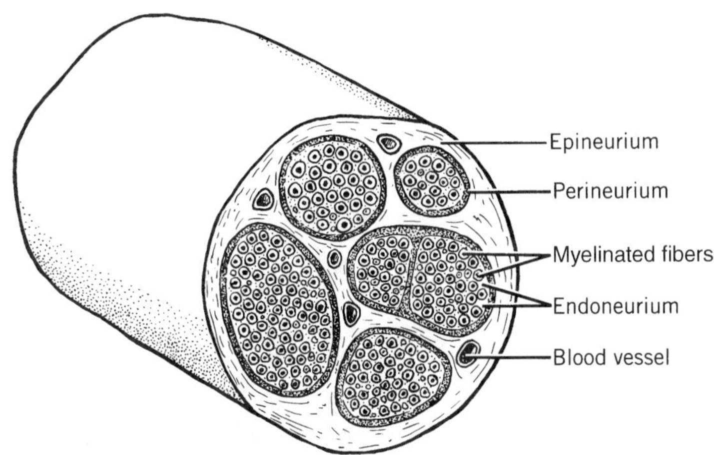

FIGURE 17.6 CROSS SECTION OF A PERIPHERAL NERVE.

glands of the various organ systems. It is useful to designate fibers as somatic sensory, visceral sensory, somatic motor, or visceral motor, because these types of fibers tend to be structurally independent.

Most organs innervated by visceral nerves receive two complete sets of fibers that elicit opposed responses. The sets have a measure of structural and functional independence from the remainder of the peripheral nervous system and are together called the **autonomic system.**

DEVELOPMENT OF SPINAL CORD AND PERIPHERAL NERVES

Neurulation, or the process that establishes the central nervous system, was described on pp. 77 and 78. Each step occurs first at the anterior end of the embryo and progressively later and later in moving toward the posterior end. This is one of many developmental gradients of the body. In the head of the embryo, the neural tube becomes relatively large and soon is compartmentalized into vesicles that foreshadow the regions of the brain. Further discussion of the development of the brain will be deferred to the next chapter.

As seen in cross section, the embryonic neural tube forms three layers having conspicuously different staining properties. In order, from neurocoel outward, these are the **ependymal, mantle,** and **marginal layers** (Figure 17.7). Some cells of the ependymal layer remain in place to become the thin, ciliated lining of the adult central canal. Most ependymal cells migrate outward to join mantle cells in forming both neurons and (a little later) neuroglia. These will be the gray matter of the adult cord. The marginal layer forms neuroglia and is penetrated by nerve fibers growing out of the deeper layers. It becomes the white matter of the cord.

As the cord grows in diameter, it enlarges every place except at the thin roof and floor of the neurocoel. Growth, therefore, establishes longitudinal surface grooves at these places known in the adult as the **dorsal median sulcus** (= *furrow*) and **ventral median fissure** (= *a split*) (Figure 17.9).

Nerve cell bodies of the sensory fibers of spinal nerves are located in spinal ganglia located near the cord. These are derived from neural crest cells. Fibers from the developing ganglia grow outward to receptor organs and inward to penetrate the cord. Sensory fibers of cranial nerves and their ganglia develop in the same way, except that there are contributions from the ectodermal placodes that will form certain sense organs of the head (see p. 78). The autonomic system also has ganglia. These are derived in part from neural crests and probably in part from cells that migrate out of the cord. Fibers of motor nerves grow out of the mantle layer of brain and cord.

Cells from the versatile neural crests also migrate to contribute to the formation of Schwann cells, myelin, and the neurilemma.

Throughout life, the nerve cell body actively produces cytoplasm that slowly flows outward along axon and dendrites. During growth, some neurons and axonal branches degenerate in a process that helps to match the nerve tree to target structures. It was long

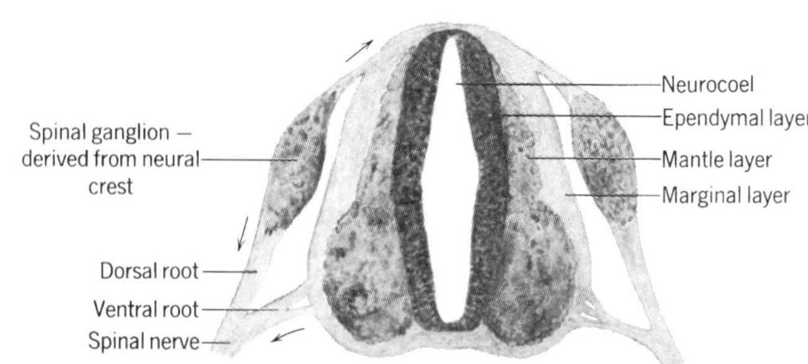

FIGURE 17.7
STAGE IN THE DEVELOPMENT OF THE MAMMALIAN SPINAL CORD, SPINAL NERVES, AND GANGLIA. Cross section. Arrows show directions of fiber growth.

thought that there is little plasticity in adult nervous systems and that adult neurons do not divide. It is now known that there is turnover of neurons in the brains of various vertebrates and that, following injury at least, there is commonly a proliferation of axon collaterals. If a nerve is cut, fibers severed from their nerve cell bodies degenerate whereas fibers that are still nourished slowly regenerate. They tend to feel their way along the empty sheath tubes and may reach their accustomed destinations. However, if the tubes are also destroyed, the regenerating fibers do not find their original end organs. In some instances (but seemingly not in others) anamniotes can then retrain the central nervous system to respond correctly to the new stimuli, but amniotes cannot.

SPINAL CORD

Function and Structure In the simplest possible reflex arc, messages from receptor organs are transferred within the cord directly from afferent fibers to efferent fibers, which then send appropriate messages to effector organs. Nearly always, however, one or more interneurons are interposed between afferent and efferent neurons (Figure 17.8). Since each afferent fiber synapses with many interneurons, possible pathways run to any of very many effector fibers. These may be on the same side of the cord as the afferent fiber or on the opposite side, at the same level of the cord or at a different level, or in the brain. Thus the function of the cord is to receive incoming impulses, integrate and coordinate them, transmit them wherever they should go within the central nervous system, and send responses to the peripheral nervous system as appropriate.

The general structure of the spinal cord is best exemplified by a cross section of the cord of an amniote. The gray matter is internal and has an irregular shape resembling the letter H (Figure 17.9). The upper arms of the H are the **dorsal gray columns** (or horns) and the shorter, broader, lower arms the **ventral gray columns** (or horns). Nerve cell bodies of interneurons that synapse with somatic sensory fibers are on the medial side of a dorsal column. Cell bodies of interneurons synapsing with visceral sensory fibers are in a smaller, lateral, and slightly more ventral part of the dorsal column. Nerve cell bodies of somatic motor neurons fill the ventral columns. Nerve cell bodies of visceral motor neurons are in a small, intermediate, and lateral position. The **gray commissure,** just above and below the central canal, makes up the cross arm of the H and transmits fibers from one side of the cord to the other.

The external white matter is divided into right and left sides by the dorsal median sulcus and ventral fissure of the cord. Each half is further divided by the gray columns into three funiculi. The **dorsal funiculus** is between the dorsal column and the dorsal median sulcus. It transmits axons toward the brain. The **ventral funiculus** is between the ventral fissure and the

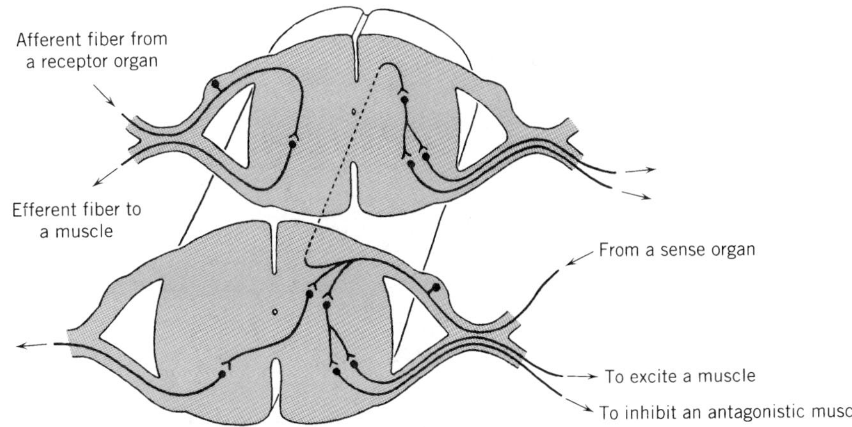

FIGURE 17.8 EXAMPLES OF REFLEX ARCS. Above left, a two-neuron arc; elsewhere, arcs including interneurons.

COMMENT 17.1

ON THE TRAIL OF EVOLUTIONARY TRENDS IN THE SPINAL CORD

Studies of the comparative morphology of the vertebrate nervous system benefit immensely by their placement in a phylogenetic framework. This is illustrated by a series of studies done by Fetcho and colleagues. We learned in Chapter 10 that the axial musculature consists of a series of blocks, or myomeres, lying along each side of the vertebral column and that these have epaxial and hypaxial divisions (somewhat obscured in adult birds and mammals). Within the ventral gray columns of the spinal cord may be either of two arrangements of the motor neurons that innervate these myomeres. In lampreys, goldfish, and mudpuppies (a fully aquatic salamander), neurons extending to epaxial and hypaxial muscles are loosely intermingled. In representative snakes, rats, and monkeys, in contrast, epaxial motor neurons are spatially segregated from hypaxial motor neurons, each being located in a specific part of the gray columns. Thus, during evolution there was not only sorting of gray matter from white (Figure 17.10), but also sorting within the gray matter.

When Fetcho and colleagues mapped their data on a cladogram, it became evident that all species studied that had unsegregated motor neurons are anamniotes, whereas all with segregated neurons are amniotes. It was also noted, however, that the group with unsegregated motor neurons are fully aquatic and the group with segregated neurons terrestrial. Was the difference in neural structure primarily evolutionary, or primarily the consequence of mode of locomotion? To answer this question the researchers studied the ventral gray columns of the tiger salamander, a permanently terrestrial anamniote. The result: unsegregated epaxial and hypaxial neurons. It appears that the segregation of motor neurons may have evolved in conjunction with the origin of amniotes; confirmation from a wider database is desirable.

---

ventral column. It transmits axons away from the brain. The **lateral funiculus** is between the dorsal and ventral columns and transmits fibers in both directions; those going toward the brain tend to be more superficial. The positions of the specific pathways vary among the vertebrates. Many of the fibers to and from the brain cross from one side to the other; the change is sometimes in the cord and sometimes in the brainstem. The reason for this crossing is not known.

There are within the spinal cord of some vertebrates (and perhaps all) neural networks capable of generating, for each appendage, cyclic outflow to locomotor muscles, as for rhythmic undulations in swimming or routine movement of limbs. These **central pattern generators** are influenced by input from the brain and sensory receptors, yet continue their function when such input is blocked.

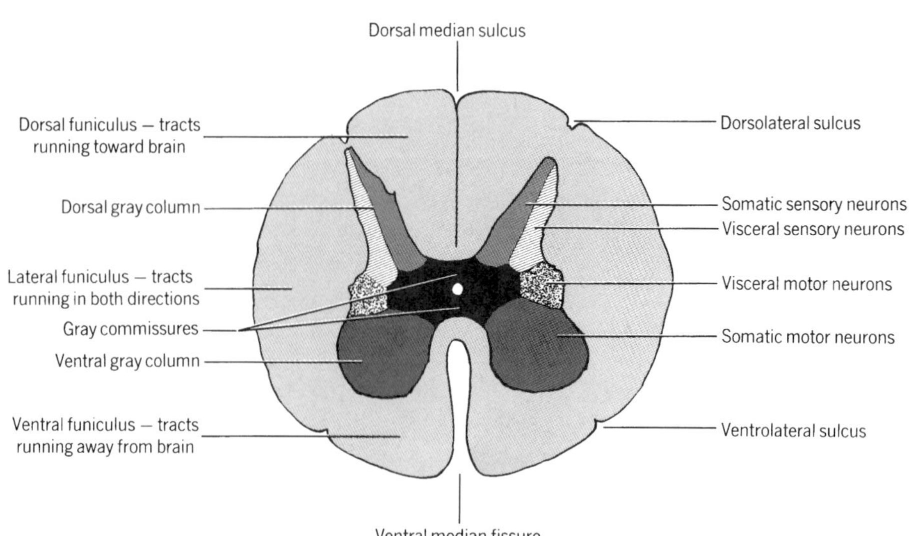

FIGURE 17.9
STRUCTURE OF THE MAMMALIAN SPINAL CORD.

The spinal cord is supported and protected by one or more layers of tissue called **meninges.** These are similar to, and continuous with, meninges of the brain and will be described in the next chapter.

Evolution of the Spinal Cord The spinal cord of AMPHIOXUS is seemingly primitive in that the embryonic neural folds do not completely fuse (Figure 17.10). Accordingly, the adult neural canal communicates with the space around the cord by a slitlike groove. Gray and white matter cannot be distinguished because nerve fibers of amphioxus are not myelinated.

CYCLOSTOMES, like other vertebrates, complete the neurulation process to enclose a central canal. The boundary between gray and white matter remains indistinct. The cord is wide and its ventral surface is concave where it fits against the notochord.

The cord of FISHES and AMPHIBIANS is nearly circular in cross section. Gray and white matter have become distinct. The configuration of the gray matter is various; ventral gray columns are usually evident. Dorsal median sulcus and ventral median fissure are making their appearance among these animals as increasing complexity of nuclei and tracts cause the cord to enlarge. The diameter of the cord enlarges moderately opposite the appendages, where there is greater need for nervous integration.

AMNIOTES have a deep sulcus and fissure. The cord again enlarges opposite the appendages, the **cervical enlargement** being the more pronounced if the pectoral appendages are emphasized (bats, apes) and the **lumbar enlargement** more pronounced if the pelvic appendages are emphasized (ostrich, bipedal dinosaurs). In cross section, the gray matter now is shaped like an H with thick arms.

A distinctive feature of BIRDS is the **glycogen body.** This is a conspicuous glandlike mass of tissue wedged into the cord of the lumbar area where the dorsal median sulcus opens to receive it. Its function has been debated but is not surely known.

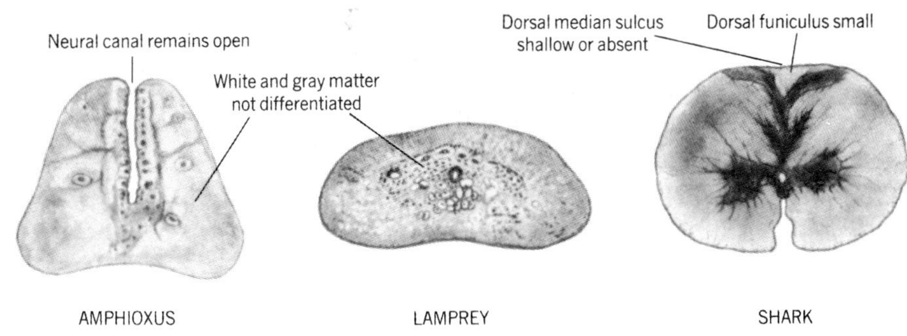

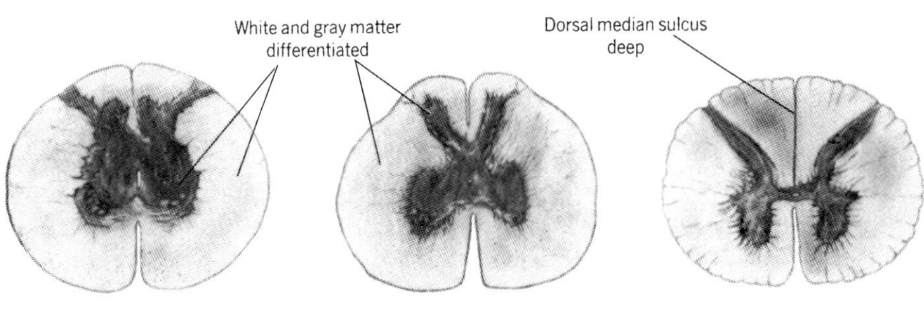

FIGURE 17.10
COMPARATIVE ANATOMY OF THE SPINAL CORD.

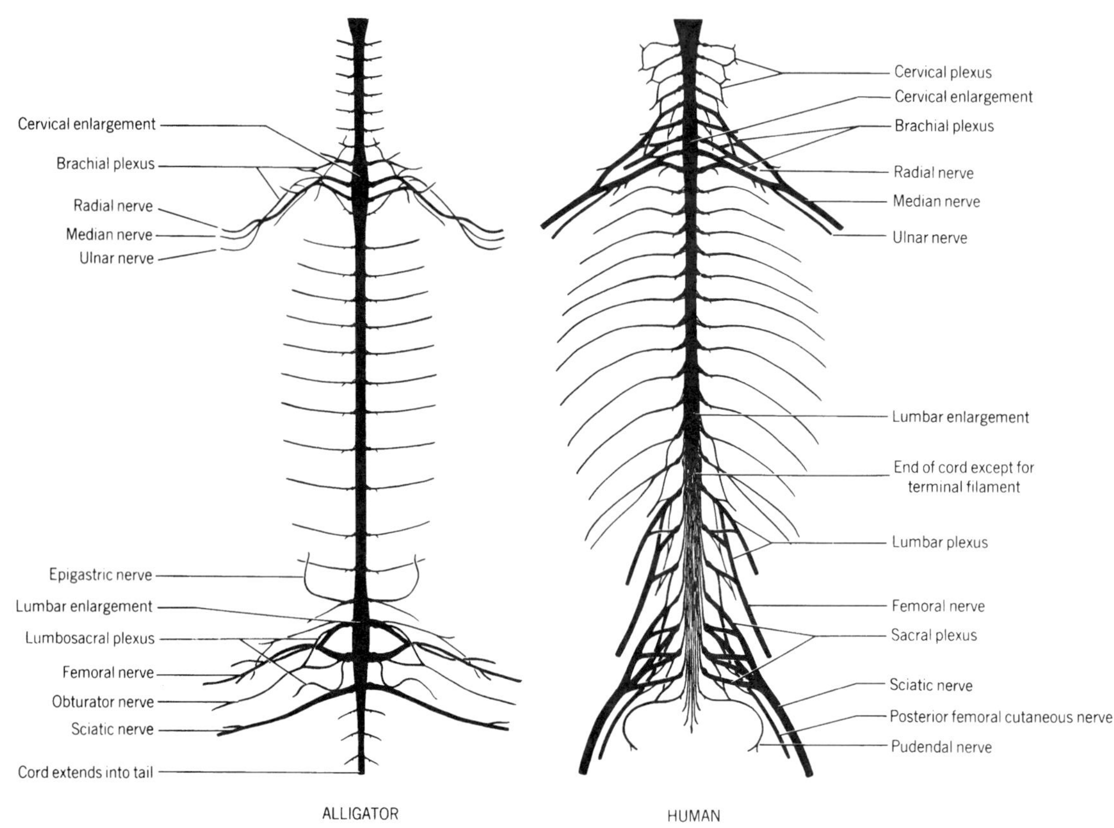

FIGURE 17.11 SPINAL CORD, PRINCIPAL NERVES, AND PLEXUSES OF A REPTILE AND A MAMMAL. Dorsal views.

Spinal cords of MAMMALS frequently have **dorsolateral** and **ventrolateral sulci** (Figure 17.9). The dorsal and ventral spinal nerve roots, respectively, join the cord along these grooves. The cords of anurans and some fishes are shorter than the canal within the vertebral column. This condition is not seen in reptiles and birds, but appears again in mammals (except monotremes). The embryonic cord fills its bony housing but grows more slowly than the spine so that the adult cord ends in the lumbar region (except for a terminal filament), and the more posterior nerves must angle back to reach their destinations (Figure 17.11).

EVOLUTION OF SPINAL NERVES

AMPHIOXUS has a series of paired "dorsal" spinal nerves that contain three kinds of fibers: somatic sensory from skin and muscle, visceral sensory from internal organs, and visceral motor. These nerves are intersegmental and run with the myosepta between muscle segments. All nerve cell bodies of sensory neurons are located within the cord. Consequently, there are no ganglia on the spinal nerves. The structures formerly called ventral spinal nerves are really segmentally arranged specialized muscle fibers that run to the surface of the spinal cord where their motor end plates are located. In this amphioxus is specialized.

The further evolution of spinal nerves follows a relatively simple and straightforward progression. LAMPREYS have intersegmental dorsal spinal nerves that are like the spinal nerves of amphioxus except that some of the sensory neurons have cell bodies outside the cord (Figure 17.12). Lampreys also have segmental ventral spinal nerves that contain only

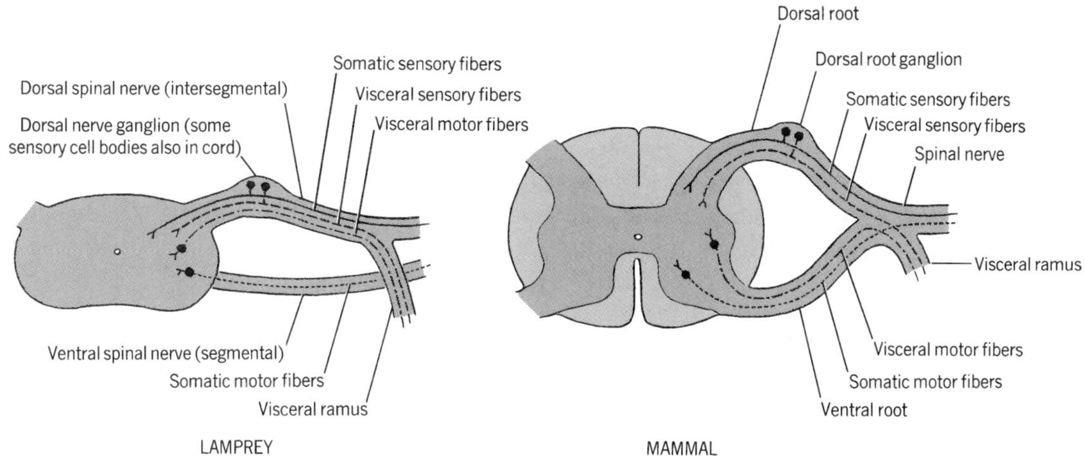

FIGURE 17.12 DISTRIBUTION OF NERVE COMPONENTS IN SPINAL CORD AND NERVES AT TWO EVOLUTIONARY LEVELS. More detail for the branching of spinal nerves and the distribution of visceral motor fibers is shown in Figure 17.15. Size of the nerves is exaggerated.

somatic motor fibers. Each ventral nerve joins the cord anterior to the corresponding dorsal nerve. Dorsal and ventral spinal nerves do not join in lampreys (or cephalaspids) but run independently to their destinations.

The HAGFISH, FISHES, and AMPHIBIANS illustrate the next structural level. The dorsal and ventral nerves of each body segment join outside the vertebral column. This establishes a single spinal nerve per segment on each side of the body. The nerve joins the cord by separate **dorsal** and **ventral roots.** Each root tends to emerge from the cord as a row of adjacent twigs. Close beyond the union of the roots, the spinal nerve divides into a **dorsal ramus** that goes to structures of epaxial origin, a relatively large **ventral ramus** that goes to the appendages and structures of hypaxial origin, and a **visceral ramus** that goes to structures derived from the hypomere (visceral and branchial muscles and glands). At the levels of the paired appendages the ventral rami form plexuses of varying complexity.

Nerve cell bodies of sensory neurons are now all located in the **dorsal root ganglion** of each nerve. The distribution of fibers in the dorsal and ventral roots is the same as for the dorsal and ventral nerves of the lamprey except that some visceral motor fibers exit in each root.

Several further advances are found in AMNIOTES. Dorsal and ventral roots of spinal nerves join inside the vertebral column. Each dorsal root joins the cord at the same level as the corresponding ventral root rather than posterior to it. Usually all visceral motor fibers exit from the cord in the ventral root. This completes the gradual shift of these fibers from the dorsal spinal nerve and leaves the dorsal root with only sensory neurons. **Brachial** and **lumbosacral plexuses** tend to be more complex than for anamniotes, but the patterns of interweaving are various (Figure 17.11).

CRANIAL NERVES

Origin and Nature Spinal nerves are conveniently uniform in regard to occurrence, configuration of roots and branches, nerve fiber components, and relation to the central nervous system. This is not true of cranial nerves, which emerge directly from the brain. Although generally consistent in number, there is some variation: A cranial nerve may be present in some vertebrates and missing in others (e.g., terminal nerve). A nerve may split in the course of evolution to become two (spinal accessory nerve from vagus nerve). Conversely, two nerves may fuse to become one (evolution of amniote trigeminal nerve). The same nerve that is a cervical spinal nerve of one vertebrate may be a cranial nerve of another (hypoglossal nerve). Also, fiber components believed to have been present in an ancestral nerve may become lost.

A further difference between spinal and cranial nerves stems from the nature of the segmentation of the central nervous system. The somites are already segmented when they first appear in the embryo, but the central nervous system is not. Motor nerves grow out of the cord at intervals to penetrate segmented somite derivatives. Sensory nerves and ganglia are derived from neural crests that become intersegmental as they are squeezed between the bulging somites. Thus, the segmentation of spinal nerves is regular but seems to be secondary. In the head, the transitory nature of somitomeres, their questionable relation to neuromeres, and the varying patterns of migration of cranial neural crests make it evident that the serial nature of cranial nerves is both secondary (to structures served) and irregular. Furthermore, the head is in part segmented ventrally in an independent series related to the visceral arches (see additional coverage of head segmentation on p. 116 and Figure 8.2). In consequence, serial homology is less evident in cranial than in spinal nerves. There are at least three series of cranial nerves that reflect the developmental contributions of the neural crests, neuromeres, and placodal and somitomeric tissue that contribute to the development of the head (see Northcutt in References). Cranial nerves are numbered in spatial sequence, but it should be understood that in terms of fundamental nature the assignment of numbers is rather arbitrary.

These difficulties were a welcome challenge to morphologists in the late 1800s and early 1900s, when much attention was given to the analysis of cranial nerves and concepts of their organization and evolution were formulated. There even developed a general consensus about the evolution of cranial nerves, but with an increase in the sophistication of functional, developmental, and anatomical studies, this consensus is being challenged and is under review (see Boord in References). Various kinds of clues were used. For example, comparative anatomical evidence shows that the trigeminal nerve of amniotes is a composite nerve: One of its principal branches joins the brain independently in some fishes. Developmental studies in these forms also reveal the spinal accessory nerve to be a composite nerve: One component of its motor nerves seems to be derived from the vagus, whereas a second, much older component is not (see Wake in References). Physiological evidence shows the unique nature of some nerves (only aquatic vertebrates have a lateral line system, and it is served only by cranial nerves).

As a way of organizing the cranial nerves (it should be kept in mind that this scheme is under scrutiny), one can place them in three general categories. First there are seven nerves (numbers 0, V in two or three parts, VII, IX, X, XI) that appear to be in series with dorsal roots of spinal nerves or, more closely, with the dorsal spinal nerves of lampreys, which, as described in the previous section, do not join their respective ventral spinal nerves (Figure 17.13). These nerves all join the brainstem at a lateral (not ventral) level. It is postulated that in the ancestral condition, each nerve carried the same components as dorsal spinal nerves: somatic sensory, visceral sensory, and visceral motor. A sensory ganglion was present close to the entrance of each nerve into the brain. Furthermore, it is postulated that in the remote ancestral vertebrate, these nerves served branchial and pharyngeal areas: Each nerve had a major branch to each hemibranch bordering a gill slit, and minor branches to the adjacent pharyngeal wall and skin.

The second category includes four cranial nerves (III, IV, VI, XII) that appear to be in series with ventral spinal nerves. All but one join the brainstem at the expected ventral level. [The exception is only partial; the motor neuron cell bodies of the trochlear nerve (IV) are ventral in the brainstem, but most of the fibers arch over the brain to emerge on

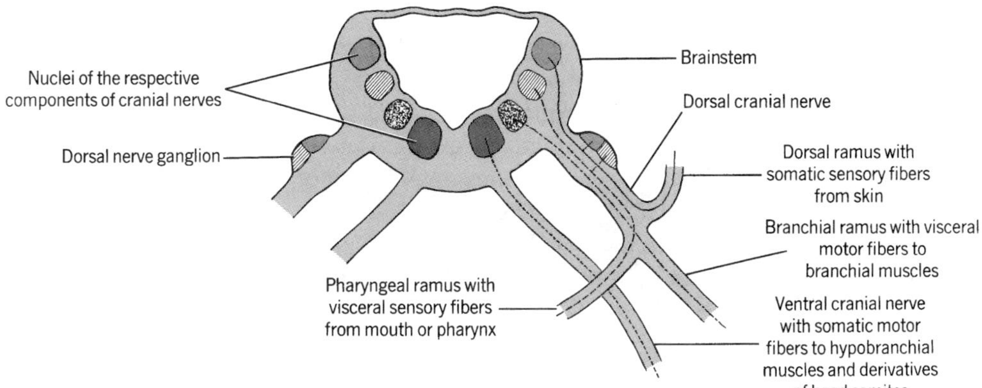

FIGURE 17.13 POSTULATED ANCESTRAL DISTRIBUTION OF NERVE COMPONENTS IN THE BRAINSTEM AND IN CRANIAL NERVES OF THE DORSAL AND VENTRAL SERIES. Shading of nuclei as in Figure 17.9.

the opposite side.] These nerves carry only somatic motor fibers. Ganglia are generally absent. The nerves innervate hypobranchial muscles and derivatives of head somites or somitomeres.

The third category has no counterpart in the spinal series because its nerves serve structures that are peculiar to, or centered on, the head: the nose, eye, ear, and lateral line system. Nerves I, II, VII, VIII, IX, and X are in this category. These nerves are sensory. Their ganglia, unlike those of other sensory nerves are derived, at least in part, from ectodermal placodes. These nerves are usually considered to be somatic, though the designation is somewhat arbitrary and perhaps superfluous. Common practice is to call them "special" sensory nerves in recognition of their distinctive nature (but see Finger in References). To a degree, this is a category of leftovers because the nerves of the nose and eye are not serially homologous either with one another or with the other nerves of the category.

There is one further complication. The visceral motor nerves of the body are so distinctive in various ways that together they make up the autonomic system, which is described at the end of this chapter. Even at levels of the cord, it was the visceral motor fibers that were least stable, emerging sometimes in the dorsal root and sometimes in the ventral root. In the head, visceral motor fibers of the autonomic system join four cranial nerves (III, VII, IX, X). These include nerves of both the dorsal and ventral categories.

Counting 7 nerves in the dorsal series, 4 in the ventral series, 6 in the special series, and 4 with autonomic fibers, and knowing that there are 13 cranial nerves in all, it is evident that there is some doubling up. Thus, the oculomotor nerve is in the ventral series but has autonomic fibers tagging along; the facial nerve is in the dorsal series but may be augmented by both special and autonomic components.

This general information provides the basis for a more specific account of the evolution of each cranial nerve. The following account is clarified by Figure 17.14 and also figures on pp. 138 and 326.

Structure and Evolution of Cranial Nerves The TERMINAL nerve was not discovered until the other cranial nerves had been given numbers from I through XII. Hence, it has the number 0. Regarded as being in the dorsal series, it originates in nasal epithelium and sometimes in the vomeronasal organ. It seems to mediate responses to sex pheromones. Having one or more ganglia, it is classed as a somatic sensory nerve, any ancestral visceral fibers having been lost. The nerve is present in all vertebrates except cyclostomes, *Latimeria,* birds, and some mammals (including humans). It is largest in elasmobranchs.

The OLFACTORY nerve (number I) is in the special series. It runs from the olfactory epithelium and vomeronasal organ (if present) to the olfactory bulb of the brain. It is

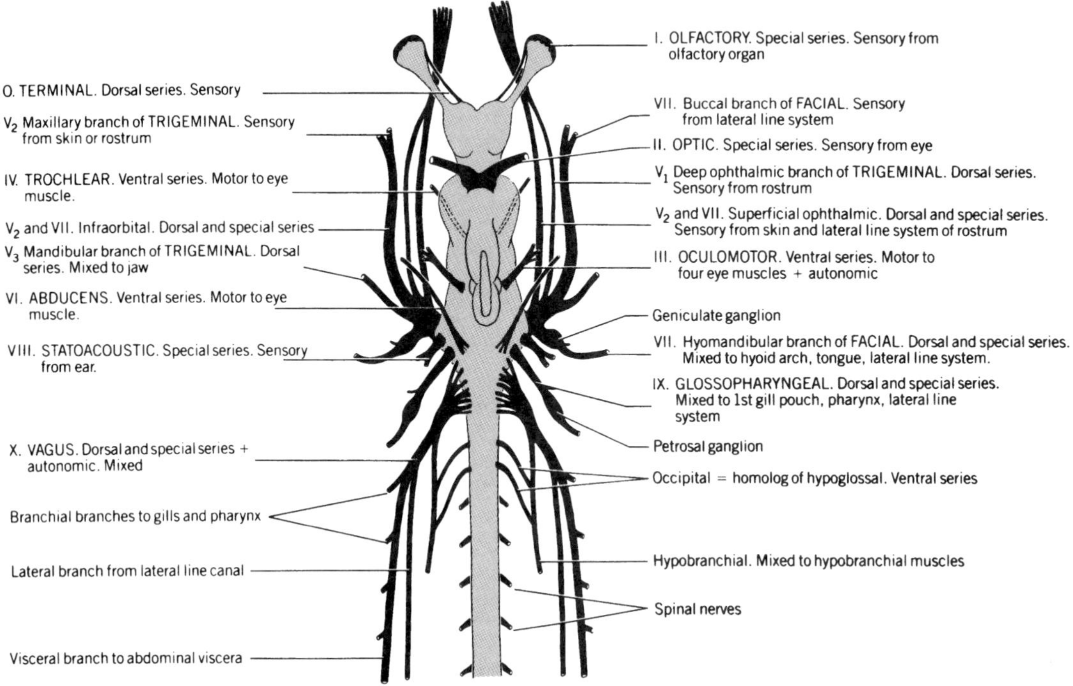

FIGURE 17.14 CRANIAL NERVES OF THE SHARK, *Squalus.* Ventral view. (Trigeminal and facial nerves are more independent in most other vertebrates.)

unique in that its fibers are extensions of the receptor cells. Consequently, it has no ganglion even though it is sensory. The nerve is present in all vertebrates, its size relating to the excellence of the olfactory sense. It is long if the rostrum is long and the olfactory tracts of the brain are short; otherwise the nerve is short. Often it is divided into many twigs. It is paired in cyclostomes in spite of the unpaired nature of that animal's nasal pouch.

The OPTIC nerve (II) is also in the special series. It runs from the eye to the brain and is unique in that it develops as a tract of the embryonic brain (see p. 362). The associated ganglion cells are in the retina. The nerve is constant in all vertebrates. The two optic nerves may completely cross under the brain (teleosts, birds, and some other vertebrates), but often some fibers cross and some do not. In mammals, half the fibers of each nerve cross to the other side, an arrangement that in this class probably contributes to the coordination of eye movements.

The OCULOMOTOR (III), TROCHLEAR (IV), and ABDUCENS (VI) nerves are in the ventral series. They innervate the extrinsic muscles of the eye. All are somatic motor nerves, though the oculomotor is joined by autonomic fibers passing to muscles of the iris and ciliary apparatus of the eye. These nerves are constant in virtually all vertebrates.

The DEEP OPHTHALMIC (or profundus) nerve (V_1) is the second nerve of the dorsal series, though like the terminal nerve it does not relate to a gill slit in any surviving vertebrate. It runs to the skin of the rostrum. It is exclusively somatic sensory; any visceral fibers it may once have had were lost with the associated gill slit. The nerve is represented in all vertebrates but is an independent nerve with its own ganglion only in ostracoderms, placoderms, and some primitive bony fishes. In other vertebrates it becomes a branch of the next nerve to be described.

The large TRIGEMINAL nerve (V) has two branches in those vertebrates having an independent deep ophthalmic nerve. In other vertebrates the deep ophthalmic (in mammals simply called ophthalmic) V_1, maxillary (V_2), and mandibular (V_3) nerves are the three branches that give the trigeminal nerve its name. (A superficial ophthalmic branch of the trigeminal may also be present, and unfortunately the facial nerve has a branch with the same name.) The maxillary and mandibular branches may represent a branchial nerve that once served the premandibular gill slit. The maxillary branch retains only somatic sensory fibers from the teeth, gums, and skin of the upper jaw. The mandibular branch similarly serves the lower jaw. It also has motor fibers to the various jaw muscles derived from the mandibular arch. These are striated, voluntary muscles, but since they relate phylogenetically to the pharynx (see pp. 178 and 179), the associated nerves are designated as visceral. The large **semilunar** (or Gasserian) **ganglion** is located where these branches all merge before entering the brain.

The FACIAL nerve (VII) is in part the nerve in the dorsal series that is associated with the spiracular cleft and derivatives of the hyoid arch. It is the first nerve of this series to retain all the components of the ancestral dorsal nerves: somatic sensory to related areas of the skin, visceral sensory to much of the mouth and taste buds, and visceral motor to all muscles derived from the hyoidean arch. Included among the latter are the muscles of facial expression of humans—hence the name of the nerve. The facial nerve of fishes also has a large component from the special series. This consists of the sensory fibers from the cranial part of the lateral line system. Finally, there is a component of the autonomic system serving tear glands and several salivary glands. The composite nerve is large. Its ganglion is the **geniculate ganglion.**

The STATOACOUSTIC nerve (VIII) (also called vestibulocochlear or auditory) serves the inner ear. It is therefore in the special series and relates, developmentally and phylogenetically, to the lateral line components of nerves VII, IX, and X. It always has two principal branches. In most vertebrates the more anterior branch serves most of the organ of equilibrium, and the posterior branch serves both the organs of equilibrium and hearing. In eutherian mammals the posterior branch functions only in hearing.

The GLOSSOPHARYNGEAL nerve (IX) is in part the nerve of the dorsal series associated with the first branchial gill slit and arch of fishes. Somatic sensory fibers are present in some vertebrates. Visceral sensory fibers run to part of the pharynx and some taste buds. Visceral motor fibers serve some small muscles. As for the previous nerve, there is a component from the special series of nerves innervating the lateral line system, this time at the back of the head, and another from the autonomic system, this time to a salivary gland. The composite nerve is usually small. Any somatic sensory fibers present have a **superior ganglion;** visceral sensory fibers have a larger **petrosal ganglion.**

The VAGUS (X) and ACCESSORY (XI) nerves are best listed together because the latter, identified in amniotes and some salamanders, is derived, at least in part, by splitting away from the original vagus. (Another part is of spinal origin, in various amphibians.) The vagus spans the levels of several ancestral head somites and joins the brainstem in a linear series of twigs. These nerves are in part the last of the dorsal series, and as such are associated with all remaining branchial structures. A small somatic sensory branch of the vagus serves skin in the gill and ear region. Visceral sensory fibers come from posterior taste buds and the pharynx. Sensory fibers of the accessory nerve do not mature. Visceral motor branches serve muscles of the branchial arches and their derivatives (including some muscles of the shoulder). A large component of the vagus nerve of fishes innervates the part of the lateral line system that is on the body, and hence is from the special series of nerves. Finally, the large, long, and important **visceral branch** of the vagus is the autonomic component of that nerve. Its fibers serve the heart, lungs (if present), and gut.

Somatic sensory fibers have a **jugular ganglion;** visceral sensory fibers have a large **nodose ganglion.**

The HYPOGLOSSAL nerve (XII) is in the ventral series of nerves. It is an exclusively somatic motor nerve and innervates the hypobranchial muscles of throat and tongue. It is a cranial nerve in amniotes and some labyrinthodonts, and a cervical nerve (called the hypobranchial nerve) in cyclostomes and fishes. It is derived from several postotic somites (the same for which the vagus was the dorsal nerve) and joins the central nervous system by a linear series of twigs.

AUTONOMIC SYSTEM

The autonomic system is not isolated, structurally or functionally, from either the central or peripheral nervous systems. Hence, it is difficult to set its limits. It relates exclusively to involuntary functions of the body. Accordingly, the system includes only visceral fibers, and the visceral fibers serving the striated branchial muscles are excluded. Structures innervated by autonomic nerves are, therefore, the heart and vessels, some respiratory organs, glands, gut tube, urogenital organs, pigment cells, fat tissue, and intrinsic muscles of eye and skin. These structures are, of course, served by both sensory and motor fibers. There is nothing very remarkable, however, about the sensory neurons: Like somatic sensory neurons they have their nerve cell bodies in dorsal root ganglia and certain cranial ganglia. Sensory fibers are therefore excluded from the discussion.

What, then, is distinctive about the visceral motor neurons that innervate involuntary organs? First, every pathway includes a neuron having its cell body inside the central nervous system, and in addition (except in the adrenal medulla) a neuron (or in some instances several neurons) having its cell body outside the central nervous system (Figure 17.15). Cell bodies of the latter are in motor ganglia. Fibers between ganglia and the central nervous system are designated **preganglionic;** those between ganglia and end organs are designated **postganglionic.** Preganglionic fibers are myelinated; postganglionic fibers have little or no myelin.

A second distinctive feature of the autonomic system is that it is divisible into sets of fibers. For amniotes it is usual to recognize **sympathetic** and **parasympathetic divisions,** which are anatomically distinct and usually elicit antagonistic responses in the organs served. The functional distinctions between the postcranial parts of these divisions are not always sharp for amniotes, however, and break down for lower vertebrates. The relatively neglected **enteric division** of the autonomic system consists of the complex net formed by

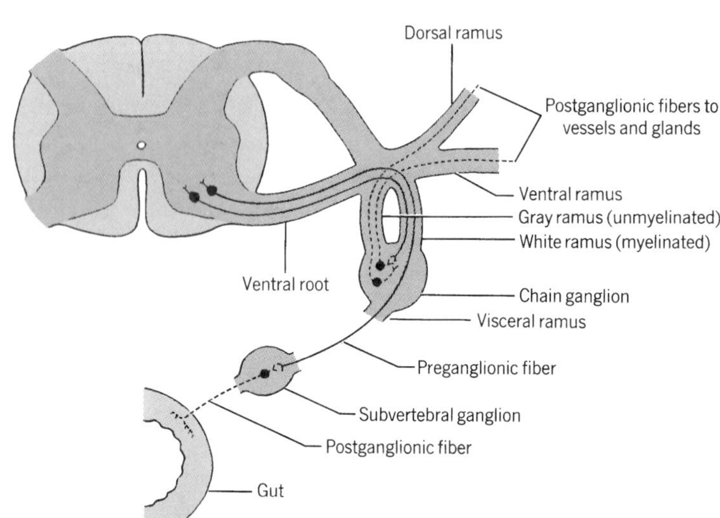

FIGURE 17.15 TYPICAL PATHWAYS OF THE SYMPATHETIC DIVISION OF THE AUTONOMIC SYSTEM OF AMNIOTES. Cross section at posterior thoracic level. Other fibers run lengthwise in the body from the chain ganglia. Size of the nerves is exaggerated.

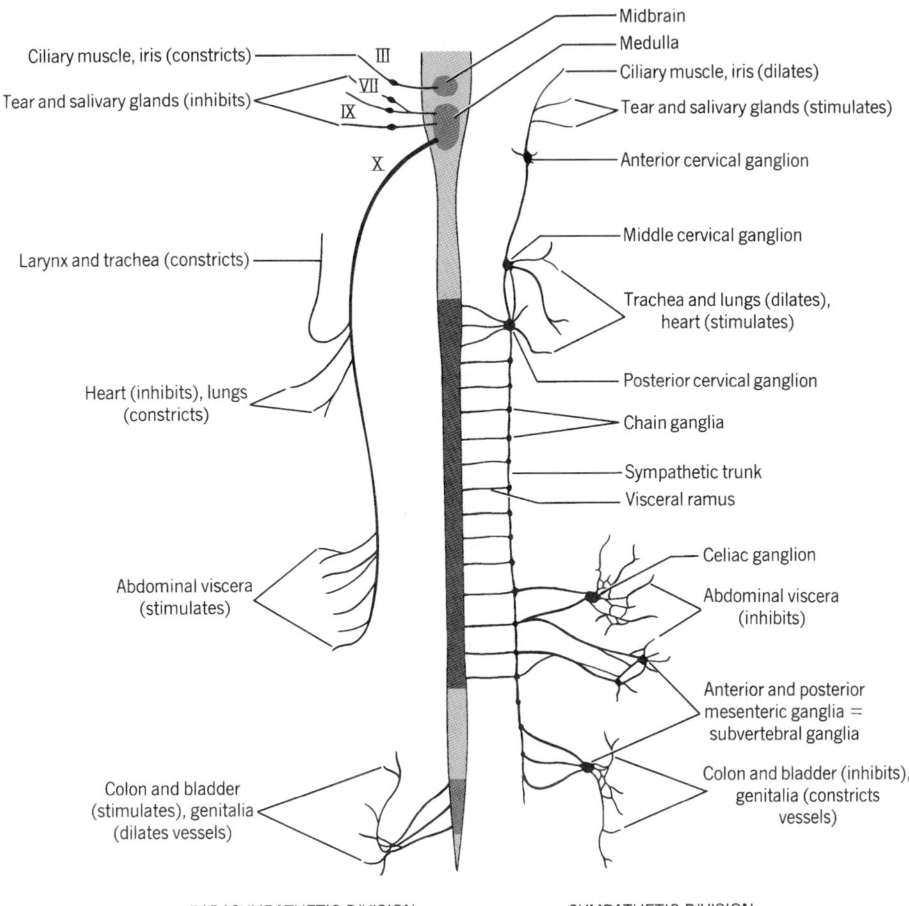

FIGURE 17.16 DIAGRAM OF THE MAMMALIAN AUTONOMIC NERVOUS SYSTEM. General distribution of fibers to vessels and skin is not shown.

the exceedingly numerous neurons located within the wall of the gut. These neurons appear to be activated directly by local physical and chemical stimuli and to mediate local reflexes.

The following descriptions of the two familiar divisions of the autonomic system are based on amniotes (particularly mammals). Comments on the system in lower vertebrates will conclude the chapter.

The SYMPATHETIC division of the system is said to have a **thoracolumbar outflow** because that term indicates the levels at which its preganglionic neurons emerge from the central nervous system in spinal nerves (Figure 17.16). Preganglionic fibers are relatively short. They reach the sympathetic ganglia by way of the visceral rami of spinal nerves (or the white part of the connection—see Figure 17.15). Most of the small but tough ganglia are **chain ganglia** arranged like two parallel strings of beads on the **sympathetic trunks** that are just ventral to the vertebral column. There are also three pairs of **cervical ganglia** against the carotid arteries in the neck, and about three unpaired **subvertebral ganglia** at the bases of the major arteries of the viscera.

A preganglionic fiber may synapse with a postganglionic fiber in the chain ganglion closest to its exit from a spinal nerve, or it may go through that ganglion to a chain ganglion on the other side of the body, a chain ganglion at another level of the spine, or a cervical or subvertebral ganglion. Postganglionic fibers are relatively long. Most of them that emerge from chain ganglia reenter a spinal nerve and run with it and its branches to their destinations. Most postganglionic fibers that emerge from cervical or subvertebral ganglia run parallel with blood vessels to their destinations.

Postganglionic fibers, like the medulla of the adrenal gland, release **noradrenalin** (and are designated as adrenergic). There are fewer terminal fibers than there are muscle and gland cells to be influenced. These nerve junctions, therefore, release relatively large quantities of noradrenalin, which then spreads to adjacent effector cells.

In general, the sympathetic division of the system elicits responses of alertness, excitement, alarm, and the expenditure of energy as necessary to meet emergencies. Vegetative functions tend to be inhibited.

The PARASYMPATHETIC division of the autonomic system is said to have a **craniosacral outflow** because it emerges from the central nervous system with cranial nerves III, VII, IX, and X and with about three sacral spinal nerves. Preganglionic fibers are relatively long. There are about four pairs of small parasympathetic ganglia in the head. These are located near the organs served (eye, salivary glands, tear glands). Elsewhere, ganglion cells are dispersed in the tissues of the viscera and are not found by dissection. Postganglionic fibers are quite short.

Postganglionic fibers of parasympathetic neurons are like most somatic fibers and preganglionic fibers of both divisions of the autonomic system in releasing acetylcholine (they were cholinergic), though again, the quantity is large. (Under study are several neurotransmitters of the autonomic system that are neither adrenergic nor cholinergic.)

The parasympathetic division of the system elicits responses appropriate to quiet, vegetative activities such as digestion and maintenance of resting levels of blood sugar.

The above account fits most AMNIOTES. The system was slow to evolve. AMPHIOXUS has visceral motor fibers running from each spinal nerve to the gut, but there are no ganglia outside the viscera. CYCLOSTOMES have autonomic fibers in the vagus nerve, but otherwise the system is rudimentary. CARTILAGINOUS FISHES send visceral motor fibers to vessels and viscera by way of the expected cranial and spinal nerves. There are autonomic ganglia under the spine, but no distinct ganglionic chains. In BONY FISHES the well-developed sympathetic ganglionic chains extend far into the head, and in AMPHIBIANS strands join the anterior chain ganglia to the autonomic ganglia of the head. Accordingly, the cranial outflow is limited in these groups.

REFERENCES

Boord, R.L. (ed.). 1993. Symposium—Structure, development, and phylogeny of cranial nerves. *Acta Anat.* 148:67–168.

Cotman, C.W., and M. Nieto-Sampedro. 1984. Cell biology of synaptic plasticity. *Science* 225:1287–1294.

Fetcho, J.R. 1987. A review of the organization and evolution of motor neurons innervating the axial musculature of vertebrates. *Brain Res. Rev.* 12:243–280.

Fetcho, J.R. 1992. The spinal motor system in early vertebrates and some of its evolutionary changes. *Brain Behav. Evol.* 40:82–97.

Fetcho, J.R., and N.T. Reich. 1992. Axial motor organization in postmetamorphic tiger salamanders (*Ambystoma tigrinum*): a segregation of epaxial and hypaxial motor pools is not necessarily associated with terrestrial locomotion. *Brain Behav. Evol.* 39:219–228.

Finger, T.E. 1993. What's so special about special visceral? *Acta Anat.* 148:132–138.

Iwaniuk, A.N., S.M. Pellis, and I.Q. Whishaw. 1999. Brain size is not correlated with forelimb dexterity in fissiped carnivores (Carnivora): a comparative test of the principle of proper mass. *Brain Behav. Evol.* 54:167–180.

Jacobson, M. 1991. *Developmental neurobiology.* 3rd ed. Plenum, New York. 776p.

Keynes, R.D., and D.J. Aidley. 1991. *Nerve and muscle.* 2nd ed. Cambridge Univ. Press, New York. 181p. The basics of the chemistry and physics of nervous activity.

Krstić, R.V. 1985. *General histology of the mammal: an atlas for students of medicine and biology.* Springer, New York. 404p. Includes 47 three-dimensional reconstructions of nervous tissue.

Nauta, W.J.H., and M. Feirtag. 1986. *Fundamental neuroanatomy.* Freeman, New York. 340p.

Nieuwenhuys, R., and H.J. Donjelaar. 1997. *The central nervous system of vertebrates.* 3 vols. Springer, New York.

2,219p. Enormous, detailed, comprehensive; excellent illustrations.

Nilsson, S. 1983. *Autonomic nerve function in the vertebrates.* Springer, New York. 253p.

Nishikawa, K.C. (ed.). 1997. Symposium—Evolution of neural ontogenies: the ontogeny and phylogeny of invertebrate and vertebrate nervous systems. *Brain Behav. Evol.* 50:5–59. Illustrates how a synthesis of comparative neuroanatomy, developmental neurobiology, and phylogenetic analysis provides insights into the evolution of neural systems.

Noback, C.R., and R.J. Demarest. 1996. *The human nervous system: introduction and review.* 5th ed. McGraw-Hill, New York. 386p. A clear, well-illustrated account of the human nervous system emphasizing pathways.

Northcutt, R.G. 1993. A reassessment of Goodrich's model of cranial nerve phylogeny. *Acta Anat.* 148:71–80.

Northcutt, R.G., and W.E. Bemis. 1993. Cranial nerves of the coelacanth, *Latimeria chalumnae* (Osteichthyes; Sarcopterygii; Actinistia), and comparisons with other craniata. *Brain Behav. Evol.* 42:1–75. An extraordinarily complete treatise and model study in comparative morphology.

Sarnat, H.B., and M.G. Netsky. 1981. *Evolution of the nervous system.* 2nd ed. Oxford Univ. Press, New York. 504p. An excellent source for comparative and functional material.

Smith, K. 1994. Are neuromotor systems conserved in evolution? *Brain Behav. Evol.* 43:293–305. Thoughtful review of the neuromotor conservation hypothesis.

Snyde, S.H., and D.S. Bredt. 1992. Biological roles of nitric oxide. *Sci. Am.* 266(5):68–77.

Wake, D.B. 1993. Brainstem organization and branchiomeric nerves. *Acta Anat.* 148:124–131.

Chapter 18

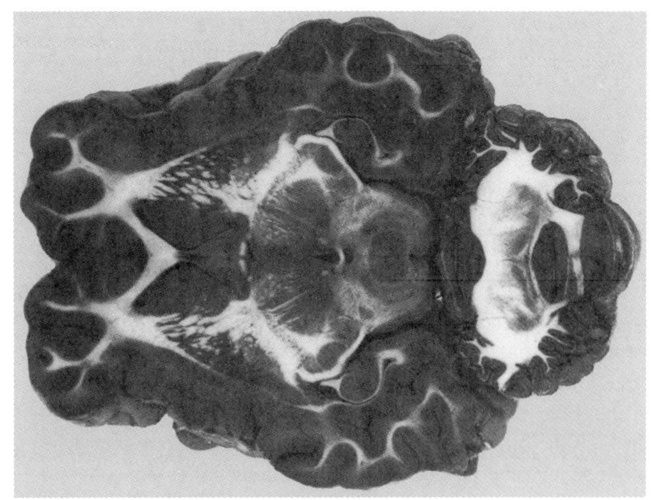

Nervous System: Brain

The brain is the most complicated and for many persons the most amazing organ of the body. It is under study by many researchers in various fields. Since understanding of its function must ultimately be based largely on its wonderfully intricate structure, the morphologist has a major role to play in the analysis of this master organ.

In order to gain an adequate background in brain structure for either teaching or advanced study, it is desirable to study the brain in enough depth to acquire basic vocabulary, learn general relationships, and identify the major functional components.

HOW THE BRAIN IS STUDIED

The brain is studied in many ways. Descriptions are made of gross structure using entire or dissected brains, of fine structure using serial sections, and of ultrastructure using electron micrographs. Thick slices of the brain are stained to distinguish, in striking color contrast, the myelinated tracts from the unmyelinated cortex and nuclei (see above). Histological techniques reveal individual neurons in detail, as do several anatomical tracer techniques. The challenge of placing the individually identified nerve cells forthcoming from anatomical tracer studies into a functional ensemble for the brain and other parts of the nervous system is being met on several fronts. The use of neurotropic viruses capable of replicating in one neuron and passing transsynaptically to infect adjacent neurons (see Card in References) and the use of intracellular calcium concentration changes to optically monitor the activity of individual neurons within the brain and spinal cord of zebrafish (see Cox and Fetcho in References) are two approaches being explored. The development of various noninvasive techniques of tomography for studies of the brain in vivo has resulted in tremendous advancement in our understanding of brain function (see Comment 18.1).

Some studies require repeated manipulation of a specific cell type or anatomically defined region of the brain under study. A device called a stereotaxic instrument is used to

Above: Stained frontal section of a sheep brain.

COMMENT 18.1

NONINVASIVE STUDIES OF THE BRAIN HOLD PROMISE

Several noninvasive techniques for the study of brain structure and function have been developed in recent years. These are based on the technique of tomography (= *cut* + *depict*), which reconstructs a two- or three-dimensional image of a structure through the use of multiple-angle pictures. Computerized (axial) tomography scanning, or CT scanning, enables researchers to obtain a reconstructed picture of a "slice" through the brain (or even hard structures; see Comment 8.2) without invasive procedures. CT scanning relies on conventional X-rays combined with computer technology: Beams of X-rays from different angles are passed through the object, and a picture based on subtle density differences of adjacent tissues is assembled.

Position emission tomography (PET) scanning works similarly, but in addition to information about brain structure, PET scans provide data about brain activity. Subjects are given radioactively labeled 2-deoxy-D-glucose (2DG), an analog of glucose. Active neurons, which require more glucose than do resting neurons, take up 2DG along with the glucose. However, the 2DG cannot be metabolized and accumulates in active neurons, thereby labeling an active brain region. Another technique is to inject a radiolabeled substance into the blood, which allows changes in blood flow through the brain to be registered. Since blood flow increases in active brain areas, changes in brain activity can be detected.

Magnetic resonance imaging (MRI) is a technique that uses the magnetic properties of the hydrogen nucleus when excited by radio frequency radiation transmitted by a coil surrounding the subject. The excited hydrogen nuclei emit a signal that is detected by a receiver coil and, as for a PET scan, is analyzed by computers to produce an image (Figure 18.1). Some tissues (myelin) contain more water (and hence hydrogen) than other tissues; accordingly, MRI provides excellent differentiation between tissues of the brain. When introduced, MRI was combined with PET to provide a structural and functional picture of brain activity. Researchers can now combine functional imaging with MRI by taking advantage of the transient local increase in the proportion of oxygenated to nonoxygenated hemoglobin that occurs in the blood in the region of metabolically active neural tissue. This altered ratio changes the magnetic properties of the region, allowing for detection of the active tissue at the same time that an image of brain structure is obtained. This combination of functional and anatomical images is known as functional MRI, or fMRI.

PET and MRI tomography is expensive, and as a result studies of human brain function dominate the literature. Nevertheless, some studies of model species of animals have been completed, and these techniques hold promise for future comparative studies.

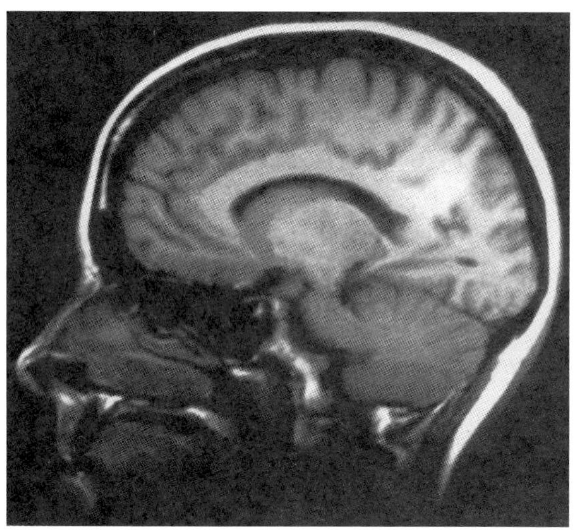

FIGURE 18.1 MAGNETIC RESONANCE IMAGE OF A HUMAN HEAD IN SAGITTAL SECTION. (Courtesy of M. Ridlen, M.D.)

position the head and brain in a precise standard position. The location of every minute part of the brain can then be expressed in terms of the three coordinates of space. Using the stereotaxic atlases that are now available for many experimental animals, the investigator can introduce microinstruments into any desired part of the living brain. Once this is done, various techniques are used: Restricted areas are destroyed with electricity, freezing, or a chemical agent, or very small bits of tissue are removed by suction. Impairment of function is then noted, and after a period of time, the animal is killed and suitable staining techniques are used to reveal on sections of the brain the paths of degenerating nerve fibers. Alternatively, normal stimuli, microelectrodes, or chemicals are used to activate specific parts of the brain, and resultant behavior is observed. Animals may be taught to stimulate their own brains for the reward of pleasant sensations, and conscious human subjects have reported sensations and thoughts accompanying local stimulation of the brain during surgery.

The impairment of function that follows accident or pathology is studied. Parts of the brains of experimental animals are severed and changes in behavior analyzed. Differences in the brains of different vertebrates are related to their widely different adaptations. Progressive maturation of the fetal brain is correlated with the onset of function. Natural electrical discharges of the organ (brain waves) are recorded during various activities. Culture techniques are used to study the growth and physiology of isolated neurons. The response of the brain to both the general and local administration of drugs is revealing. Finally, the science of cybernetics adapts mathematical models to brain function in an effort to elucidate mechanisms.

From all this come several overlapping levels of advancement in the study of this challenging organ. The first level is descriptive. Though not finished for any species and not even begun for many, this necessary level of investigation has produced hundreds of technical terms and shelves of much-labeled drawings. Study of the brain can be tedious when the emphasis is on description, but an understanding of anatomy is essential as a first step in the interpretation of how the brain evolved and how it functions. But even the most carefully prepared anatomical description of the brain's components will not provide all the answers. For some of the anatomically best-known pathways (e.g., that of vision), no part of the system is the exclusive site of any one function. Of all the organs of the vertebrate body, the brain best illustrates "emergent properties": the whole is more than the sum of its individual parts. Bullock reminds us that not only are there a limited number of detailed comparative studies of the vertebrate brain available, but our ignorance is deep when it comes to even understanding the ways in which more complex brains are different from less complex ones. The brain does not merely transmit, reject, or store the information in the 3 billion impulses that reach its 10^{10} cells every waking second. It transforms the information, adapts it, and chooses among alternative responses in ways that surpass present comprehension.

DEVELOPMENT OF THE BRAIN

By the time the neural folds close over the neurocoel in the later stages of neurulation, the future brain is already of greater diameter than the spinal cord. As soon as the tube is formed, the developing brain expands at three levels to form vesicles separated by constrictions. These **primary vesicles** are the **forebrain** or **prosencephalon, midbrain** or **mesencephalon,** and **hindbrain** or **rhombencephalon** (Figure 18.2). The prosencephalon lies anterior to the notochord; the other vesicles are dorsal to the notochord.

At the next stage of development, additional constrictions divide the brain further into five **secondary vesicles.** The anterior part of the prosencephalon becomes the **telencephalon,** largely through expansion of its lateral walls. These expansions will form the **cerebral hemispheres** of the adult. The posterior part of the prosencephalon becomes the **diencephalon.** The mesencephalon remains undivided. The rhombencephalon forms an anterior **metencephalon,** which forms the adult **cerebellum,** and a posterior **myelencephalon.** ("Encephal" = *brain;* most of the prefixes indicate position.)

Divisions between certain of these secondary vesicles are slight, and there is little functional and evolutionary basis for recognizing them. Nevertheless, it is a convenience to divide the brain into these parts for purposes of instruction. The embryonic brain vesicles are related to the basic structure of the adult amniote brain in Figure 18.3.

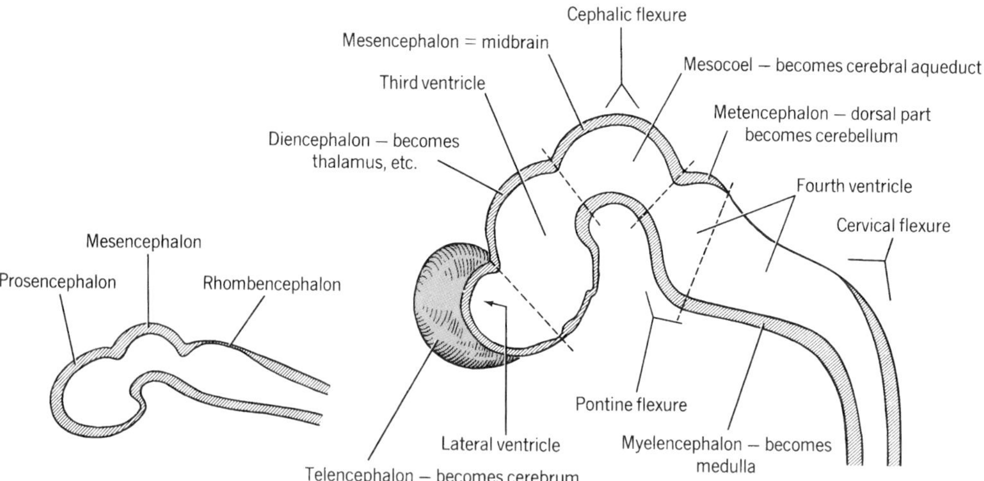

FIGURE 18.2 DEVELOPMENT OF THE MAMMALIAN BRAIN. Stage of primary vesicles on left; stage of secondary vesicles on right. Brains cut in the sagittal plane.

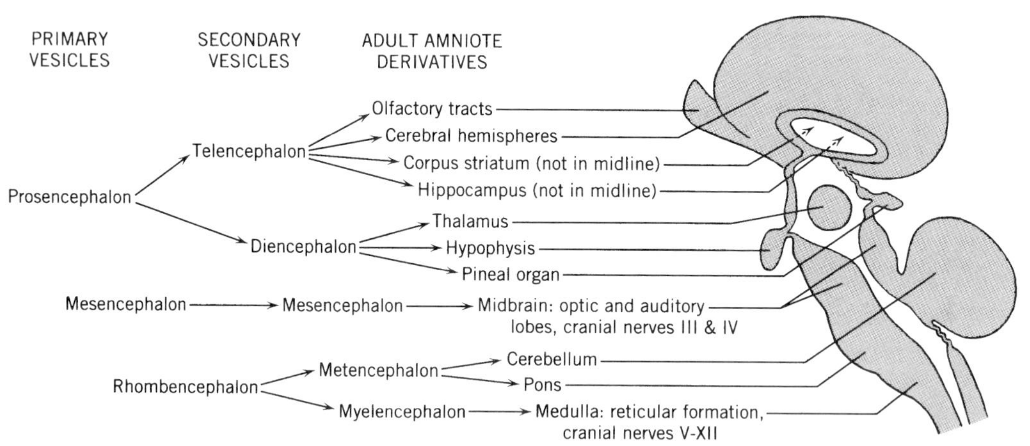

FIGURE 18.3 BASIC ORGANIZATION OF THE ADULT AMNIOTE BRAIN in relation to embryonic brain vesicles. Compare with Figures 18.7 and 18.12.

The embryonic neurocoel is larger in the brain than in the cord. Within the vesicles it forms expansions called **ventricles.** The **lateral ventricles** occupy the cerebral hemispheres. If, as in fishes, the hemispheres are partly joined, then they share a common ventricle. The **third ventricle** is in the diencephalon. The neural canal expands within the mesencephalon of most vertebrates, but it is relatively restricted and tubelike in mammals and is then called the **cerebral aqueduct.** The **fourth ventricle** is located in both the metencephalon and the myelencephalon.

Most brains have nearly straight axes. Brains of birds and mammals, however, acquire three **flexures** in the embryo. The sharpness of each flexure depends on posture and cranial architecture, and does not relate directly to function. The **cephalic flexure** is in the mesencephalon and is concave ventrally. The **pontine flexure** is in the part of the metencephalon called the pons and is concave dorsally. The **cervical flexure** is within the posterior part of the myelencephalon and again opens ventrally. This flexure is most evident in bipeds that hold the head at an angle to the neck.

It is useful to divide cross sections of the tubelike embryonic brain (and cord) into quadrants. The dorsolateral quadrants are called **alar plates** and the ventrolateral quadrants **basal plates.** In the cord, the dorsal gray columns (see figure on p. 306), with their interneurons and axons of sensory neurons, develop from the alar plates, whereas the ventral gray columns, with their motor neurons, develop from the basal plates. In the brain, alar and basal plates become discontinuous, yet (as explained further below) they form neurons corresponding to their counterparts in the cord. The basal plate terminates at the diencephalon.

The three tissue layers of the embryonic spinal cord (ependymal, mantle, and marginal) are also present in the brain. The mantle layer is thick, and in the cerebral hemispheres and cerebellum of amniotes most of its inner cells migrate peripherally into the marginal layer. The vacated region will thus become white matter, and the invaded surface of the brain will become gray matter. This is the reverse of the spatial relationship elsewhere in the central nervous system, where gray matter is internal. Myelination is a slow process. It does not start until the fiber tracts are well established and is not finished until well after birth.

As in the cord, fibers grow away from their respective cell bodies. The specialized tips, or growth cones, of developing axons recognize and follow the correct pathways guided by a variety of specific molecules provided by cells located along the way. The target organ may also release molecular cues. As networks mature, stimulation and activity are required for their normal completion.

In old age, cells of the human brain shrink or die, causing the brain to lose weight.

MORE ABOUT THE ORGANIZATION OF THE BRAIN

In the previous section the brain was divided into five regions on the basis of development. Another division into three regions—**brainstem, cerebellum,** and **cerebrum**—is also useful. The central axis of the brain is the brainstem. It is the first region to form in ontogeny, the least variable, and the most like the spinal cord in structure. It receives all cranial nerves except the atypical terminal and olfactory nerves, relays impulses to the other two regions, and independently controls various vegetative functions of the body. Part of the adult metencephalon and all of the diencephalon, mesencephalon, and myelencephalon are included in the brainstem. The adult mesencephalon is usually called the **midbrain** and the adult myelencephalon the **medulla.**

The cerebellum and related pons (if present) are the principal adult derivatives of the metencephalon. They contribute to the coordination of motor functions. The cerebellum of amniotes and of some fishes is a conspicuous appendage of the brainstem covering much of the posterior part of its dorsal surface.

The cerebrum is the adult derivative of the telencephalon. Gradually this region of the brain enlarged and added new parts and functions until, in mammals, it came to dominate the brain in both size and control.

Within the spinal cord, nerve cell bodies mass in the dorsal and ventral gray columns, and these are continuous throughout the length of the cord. In the brain, by contrast, functionally related nerve cell bodies either mass at the surface of the cerebrum or cerebellum, where they make up the **cortex** of those organs, or cluster in discontinuous masses within the brain. Such a cluster is usually called a **nucleus** but may also be termed a center or body (see Figures 18.4 and 18.12).

Similarly, within the cord, bundles of functionally related nerve fibers are usually called tracts. In the brain, such a bundle is also commonly called a tract, but unfortunately may have another name (fascicle, capsule, brachium, peduncle, lemniscus) depending on size, shape, and relationships.

Sensory fibers enter the brain from the cord and cranial nerves, and terminate in a nucleus of the brainstem or in the cortex of the cerebellum. Incoming impulses are usually passed from a first nucleus in the brainstem to one or more other nuclei or to the cortical

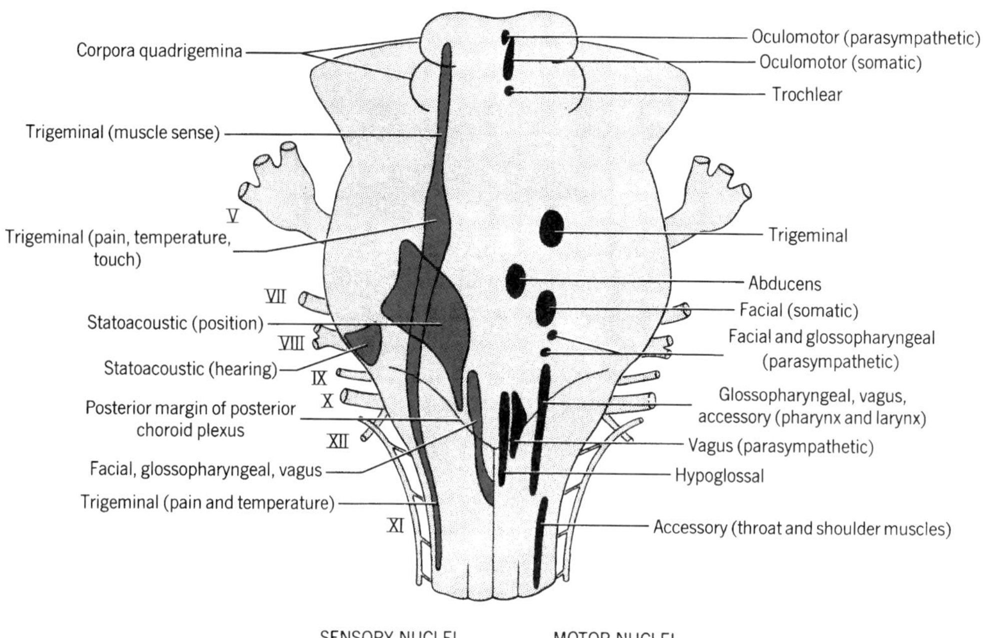

FIGURE 18.4 NUCLEI OF CRANIAL NERVES IN THE BRAINSTEM OF THE HUMAN. Dorsal view. Labels show nerve affiliations of nuclei, not their technical names.

areas (including the cerebral cortex). Each incoming impulse "comes to the attention of" many (usually thousands) of neurons in the gray matter of the brain. In this process of "considering" messages received, the nuclei are not mere relay stations for transmitting impulses; each is also an association center that processes, alters, and redistributes the messages in some way. The more complicated and the less automatic the ultimate response, the more likely that relevant impulses will reach the anterior part of the brainstem and cerebrum. Messages passing posteriorly may also be processed in "lower" nuclei before exiting from the brain in the spinal cord or cranial nerves. Furthermore, tracts called **commissures** pass from one side of the brain to *corresponding* parts on the other side. These permit the integration of sensations and learning experiences from the two sides of the body. Crossings of sensory or motor fibers from one side of the brain or cord to different parts on the other side are termed **decussations.**

POSTERIOR BRAINSTEM: MEDULLA THROUGH MIDBRAIN

Nuclei of Cranial Nerves It is again useful to refer to structure of the spinal cord as a frame of reference. There, somatic motor neurons are derived from the embryonic basal plates, whereas other neurons are from the alar plates. The same is true in the brainstem. In the cord, there is a more or less dorsal-to-ventral linear arrangement in the gray columns of somatic sensory, visceral sensory, visceral motor, and somatic motor neurons. This linear arrangement is roughly preserved in the brainstem. However, spreading of the dorsal parts of the alar plates of the medulla to accommodate the fourth ventricle causes the orientation of the linear series to change from dorsal–ventral to somewhat lateral–medial (see figure on p. 311). Also, these neurons are pushed up close to the floor of the fourth ventricle as other structures come to occupy the ventral part of the medulla. Furthermore, these components are continuous in the gray columns of the spinal cord but are broken up into discontinuous nuclei in the brain. Finally, nuclei without counterpart in the cord are present in the brainstem to relate to the "special" senses of the head. These shoulder their way into position near their nonspecial equivalents.

It follows from this general plan that each cranial nerve tends to have a nucleus in the brainstem for each kind of component fiber it carries, and that, although corresponding nuclei of different nerves are at different anterior–posterior positions in the brainstem, they tend to be at roughly equivalent lateral–medial and dorsal–ventral positions.

These general relationships are useful for interpreting the anatomy of the nuclei of cranial nerves in specific animals, but nature contrives to complicate matters somewhat: One nucleus may serve two or more adjacent nerves, and one ancestral nucleus may split into several nuclei. Some nuclei are long and some short, some large and some small. Jostled by neighboring nuclei and tracts, a nucleus may stray somewhat from its postulated ancestral position.

The oculomotor and trochlear nerves join the midbrain, and their nuclei are confined there. Nerves V to XII join the medulla, and most of them have nuclei there. Nuclei of the more anterior of these nerves commonly also (or instead) are located in the metencephalon, and a nucleus of the trigeminal nerve reaches forward to the midbrain.

To name the various nuclei of cranial nerves and to compare their numbers and positions in different vertebrates would not serve the objectives of this book. However, several representative nerve-nucleus relationships will be noted to illustrate the structural plan already presented: The oculomotor nerve has a medially situated nucleus for its somatic motor neurons and another, smaller nucleus for its autonomic (visceral motor) fibers (Figure 18.4). The single nucleus of the trochlear nerve is found, as expected, just posterior to the somatic motor nucleus of the oculomotor nerve, even though the nerve itself exits from the brain at an unaccountably dorsolateral position. The large trigeminal nerve has a long somatic sensory nucleus in the expected dorsolateral part of the medulla. The visceral motor fibers of its mandibular branch (which innervate jaw muscles of branchial origin) have a midbrain nucleus that is relatively large in animals that chew their food. The facial nerve, having several kinds of component fibers, has several nuclei in the brainstem. The statoacoustic nerve has separate "special" nuclei (or clusters of nuclei) for its vestibular and acoustic branches. The spinal accessory nerve shares one of its nuclei with the vagus, from which it is derived.

Reticular Formation The reticular formation occurs in all vertebrates. It is derived from the basal plates and is located in the central part of the brainstem from midbrain to medulla and, in reduced form, also in the anterior part of the cord (see Figure 18.12, lower right). It consists of a diffuse, interlocking mass of nerve cell bodies and fibers; hence it is a mixture of gray and white matter. Its boundaries are indistinct (Figure 18.6). More or less sharply defined clusters of nuclei tend to form in its substance at several levels of the brainstem.

The reticular formation has sensory input from virtually all parts of the body and all senses. It in turn projects to the cerebrum, cerebellum, various cranial nuclei, and the cord. It is essential for consciousness. Stimulation awakens a sleeping animal and makes an awakened animal more alert. It contributes to the activities of both voluntary and involuntary muscles by facilitating, inhibiting, screening out noise, and coordinating stimuli. It also contributes to control of the cardiovascular and respiratory systems.

Other Nuclei of the Posterior Brainstem There are numerous paired nuclei in the posterior brainstem other than those relating directly to cranial nerves. Several of the more prominent ones will be mentioned. The **olivary nuclei** arise from the alar plates, yet during development migrate down to the ventrolateral wall of the medulla. This complex of nuclei is present in all vertebrates. It is best developed in the most active representatives—mammals, birds, and some fishes. The largest component is the **inferior olive,** which may form a low bulge on the sidewall of the medulla (Figures 18.5, left, and 18.6, below). In

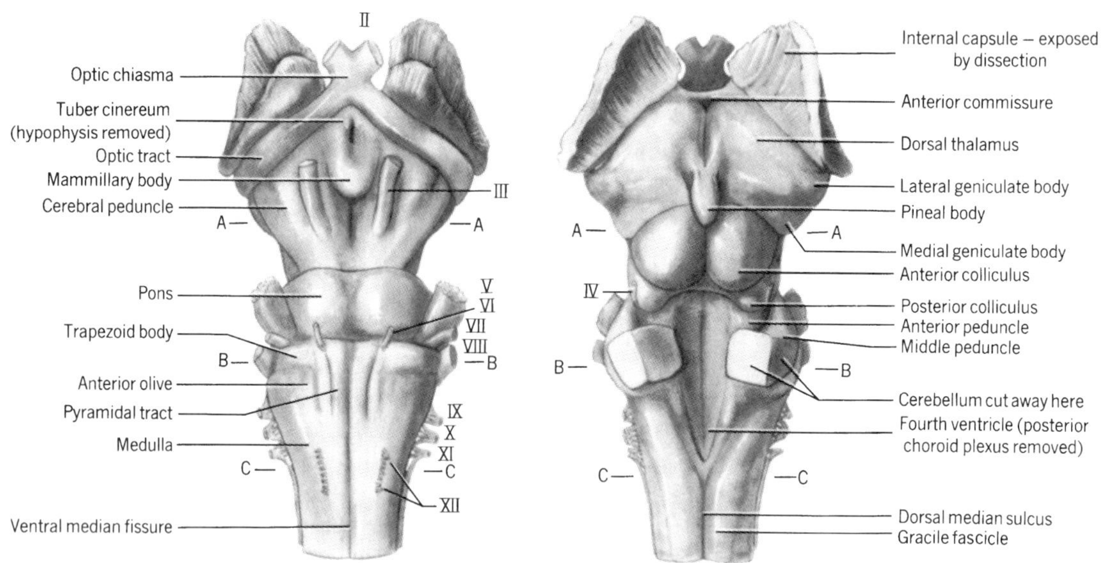

FIGURE 18.5 BRAINSTEM OF THE COW in ventral (left) and dorsal (right) views. Cerebrum and cerebellum removed by dissection. Letters show the levels of the cross sections in Figure 18.6. (The brain of the sheep, commonly studied in the laboratory, is similar to the brain of the cow.)

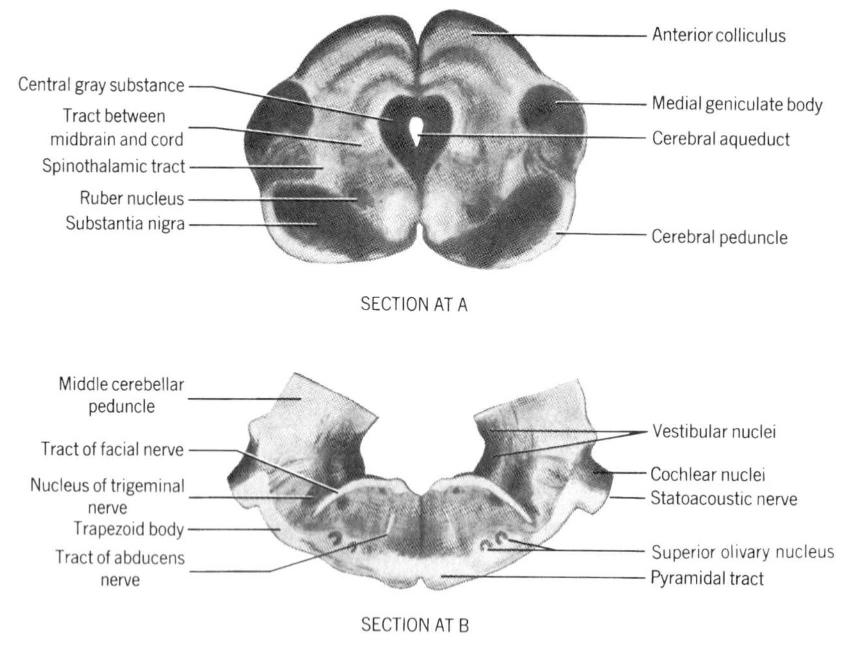

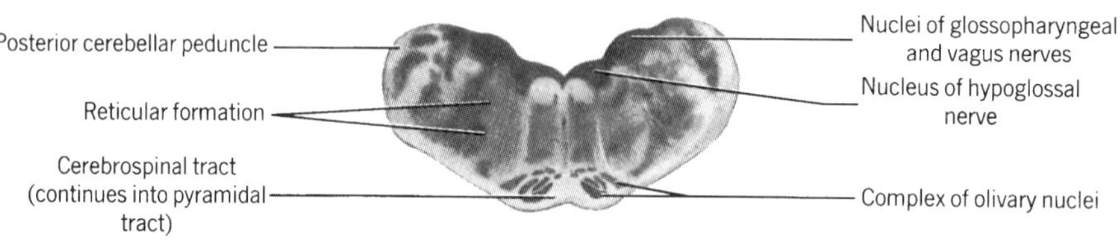

FIGURE 18.6 CROSS SECTIONS OF THE BRAINSTEM OF THE COW at the levels shown by letters in Figure 18.5. The contrast between nuclei and tracts has been enhanced by staining.

humans the inferior olive, which is a motor coordination center, has the size and crumpled shape of a raisin, receives interneurons from some other nuclei of the brainstem and proprioceptive (muscle sense) impulses from the spinal cord (and some other input), and projects dorsally to the cerebellum.

The **ruber** (or red) **nucleus** and **substantia nigra** (which is pigmented in some mammals) are located deep in the mesencephalon (and may extend into the diencephalon) (Figure 18.6, above). The ruber nucleus is present in all vertebrates but is most developed in mammals, whereas the substantia nigra first appears in reptiles and is best developed in primates. Each nucleus may have evolved from the reticular formation. In man the ruber nucleus resembles a pea and the substantia nigra a lima bean. These are relay stations between the forebrain, on one hand, and posterior brainstem and cord, on the other. The ruber nucleus plays a role in coordination of motor functions, particularly of flexor muscles. The substantia nigra is involved with the memory of learned tasks. Furthermore, death of its cells is associated with Parkinson's disease.

The roof of the midbrain is called the **tectum** (= *covering*). In all vertebrates except mammals it consists primarily of a bilateral pair of conspicuous hemispherical eminences termed **optic lobes** (Figure 18.13). With mammals again excepted, the optic lobes are the primary center for the perception of vision (though even in fishes, the retina, diencephalon, and other areas of the mesencephalon are also involved in this complicated sense). The optic lobes are distinctly striated internally, and the visual image is projected onto the lobes point for point. The lobes are the most prominent feature of the brains of fishes that locate food by sight. They remain conspicuous in reptiles and birds.

The perception of vision is largely transferred by mammals to the cerebrum, though the midbrain tectum still has an important function: It tells the mammal where in space a visual object is, whereas the cerebrum tells what the object is. The much smaller optic lobes are here called **anterior colliculi** (Figure 18.5). Behind are the **posterior colliculi,** which are possibly involved in coordination of auditory reflexes. Taken together, these four bumps are the **corpora quadrigemina.** There are commissures between the lobes of a pair.

Some of these brainstem nuclei, and others not named, are sufficient to support various vegetative functions in normal circumstances. Even with other parts of the brain removed, experimental animals maintain heart beat, respiration, swallowing, and digestion.

Additional Features of the Posterior Brainstem The brainstem transmits all impulses moving to or from other parts of the brain. Obviously, much of its substance must consist of tracts of nerve fibers. The pathways are complex yet are well known for many vertebrates. Several tracts that are large and can usually be seen on the surface of the brain will be identified here.

Particularly large paired tracts of motor fibers run directly from the cerebral cortex of mammals to the spinal cord without interruption. Similar but less prominent tracts are found in other tetrapods and sharks. In mammals they emerge at the surface of the brain on the ventrolateral wall of the midbrain, where they converge on each side to form the **cerebral peduncles** (= *small feet*, hence supports). The fibers are lost from surface view in the metencephalon but continue in most mammals as the prominent **pyramidal tracts** on the ventral surface of the medulla as they move to the lateral and ventral funiculi of the cord (Figure 18.5).

The **trapezoid body,** which may be evident externally in mammals, transmits fibers from the statoacoustic nerve to the superior olive and elsewhere. Large paired columns of sensory fibers associated with light touch and pressure enter the medulla from the dorsal funiculi of the spinal cord. These **gracile** and (more lateral) **cuneate fascicles** flank the dorsal median sulcus of cord and posterior medulla until they terminate in the **gracile** and **cuneate nuclei.** These nuclei project forward to the diencephalon.

The cerebellum, which arches over the fourth ventricle, joins the brainstem by three pairs of peduncles (or brachia) that flank the ventricle. If the cerebellum is large, the peduncles are then prominent surface features. The **posterior peduncle** carries fibers between the cerebellum and spinal cord, inferior olive, and vestibular nuclei. The **middle peduncle** of mammals is the largest and most lateral peduncle, partly hiding the others (Figure 18.5). It transmits fibers between the two sides of the cerebellum and from cerebrum to cerebellum. The **anterior peduncle** has fibers joining the cerebellum to the ruber nucleus and diencephalon.

The roof of the fourth ventricle is covered by the **posterior choroid plexus.** This is a delicate, much-pleated, highly vascular membrane to which the brain contributes only its thin ependymal layer. The function of choroid plexuses is described at the end of this chapter.

ANTERIOR BRAINSTEM: DIENCEPHALON

The evolution of the diencephalon and telencephalon continues to attract much interest. (For a recent series of papers devoted to this topic, see Braford in References.) The anterior part of the brainstem differs from the posterior part in being derived entirely from the embryonic alar plates, in having no nuclei for cranial nerves and no reticular formation, and in relating to functions that are more highly evolved. The embryonic telecephalon forms the cerebrum, which in most vertebrates is bilaterally paired and has no structures in the midline except for several commissures and the thin anterior wall of the ventricle. Consequently, the diencephalon forms the most anterior part of the brainstem. It is convenient to divide the diencephalon into three parts.

The narrow dorsal part of the diencephalon is the **epithalamus,** much of which is nonnervous in function. Most anterior is the **anterior choroid plexus** of the third ventricle (Figure 18.7). Posterior to the plexus is an evagination of similar structure (the paraphy-

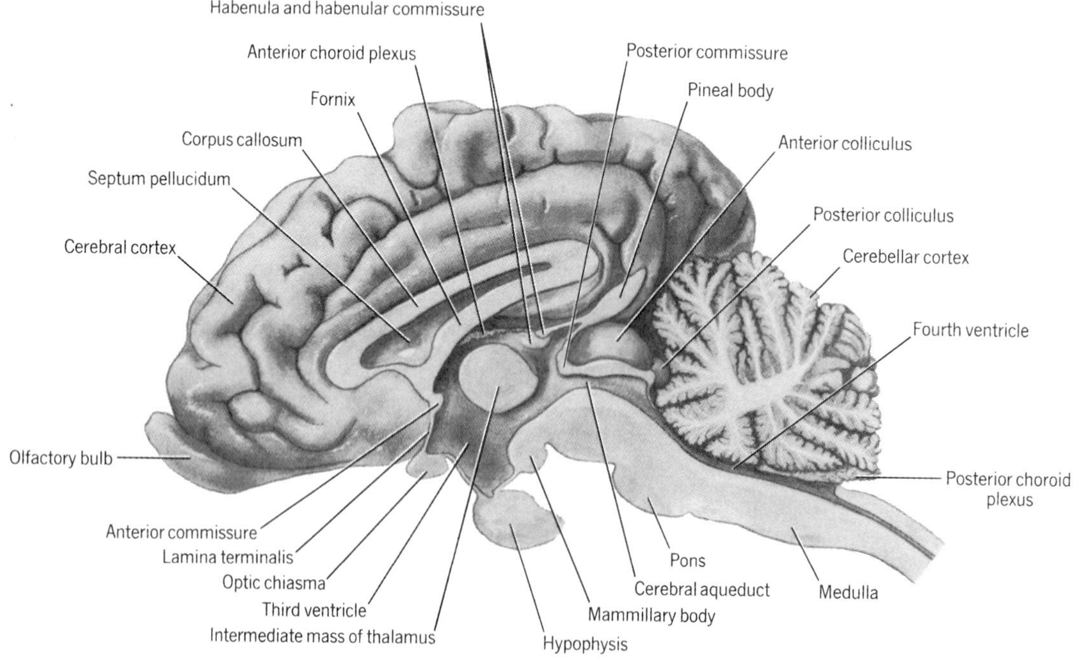

FIGURE 18.7 SAGITTAL SECTION OF THE BRAIN OF THE COW.

sis), which, however, does not mature in humans. Next in line are the **habenular nuclei** or habenulae (which are often of different size on the right and left). They are present in all vertebrates and contribute to the coordination of olfactory reflexes. Behind the habenulae are two evaginations, the **parietal organ** and **pineal body.** These structures do not lend themselves to the organization of chapters by organ system. They are mentioned here because of their origin and relationships, in Chapter 19 because each may function as a sense organ, and in Chapter 20 because one may function as an endocrine gland. The habenulae have a small commissure, and a larger **posterior commissure** marks the posterior boundary of the epithalamus. Its fibers join some nuclei of the two sides of the diencephalon and perhaps of the mesencephalon as well.

Each thick lateral wall of the diencephalon is called a **thalamus.** This is the largest part of the anterior brainstem, being 4 cm long in humans. The two thalami are bordered in part by the lateral ventricles and are separated from one another by the third ventricle, except where the large **intermediate mass** or middle commissure spans that cavity (Figures 18.7 and 18.8). Each thalamus is a compact oblong mass of many nuclei—30 or more are recognized. A ventral cluster of nuclei (**ventral thalamus**) is represented in all vertebrates and projects motor fibers posteriorly in the brain. A dorsal cluster (**dorsal thalamus**) projects sensory pathways (except olfactory pathways) to the cerebrum (to both its striatal and cortical parts, as defined below). The dorsal nuclei are most developed in tetrapods—particularly in mammals.

Two of the dorsal nuclei are of particular importance. These are located at the posterior end of the thalamus close to the tectum of the midbrain. One is the **medial geniculate body,** which forms part of the auditory pathway and is joined by a pathway to the posterior colliculi of the tectum (Figures 18.5 and 18.6). The other is the **lateral geniculate body,** which is a relay station in the primary visual pathway. Seen in cross section, the lateral geniculate bodies of some mammals (like the optic lobes of many vertebrates) are striated into about six conspicuous layers. The visual field is projected onto these layers.

The thalamus is a relay center to the cerebrum and also is the last center short of the cerebrum where bodily functions are modulated. There is evidence that in several classes of vertebrates a level of awareness, including perception of both pain and pleasure, is located in the thalamus.

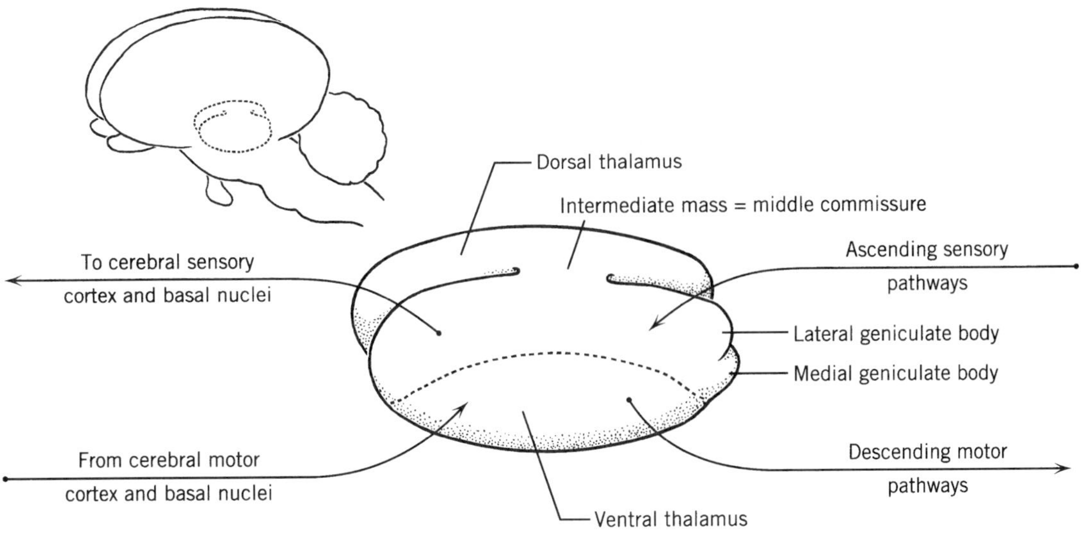

FIGURE 18.8 MAMMALIAN THALAMUS seen from the left and a little above.

The ventral part of the diencephalon is the **hypothalamus.** It contains about a dozen pairs of nuclei that together integrate and largely control the autonomic functions of the body, including water balance, temperature regulation, appetite and digestion, blood pressure, sleep and waking, sexual behavior, and emotions. Consistent with the dual nature of the autonomic system, each function has two centers in the hypothalamus—one for facilitation and one for inhibition.

On the ventral surface of the hypothalamus is the **optic chiasma,** where the optic nerves converge and cross (usually with partial decussation of their fibers) before continuing up the sides of the brain as the **optic tracts** (Figures 18.5 and 18.7). The optic tracts terminate in the lateral geniculate bodies. Just posterior to the optic chiasma is an area called the **tuber cinereum,** which encloses several nuclei of the parasympathetic system and subtends the **hypophysis.** The important hypophysis is in part of nervous origin but functions (partly under the influence of nuclei in the hypothalamus) as an endocrine gland. Accordingly, it is described in Chapter 20. Posterior to the hypophysis is a pair of small but evident lumps called the **mammillary bodies.** Within are nuclei of the same name that function in olfaction. These are present in all tetrapods and may be represented in fishes.

CEREBELLUM AND PONS

The cerebellum is an ancient part of the brain, yet has enlarged and changed in the course of evolution much more than has the brainstem. It follows that it is a variable part of the brain and that portions of the cerebellum of mammals are relatively "new." The cerebellum develops from the dorsal part of the metencephalon and lags a little behind the development of the brainstem.

The cerebellum of cyclostomes and amphibians is small and smooth. Because it is merely a thickening of the wall of the brain tube, it has no cavity. The cerebellum is usually still smooth in fishes and reptiles, yet in them it is prominent and encloses part of the fourth ventricle. In birds and mammals the organ is very large, lobed, and convoluted into tight **gyri** (convex folds) and **sulci** (concave grooves) (Figure 18.9). Its solid walls then nearly exclude the fourth ventricle.

Unlike the spinal cord and brainstem, the gray matter of the cerebellum is in a thin superficial cortex. The central white matter branches to each lobe and gyrus. Because of its appearance in longitudinal section, this branching white matter is called the **arbor vitae** (= *tree of life*).

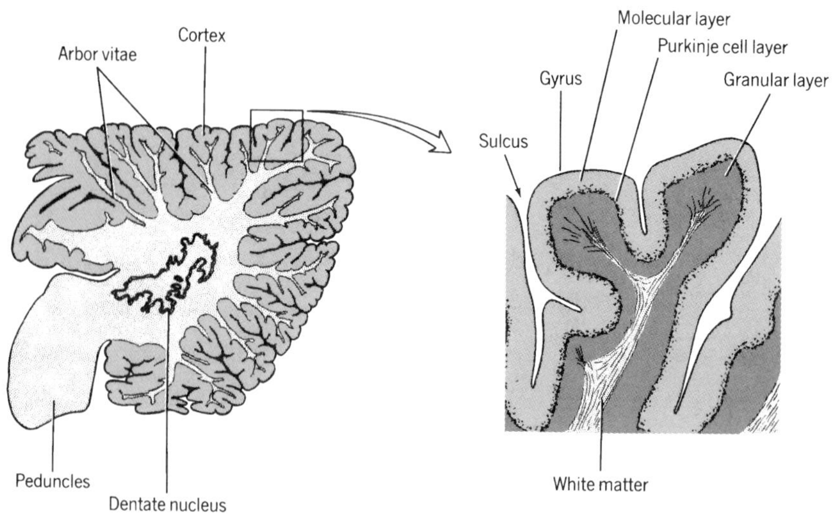

FIGURE 18.9
PARASAGITTAL SECTION OF THE HUMAN CEREBELLUM.

The cerebellar cortex of all vertebrates is divided histologically into three regions: a deep **granular layer,** which has several types of cells; a middle **Purkinje cell layer;** and a superficial **molecular layer,** which has scattered cells but consists mostly of fibers and synapses. Afferent impulses are relayed twice within the cortex before reaching the large and exceedingly branched Purkinje cells (see the most branched neuron of the figure on p. 300). The arrangement of cells in the cortex assures that impulses will spread widely.

The cerebellum functions to control motor coordination and to maintain equilibrium. It does not initiate motor activities, but processes those initiated elsewhere. It provides unconscious timing and integration of muscles that must contract together or in sequence, and of antagonistic muscles that must simultaneously relax. If the cerebellum is damaged, muscle activity loses control and precision. Some memory trace circuits may also be located in the cerebellum.

To do its job, the cerebellum needs extensive input from the sensory system. In primitive vertebrates the cerebellum has two parts. One is the **archicerebellum.** Its principal input comes from the labyrinth of the ear and the lateral line system by way of associated nuclei in the brainstem, though pathways from the cord may also exist. The archicerebellum is retained in amniotes as the auricular, or floccular, lobes of the more evolved organ (Figure 18.15). The other part is the **paleocerebellum,** which is located in the midline of the organ. Its input is largely from proprioceptor (muscle sense) organs of the trunk and is relayed by the spinal cord and inferior olivary nucleus. This part of the cerebellum is important for the control of swimming in fishes. Finally, when appendicular muscles became more prominent with the evolution of amniotes, and the cerebral cortex became larger and more dominating, the cerebellum responded by evolving an associated **neocerebellum.** Though not sharply set off, the neocerebellum forms the hemispheres of the organ. There is also input to the cerebellum from the senses of touch, sight, and hearing, and from the reticular formation.

The cerebellum has fewer efferent than afferent fibers. Outgoing impulses originate in the Purkinje cells and are relayed by **cerebellar nuclei.** The location of these nuclei is varied and complex in anamniotes. They are located in the base of the larger cerebellum of amniotes, and in mammals split to become three or (in primates) four separate pairs of nuclei. The largest and most distinctive is the **dentate nucleus,** which (like the inferior olive of the medulla) has a crumpled contour. There is a topographic relationship between the cerebellar nuclei and the regions of the cerebellar cortex. Efferent pathways run from these nuclei to the vestibular nuclei, reticular formation, ruber nucleus, and dorsal thalamus.

Afferent and efferent pathways between the cerebellum and cerebral cortex evolve as the newer, more dominant parts of the cerebral cortex evolve. Associated with this change is the appearance in some birds, and the prominence in mammals, of the **pons.** This addition to the ventral part of the brainstem at the level of the cerebellum receives in its **pontine nuclei** fibers from the cerebral cortex and relays impulses to the cerebellum through the middle peduncles. (The peduncles supporting the cerebellum were described above with the posterior brainstem.) The pons also transmits impulses from one side of the cerebellum to the other.

CEREBRUM

General Structure The cerebrum is the adult telencephalon. The most anterior part of the cerebrum is always bilaterally divided. Nearly all of the organ is divided in amniotes, and each half is then called a cerebral hemisphere.

The cerebrum is like the cerebellum in that it is of ancient origin yet enlarges and changes in the course of evolution. However, the changes are here more pronounced and do not form a single progression but instead follow separate trends leading to different end points in teleosts, birds, and mammals. The phylogeny of the cerebrum is one of the most striking in vertebrate evolution.

At the anterior end of each hemisphere is the **olfactory bulb,** which receives the olfactory nerves. The remainder of each hemisphere is divided into two principal parts: The **corpus striatum** is in a ventral position. It is composed of various prominent nuclei and often swells to become large and globular. The **cortex,** or pallium, forms the roof and sidewalls of the cerebrum. Its association centers tend to be spread out into a sheet. In vertebrates other than actinopterygians, the corpus striatum and cortex are in part separated by the lateral ventricles. Where cortex and corpus striatum merge, the boundary may be indistinct. Homologies of the various nuclei are sought using structural, comparative, functional, and histochemical evidence. The cerebrum is completed by various tracts and commissures. These parts will be discussed in order.

Olfactory Bulb and Tract Axons of the conductive receptor cells of the nasal epithelium enter the olfactory bulbs (Figure 18.7). There they converge to synapse with two kinds of neurons within ball-like tangles of nerve endings called **glomeruli.** Axons of these second-level neurons enter the olfactory tracts, which conduct impulses out of the bulbs. Another kind of neuron provides local feedback circuits within the bulbs. There are fewer fibers in the outgoing tracts than in the incoming nerves, so the bulbs accomplish a summation of impulses. Third- and fourth-level neurons elsewhere in the brain are also involved in the complicated processing of olfactory input.

The size of these structures relative to the remainder of the brain varies over a wide range according to the importance of olfaction in the life of the animal (Figure 18.15).

Corpus Striatum and Basal Nuclei The corpus striatum (= *body* + *striped*) is named for the appearance of part of its substance in some vertebrates as seen in section. For mammals, **basal nuclei** is an approximate synonym. The term **basal ganglia** is also used. The corpus striatum has three principal parts, which are named according to function. It should be recognized, however, that historical misconceptions about sequence in their evolution have given emphasis to their distinctions. Also, the homologies assigned to the parts among the different classes of vertebrates are in some instances tentative.

One part of the striatum is called the **archistriatum** (Figure 18.10). It integrates olfactory and general somatic senses. The archistriatum of fishes consists of several indistinctly segregated nuclei called the **amygdaloid** (= *almond-shaped*) **complex.** Tetrapods retain the structure, and in mammals the corresponding amygdala is a globular mass that tends to be ventral to the other basal nuclei. Even in mammals it remains in part an association center for olfactory input, but also contributes to food intake, arousal, and emotions (including fear) and emotional aspects of memory.

The **paleostriatum** is represented in all vertebrates, with various names assigned to its parts. The homologous basal nucleus of primates is the **globus pallidus** (= *ball* + *pale*), the function of which is noted below.

The name **neostriatum** reflects the long-held belief that this part of the striatum evolved only in amniotes—a view that is no longer held. Some structures to which this name has been assigned may prove not to be homologs (particularly in birds), but the derivative mammalian basal nuclei are the long **caudate** (= *tailed*) **nucleus** (Figure 18.11), which arches over the others, and the **putamen** (= *shell*).

The corpus striatum is the highest integrative center of fishes (though, as we shall see, the cortex should not be discounted). The striatum appears to be enormous in reptiles and is usually considered to be their highest integrative center, though it now seems probable that part of the cortex is in fact incorporated into the "striatum." Paleostriatum and neostri-

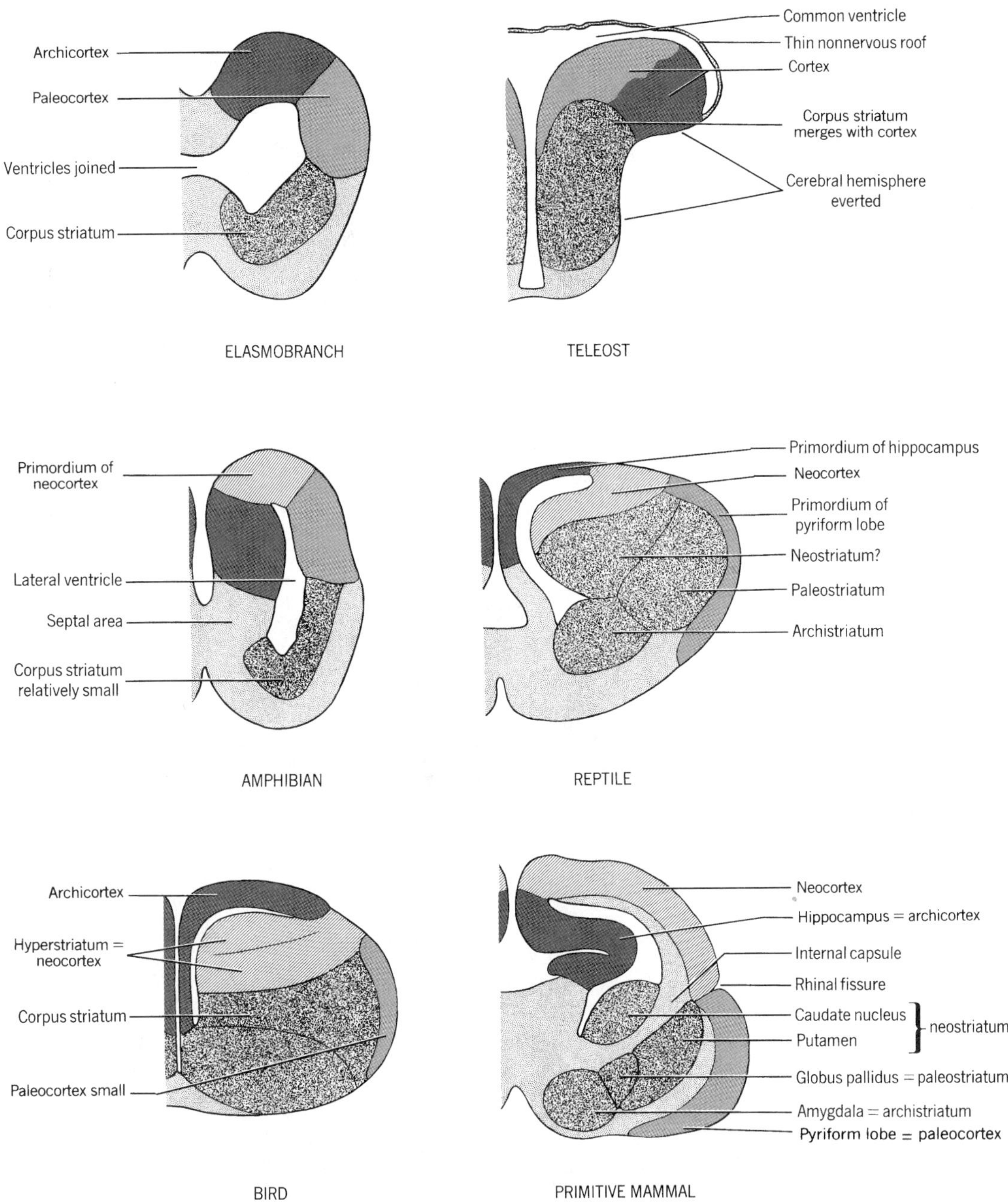

FIGURE 18.10 DIAGRAMMATIC CROSS SECTIONS OF CEREBRAL HEMISPHERES SHOWING THE COMPARATIVE STRUCTURE OF THE CORPUS STRIATUM AND CORTEX.

atum are also large in birds and over these is a unique, thick, four-layered **hyperstriatum.** This is again derived from the cortex.

Considering their prominence, the function of the basal nuclei is poorly known for mammals. The caudate nucleus, putamen, and globus pallidus are here dominated by the cerebral cortex. They receive fibers from the cortex and thalamus. They project fibers to each other and to the thalamus and substantia nigra. Together they are thought to contribute to the function of muscle masses, as opposed to delicate control of individual muscles.

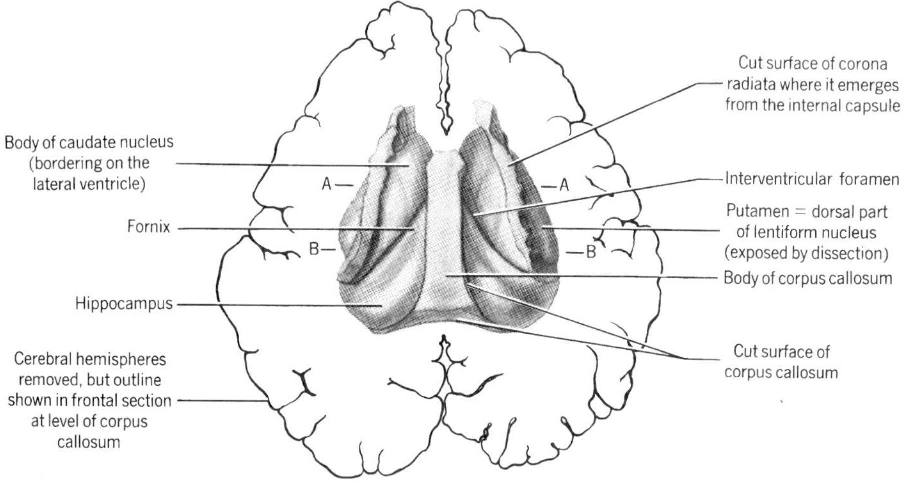

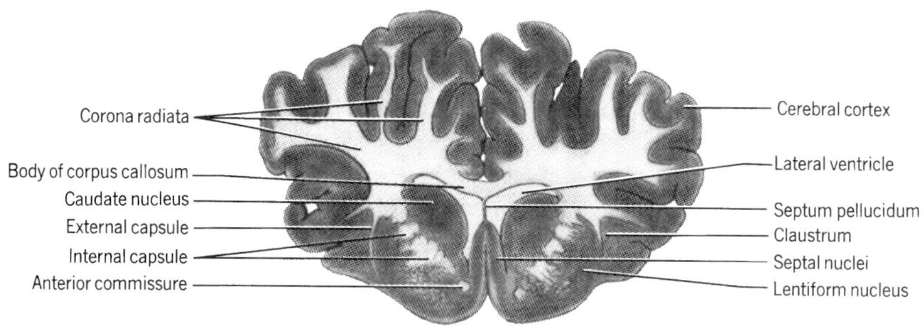

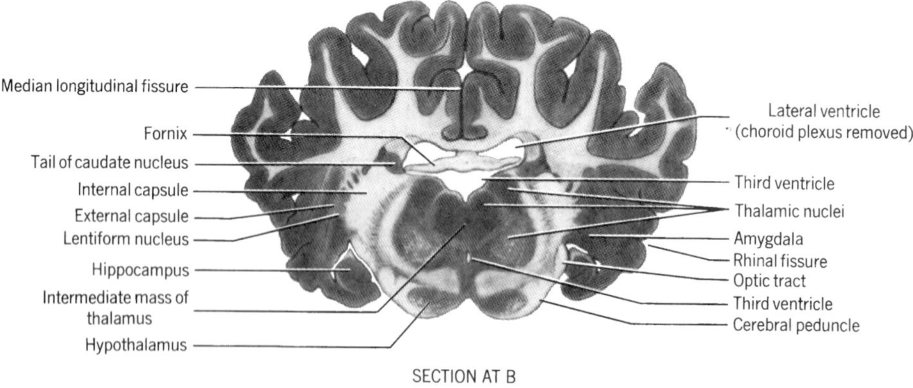

FIGURE 18.11 FOREBRAIN AND THALAMUS OF THE COW shown by a dissection seen in dorsal view (above) and by cross sections (below) made at the levels indicated by letters. On the sections, contrast between nuclei and tracts has been enhanced by staining.

Lesions or deterioration of one or more of the basal ganglia result in specific motor disturbances, recognizable as different from those resulting from lesions of cerebellum or cerebral cortex. Furthermore, the utilization of glucose by the basal nuclei when the eye is stimulated has implicated the nuclei with vision in a monkey.

[Two additional nuclei are commonly named in reports about the mammalian corpus striatum. Although phylogenetically and functionally more closely related to the caudate nucleus, the putamen may merge structurally with the globus pallidus to form the lentiform (= *lentil-shaped*) nucleus. Also classed as a basal nucleus is the flat, laterally placed claustrum. It may in fact be derived from the cortex rather than the corpus striatum. Although it is present in reptiles as well as mammals, its function is not clear.]

Cortex Three principal parts of the cortex are recognized: **paleocortex, archicortex,** and **neocortex.** Except in actinopterygians and mammals, the paleocortex is lateral to the ventricle, and the archicortex is dorsal or median to the ventricle. The neocortex may be between the other parts or, apparently in reptiles and birds, ventral or lateral to the ventricle in association with the striatum.

The paleocortex and archicortex were once thought to relate only to olfaction and to be the only parts of the cortex of fishes. It is now known that much of the cortex of many fishes is *not* olfactory in function. Experiments have shown that parts of the cortex contribute to schooling, aggressive, and reproductive behavior and make learned responses possible. Accordingly, the neocortex is probably also ancient, though its limits are not clear at the fish level of evolution.

The architecture of the cerebrum of actinopterygians is unique. Cortex and corpus striatum are thick, merge together, and lie lateral and ventral to a common ventricle. Paleocortex and archicortex are certainly present, as is probably also the neocortex.

In reptiles and birds a thin cortex is stretched wide to arch over the enlarged corpus striatum. As noted above, however, it now seems likely that part of the large striatum (the hyperstriatum portion in birds) is really neocortex (see Medina and Reiner in References). Birds do not seem to be seriously inconvenienced by the surgical removal of their thin superficial cortex, though there are indications that some capacity for memory is sacrificed. Contrary to common belief, some birds are more intelligent than many nonprimate mammals, and it has been proven that their ability to solve problems, and to remember how to solve new problems of a related kind, resides in the hyperstriatum.

In mammals the paleocortex and archicortex are wedged apart as the evolving neocortex enlarges between them. Anteriorly, the paleocortex includes the olfactory tracts. Posteriorly it is pushed down around the sidewall of the hemisphere to the position flanking the anterior brainstem. The resulting **pyriform** (= *pear-shaped*) **lobe** is prominent in mammals having a keen sense of smell. It is separated from the neocortex above by the **rhinal fissure.** The pyriform lobes are the olfactory cortex.

The archicortex is pushed by the neocortex in the other direction to the crown of the brain near the midline (reptiles, monotremes, marsupials) or over the edge onto the medial wall of the hemisphere (other mammals). As it moves, it rolls on itself lengthwise and sinks largely below the surface, thus forming a long arching band that impinges on the lateral ventricle. Its name, **hippocampus** (= *sea horse*), is suggested by its rolled appearance as seen in cross section (Figures 18.10 and 18.11). The hippocampus of mammals is required for memory of spatial relationships. The disorder of the memory associated with Alzheimer's disease may result from pathology that isolates the hippocampus from other parts of the brain.

Together with the hypothalamus, amygdala, habenula, and mammillary bodies, the hippocampus contributes to the **limbic system,** which functions in aspects of sexual and emotional behavior, memory, learning, and motivation. It seems to regulate more primitive parts of the brain by inhibiting stereotyped behavior, thus permitting accommodation to new circumstances.

All large mammals and some smallish ones have a convoluted neocortex (or cerebral cortex, or simply cortex); the gyri and deep sulci greatly increase total surface area, and hence the number of cell bodies and synapses. Intelligence relates to absolute brain size, relative brain size, and convolutions, but only in a general way; the human brain is not the largest, nor the largest in relation to body weight, nor even the most convoluted.

Gray matter is internal in the cortex of anamniotes but external in that of amniotes. The various parts of the cortex can be distinguished histologically. Characteristic cell layers differ in the density, size, configuration, connections, and staining properties of their neurons. Cells in the six successive layers of the cortex of primates tend to be arranged in vertical columns, each having as many as 100 cells. Small mammals have more cells per unit volume of brain than large mammals.

The primate cortex is divided into regions for ease of reference (i.e., frontal, temporal, parietal, and occipital), and specialists have a standardized way of indicating subregions. In a band across the crown of the human brain, at the posterior margin of the frontal region, is the primary somatic motor cortex (Figure 18.12). Ultimate control of voluntary

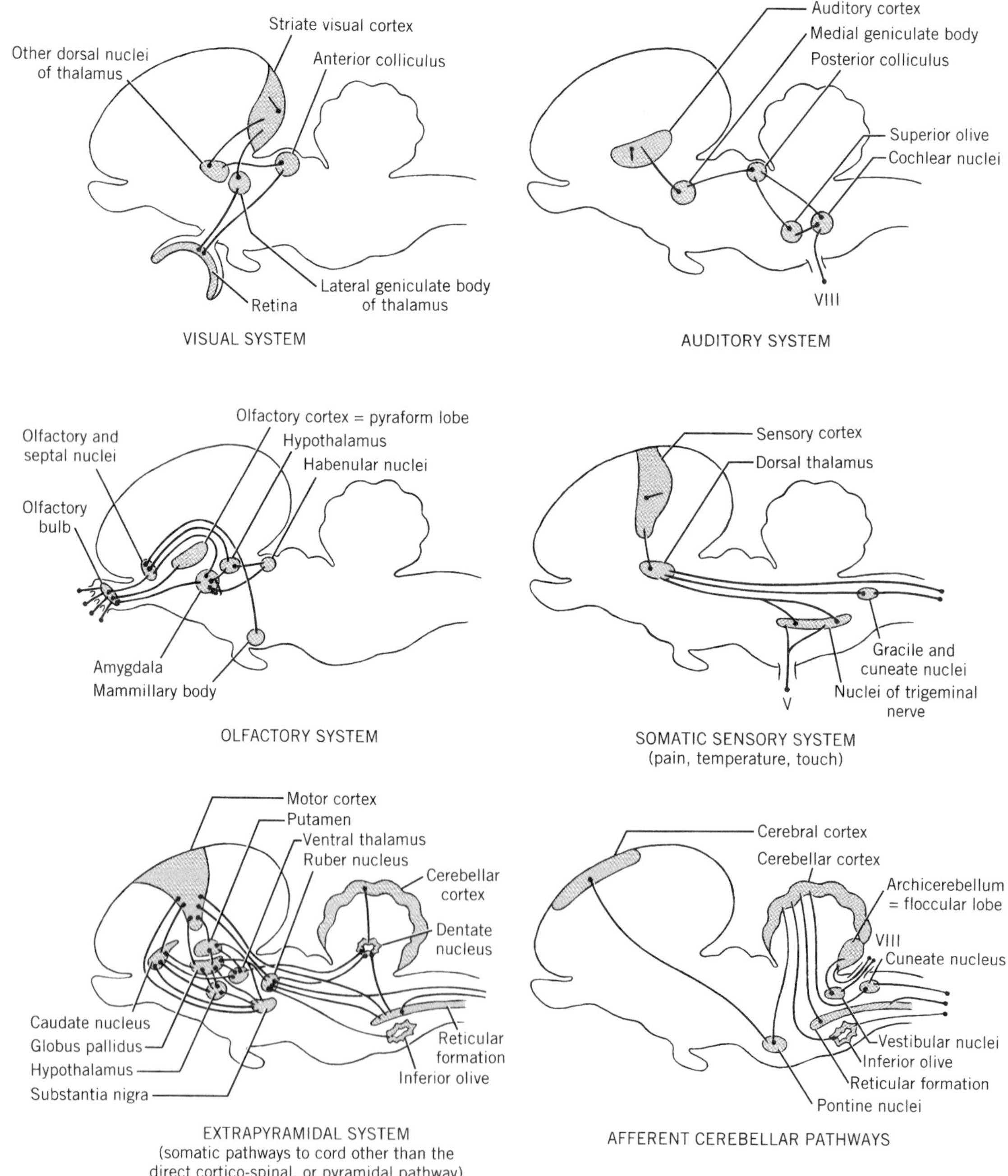

FIGURE 18.12 SCHEMETIC DIAGRAMS OF SOME PRINCIPAL COMPONENTS OF SOME FUNCTIONAL SYSTEMS OF THE MAMMALIAN BRAIN.

motor activity rests here, though this part of the cortex is "coached" and "advised" by other parts of the cortex and by the cerebellum, reticular formation, basal nuclei, ruber nucleus, and other centers.

Just behind the somatic motor cortex, at the anterior margin of the parietal region, is the primary somatic sensory cortex. Similarly, the primary auditory cortex is at the top of the temporal lobe, and the primary visual cortex (which develops in mammals as the optic lobes of the midbrain relinquish most of their function) is in the occipital region. These areas have a point-for-point relationship with function (though in some mammals there is overlap among adjacent points): Motor control of the thumb is at a specific place just below the forefinger point and above the neck point, tones are perceived at given points in treble-to-bass sequence, and sight points are arranged in relation to the visual field. If the face is relatively sensitive (as in humans), the corresponding areas of the cortex are relatively large; if the hands are relatively sensitive (raccoon), the hand areas are large.

Although the above is true, it does not go far enough to indicate the complexity of cortical function. For unknown reasons, the two hemispheres in human beings are not symmetrical, either in structure or in their control of various functions. For example, speech, reading, writing, and rational and practical behavior are vested primarily in the left hemisphere (of right-handed persons), whereas artistic ability, intuition, spatial recognition, and body awareness are dominated by the right hemisphere. The primary cortical areas noted above account for most of the cortex of lower mammals (marsupials, insectivores) but for only one-quarter of the human cortex. In the parietal and occipital regions there are secondary and tertiary projection areas. These are not related to function on a point-to-point basis and have fuzzy boundaries. They code messages, store information, combine input, and provide spatial orientation. If they are impaired, a person might see well yet confuse right and left, might walk well yet get lost in a familiar place. Finally, the frontal region of the cortex relates to programs, intentions, orientation to goals, and sequence in the performance of activities. The entire cortex of experimental animals increases in mass if the animals live in a relatively diverse, stimulating, and "enriched" environment.

Some Other Features of the Cerebrum The tracts of the cerebrum are at least as constant as the nuclei. Interneurons, short and long, loop between the different parts of the cortex. The olfactory tracts divide into several paths as they merge into the brain: One crosses to the other olfactory bulb. Another stops in **olfactory nuclei,** which send third-level neurons to the habenula and mammillary bodies of the diencephalon. Still another runs to the amygdaloid nuclei and pyriform lobe. All these structures function in the perception and integration of olfactory stimuli (Figure 18.12).

The conspicuous motor tract that exits from the primary motor cortex runs right through the mammalian corpus striatum as a broad white band called the **internal capsule** (Figures 18.10 and 18.11). It separates the caudate nucleus from the functionally related putamen. It is this band that continues, as the cerebral peduncles and subsequently as the pyramidal tracts, directly into the spinal cord. Collectively this is the **pyramidal system,** as contrasted to the **extrapyramidal system,** which includes the same cortical areas but also other parts of the cortex and efferent relay centers in the corpus striatum, thalamus, and elsewhere.

The cerebrum has several commissures. The small **habenular** and larger **anterior commissures** are derived from the archicortex and appear in the lower classes. Therian mammals have evolved a commissure, the **corpus callosum,** which in all but marsupials is the largest of all (Figures 18.7 and 18.11). It joins one neocortex to the other and is seen in the dorsal midline as a horizontal shelf when the hemispheres are pulled apart. If the corpus callosum is cut, something learned or experienced on one side of the body cannot be adequately used or acted upon by the other. (Curiously, the human corpus callosum varies in size both with gender and handedness.) The **fornix** is a conspicuous arching pathway positioned ventral to the corpus callosum anteriorly and lateral to the hippocampus posteriorly. It transmits fibers between the hippocampus and the hypothalamus.

Each lateral ventricle encloses a large choroid plexus that is rooted at the roof of the diencephalon with the smaller plexus of the third ventricle.

CIRCUITS, VERSATILITY, AND MEMORY

In presenting the basic structure of the brain the preceding sections indicated some functional relationships and also contrasted structure among the vertebrate classes. Nevertheless, the organization of the chapter so far has been regional. Certain material will now be reviewed, and somewhat extended, by identifying various functional pathways of the brain and (in the next section) by reviewing structure by taxon.

The cortical areas and nuclei of the brain are interconnected by fiber tracts that have been traced in detail for mammals and are becoming better known for the other classes. Accordingly, neurologists have made wiring diagrams for the various functions of the body. Figure 18.12 shows six examples. Such diagrams depict reality: Injury, disease, or experimental intervention at any part of a circuit results in predictable changes in function.

Having said this, however, it is necessary to make important qualifications. It is probable that, in appropriate circumstances, any neuron can interact, however indirectly, with most other neurons of the brain. There are "explosions" of activity in the cerebral cortex. Therefore, any diagram could (if we had the knowledge) be much elaborated. Also, if one neuron is removed, another takes over its function, and the details of circuits may be altered by conditioning. Hence, they vary between individuals, and in the same individual at different times. There is no ultimate specificity of brain circuits; far from being "hard-wired," they are variable and plastic. The brain selects from a vast repertoire of alternatives and makes generalizations. Perception is the consequence of the simultaneous, cooperative activity of millions of widely distributed neurons that reach a consensus response.

Memory remains mysterious, but less so than formerly. It characterizes all vertebrates, and humans in particular. We recall skills, sense perceptions, habits, facts, events, and thoughts. We have distinct short-term and long-term memories. The mechanism involves changes, elicited by conditioning, in target neurons and in synaptic resistances. All of the brain undergoes learning-dependent changes. Certain regions, however, are directly involved, although in this instance the related circuits are not clear. The limbic system (hippocampus, hypothalamus, amygdala, mammillary bodies, habenula) functions in memory. Pyramidal cells of the hippocampus bring together independent relevant stimuli and function in the short-term storage of simple information. The medial temporal lobe of the cortex is a site of convergence and plays an enabling role in establishing long-term memory. The neocortex, however, is the ultimate storage place, and recall fades elsewhere.

EVOLUTION OF THE BRAIN

How did the vertebrate brain evolve? Until recently, we could address this question only in a most general way because we had little data upon which to construct an informed hypothesis. With an understanding of *Hox* genes, realization of the power of phylogenetic analyses, and a better understanding of development, testable hypotheses are emerging. Two recent reviews of the evolution of the vertebrate brain from that of cephalochordates (amphioxus) are in general agreement regarding the major events (see Butler and Northcutt in References). There is consensus that craniate brains are an elaboration of a cephalic tube similar to that amphioxus, but with new levels of neural organization providing for evolutionary advancements. The simplest explanation for the origin of the craniate brain is that the duplication of several different homeobox genes in ancestral forms allowed far more complex tissue-tissue interactions. Recall that the role of *Hox* genes in specifying brain organization is best documented for hindbrain segmentation (Comment

5.1 and figure on p. 116). *Hox* duplication resulted in the genesis of neural crests and neurogenic placodes (Table 5.1), two new features fundamental to the development of the skeletal elements of a new pharyngeal pump and, as particularly stressed by Northcutt, new receptors advantageous to a more active lifestyle. These changes are consistent with the belief that the origin of vertebrates represents an ecological shift from suspension feeding to active predation. It appears that the ancestral vertebrate possessed a brain that exhibited paired cerebral hemispheres, a diencephalon, an optic tectum, and a medulla, structures observed in both hagfishes and lampreys but not in amphioxus.

The brain of CYCLOSTOMES is primitive, but also in some ways specialized and degenerate. The anterior part of the brain is foreshortened in response to crowding by the terminal mouth and dorsal nasal chamber (Figure 18.13). The large olfactory bulbs are separated by only a shallow constriction from thick cerebral hemispheres. Corpus striatum, paleocortex, and archicortex are somewhat vaguely delimited. Parietal organ, pineal body, and large habenula are all visible on the roof of the diencephalon. Optic lobes are evident in lampreys and small in blind hagfishes. The medulla is relatively large and supports a large everted posterior choriod plexus. The cerebellum is rudimentary, as would be expected for these sluggish creatures.

The brain of ELASMOBRANCHS is well developed and large, the ratio of brain weight to body weight overlapping more with that of birds and mammals than with that of bony fishes. Olfactory bulbs are large, are widely separated, and enclose extensions of the

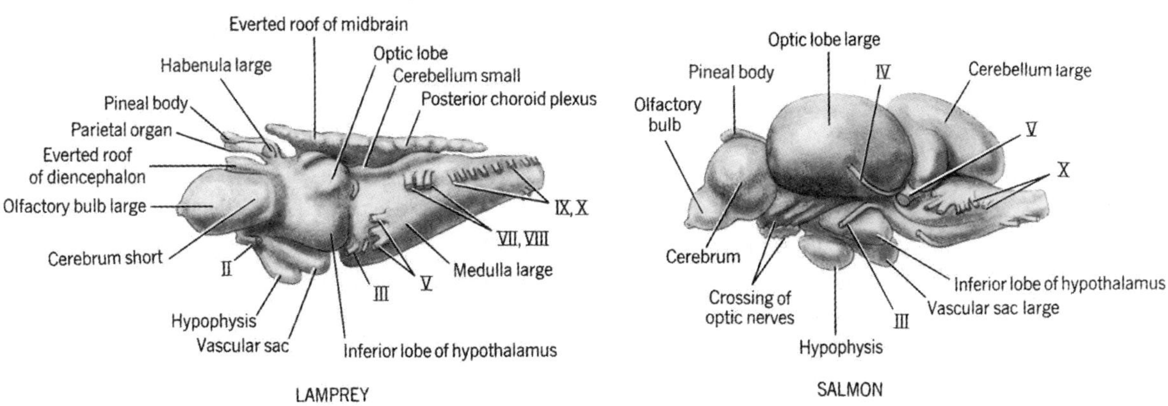

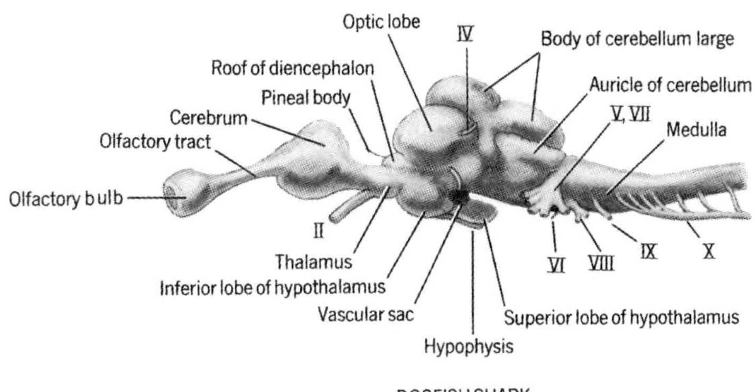

FIGURE 18.13 BRAINS OF REPRESENTATIVES OF THREE CLASSES OF FISHES.

lateral ventricles. Olfactory tracts are in evidence. Cerebral hemispheres are distinctive in that they are broadly joined in the midline and share a common ventricle. Corpus striatum and cortical areas are well established. Present in the diencephalon are pineal body (but not parietal organ), choroid plexus, habenula, lateral geniculate body, thalamus, and large hypothalamus. The hypothalamus has paired **inferior lobes** flanking the hypophysis, and a thin **vascular sac** that functions in depth perception and correlated behavior. These two structures are characteristic of fishes in general, particularly of deep-sea fishes, and are absent in tetrapods. Optic lobes are usually prominent and surround an expansion of the midbrain ventricle. The ruber nucleus has evolved. The cerebellum is large in active species, median in position, somewhat fissured in large sharks, and contains an expansion of the fourth ventricle. Archicerebellum and paleocerebellum are distinguished, and the cortex has differentiated into the three layers characteristic of higher vertebrates. Olivary nuclei and reticular formation are present in the medulla.

Brains of BONY FISHES are diverse. Those of Dipnoi resemble those of Elasmobranchii, whereas those of Actinopterygii have forebrain architechture shared by no other vertebrates. The brain of the surviving representative of the Crossopterygii is intermediate.

The cerebral hemispheres of dipnoans evaginate in the usual manner, each becoming convex on its outer surface. Those of actinopterygians evert instead: The dorsal lip of the hemisphere curls outward, so the hemisphere becomes concave on its outer surface. Cortex and corpus striatum are contiguous and thick. The homologies of the parts of the cortex are uncertain.

Conspicuous inferior lobes and vascular sac are present under the diencephalon as for cartilaginous fishes. Optic nerves pass one another at the optic chiasma or weave through one another in various patterns. Bulging optic lobes are usually the most prominent part of the brain of bony fishes. The cerebellum is smooth but usually large. It is more solid than in cartilaginous fishes. The cerebellum of actinopterygians has a distinctive anterior projection called the **valvula,** which pushes into the ventricle of the midbrain, contributing to the separation of the optic lobes.

Large pathways relating to taste are present in the medulla. If food is tasted primarily on barbels and lips, a median **facial lobe** forms behind the cerebellum to house nuclei of the facial nerves. If food is tasted primarily within the mouth and pharynx, lateral **vagal lobes** form to house nuclei of the glossopharyngeal and vagus nerves. In the medulla of actinopterygians (and urodeles) there is a single pair of giant neurons called **Mauthner cells** that relate to the ear and lateral line system. They mediate escape reflexes to muscles used in swimming. The choroid plexuses of fishes are everted.

The brain of AMPHIBIANS is remarkably unspecialized (particularly in urodeles) and is scarcely more advanced than that of cartilaginous fishes and dipnoans (Figure 18.14) (see Roth et al. in References). The cerebral hemispheres are more separate from one another than in fishes, so they share little common ventricle. Primitive hippocampal and pyriform areas have formed, respectively, from the archicortex and paleocortex. The corpus striatum is small. The pineal body is well developed in anurans. The dorsal thalmus is beginning to enlarge. Incipient mammillary bodies are present in the hypothalamus. Optic lobes are of moderate (anurans) or small (urodeles) size. The cerebellum is rudimentary.

The brain of REPTILES is narrow, elongate, and nearly straight. Olfactory bulbs tend to be smaller than for fishes. Olfactory tracts are long. The cerebrum is large because of the expansion of the corpus striatum and associated neocortex. The superficial parts of the cortex are thin, and gray matter has become external. Relative sizes and positions of the divisions of cortex and corpus striatum indicate that there have been two trends in the evolution of the reptilian forebrain: one line (represented by turtles) in the direction of mammals and another (represented by crocodilians) in the direction of birds.

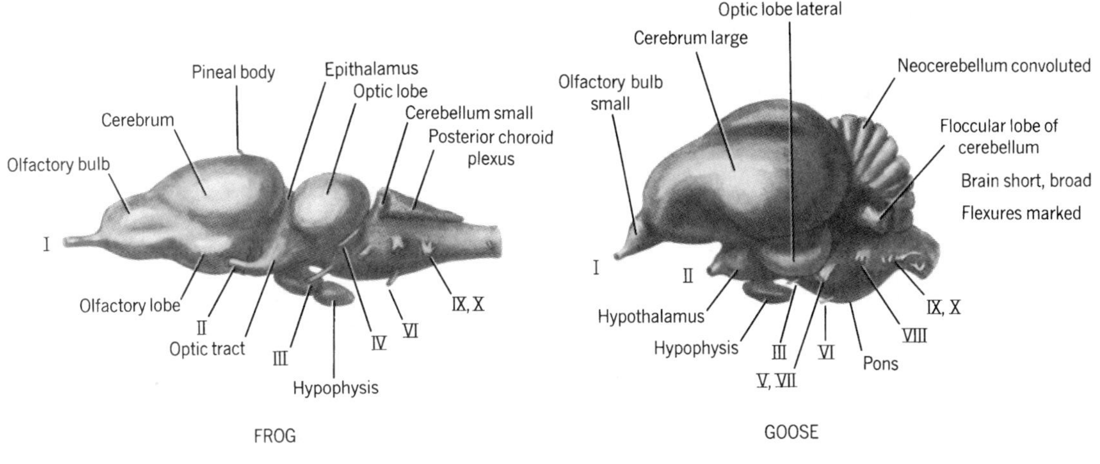

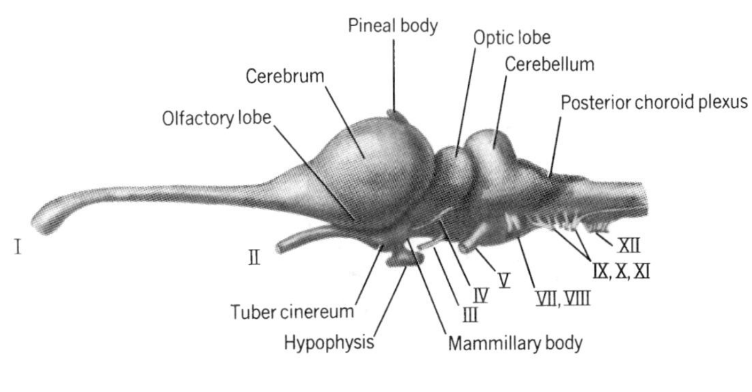

FIGURE 18.14 BRAINS OF REPRESENTATIVES OF THREE CLASSES OF TETRAPODS.

The parietal organ is functional in lizards (see Chapter 19). The dorsal thalamus is larger and more complex than in the lower classes, and the ventral thalamus has all the nuclei regularly recognized in mammals. Since most reptiles have excellent vision, the optic lobes are conspicuous. The midbrain still encloses an expanded ventricle as in anamniotes.

The reptilian cerebellum is smooth. It is largest in swimmers and rudimentary in snakes. Cerebellar nuclei are now within rather than below the organ. Choroid plexuses are inverted.

The brains of BIRDS are relatively large, uniform, and distinctive. The organ is short and broad. There are marked cranial, pontine, and cervical flexures. Olfactory bulbs and tracts are evident in the scavengers but in general are smaller than in other vertebrates. The avian cerebral hemisphere is surpassed in size only by that of some mammals. This is because of the enormous development of the corpus striatum with its associated “hyperstriatum” or neocortex. The parts of the cerebrum have a different configuration in birds that emphasize hearing (owl) than in birds that emphasize touch and manipulation with the beak (duck, snipe, parrot). The superficial parts of the cortex are exceptionally thin and have little function.

The dorsal thalamus is even more developed than in reptiles. Optic nerves, chiasma, and tracts are large. Optic lobes are particularly large and are layered within. They have connections from all sense organs and with the cerebrum. Squeezed between the cerebrum and cerebellum, the optic lobes have a uniquely lateral position.

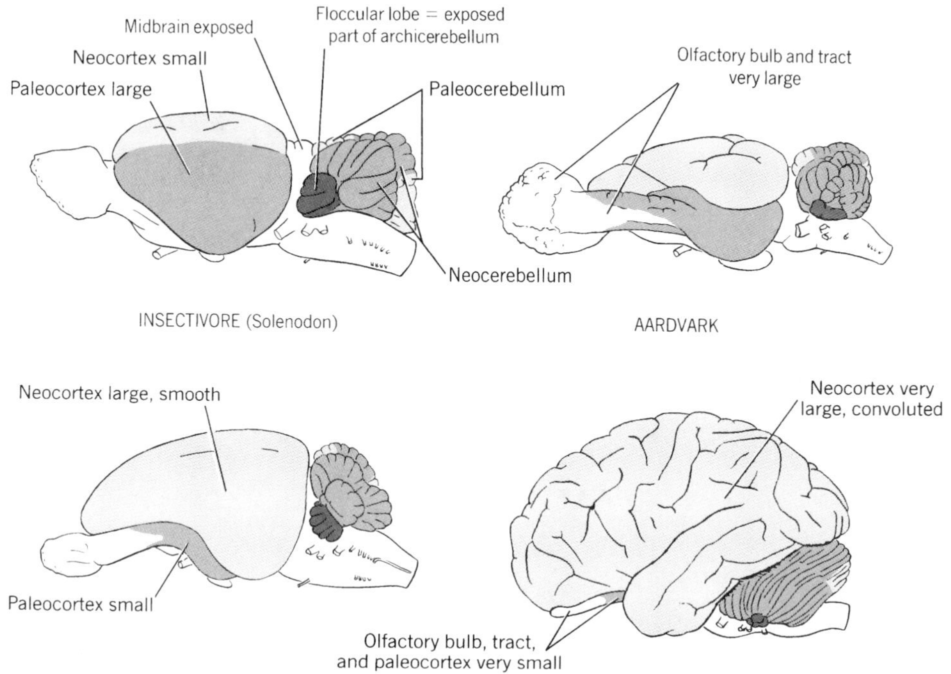

FIGURE 18.15 VARIATION IN THE STRUCTURE OF THE MAMMALIAN BRAIN. (Derivatives of the archicortex are hidden from lateral view.)

The cerebellum is larger than in other vertebrates except some mammals. It is tightly convoluted, and the organ is high and narrow. Related to the marked development of the cerebellum are the appearance of the pons under the brainstem, and enlargement of the olivary nuclei within the broad medulla.

The olfactory bulbs and tracts of MAMMALS range from huge (aardvark, armadillo, anteater) to very small (primates) (Figure 18.15). Although relatively smaller than in reptiles and birds, the corpus striatum is prominent. It is represented by the basal nuclei, of which the caudate and putamen, derived from the neocortex, are relatively large. The extensive neocortex is the hallmark of the class. It dominates the entire brain both structurally and functionally. The hemispheres are smooth in most small mammals and convoluted in most large mammals. A new commissure, the corpus callosum, joins the hemispheres of therian mammals. The archicortex is represented by the large hippocampus. Pyriform lobes are lateral or ventral in position. They are extensive if olfaction is acute; otherwise they are restricted.

Thalamus and hypothalamus are highly differentiated. The midbrain is exposed in only a few mammals. Optic lobes, now called anterior colliculi, are small because the cerebral cortex has taken over much of their function. Posterior colliculi are present to complete the corpora quadirigemina of the midbrain tectum. The ventricle of the midbrain is restricted to a narrow cerebral aqueduct.

The mammalian cerebellum is large, much convoluted, and relatively broad. The cerebellar nuclear complex has differentiated into three or four distinct pairs of nuclei. A pons is prominent.

SUPPORT AND NOURISHMENT OF THE CENTRAL NERVOUS SYSTEM

The central nervous system of cyclostomes and fishes is loosely surrounded by a protective fibrous envelope called the **primitive meninx.** Amphibians and reptiles have two envelopes (or meninges), an outer tougher **dura** and an inner **pia-arachnoid.** Birds are structurally intermediate between reptiles and mammals in having the beginnings of a third layer.

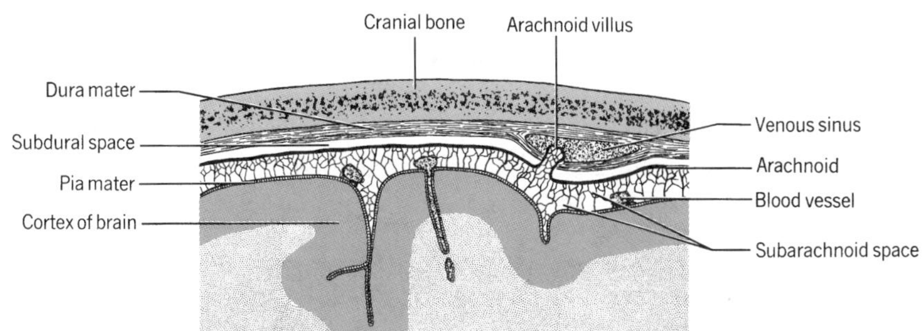

FIGURE 18.16 MENINGES OF THE MAMMALIAN BRAIN.

Mammals have three meninges (Figure 18.16). The strong **dura mater** (= *hard* + *mother*) is outermost. It is of mesodermal origin. In the vertebral column it is separated from the vertebrae by fat; in the cranium it adheres directly to the bones. The innermost meninx is the thin vascular **pia mater** (= *tender* + *mother*), which adheres to the nervous tissue, following contours into every fissure and sulcus. Between the dura and pia is the nonvascular **arachnoid** (= *spiderlike*), which is delicately fibrous and sends strands to the pia mater. Pia and arachnoid are both derived in part from neural crests and in part from mesenchyme. Between the dura and arachnoid is a shallow, fluid-filled, **subdural space.** Between arachnoid and pia is a deeper **subarachnoid space.**

The central canal of the spinal cord, the ventricles of the brain, and (in mammals) the subarachnoid space are filled with a considerable quantity of **cerebrospinal fluid.** This fluid is clear, colorless, and similar to blood plasma except that it contains more chloride and virtually no protein. It flows slowly by secretion pressure and also by action of ciliated ependymal cells. Motion is toward the fourth ventricle from both the anterior ventricles and the central canal of the cord. The fluid escapes through holes in the roof of the medulla to the subarachnoid space. From there it reenters the bloodstream in various places, particularly through venous sinuses near the brain. Some cerebrospinal fluid also drains into lymphatics.

The pattern and extent of this circulation must be somewhat different in nonmammals (there being no subarachnoid space), but such vertebrates also have some cerebrospinal fluid outside the central nervous system. (The choroid plexuses are usually everted into thin sacs of the fluid instead of inverted into their respective ventricles.)

Recall that in the lateral ventricles, and on the roof of the fourth ventricle, there are choroid plexuses. These consist of a richly vascular and convoluted pia mater, together with the adherent ependymal layer of the brain vesicles. The plexuses rapidly produce most of the cerebrospinal fluid (a small amount is probably furnished by ependymal cells of the cord). Bathing and buoying the central nervous system as it does, the fluid protects it from injury when there is a blow to the head or back. Together, plexus and cerebrospinal fluid also function importantly in maintaining the chemical stability of the central nervous system.

Nourishment is provided in two ways. First, where capillaries come in direct contact with nervous tissue (called the **blood-brain barrier**), many materials, including large molecules and drugs, are prevented from crossing, but there is rapid transport into the brain of substances that are quickly consumed (e.g., respiratory gases, glucose, amino acids, lactate). Transport is by facilitated diffusion, which uses no energy. Second, each choroid plexus forms a selective barrier between the blood and cerebrospinal fluid. Various nutrients (including vitamins C and B) are moved into the fluid by active transport, which uses energy. The cerebrospinal fluid, in turn, has free access to the interstitial fluid of the brain through the pia mater. The plexuses also serve as kidneys for the brain, moving the waste products of metabolism from the cerebrospinal fluid to the blood.

The structure and function of the circulatory system ensure a rich blood supply to the central nervous system. Associated vessels are usually large and numerous, collateral systems are present, and in time of stress this system is given priority over others. Brain and cord need much blood for two reasons. First, their metabolic rate is high and their need for oxygen is constant and great. Second, the exchange between blood and tissues there may be less efficient than elsewhere. Nervous tissue is dense, interstitial spaces are minute, and there are no lymphatic vessels.

REFERENCES

Alkon, D.L. 1989. Memory storage and neural systems. *Sci. Am.* 26(1):42–50.

Braford, M.R. (ed.). 1995. Evolution of the forebrain. *Brain Behav. Evol.* 46:181–338 (entire issue).

Bullock, T.H. 1993. How are more complex brains different? *Brain Behav. Evol.* 41:88–96.

Burt, A.M. 1993. *Textbook of neuroanatomy.* Saunders, New York 541p. Provides basic overview with emphasis on humans.

Butler, A.B. 2000. Chordate evolution and the origin of craniates: an old brain in a new head. *Anat. Rec.* (*New Anat.*) 261:111–125. Provides a model for the evolution of the brain by sequential steps caused by shifts in expression of the genes that specify nervous system development from the ectoderm.

Butler, A.B., and W. Hodos. 1996. *Comparative vertebrate neuroanatomy: evolution and adaption.* Wiley-Liss, New York. 514p. Informative, detailed account of chordate neuroanatomy discussed in an evolutionary context.

Card, J.P. 1998. Exploring brain circuitry with neurotropic viruses: new horizons in neuroanatomy. *Anat. Rec.* (*New Anat.*) 175:176–185.

Cox, J.A., and J.R. Fetcho. 1996. Labeling blastomeres with a calcium indicator: a noninvasive method of visualizing neuronal activity in zebrafish. *J. Neurosci. Methods* 68:185–191.

Deacon, T.W. 1990. Rethinking mammalian brain evolution. *Am. Zool.* 30:629–705.

Delcomyn, F. 1996. *Foundations of neurobiology.* Freeman, New York. 648p. Neuroanatomy presented and discussed in the context of physiology and behavior. Excellent treatise.

Freeman, W.J. 1991. The physiology of perception. *Sci. Am.* 264(2): 78–85.

Isaacson, R.L. 1982. *The lymbic system.* 2nd ed. Plenum, New York. 327p.

Kandel, E.R., and R.D. Hawkins. 1992. The biological basis of learning and individuality. *Sci. Am.* 267(3):78–86.

Kimelberg, H.K., and M.D. Norenberg. 1989. Astrocytes. *Sci. Am.* 260(4):66–76.

Kimura, D. 1992. Sex differences in the brain. *Sci. Am.* 267(3):118–125.

Krubitzer, L. 1995. The organization of neocortex in mammals: are species differences really so different? *Trends Neurosci.* 18:408–417. Critical analysis is presented, based on studies of the sensory cortex of diverse mammals, for the origin and evolution of the brain's cortical fields.

Medina, L. and A. Reiner. 2000. Do birds possess homologues of mammalian primary visual, somatosensory and motor cortices? *Trends Neurosci.* 23:1–12. Evidence that some important structural and functional components of the avian and mammalian brain long thought to be similar, may have evolved independently.

Nieuwenhuys, R., H.J. ten Donkelaar, and C. Nicholson (eds.). 1998. *The central nervous system of vertebrates.* 3 vols. Springer, New York. 2,219p. Enormous, detailed, comprehensive; excellent illustrations.

Nieuwenhuys, R., J. Voogd, and C. van Huijzen. 1998. *The human central nervous system: a synopsis and atlas.* 3rd ed. Springer-Verlag, New York. 440p. Superb illustrations of structure and pathways.

Northcutt, G.R. 1996. The agnathan ark: the origin of craniate brains. *Brain Behav. Evol.* 48:237–247.

Paulin, M.G. 1993. The role of the cerebellum in motor control and perception. *Brain Behav. Evol.* 41:39–50. General review for cyclostomes, elasmobranchs, amphibians, reptiles, birds, and mammals; concludes that in addition to tracking the movements made by the animal itself, the cerebellum is involved in tracking objects moving around the animal as well.

Roth, G., et al. 1993. Paedomorphosis and simplification in the nervous system of salamanders. *Brain Behav. Evol.* 42:137–170. A comprehensive analysis of the apparent paradox that the salamander brain is less complex than is expected based on its phylogenetic position.

Selkoe, D.J. 1992. Aging brain, aging mind. *Sci. Am.* 267(3):134–142.

Spector, R., and C.E. Johanson. 1989. The mammalian choroid plexus. *Sci. Am.* 261(5):68–74.

Wake, D.B. 1994. Brainstem organization and branchiomeric nerves. *Acta. Anat.* 148(2–3):124–131.

Chapter 19

Sense Organs

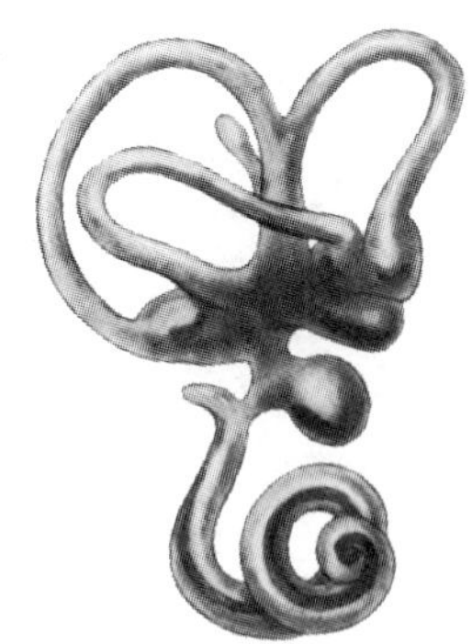

All cells of the body are responsive to their environments. Certain cells and organs, however, are specialized for monitoring the environment, and these compose the sensory system. They stimulate the central nervous system, and it is there that sensation is perceived and integrated and where action, if any, is initiated.

Receptors range from mere nerve endings, through simple microscopic capsules, to the large and complex eye. Various classifications of these diverse organs are used: (1) general (widely distributed like pressure receptors) versus special (localized like the ear), (2) somatic (conscious reception of relatively superficial stimuli) versus visceral (unconscious reception of deep stimuli), (3) stimulated from internal sources (muscle tone, balance) versus external sources (cold, light), and (4) responsive to mechanical stimuli (touch, sound) versus electromagnetic stimuli (heat, light) versus chemical stimuli (taste, smell). None of these schemes is entirely satisfactory for our purpose of analyzing the major advances of vertebrate structure. The more complicated sense organs contribute interesting histories to the story of evolution and will be presented in approximate order of increasing complexity after brief attention is given to the relatively simple receptors.

SOME MISCELLANEOUS SMALL RECEPTORS

The structural bases for the perception of such human sensations as hunger, fatigue, sex drive, and anxiety are unknown, and many vertebrates have sense organs, not shared by humans, for which functions can only be approximated. We can never know what it feels like to be a fish, and much remains to be learned about sense reception—particularly in lower vertebrates.

Various sensations, including touch, pressure, stretch, heat, cold, and general chemical sense, are at least sometimes received by naked nerve endings. There can be little to say about the phylogeny of these. Numerous kinds of **sense capsules** are found in the epithelial and connective tissues of tetrapods. These consist of variously modified nerve endings surrounded by small, simple capsules having many configurations. They appear to monitor heat, cold, touch, pressure, and pain. It is difficult to identify the function of each kind of

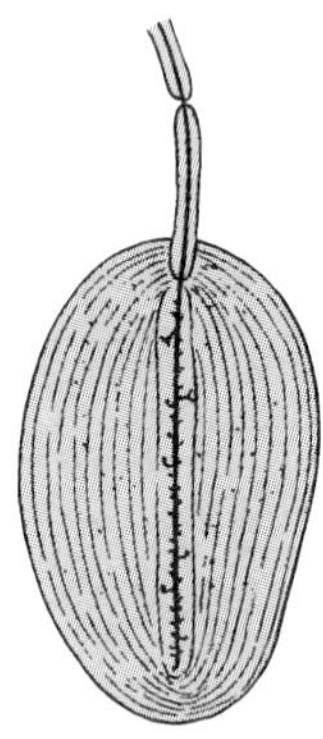

FIGURE 19.1 PACINIAN CORPUSCLE, a pressure transducer.

sense capsule, and it is possible that some kinds respond to more than one stimulus. Virtually nothing is known of their phylogeny, and homologies can rarely be made. The best-known is the **pacinian corpuscle,** which is a marvelously effective tiny pressure transducer (Figure 19.1). Slight distortion of the 30 to 50 onionlike layers of the capsule sets up a **generator potential** in the nerve ending within the organ. If the potential reaches threshold intensity, the capsule greatly steps up the impulse at the first node of Ranvier. The resulting **action potential** is then transmitted along an axon.

The **proprioceptive senses** apprise an animal of the relative positions of the parts of its body. Half awakening from deep sleep, one may not know for an instant how the legs are flexed or the arms placed, but the slightest tensing of the muscles makes one's position known. Three kinds of proprioceptors are involved, all of which respond to tension and contribute to postural reflexes. They appear to be present in all tetrapods. **Joint receptors** have complex nerve endings within the connective tissue of joint capsules. Two kinds of muscle receptors, best known from studies of mammals, are **tendon organs** and **muscle spindles** (Figure 19.2). These receptors monitor muscle length and force. Both are plentiful in striated muscles that develop relatively substantial force and are subject to unpredictable loads. Tendon organs are located along the aponeuroses of tendons of both origin and insertion, and aponeuroses that extend into the belly of the muscle (but generally not in the tendons proper). Tendon organs possess one sensory afferent neuron and are structurally simple. Spindles, in contrast, may be relatively simple in structure, but are considered by some to be complex. Spindles are located throughout the muscle belly and near aponeuroses of origin and insertion. A typical muscle spindle in mammalian muscle is composed of two types of modified muscle fibers, a capsule, and both afferent (sensory) and efferent (motor) neurons. Spindles and, to a lesser extent, tendon organs have been intensely studied, but their precise role in locomotor control remains elusive (see Stuart and McDonagh in References).

Several dissimilar kinds of sense organs are macroscopic yet relatively simple in structure. The **carotid body** is a small mass of cells of disputed origin that lie in the fork of the internal and external carotid arteries of tetrapods. Richly supplied with both nerves and blood, it detects fluctuations in the concentrations of carbon dioxide and oxygen in the blood stream and sends impulses via the ninth cranial nerve to the parts of the brain controlling circulation and respiration. **Otoliths** are hard objects within the internal ear of

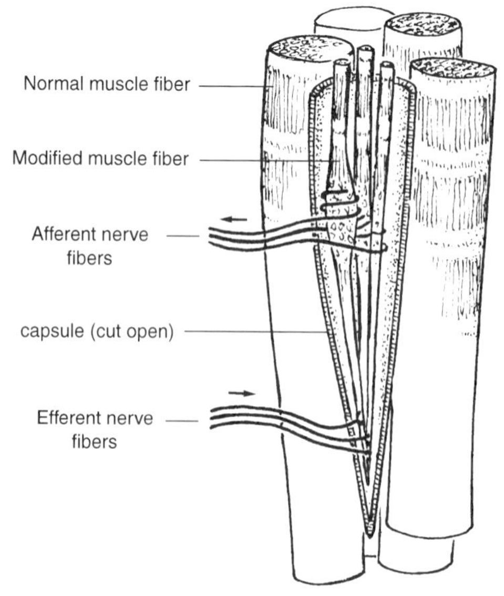

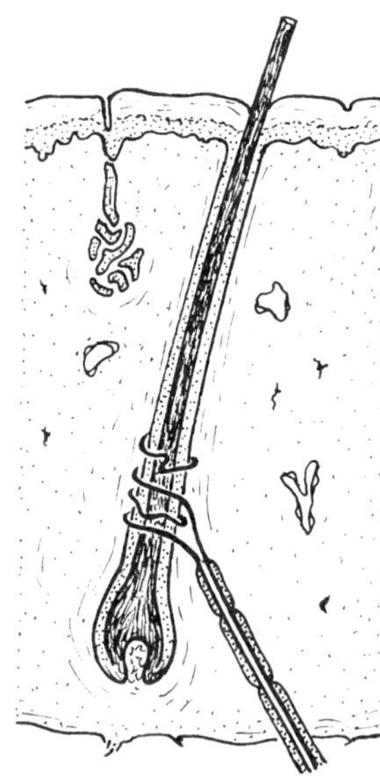

FIGURE 19.2 TWO MECHANORECEPTORS: MUSCLE SPINDLE (left) and MOTION SENSOR OF A HAIR FOLLICLE (right).

some vertebrates. In addition to their function in equilibrium (see below), they probably act as accelerometers. Also, the large otoliths of several teleost fishes have been shown to have piezoelectric properties (i.e., converting force to electric potential); theoretically they could function as depth registers. **Pit organs** give their name to the American snakes called pit vipers. The conspicuous pits are located between the nostril and the eye. The floor of the blind pit is vascular and rich in superficial nerve endings of cranial nerve V. The organ is a remarkably sensitive thermoreceptor. The background discharge is modified if the temperature increases or decreases by as little as 0.003°C. This enables the snake to detect, in a fraction of a second, the presence and exact position of warm-blooded prey that comes within striking distance. Pythons and boas either have series of similar, but smaller and less sensitive, pits in the scales surrounding the mouth, or they have heat-sensitive skin areas.

Behavioral experiments have demonstrated that many vertebrates sense the earth's magnetic field to orient themselves while migrating or homing. What has not been demonstrated, however, are the physiological mechanisms by which a **magnetic sense** works, nor have the anatomical receptors responsible been identified. To demonstrate the biological basis for a magnetic sense remains a fascinating challenge (see Lohmann and Johnsen in References). Several mechanisms, some with strong theoretical support but little empirical evidence, have been proposed for how magnetic fields are perceived and their information transmitted. They include electromagnetic induction and sensitivity to minute movement of magnetic crystals. One, or more than one, may be used by an individual species. Electromagnetic induction has been implicated as the means by which elasmobranchs use the ampullae of Lorenzini, in combination with electroreceptors on their skin, to detect voltage changes generated by shifts of the body's orientation as it moves through the earth's magnetic field. Hatchling sea turtles initially find their way from beach to water by moving toward the light of stars or moon reflected from the ocean. They swim against the incoming waves to get offshore. By the time they reach the needed ocean current, their magnetic compasses have been set. Apparently they determine their latitude by sensing the angle of the earth's magnetic field (which becomes vertical at the poles and horizontal near the equator).

The mineral magnetite has been detected in trout and salmon, birds, sea turtles, and some other animals known to orient to the earth's magnetic field. In trout, for example, it has been suggested that 50 nm magnetite particles located in the snout exert pressure or torque on secondary receptors (stretch receptors, hair cells, or mechanoreceptors) as the particles attempt to align with the geomagnetic field. Or the movement of the crystals might serve to open ion channels directly.

The remarkable homing ability of pigeons appears to result from several methods of orientation. The birds use vision, and possibly also hearing and smell, to construct a landscape map that is most detailed near the home roost, but may extend out to 1000 km. This is supplemented by a time-corrected sun compass based on either the sun disc or sky polarization. Experiments show that should conditions make these methods inadequate, the pigeon then uses a magnetic compass. The flight of migrating birds is disturbed as they fly over magnetic anomalies in the earth's crust.

ORGANS OF CHEMORECEPTION

Olfactory Organs Seemingly unspecialized nerve endings at various places on the surface of the body of many vertebrates are sensitive to the chemical environment. Response of the human eye to smog and onions is an example. The principal chemical senses are olfaction, pheromone reception, and taste. Olfaction is a primitive sense. The earliest vertebrates appear to have had a keen sense of smell, and fossils suggest that the forebrain was devoted to olfactory signals.

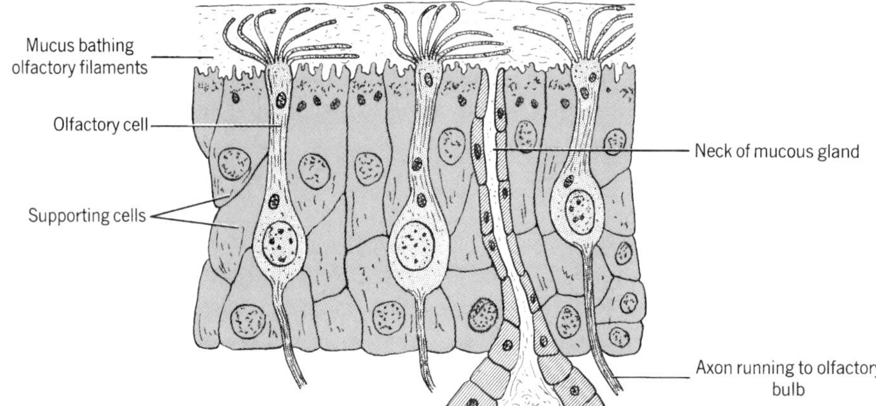

FIGURE 19.3
SECTION OF THE OLFACTORY EPITHELIUM OF A TETRAPOD.

Olfactory epithelium is localized in the nasal pits of fishes and in protected outpocketings of the respiratory passages in air breathers. Only part, and usually a small internal part, of the epithelium lining the nasal pit or passage contains olfactory cells. These are columnar cells each of which has about eight hairlike filaments on its free surface (Figure 19.3). These cells are unique in that they continue as axons into the olfactory bulb of the central nervous system. They are also unusual in that they can be replaced. Supporting cells are dispersed among the thousands or millions of olfactory cells.

The combined area of all the filaments of the olfactory epithelium may exceed the surface area of the body, and it is on these filaments that dissolved chemicals are detected. The sense of smell is so wonderfully acute that the theoretical limit of sensitivity is reached by many animals: Each filament can respond to a single molecule of certain odoriferous materials. Each kind of animal has its own spectrum of sensitivity, being very sensitive to some odors and little sensitive to others. Smells are detected when odorants (small, volatile, lipid-soluble molecules) bind to proteins on the filaments of the receptor neurons. There are about 1000 kinds of receptors. (This was established by finding the genes that encode them.) Mammals can distinguish about 10,000 odors, so each receptor responds to several odor molecules. The kinds of receptor neurons are randomly distributed in each region of the nasal epithelium, but the axons of like kinds converge to particular glomeruli in the olfactory bulb (see p. 332). Like kinds of glomeruli form a pattern in the olfactory bulb and hence project onto the olfactory cortex a two-dimensional representation of odors detected.

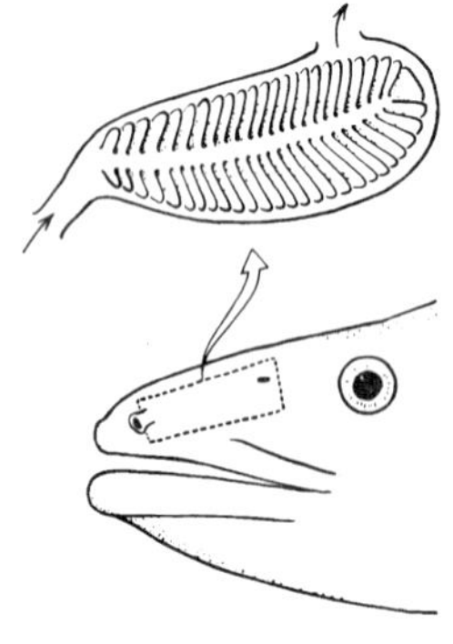

FIGURE 19.4
DISSECTION OF THE NASAL PIT OF AN EEL, *Anguilla*.

Olfaction has not been detected with certainty in amphioxus. CYCLOSTOMES have a median olfactory sac that is ventilated by water passing in and out of the single nasal pouch. However, it is probable that the olfactory structures of the ancestors of these animals, and doubtless also of anaspids and cephalaspids, were paired as for other vertebrates. This is indicated by the bilobed nature of the sac, its innervation, and its ontogeny.

FISHES admit water into each nasal pit by one compressed opening (Selachii) or by two openings (most Teleostei) constructed so that a continuous stream passes over the olfactory epithelium. The nasal pit of cartilaginous fishes lies in the respiratory current, thus ensuring ventilation even when the fish is stationary. The internal nostrils of Sarcopterygii may have evolved partly to benefit smelling in water (through increased ventilation) rather than breathing in air. The lining of each pit is pleated (Figure 19.4). The number and shape of the pleats relates to both species and age. The olfactory epithelium is largely confined to the clefts between the pleats.

AIR-BREATHING VERTEBRATES add mucous cells to the olfactory epithelium. This is necessary both to dissolve the particles to be smelled and to wash away material that has already been detected so that fresh samples of air can be examined.

The size of the nasal chamber increases, and its structure becomes more complex, as the secondary palate evolves. Most amniotes increase the surface area of the nasal epithelium by folding it over several pairs of turbinates (or conchae), or scrolls of bone, which curl into the nasal chamber from its lateral walls. Turbinates are relatively simple in most vertebrates but exceedingly complex in some mammals. The area of the olfactory epithelium may be increased in this way, but the primary function of the folded nasal epithelium is to clean, moisten, and sometimes to alter the temperature of respired air before it reaches the lungs. The olfactory sense was formerly thought to be rudimentary in birds, but it is now known that most birds respond to odors. Olfaction contributes to the recognition of familiar surroundings by homing pigeons, the avoidance of noxious insects by chickens, the selection of nest-building materials by starlings, the location of pelagic food by petrels, and the finding of carrion by vultures. Olfaction is relatively weak in aquatic mammals, which, unlike fishes, cannot ventilate the olfactory epithelium while submerged.

Pheromone Receptors Pheromones are chemicals released by an animal that elicit behavioral responses in conspecifics. The chemical signals are sensed in the **vomeronasal organ,** located at the base of the nasal septum. It is confined to tetrapods and is rudimentary or absent in turtles, crocodilians, birds, aquatic mammals, and higher primates, including humans. In mammals the complex response mechanism involves hormones produced elsewhere, and functions to modify sexual and social behaviors. In snakes the vomeronasal organ serves in identifying prey and in following their trails. This enigmatic little organ is described further in Comment 19.1.

Taste Organs Taste organs, a third chemosensory system, are similar to olfactory organs in being chemoreceptors having epithelial hair cells. They differ from olfactory organs in being less sensitive by about four orders of magnitude, having more restricted response (there are fewer different tastes than odors), relating to different parts of the brain, being of endodermal instead of ectodermal origin, and having the receptor cells aggregated into groups, or **taste buds.**

Each taste bud consists of supportive cells and 30 to 40 columnar taste cells all arranged in a barrel-shaped cluster (Figure 19.6). The buds may be dispersed or grouped on little hummocks of the epithelium called papillae. Papillae are of various sizes and shapes and may be arranged in various patterns—all of unknown functional significance. Taste buds, unlike olfactory epithelium, are exposed and subject to wear. Accordingly, their cells are short-lived and are constantly replaced. Each bud is most sensitive to one of the four basic tastes: salt, sour, sweet, or bitter. Nevertheless, all taste buds look alike, and the different vertebrates are not equally sensitive to these tastes. Being derived from epithelium at the levels of several visceral arches, taste buds are innervated by several cranial nerves: the seventh, ninth, and tenth.

Taste buds are abundant in the mouth and pharynx of cyclostomes and fishes, but they also may be located on the surface of the body, particularly on the heads and oral feelers of fishes that find food in sand, mud, or murky water. Amphibians have taste buds on the tongue, pharynx, and skin. Frogs have taste buds positioned to taste bits of tissue abraded by the palatal teeth and dissolved by enzymatic action of an oral secretion. Reptiles and birds, having dry skins and usually somewhat keratinous tongues, distribute most of their taste buds in the pharynx. These animals, especially birds, have a relatively poor sense of

COMMENT 19.1

VOMERONASAL ORGAN: NO LONGER A MYSTERY, BUT QUESTIONS REMAIN

The **vomeronasal organ** was described in 1813 by the Danish anatomist L. Jacobson, but remained puzzling for 150 years. It was long considered to be a somewhat distinct part of the olfactory epithelium, the reason for its separation being elusive. However, the vomeronasal organ differs from the olfactory organ in fine structure; in the molecules to which it responds; in having its own nerves, which terminate at a bulb that is adjacent to but separate from the olfactory bulb; and in projecting independently to the amygdala and the neuroendocrine hypothalamus. In mammals, stimulation of the organ triggers the release of hormones relating to luteinizing hormone and prolactin, and ultimately affects courtship, mating, maternal care, and aggression. Snakes, amphisbaenians, and many lizards have a forked tongue that is flicked in and out and up and down, often touching the substrate. The widely spaced tips of the tongue sample the chemical environment at different points and deliver their stimuli independently to the front of the mouth, whence they reach the paired vomeronasal organ. This enables the animal to locate the margins of, and therefore follow, a scent trail laid down by prey or by another of its kind.

The organ lies at the base of the nasal septum, right and left, in a tunnel in the vomer bone or cartilage (Figure 19.5). Each side has a crescent-shaped lumen. The sensory epithelium is on the concave side of the lumen; it is distinctive in having microvilli. The nonsensory epithelium on the convex side is ciliated. Each lumen is blind at its posterior end and opens anteriorly, by way of small ducts, into the nasal chamber, oral chamber, or both. The stimulus (which for mammals usually emanates from urine or vaginal secretions) is trapped by nasal or oral fluids and then travels to the vomeronasal organ. The epithelium is bordered by venous sinuses surrounded by smooth muscle. Vessels and muscle are richly supplied with autonomic nerves. Alternating vasoconstriction plus contraction with dilation plus relaxation apparently creates a pump that circulates fluid in and out of the organ.

The nature of the stimulants needs clarification. How fast can the gland be emptied and restimulated? The mechanism of reception and transduction is not clear. There are two classes of receptor cells that are intermixed in the epithelium but are sorted out to different parts of the vomeronasal bulb. Why? Many mammals lift the head, curl the upper lip, and interrupt breathing during sensing (the Flehman behavior). How does that help? Is reception or response conscious?

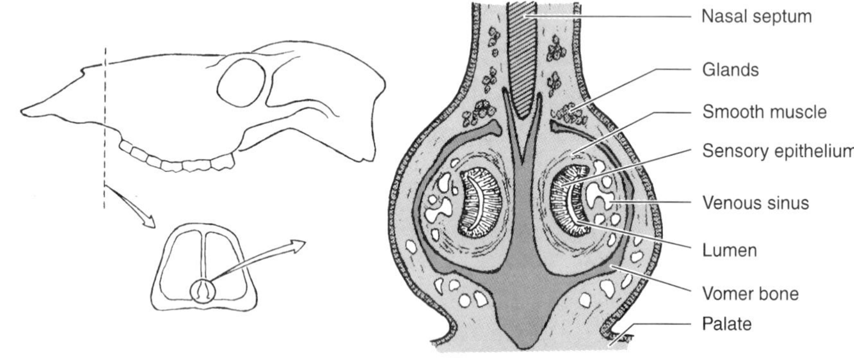

FIGURE 19.5 CROSS SECTION OF THE VOMERONASAL ORGAN OF A MAMMAL. (Turbinate bones of the nasal cavity not shown.)

taste. Mammals also have taste buds in the mouth generally and on the pharynx, but concentrate them on the fleshy tongue.

LATERAL LINE SYSTEM; ELECTRORECEPTION

The lateral line system and internal ear are so related by structure, function, and ontogeny that together they are called the **octavolateralis system** (or acousticolateralis system). In most fishes other than teleosts the lateral line system, in turn, has two components. One consists of electroreceptors and the other of mechanoreceptors that measure water displacement.

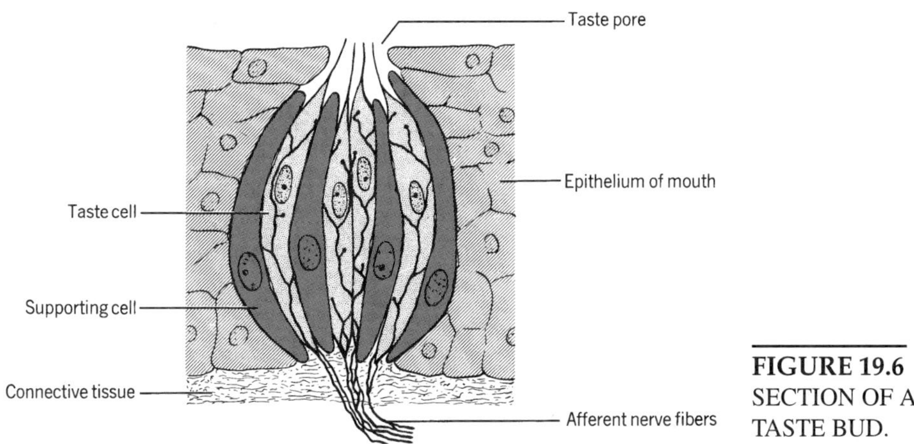

FIGURE 19.6
SECTION OF A TASTE BUD.

The two components of the lateral line system have a similar developmental origin. Several ectodermal placodes form on each side of the head and neck area of the embryo. Cells from neural crests may join the placodes or influence their development. One placode on each side forms the internal ear. Cells move away from the other placodes to become the lateral line system and the ganglion cells of the nerves that will innervate that system. These are branches of the seventh cranial nerve, which serves most of the head; the ninth nerve, serving a small area at the back of the head; and the tenth nerve, serving the remainder of the system.

The MECHANORECEPTIVE COMPONENT of the system is present in all fishes and in both larval and aquatic adult amphibians. It consists of thousands of microscopic organs called **neuromasts.** These are always freely distributed at the surface of the skin and usually are also located in shallow surface pits, along canals in the skin (Figure 19.7), or along horizontal tubes tunneling under the skin and even through dermal bones. Such tunnels always communicate with the surface by pores spaced along their length. The longest and most constant tube or canal is the **lateral line canal**—hence the name of the system (**sensory canal system** is also used). These ways of distributing and protecting neuromasts intergrade. Elasmobranchs may also have neuromasts on an organ within the spiracle.

A neuromast looks somewhat like a taste bud. Found only in vertebrates, it is a clump of sensory hair cells and supportive cells. Each sensory cell has on its exposed surface a bundle of 20 to 50 sensory hairs called **stereocilia,** which are graded in length, and usually also one larger **kinocilium,** which stands at one edge of the stereocilia. Each neuromast that is free at the surface of the skin is capped by a tall, fragile, and transparent dome of extracellular material that protrudes into the water. This structure is the **cupula.** All sensory hairs

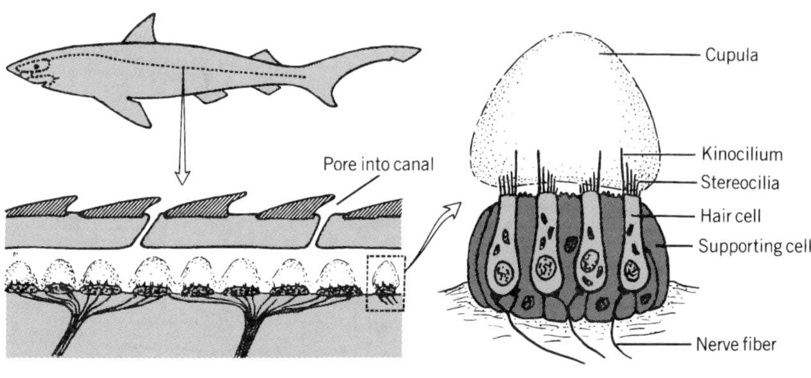

FIGURE 19.7
POSITION OF LATERAL LINE SYSTEM BELOW SKIN, LONGITUDINAL SECTION OF A CANAL, AND STRUCTURE OF A NEUROMAST.

of the neuromast are embedded in the base of the cupula. Neuromasts opening into subsurface canals usually have cupulas also, but some are instead covered by a jelly that fills the surrounding space. Any high-frequency water motion, such as might be generated by potential prey, causes the cupula to pivot, thus tilting the embedded sensory hairs in the direction of motion. Each hair cell is a pressure transducer that discharges electrical signals at a basic frequency that increases when displacement of the hairs is in one direction and decreases when displacement is in the opposite direction. There are two sets of hair cells, one positively activated by motion of fluid along the canal in one direction and the other by motion in the opposite direction.

Collectively, free neuromasts and those of canals and pits form a system of distant touch that is so well coordinated that it gives both spatial and temporal information. The system contributes to obstacle avoidance and orientation of the body relative to currents and eddies. Even when blinded, fishes can determine the position and motions of nearby prey and of other fishes. They can also sense the relative motion of their own bodies and the water. A schooling fish can make its precise, coordinated motions when deprived of either its lateral line system or its eyes, but requires each to maintain normal spacing with other fishes. Sound waves from a distant source are not sensed by the lateral line system. However, low-frequency sound waves from a near source cause relatively great displacement of the water and hence are probably heard (whatever they may "sound" like to a fish).

Cyclostomes and amphibians have no sensory canals on the head or body. Their neuromasts are free or in pits that tend to be linearly arranged. The lateral line system is well known from the armor of ostracoderms and placoderms. Fishes have lateral line canals (sometimes doubled or branched) and a complicated but rather stable system of head canals. The canals are best developed in active fishes and least protected from the surface in fishes inhabiting quiet water. Most of the canals of chimaeras are open to the surface (see figure on p. 44).

ELECTRORECEPTION in fishes is accomplished by neuromasts usually located at the bottoms of **ampullary organs,** which are deep, tubelike pits located in the skin and subcutaneous tissue of the head. Cupulas are absent; the neuromasts are covered with jelly. The system detects weak electric stimuli. These may have been reflected by nearby objects, the original pulses sometimes having been generated, like radar, by the fish itself (see pp. 191 and 192). The system may also sense the muscle potentials of nearby fish and thermal, mechanical, or magnetic stimuli. The system is present in lampreys, cartilaginous fishes, *Polypterus,* Chondrostei, and some amphibians. A corresponding system is present in a few teleosts. Of entirely independent origin is the electroreceptive system found in the bill of the platypus, a diving, prototherian mammal. The sensors are innervated by the trigeminal nerve and surround the pores of mucous glands. They detect weak electric fields generated by the activities of invertebrates on which the platypus feeds.

EAR

Organ of Equilibrium Details of the origin of the ear are lost in antiquity, but the story can be surmised with reasonable confidence. It is probable that the lateral line system evolved before the ear, and that of the two basic functions of the ear, equilibration and hearing, the former was perfected before the latter was well-initiated. It is desirable for an organ of equilibration to be on the head, where it is close to the feeding apparatus and other special senses, and away from the oscillating locomotor mechanism. Furthermore, it is desirable for the organ to be away from the surface of the body, where signals of the type monitored by the lateral line system would be distracting. Evidently a portion of the lateral line system at the side of the head sank below the skin and was modified to become the new organ.

This probable phylogeny appears to be recapitulated by embryos in that the internal ear forms from an ectodermal placode that, in fishes, is in the middle of the series of placodes that form the lateral line system. This otic (or auditory) placode develops under the inductive influence of the hindbrain, thus ensuring its position at the back of the head. The placode sinks in, forming a pit, that then pinches away from the skin ectoderm to become the hollow **otic vesicle.** Gradually, the vesicle assumes the complicated shape that makes appropriate its adult name of **labyrinth** (Figure 19.8). Ganglion cells of the associated eighth cranial nerve are also derived from the vesicle.

The body of the labyrinth becomes more or less divided into a dorsal chamber, or **utricle,** and a ventral chamber, or **saccule.** A diverticulum from the utricle is called the **endolymphatic sac,** and another from the saccule is called the **lagena** (which becomes the mammalian cochlea). The labyrinth is completed by three looping **semicircular canals,** each of which joins the utricle at both ends. The labyrinth and its canals are filled with the fluid **endolymph.**

In both utricle and saccule there is at least one patch of sensory and supportive cells. Each patch, called a **macula,** resembles a neuromast of the lateral line system but is somewhat larger. Each sensory cell has the familiar clump of stereocilia and single, asymmetrically placed kinocilium. The hairs of each macula are embedded in a modified cupula that is made heavy by deposits of calcium. These are in the form of crystals in some groups and in the form of solid masses called **otoliths** in others. The size and shape of otoliths are distinctive for many vertebrates.

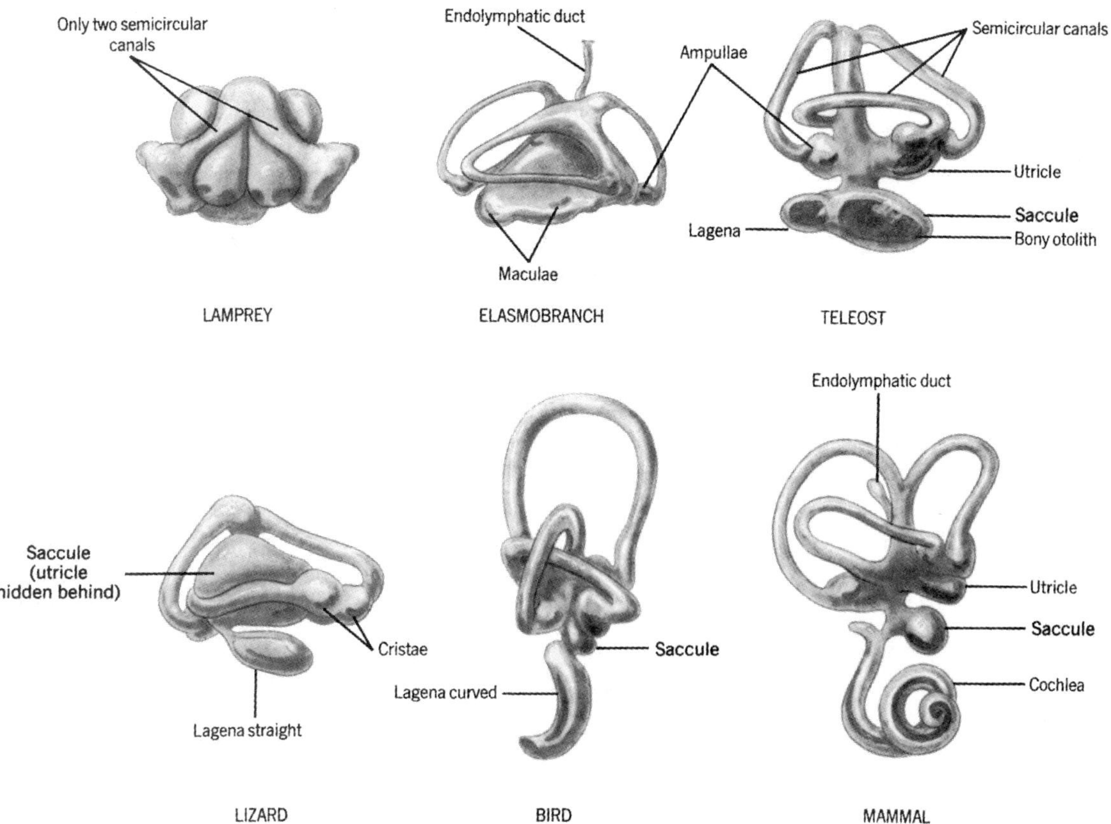

FIGURE 19.8 COMPARATIVE ANATOMY OF THE LABYRINTH. Lateral view of the right organ.

When the head of the animal moves to a new position, the cupulas or otoliths tend to slide over the maculae in response to gravity, thus establishing the shearing force to which hair cells are sensitive. This makes the labyrinth an excellent position register.

The labyrinth evolved a related function, which is the detection of angular acceleration, or rotation of the head. To understand the mechanism involved, imagine a straight garden hose filled with water. If one jerked the hose in the direction of its length, the inertia of the long column of contained water would cause it to tend to remain stationary while the hose itself moved. As a result, pressure would increase at the end of the hose away from the motion and water would tend to escape there. (If the hose were jerked sideways, there would be little pressure on any area equal to that of the end of the hose because there would be no long column of water in that plane.) Similarly, when a semicircular canal is rotated in its own plane, the canal moves slightly past the contained endolymph. At one end of each canal is a swelling, the **ampulla,** within which is a patch of sensory cells, here called a **crista,** which responds to the motion. Cristae have high, domed cupulas that extend across the canals, completely occluding the lumens. The hair cells are all oriented so as to respond to motion in the appropriate direction. (Hair cells of maculae are oriented in various directions.) Since one of the three semicircular canals lies in each plane of space, every rotational motion is registered.

In order to protect the delicate labyrinth, and to mask out distortions from irrelevant sources, the entire structure is surrounded by cartilage or bone. However, the labyrinth does not fit quite snugly inside its bony housing. The thin space between labyrinth and skeleton is filled (at least in places) with the fluid **perilymph,** which cushions the organ against harm and plays a role in the hearing function of the ear of terrestrial vertebrates.

Organ of Hearing Hearing in vertebrates is the response to sound vibrations by the ear. It includes both pressure waves and (in water) displacement of small particles, but excludes response to pain, touch, and currents, and also to vibrations that do not activate acoustic receptors.

As sound waves in water leave their source, their amplitude initially falls off very fast with distance, then more gradually. Close, or *near-field* signals are therefore easier to detect, but most sources are farther away, or *far-field.* Fishes having no gas chamber can detect only particle motion, not pressure waves. Far-field sound waves that are carried underwater cannot be detected by an animal of uniform density because the waves simply pass through the body; no tissue moves relative to another. However, hair cells of the internal ear might move relative to a large, dense, bony otolith because the latter would have more inertia. This appears to be the basis for hearing, including sound localization, in many fishes. Sound can also be heard underwater if the pressure waves are received and amplified by a confined compressible medium (that is, a gas) and are then translated to motion changes in the endolymph bathing hair cells. The gas bladder of bony fishes was a ready-made resonator, and two groups of fishes independently evolved mechanisms for translating pressure waves from bladder to labyrinth, as described further in the next section. These fishes are known to hear well at frequencies below 1000 Hz.

Most airborne sound waves do not strike the body with sufficient force to carry through soft tissues to a deep resonator such as the gas bladder (much of the force is simply reflected away). Consequently, in order for an animal to hear when out of water, there must first be a mechanism for receiving sound waves. Most have a thin surface membrane, the **tympanum,** or eardrum, that is free to vibrate against the low resistance of an air chamber, the **middle ear,** on its inner side. However, the evolutionary story is complex: A tympanum and middle ear evolved many times, and numerous small, ground-living amphibians and reptiles hear well in spite of not having tympanic middle ears at all. For them, sound waves are received by the body wall and skeleton.

There is a second requirement, which is to translate the motion of the sound waves to part of the labyrinth. Even if there is a middle ear cavity (the usual circumstance), merely to have the cavity impinge on the labyrinth would not be adequate, because sound waves would be damped by the soft walls of the eustachian tube, and most of the energy of the waves would be reflected back into the air in the middle ear at its interface with the more dense liquid in the labyrinth. Accordingly, one or more bony ossicles mechanically transmit the waves across the cavity. The footplate of the ear ossicle (or innermost ossicle, if there are three) vibrates against the **oval window,** which is an opening in the bony housing of the labyrinth (Figure 19.9). The area of the oval window is always much smaller than the area of the tympanum, thus increasing the pressure, or force of vibration per unit area, by the 14 to 60 times (for various mammals) needed to create adequate displacement of fluid. Furthermore, the outermost ossicles of mammals (incus and malleus) are constructed so as to decrease amplitude but increase force as they pivot. In the absence of a middle ear cavity, sound waves may be conducted directly to the labyrinth by some combination of bone, ligament, and muscle.

What remains is to translate motion of the innermost ossicle to shearing force over hair cells. When the footplate vibrates against the oval window, it creates compressional waves in the perilymph beyond. This perilymph is in a tube, called the **scala vestibuli,** that extends away from the oval window and then turns sharply on itself to return as a parallel tube, the **scala tympani.** Between these two tubes is an extension of the lagena called the **scala media.** The three adherent channels lengthen and coil—slightly in birds and markedly in mammals—and are then called the **cochlea** (= *snail shell*) (Figures 19.8 and 19.10). Within the scala media is a special auditory macula called the **basilar papilla** or, if much lengthened in a cochlea, its derivative, the **organ of Corti.**

The basilar papilla or organ of Corti rests on the **basilar membrane** separating scala media from scala tympani. This organ has several to many rows of hair cells, each having dozens of stereocilia of graded length, but no kinocilium. Fine extracellular tip links slant upward from the tip of each stereocilium to the wall of its next longer neighbor. The stereocilia are embedded in a derivative of the ancestral cupula now called the **tectorial membrane.** Vibrations at the oval window cause traveling waves in the scala vestibuli, which

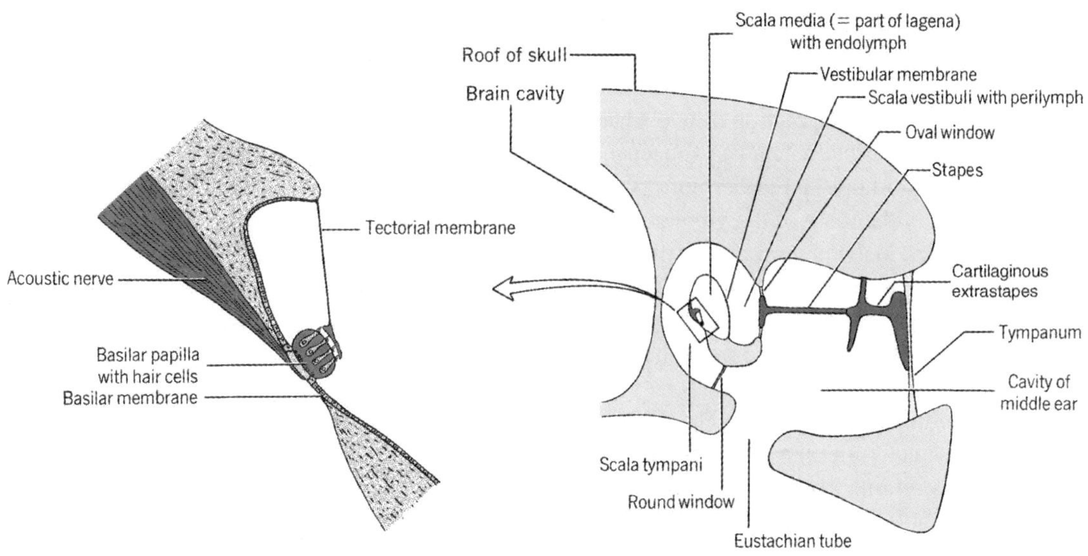

FIGURE 19.9 DIAGRAMMATIC SECTION THROUGH THE AUDITORY APPARATUS OF A LIZARD.

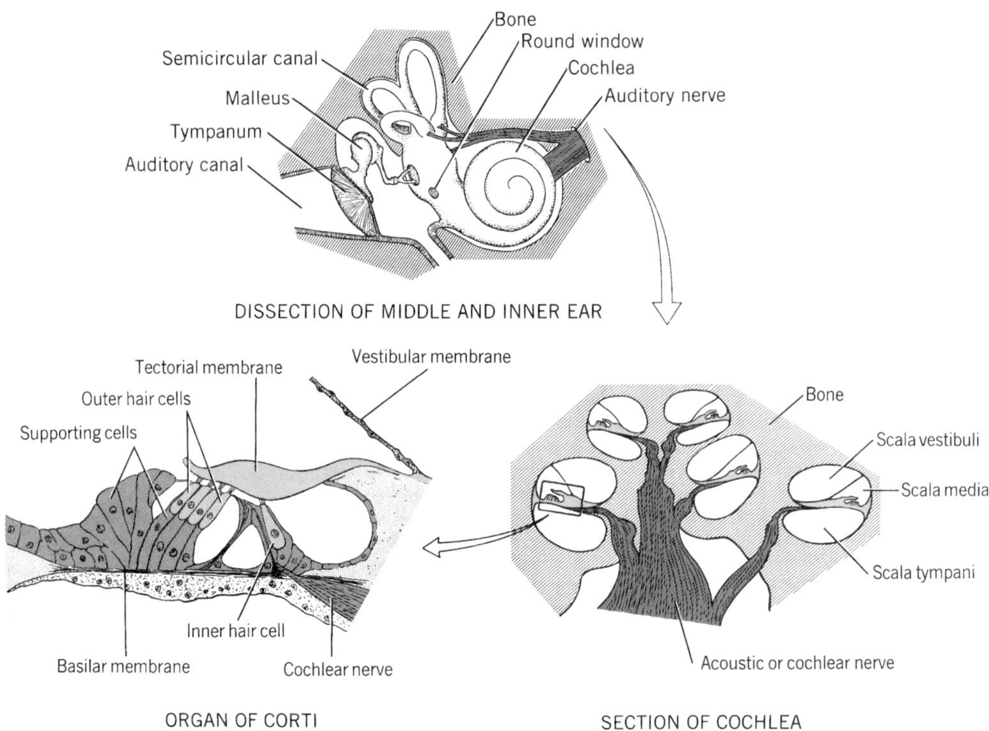

FIGURE 19.10 STRUCTURE OF THE MAMMALIAN EAR.

are translated across the thin vestibular membrane to the endolymph of the scala media, and thence across the basilar membrane to the scala tympani. These incompressible fluids can move within their bony walls because the scala tympani meets the cavity of the middle ear at a thin membrane covering the **round window.** Thus, when the tympanum and oval window move inward, the round window moves outward. (*Impedence matching* is the term given to the adjustment of airborne sound vibrations at the tympanum to the vibrations needed in the stiffer, denser inner ear.)

The basilar membrane and tectorial membrane pivot at different points, so their motions create shearing force between them, which activates the hair cells. It is thought that the slightest relative motion of two adjacent stereocilia, in the appropriate direction, tugs on their mutual tip link, which has a mechanically activated gate at each end and somehow initiates a sensory impulse. Furthermore, in some way that is unknown but under study, hair cells amplify the force of the vibration. There is a systematic change along the length of the organ of Corti in the pattern of hair cells, arrangement of stereocilia, and breadth of the organ. This is the structural basis for the differential response at different levels along the organ; high frequencies are sensed at the base of the cochlea and low frequencies toward the tip. All in all, this is an amazingly complex but effective structure.

Evolution of the Ear The labyrinth, or inner ear, evolved very early in vertebrate history and, with many variations in configuration but none of basic design and function, has been retained by all vertebrates. The middle ear evolved as tetrapods evolved, and the external ear is scarcely found except in mammals. There is fossil evidence that the first true tympanic ear occurs in the Triassic period and may have been related to the evolution of sound production by insects.

CYCLOSTOMES, and at least some ostracoderms, have fewer than three semicircular canals (Figure 19.8). It is not known if this is a primitive or degenerate condition. Utricle and saccule are not set apart, and the organ is compact.

JAWED FISHES have all three semicircular canals plus utricle, saccule, and lagena. Elasmobranchs may have, in the organ of equilibrium, fine sand grains that enter by an endolymphatic duct (where the embryonic otic vesicle pulled away from the skin ectoderm) (see Figure 19.8). Bony fishes have instead hard calcareous otoliths that rest on the maculae, of which there may be three or four. The otolith of the saccule ranges from small to so large that it nearly fills the chamber. Most fishes can hear, and some hear very well. Two groups of bony fishes have achieved far-field hearing of low-frequency sounds by adapting the gas bladder as a resonator. One group (cods, herrings) have a pair of long extensions of the gas bladder that penetrate the brain cavity to impinge on extensions of the endolymphatic sacs of the labyrinths. The other group (minnows, catfishes, goldfishes) has modified processes of adjacent vertebrae into paired chains of three or four ossicles that pivot on the spine to convey vibrations from gas bladder to endolymphatic sac. There is experimental evidence that many fishes have directional hearing as a consequence of the transfer of particle motion from water to tissue to ear. The exact mechanism remains obscure. The velocity of sound is 4.5 times greater in water than in air, and the wavelength of sound is much greater. Accordingly, time difference in reception at the two ears is not a basis for directional hearing. Rain, waves, wind, and the grating of material over the substrate can make the underwater environment noisy. Fishes must discriminate the sounds for which they are listening.

The ears of LIVING AMPHIBIANS are off the main line of descent. Caecilians, urodeles, and some anurans lack a tympanum and middle ear cavity, but have a stapes, which is attached to the shoulder girdle or skin. These ears are particularly suited for hearing low-frequency ground vibrations. Some salamanders seem to hear well both in air and water. Most adult anurans do have tympanic middle ears and, being highly vocal, hear well in air. Nevertheless, their ears are distinctive. In addition to a stapes (which is in contact with the large tympanum) there is a second ossicle (the operculum), which, with the help of a small muscle, joins the pectoral girdle to the oval window. The stapes conducts high-frequency sound and the operculum (particularly in small frogs) conducts low-frequency sound. In addition to the basilar papilla there is a second, larger **amphibian papilla,** which is unique to the class, though (according to Fritzsch and Wake) it may have had a common origin. The former papilla is sensitive to frequencies of about 1200–1600 Hz, the latter to frequencies in the range of 200–800 Hz.

Most REPTILES (and perhaps the large extinct amphibians) have a large tympanum that is either flush with the surface of the head or protected by a flap of skin (Figure 19.9). The eustachian tube may be broadly open to the pharynx. The single ossicle, or stapes, is commonly a slender but complicated structure having a cartilaginous portion with arms (the extrastapes) at its outer end. The lagena is somewhat lengthened but coils only in mammal-like reptiles. The basilar papilla is elongate and, like the tectorial membrane, is constructed in a variety of ways often departing considerably from the mammalian condition. The number of hair cells along the papilla gives some indication of the range of frequencies that can be detected and of the general sensitivity of the ear. Snakes, amphisbaenians, and some lizards lack a tympanum and middle ear. Their stapes joins the mandible or quadrate, which facilitates reception of vibrations from the substrate.

The vestibular membrane is delicate in MAMMALS but heavy in BIRDS. The basilar membrane of birds is much shorter and broader than that of mammals. The organ of Corti has one inner, and three or four outer rows of hair cells in mammals, but about 10 times as many in birds. A coiled (hence lengthened) cochlea is present in each class

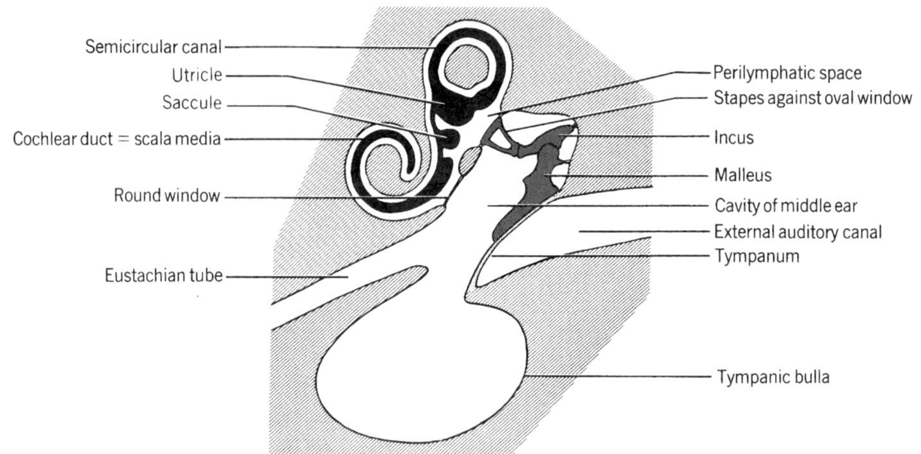

FIGURE 19.11
DIAGRAMMATIC SECTION THROUGH THE AUDITORY APPARATUS OF A MAMMAL. Somewhat distorted for two-dimensional representation. Bulla may be relatively much larger.

(except multituberculates and monotremes), making as many as five turns in mammals. Ears of mammals are distinctive in several ways: An external ear, or pinna, is usually present. This structure helps to funnel sound waves toward the tympanum, but may have other functions (cooling, recognition) and can be an inefficient funnel. The middle ear is housed in a bony tympanic bulla, and extensions of the middle ear cavity may penetrate adjacent bones as mastoid sinuses.

The bullae may be free from the remainder of the skull to limit bone conduction from extraneous sources (as from the larynx of echolocating bats) or to allow sampling of sound intensity at two independent ears (perhaps the basis for directional hearing of cetaceans). Large inflated bullae increase responsiveness to low-frequency sounds (as for kangaroo rats underground) by reducing the stiffness of the system. There are, of course, three ossicles (Figure 19.11). Mammals commonly hear sound over the wide range of 20 to 20,000 Hz. Dogs hear at 40,000 Hz, and some volant and aquatic species emit and receive sound at much higher frequencies as a means of echolocation. At the other extreme, elephants, rhinos, and some whales can hear sounds as low as 12 Hz. The ossicles of cetaceans are massive. Vibration may reach them by way of the auditory canal (usually curved and occluded) or a thin region of the jawbone coupled with a pad of fat. Loudness of sound is detected over about 10 orders of magnitude. Very strong vibrations are damped to prevent injury by tiny muscles and by buckling of the chain of ossicles. Birds and mammals have remarkable ability to locate the source of a sound—even to within 2° or 3° in owls and bats. Several cues are used to contrast the acoustic waveform at the two ears. Birds are probably assisted in this by broad communication between the two middle ears, and some mammals by the structure and function of their external ears.

EYES

All vertebrates have a pair of lateral eyes, and some also have an unpaired dorsal eye. The lateral eye is a complex precision instrument. It is unmatched for range of adaptation, being modified to function in water or air, by day or night, at short range or far, and in habitats ranging from the sky to the depths of the ocean. From the eye alone, morphologists can determine much about the habits of any vertebrate. Evolutionary lineages are less easily established from eye structure, though some trends are evident. In several instances, prior knowledge of phylogeny helps to explain otherwise puzzling characteristics of the eye.

Structure and Function Structure and function can profitably be studied in terms of the requirements of the eye as an optical instrument. A first requirement is for a firm housing;

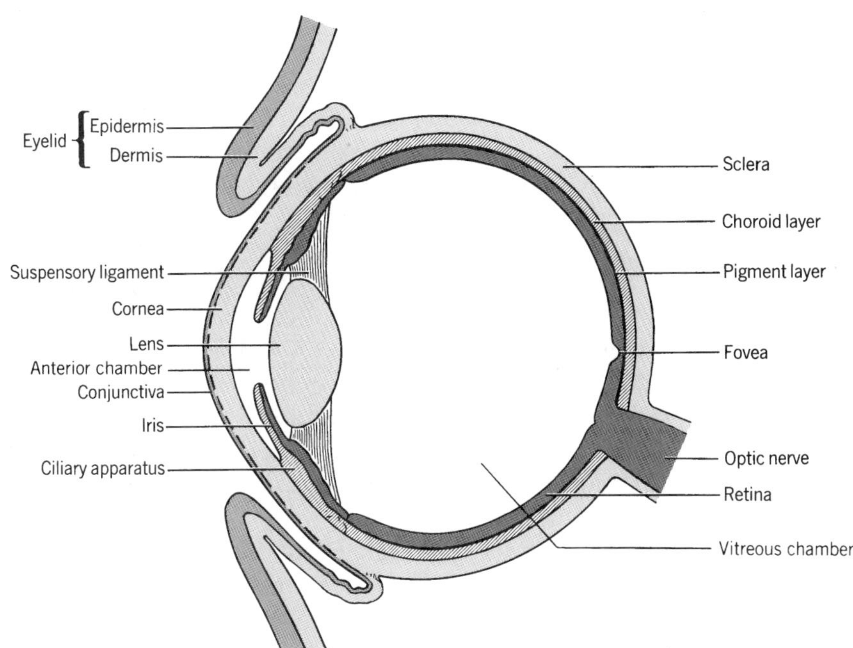

FIGURE 19.12 GENERAL STRUCTURE OF THE EYE OF A DIURNAL MAMMAL.

sharp images would not be possible on the filmplate of a camera having soft walls. The capsule of the eye is kept rigid by its outermost envelope, the **sclera,** which is stiffened by cartilage, or bone, or a tough mesh of collagenous fibers, or by a combination of these (Figure 19.12). Pressure of the fluid within the eye also contributes to keeping the envelope distended and firm.

Another requirement is for control of the amount of light entering the eye. This is accomplished by a curtain, called the **iris,** which surrounds the light window, or **pupil.** Delicate muscles within the iris cause it to pull back and enlarge the pupil when light is dim, and extend to narrow the pupil to a small circle or slit when light is bright.

There must be no random reflections of light within the eye, so the capsule is usually lined by a nonreflective black pigment. This **pigment layer** is just behind the light-sensitive cells, where it can, in some vertebrates, temporarily surround the ends of those cells to screen them from excessively bright light. (There is directed reflection of light by some eyes—see below.)

An important requirement is for the formation of an image. When a ray of light passes from one material into another having a different refractive index, it changes direction at the interface. The direction and amount of the change depend on the materials and on the angle the ray of light makes with the interface. A **lens** is a transparent object shaped so as to bend light into an undistorted image. The image is "upside down" both on the filmplate and the retina (Figure 19.13). The brain somehow turns the image over, or so it seems.

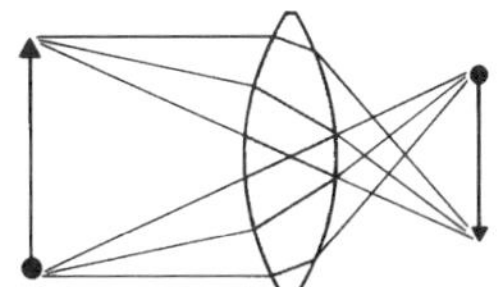

FIGURE 19.13 FORMATION OF AN IMAGE BY A LENS.

Over the front of the eye, the sclera merges into the transparent **cornea.** The cornea has about the same refractive index as water. Accordingly, corneas of eyes that function underwater do not bend light and can be of any shape. They are usually flat or streamlined. Rays of light entering the cornea from air are bent sharply. Hence, the cornea of terrestrial animals must be accurately curved to avoid the distortion of the image called astigmatism. The lens then completes the image formation initiated by the cornea.

When passing through an ordinary lens, central and marginal rays have slightly different focal points, thus causing spherical aberration. The lensmaker overcomes this problem by building fine lenses from several apposed elements. The eye overcomes it by having the cornea curve slightly less at its margins, and by having the lens more dense at its core. Furthermore, light rays of different wavelengths bend slightly differently when passing through an ordinary lens, thus causing chromatic aberration. The maker of fine lenses combines crown and flint glass, which have compensating properties. The eye that requires a sharp color image screens out light having the shortest wavelengths (violet and blue) by using a color filter. Yellow pigment is added to the lens or retina, or oil droplets of yellow or red color are distributed in the photoreceptor cells.

Provision must be made for focusing the image where it will be recorded. In the eye, this is called **accommodation.** Some vertebrates (like the photographer) move the lens outward from a farsighted resting position to focus on near objects. Others move it inward from a nearsighted resting position to focus on far objects. Amniotes instead focus by changing the shape of a stationary lens, causing it to bulge for near vision and flatten for far vision. This is done by changing the tension in the **ciliary apparatus** that suspends the lens. Different amniotes use different mechanisms described in a later section. The very small lenses of small vertebrates naturally have relatively great depth of focus and close near points, so little accommodation is needed.

The next requirement is for perception of the image. Photoreceptor cells are of two kinds, **rods** and **cones.** Each has a nervelike base, an inner segment with nucleus and mitochondria, and an outer segment that is long and cylindrical for rods but shorter and conical for cones (Figure 19.14). The outer segment, which is narrowly connected to the nourishing inner segment, is intricately folded into a stack of hundreds of lamellae.

On these lamellae is **visual pigment,** which is instantly altered chemically in the presence of light. This change is translated to nervous impulses in a manner as yet not understood. Rods are all alike. Their pigment is **rhodopsin** (or, in freshwater fishes and tadpoles, a related pigment), each molecule of which responds to a single quantum of light energy—the theoretical limit. Cones are of three classes having different pigments and responding to light of different wavelengths. Thus, cones are the basis for color vision. Some vertebrates have double cones, which have different pigments, and some have twin cones, which have the same pigment. All types of cones are commonly distributed over the retina in a regular mosaic pattern according to species. Rods are more sensitive than cones by about two orders of magnitude but do not record color. Nocturnal animals have only

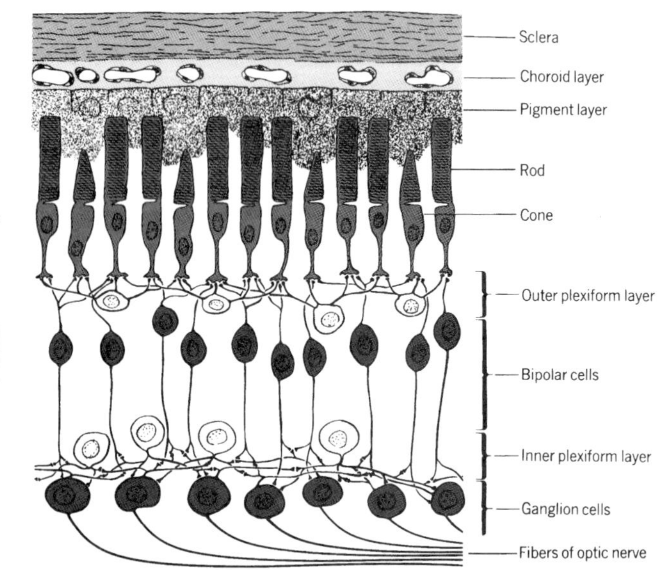

FIGURE 19.14 HISTOLOGY OF A MAMMALIAN RETINA seen in cross section.

rods. Some fishes, turtles, birds, and rodents have visual pigments that are appropriate for sensing ultraviolet light.

Beyond the rods and cones (in the direction of the inside of the eyeball) are **bipolar cells** and, beyond these, **ganglion cells,** whose axons extend through the optic nerve. There are also two plexiform layers of nerve cells, one meshed into the system at each end of the bipolar cells. This complex of nerve cells accomplishes summation of stimuli from different photoreceptors (particularly rods) and links cones in ways that are essential for integrated color vision.

In order for the eye to have acuity of vision—as is usual for diurnal animals that rely on vision in feeding, moving about, or avoiding danger—the retina must record the image with a fine grain; photoreceptors must be tightly packed so different cells can respond differently to adjacent parts of the small image. In most diurnal vertebrates having color vision (most teleosts, frogs, most reptiles and birds, some mammals), the sharpest image is perceived at a place on the retina called the **area centralis** where the cells are most slender and closely packed (as many as $10^6/mm^2$), where there are only cone cells, and where many animals have a pit called the **fovea,** which bends light rays to enlarge the image by as much as 30 percent. Humans have a "good" area centralis (we move our eyes when we read to keep the print focused there), but not the best: We need a low-power binocular to gain the acuity of vision of a hawk!

If the eyes move independently in their orbits, their fields of vision may be independent or may at least partly overlap. If they have overlapping fields of vision, depth perception, or stereoscopic vision, is achieved. In mammals, nervous coordination prevents independent movement of the eyes.

Finally, the eye requires some structures that are accessory to its optical functions. A vascular **choroid layer** nourishes the eye. Extrinsic muscles turn the eyeball in its socket (see figure on p. 184). Glands and lids (sometimes including a third eyelid, or **nictitating** membrane) moisten and protect the eye in air. The eye appears to look through a hole in the skin, but the edges of the lids are in fact folds in the skin, not breaks, and the skin continues over the cornea as the thin, transparent **conjunctiva.**

This book cannot cover many of the specializations of the vertebrate eye, but two principal adaptations will be noted: The general structure of eyes adapted for vision underwater is presented in Chapter 27, and adaptations for vision when there is little light are presented here.

The many amphibians and mammals, and few reptiles and birds that are active at night, some fishes that frequent murky water, and many of the fishes that live deep in the ocean (where sunlight does not penetrate but where animals are luminescent) must be able to see in dim light. Their eyes are very large in front to gather in as much light as possible (Figure 19.15). The pupil opens very wide (though it may also close to a narrow slit by day). The lens is large, spherical, and placed well back in the eye to be close to the retina. The retina is relatively small, so the entire eye is deep in the optical axis, or even tubular. This arrangement mimics the slide projectionist who gains a bright image by placing his projector close to a small screen.

Cone cells are few or absent. The more sensitive rod cells are slender and closely packed. There is so much summation by the ganglion cells that light striking a thousand or more rods may trigger an impulse in a single nerve fiber. This enormously increases sensitivity to dim light, but slightly blurs the image. An area centralis is often lacking so that the eye, which may be too large to turn much in its socket, need not be oriented exactly toward the objects perceived. The entire head is turned if need be—some owls can rotate the head 270°!

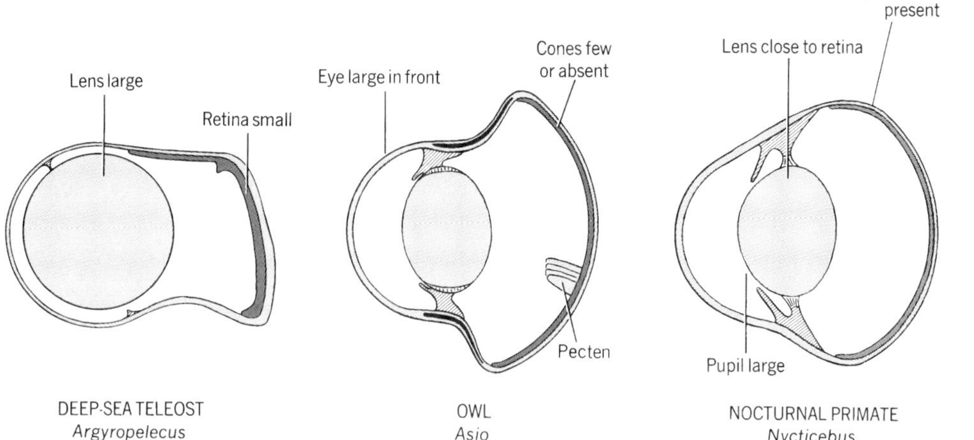

FIGURE 19.15 ADAPTATIONS OF THE EYE FOR VISION IN DIM LIGHT.

Finally, most nocturnal vertebrates have a mirror, or **tapetum lucidum,** behind the rods that reflects light back out of the eye, so light passes through the rods twice instead of once. Analogous mirrors have evolved several times; reflection may be from crystals of guanine, lipid particles, or other pigmented layer, or supportive tissues.

Development and Origin Developmentally, the eye has three principal parts: retina and pigment layer, lens, and supportive tissues. First to appear is the primordium of the retina, which is an evagination of the sidewall of the diencephalon. This expansion, or **optic vesicle,** pushes toward the skin ectoderm, trailing an **optic stalk** behind (Figure 19.16). Next, the lateral surface of the vesicle sinks into the cavity of the vesicle, thus forming a double-

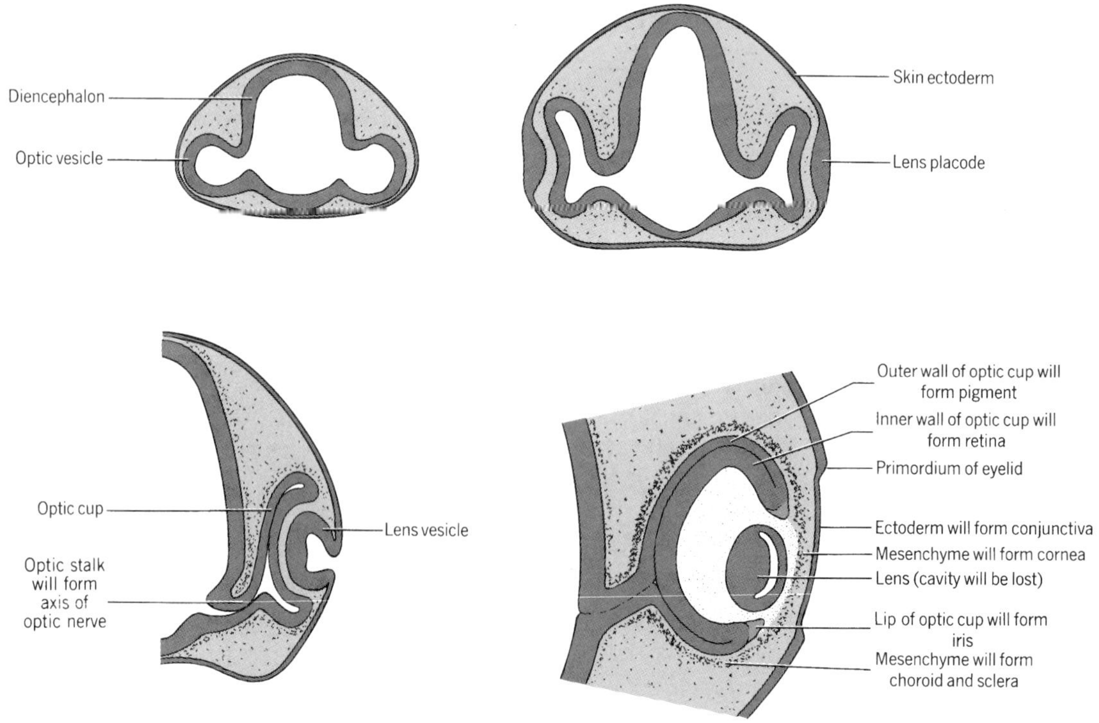

FIGURE 19.16 DEVELOPMENT OF THE EYE. Cross sections of the head.

walled cup resembling the bulb of a rubber syringe indented on one side by the thumb. The outer wall of the cup becomes the pigment layer and the inner wall the retina. The lip of the cup forms the iris. The optic stalk becomes the optic nerve as axons from ganglion cells in the retina extend down its length. In a sense, the retina is a nucleus of the brain and the optic nerve a tract of the brain.

The lens forms from a placode of the skin ectoderm that develops only under the inductive influence of the underlying optic vesicle. The **lens placode** becomes a hollow **lens vesicle,** but the cavity is soon diminished and then obliterated by the thickening of the inner wall of the vesicle.

Mesenchyme surrounding the optic cup differentiates into choroid, sclera, cornea, and ciliary apparatus. Developmentally, these are continuous with, and apparently also homologous with, the meninges of the brain.

The phylogenetic origin of the vertebrate eye is unknown; the first known vertebrates had already perfected the organ. This has not discouraged morphologists from speculating, however, and nearly a dozen theories have been proposed since the 1870s. The most plausible theory is strongly supported by developmental, and weakly supported by histological and comparative anatomical clues. The neurocoel and ventricles of the brain are lined with ependymal cells that are usually ciliated. In the head region of many chordates, from amphioxus to birds, these cells are light-sensitive. If phylogeny is repeated by ontogeny, then rods and cones evolved from ependymal cells of a part of the forebrain, and the eye could have been functional throughout its evolution. The outer segments of rods and cones are considered to be modified cilia.

(A dermal light sense is common in cyclostomes, fishes, and amphibians. In some instances the receptor is unknown, but light-sensitive cells are identified in the skin of some species. This is probably an ancient condition.)

Evolution of Lateral Eyes The hagfish, a scavenger in deep water, has degenerate eyes and is blind. The well-developed eyes of the LAMPREY are primitive in two respects: The conjuctiva is not fused to the cornea, and the ependymal layer is retained in the core of the optic nerve. Other distinctive features may or may not be primitive: The size of the pupil is fixed; the sclera is not stiffened by cartilage or bone; accommodation results when an extrinsic muscle pulls against the cornea, thus pressing the lens inward; and the lens is held in position by pressure only, not by a suspensory apparatus. As for primary swimmers in general, the eye is large and shallow along its optical axis, the lens is large and spherical, eyelids and glands are absent, and the extrinsic oblique muscles rotate the eyeball around its optical axis.

There is a wide range of eye structure within the large assemblage of fishes, but some generalizations can be made. ELASMOBRANCHS stiffen the sclera with cartilage (Figure 19.17). They accommodate somewhat by pulling the lens forward from its resting position with a small intrinsic muscle of ectodermal origin. A unique cartilaginous pedicel props the eyeball away from the back of the orbit; its function is not clear. Cones are few or absent. An area centralis is present. Crystals of guanine in the choroid cause it to reflect light as a tapetum lucidum. The chamber between lens and cornea is small.

BONY FISHES stiffen the sclera with cartilage and, in most teleosts, by several bony plates. The cornea is flat or streamlined. Cones are usually present, so (as people who enjoy fly-fishing should know) color vision is typical. (Pigment cells have been identified in the armor of a placoderm, suggesting that color vision had already evolved in the Devonian.) Usually, an intrinsic mesodermal muscle pulls the lens inward to accommodate for far vision (chondrosteans, and some other fishes, have no accommodation). Teleosts have a nutritive structure derived from the choroid, called the **falciform process,** which projects into the cavity of the eyeball. Many teleosts have an area centralis, and some have a fovea.

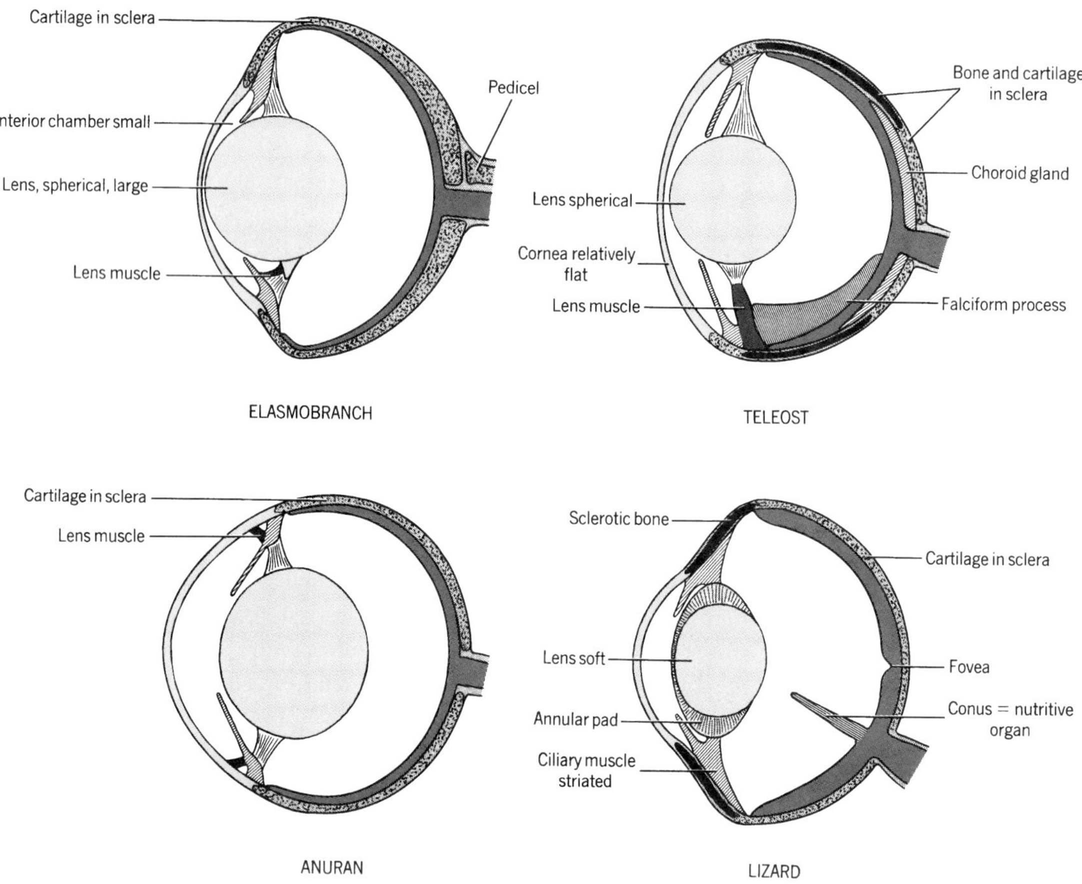

FIGURE 19.17 COMPARATIVE ANATOMY OF THE EYE.

As larvae, AMPHIBIANS have fishlike eyes. At metamorphosis they develop eyelids and glands and an evenly curved cornea. A small mesodermal muscle within the eyeball moves the lens forward for near vision. The eyes of anurans are better developed than those of urodeles. Some have color vision, as, in all probability, did many labyrinthodonts. (Caecilians and some salamanders are blind.)

REPTILES and BIRDS, most of which are diurnal, have the finest visual acuity and accommodation of all vertebrates. The extent to which the binocular vision of birds (overlap of the visual fields of the two eyes) results in depth perception has been the subject of research and discussion (see Martin and Katzir in References). The eye is large, particularly in birds; eyes of some hawks are larger than those of humans. The sclera is stiffened by a cartilaginous cup behind and by a ring of about 15 small, overlapping bones on the forward wall, where the eyeball might otherwise be distorted by the ciliary muscles. These muscles are striated (though of ectodermal origin) and, hence, faster in action than those of other vertebrates. The small, soft, somewhat flattened lens has a peripheral **annular pad** that makes firm contact with the ciliary apparatus. Accommodation is active and instant; muscular contraction causes the lens to bulge in front. Cones are usually numerous and color vision excellent (except in snakes and crocodilians). An area centralis is present and there is one or, in many birds and some lizards, two foveas. There is a vascular projection into the cavity of the eyeball. This is particularly large and complexly folded in birds, where it is called the **pecten** (= *comb*) (Figure 19.15). Of the many functions postulated for this structure,

COMMENT 19.2

CONSERVATION AND INNOVATION: SIGNPOSTS FOR FUTURE STUDIES

Discovery of the basis for the evolutionary diversity of vertebrate form is, of course, a fundamental goal of comparative morphology. The sensory systems are tractable for pursuit of this goal because at one level of structural and functional organization, they illustrate much innovation and structural diversity, but on another they have remained conservative throughout vertebrate history (see Fay and Popper in References). Comparative studies illustrate, for example, that sensory end organs, the peripheral sensory components responsible for acquiring and transducing various stimuli, often demonstrate a wide diversity of morphologies among members of the same class, family, and even order. The central components responsible for receptor coding and neural processing, in contrast, appear quite similar across vertebrates, and thus reflect a phylogenetic conservatism. The latter was illustrated for olfaction, vision, taste, and the vestibular and mechanosensory lateral line systems when we considered the cranial nerves: the central pathways are quite similar among vertebrates.

Hodos and Butler point out that often, with the appearance of novel sensory systems such as electroreception or infrared detection, novel nuclei also appear in the central pathways. In addition, the evolution of sensory specialists, forms that can detect an exceptionally broad range of stimuli and can make exceptionally fine distinctions between different sensory qualities (e.g., bats for hearing, birds for color and visual discrimination), also demonstrate correlative changes in the central nervous system.

Also evident has been the development of novel sensory receptors that appear to be the result of a modification of existing receptors. An example is electroreception, a sense that has been lost and then "re-evolved" a number of times in various taxa (see New in References). The myxinoids (hagfishes), the sister group to the "remaining vertebrates" (an awkward term in this context, see p. 33), do not appear to possess an electrosense, but they do possess a mechanosensory lateral line system, the system from which electrosense is thought to have evolved. Lampreys (petromyzontids), the most primitive of the remaining vertebrates, possess both. Finally, all teleost fishes possess a mechanosensory system, but only a few possess an electrosensory system and, among those, two distinct morphologies in the receptor organs are observed. When these novel and conserved features of receptor mechanisms and of the peripheral and central nervous systems are mapped on a phylogenetic construct, one is able to generate testable hypotheses concerning the evolution of specific morphologies and of their evolutionary history. "How do new sensory nuclei develop?" "How do new sensory systems evolve?" (As most of the examples given here are based on limited studies of only a few organisms, much remains to be done.)

that of nutrition is the most probable. Lacrimal gland and nictitating membrane are present. Iris musculature is striated. A tapetum lucidum is rare. Some binocular vision is typical, particularly in predatory birds. (The eyes of snakes are in some ways atypical because of derivation from burrowing ancestors for which vision was unimportant. Amphisbaenians, some lizards, and some snakes are blind.)

MAMMALS first evolved as small nocturnal creatures, and at that stage lost the perfection of eye structure of their reptilian ancestors. Some of the loss was later regained, but not all. There is no cartilage or bone in the sclera. Muscles of the ciliary apparatus and iris are smooth and, therefore, relatively slow. The shape of the lens is adjusted to focus the image, but the mechanism is inferior to that used by reptiles and birds; contraction of the ciliary muscles relieves tension on the suspensory apparatus, allowing the lens to bulge of its inherent elasticity. This process is relatively slow, particularly in old age, when it may also become incomplete, making near vision impossible (without eyeglasses). There is no pecten or corresponding nutritive organ. A tapetum lucidum is confined to several orders; a nictitating membrane is rare. Color vision has been at least partially regained in diurnal species of various orders and is highly evolved in primates and some rodents. An area centralis is present in some orders; a fovea is present only in higher primates. Binocular vision is common and coordination of the eyes is superior. (The eyes of monotremes are in some respects atypical of the class.)

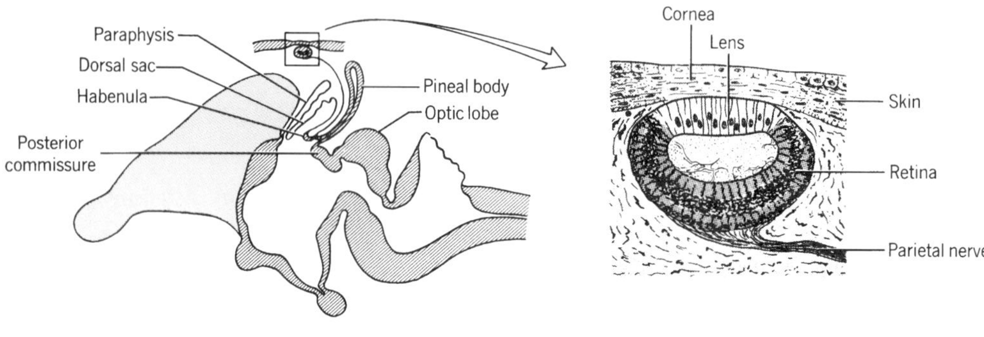

FIGURE 19.18
DORSAL EYE OF A LIZARD.

Dorsal Eyes On top of the diencephalon there are, in the midline, two small evaginations: an anterior **parietal organ** (or parapineal) and a posterior **pineal body** (or epiphysis). One or both of these structures may be photoreceptive, and it is then termed the third eye.

Asymmetry of these structures in adult lampreys and embryo lizards, and the conformation of ostracoderm head armor, indicate that parietal and pineal organs may be derived phylogenetically from a bilateral pair of organs, the left member of which shifted forward to the midline while the right member slipped behind.

The pineal organ is present in nearly all vertebrates. It has photoreceptive cells in lampreys, and endocrine properties are known for various vertebrates. Initially, at least, the endocrine functions seem to have been light-related (see p. 377). The parietal organ also has photoreceptor cells in lampreys, but in these animals it is subordinate to the pineal eye. The parietal eye is functional in tadpoles and salamander larvae and in many lizards, where it may have a lens, retina, and tiny nerve (Figure 19.18). Histologically, its photoreceptive cells are closely similar to the cones of lateral eyes. Such light-associated behavior as thermoregulation and activity rhythms is influenced by this organ in some lizards.

It is evident from the widespread presence of a parietal (or pineal) foramen in skulls of ostracoderms, placoderms, and early representatives of bony fishes, labyrinthodonts, and reptiles that a third eye has been persistent in the evolution of vertebrates.

REFERENCES

Able, K.P. 1991. Common themes and variations in animal orientation systems. *Am. Zool.* 31:157–167.

Ali, M.A., and M.A. Klyne. 1985. *Vision in vertebrates.* Plenum, New York. 272p.

Axel, R. 1995. The molecular logic of smell. *Sci. Am.* 273(4):154–159.

Barinaga, M. 1991. How the nose knows: olfactory receptor cloned. *Science* 252:209–210.

Bertmar, G. 1981. Evolution of vomeronasal organs in vertebrates. *Evolution* 35:359–366.

Blaxter, J.H.S. 1987. Structure and development of the lateral line. *Biol. Rev.* 62:471–514.

Borg, E., and S.A. Counter. 1989. The middle-ear muscles. *Sci. Am.* 261(2):74–80.

Clack, J.A. 1997. The evolution of tetrapod ears and the fossil record. *Brain Behav. Evol.* 50:198–212.

Crescitelli, F. 1977. *The visual system in vertebrates.* Springer, New York. 813p. Includes chapters on adaptations to the deep sea, vision in turtles, the pineal system, the avian eye, and comparative optics in mammals.

Doving, K.B., and D. Trotier. 1998. Structure and function of the vomeronasal organ. *J. Exp. Biol.* 201:2913–2925. A review article.

Eakin, R.M. 1970. A third eye. *Am. Sci.* 58:73–80.

Fay, R.R., and A.N. Popper. 1985. The octavolateralis system, pp. 291–316. *In* M. Hildebrand et al. (eds.), *Functional vertebrate morphology.* Harvard Univ. Press, Cambridge, MA.

Fay, R.R., and A.N. Popper (eds.). 1997. Symposium: evolution of vertebrate sensory systems. *Brain Behav. Evol.* 50:187–259.

Fay, R.R., and A.N. Popper (eds.). 1998. Comparatative hearing: fish and amphibians. Springer, New York. 438p. Each chapter is a review article.

Fritzsch, B., and M.H. Wake. 1988. The inner ear of gymnophione amphibians and its nerve supply: a comparative study of regressive events in a complex system (Amphibia, Gymnophiona). *Zoomorphology* 108:201–217.

Gamow, R.I., and J.F. Harris. 1973. The infrared receptors of snakes. *Sci. Am.* 228(5):94–100.

Halpern, M. 1987. The organization and function of the vomeronasal system (review). *Annu. Rev. Neurosci.* 10:325–362.

Halpern, M., and D. Holtzman (eds.). 1993. Symposium: chemosensing and chemosignaling in reptiles. *Brain Behav. Evol.* 41:119–268. In-depth reviews of olfaction, vomeronasal organs, and other chemical sensing systems.

Hodos, W., and A.B. Butler. 1997. Evolution of sensory pathways in vertebrates. *Brain Behav. Evol.* 50:189–197.

Hudspeth, A.J. 1985. The cellular basis of hearing: the biophysics of hair cells. *Science* 230(4727):745–752.

Jacobs, G.H. 1992. Ultraviolet vision in vertebrates. *Am. Zool.* 32:544–554.

Knudsen, E.I. 1981. The hearing of the barn owl. *Sci. Am.* 245(6):113–125.

Land, M.F., and R.D. Fernald. 1992. The evolution of eyes. *Annu. Rev. Neurosci.* 15:1–30.

Levine, J.S. 1985. The vertebrate eye, pp. 317–337. *In* M. Hildebrand et al. (eds.), *Functional vertebrate morphology.* Harvard Univ. Press, Cambridge, MA.

Lohmann, K.J., and S. Johnsen. 2000. The neurobiology of magnetoreception in vertebrate animals. *Trends Neurosci.* 23:153–159.

Lombard, R.E. 1991. Experiment and comprehending the evolution of function. *Am. Zool.* 31:743–756. About the tetrapod ear.

Lombard, R.E., and T.F. Hetherington. 1993. Structural basis of hearing and sound transmission, vol. 3, pp. 241–302. *In* J. Hanken and B.K. Hall (eds.), *The skull.* Univ. Chicago Press, Chicago.

Martin, G.R., and G. Katzir. 1999. Visual fields in short-toed eagles. *Circaetus gallicus* (Accipitridae), and the function of binocularity in birds. *Brain Behav. Evol.* 53:55–56. Discussion and references serve as an introduction to controversy as to whether or not birds and mammals with binocular vision possess three-dimensional discrimination.

New, J.G. 1997. The evolution of vertebrate electrosensory systems. *Brain Behav. Evol.* 50:244–252.

Newman, E.A., and P.H. Hartline. 1982. The infrared "vision" of snakes. *Sci. Am.* 246(3):116–127.

Northcutt, R.G. 1992. Distribution and innervation of lateral line organs in the axolotl. *J. Comp. Neurol.* 325:95–123.

Popper, A.N., and R.R. Fay. 1993. Sound detection and processing by fish: critical review and major research questions. *Brain Behav. Evol.* 41:14–38. Strong message that studies from comparative morphology are critically needed.

Schnapf, J.L., and D.A. Baylor. 1987. How photoreceptor cells respond to light. *Sci. Am.* 256(4):40–47.

Schwentk, K. 1994. Why snakes have forked tongues. *Science* 263:1573–1577.

Stuart, D.G., and J.C. McDonagh. 1997. Muscle receptors, mammalian, vols. 1–2, pp. 1249–1253. *In* G. Adelman and B. Smith (eds.), *Encyclopedia of neuroscience,* Elsevier Science, New York.

von der Emde, G. 1999. Active electrolocation of objects in weakly electric fish. *J. Exp. Biol.* 202:1205–1215. Illustrates through a combination of behavioral, anatomical, and physiological studies of isolated neuromasts the differences in the neural mechanisms for detection of two groups of weakly electric fish.

Webster, D.B., R.R. Fay, and A.N. Popper (eds.). 1992. *The evolutionary biology of hearing.* Springer, New York. 859p. A reference of 36 chapters by 78 contributors. Indexed. Fully referenced.

Wever, E.G. 1985. *The amphibian ear.* Princeton Univ. Press, Princeton, NJ. 488p.

Zeigler, H.P., and H.-J. Bischof (eds.). 1993. *Vision, brain, and behavior in birds.* MIT Press, Cambridge, MA. 415p.

Chapter 20

Endocrine Glands

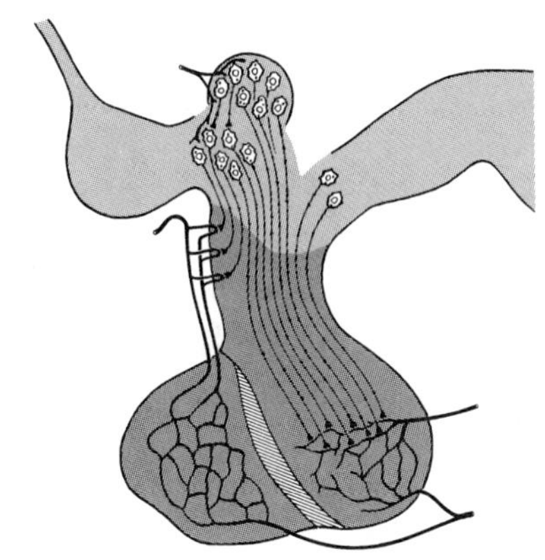

CHEMICAL MEDIATION

When a chemical released at one place in the body influences biological activity at another place, chemical mediation is achieved. This process is so ubiquitous that it is useful to sort out some of the variations. A chemical (e.g., a metabolite) produced by a cell may directly affect the subsequent activity of that same cell (a pathway termed intracrine), or may first diffuse out of the cell and then reenter and alter activity, perhaps of the cell membrane (the autocrine pathway). Also, a chemical may diffuse out of one cell and then alter the activity of another cell (paracrine pathway). The cells may be of the same kind; neurotransmission across a synapse is of this sort. The cells may also be of different kinds; embryonic induction (as of the neural plate by adjacent chordamesoderm) is of this sort. From there it is only a small step to the *endocrine* pathway.

GENERAL NATURE OF ENDOCRINE GLANDS

The **endocrine glands** are ductless glands having secretions, called **hormones,** that are discharged into the blood (or in some instances, lymph or cerebrospinal fluid) for distribution to responsive tissues elsewhere. Hormones may influence much of the body (e.g., growth hormone, thyroid hormones), but usually only certain target tissues are responsive to each hormone. Response may be morphologic, as for sex hormones that influence the development of secondary sexual characteristics, or physiologic, as for an adrenal hormone that influences kidney function. Many hormones act directly on tissues that are receptive to their "messages." Often, however, the action is indirect. Thus, a hormone secreted by the hypothalamus of the brain and released by the hypophysis causes the ovary to secrete a hormone to which the lining of the uterus responds.

Taken together, the endocrine glands function like the nervous system in several respects: Each controls and integrates bodily functions, each mediates control through the release of chemicals (often the same chemicals), and each may accomplish interaction within its own system to coordinate its activities. Endocrine control differs from nervous control in tending to be slower and more sustained; however, the two systems merge.

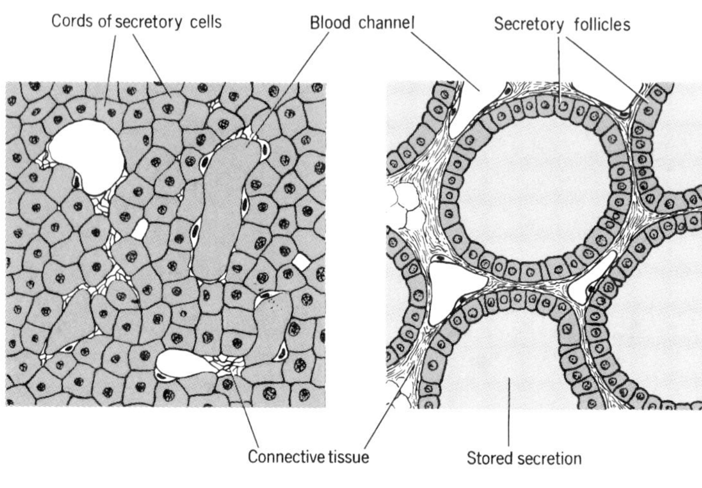

FIGURE 20.1 TWO HISTOLOGICAL TYPES OF ENDOCRINE GLAND.

All endocrine glands are small and highly vascular. They are often diffuse in lower vertebrates, but tend to be discrete in tetrapods. Most are constructed of cords of more or less cuboidal cells arranged among sinusoids and supported by a matrix of connective tissue (Figure 20.1). Several endocrine glands (neurohypophysis, urophysis) are instead constructed of thin attenuated cells, and one (thyroid) is constructed of follicles. The hormones of glands of mesodermal origin (gonads, adrenal cortex, placenta) are steroids, whereas hormones of glands of ectodermal or endodermal origin are proteins, peptides, or other derivatives of amino acids.

Since endocrine glands utilize the circulatory system to transmit signals, their shape is of little importance, and a gland may usually be single, multiple, or diffuse without relation to function. Only the aggregate volume of cells is critical. The system being ancient (the endocrine and nervous systems probably evolved together), its secretions have been remarkably constant over the ages, though there are some exceptions, and responses to the secretions are various. Endocrine glands have diverse developmental origins, are usually unrelated to one another in space, and are incompletely related in function. They do not compose an organ system in the usual sense.

Although generalities about endocrine organization and hormone function have remained somewhat stable over the past several years, keep in mind that new hormones, pathways, and subtleties of cell interaction are being discovered and described at a rapid pace. The use of immunohistochemistry and radioimmunoassays as techniques have contributed to these advancements, as has the recognition that the endocrine and nervous systems are tightly integrated. But even in this time of accelerated learning, there remains much comparative work to be done because the number of vertebrate species examined to date is small.

STRUCTURE, FUNCTION, AND EVOLUTION OF THE GLANDS

Hypophysis The **hypophysis,** or **pituitary gland,** is located under the hypothalamus of the brain (see figures on pp. 328, 339, and 341). In mammals it is housed in a bony pocket of the basisphenoid bone. Although it is small, this gland is, in both structure and function, one of the most complicated organs of the body. Developmentally it has a surprising dual origin: The adult portion called the **neurohypophysis** forms from the part of the floor of the embryonic diencephalon termed the **infundibulum.** This structure may evaginate (most tetrapods) or remain nearly unfolded (amphibians, some fishes). The remainder of the gland, or **adenohypophysis,** forms from an evagination of the ectodermal part of the embryonic mouth cavity (the stomodaeum) called the **hypophyseal pouch,** or Rathke's

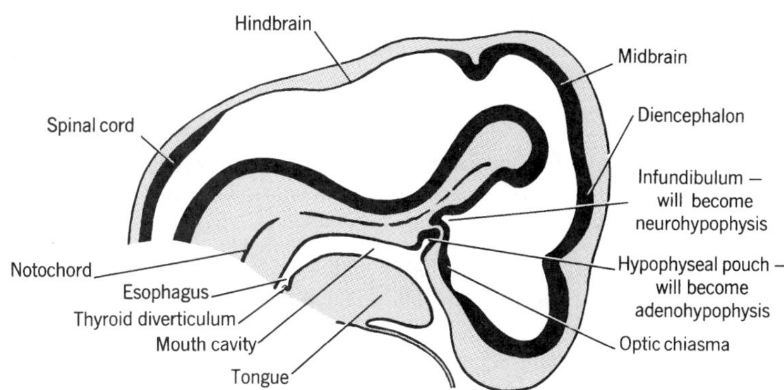

FIGURE 20.2 EMBRYONIC ORIGIN OF THE HYPOPHYSIS shown by a sagittal section of the head of a mammalian embryo.

pouch (Figure 20.2). This pouch and its derivatives are variously lobed in the different vertebrates. Its connection to the mouth is usually lost during maturation.

In general, the *neurohypophysis* has an anterior subdivision, the **median eminence** (not identified in forms below lungfishes), and a posterior or ventral subdivision, the **pars nervosa,** or posterior lobe of the gland (Figure 20.3). The *adenohypophysis* has several

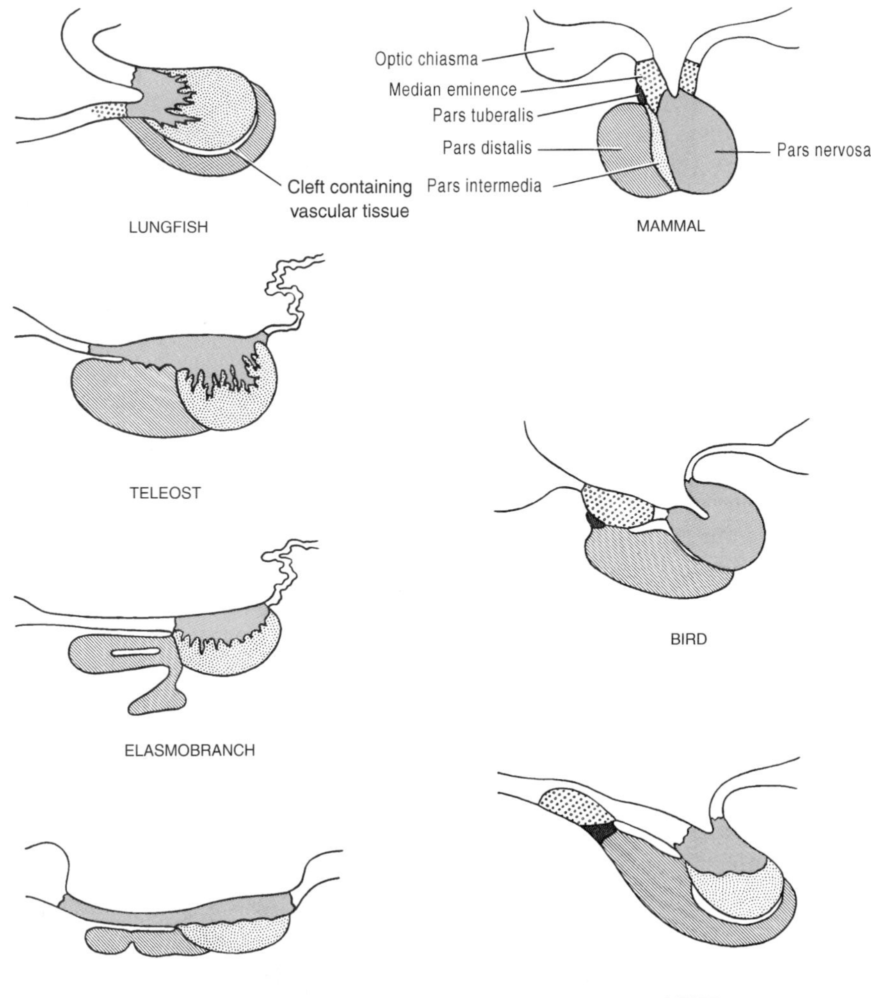

FIGURE 20.3
COMPARATIVE ANATOMY OF THE HYPOPHYSIS as seen in sagittal section. Anterior is to the left.

parts: The largest, most constant, and most active is the **pars distalis** (or anterior lobe). A **pars intermedia** is usually present but is lacking in birds. Tetrapods have a thin **pars tuberalis** between the median eminence and pars distalis. Its function is not clear; it may relate to reproduction. The parts of the pituitary differ widely among the vertebrates, and homologies can be made, if at all, only by combining clues from embryology and histochemistry. The median eminence and pars distalis have a common blood supply; the pars nervosa has an independent blood supply.

The neurohypophysis is atypical of endocrine glands in that it is constructed largely of long parallel nerve fibers originating in the hypothalamus of the brain (Figure 20.4). Indeed, this part of the hypophysis functions by storing and releasing into the bloodstream hormones elaborated in the hypothalamus and transferred to the neurohypophysis by neurosecretion (of which more is discussed below). Very different is the adenohypophysis, which has cords of secretory cells of various kinds that branch without pattern among sinusoids.

The *pars intermedia* secretes one hormone, **melanocyte-stimulating hormone,** which causes the melanin in pigment cells to disperse, thus making the skin darker. The *pars nervosa* releases the polypeptide **antidiuretic hormone** (vasopressin), which contributes to osmoregulation by causing the kidney to hold back fluid. It also causes vasoconstriction. In mammals the pars nervosa produces **oxytocin,** which initiates the letdown of milk and the contraction of the uterus.

The *pars distalis* secretes at least six hormones, four of which function by stimulating other endocrine glands: **thyroid-stimulating hormone** (thyrotropin) acts on the thyroid, **adrenal corticotropic hormone** acts on the adrenal cortex, **follicle-stimulating hormone** stimulates the ovary during the ripening of eggs, and **luteinizing hormone** stimulates the ovary during the formation of the corpus luteum. The hormones produced by these target organs, in turn, suppress the production of these pituitary hormones. The pars distalis also secretes **growth hormone** (or somatotropin), which greatly influences growth and may influence fat metabolism, and **prolactin** (or lactogenic, or luteotropic hormone), which is needed for lactation in mammals but has a wide range of functions in other vertebrates.

It has been postulated that any of several structures of amphioxus may be homologous with the vertebrate hypophysis, but no conclusions can be drawn. In cyclostomes and

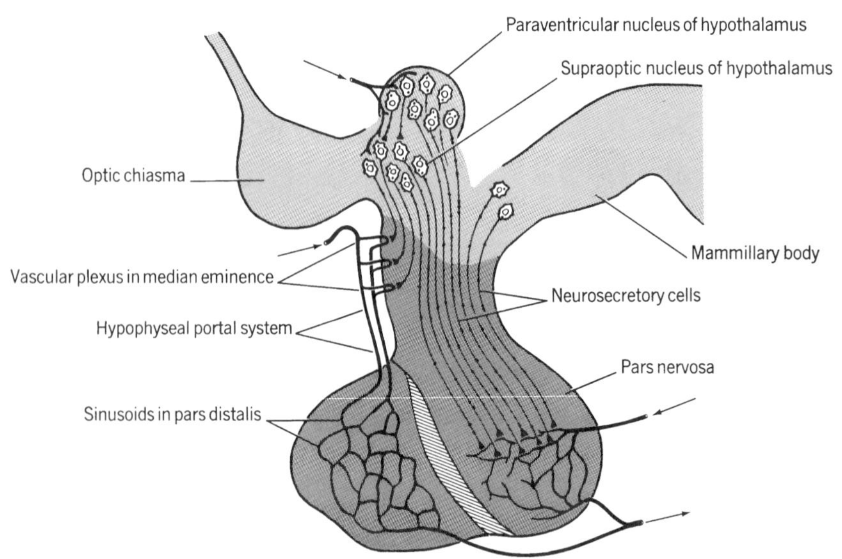

FIGURE 20.4 RELATIONSHIPS OF THE NEUROSECRETORY CELLS OF THE HYPOPHYSIS. Anterior is to the left.

fishes other than Sarcopterygii, the neurohypophysis is the more or less flat floor of the brain above other parts of the gland; a pars nervosa cannot be clearly distinguished. The atypical adenohypophysis of hagfishes is scattered as islets among other tissues. The adenohypophysis of Selachii is unique in having a ventral lobe of unknown homology. Fishes, and particularly ray-finned fishes, are distinctive for the way that the adenohypophysis and neurohypophysis interdigitate over a broad area.

In dipnoans and tetrapods, interdigitation between the neurohypophysis and adenohypophysis is much reduced or, more often, entirely lost. A pars nervosa forms from all or part of the infundibulum, which is now usually evaginated from the floor of the brain. This causes the gland to subtend more from the brain. The pars intermedia is large in reptiles, small or absent in mammals, and absent in birds. When absent, its hormone, melanocyte-stimulating hormone, may be produced by the pars distalis.

Thyroid The thyroid gland is located in the throat. Its secretory cells are derived from a midventral evagination of the endoderm of the embryonic pharynx at about the level of the second pharyngeal pouch (Figure 20.2). Surrounding mesenchyme contributes supportive tissues. The gland always consists of a cluster of rounded follicles, and each follicle is lined by a single layer of cells that are usually cuboidal (sometimes columnar when highly active) and have microvilli on their free surfaces (Figure 20.1). The thyroid has an exceedingly rich blood supply for its size. It is the only endocrine gland to have extracellular storage of its secretion. This secretion, called **colloid,** fills the follicles and contains the glycoprotein **thyroglobulin.** This iodine-rich protein is converted by hydrolysis to either of two hormones, **thyroxine** (T_4) and (in much lesser quantity) **triiodothyronine** (T_3).

The gland begins to function early in ontogeny, contributing to the control of differentiation, growth, metamorphosis, the distribution of pigment, and sexual development. It has a profound effect on metabolic rate and may influence molt (amphibians and reptiles), feather shape, body temperature, and functions of the nervous, digestive, and excretory systems. The thyroid interacts with the hypophysis and in at least some instances (anurans) with the hypothalamus. The gland enlarges when diseased.

All vertebrates have a thyroid, and its origin traces back to cephalochordates. The endostyle on the floor of the pharynx of amphioxus, and also of the larval lamprey, ammocoetes (see figure on p. 38), functions in the production and movement of mucus for filter feeding. Nevertheless, there is evidence that part of the endostyle is the phylogenetic precursor of the thyroid: (1) The endostyle, like the thyroid, forms from a midventral evagination of the pharynx. (2) Part of it, like the thyroid, concentrates iodine from the blood. (3) At metamorphosis the endostyle of ammocoetes is partly converted to adult thyroid.

In cyclostomes and many teleosts the thyroid is relatively diffuse and is variously distributed near the ventral aorta, branchial afferent arteries, heart, gills, head kidney, spleen, brain, or eye. It is more discrete in other vertebrates but may be paired (amphibians, lizards, birds), bilobed (dipnoans, many mammals), or single (cartilaginous fishes, most reptiles) (see Figure 20.5 and figures on pp. 183 and 258). The thyroid of tetrapods is usually near the larynx, trachea, or bronchi.

Parathyroid The secretory portion of the parathyroid glands differentiates from the epithelium of the third and fourth (and in reptiles also the second) pharyngeal pouches. Curiously it is the dorsal wings of the pouches that contribute in mammals, but the ventral wings in other vertebrates. It is clear from this and other evidence that the pharyngeal pouches share the potential for forming glandular tissue, yet the kind of gland formed by a specific region is not constant.

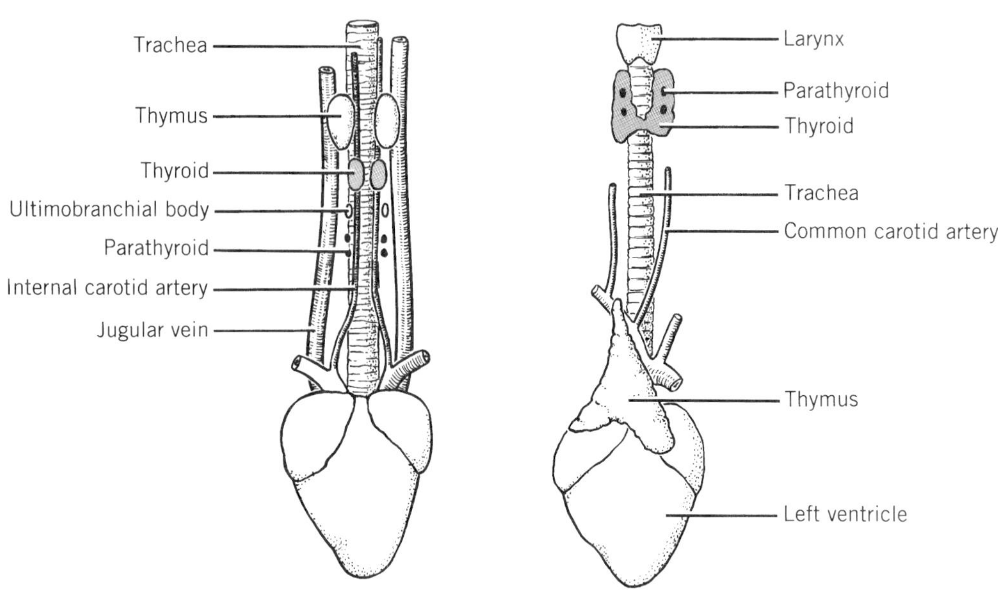

FIGURE 20.5 SOME ANTERIOR ENDOCRINE GLANDS AND THE THYMUS of a bird (left) and mammal (right). Ventral views. (Configurations may vary among genera.)

The gland consists of densely packed cells arranged in cords and clumps. The **parathyroid hormone** (parathormone) is a polypeptide. It affects the level of calcium, and less directly the level of phosphorus, in the blood. In the absence of the hormone, calcium disappears from the blood in a matter of hours, tetanus occurs in muscles, and death follows. Deficiency of the hormone leads to abnormalities of bones and teeth.

Glandular tissue of unknown function has been identified in cyclostomes and fishes that may be homologous with the parathyroid. However, the gland is known with certainty only in tetrapods, and hence evolved relatively late. It is usually divided into one or two pairs of small glands. These may be linear (birds) but are usually more or less globular. They are located in the throat area, commonly near, or even embedded in, the thymus or thyroid (Figure 20.5).

Ultimobranchial Bodies and Parafollicular Cells All vertebrates except mammals and cyclostomes have **ultimobranchial bodies.** They develop from the most posterior pair of pharyngeal pouches (although there is experimental evidence that head mesenchyme of neural crest origin contributes in quail). Ultimobranchial bodies may be single or paired, and are located near the esophagus (Figure 20.5). In reptiles and birds, part of their tissue may also be within the thyroid, parathyroids, or thymus. In mammals there are no discrete glands, but **parafollicular cells** within the thyroid are homologous with ultimobranchial bodies. The hormone of this tissue, wherever located, is **calcitonin,** which affects calcium metabolism in a manner similar to, but more rapid than, parathyroid hormone.

Interrenal Organ and Adrenal Cortex The adrenal glands of amniotes are located adjacent to the kidneys. There are two kinds of adrenal tissues. These are usually intermixed, but may be separate in elasmobranchs, and in mammals are segregated into the **cortex** and **medulla** of the adrenal glands. The two tissues are different in function and embryonic origin. The cortical type of tissue is discussed here, and the medullary tissue is discussed in the next section. In bony fishes the tissue that is equivalent to the adrenal cortex is called **interrenal tissue.**

These structures are similar to the gonads in the steroid nature of their hormones and also in their embryonic origin. Cortical tissue is derived from the mesodermal lining of the coelomic cavity close to the place of origin of the genital ridges.

The secretory cells of cortical and interrenal tissues form cords that are arranged in three layers in mammals but have little or no organization in other vertebrates. Many cortical hormones are known in mammals, but some of these are also produced elsewhere in the body, some are readily converted to others, and only about a dozen are known to be physiologically active. Together these hormones are called **adrenocorticosteroids.** They are classed according to chemical structure and general function into four groups. One group, called glucocorticoids (including cortisone, corticosterone, and cortisol), is antiinflammatory and functions in carbohydrate and protein metabolism. In its absence the liver is unable to synthesize carbohydrate. A second group (including deoxycorticosterone) affects salt and water metabolism. In its absence there is dehydration, lowering of blood pressure, and death. The third group, called mineralocorticoids (including aldosterone), is typical only of mammals and relates to sodium and potassium metabolism. The last group (including adrenosterone) resembles the male sex hormones. Most of these hormones interact with the pituitary, and many are involved in responses to stress.

Interrenal tissue of cyclostomes is scattered along the posterior cardinal veins and other vessels. In teleost fishes, interrenal tissue may be diffuse or discrete, but usually forms numerous small flecks located near or within the head kidneys. The gland is characteristically elongate and between the kidneys in cartilaginous fishes (Figure 20.6), elongate and adherent to the kidneys in anurans, and diffuse and adherent to the kidneys in urodeles. The cortical tissue of amniotes forms a pair of compact bodies located on or near the anterior ends of the kidneys (see figures on pp. 290 and 291).

Chromaffin Bodies and Adrenal Medulla Tissue corresponding to the adrenal medulla tends to be much scattered in some vertebrates and is then termed **chromaffin tissue** because of its staining properties. Chromaffin and medullary tissue is innervated by preganglionic neurons of the autonomic nervous system. These nerves and glands are all derived from the ectodermal neural crests of the embryo, and all secrete the catecholamines **epinephrine** (= *adrenalin*) and **norepinephrine,** though the glands produce much more (particularly of epinephrine) than does the nervous system. After the production,

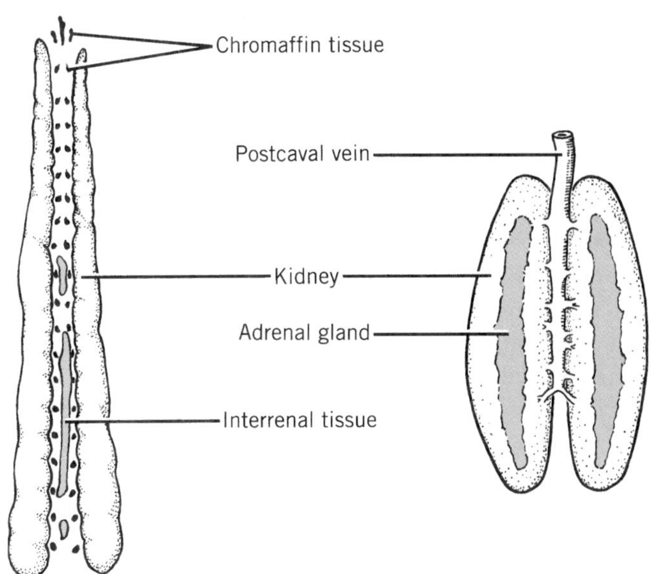

FIGURE 20.6
ADRENAL GLANDS of an elasmobranch (left) and frog (right). Ventral views. (Configurations may vary among genera.)

these hormones are stored in granules until needed; they are not released constantly, as are so many other hormones. The body responds to these hormones in many ways that better enable it to meet sudden emergencies (e.g., increased blood sugar and blood pressure, inhibition of smooth muscles).

The distribution of chromaffin tissue in the body corresponds to that of interrenal tissue but tends to be even more diffuse, particularly in fishes, where it may occur along the postcardinal veins as well as near, on, or in the kidneys. Chromaffin tissue may lie near, but separate from, interrenal tissue (some fishes and some lepidosaurs) (Figure 20.6), may be intermingled with interrenal or cortical tissue (some fishes, most amphibians and reptiles, birds), or may lie as a medulla within a covering of cortical tissue (most mammals). Even mammals, however, have chromaffin bodies or **paraganglia** associated with some sympathetic ganglia.

Gonads and Placenta The development and structure of the gonads were presented in Chapter 16, but these organs should be mentioned again as endocrine glands. The ovary produces several **estrogens** (the principal ones are estradiol and estrone), **progesterones,** and in mammals, **relaxin.** Estrogens control the growth and development of the female genital duct system and are essential for reproduction. They also initiate and maintain secondary sexual characteristics, which are marked in some vertebrates and inconspicuous in others. The site of estrogen secretion appears to be the internal theca in mammals, and probably also the stratum granulosum (see figure on p. 286), but is in doubt for vertebrates having virtually no theca.

Following ovulation, the ruptured mammalian follicle is transformed into a temporary but pronounced gland termed the **corpus luteum.** This structure then secretes progesterone, a hormone that is essential for the final differentiation of the female reproductive tract in preparation for fertilization and pregnancy, and also for maintaining pregnancy. Structures similar to the mammalian corpus luteum form in the ovaries of various other vertebrates (sharks, teleosts, urodeles, birds, some reptiles) but seemingly not in others. *All* vertebrates tested, however, produce one or more progesterones, which is puzzling.

Relaxin is the only gonadal hormone that is not a steroid. In some mammals, at least, it acts on the pelvic symphysis, mammary glands, and genitalia, readying them for their functions at delivery and thereafter.

The interstitial cells of the testis (see figure on p. 287) produce male hormones collectively called **androgens.** The principal androgens are **testosterone** and **androstenedione.** Androgens are required for the growth, differentiation, and function of the male genital ducts and copulatory organ (if present) and for control of secondary sexual characteristics and sexual behavior. All vertebrates have androgens. Interstitial tissue has not been identified in all, but homologous cells may occur in the seminiferous tubules or bounding the seminiferous ampullae. To complicate the situation, males produce estrogens and females produce androgens, all from inadequately known sources, though the adrenals are probably involved. Supportive (or Sertoli) cells within seminiferous tubules may produce a hormone necessary for sperm maturation.

The mammalian **placenta** is a rich source not only of estrogen and progesterone, but also of a gonadotropin and of prolactin, which is otherwise produced by the adenohypophysis. Prolactin is needed for milk production in mammals but is also present in lower vertebrates, where it has varied functions. The placenta, like the ovary, also produces relaxin.

All the gonadal hormones have complex interactions with the hypophysis; some relate to interrenal or cortical function or to the activities of the thyroid or pineal. Gonadal function is cyclic—usually seasonal, the ultimate control often being photoperiod as mediated by the hypothalamus. The pathways are poorly known for fishes.

Miscellaneous, Possible, and Near Endocrine Glands The substance **secretin,** produced by the duodenum in response to stomach acid, inhibits motility of most of the gastrointestinal tract and was the first described "hormone" by Bayless and Starling in the early 1900s. Since that time, a number of chemical mediators known as **gastroenteropancreatic hormones,** produced by cells of the stomach, intestine, or pancreas, have been isolated. The stomach also produces gastrin and the jejunum cholecystokinin, two additional hormones that act to coordinate the digestive process. Surprisingly, many of the gastroenteropancreatic hormones are also secreted by cells of the nervous system, which in turn, act on target cells that are either distant or near to the site of production. Most of what we know about this complex of hormones comes from studies of mammals (see Sheridan and Sower in References). However, Viga provides evidence that cholecystokinin evolved in chordate ancestors and that gastrinlike peptides evolved from an ancestral cholecystokinin at the level of the divergence of tetrapods from fishes.

The origin and exocrine nature of the pancreas were noted in Chapter 12. Its thousands of islets (see figure on p. 216) secrete two very important hormones: **insulin,** which controls the deposit of glycogen in the tissues, and **glucagon,** which controls its release. Among fishes there may be fewer, larger islets located within the pancreas or along the bile duct. There is great variation in response to these hormones; protein metabolism and solutes of the blood may be involved.

The tiny **pineal** organ atop the brain (see figure on p. 366) is a photoreceptor in the lower classes. In tetrapods it assumes endocrine functions that seem usually to involve the relations between temperature control or reproduction and illumination. In most tetrapods the gland secretes **melatonin** under chemical signals from nerves. Its production has a daily rhythm (stimulated by darkness and inhibited by light) and can trigger seasonal breeding or migratory activities. Melatonin affects melanophores of frogs and suppresses the gonadal activity in mammals. Animals living in the arctic are subject to 10-week periods of total darkness and 10 weeks of "midnight sun." Researchers take advantage of these climate extremes to study the daily and seasonal patterns of a number of hormones including melatonin (see Deviche and Barnes in References). Studies of melatonin in reindeer in northern Finland, for example, reveal a remarkably tight correlation between levels of melatonin and changing day length (Eloranta et al.). Thus in species such as reindeer, the external photoperiodic environment translates into a chemical signal (melatonin), which in turn regulates reproductive functions. A binding site for melatonin is the nucleus of the hypothalamus (called suprachiasmatic) that is the location of the biological clock. (A second circadian clock related to the production of melatonin may be present in the eyes of some rodents.)

The atria of mammals are stimulated by stretching to produce the peptide hormone **atrial natriuretic factor,** which is stored in granules in the cardiac cells, and which interacts with blood vessels, kidneys, adrenals, and brain in the complex regulation of blood pressure and the excretion of water and sodium.

Fishes, but not tetrapods, have neurosecretory cells in the caudal part of the spinal cord. In teleosts these relate to a vascular, ventral swelling termed the **urophyis.** The hormones produced by the nerves, and perhaps stored in the swelling, are **urotensins,** which cause contraction of smooth muscle, particularly in the urinary bladder, and may contribute to osmoregulation.

Prior to the discovery of naturally occurring analgesic (pain-killing) hormones in the nervous system, it was known that the opiate morphine reduced pain. After an extensive search, two major groups of opiate-like peptides have been isolated: the endogenous morphines, or **endorphins** (= *from within*), and the **enkephalins** (= *in the head*). These are produced in the pituitary and CNS (primarily in the medulla) in response to pain stimuli and are the focus of intense research. These are thought to relate to the "runner's high."

In response to a drop in blood pressure, **juxtaglomerular cells** in the afferent arterioles of glomeruli of the kidney release **renin.** Renin is an enzyme, not a vasoactive substance itself, that acts on a second plasma protein, angiotensin. Renin converts angiotensin to **angiotensin I,** which, in the lungs and in the presence of *converting enzyme,* is converted to **angiotensin II.** Angiotensin II is a powerful vasoconstrictor that, by constricting peripheral arterioles, increases peripheral resistance to raise the arterial pressure back toward normal. The most commonly used drug to control high blood pressure in humans acts to inhibit converting enzyme.

The **thymus gland** has been suspected of endocrine function, but none has been definitively established.

Not only are there these apparent endocrine glands of unknown function, but there are apparent "hormones" known not to be produced by glands. For example, urea and carbon dioxide, derived from nonendocrine tissues, are distributed by the blood and convey "messages" to organs distant from their place of release. In order not to confuse the definition of hormone, these and similar substances are called **parahormones.**

NEUROSECRETION

Reemphasis should be given to the close relationship noted above between certain parts of the nervous system and certain endocrine organs. Neurons in two or more nuclei of the hypothalamus actually extend into and constitute the neurohypophysis. Those reaching the pars nervosa elaborate the hormones of that part of the gland. These hormones are released into the general circulation. Those neurons reaching the median eminence release hormones that enter a miniature portal system running from the median eminence to the adenohypophysis (Figure 20.4). This portal system is lacking in fishes, but the interdigitation of neurohypophysis and adenohypophysis in many fishes may accomplish the same interaction.

Similarly, nerves of the teleost spinal cord extend into the urophysis, where they release hormones. The adrenal medulla and other chromaffin tissues do not resemble nervous tissue histologically, yet they function as though they had evolved from postganglionic fibers of the autonomic nervous system. Nervous stimulation causes the medulla to secrete. The pineal gland of at least some mammals secretes its hormone upon stimulation of the autonomic nervous system. Neurosecretion is also known in crustacea, insects, and other invertebrates.

These observations make it clear that the distinction between the nervous system and endocrine glands is not sharp—and that much remains to be learned about the relationship.

REFERENCES

Bentley, P.J. 1998. *Comparative vertebrate endocrinology.* 3rd ed. Cambridge Univ. Press, New York. 526p. Broad and somewhat detailed account organized to illustrate patterns of endocrine function in response to environmental demands.

Bern, H.A. 1985. The elusive urophysis: twenty-five years in pursuit of caudal neurohormones. *Am. Zool.* 25:763–769.

Bern, H.A. 1990. The "new" endocrinology: its scope and its impact. *Am. Zool.* 30:877–885.

Callard, I., and G. Callard. 1990. Symposium: unconventional vertebrates as models in endocrine research. *J. Exp. Zool.* 4(2–5):1–218. Channel catfish, clawed frogs, and musk shrews—an instructive series of papers that demonstrate clearly the concept of "model" species and their value for the pursuit or specific questions.

Deviche, P., and B.M. Barnes. 1995. Introduction to the symposium: endocrinology of arctic birds and mammals. *Am. Zool.* 35(3):189–190. Diverse papers that discuss the endocrine challenges and responses that vertebrates face with large seasonal changes in photoperiod and temperature.

Eloranta, E., et al. 1995. Seasonal onset and disappearance of diurnal rhythmicity in melatonin secretion in female reindeer. *Am. Zool.* 35:203–214.

Gorbman, A., et al. 1983. *Comparative endocrinology.* Wiley, New York. 572p.

Guyton, A.C., and J.E. Hall. 1996. *Textbook of medical physiology.* 9th ed. Saunders, Philadelphia. 1148p.

Matsumoto, A., and S. Ishii (eds.). 1992. *Atlas of endocrine organs: vertebrate and invertebrate.* Springer, New York. 307p. Beautiful and informative.

Norris, D.O. 1997. *Vertebrate endocrinology.* 3rd ed. Academic Press, New York. 634p. Broad and instructive phylogenetic approach to comparative endocrinology, including a chapter on approaches to endocrine research. Informative illustrations, chapter summaries.

Pang, P.K.T., and M.P. Schreibman. 1986. *Vertebrate endocrinology: fundamentals and biomedical implications.* Academic Press, New York. 496p.

Reiter, R.J. 1981. The mammalian pineal gland: structure and function. *Am. J. Anat.* 162:287–323.

Sheridan, M.A., and S.A. Sower. 2000. A tribute to Erika M. Plisetskaya: new insights on the function and evolution of gastroenteropancreatic hormones. *Am. Zool.* 40:161–308.

Sower, S.A., S. Kunimasa, and K.L. Reed. 2000. Perspective: research activity of enteropancreatic and brain/central nervous system hormones across invertebrates and vertebrates. *Am. Zool.* 40:165–178.

Viga, S.R. 2000. Evolution of the cholecystokinin and gastrin peptides and receptors. *Am. Zool.* 40:287–295.

Part Three

Structural Adaptation:

Evolution in Relation to Habit and Habitat

Chapter 21

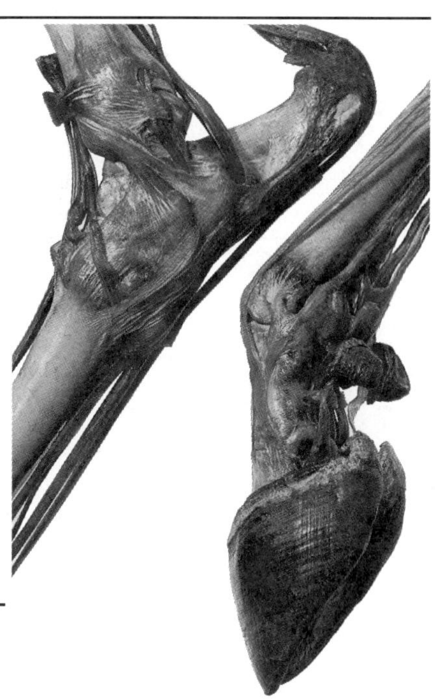

Structural Elements of the Body

ANIMALS AS SPECIALISTS

Part II of this book includes much functional interpretation of structure, particularly for the respiratory, circulatory, and excretory systems and for the eye and brain. Emphasis, however, is on analysis of structure in relation to the long sweep of phylogeny: on conservative evolutionary changes common to all animals in such large taxa as classes and subclasses. Primitive and unspecialized characters are featured. Structures that are useful to animals of varied habits are stressed: Jaws are generally advantageous; two pairs of appendages proved to be a good general plan; a circulation divided into pulmonary and systemic circuits is superior for tetrapods.

Part III deals with the parallel or convergent influence of functional adaptation on different vertebrates; particular attention is paid to locomotor and feeding mechanisms, in which parallels are seen most clearly. Animals with similar specialties are found scattered among the systematic categories, but are here brought together, their common problems are identified, and their various adaptations are interpreted on the basis of functional morphology.

Analysis is complicated by several factors. Different animals may do similar things in different ways: Squirrels and pottos both climb, but pottos slowly grasp the branches whereas squirrels run on the limbs or cling to them with sharp claws. No one kind of animal has all the structural modifications that are associated with its general habit. Furthermore, one kind of animal may have several specialties: Frogs jump and swim, flying squirrels climb and glide, cormorants fly, swim, and dive. Also, the activities of animals are determined not only by structural features but also by behavioral factors. Thus, gray foxes climb trees whereas red foxes do not. One cannot learn by dissection which climbs, or, indeed, that either climbs. Even without special structural adaptedness, many animals can run, swim, climb, and dig somewhat. Conversely, animals may fail to move in ways for which they seem, on the basis of morphology, to be adapted. Thus, the adult gorilla has

Page 381: Air-dried dissection of the arm of a harbor seal.
Above: Air-dried dissection of the right tibiotarsal joint and left foredigits of a cow.

the structure that correlates with climbing by arm-swinging under the branches, but because it is very large, it seldom does so.

In spite of these complications, however, it is rarely difficult to determine the principal habits of a vertebrate animal from its structure. Some clues are subtle and some obvious, but all make sense. Their identification and interpretation can be very engaging. Animals are so good at their specialties!

This part of the book begins with three chapters presenting basic mechanical principles that relate to feeding, posture, and locomotion in general. Subsequent chapters analyze specific adaptations.

Although all of the principal locomotor and feeding adaptations are presented, no book (or student or professor) can completely cover such broad topics. Even in areas selected for emphasis it is necessary to summarize, and ancillary topics are omitted. Therefore, parenthetically, we mention some adaptations, not included in the following pages, that could be subjects for supplemental study or special reports.

There are structural as well as physiological and behavioral adaptations for living at high elevations, deep in the sea, and in caves. Some vertebrates are modified for walking on snow, mud, or floating vegetation. Adaptations for prenatal development and parturition include placental morphology, length and structure of the umbilical cord, and shape of the uterine cavity in relation to the shape, flexibility, and densities of different parts of the fetus. The larynx, syrinx, or other sound-producing mechanism exhibits marked specializations, as does the ear. Provisions for defense and escape include a diversity of structures and functions, among which are armor, teeth, claws, horns, antlers, tusks, beaks, spines, quills, camouflages, and mechanisms for stinging, poisoning, crushing, inflating the body, dropping the tail, and producing slime, irritants, or noxious odors.

We humans do not receive special attention in this book, but we also have distinctive locomotor skills. Even if other animals could have the incentive to try and the patience to learn such "artificial" activities as gymnastics, diving, skiing, figure skating, and pole vaulting, none could match human athletes. Also, humans are not only uniquely expert but also impressively skilled at throwing.

PROPERTIES OF SUPPORTIVE MATERIALS

The materials of the body that accomplish support and movement are bone, cartilage, muscle, tendon, and ligament. (Soft organs are further supported by meshworks of collagenous fibers.) The suitability of these materials for the various requirements of the body depends on their properties. (Hydrostatic systems are also identified, as in some tongues and genitalia, and are the basis for support and locomotion in many invertebrates.)

Three important properties of living supportive tissues are not shared by any material available to architect or engineer: First, all display **growth** without interruption of function. They have remarkable capacity for repair of both major breaks and minor damage. This property protects these tissues from fatigue, or loss of strength with repeated loading, which is characteristic of nonliving supportive materials, and minimizes the accumulation of injuries as time passes. The rate of repair is faster for muscle and bone than for cartilage, and in all these tissues the capacity diminishes with age. Second, all have amazing **capacity to adjust to circumstance,** slowly altering their substance and configuration in response to demand. It is common knowledge that muscular strength increases with exercise; the adaptability of other supportive tissues is exemplified later in this chapter. Finally, these properties taken together ensure adequate **durability** for a lifetime of constant use. No man-made apparatus having even remotely comparable complexity of moving parts approaches the body in this regard.

The strengthening of supportive materials that results from their heterogeneity and the presence within them of microscopic lacunae was noted on p. 114. Before considering further the strength of the structural elements of the body it is necessary to review or introduce several concepts and terms. The important concept of **force** was presented on pp. 174 and 175, and should be reviewed there now. The weight of an animal pressing on the ground is a force, the pull of a muscle on its insertion is a force, and the push of a fish tail against the water is a force. Since forces of the body are concentrated at such places as insertions of tendons and contacts between bones, it is useful to consider force per unit area which, strictly speaking, is called **stress** if the force is in one direction (those tendons and bones) and **pressure** if the force is in all directions (gas within a lung). In common practice, however, the word "pressure" is also used for stresses against surfaces (teeth, bones, mud, snow). Stress and pressure are expressed as kilograms per square centimeter (or newtons per square meter, or pascals).

Load is a general term referring to any force that is applied to a solid object. Adjacent bones of the legs and spine load each other; active muscles load related bones. In order for loaded objects to remain in equilibrium, equal forces must operate in opposite directions (this is an application of Newton's third law). Thus, as a tetrapod stands at rest, the downward force of its weight is opposed by an equal upward force from the ground. The weight is transmitted through the bones of the legs to the ground. When objects transmit loads, there are internal forces of one part of the object acting on adjacent parts. The internal transmission of a load is, again, stress. External loads cause internal stresses.

When any load is applied to any object, deformation occurs; there is a change in length, volume, or angle. Relative deformation is called "strain." Thus, for length, **strain** = change in length/original length. Because strain is a ratio, it has no units. For hard materials, such as bone, strain is directly proportional to stress: if one is doubled, the other is doubled. (They are said to follow Hooke's law.) The regression line expressing the relationship (short of damage to the material) is linear (Figure 21.1, left). Furthermore, the curve is the same for loading and unloading. The area under the curve is proportional to the work required to produce the strain. This **strain energy** is stored for as long as the load is applied and is released when the load is removed (as for a spring).

For tendons and ligaments (in contrast to steel and bone), stress and strain depart from direct proportionality (i.e., Hooke's law does not apply). The regression is not linear; it

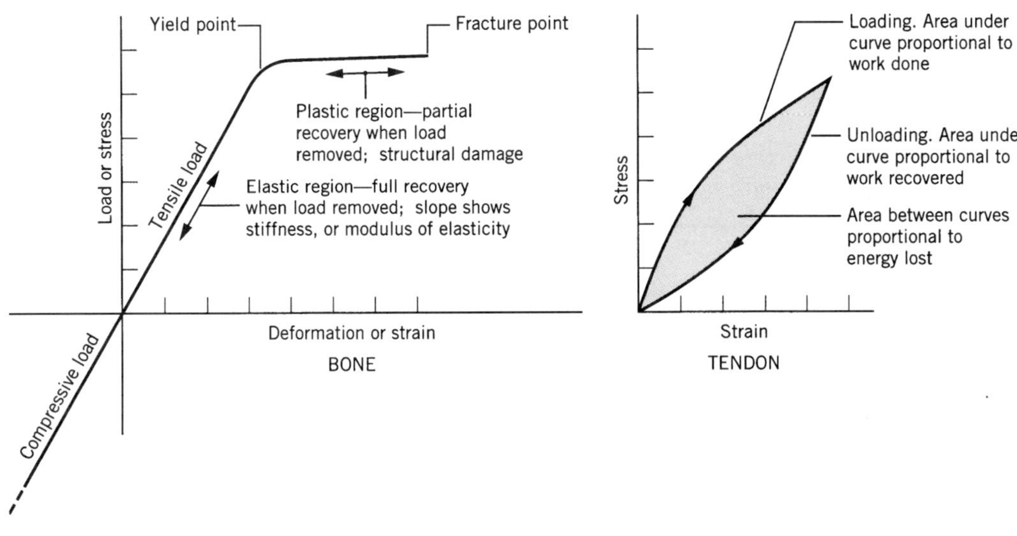

FIGURE 21.1 LOAD-DEFORMATION CURVES.

arches upward during loading and downward during unloading (forming what is called a hysteresis loop) (Figure 21.1, right). The work done in stretching a tendon, and the work recovered as it is released, are again each proportional to the area under the respective curves, but the curves are not the same. The area between the curves is proportional to the strain energy lost as heat. As we shall see, moving vertebrates use tendons as springs. This analysis shows why some energy is always lost.

Deformation may be permanent or temporary. Even moderate loads cause permanent deformation of modeling clay. Such materials could hardly support the animal body. If deformation is temporary, recovery may be almost immediate, as for bone, or somewhat slower, as for cartilage, tendon, and ligament. The capacity of a material to return completely to its original shape after a load is removed is called **elasticity.** (Note that this is not the same as the layman's use of the word to mean "stretchability.") The structural materials of the body have virtually perfect elasticity within usual load limits. Some elastic materials, like rubber, are much deformed by moderate force, whereas others, like steel and bone, are only slightly deformed by great force. Tendons are of intermediate stiffness. The ratio of stress to strain, and hence also the slope of their regression line, is a measure of this stiffness and is called the **modulus of elasticity,** or Young's modulus. It has the same units as stress, or force per unit area.

The important property of **strength,** as applied to supportive materials, is the capacity to resist force without breakage or permanent deformation. Strength varies, of course, with material, and is proportional to the cross-sectional area of the object (i.e., the more bone, the greater the strength). As noted on pp. 114 and 115, strength of heterogeneous materials varies with relative orientation of force to grain. Strength also varies importantly according to the direction of the applied force in relation to a surface; that is, the interaction between adjacent objects differs depending on the direction of the forces acting between them. This is true both for a load applied to an actual external surface of an object and for a stress applied to an imaginary internal surface. Forces are of only two kinds: perpendicular to a surface or parallel to a surface. They can, of course, be applied at intermediate angles, but analysis of a kind presented in Chapter 22 shows that such forces can always be broken down into a perpendicular component and a parallel component.

Perpendicular forces, in turn, are of two kinds: **Compression** results from force directed *toward* an object. It tends to make the object shorter in the direction of the applied force (strain is then said to be negative) (Figure 21.2). **Tension** results from a force directed *away* from an object. It tends to make the object longer (strain is positive). Columns and pillars withstand compressive force; guy wires and cords that suspend objects withstand tensile forces. Compression and tension can occur together, but at right angles to one another.

Forces applied parallel to a surface but in opposite directions cause **shear.** Shear slides one part of a material crosswise to adjacent parts. Scissors cut by shearing. If a closed

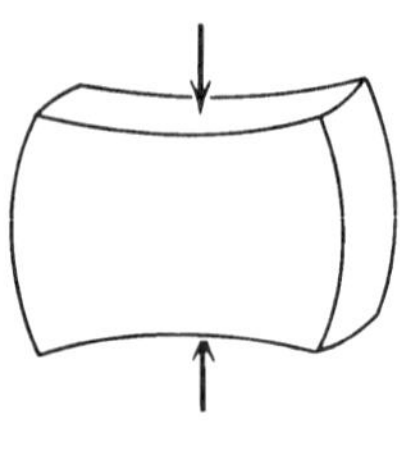

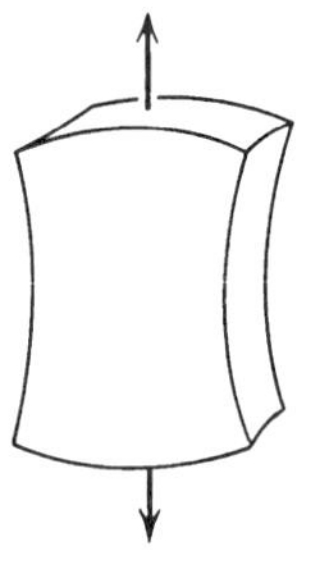

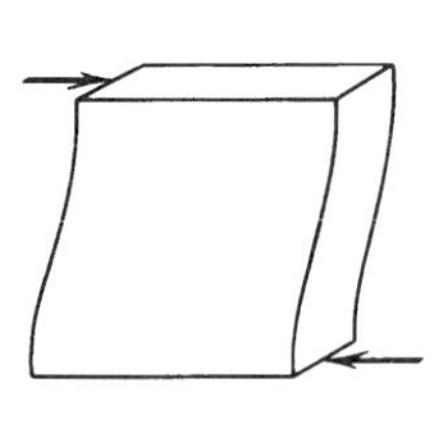

FIGURE 21.2 THE THREE PRINCIPAL KINDS OF FORCES and the distortions they tend to cause in solid objects.

book is held between the palms of the hands and one cover is pushed or twisted relative to the other, the book is distorted by shear as the pages slip over one another.

Equipped with these concepts, let us now consider the strength of the supportive materials of the body. Fresh compact bone (not dry or embalmed or cancellous bone) loaded parallel to its grain (as determined by the orientation of osteons) has a compressive strength of 1330 to 2100 kg/cm^2 (the international unit of compression is the pascal, or newton/square meter). About 170 students would somehow have to stand on a single 1 in. cube of compact bone in order to crush it! Values for cartilage vary, but are lower than those for bone. Tendons and ligaments, like string, merely crumple when compressed lengthwise.

Fresh compact bone loaded parallel to its grain has a tensile strength of 620 to 1050 kg/cm^2, or about half its compressive strength. The tensile strength of cartilage is again less than that of bone. Tendon and ligament, however, although softer and lighter materials, have about the same tensile strength as bone.

The resistance of compact bone to shear may be as low as 500 kg/cm^2 if stressed parallel to the grain, and as high as 1176 kg/cm^2 if stressed crosswise to the grain. Cartilage, tendon, and ligament have less resistance to shear.

It might seem that resistance of tendon and ligament to tension, and of bone to all forces, are far in excess of demand. So they are for maintaining posture and engaging in moderate activity. In strenuous activity, however, forces exerted on the skeleton by individual tendons of human-size animals may reach several hundred kilograms, and excessive loads, as from a fall, occasionally cause tearing or breakage. It is clear from the above figures that shearing forces would be limiting in the body if they approximated usual compressive and tensile forces. Actually, pure shear is unusual, but bones may be sheared by twisting (i.e., rotating) at the same time they are compressed, and bending (i.e., bowing, or curving) forces, which are very common in the skeleton, combine shear, compression, and tension. The relative magnitudes of the kinds of stresses in the skeleton seem usually to be in proportion to the capacity of bone to withstand them: Compressive forces are largest and shearing forces are smallest. When bones do fail, any of the types of forces may have been responsible, though compressive fractures are least common.

STRESS AND STRESS LINES

It will be easier to understand how the structural elements of the body are constructed for maximum effectiveness after considering the transmission of forces within homogeneous objects. When a solid cylinder resting on the ground is compressed by a uniform load, the downward force of the load is opposed by an equal and opposite upward force at the ground. These forces are represented by large arrows in Figure 21.3A. If the opposed forces are depicted instead by many arrows, each representing a unit of force, more information is included because even spacing of the arrows then shows that pressure is uniform over the ends of the cylinder (part B). Within the cylinder, units of stress have the same magnitude and direction as the external pressure, so at any arbitrary plane they can also be represented by arrows, the number of arrows being proportional to the area taken. The lines we draw to represent the paths followed by units of force as they pass through an object are called **stress lines.** In this example the lines are straight and evenly spaced because loading is uniform (part C). They represent only compression and can also be called **compression lines.** The magnitude of compressive stress in the plane of the illustration is proportional to the height of the shaded rectangle (the units being arbitrary). At surfaces of the cylinder, and at vertical planes within, there is no tension or shear. (In this and the following examples, stress resulting from weight of the object itself is ignored for the

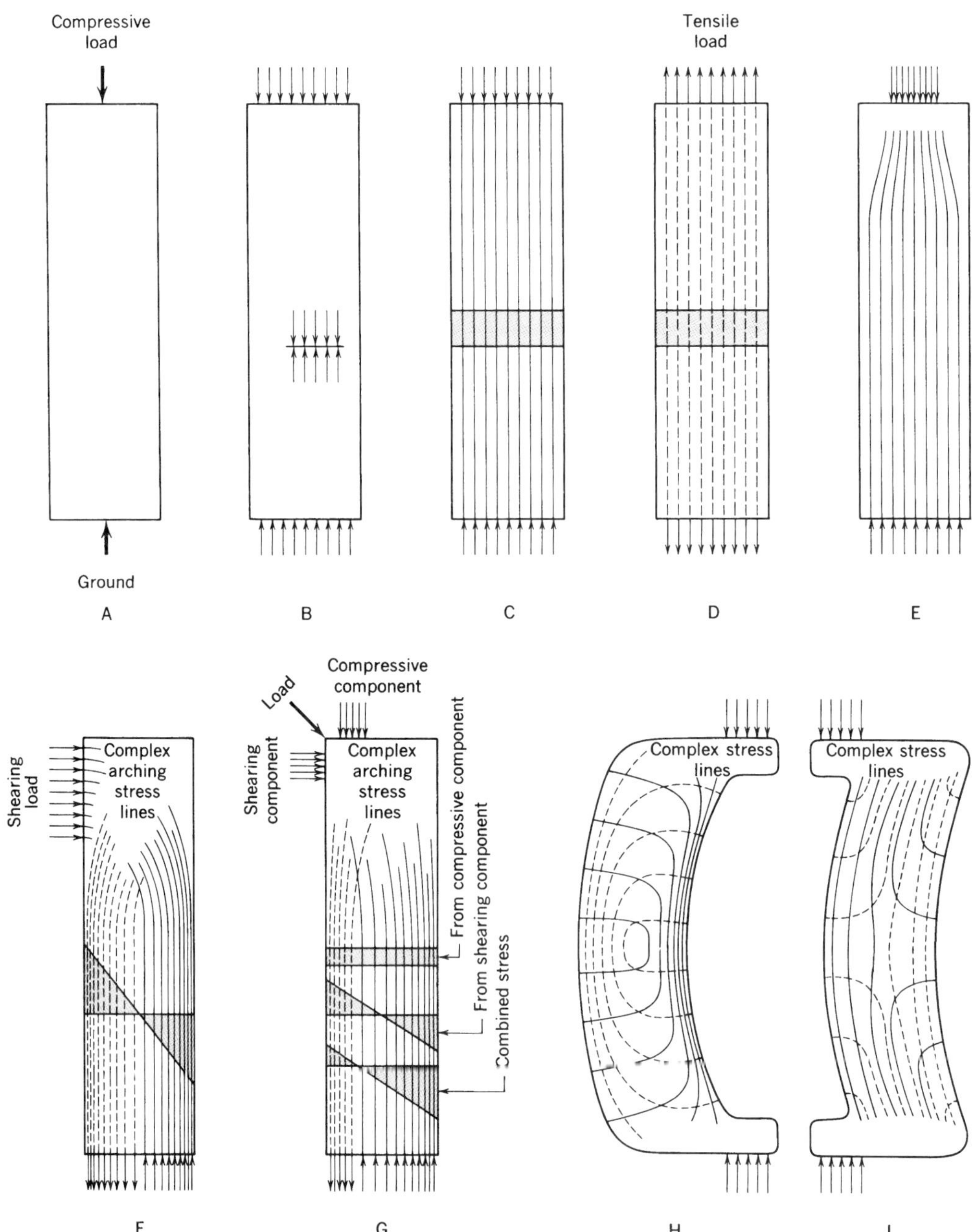

FIGURE 21.3 DIAGRAMS OF STRESS LINES WITHIN CYLINDRICAL OBJECTS.

sake of simplicity. Also, it is assumed here that loading is insufficient to cause the column to bend or buckle.)

Tensile force applied uniformly to one end of a cylinder or rod, and that which opposes the load at the other end, can also be depicted by arrows representing units of force. Again, straight lines show the paths of the forces within the object, but this time they are **tension lines** (represented by dashed lines in part D). The magnitude of tensile stress is again proportional to the height of the shaded rectangle. This model is closely approximated by stressed tendons and ligaments.

Bones are never cylinders evenly compressed over their ends in the direction of their long axes. If the end of a cylinder is compressed over a restricted area, then the adjacent stress is great and is represented by compression lines that are close together (part E). As the lines pass away from the point of application of the load, however, they spread out until they are evenly distributed. The upper, outer parts of the cylinder are devoid of stress, but at the boundaries between stressed and unstressed areas, and also immediately under the load, stresses are complicated (and not illustrated). This model was approximated by the nearly solid, cylindrical, 2 m long femur of the great 54,500 kg (120,000 lb) dinosaur *Apatosaurus,* but for reasons given below, the long bones of tetrapods are rarely solid cylinders. We progress, therefore, to other applications of the concept of stress lines.

If a load is not applied perpendicularly to the end of a cylinder (as in the above examples) but is instead applied along an upper edge perpendicular to the axis of the cylinder, then the resultant stress must resist bending. Compression lines arch away from the load and come to run lengthwise in the opposite side of the cylinder at some distance away from the load (part F). Tension lines run lengthwise in the side of the cylinder near the load. The configurations of compression and tension lines near the load are complicated and somewhat dependent on the material, and hence are not illustrated. Shear is also present. Note that compression and tension are each greatest at their respective edges of the cylinder (stress lines are closest together there) and each diminishes to zero at the central axis of the cylinder. The stress at any intermediate point between edge and center is proportional to the height at that point of the relevant triangle, as shown in part F. If a tension load replaces the compression load, the pattern of stress remains the same, but the kinds of stresses are reversed. These models are approximated by the force of food against the jaw (compressive load) and by the forces of muscles on opposite edges of the summits of vertical neural spines (tensile load), though for reasons explained in the next section these bones are not cylindrical.

Most loads applied to long bones are neither parallel nor perpendicular to their long axes, but instead are at an intermediate angle, as when one bone loads another across a flexed joint, or a tendon inserts obliquely onto a bone. Such loads can be converted, however, to longitudinal (compressive or tensile) and transverse (shearing) stresses (see p. 407). Part G represents this more general situation when both compression and shear are present. The magnitudes of these stresses in the plane illustrated can be independently represented, respectively, by the heights of a rectangle and a pair of congruous triangles as before. Total stress can be represented by combining these figures as shown. It is seen that compression exceeds tension and that the axis of zero stress is no longer the central axis of the cylinder.

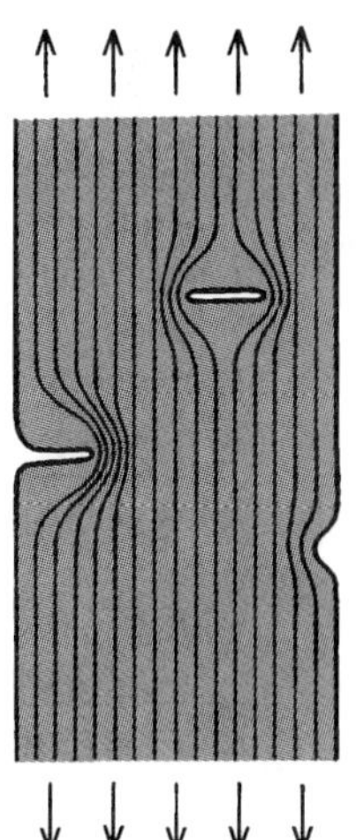

FIGURE 21.4 STRESS LINES WITHIN AN OBJECT ARE CONCENTRATED BY CRACKS AND IRREGULARITIES.

Finally, although the cylinder has provided a conveniently simple model thus far, few bones closely approach cylindrical shape with straight parallel sides. Most large bones have somewhat enlarged ends and curved shafts. Solid models of this general shape can be loaded either by compression or tension tending to bend them further (see part H) or straighten them (part I). The patterns of stress lines in the curved shafts are shown; those near the applications of the loads are intricate and are omitted. Note that stress is distributed throughout each shaft (except for one central focal point in each), but is greatest near the convex and concave edges of the shafts.

Small cavities, notches, and channels all weaken materials by causing local concentrations of stress (Figure 21.4). Bones are constructed to minimize such loss of strength. It is the shafts, not the ends, that have the greatest strain, and these tend to be very smooth. Canals for blood vessels usually run at an angle to the long axis of a bone, and the lacunae housing osteocytes have their shortest axes at right angles to the long axis of the bone. These configurations reduce the concentration of stress. (The presence of lacunae in cellular bone is also

COMMENT 21.1

BENDING STRESS CAN BE COMPLICATED

As shown in Figure 21.3F and G, when a bending force is applied near the end of a smooth cylinder made of a stiff material a *neutral axis* runs lengthwise within the cylinder (at its center or somewhat offset) where there is neither compression nor tension. Let y be the perpendicular distance from the neutral axis to the point on the cylinder at which we wish to consider bending stress (the radius of the cylinder when the neutral axis is central), and let M_y be the moment of the applied force around the neutral axis. The bending stress equals M_y / I, where I is the **moment of inertia** or **second moment of area** of the cross section of the cylinder. $I = \int (y^2\, dA)$, dA being an increment of the area. Thus, I is a measure of the distribution of material on the cross section in relation to the designated axis. As it is the square of y that is multiplied by each increment of area, the value of I increases rapidly as the material is moved away from the neutral axis, thus supporting the inference made in the main text. For round and oval cross sections, solid and hollow, the geometries of their areas provide reasonably simple formulas for calculating I. (For examples and further analysis, see the book chapter by Sharon Swartz listed in References.)

To turn from a perfect cylinder to a living limb bone introduces complications. There are usually multiple applied forces, probably acting in different planes and usually changing over time. The neutral axis is unlikely to be exactly at the centroid (center of mass) of the section and could be shifting. Cross sections of the bone are somewhat irregular in outline and thickness. To calculate such cross-sectional areas, it is best if the shapes are digitized and their properties determined using computer algorithms that sum elements of the area with respect to the chosen axis (e.g., anteroposterior, mesolateral, or that giving the maximum value of I). (Instructions, and much related information, are found in the book edited by Andrew Biewener listed in References.)

Experimentalists are likely to cut corners by using strain gauges *in vivo* to measure stress directly; the rest of us may find that the general formula shown above at least helps to clarify the variables.

---

advantageous for stopping the spread of microfractures, and microfractures are removed by a constant turnover of bone tissue.) Furthermore, bones tend to have at least small elevations or crests where tendons join them. This causes less concentration of stress within the body of the bone than would otherwise occur.

USE AND DESIGN OF STRUCTURAL ELEMENTS

Tendons, Ligaments, and Cartilages TENDONS transmit the pull of muscles to bones, a function for which their flexibility and tremendous resistance to tension well suit them (Figure 21.5). They consist of tightly packed parallel bundles of collagenous fibers. Their properties vary somewhat according to function: flexor tendons tend to be stronger and stiffer than extensor tendons. Tendons that must move appreciably relative to adjacent tissues have sheaths, and in some instances slip through lubricated channels resembling the spaces around movable joints of the skeleton.

The force of contraction of even a large muscle is usually concentrated by its tendon on a small area of the skeleton. This contributes to precision of movement and allows several muscles to act in different ways at about the same place. It is important that tendons enable muscles to move skeletal parts at a distance from their own positions. Weight distribution and body contours are thus controlled in ways that contribute to speed, endurance, and agility (discussed further in subsequent chapters). Human fingers would be useless if encumbered by all of their own muscles.

Some tendons transmit tension around corners at movable joints in the manner of cords passing through pulleys. The living pulley may be a tunnel of bone (as where tendons of digital flexors pass around the proximal end of the tarsometatarsus of some birds), a bony projection forming a channel (as where tendons of abductors of the foot angle at the outside of the ankle of many mammals), or a ligamentous loop (as where tendons of digital extensors

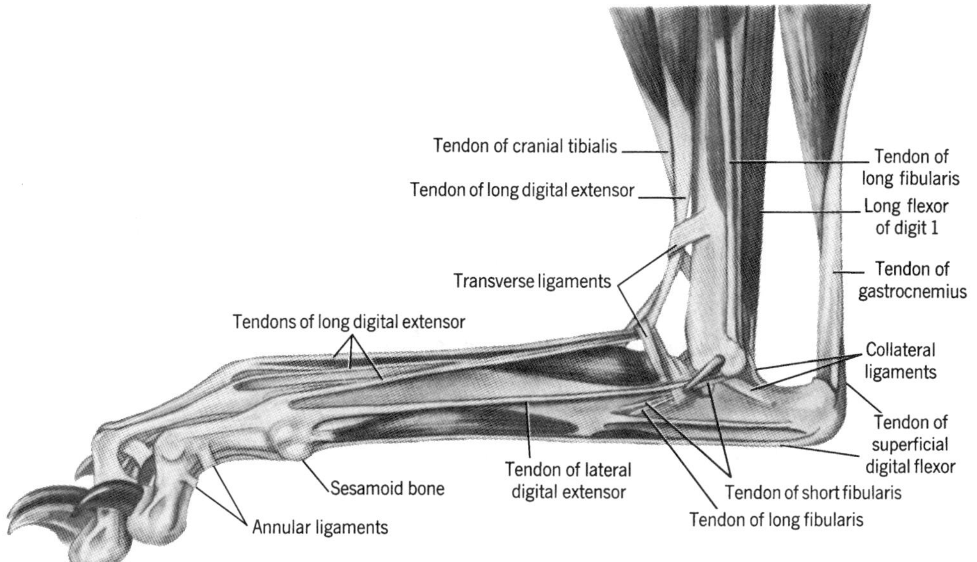

FIGURE 21.5 EXAMPLE OF RELATIONSHIPS AMONG TENDONS, LIGAMENTS, AND BONES shown by a lateral view of the left ankle and foot of the gray fox, *Urocyon.* (Drawn from a freeze-dried dissection and hence slightly shrunken.)

turn in front of the ankle). Loaded tendons that are straight sustain only tensile forces, but where a tendon bends, shearing forces also occur. Accordingly, tendons compensate by becoming thicker at such places, and by incorporating some fibrocartilage, which withstands compression. If forces other than tension are particularly severe (as where the tendon of the quadriceps muscle passes in front of the knee), small bones, which are better able to withstand such forces, interrupt the tendons. These are called **sesamoid bones.**

Some LIGAMENTS have virtually the same structure and properties as tendon, whereas others have less regularly oriented collagenous fibers and contain elastic fibers in various proportions. Because tendons always relate to muscles, they function only when muscles function. Ligaments, by contrast, function passively, using no energy, and are therefore superior where constant tension is required. In the manner of lashings, ties, and elastics, they bind the skeleton together, limit the motion of some joints, and contribute to antigravity mechanisms.

Some ligaments are merely thickened portions of the capsules around movable joints. These merge with adjacent connective tissue and have indistinct margins. Other ligaments that also bind movable joints are tough and prominent. Ligamentous loops and sleeves guide tendons, particularly at joints. Some of these merge into the sheaths of the tendons.

The **nuchal ligament** is an example of an antigravity mechanism. This strong and extensible ligament (i.e., having a low modulus of elasticity) is prominent in large mammals with heavy heads and long necks (Figure 21.6) and, judging from scars on neural spines of vertebrae, also in long-necked sauropod dinosaurs. It extends from the summits of anterior thoracic neural spines to the back of the skull and neural spines of anterior cervical vertebrae. The head and neck are held in normal resting posture without muscular effort. A small muscular tug depresses the head to the ground, simultaneously stretching the ligament. When the muscles relax, the ligament shortens, thus elevating the head.

Ungulates have a suspensory mechanism that cushions their footfalls. Each foot is supported by an elastic ligamentous sling. This mechanism is described on p. 551. The claws of cats are passively retracted by elastic ligaments.

CARTILAGE is found where moderate resistance to compression, tension, and shear (in different combinations according to circumstance) must be combined with firmness and some flexibility. It has the advantage over bone that it is lighter. The various types of

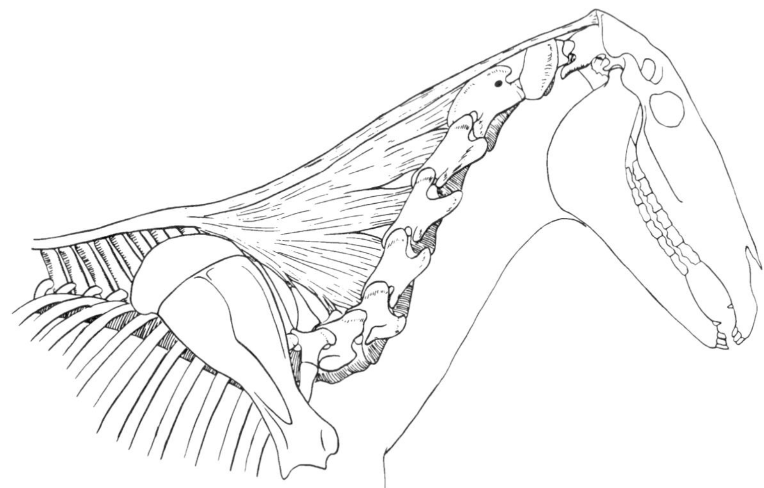

FIGURE 21.6 NUCHAL LIGAMENT OF THE HORSE, an antigravity mechanism.

cartilage (see p. 114) can be distorted in varying degrees, but all return to their original form when released. Elastic cartilage, the most flexible type, supports the external ear, nose, and epiglottis of mammals. Fibrous cartilage provides a tough, but somewhat flexible, cushion. It forms intervertebral disks and the pelvic symphysis of some tetrapods. It is also found at the insertions of some tendons. Hyaline cartilage forms the skeletons of embryos and elasmobranchs and parts of the skeletons of adults of many vertebrates. In tetrapods, it may substitute for bone where the greater strength of bone is not needed (e.g., in the carpus and tarsus of salamanders). It also stiffens the trachea and covers the articular surfaces of movable joints, where its hardness, smoothness, and release of water under pressure reduce friction and benefit lubrication. The principal kinds of cartilage may intergrade. Varying degrees of calcification may harden the hyaline cartilage of the sternal ribs of mammals, the epiphyses of amphibians, and the skeletons of elasmobranchs.

Bones that Resist Compression or Tension In order to save on weight, bulk, and metabolic requirements, the supportive elements of the body are designed to provide adequate strength with minimum material. This principle is important for analysis of the skeleton.

Adequate strength to resist all stresses is provided to a mouse by even a slender skeleton. For reasons explained in Chapter 23, however, very large tetrapods would be unable to sustain the resultant loads if they were proportioned like enormous mice. Elephants, various extinct mammalian giants, and many dinosaurs are (or were) obliged to modify structure, posture, and behavior to minimize stresses on the skeleton and to resist the remaining stresses effectively. Figures given above show that bone can sustain more compressive force per unit of cross-sectional area than any other kind of force. An animal could support itself with the least bone (and weight and bulk) if it could limit all loads to compression. This is not even possible for a static table, let alone a moving animal, yet the largest land animals do minimize stresses other than compression in ways that can be predicted from study of Figure 21.3. The columnlike limbs of such animals have bones that are relatively cylindrical with heads nearly in line with their shafts, thus reducing bending forces (see figure on p. 428).

In Chapter 22 it is explained that the vertebral centra of most land tetrapods are subjected primarily to compressive forces acting on their opposing ends. If they were large enough for the necessary muscle attachments and leverages and also solid, they would have to be much stronger (and heavier) than needed. They are not solid, but neither are they hollow. Spicules of bone called **trabeculae** brace the inside of each centrum, and

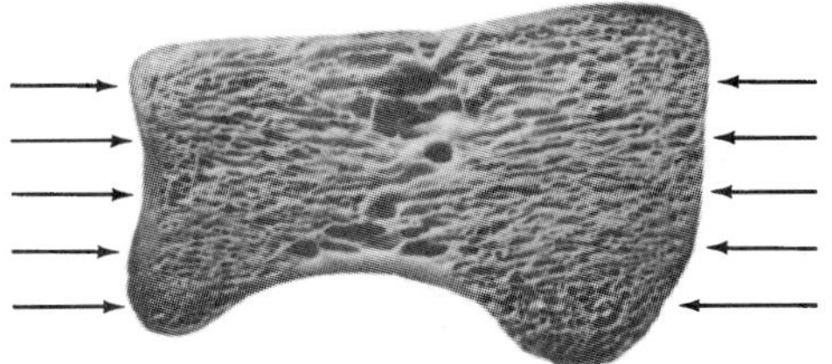

FIGURE 21.7 LONGITUDINAL SECTION OF A CENTRUM SHOWING ORIENTATION OF TRABECULAE PARALLEL TO COMPRESSIVE FORCES. The specimen is a lumbar vertebra of a caribou, *Rangifer.*

these are largely oriented lengthwise in the direction of the predicted compression lines (see Figure 21.7).

The arm bones of gibbons are stressed primarily by tensile forces as the animals swing under tree branches. Few bones of few animals, however, are loaded primarily by tension; if constant or frequent tension must be withstood, a ligament, being as strong for its weight and less likely than a thin bone to fracture if bent sideways, is substituted. An example is the sacrotuberous ligament shown in Figure 21.8. Bones are heavily stressed locally by tensile forces where tendons insert on them. This most often occurs near the ends of bones where bending forces are also common.

Bones that Resist Bending in One Plane When a solid cylinder resists a bending force applied in one plane (e.g., the plane of the paper in parts F and G of Figure 21.3), the resultant stresses are concentrated in that plane and are greatest at the surface of the cylinder. Accordingly, although a cylinder is useful for sustaining compression alone, or tension alone, it is not economical of material for resisting bending in one plane: Too much of the material is not stressed and hence is wasted. The engineer uses instead an "I-beam" as a girder. This beam has upper and lower bars of steel that are stressed when loaded, and a wall to hold the bars apart. (Which bar is compressed and which is tensed depends on the relation of the load to the support—see Figure 21.9.) Bones are never designed as simple I-beams, yet the same principle of construction explains the dumbbell-like distribution of material sometimes seen in cross sections of bones.

The carpenter's beam and joist provide another useful analogy. Lumber is used that has rectangular, but not square, cross section, and is always oriented so that the longer dimension is parallel to the load (i.e., usually is vertical). The reason for this is that resistance to bending is proportional to the width of the beam (dimension transverse to the

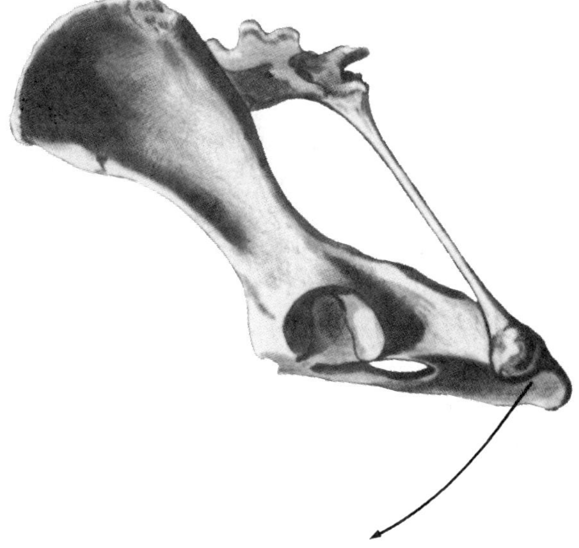

FIGURE 21.8 AN EXAMPLE OF A LIGAMENT THAT WITHSTANDS FREQUENT TENSION is the sacrotuberous ligament of the dog, which resists the tendency of the innominate bone to rotate on the sacrum in the direction shown by the arrow when muscles that swing the leg to the rear pull on the ischium.

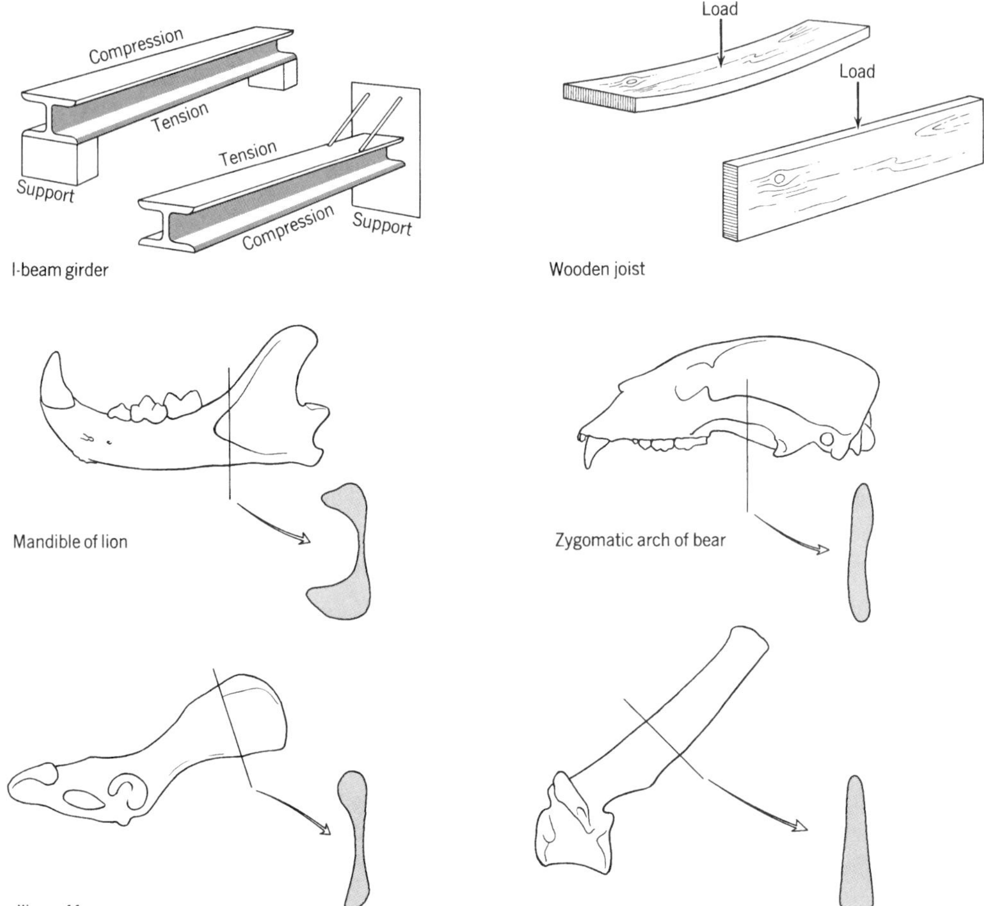

FIGURE 21.9
BONE STRUCTURE ANALOGOUS TO THE I-BEAM (left) AND JOIST (right).

load) times the square of its height (dimension in line with the load). If one dimension is twice the other, then the beam is about twice as strong on edge as it is flat; if one dimension is three times the other, the beam is three times stronger when on edge. Clearly, the animal body should "know about" this, and it does. The zygomatic arch is a bony beam turned on edge to the muscles acting on it. The same is true of most of the neural spines. The pygostyle of birds is a blade of bone oriented parallel to the air resistance transmitted to it by the tail feathers. The lower jaw at the level of the teeth is a modified beam of expected orientation in relation to loads imparted by teeth and muscles.

Resistance of a beam to bending also varies inversely as the square of its length. For this reason, bony beams are not long; other kinds of construction resist loads that must be held at some distance from the support of the mechanism. The nuchal ligament is one example; others will be described later.

Bones that Resist Bending in Several Planes We have seen that a flat beam effectively resists bending when loaded edge on, but is weak when loaded flat-side on. The long bones of the appendages of tetrapods must resist bending in many directions; hence, they cannot be flat. The cylinder, discarded as wasteful of material when bending forces are in one plane, is here an effective model because compressive and tensile forces can concentrate at opposite edges of the cylinder no matter what the direction of loading. It is still true, however, that stresses are least toward the central axis of the cylinder (see again parts

F–I of Figure 21.3). Therefore, the most strength for the least material is achieved by a hollow cylinder. This explains the hollow shafts of long bones.

Controlled bending of the vertebral column of a fish occurs at the intervertebral joints. The centra themselves must resist the bending forces produced by axial muscles. The spine as a whole functions as a somewhat flexible bony tube, the discontinuous cavity of which is the spaces between the markedly amphicoelous centra.

Returning to tetrapods, since resistance of a tube to bending varies inversely as the square of its length, and since the long bones are stabilized by muscles acting at their ends, they would be most subject to fracture at the centers of their shafts if there were no provision to the contrary. The shafts of such bones compensate by being a little thicker at midlength, by increasing their diameter, or both (see figure on p. 161).

Although long bones are subject to bending forces in various directions, force may be greatest in one plane. The bone may then compromise between the beam and tube by becoming oval in cross section, with the long axis of the oval in the direction of the dominant load and with thicker walls on the sides toward and away from the load. The phalanges of bats and pterosaurs tend to have this configuration (Figure 21.10). It sometimes happens that stress is greatest along the concave side of a curved bone (Figure 21.3, part H), and the wall of the bone may then be thickest there.

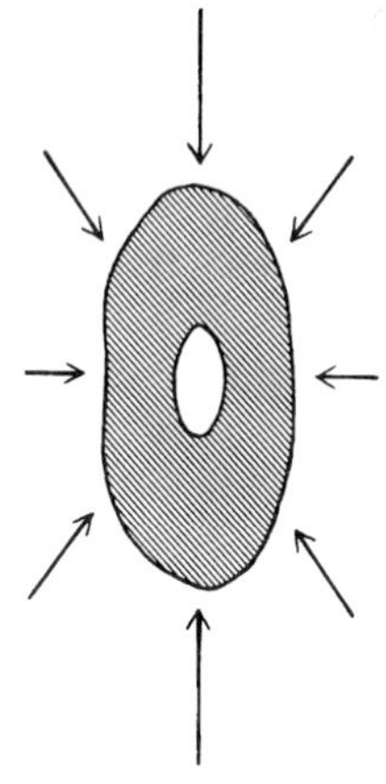

FIGURE 21.10
CROSS SECTION OF A PHALANX OF A BAT SHOWING STRUCTURE IN RELATION TO USUAL FORCES.

It is evident from parts F to I of Figure 21.3 that since most muscles insert near the ends of the long bones, and since forces transmitted by one bone to another across a flexed joint are rarely parallel to the shaft of either, stress lines arc across the ends of those bones. It follows that the ends of long bones should not be tubes. Since solid ends would be stronger (and heavier) than needed, the most economical design is a network of interconnecting trabeculae and thin sheets of bone that follow the stress lines, and this is what we find. Stress lines change somewhat as loads change so the body adopts lines that are a compromise of the more usual loads. (The arching of the trabeculae is usually evident, as for the lower trabeculae in Figure 21.11, but rarely is as regular as in the head of the human

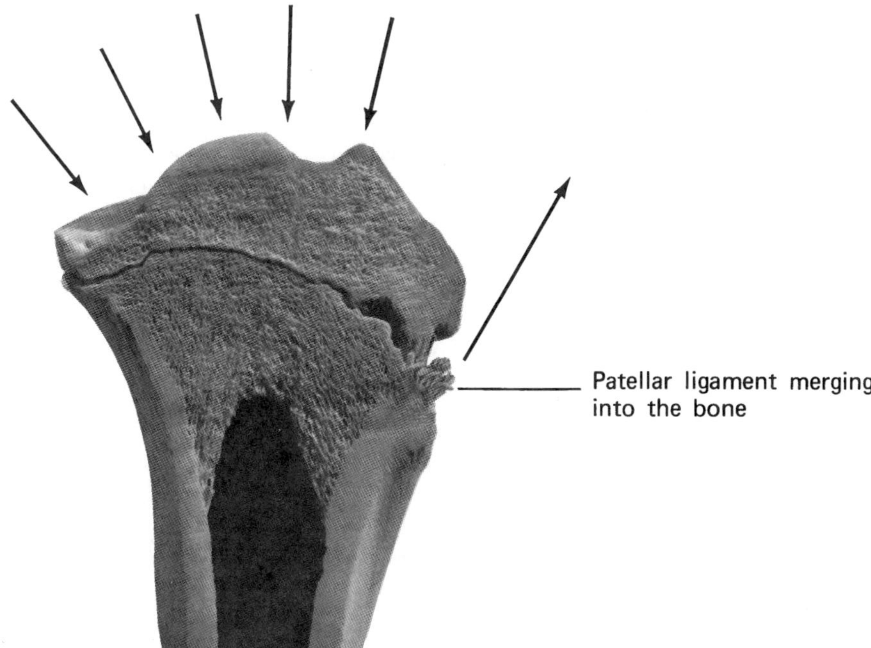

FIGURE 21.11
LONGITUDINAL SECTION OF THE PROXIMAL END OF A COW TIBIA SHOWING ORIENTATION OF TRABECULAE IN RELATION TO SOME OF THE FORCES ACTING ON THE BONE.

COMMENT 21.2

REMODELING OF BONE: PROCESS AND STIMULUS

Remodeling of bone is the process of removing bone in some places while adding it elsewhere. Remodeling is essential for growth (see figure on p. 162), healing of microcracks, repair of breaks, and adaptation to loading. It is episodic at any given locus, but overall is a constant and life-long process.

Remodeling is accomplished by *basic multicellular units*, which are teams of bone-resorbing cells (*osteoclasts*) and bone-forming cells (*osteoblasts*). The former excavate tunnels in existing bone (removing any damage); the latter follow behind laying down new bone until the tunnels are only large enough (the diameter of a human hair) to house the Haversian canals of new osteons. Osteoblasts are trapped here and there in the matrix to become new osteocytes.

It is likely that the signal to initiate remodeling can be chemical, mechanical, or electrical in nature. The last has been the most studied and probably guides remodeling that alters bone shape in response to loading. Deformation causes stress-generated polarity in two ways. *Piezoelectric potential* results when pressure is applied to a crystalline substance. *Streaming potential* (which is probably the dominant signal) results when an ionized fluid flows in channels with charged surfaces. The effective channels are apparently the canaliculi radiating from the lacunae of the bone. Flow is proportional to strain. Compression (which squeezes fluid away) produces negative charges, resulting in a net gain of bone tissue. Tension (sucking fluid in) produces positive charges, resulting in loss of bone. Cyclic loading is a much more effective stimulus than constant loading. These responses cause curved bones to straighten (as where a fracture has healed at an angle), but most bones are normally curved and do *not* straighten. Seemingly it is only strains that are above threshold levels that tend to alter normal curvature. (This general process explains how dental braces gradually move teeth.)

Clearly there remains much to learn about the intricacies of this dynamic system. (See, in References, the book by Martin, Burr, and Sharkey.)

---

femur, which is commonly used to illustrate this phenomenon, or the bones of Figure 21.13, which are diagrammatic.) The spongy nature of the ends of long bones also provides that they can function as shock absorbers.

Where several bones function as a unit in sustaining usual loads, stress lines, and hence trabeculae, also traverse those bones as a unit. The human tarsus is an example. It functions as a beam loaded in the middle by the tibia (though we shall see later that this is not all of the story).

We have seen that the shapes of bones are usually adaptive: Beamlike bones withstand bending in one plane; hollow cylindrical bones withstand bending in several planes; internal trabeculae are oriented along stress lines. Each animal inherits the general form of its skeleton, but the detailed form is determined by use. The configuration and thickness of bones and the patterns of their trabeculae are established only as the young animal moves about and matures, and they are modified if changes in the distribution of mass or in behavior (including any resulting from injury) alter usual loads. Adaptability requires the remodeling of existing bone, as described further in Comment 21.2.

UNIONS OF STRUCTURAL ELEMENTS

Tendon to Muscle; Tendon and Ligament to Bone The tensile strength of tendons is roughly four times the maximum loads delivered to them by their respective muscles. The union of tendon to muscle is sometimes a little less strong than the muscle, but only a little. The tendon may appear to end where the muscle begins, but it branches and pervades the muscle, its fibers merging with those of the perimysium and endomysium. In pulling on its own fibrous framework the muscle also pulls on its tendon.

Muscles that take origin from large areas of bone (e.g., supraspinatus) may gain sufficiently firm attachment by merging their connective tissue with the periosteum of the bone. Tendons and ligaments, however, concentrate so much force on such small areas that a stronger attachment is needed. Imagine the difficulty of joining with glue the end of a flexible cord to a hard smooth material with enough strength to sustain loads of 900 kg/cm^2 even though the angle of attachment changes! Insertions of ligaments and tendons do tend to be weak points in the bone-muscle system, yet the body surpasses human technology in solving the problem.

The collagenous fibers of tendons are not attached to bone; they merge into it (Figure 21.11). Fibrous bone forms all of the skeleton of small animals. Large animals have osteons throughout most of the skeleton, where compressive forces dominate, but retain fibrous bone at the insertions of tendons. Fibers of tendons penetrate the bone and there become indistinguishable from its fibers. (Fibrous cartilage may intervene. Also, calcification may merge into a tendon at the tendon-bone junction.)

There remains an apparent source of weakness. Consider a large tendon of circular cross section that inserts at right angles to the surface of a bone. If all bundles of the tendon are of equal length, they share the load equally when tension occurs. If the angle of insertion changes, as would be expected if the tension causes motion, the relative distance from muscle to bone decreases on the side of the tendon now forming an acute angle with the bone and increases on the side forming an obtuse angle. One might expect that fibers on the long side of the tendon would carry all the stress and would give way one by one. This does not happen, primarily because the collagenous fibers of a tendon, although parallel in the body of the tendon, weave at its insertion, thus distributing the load throughout the insertion. Furthermore, the strength of a tendon usually has a large safety factor; not all of it need support the pull of its associated muscle. Fibers of a resting tendon are a little wavy; a pull that straightens them on the long side transmits some tension also to the short side if the difference in length is slight. Moreover, the angle of insertion rarely changes very much.

The elastic fibers of ligaments are largely replaced by collagenous fibers at their insertions. Thus, the insertions of ligaments are like those of tendons, though details for each differ according to size of animal, general angle of insertion, and specific location.

Kinds and Functions of Joints Joints between bones are classified on the basis of both structure and function, though the two are, of course, related. A first structural category is the immovable joint or **synarthrosis** (= *together* + *joint*). The bones may be joined only by connective tissue, which is the rule for membrane bones (e.g., on the roof of the skull), or only by cartilage, which is usual for replacement bones (e.g., at the base of the skull and between shafts and epiphyses of long bones). The cracks between bones joined by synarthroses are called **sutures.** Although "immovable," some sutures are flexible enough to provide some shock absorption, particularly in response to tension. These joints are places of growth; sutures must remain open for growth to occur. When the growth period terminates, synarthroses of birds and mammals tend to ossify and thus become obliterated one by one on a schedule characteristic of each species. Most sutures of marsupials, however, and some sutures of many other mammals, remain open for life.

Synarthroses are further characterized by the configuration of the suture, and this relates to function. If the suture is approximately straight and the bones have nearly squared-off edges, then a **butt joint** is formed, as between the two nasal bones, and between bones of the basicranium of most mammals (Figure 21.12). Butt joints can withstand compression but little shearing or bending.

If the same square-edged bones were instead joined by overlapping, the union, a **lap joint,** might be somewhat stronger. However, as human builders have learned, when a lap joint is compressed or tensed, the area of contact is not evenly stressed. When glue is

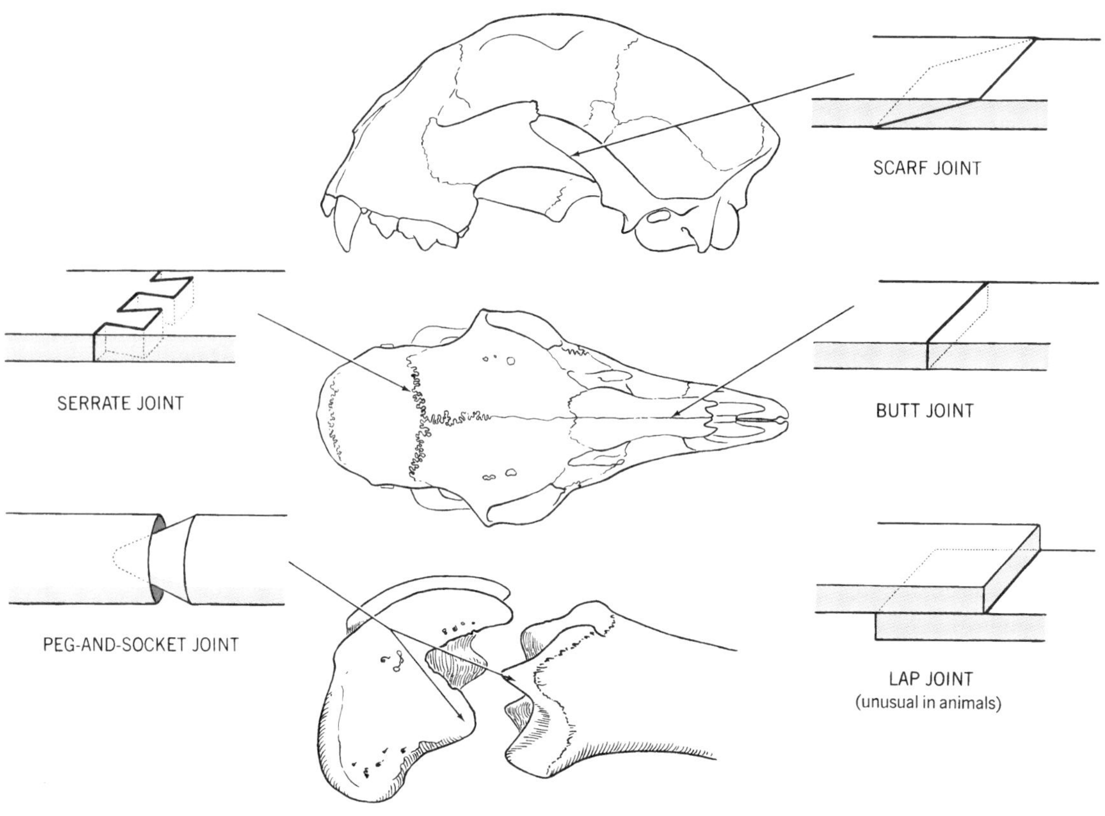

FIGURE 21.12 SOME KINDS OF SYNARTHROSES. Top, skull of a cheetah, *Acinonyx;* middle, skull of a deer, *Odocoileus;* bottom, distal epiphysis and shaft of the femur of a young wolf, *Canis.*

used, it tends to give way at the leading edges and to break toward the middle. If the overlapping edges taper instead so that the two members remain in line, the union is called a **scarf joint.** Here the entire area of contact is evenly stressed by most loads and strength is much improved. Scarf joints (also called squamous joints) often join thin flat bones. They occur between some bones of the mandible of reptiles, and in most vertebrates join various of the cranial bones.

A synarthrosis that is very effective for withstanding compression and shear between hard structures that are not thin and flat is the **peg-and-socket** (also called gomphosis). Such joints join thecodont teeth to the jaw bones and often the jugal to the maxilla. Most epiphyses of long bones join their shafts by complex joints including several pegs and sockets, sometimes relatively deep (distal end of femur of mammals) and sometimes shallow (proximal end of tibia). Another synarthrosis is the **serrate joint,** which has such an irregular suture that the adjoining bones interlock repeatedly throughout the union. This firm type of joint is found between roofing bones of the cranium of some tetrapods, particularly of amphisbaenians (which dig with the head) and artiodactyls (which support horns or antlers). The configuration of the interlocking surfaces relates to the type of force that is resisted. Such joints are effective for absorbing energy; the more a suture is loaded, the more interdigitation occurs and the later it ossifies.

A second, and intermediate, structural category is the **amphiarthrosis** (= *both* + *joint*), which allows some motion in response to compression, tension, or twisting, yet is tough. The surfaces of the adjoining bones may be covered by hyaline cartilages which, in turn, are joined by a pad of collagenous fibers or by fibrous cartilage. The union between the bones of such a joint is called a **symphysis** instead of a suture. Examples are the mandibular symphysis of many vertebrates, the pelvic symphysis (which allows more motion in females toward the end of pregnancy than in males), and the joints between most vertebral centra (which provide for motions of the spine). The function of joints between centra is conditioned by configuration (for terminology, see p. 142). Joints between procoelous and between opisthocoelous centra allow adequate motion in any direction, withstand compression, and resist dislocation better than platyan centra. Procoelous and opisthocoelous vertebrae are, therefore, common in the necks of tetrapods and in their tails if the tail is strong. Platyan centra are usually restricted to the trunk, where shearing is minimal.

In another kind of amphiarthrosis called a **syndesmosis,** the bones are joined by moderately thick zones of collagenous fibers or by ligaments, and somewhat greater motion is allowed. Examples are the unions of radius to ulna and of fibula to tibia in certain mammals having some play between these pairs of bones. However, syndesmoses are more characteristic of various other classes. Thus, such joints are common among bones of the protrusible upper jaws and movable opercula of bony fishes.

The last general structural category of joints is the freely movable joint or **diarthrosis** (= *two + joint*). The articulating surfaces of the bones are covered by smooth hyaline cartilage (Figure 21.13). In fetal life the **joint cavity** develops, which is necessary for movement of one bone on the other. Where not bordered by the cartilage-covered bones, the cavity is enclosed by a **joint capsule.** The capsule may be thin and membranous but is usually at least partly tough and fibrous, containing both collagenous and elastic fibers. The capsule is lined by a cellular **synovial membrane,** which is more or less folded. It contains fat cells and, in some joints, fat pads that encroach on the cavity and help cushion its changing configuration as the joint moves. Ligaments binding a diarthrosis may be within or partly outside the capsule, or may be inside the joint cavity, as at the hip and knee of mammals. Tough pads of fibrous cartilage called **menisci** (singular, meniscus) are anchored inside the cavities of several joints. They may guide the moving bones where the bony surfaces otherwise have

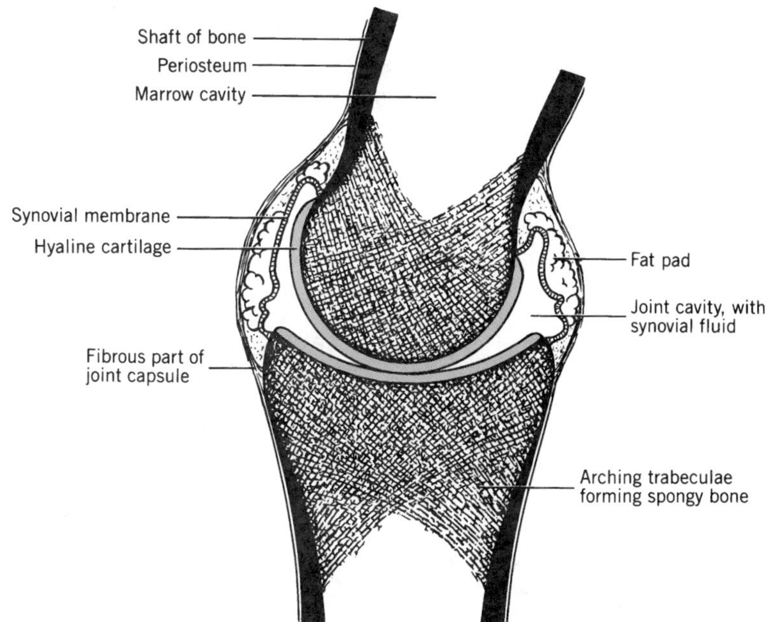

FIGURE 21.13
STRUCTURE OF A DIARTHROSIS.

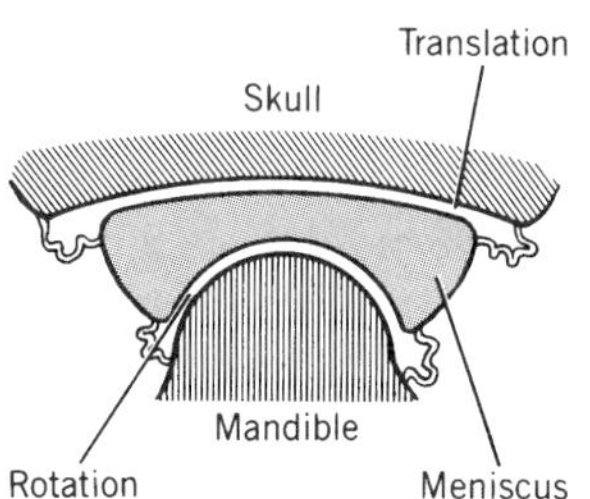

FIGURE 21.14 DIAGRAM OF A TEMPOROMANDIBULAR JOINT showing, in lateral view, a probable function of the meniscus.

a poor fit, as at the knee. At some joints a meniscus increases the kinds of motion possible. Thus, in some mammals there is rotation between the mandible and a fibrous pad, but translation between the pad and the skull (Figure 21.14). By implanting a variable-reluctance transducer into menisci of miniature pigs, researchers at the University of Washington learned the displacement of the pad during various jaw motions and showed that it is under tension or compression according to function.

Bathing joint cavities is a small quantity of **synovial fluid,** apparently produced by the synovial membrane. This clear or yellowish fluid is similar to tissue fluids and contains mucin. It is more or less viscid according to the joint. Its function is to nourish the hyaline cartilage (which is devoid of blood vessels) and, importantly, to lubricate the joint. White cells in the synovial fluid remove from the joint capsule any microscopic fragments of cartilage that result from the contacts between surfaces.

No matter how congruously two surfaces may be shaped, and how carefully polished (e.g., two flat pieces of glass), when they rest together, they touch only at microscopic elevations. When the dry surfaces slide over one another, the microelevations grate, thus producing heat, friction, and wear. If a lubricant is introduced between the surfaces, fewer solid-to-solid contacts are made, and the pressure on them is less. Much of the shearing force takes place between molecules of the lubricant, so friction and wear are reduced. This kind of lubrication, common in man-made machinery, is called **boundary lubrication.** When a lubricant can be kept thick enough to hold two surfaces completely apart, even though they are under considerable pressure, then as they slide over one another the only shearing is within the lubricant and there is no wear. This is **fluid film lubrication.** To achieve it (without an external pump), the surfaces must (like animal joints) not be quite congruous; therefore, the lubricating surface is shaped like a wedge. As the joint turns, the surfaces roll onto a film of lubricant of ever-increasing thickness, thus compensating for lubricant that is squeezed away.

Some investigators have thought that the lubrication of living joints is of the fluid film kind, except when motion starts and stops. However, it now appears that the exceedingly low friction of living joints results primarily from a combination of boundary lubrication and another kind called **weeping lubrication.** Joint cartilage consists in part (20–40%) of collagen (which is stiff in tension), proteoglycans (stiff in compression), cells, lipids, and other protein. These form a spongelike matrix. The remainder of the cartilage is synovial fluid, which fills the sponge. Under pressure, the fluid is squeezed through the pores (thus dissipating stress) and out of the cartilage into the loaded part of the joint, where it lubricates. When load is removed from a part of a joint, as usually occurs during turning, fluid is sucked back into the cartilage by the elasticity of the matrix and by electrical bonding of water to proteoglycans. The imperfect fit of the joint surfaces facilitates this recharging process.

The lubrication of living joints is remarkably effective. As explained further on p. 480, friction at a joint equals the force crossing the joint times a coefficient of friction. For a

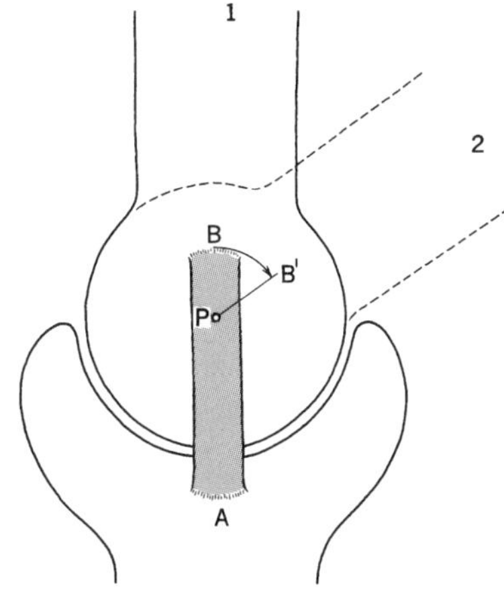

FIGURE 21.15 MECHANICS OF A SNAP JOINT. As the upper bone rotates, hingelike, from position 1 to position 2 on the lower bone, which is fixed, insertion B of ligament AB moves to B′ in an arc around the pivot of motion, P. Since distance AB′ is shorter than AB, the joint is unstable in position 1.

waxed ski sliding on dry snow the coefficient is about 0.18, for ice sliding on ice it is about 0.02, and for movable joints common values are 0.002 to 0.04.

Diarthroses are classified according to both function and shapes of articulating surfaces. Function and shape are correlated, but several shapes may serve similar functions, and one general shape may serve several functions. Thus, terminologies tend to be either inadequate or inconsistent. Furthermore, there are intergrades and combinations. Nomenclature does not substitute for interpretation.

A **hinge joint** has a more or less cylindrical head that rotates in a corresponding socket. Motion is primarily around one axis, as for a door hinge. The articulation of the mandible to the skull of carnivores is a simple hinge joint. The head of a hinge joint may have splines, flanges, or bulges that mesh with corresponding grooves or waists in the socket (see figure on p. 445). These devices further limit motion to one plane and resist dislocation. Examples are the elbow joint, ankle joint of many mammals, and joints between phalanges, particularly of runners.

Some hinge joints can further be designated **snap joints** (Figure 21.15). These are stabilized by their ligaments in the open and closed positions and are unstable in intermediate positions. Any hinge joint revolves around an axis that lies within the convex member and is transverse to the plane of motion. Ligaments holding the joint together are lateral in position. If one end of such a ligament inserts exactly at the pivot, then its length remains constant as the joint moves and tension does not change. If, however, a ligament crosses over the pivot to insert slightly beyond, then its length and tension are reduced as the joint moves in either direction from an intermediate position. The joint then snaps into the open or closed positions. Snap joints are found at the elbow and ankle (hock) of various large mammals. The mechanism provides some passive support in the standing posture. In order to better secure the union, snap joints usually have at each side of the hinge either two ligaments that cross or a single broad ligament that twists. (Can you say how the ligaments of a hinge joint could be arranged to snap the joint into a single given position, open, closed, or intermediate?)

A **ball-and-socket joint** has a hemispherical head that turns in a nearly congruous socket. A wide range of motions, including rotation, is implied, and the shoulder and hip joints are the usual examples (see figures on pp. 165 and 166). The union of the occipital condyle to the atlas of archosaurs can also be cited. A shallow socket (shoulder) allows more excursion than a deep socket (hip). Joints between procoelous and between opisthocoelous centra are ball-and-socket in structure, but they have less range of motion and are usually not so designated, even if the joint is a synarthrosis. A modification of the ball-and-socket joint is the peg-and-socket diarthrosis that allows only rotation around the long axis of the peg. This unusual kind of joint joins the avian quadratojugal to the quadrate and functions with the mechanism that moves the upper part of the bill (see p. 562).

Somewhat similar to the last named joint in function is the **pivot joint,** which allows rotation of one bone around its own long axis. During pronation and supination of the manus, the proximal end of the radius pivots on the ulna; its disk-shaped head revolves in the radial notch. Likewise, the manus pivots on the styloid process of the ulna.

If the convex head of a bone is biaxial instead of hemispherical, and fits into a biconcave socket, an **ellipsoid joint** results. Motion is around two axes (e.g., flexion-extension and adduction-abduction), and motion around the third axis is prevented—very different from a pivot joint. The human radius-to-carpus joint is an example. The same in function, though different in structure, is the **saddle joint** found between the heterocoelous cervical vertebrae of birds (see the pelican in the figure on p. 143). The articulatory surface of the anterior vertebra (i.e., at the posterior end of the centrum) is convex horizontally and concave vertically, whereas that of the posterior vertebra is concave horizontally and convex vertically.

Another family of joints has the name **plane joint.** The articulating surfaces are more or less flat and permit various motions depending largely on the nature of associated ligaments. Contact between the bones may be maintained if they glide over one another as do the pre- and postzygapophyses of vertebrae. Usually the articulating surfaces are not quite flat, so that gliding motions force them apart. Some motions separate flat-ended bones to a surprising degree; the joints between the carpal bones of large mammals are a striking example (see the vicuna in the figure on p. 446).

The patella has a curved surface that slides in the patellar groove of the femur. Lumbar vertebrae of artiodactyls have postzygapophyses that are rolled into scrolls, and prezygapophyses that are trough-shaped. These joints limit some kinds of motions. They are no longer so "plain," yet have no special name.

Still other kinds of joints defy current terminology yet invite attention. Thus, nature has designed the mammalian knee joint without regard for orderly classification: The femoral condyles largely rotate on the platform of the head of the tibia, but they also roll over it. Motion is mostly around one axis, as for a hinge joint, but not entirely so, and there is also slight rotation of the tibia around its axis. Many birds can move the upper bill on the braincase (discussed further in Chapter 30). The joint usually consists merely of a zone of thin flexible bone—a type not named in our anatomy texts. Anurans have either a syndesmosis or diarthrosis between the sacral vertebra and the arms of the pelvic girdle. The joint may function as a hinge in the vertical plane or may allow side to side bending. In some frogs these motions are possible to a degree, and the vertebral column can also telescope forward and backward on the pelvic girdle under the control of apposed sets of muscles.

GENERAL REFERENCES FOR PART III

Alexander, R. McN. 1967. *Functional design in fishes.* Hutchinson, London. 160p. Discusses swimming, buoyancy, respiration, feeding, and sense organs.

Alexander, R. McN. 1983. *Animal mechanics.* 2nd ed. Blackwell Scientific Publications, Boston. 301p. Far-ranging application of mechanical principles to animal functions.

Alexander, R. McN., and G. Goldspink (eds.). 1977. *Mechanics and energetics of animal locomotion.* Wiley, New York. 346p.

Biewener, A.A. (ed.). 1992. *Biomechanics (structures & systems): a practical approach.* Oxford Univ. Press, New York. 290p.

Feder, M.E., A.F. Bennett, W.W. Burggren, and R.B. Huey (eds.). 1987. *New directions in ecological physiology.* Cambridge Univ. Press, New York. 364p.

Gans, C. 1974. *Biomechanics: an approach to vertebrate biology.* Lippincott, Philadelphia. 259p.

Gans, C.L., A.S. Gaunt, and P.W. Webb. 1997. Vertebrate locomotion, pp. 56–213. *In* W.H. Dantzler (ed.), *Handbook of physiology,* Sec 13: Comparative physiology. American Physiological Society, Oxford Univ. Press, New York. Comprehensive overview of all groups; 2000+ references.

Gordon, J.E. 1976. *The new science of strong materials, or why you don't fall through the floor.* Penguin, New York. 287p.

Gordon, J.E. 1978. *Structures, or why things don't fall down.* Plenum, New York. 395p.

Hertel, H. 1963. *Structure, form, and movement.* Otto Krausskopf-Verlag, Germany. English edition, 1966, Reinhold, New York. 251p. An engineer's analysis of body mechanics. Emphasis on swimming and flying.

Hildebrand, M., D.M. Bramble, K.F. Liem, and D.B. Wake (eds.). 1985. *Functional vertebrate morphology.* Harvard Univ. Press, Cambridge, MA. 430p. An outstanding source at a next level of advancement beyond this book.

Nowak, R.M. 1999. *Walker's mammals of the world.* 6th ed. Johns Hopkins, Baltimore. 2 vols., 1936p. All genera are described and illustrated. Valuable for visualizing body form.

Rayner, J.M.V., and R.J. Wootton (eds.). 1991. *Biomechanics in evolution.* Cambridge Univ. Press, New York. 273p.

Thomason, J.J. (ed.). 1995. *Functional morphology in vertebrate paleontology.* Cambridge Univ. Press, Cambridge, UK. 277p.

Vogel, S. 1988. *Life's devices.* Princeton Univ. Press, Princeton, NJ. 367p.

Wainwright, P.C., and S.M. Reilly. 1994. *Ecological morphology. Integrative organismal biology.* Univ. Chicago Press. 367p.

Wainwright, S.A., W.D. Biggs, J.D. Currey, and J.M. Gosline. 1976. *Mechanical design in organisms.* Wiley, New York. 423p.

Weibel, E.R., C.R. Taylor, and L. Bolis. 1998. *Principles of animal design.* Cambridge Univ. Press, New York. 314p.

REFERENCES FOR CHAPTER 21

Alexander, R. McN., and M.B. Bennett. 1987. Some principles of ligament function, with examples from the tarsal joints of the sheep (*Ovis aries*). *J. Zool. Lond.* 211:487–504.

Alexander, R. McN., and N.J. Dimery. 1985. The significance of sesamoids and retro-articular processes for the mechanics of joints. *J. Zool. Lond.* 205:357–371.

Biewener, A.A. (ed.). 1992. *Biomechanics—structures and systems: a practical approach.* Oxford Univ. Press, New York. 290p.

Bennett, M.B., R.F. Ker, N.J. Dimery, and R. McN. Alexander. 1986. Mechanical properties of various mammalian tendons. *J. Zool. Lond.* 209:537–548.

Bock, W.J., and B. Kummer. 1968. The avian mandible as a structural girder. *J. Biomechanics* 1:89–96.

Currey, J. 1984. *The mechanical adaptations of bones.* Princeton Univ. Press, Princeton, NJ. 294p.

Currey, J.D. 1999. The design of mineralised hard tissues for their mechanical functions. *J. Exp. Biol.* 202:3285–3294. Thoughtful discussion of the "evolutionary view" of mineralized hard tissues of invertebrates and vertebrates.

Dimery, N.J., R. McN. Alexander, and K.A. Deyst. 1985. Mechanics of the ligamentum nuchae of some artiodactyls. *J. Zool. Lond.* 206:341–351.

Gardner, E. 1950. Physiology of movable joints. *Physiol. Rev.* 30:127–176. Review article with extensive bibliography.

Hall, M.C. 1966. The architecture of bone. Thomas, Springfield, IL. 346p. Illustrates bone sections showing internal structure.

Hermanson, J.W., and B.J. Macfadden. 1992. Evolutionary and functional morphology of the shoulder region and stay-apparatus in fossil and extant horses (*Equidae*). *J. Vert. Paleontol.* 12:377–386.

Hermanson, J.W., and B.J. Macfadden. 1996. Evolutionary and functional morphology of the knee in fossil and extant horses (*Equidae*). *J. Vert. Paleontol.* 16:349–357.

Herring, S.W. 1972. Sutures—a tool in functional cranial analysis. *Acta Anat.* 83:222–247.

Ker, R.F., and P. Zioupos. 1997. Creep and fatigue damage of mammalian tendon and bone. *Comments Theor. Biol.* 4:151–181.

Kier, W.M. 1985. Tongues, tentacles and trunks: the biomechanics of movement in muscular-hydrostats. *Zool. J. Linn. Soc.* 83:307–324.

Kummer, B. 1976. Biomechanics of the mammalian skeleton. Problems of static stress. *Fortschritte der Zool.* 24(2/3):57–73.

Lanyon, L.E., and C.T. Rubin. 1985. Functional adaptation in skeletal structures, pp. 1–25. *In* M. Hildebrand et al. (eds.), *Functional vertebrate morphology.* Harvard Univ. Press, Cambridge, MA.

Martin, R.B., D.B. Burr, and N.A. Sharkey. 1998. *Skeletal tissue mechanics.* Springer, New York. 392p.

McCutchen, C.W. 1983. Lubrication of and by articular cartilage, vol. 2, pp. 87–107. *In* B.K. Hall (ed.), *Cartilage.* Academic Press, New York.

Myers, E.R. 1983. Biomechanics of cartilage and its response to biomechanical stimuli, vol. 1, pp. 313–341. *In* B.K. Hall (ed.), *Cartilage.* Academic Press, New York.

Swartz, S.M. 1991. Strain analysis as a tool for functional morphology. *Am. Zool.* 31:655–669.

Swartz, S.M. 1993. Biomechanics of primate limbs, pp. 5–42. *In* D.L. Gebo (ed.), *Postcranial adaptation in nonhuman primates.* N. Illinois Univ. Press, DeKalb.

Vis, J.H. 1957. Histological investigations into the attachment of tendons and ligaments to the mammalian skeleton. *Koninkl. Ned. Akad. van Wetenschappen, Proc. Ser. C,* 60:147–157.

Weiner, S., and H.D. Wagner. 1998. The material bone: structure-mechanical function relations. *Annu. Rev. Mat. Sci.* 28:271–298. The authors present bone as a composite tissue made up of seven hierarchial levels, the proportion of which determines its mechanical properties.

Chapter 22

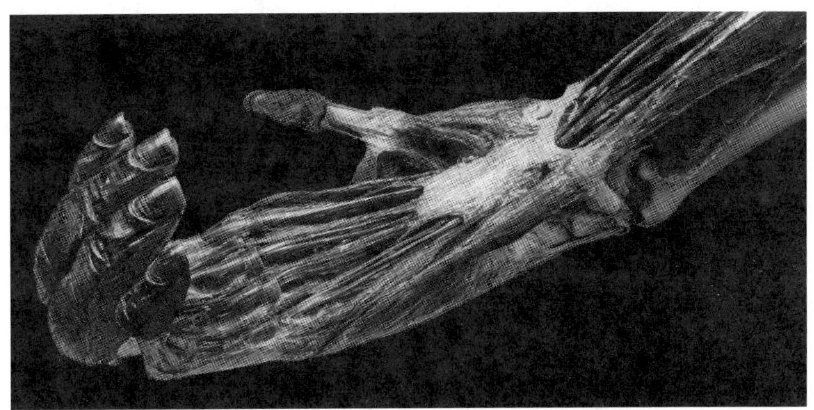

Mechanics of Support and Movement

MAGNITUDE AND DIRECTION OF FORCES

Force Vectors and Their Resolution Force was defined on p. 385. Forces are **vector quantities;** that is, they have both magnitude and direction. Each property is important to the analysis of bone-muscle systems, and each can be represented graphically by an arrow called a **vector.** The arrow is usually placed so that its tail is at the point of application of the force, for example, the insertion of a tendon (alternatively, the head of the arrow could be placed at the insertion). The orientation of the arrow represents the direction of the force, and its length represents the magnitude of the force according to some arbitrary scale (e.g., 1 cm = 10 N). In Figure 22.1, part A shows the long head of the triceps muscle inserting on the olecranon process of the mammalian ulna. If the force of contraction is approximated from the cross-sectional area of the muscle, or better, by direct measurement from the live muscle, then the force of contraction can be represented by the vector $\mathbf{F}_1$ in part B.

The medial head of the triceps (part C) inserts at the same place by the same tendon and can similarly be represented by the vector $\mathbf{F}_2$ of part D. (If the second muscle inserted near the first but not by a common tendon, the forces could still be considered to act at a common point if extensions of their lines of action intersected.)

With two muscles pulling on the same point but in different directions and with different tensions, it becomes important to learn the magnitude and direction of their common, or net, effect. What, we ask, is the vector of the single hypothetical muscle that, acting alone, would load its insertion in just the same way as the actual muscles acting together? The desired force is called the **resultant** of the given forces, and its derivation is called the adding of forces. The resultant is usually determined graphically by drawing a **parallelogram of forces,** as shown in part E: $\mathbf{F}_1$ and $\mathbf{F}_2$ form two sides of a parallelogram that is completed by drawing dotted lines. The diagonal, $\mathbf{R}_1$, is the desired vector of the resultant force.

Above: Air-dried dissection of a chimpanzee hand.

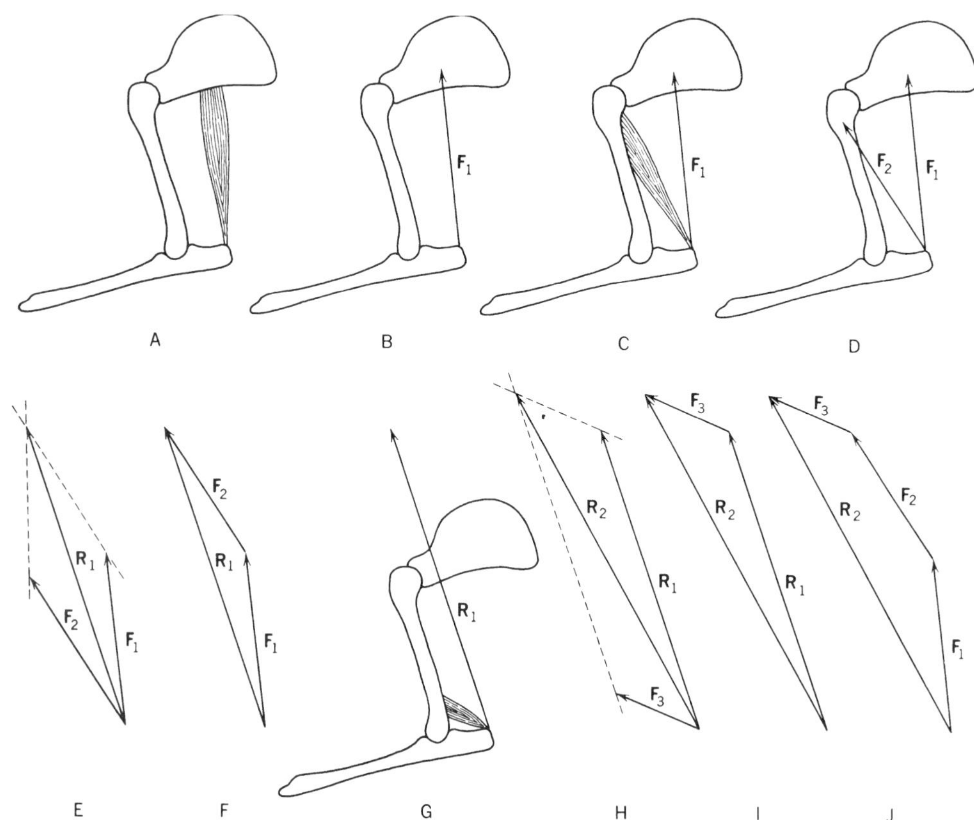

FIGURE 22.1
FORCE VECTORS AND THE ADDING OF FORCES.

Since opposite sides of a parallelogram are equal, an alternative graphical solution is to place $\mathbf{F}_2$ so that its tail is at the head of $\mathbf{F}_1$ (or vice versa), and then draw the line that makes the third side of a triangle (i.e., the triangle that is half of the parallelogram constructed above). That third side is again the desired vector, $\mathbf{R}_1$ (see part F).

A third muscle, the anconeus, also inserts at the same place (part G), its vector being $\mathbf{F}_3$. To determine the resultant force, $\mathbf{R}_2$, of all three muscles contracting simultaneously, either one can draw the diagonal of the parallelogram having vectors $\mathbf{R}_1$ and $\mathbf{F}_3$ as sides, or one can close the triangle having $\mathbf{R}_1$ and $\mathbf{F}_3$ as sides (parts H and I). Either method is a two-step solution, because $\mathbf{R}_1$ first has to be determined from $\mathbf{F}_1$ and $\mathbf{F}_2$. The problem can also be done in one step by joining vectors $\mathbf{F}_1$, $\mathbf{F}_2$, and $\mathbf{F}_3$ tail to head (in any sequence) and then closing the polygon (part J). The closing line is again $\mathbf{R}_2$, or the vector of the resultant force of all three muscles acting together.

The magnitude and direction of a resultant force can be calculated somewhat more precisely, if need be, by trigonometric methods.

The determination of the magnitude and direction of the force of contraction of a pinnate muscle provides an application of the adding of forces. Consider the stylized, flat pinnate muscle shown in Figure 22.2A. It can be regarded as two muscles pulling in different directions on a common central tendon. The force of contraction of each side taken alone is in the direction of the fibers of that side and has a magnitude that is roughly proportional to the cross-sectional area of all the fibers of the side (or to the distance AB or AC if the muscle is of uniform thickness). Vectors $\mathbf{F}_R$ and $\mathbf{F}_L$ of part B represent the forces of the right and left sides of the muscle. Completing the parallelogram of forces, $\mathbf{R}$ is found to be the vector of the resultant force. (Alternatively, $\mathbf{R} = 2\mathbf{F} \cos \theta$.) The value of $\mathbf{R}$ is greatest when the muscle fibers of each side insert on the central tendon at an angle

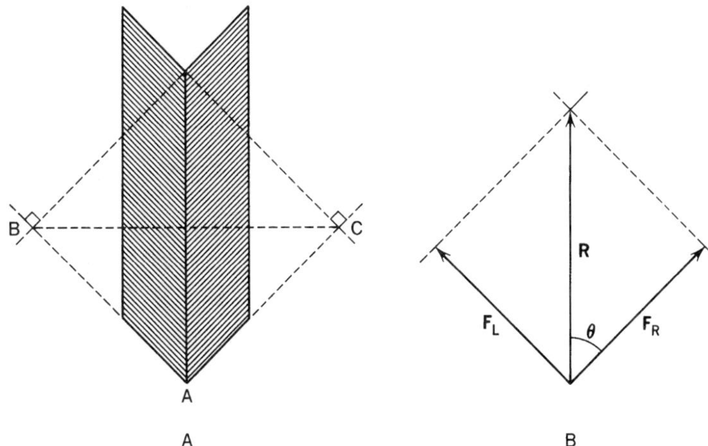

FIGURE 22.2 APPLICATION OF THE BREAKING DOWN OF FORCES TO A STYLIZED, FLAT PINNATE MUSCLE.

of about 45°. Note, however, that the angle of insertion must change as the muscle shortens.

In order to have the same force of contraction as this pinnate muscle, a straplike muscle with parallel fibers would need to have the much greater width BC (Figure 22.2A). Such a muscle having the same overall length as the pinnate muscle would have much longer fibers, however, and could move its insertion farther. Clearly, a pinnate muscle can develop great force for its overall width, but has a short contraction distance. Pinnate muscles, unlike muscles with fibers parallel to their tendons, do not become wider during contraction—an advantage if the muscle must function in a confined space. Furthermore, pinnation allows a muscle to have an irregular shape.

Among mammals, pinnation is seen in most of the flexors of the limbs and in the mylohyoid. The human deltoid is a complexly pinnate muscle, as is the subscapularis of some mammals (see figure on p. 465). The central tendons of some pinnate muscles are stiffened by ossification; examples are found in the "drumstick" of the turkey. Similar splints of bone are found in the neck and jaw muscles of certain birds and were present in the backs of some large dinosaurs.

Components of Forces Just as two or more forces can be combined into one resultant force, so a given force can be broken down into two or more components. And just as an infinite series of pairs of forces can have the same resultant force, so conversely can a given force have an infinite number of pairs of components. However, it is usually desirable to specify the direction of each component, and there is then only one solution.

Consider the vector of the mammalian triceps muscle (one head or all heads in combination) shown in Figure 22.3A. When the muscle contracts, the ulna turns counterclockwise on the humerus. The insertion of the muscle swings in an arc around the pivot of motion. The radius of the arc is the distance from the pivot to the insertion. At any instant in time the direction of motion of the insertion is in the direction of the tangent to the arc that passes through the point of insertion. (A tangent of a circle touches the circumference of the circle at one point and is perpendicular to a radius drawn to that point. At successive instants in time the tangent will change as the joint moves because the insertion will move along the arc.)

Since the insertion usually does not move exactly in the direction of the pull of the muscle, it is important to learn the magnitude of the part, or component, of the pull that *is* in the direction of motion. Two components must be selected such that their resultant is the given force of the triceps. One component will be selected to include all the force in

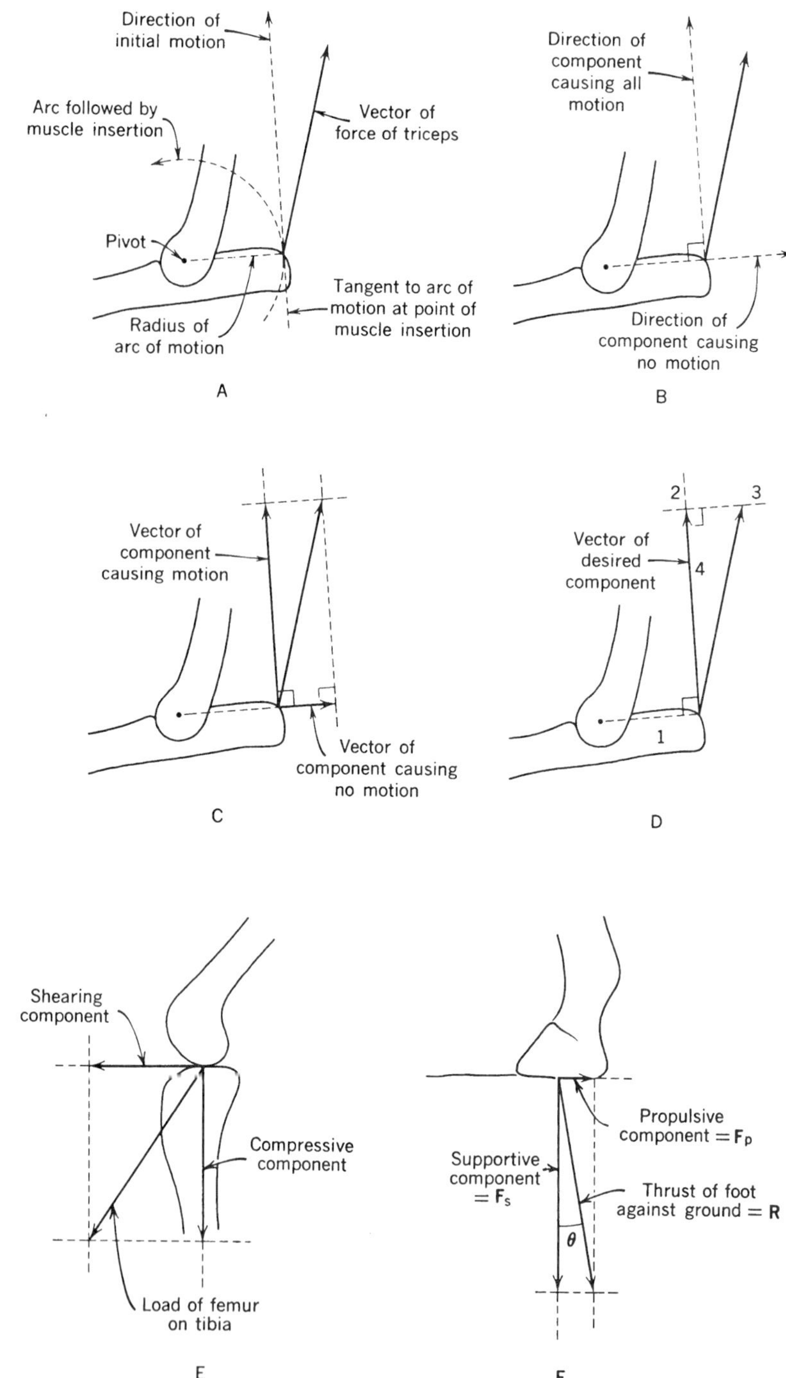

FIGURE 22.3 DETERMINATION OF A DESIRED COMPONENT OF A GIVEN FORCE.

the direction of motion, and, as just explained, this must be along the tangent passing through the insertion. The other component must be selected so as to cause no motion at all, either clockwise or counterclockwise. Only one direction fits this requirement, and that is normal (or perpendicular) to the arc of motion at the point of insertion (see part B). (The vector of this component is an extension of the radius of the arc at the insertion.) Given the directions of the two components, it is simple to derive their magnitudes graphically by completing the parallelogram of forces that has the vector of the force of the triceps as a diagonal. Since the components have been selected to be perpendicular to one another, the parallelogram is a rectangle (part C).

Because the component in the direction of motion is the only one of concern in this instance, and because the desired parallelogram has only right angles, the problem can be solved more directly as follows (part D): (1) Draw a line from the estimated pivot of motion to the insertion of the muscle. This is the radius of the arc of motion. (2) Draw the line perpendicular to this radius that passes through the point of insertion. This is the tangent giving the direction of the desired component. (3) Draw the line perpendicular to the tangent that extends from the tangent to the tip of the vector of the given force. This line is the side of the rectangle of forces that is opposite to (and therefore equal to) the component of force selected to cause no rotation of the ulna on the humerus. (4) The vector of the desired component in the direction of motion is now that part of the tangent between the point of insertion of the muscle and the intersection with the perpendicular drawn in the previous step. It only sounds complicated; one can draw the diagram in less time than it takes to read about it.

In Chapter 21, it was noted that a load applied diagonally to the end of a long bone can be converted to a longitudinal compressive component and a transverse shearing component that tends to bend the bone. The directions of the desired components are determined by the nature of the problem and are at right angles to one another. Completion of a rectangle of forces establishes their magnitude (part E).

When the foot of a running animal thrusts against the ground, it both supports and propels the animal. The direction of the supportive component is vertical (in opposition to the pull of gravity) and the direction of the propulsive component is horizontal in the line of travel. If the thrust of the foot at a given instant can be determined, then it is easy to solve for the magnitudes of the components (part F). Similarly, one can calculate the components of an applied force that are necessary for analyzing centrifugal force (see figure on p. 447), lateral undulation of snakes (see figure on p. 469), friction (see figure on p. 480), and the forward and lateral components of the diagonal thrust of the tail of a fish against the water (see figure on p. 502).

Components of forces can be determined trigonometrically as well as graphically. In this instance the calculations are more simple and direct than for the resolution of forces because only right triangles need be used. Thus, in Figure 22.3F, $\mathbf{F}_p = \mathbf{R} \sin \theta$ and $\mathbf{F}_s = \mathbf{R} \cos \theta$.

Note that as the angle between the given force and a component increases from 0 to 90° the magnitude of the component decreases from that of the given force to zero. In terms of mechanics, muscles are most effective for pivoting bones when they pull in the direction of motion. By inserting onto its central tendon at an angle, a pinnate muscle increases the number of its fibers and hence its force of contraction. However, as the angle of insertion increases, the effective component of the force decreases. The optimum angle (usually a little less than 45°) represents a compromise between these opposing factors.

BONE-MUSCLE SYSTEMS AS MACHINES

A **machine** is a mechanism that transmits force from one place to another, usually also changing its magnitude. Thus, when a screwdriver is used to pry the lid off a can, moderate downward force applied to the handle produces great upward force at the tip of the tool against the lid. Similarly, when the triceps muscle pulls up on the olecranon process, a downward force is produced at the forefoot (Figure 22.4, parts A and B). All bone-muscle systems are machines. It is useful to designate any input force applied to a machine as an **in-force** ($\mathbf{F}_i$) and any output force derived from a machine as an **out-force** ($\mathbf{F}_o$). In the body, in-forces are applied by the pull of tendons or tensed ligaments, by gravity, and by external loads; useful out-forces are ultimately derived at the teeth, feet,

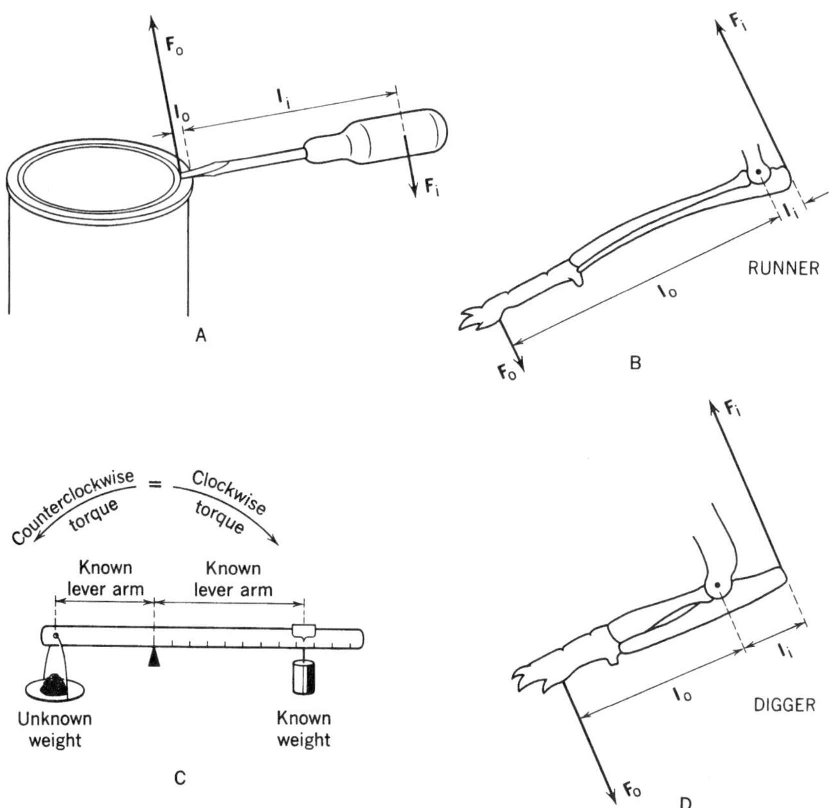

FIGURE 22.4 PRINCIPLES OF IN-FORCES AND OUT-FORCES, LEVER ARMS, AND TORQUE. (In- and out-torques are not adjusted to be in equilibrium.)

digits, and elsewhere. For now, we will consider only simple machines having one in-force and one out-force.

Lever Arms and Torques An in-force may be transmitted to an out-force by a crankshaft, hydraulic device, pulley, lever, or other mechanism. Most feeding and locomotor systems of the body transmit forces by levers, and only these will be considered here. A **lever** is a rigid structure, such as a crowbar or bone, that transmits forces by turning (or tending to turn) at a pivot. Each force is spaced from the pivot by a segment of the lever called a **lever arm;** the **in-lever arm** ($\mathbf{l}_i$) (or power arm) extends from the in-force to the pivot, and the **out-lever arm** ($\mathbf{l}_o$) (or load arm) extends from the pivot to the out-force. In the example of the screwdriver used to pry open a lid, $\mathbf{l}_i$ extends from the hand on the handle to the lip of the can, and $\mathbf{l}_o$ extends from the lip to the tip of the tool pressing on the lid. In the other example, $\mathbf{l}_i$ is the olecranon process and $\mathbf{l}_o$ the forearm from elbow joint to forefoot.

The product of a force times its lever arm is called a turning force, or moment, or **torque** (τ). Every functioning lever includes at least two torques, one for the in-system and one for the out-system. Thus (as qualified below), $\tau_i = \mathbf{F}_i\mathbf{l}_i$ and $\tau_o = \mathbf{F}_o\mathbf{l}_o$. Torques are expressed in dyne-centimeters, or equivalent units (or in gram-centimeters if one substitutes weight units for the more precise force units). When $\mathbf{F}_i\mathbf{l}_i > \mathbf{F}_o\mathbf{l}_o$, the lever rotates in the direction of $\mathbf{F}_i$; when $\mathbf{F}_i\mathbf{l}_i < \mathbf{F}_o\mathbf{l}_o$, the lever rotates in the direction of $\mathbf{F}_o$; when $\mathbf{F}_i\mathbf{l}_i = \mathbf{F}_o\mathbf{l}_o$, the system is in equilibrium and there is no motion.

When a lever system is in equilibrium, any of the variables can, of course, be easily obtained if the other variables are known. This is the principle of a beam balance: The product of a known weight times its lever arm (the calibrated scale) is adjusted until it equals the product of the unknown weight times a fixed lever arm (Figure 22.4C). Likewise, when tetrapods stand, the forces of all postural muscles are adjusted so that in-torques equal out-torques and support is maintained without motion. Other examples are given later in this chapter.

It is important that the student of body mechanics be able to solve the equation $\mathbf{F}_i\mathbf{l}_i = \mathbf{F}_o\mathbf{l}_o$ for any of the variables and to understand the relation of each to the others. Thus, if it is desirable for a mammal that digs to produce a large out-force at the forefoot when the triceps contracts, then since $\mathbf{F}_o = \mathbf{F}_i\mathbf{l}_i/\mathbf{l}_o$, it is seen that the animal can increase the out-force by increasing $\mathbf{F}_i$ or $\mathbf{l}_i$, or by decreasing $\mathbf{l}_o$. The adaptations of many diggers include all of these (compare parts B and D in Figure 22.4).

Actual versus Effective Forces and Lever Arms There is one further qualification. The product of force times lever arm is equal to the torque of the system only when the force and lever arm are at right angles to one another. There are two ways to ensure that this condition is met. Each is simple and either may be the easier to apply to a dissection or experimental animal, so each will be presented.

The actual length of the lever arm of an in-force is the straight-line distance from the insertion of the relevant muscle to the pivot of the motion caused by contraction of the muscle. This is called the **actual lever arm** ($\mathbf{l}_a$). As noted in the section on components of forces, this line (a radius of the arc traveled by the insertion as it moves) is at right angles to the effective component of force that is in the direction of motion. If we call the force of the muscle the **actual force** ($\mathbf{F}_a$) and the force of the effective component of the actual force the **effective force** ($\mathbf{F}_e$), then the required condition is met when $\tau = \mathbf{F}_e\mathbf{l}_e$ (Figure 22.5A). (To indicate that these variables relate to the in-torque, we can write $\tau_i = \mathbf{F}_{ie}\mathbf{l}_{ia}$.) Using this method, one must always calculate the effective force before calculating the torque.

The alternative method uses instead the actual force, regardless of the relation of its line of action to that of its effective component. The appropriate, or **effective lever arm**

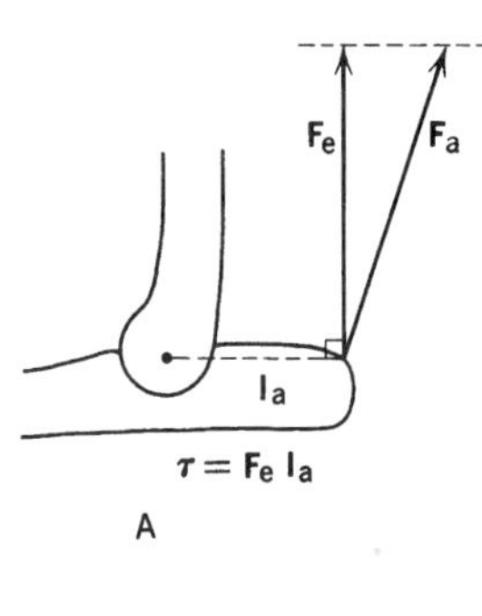

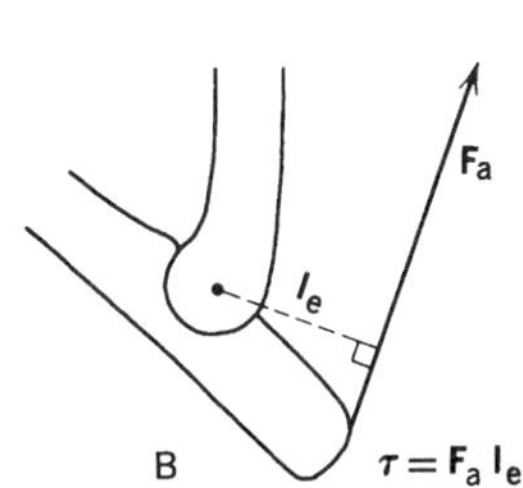

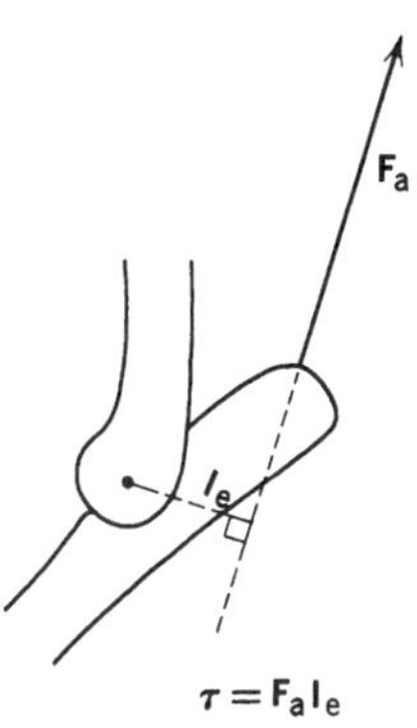

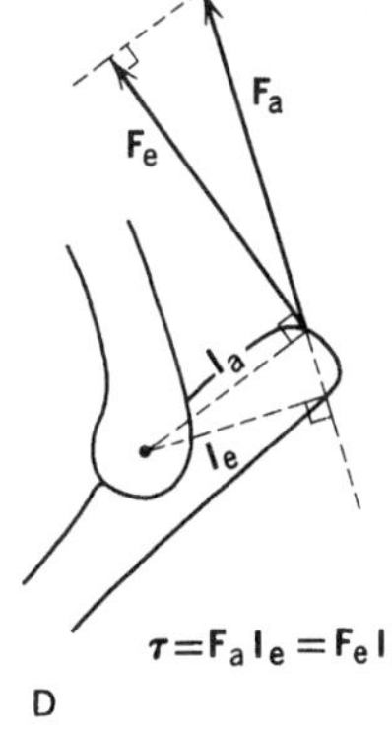

FIGURE 22.5 USE OF ACTUAL AND EFFECTIVE FORCES AND LEVER ARMS IN CALCULATING TORQUE.

(l_e) is then the perpendicular extending from the line of action of the muscle to the pivot. This perpendicular strikes $\mathbf{F}_a$ short of the insertion when the angle between $\mathbf{F}_a$ and $\mathbf{l}_a$ is acute; it strikes a projection of $\mathbf{F}_a$ beyond the insertion when the angle between $\mathbf{F}_a$ and $\mathbf{l}_a$ is obtuse (contrast parts B and C of Figure 22.5). Now $\tau = \mathbf{F}_a\mathbf{l}_e$, and also for any given position of the joint, $\mathbf{F}_e\mathbf{l}_a = \mathbf{F}_a\mathbf{l}_e$. The two methods are compared in part D. Can you position the forces so that $\mathbf{F}_a = \mathbf{F}_e$ and $\mathbf{l}_a = \mathbf{l}_e$?

Relations of In-Force to Out-Force Thus far our examples have shown the pivot to be located between the in-force and out-force. The torques are then in different directions, one clockwise and one counterclockwise around the pivot. This relationship is common in the body and is usual with extensor muscles (see Figure 22.6A). Two other arrangements are possible, each having the in-force and out-force on the same side of the pivot and turning in the same direction: Either the out-force can be closer to the pivot than the in-force (a less common arrangement, but two examples are depicted), or it can be farther from the pivot than the in-force (a usual arrangement with flexor muscles as shown by part C). Note that in the first arrangement (or first-order lever), the lever arms are independent and either arm can be longer, though in the body the out-lever arm is usually much longer. In the other arrangements the longer lever arm (which is the in-lever arm for second-order levers and the out-lever arm for third-order levers) includes all of the shorter lever arm—they "share" part of the lever.

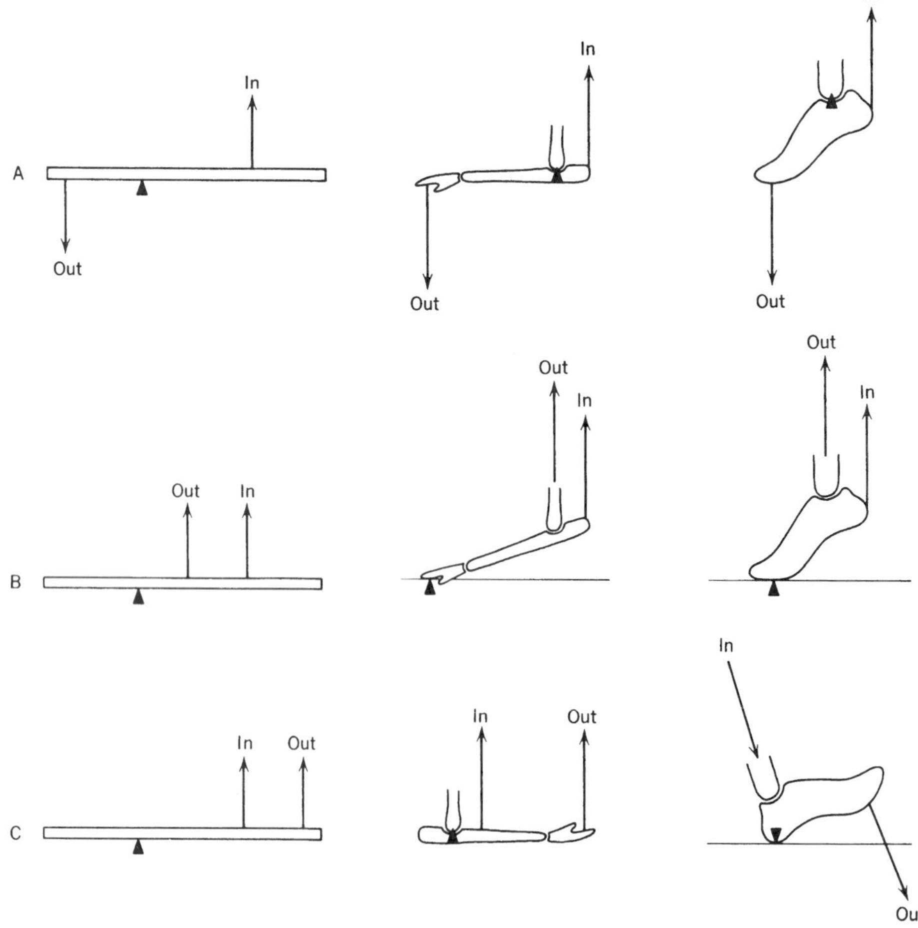

FIGURE 22.6 THREE WAYS THAT IN-FORCES AND OUT-FORCES CAN BE ARRANGED IN LEVER SYSTEMS.

Figure 22.6 shows that one joint and one muscle can function as different kinds of levers. When one walks on dry sand, the foot (to one's distress) even functions as two kinds of levers at the same time. The important thing is not to memorize diagrams but to learn to identify the pivot, in- and out-forces, and in- and out-levers in any specific bone-muscle system.

Summation of Torques; Two-Joint Systems More than one in-force may tend to turn the same lever. In order to determine the net effect, one determines the torque of each force independently, adds all that tend to turn the lever in one direction, and subtracts the sum of all (if there are such) that tend to turn it in the other direction. Multiple out-forces, as when several teeth simultaneously crush food, are treated in the same way.

If the weight of a part of a living machine is itself to be considered, as it should be in making an accurate calculation, then in addition to torques resulting from the action of muscles, one must allow for the torque from the pull of gravity on the lever (e.g., forearm or thigh). The lever responds to gravity as though all its weight were concentrated at its center of mass. The weight of the lever times its effective lever arm, calculated using the center of mass, is the desired torque. The center of mass of an irregularly shaped object is difficult to locate by calculation but is easy to locate experimentally if the object can be isolated: The object is suspended (e.g., by a string) successively from two points on its surface, and the point where extensions of the lines of support intersect is the center of mass.

Two-joint muscles, which pass over two diarthroses, are common (e.g., gastrocnemius, gracilis, biceps, long head of triceps). Each muscle can move either joint, both, or neither, depending on the actions of other muscles and loads. In any event, contraction moves, or tends to move, both joints, so the muscle simultaneously provides in-force to two lever arms. Tendons of some digital flexors pass over three or more joints. Such systems are virtually impossible to analyze in detail, either theoretically or experimentally, even for one instant in time. However, practical approximations of the mechanics of a unit of activity (a digging stroke, a running stride) have been undertaken by monitoring the principal out-forces, learning the approximate strengths and activities of relevant muscles (by electromyography, by preparing length-tension curves, and in other ways), determining usual postures (by various methods including X-ray cinematography), and measuring lever arms. We should not expect anything so wonderfully complex as the moving body to lend itself to simple analysis.

MECHANICS OF BODY SUPPORT

Balance and Counterbalance The trunk of tetrapods is supported according to several mechanical principles. The spines of very small mammals, such as mice and shrews, are strongly arched and may provide some support in the manner of an arched bridge. A truss bridge instead has a superstructure (or substructure) of braces that is suggestive of the rigid body of birds. Also like a truss are lengthened neural spines, found above the shoulders of many mammals, together with the ligament that joins their summits (Figure 22.9).

A more generally applicable analogy is the cantilever bridge (Figure 22.7). The roadbed is equated to the spine (each is under compression), and the top profile of the bridge is

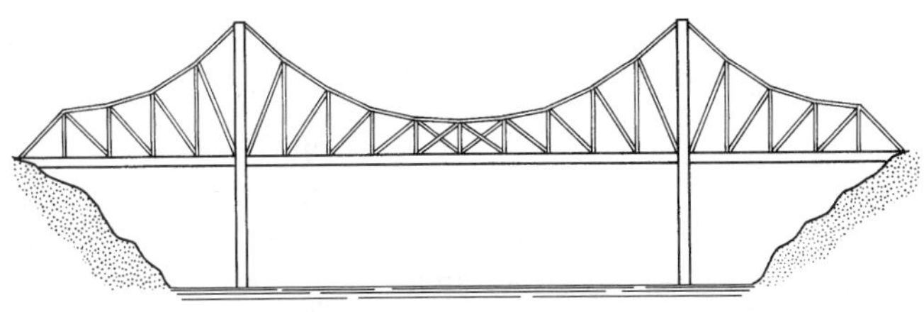

FIGURE 22.7
THE CANTILEVER BRIDGE IS AN ANALOG FOR ONE MECHANISM OF BODY SUPPORT.

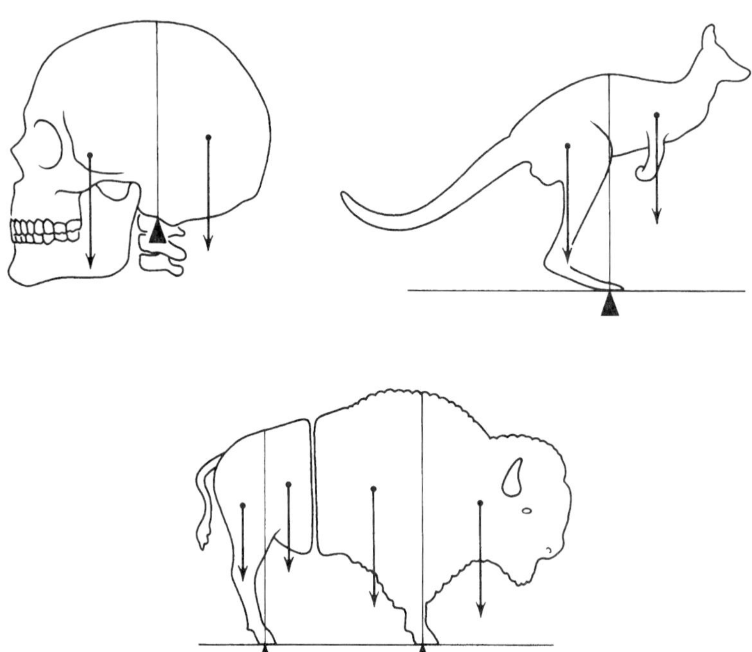

FIGURE 22.8 THE PRINCIPLE OF BALANCE AND COUNTERBALANCE IN THE SUPPORT OF THE BODY.

equated to the long epaxial muscles. The crisscrossing braces of the bridge are equated to the neural spines and short epaxial muscles. It is here that the analogy partly fails, because the appropriate structures are not constructed to provide enough rigidity for constant support. It does not follow that no support can be provided in this way when muscular activity contributes tension.

Another characteristic of the cantilever bridge does apply to the support of the body of many tetrapods. The functional unit of a cantilever bridge is from midspan to midspan. During construction, the overhanging cantilevers are built out equally on each side of each tower. (A cantilever is a projecting member that is supported at only one end.) Thus, the load on one side is counterbalanced by the load on the other. In different terms, the mass of one cantilever times the distance of its center of mass from the tower is a torque that is equal in magnitude, but opposite in direction, to the similar torque of the cantilever on the other side of the tower; hence there is equilibrium and the tower does not fall. The principle of balance and counterbalance is widely used by tetrapods (Figure 22.8).

Bows and Bowstrings from Bones and Fibers Another analogy for the support of the body is more valid than any of the bridge analogies. It equates the trunk vertebrae to a bow, which may be either curved like the archer's bow (many small mammals) or nearly straight with the ends bent down like the violinist's bow (salamanders, crocodilians, lizards, many large mammals) (see Figure 22.9, parts A and B). The principal string of this living bow is the ventral abdominal musculature, scalenus and related muscles, and intervening sternum. The psoas and quadratus muscles under the lumbar vertebrae form a short secondary bowstring. The spine cannot sag if tension is maintained in these bowstrings.

The cervical vertebrae of the larger mammals form another bow, this time inverted. The strands of the nuchal ligament form multiple bowstrings. The ligament can be stretched by muscles, thus decreasing the arch of the bow to depress the head.

All the analogs proposed have the shortcoming that they are passive, static systems, whereas the body is active and dynamic. The trunk cannot remain balanced on the limbs as

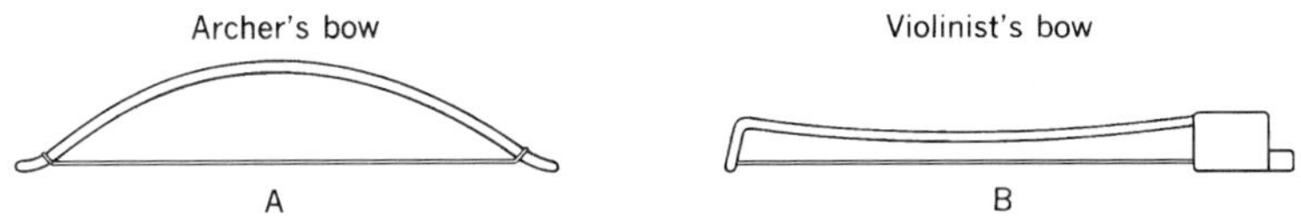

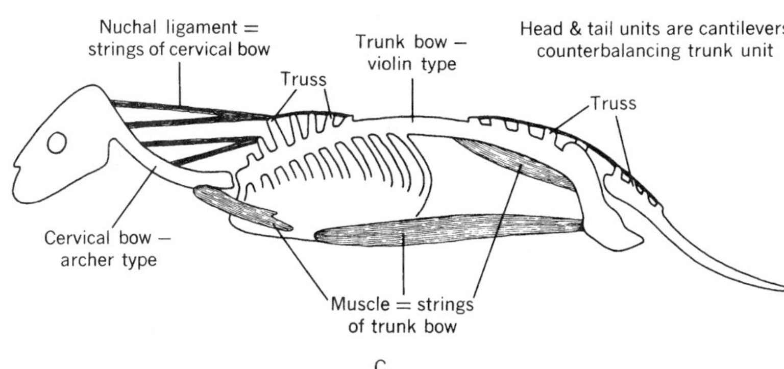

FIGURE 22.9 BOW ANALOGS AND COMBINED FACTORS IN THE SUPPORT OF THE BODY.

a bridge is balanced on piers; the length and orientation of animal cantilevers are subject to change; the loads on animal bows and bowstrings shift in magnitude and direction.

Furthermore, it is better to combine analogs than to choose among them. The trunk can be regarded as having a bow, the spine, which is prevented from collapsing by muscular bowstrings, which in turn may be relieved in part by the counterbalance of a trussed cantilever (a heavy, extended tail) and in part by the counterbalance of an inverted-bow cantilever (the neck) loaded at its extremity by the head (Figure 22.9C). We may assume that when a tetrapod stands at rest, this complicated system remains in equilibrium with only slight sustained tension in a few muscles and intermittent corrective tension in some others.

On p. 396 the human foot was likened to a beam loaded at the center. So it is, but it is also a bow, touching the ground at the ball of the foot and the heel, that is prevented from collapsing by muscles and by a bowstring (the plantar aponeurosis) that can be tightened by lifting the toes. On p. 394 the zygomatic arch was likened to a beam, and so it is, but it is often also an arch, firmly anchored at each end and strengthened by a bony string, the basicranial axis, and sometimes also by a postorbital ligament that suspends the arch from the postorbital process above.

Stops, Slings, and Locks The factors contributing to support that have thus far been presented relate primarily to the axial part of the body. Since the legs are not rigid vertical columns, but instead are hinged and usually bent struts, it is important to learn how they can support the body. During periods of activity, muscles provide support as well as motion. During periods of inactivity the animal may avoid a support role for the legs by resting sprawled on its ventral surface (amphibians, reptiles), by crouching (rodents, rabbits), by sitting (primates), or by lying down (carnivores, many artiodactyls). Nevertheless, some large tetrapods stand for long periods; their limbs must then support them without excessive muscular effort. Various factors contribute.

As described in Chapter 21, joints, together with their ligaments, are constructed to limit both the kind and extent of motion that can occur. In large mammals, no joint distal to shoulder and hip (and these to only a limited degree) can flex in the transverse plane; passive support is provided against lateral bending. Also, hyperextension in the sagittal plane is prevented by either ligamentous or bony stops: Elbows, knees, hocks, and digits cannot collapse by "bending backward."

The tendency to collapse by flexion may be reduced or avoided in any of several ways other than muscular contraction. The largest tetrapods (living and extinct) stand (or stood) with limb segments vertically aligned, or nearly so, balancing one bone on the next and reducing bending torques (more on this in the next chapter).

The human knee (and probably some upright limb joints of other animals) can be extended just beyond vertical. When weighted, there is a slight torque in a direction in which the joint cannot bend further. This prevents bending in the opposite direction and provides the standing body with a lock against flexion. Some resistance to flexion is also provided by snap joints (see p. 401 for a description). This resistance may help to prevent collapse of a joint that is otherwise nearly in equilibrium.

Ungulates also have a sling mechanism that prevents collapse of the fetlock joint between metapodial (or cannon bone) and proximal phalanx. This joint is always bent when weighted; the angulation depends on the load. It is thus an important shock absorber. Because it is bent, it must be supported. This is accomplished by a remarkable ligamentous sling that extends from the posterior surface of the proximal end of the metapodial, down under the fetlock joint (where it is anchored to sesamoid bones), and then around the proximal phalanx to insert on the anterior surfaces of the distal phalanges (Figure 22.10). As the foot is heavily loaded by shifting weight when at ease, or by impact when moving, the ligament stretches, allowing the joint to flex sharply. In doing so, the ligament stores potential energy that is released by returning the joint to a neutral angulation when the load is diminished (see also figure on p. 551).

Ungulates, unlike elephants, usually have flexed stifle joints (corresponding to the human knee) and moderately flexed hock (or heel) joints. Passive support to relieve exten-

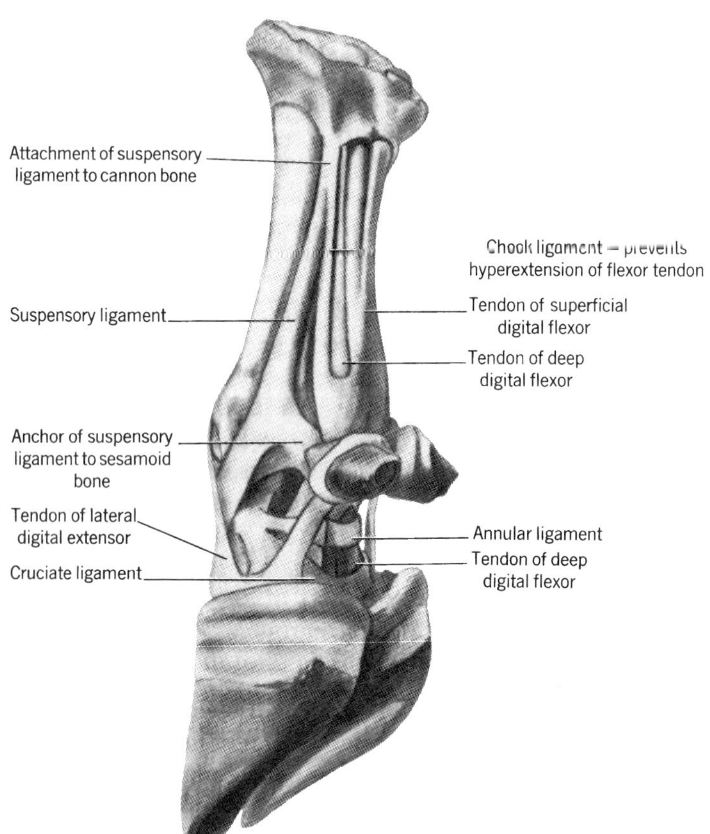

FIGURE 22.10 STRUCTURE OF THE RIGHT FORE-FOOT OF THE COW INCLUDING THE SUSPENSORY MECHANISM. (Drawn from a freeze-dried dissection and hence slightly shrunken.)

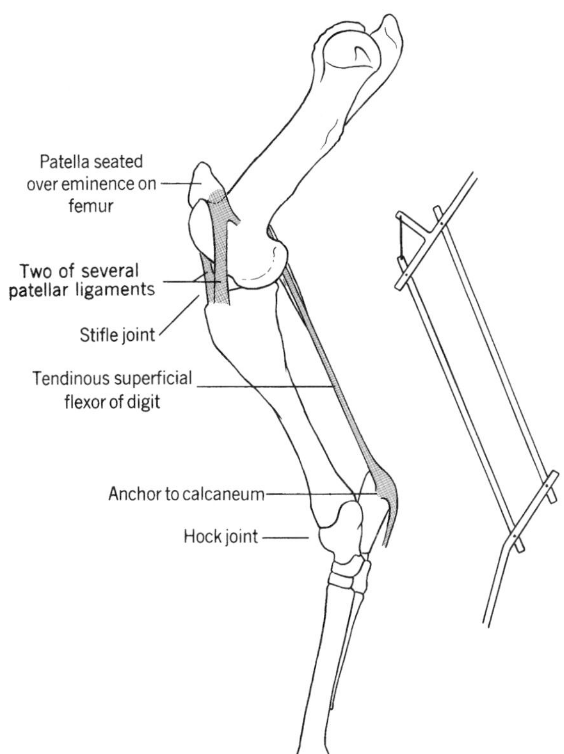

FIGURE 22.11 SUPPORT MECHANISM OF THE HIND LEG OF MANY UNGULATES shown by a medial view of the right leg of a horse.

sor muscles is provided to perissodactyls and some artiodactyls by a mechanism that is effective, simple, and subject to voluntary control. The tibia on one side, and the almost completely tendinous superficial flexor muscle of the digit (or digits) on the other side, form the long arms of a parallelogram that is completed above by the distal end of the femur and below by the calcaneum and other tarsal bones (Figure 22.11). All angles of this parallelogram must change simultaneously. The stifle and hock cannot flex independently, and if any of the angles is prevented from changing, the entire system remains rigid. This is accomplished by a locking device at the stifle. The ridge flanking the patellar groove on the medial side is enlarged and ends proximally in an eminence. The patella is anchored to the tibia by several strong ligaments that just permit the patella to be pulled (by the quadriceps muscle) behind this eminence when the joint is extended. When the patella is thus seated, the ligaments prevent flexion of the joint; the patella must be pulled sideways and down into the patellar groove before the limb can be flexed.

These support mechanisms are representative. Numerous others are known (some that relate to specific adaptations will be described in subsequent chapters), and doubtless still more await identification.

MECHANICS OF MOTION

Velocities and Lever Arms Movement is change of position. Speed is the rate of change of position. **Velocity** (**v**) is speed in a given direction. An animal is said to be able to swim or run at a stated speed because it can move in any direction; muscles and bones, however, are said to attain certain velocities because their directions of motion relative to associated structures are always restricted. Velocity is expressed in centimeters per second, or equivalent units, and direction is stated or implied. Velocity, like force, is therefore a vector quantity and can be represented by an arrow (see p. 406).

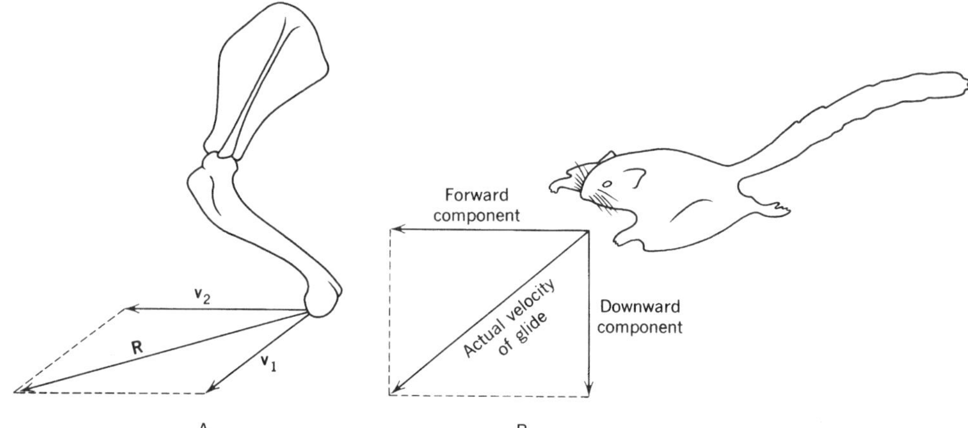

FIGURE 22.12
THE CALCULATION OF RESULTANTS AND COMPONENTS OF VELOCITIES.

Just as parallelograms can be drawn to add several forces acting together, so they can be drawn to add several constant or instantaneous velocities acting together. When a person swims straight out from the bank of a river, his velocity relative to the water and the velocity of the current relative to the bank combine to move him diagonally downstream at a rate exceeding either velocity taken alone. Similarly, if the distal end of the humerus of a running animal is extended on the scapula at the instantaneous velocity $\mathbf{v}_1$ (Figure 22.12A) as the body is moving relative to the ground at velocity $\mathbf{v}_2$, then **R,** the diagonal of the parallelogram, is the resultant velocity of the distal end of the humerus relative to the ground.

Likewise, a velocity can be broken down into components. As for forces, a useful application establishes components at right angles to one another by constructing a rectangle of velocities as in part B, which shows the glide of a pygmy scaly-tailed squirrel.

It is often desirable to combine forces (or torques) that act independently on the same lever; two muscles can pull harder than one. Velocities, by contrast, cannot be applied independently to different parts of the same lever; disregarding the inertia of the system (which can be an oversimplification), two muscles cannot pull faster than one. The velocity of any point on a lever is determined only by its distance from the pivot and the angular velocity or rate of turning of the lever as a unit. Thus, when loads are small, the relative velocities of different insertions acting together on one bony lever are determined primarily by their positions and not by inherent differences in rate of shortening of associated muscles. (Recall, however, that the velocity of shortening decreases with load, and reducing a muscle's lever arm increases its load.)

We have learned that the relation of in- and out-forces to their respective lever arms is expressed by the formula $\mathbf{F}_i\mathbf{l}_i = \mathbf{F}_o\mathbf{l}_o$. Therefore, $\mathbf{F}_i = \mathbf{F}_o\mathbf{l}_o/\mathbf{l}_i$ and $\mathbf{F}_o = \mathbf{F}_i\mathbf{l}_i/\mathbf{l}_o$. Out-force increases with the length of the in-lever and decreases with the length of the out-lever: The digger needs a long olecranon and short forearm. It follows from the previous paragraph that **in-** and **out-velocities** ($\mathbf{v}_i$ and $\mathbf{v}_o$) are also related to their respective lever arms, but in the reverse way: $\mathbf{v}_i\mathbf{l}_o = \mathbf{v}_o\mathbf{l}_i$, so $\mathbf{v}_i = \mathbf{v}_o\mathbf{l}_i/\mathbf{l}_o$ and $\mathbf{v}_o = \mathbf{v}_i\mathbf{l}_o/\mathbf{l}_i$. For high velocity at the foot, the olecranon should be short and the forearm long (see Figure 22.4). The same lever cannot enhance both the force and the velocity of the same muscle. The problem of providing an animal with both low and high "gear systems" is discussed on page 439, and in Comment 24.1 on p. 442. (Optimum structure is determined not only by these mechanical factors but also by related physiological factors.)

It is often desirable to know the speed of an animal as a whole. It may also be useful to know the maximum velocity of a part of the body (as of the center of mass of a jumper at takeoff) or the velocity at a given instant (as of the foot of a runner when it strikes the ground), but the parts of moving animals seldom attain constant velocities, and average velocities are not often useful. Therefore, other variables must also be considered.

Mass and Acceleration Objects remain at rest or in uniform motion unless acted upon by external forces. (This is Newton's First Law.) They have the inherent tendency to remain at rest or in motion at constant velocity. This tendency is called **inertia.** It is not expressed in numerical terms, but experience tells us that heavy objects have more inertia than light objects. Actually, the critical factor is not weight, which is determined by the pull of gravity, but **mass** (m), which represents quantity of matter and remains the same in space as on earth. The vertebrates are on the earth, where mass is measured by weighing, so the difference is not of practical importance; nevertheless, it is preferable to use the concept of mass.

The capacity of an object moving in a straight line to overcome resistance is called **linear momentum** (M) and is equal to $m\mathbf{v}$. Momentum is conserved: When one system loses momentum, another system gains an equal amount. Thus, when a bird flies to a perch, the momentum lost by the bird is gained by the perch, which may cause it to sway.

The relation of mass (m) to force ($\mathbf{F}$) and change in velocity per unit time, or acceleration (a), is stated by Newton's Second Law and is simply $\mathbf{F} = ma$, when appropriate units are selected. Acceleration is expressed as centimeters per second per second, or equivalent units, and obviously relates to velocity ($\mathbf{v}$), time (t), and distance (s). If an object is at rest when acted on, then the basic relationships are $\mathbf{v} = at$, $s = \mathbf{v}t$, and $\mathbf{F}t = m\mathbf{v}$. Other equations can be derived by substitution. If an object is already in motion when acted on, then the equations must be modified somewhat. Derivations and interpretations are found in texts on elementary mechanics.

These relationships show that if an object is to be accelerated or decelerated rapidly (e.g., the body of a bird at takeoff or the tongue of a feeding chameleon), it should be light in order to avoid excessive forces; if an object is heavy (body of an elephant), relatively more time is needed to achieve maximum velocity; greater velocity is attained when a force acts on an object for a relatively long time (a reason for the very long hind legs of the jumping frog and tarsier).

These formulas and examples apply only to rectilinear motion. Since animals and their moving parts often do not travel in straight lines, other factors must now be considered.

Curvilinear and Rotational Motion When a tetrapod jumps, or momentarily lifts all feet from the ground while running, or folds its wings for an instant in flight, then (disregarding wind resistance) its center of mass tends to continue to move with the initial direction and velocity (according to Newton's First Law) but is simultaneously accelerated in a different direction by gravity. One component of its motion, the rate of fall, is not constant. Hence, the resultant motion is not rectilinear. While unsupported, the jumping body as a whole moves in a parabolic curve regardless of any motions of its appendages.

When a running animal turns sharply, its velocity also changes, even though its speed may remain constant (because velocity has both magnitude and direction). There is then a sideways, or centrifugal, force out of the curve of the turn that must be balanced by an inward centripetal force in order to prevent skidding.

These are examples of curvilinear motion. Such motion relates particularly to jumping, dodging, turning in flight, and throwing and will be noted further in subsequent chapters.

Having no wheels, vertebrates use moving parts (legs, fins) to support and propel themselves. Such parts may be thought of as segments of wheels that constantly reverse their direction of rotation. This is inefficient because the appendages must constantly be accelerated and decelerated. However, oscillating parts do have the advantage of versatility: The functional length and period of oscillation of a limb can be modified at will to adapt to rough ground or angle of terrain.

Turning wheels and the oscillating levers of moving animals rotate around axes; the mechanics of rotational motion apply. Consider an extended limb that is rapidly pivoting at the hip or shoulder during the supportive phase of a running stride. Its **angular velocity** (ω) is the rate of rotation, or degrees turned per unit of time. Its **angular acceleration** (α) is the rate of change in angular velocity. Its **moment of inertia** (I), which corresponds to mass in rectilinear motion, is an index of the resistance of the limb to acceleration. This equals the mass (m) of the limb times the square of a constant (k) called the radius of gyration, the value of which depends on the distribution of the mass of the limb. The value is greater for long limbs than short, and greater for limbs that are heavy distally than for limbs that are heavy proximally. The turning force, or torque (τ), of the limb is $I\alpha$. **Angular momentum** (L) equals $I\omega$, and like linear momentum it is conserved. By flexing a limb, or spine, or tail so as to concentrate its mass closer to its axis of rotation, k, and hence I, are decreased. Because L does not change, ω increases. Human divers and gymnasts learn to control the spin of their bodies in this way; cats, jerboas, and gibbons do likewise.

Complicating factors make it extremely difficult to apply these and related formulas to the quantitative analysis of actual animal movements: Articulated segments of the body commonly rotate around independent axes at the same time that they turn as a unit. Each segment then has its own values for each variable, and the values of the variables for the segments taken together change constantly. Furthermore, one segment may rotate simultaneously around more than one axis (e.g., extension with supination, or flexion with adduction).

In general, however, study of the formulas $I = mk^2$, $\tau = I\alpha$, and $L = I\omega$ makes it easy to interpret many morphological adaptations. It is seen that the mass and distribution of mass of an oscillating structure are critical to its function unless the part has little mass (legs and jaws of very small tetrapods), or moves slowly (legs of the sloth), or overcomes external resistance far in excess of resistance offered by its own mass (foreleg of a burrowing mammal). If an oscillating structure moves fast, then the muscles that move it, being heavy, should not be located in the fastest-moving part of the structure (distal part of leg or wing) but instead should transmit their forces by levers or long tendons from regions of relatively slow motion (shoulder of antelope, breast of bird). Forces can be reduced and angular velocities increased if the distal parts of oscillating structures are light and slender. Very large animals eliminate such oscillations as can be avoided (e.g., flexion and extension of the spine).

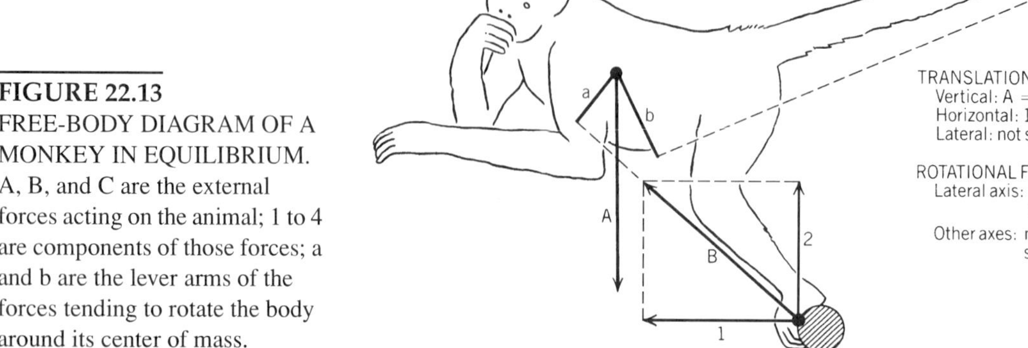

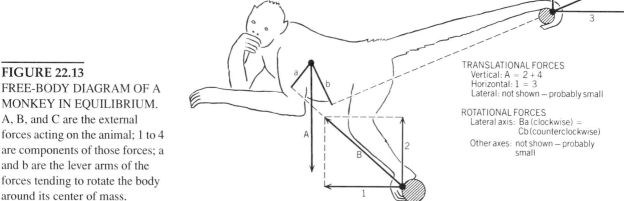

FIGURE 22.13
FREE-BODY DIAGRAM OF A MONKEY IN EQUILIBRIUM.
A, B, and C are the external forces acting on the animal; 1 to 4 are components of those forces; a and b are the lever arms of the forces tending to rotate the body around its center of mass.

FREE-BODY DIAGRAMS

In solving problems in functional morphology it is often helpful to use a **free-body diagram.** One isolates the system under study, locates its center of mass, measures all external forces acting on the system, then sketches the system and represents all of the forces by vectors (Figure 22.13). There can be a maximum of six external forces: a translational force in each plane of space (tending to shift the entire system as a unit) and a rotational force around each axis in space (tending to revolve the system around its center of mass). When the system is at rest, the sum of all translational forces must be zero, and likewise the sum of all rotational forces must be zero. This is a powerful tool for learning if all forces have been identified and none overrepresented.

The isolated mechanical unit need not be an entire animal, as in the example shown, but can instead be a limb or limb segment, a jaw, a tooth, or other structural complex. It is also possible (when the variables can be identified and measured) to construct a dynamic free-body diagram for a moving system at an instant in time (as of a human diver). However, this is a more difficult task.

REFERENCES

Alexander, R. McN. 1983. *Animal mechanics.* 2nd ed. Blackwell Scientific, Boston. 301p.

Altringam, J.D. 1994. How do fish use their myotomal muscle to swim? *In vitro* simulations of *in vivo* activity patterns, pp. 99–110. *In* L. Maddock, Q. Bone, and J.M.V. Rayner (eds.) *Symposium on mechanics and physiology of animal swimming.* Cambridge Univ. Press, New York.

Altringham, J.D., and D.J. Ellerby. 2000. Fish swimming: patterns in muscle function. *J. Exp. Biol.* 202:3397–3403. Good review of studies to date concerning how muscles power swimming movements in fish; future directions proposed.

Biewener, A.A. 1998. Muscle function *in vivo:* a comparison of muscles used for elastic energy savings versus muscles used to generate mechanical power. *Am. Zool.* 38:703–717.

Dempster, W.T., and R.A. Duddles. 1964. Tooth statics: equilibrium of a free-body. *J. Am. Dental Assoc.* 68:652–666. Explanation and application of the use of free-body diagrams.

Frohlich, C. 1980. The physics of somersaulting and twisting. *Sci. Am.* 242(3):155–164.

Girgenrath, M., and R.L. Marsh. 1999. Power output of sound-producing muscles in the tree frogs *Hyla versicolor* and *Hyla chrysoscelis. J. Exp. Biol.* 202:3225–3237.

Hill, A.V. 1950. The dimensions of animals and their muscular dynamic. *Sci. Prog.* 38(150):209–230.

Kreighbaum, E., and K.M. Barthels. 1996. *Biomechanics: a qualitative approach for studying human movement.* 4th ed. Burgess, Minneapolis. 619p.

Kummer, B. 1959. *Bauprinzipien des säugerskeletes.* Georg Thieme Verlag, Stuttgart. 235p. Mechanics of support of the static skeleton with emphasis on the spine and femur.

Marsh, R.L. 2000. How muscles deal with real-world loads: the influence of length trajectory on muscle performance. *J. Exp. Biol.* 202:3377–3385. Thoughtful discussion of the complexities encountered when muscle length and environmental loads (resistance) are (or are not) factored into muscle performance.

Slijper, E.J. 1946. *Comparative biologic-anatomical investigations on the vertebral column and spinal musculature of mammals.* Akad. van Wetenschappen, Afd. Natuurkunde, Tweede sectië, 42(5). 128p.

Swartz, S.M. 1993. Biomechanics of primate limbs, pp. 5–42. *In* D.L. Gebo (ed.), *Postcranial adaptation in nonhuman primates.* N. Illinois Univ. Press, DeKalb.

Tricker, R.A.R., and B.J.K. Tricker. 1967. *The science of movement.* American Elsevier, New York. 284p. Simple presentation of principles of mechanics as applied to athletics.

Chapter 23

Form, Function, and Body Size

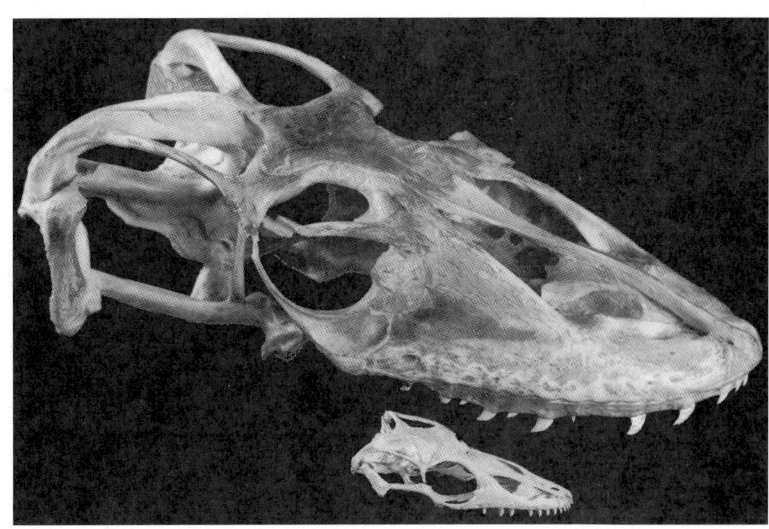

The largest land vertebrate is about one million times heavier than the smallest, and vertebrates of the same family commonly vary in size by an order of magnitude. It is obvious that animals of widely different sizes must have different requirements for shelter, feeding, and protection, even though they are related: The tiny pudu deer cannot graze under a meter of water like the moose; the hatchling monitor lizard can hide where the three-meter adult cannot and need not. It may be less evident that size also has a profound influence on form and function. This is because surface and volume do not increase equally as linear dimensions increase, and many functions of the body depend on the ratio of surface to volume. **Scaling** is the relationship between body proportion and body size among related and similarly shaped animals.

PROPORTIONATE GROWTH AND SURFACE-TO-VOLUME RATIO

To start with the simplest case, consider that as an animal, or animal lineage, gets larger, all growth is **isometric** (= *equal* + *measure*). The large and small animals then have **geometric similarity.** The small animal is made equal to the large one by multiplying all linear dimensions by the same factor. One looks like a photographic enlargement of the other (Figure 23.1). To simplify further, let our small "animal" be represented by the shaded cube in Figure 23.2.

If each of its edges has length of one unit (e.g., 1 cm), then each of its sides has a surface area of 1×1 unit^2, and, as the cube has 6 sides, its surface area is $1 \times 1 \times 6 = 6$ units^2. Its volume is $1 \times 1 \times 1 = 1$ unit^3. The important ratio of surface to volume is 6:1. The larger cube in Figure 23.2, having twice the linear dimension (L), has a surface area (SA) of $2 \times 2 \times 6 = 24$ units^2, a volume (or mass, M) of $2 \times 2 \times 2 = 8$ units^3, and a reduced ratio SA to M of 3:1. Similarly, if L increases 5 times (house cat to lion), SA goes up to 150 units^2, M jumps to 125 units^3, and the ratio SA:M falls to 1.2:1. If L increases 10

Above: Skulls of one genus: Komodo dragon and mangrove monitor.

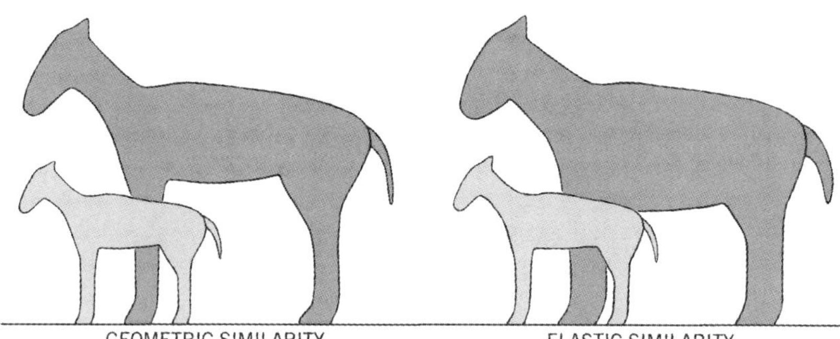

FIGURE 23.1 TWO KINDS OF BODY SCALING.

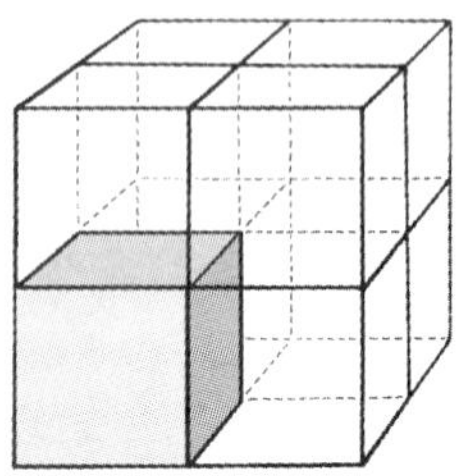

FIGURE 23.2 CUBES SHOWING THE RELATION BETWEEN SURFACE AND VOLUME AS SIZE INCREASES.

times, SA = 600 units^2, M = 1000 units^3, and SA:M is a mere 0.6:1. Even larger variations in size are relevant: the range in weight of varanid lizards (which have approximate, but not exact, geometric similarity) is about 100,000. (The photo opener for this chapter illustrates part of this range.)

One must be less precise with animals than with uniform cubes. Human swimmers learn that M varies not only with L, but also with fat and bone content (some float, some do not), and the ratio SA:M varies with shape; it is higher for slender animals with long appendages (spider monkey) than for compact animals of the same size (miniature pig). For animals, the relationships are better presented with the sign ~ (meaning similar, or proportional to) than with the equals sign. In general terms, $M \sim L^3$, and SA $\sim L^2$.

The consequences of these relationships are far-reaching for vertebrate design and function. A mammal generates metabolic heat in proportion to the number of active cells in its body—that is, in proportion to volume. It loses heat at a rate proportional instead to total surface area. Consequently, unless compensations can be made (insulation, activity, metabolic rate, etc.), small mammals may have difficulty staying warm and large ones, having low SA to M ratios, difficulty staying cool. (We return to this point in the Comment on p. 429.) Further examples: The forces that muscles can exert, and the loads that bones can support, are roughly proportional to their cross-sectional areas (a surface area dimension) (see p. 176); nevertheless muscles and bone move and support the mass of the body. As body dimensions increase isometrically by a factor of χ, loads on the musculoskeletal system outstrip strength by $\chi^3/\chi^2 = \chi$ times. Unless form and function are modified (as described and illustrated below for elephants and other giants), the body loses capacity for support and movement. Similarly, oxygen is absorbed through surfaces, but is used by the entire volume of the body; rough food is ground by the surfaces of teeth and absorbed through the surface of the gut, yet nourishes all the body; the surface of a flyer's wings must support all of its weight. Geometric similarity is widely observed in ontogeny and in the evolution of phyletic lines, provided that size differences remain moderate. Ultimately, however, as size increases, adaptations must occur in form and function to sustain body support and metabolic requirements.

DISPROPORTIONATE GROWTH

One way to adapt as size increases is to reduce the need for surface-related functions. The rate of gaseous exchange in the lungs, of absorption from the gut, and of excretion in the kidney all depend on the rate of metabolism. It is not surprising, therefore, that the basal metabolic rate of vertebrates does not scale isometrically but slows down with increased body size. It is roughly proportional to the three-quarters power of body weight.

Another way large animals adapt is by making structural modifications that increase surface areas disproportionately so that they can "keep up" with related volumes as overall size increases. Removal of wastes from the bloodstream occurs in the surface layer, or cortex, of the mammalian kidney. In order to provide enough cortex to remove enough wastes, the kidneys of large mammals are compound, each resembling a large cluster of small, smooth kidneys. Similarly, the gray matter of the mammalian forebrain (where neurons and synapses are concentrated) is on the surface of that organ. In order to have enough gray matter, the large mammal has a highly convoluted forebrain. The occlusal surfaces of the cheek teeth of large herbivores are disproportionately large and they have particularly intricate infolding of the enamel.

We have seen that geometric similarity of the support system cannot pertain over a wide range of body size; the bones and muscles of the larger animals would fail under their heavy loads. This would be prevented if the supporting members enlarged in proportion to their loads (Figure 23.1). Such animals would have **elastic similarity** (a principle developed by T.A. McMahon of Harvard University). Some of the required relationships become $L \sim M^{1/4}$, diameters of limbs and of bones $\sim M^{3/8}$, and maximum muscle force $\sim M^{7/8}$. (The derivations of these proportionalities are too complex to present here. See the summary in Alexander cited in the references and the sources he cites.) This principle also places an upper limit on body size; very large animals would have to have such stout legs that the movement of their joints would be restricted. As a design principle elastic similarity is subject to the criticism that it relates to static conditions, whereas loads (and the structure needed to sustain them) are greater during activity.

The dimensions of bones need not scale according to one principle: For some groups of mammals the width of the lumbar centrum maintains geometric similarity whereas the height has elastic similarity. Furthermore, and importantly, numerous groups of mammals adjust skeletal dimensions only a little to compensate for large body size, but instead adjust the peak forces applied to the skeleton. This results in the loss of agility observed in large vertebrates. They do not merely have skeletons that have geometric similarity (too weak) or elastic similarity (too cumbersome) to skeletons of smaller relatives; they modify their postures and activities (as described by Biewener) to ensure safety factors (commonly between 2 and 4) similar to those of smaller animals. This is called **dynamic strain similarity,** or stress similarity.

Kinematic similarity also relates to motion: Movements of small animals are made similar to those of large animals by multiplying the linear dimensions of legs and other oscillating parts by one constant and time factors by another. Thus, as legs get longer, their periods of oscillation get slower, and the joint angles described by the moving parts remain the same.

It is evident that form, function, and size are not related by any one principle. Study of the alternatives gives insight into the factors among which nature seeks compromise.

ALLOMETRY

Disproportionate growth may be termed **allometric** (= *unequal* + *measure*). In practice, however, one does not know if the relative growth of two parts is isometric or allometric until analysis is made, and one method is used in all instances. **Allometry** is the study of the correlation between form and size. It is the method for analyzing scaling.

If two structures grow at different rates, then, as size changes, body proportions also change. Allometry relates both to ontogeny (colts have relatively longer legs than adult horses) and to phylogeny (modern horses have relatively longer legs than

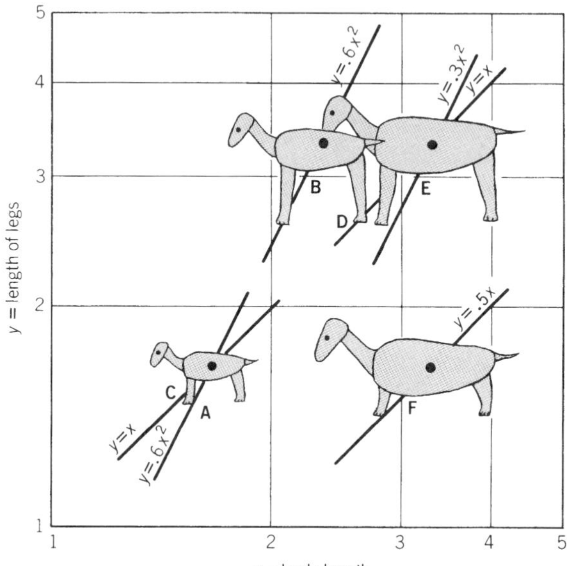

FIGURE 23.3 RELATION BETWEEN BODY SIZE AND BODY PROPORTIONS shown by graphing leg length against body length on a logarithmic grid. Each regression line represents many plotted points (not included) of one population. Sketches show body size and proportions of animals near the centers of the respective distributions.

their smaller remote ancestors). Either the altered size, or changed form, or both, are considered to be adaptive, so allometry is important for helping the morphologist interpret structure.

If the length of structure x (e.g., the legs) is plotted against the length of structure y (e.g., the spine) for many animals of a kind, the individual plots distribute themselves on the graph in such a way that a regression line, or line of best fit, can be drawn to represent their relationship. The equation $y = bx^a$ gives an adequate fit, is simple, and provides a straight regression line on a logarithmic grid. Change in the value of b shifts the line parallel to itself. Change in the value of a alters the slope of the line. When x and y are both linear measures (or both surfaces, or both masses) then, if the relative growth is isometric, $a = 1$ and the slope of the line is 45°. If x (plotted on the horizontal axis) grows slower than y (plotted on the vertical axis), then $a > 1$ and the slope of the line is $> 45°$. If y grows slower than x, then $a < 1$ and the slope $< 45°$.

In Figure 23.3, the six regression lines represent allometric equations for six populations that are distinct in age or ancestry. Populations A and C have about the same average body size and relative leg length, but in C, legs and spine are growing at the same rate, whereas in A the legs are growing faster. The analysis shows that, in terms of evolution, the populations are less alike than they appear. The same can be said of populations D and E. These animals are larger than A and C, yet have the same proportions. However, the analysis shows that C differs more fundamentally from E than from D, because C and E differ also in the relative growth rate of legs and spine. How would you interpret the relationship between A and B, or B and C, or C and F?

In general, skeletal muscle, blood volume, heart, and lungs of mammals scale approximately isometrically, whereas skeleton and fat contribute proportionately more to body mass in large mammals, and skin and brain proportionately less. Limb proportions scale variously according to function.

MINIATURES AND GIANTS

Small vertebrates tend to burn their fuel faster, respire faster, have faster heart rates, and mature and die younger than their larger cousins. The relation between body mass and energetics is complex, however, and clearly there are selective advantages to small size that may offset the energetic disadvantages. Miniature vertebrates can exploit extreme niches, finding microclimates, safe shelters, and food sources not otherwise available. In most instances the body form of diminutive vertebrates differs in no notable way from that of related larger animals.

Miniaturization is common in teleosts; many weigh <5 g and some <0.01 g. They tend to have relatively large gills (seemingly paedomorphic, see p. 70) and large ratios of body surface to mass (increasing drag). The smallest amphibians include caecilians, salamanders (down to 17 mm snout-vent length), and frogs (to 11 mm). They have advantages in concealment, water conservation, and ease of entering crevices or digging. The smallest reptiles are 1 g nocturnal lizards (geckos).

Hummingbirds weigh 2–10 g; the smallest among other birds weigh 4 g. Small size in birds may be constrained by wing-beat frequency and lift generation. Most bats (excluding fruit bats) weigh 5–10 g. Bats probably need to be small to produce ultrasound for echolocation. The smallest bats call at the highest frequencies and feed on the smallest insects. About 8 percent of terrestrial mammals weigh <15 g. Most of the miniatures are shrews (1.5–6 g) or mice. They tend to be tropical or subtropical, feeding on insects or seeds.

An evolutionary trend toward large body size has been common within vertebrate lineages, particularly for those having a moderately large ancestor. As the body weight of a terrestrial animal approaches 900 kg (2000 lb), marked adaptations for support become necessary. Beasts having such adaptations are said to be **graviportal** (= *heavy* + *to carry*).

The selective advantages of large size include (1) virtual freedom from predation (except, alas, by man); (2) the ability to roam over large areas in search of food, water, shelter, or breeding grounds; (3) the capacity to use and produce energy more slowly than small animals, so that relatively little food is required per unit of body weight and food of low nutritive quality may be sufficient; and (4) a low surface-to-volume ratio and a high capacity for heat production, enabling the larger animals to heat up and cool off slowly.

Gigantism has not characterized the amphibians, but the largest labyrinthodont was thickset and about 4.5 m long. There have been giants in many reptilian lineages: An extinct marine lizard (a mosasaur) was 10 m long. The longest snakes probably reached 12 m. A crocodile attained about 13 m, and many dinosaurs of the order Saurischia were gigantic: estimates of length range up to about 35 m, and estimates of weight to more than 80,000 kg.

Four extinct mammalian orders (Amblypoda, Embrithopoda, Notoungulata, Astrapotheria) included bulky land herbivores. The hippopotamus rarely attains 4500 kg. The living Indian rhinoceros reaches 4000 kg, and an extinct, hornless rhinoceros was the largest of all land mammals—it stood about 5 m high at the shoulder. The bull African elephant, largest of living land animals, stands 4 m high and weighs as much as 7000 kg (7.7 tons). Human greed has nearly exterminated the largest of all vertebrates, the blue whale, and few large specimens remain. This magnificent creature reached 23 to 30 m (75 to 100 ft). A 27 m specimen weighed 136,400 kg (150 tons).

Aquatic giants support their bodies effortlessly by flotation. Terrestrial giants reduce their requirements for support by avoiding unnecessary oscillations (the spine is usually stiff), jolting (they rarely jump or gallop), and exertion (the limbs are flexed little in locomotion). Furthermore, they reduce demands on their skeletons by modifying structure and posture so that compressive forces, which can be withstood with minimal tissue, are increased, and bending forces are decreased. Limb bones are oriented

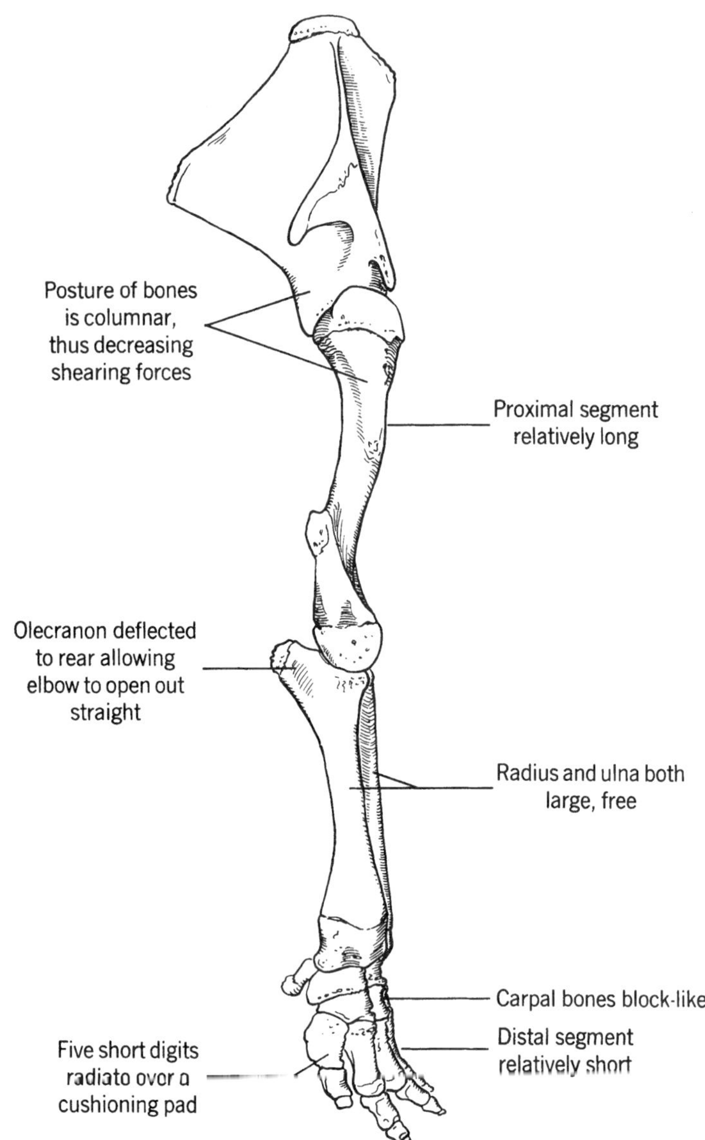

FIGURE 23.4
SOME GRAVIPORTAL ADAPTATIONS of the right foreleg of a 12-year-old Indian elephant, *Elephas*.

vertically (Figure 23.4). Their columnar shafts are straight, and joint surfaces are in line with the shafts. The olecranon process is deflected backward so that the elbow joint can open completely. The heads of the humerus and femur face more nearly upward than for smaller tetrapods, and the acetabulum faces more nearly downward. Even the scapula and ilium tend to be broad and vertically oriented. Proximal limb segments are long; distal segments are short. Radius and fibula are large and free. The feet are broad. Usually, all five toes are retained, are about equal in length, and radiate over a pad that cushions and evenly distributes the great weight. These modifications shift part of the support role from muscle to bone. Ligamentous slings and guys, and modifications of muscle mechanics to favor force and reduce velocity, contribute further to economy of effort.

For speculation about some physiological consequences of being a large dinosaur, see the accompanying Comment.

COMMENT 23.1

THE DEBATE OVER DINOSAUR METABOLISM

Animals that derive their body heat primarily from the external environment are called **ectotherms** or "cold bloods." They have relatively low metabolic rates. Their behavior may warm them somewhat (exercise, sunbathing), but their body temperature varies with ambient temperature; thus they are usually *poikilotherms* (= *varied* + *heat*). Animals that maintain an elevated body temperature by having a relatively high metabolic rate coupled with control over heat loss (insulation, sweating) are **endotherms** or "warm bloods." Their body temperature is usually nearly constant; they are usually **homeotherms** (= *equal* + *heat*). ("Usually" because there are other possibilities: A benthic fish is both ectotherm and homeotherm; a bird or mammal having seasonal or daily periods of torpor is both endotherm and facultative poikilotherm.) Where did dinosaurs fit in?

It had long been assumed that dinosaurs were ectotherms, like modern reptiles, and slow-moving except for short bursts of activity to make a capture or an escape. Then Robert Bakker (see References) argued that dinosaurs were endotherms, an idea that pleased illustrators of dinosaur books, who started depicting them as running, rearing, raging giants. Bakker cites their upright stance, wide distribution, rapid growth rate, and the microscopic structure of their bones, which rarely (and at most faintly) show the seasonal growth rings typical of the bones of living ectotherms. Of particular interest is his careful analysis of predator-prey ratios. Endothermic predators require 15 to 30 times as much food (mass for mass) as ectothermic predators. It is difficult to establish population densities for extinct animals, but Bakker believes that the ratio of carnivorous to herbivorous dinosaurs supports the hypothesis that they were endotherms.

Others (e.g., Alexander, McGowan, and O'Connor and Dodson, cited in References) question or qualify this hypothesis. If dinosaurs had mammal-like metabolism, then, it is reasoned, small dinosaurs, having no insulation, could not have kept warm. Large dinosaurs, by contrast, would not have been able to dissipate enough metabolic heat to avoid cooking themselves. If, instead, dinosaurs had the metabolic rate of crocodiles, then those of all sizes could survive, but the larger ones would have had elevated and nearly constant body temperatures in spite of low metabolic rate. It would have taken days for such creatures to overheat or cool off as a consequence of changes in ambient temperature, and overheating from physical activity would have been controlled by living leisurely lives. And speaking of leisure, an elephant must eat about 75 percent of the time to sustain its bulk. A large herbivorous dinosaur with an elephant's metabolism would require four to eight times as much food (unless that food were much more nutritious or more efficiently processed). Could it have eaten enough? One more point: birds and mammals have turbinates (scrolls of bone and epithelium in the nasal passages) that warm air on its way to the lungs. These seeming indicators of endothermy have not been found in dinosaurs.

Perhaps small dinosaurs, like most reptiles, controlled their body temperatures primarily by behavioral adaptations, but gave it a temporary metabolic boost as required. Perhaps some medium-sized dinosaurs metabolically elevated body temperature, at least to medium levels, at least part of the time.

REFERENCES

Alexander, R. McN. 1985. Body support, scaling, and allometry, pp. 26–37. *In* M. Hildebrand et al. (eds.), *Functional vertebrate morphology.* Harvard Univ. Press, Cambridge, MA.

Alexander, R. McN. 1989. *Dynamics of dinosaurs and other extinct giants.* Columbia Univ. Press, New York. 167p.

Bakker, R.T. 1986. *The dinosaur heresies: new theories unlocking the mystery of the dinosaurs and their extinction.* William Morrow, New York. 482p. Spectacular illustrations by the author.

Biewener, A.A. 1989. Scaling body support in mammals: limb posture and muscle mechanics. *Science* 245: 45–48.

Calder, T.J. 1984. *Size, function and life history.* Harvard Univ. Press, Cambridge, MA. 431p.

Christiansen, P. 1999. Scaling of the long bones to body mass in terrestrial mammals. *J. Morphol.* 239:167–190.

Emerson, S.B., and D.M. Bramble. 1993. Scaling, allometry, and skull design, vol. 3, pp. 384–421. *In* J. Hanken and B.K. Hall (eds.), *The skull.* Univ. Chicago Press, Chicago.

Gregory, W.K. 1912. Notes on the principles of quadrupedal locomotion and on the mechanism of the limbs of hoofed animals. *NY Acad. Sci. Ann.* 22:267–294. Excellent for graviportal adaptations.

McGowan, C. 1994. *Diatoms to dinosaurs: the size and scale of living things.* Island Press, Washington, DC. 288p. A popular review in narrative style.

Miller, P.J. (ed.). 1996. *Miniature vertebrates: the implications of small body size.* Oxford Univ. Press, New York. 328p.

O'Connor, M.P., and P. Dodson. 1999. Biophysical constraints on the thermal ecology of dinosaurs. *Paleobiology* 25(3):341–368.

Peters, R.H. 1983. *The ecological implications of body size.* Cambridge Univ. Press, London. 324p.

Rubin, C.T., and L.E. Lanyon. 1984. Dynamic strain similarity in vertebrates: an alternative to allometric limb bone scaling. *J. Theor. Biol.* 107:321–327.

Schmidt-Nielsen, K. 1984. *Scaling: why is animal size so important?* Cambridge Univ. Press, London. 241p.

West, G.B., J.H. Brown, and B.J. Enquist. 1997. A general model for the origin of allometric scaling laws in biology. *Science.* 276:122–126. Technical, but see also the review on p. 34 of this issue.

Chapter 24

Running and Jumping

Vertebrates that commonly run on the ground and are structurally modified to enhance speed or endurance are said to be **cursorial.** Most cursors are either predators or medium- to large-sized herbivores. Noncursors include the many animals that walk freely but run infrequently, as well as those small species that do run but have few postural or structural modifications associated with speed (such as various lizards, insectivores, and rodents). (Some definitions include only quadrupeds among cursors, excluding such fast and structurally modified runners as ostriches and certain dinosaurs. Another definition equates cursors with vertebrates having vertically oriented limbs. All cursors have an upright stance, but this definition also includes graviportal tetrapods, which, as described in the previous chapter, are modified primarily for the support of great weight.)

Animals that jump or hop are said to be **saltatorial.** Saltators are often bipedal, and if the hind legs are used in unison for a succession of jumps, kangaroo-fashion, the gait is called a **ricochet.** Saltators have evolved many times among small vegetarian tetrapods living in relatively open habitats, and among small arboreal vertebrates. Most cursors are at least fair jumpers. Saltators either move easily on the ground or are excellent climbers.

Adaptations for running and jumping can be approached from a number of perspectives. We might be interested in maximum speed, endurance, and maneuverability; the force profile of the foot as the animal bears weight and thrusts forward; the evolution and efficiency of running gaits; or the neural control of the muscles.

A first step for most studies, and a common entry point for beginning students, is an analysis of the kinematics (i.e., movements) of the limbs. We have come far in our ability to photographically "freeze for analysis" a sequence of locomotion since the late 1800s, when Eadweard Muybridge, using a series of 12 cameras linked by a clock to shutter in rapid sequence, captured the movements of galloping horses for Mr. Leland Stanford on a ranch in Palo Alto, California. For relatively rapid movements of limbs (or jaws), conventional or high-speed video or photographic equipment may be required. The specific structure or event to be measured (e.g., joint angle, speed of trajectory, time of contact, muscle activity pattern) depends on the hypothesis being tested. But sometimes it is instructive (and satisfying) to make preliminary recordings of a behavior of interest, and let what has been observed serve as inspiration for the formulation of questions to be addressed.

The advent of computers and digital cameras has greatly enhanced our ability to manipulate images of animals in motion. Just a few years ago, we were limited to "digitizing" locations of bony landmarks or other recognizable features of the image by hand for the construction of simple stick figures to represent an animal's movements. But with each generation of computers providing increased processing speed, memory, and storage capacity, our ability to depict and reconstruct complex animal movements in two and three dimensions becomes easier. We can also link the dynamics of muscle activity with the visual record provided by videotape. This approach enables the study not only of contemporary forms but, through modeling, of extinct forms as well. Once the movements of the skeletal elements during a single stride of a contemporary form are known, the limbs of extinct forms, such as dinosaurs, can be appropriately scaled and set in motion.

ADVANTAGES OF SPEED AND ENDURANCE

Cursorial and saltatorial animals have a number of selective advantages:

1. Cursors are able to forage over large areas. A pack of African hunting dogs may range over 3800 km^2, and a mountain lion works a circuit some 160 km long.

2. Cursors can seek new sources of food and water when familiar supplies fail. Africa's big game animals may travel great distances, and individual arctic foxes have wandered 1300 km.

3. Cursors can take advantage of seasonal variation of climate and food sources. Some herds of caribou migrate 2500 km each year.

4. Predators run to overtake prey, exploiting superior speed or relay tactics according to their habits.

5. Prey species run to escape predators. They are commonly about as swift as their pursuers (the latter rely partly on surprise) and may have superior endurance. Small prey species may be master dodgers: When chased, a kangaroo rat bounces in a different direction on nearly every hop. When startled on land, a frog can reach the safety of the water with several jumps.

6. Animals may leap to clear obstacles or to see over obstructions. The long, springy legs of the Peruvian maned wolf are said to enable it to keep mice in view in tall pampas grass. Rabbits make "spy hops" to check up on pursuers.

7. Arboreal saltators jump to climb and are able to move from branch to branch with great agility.

CURSORS, SALTATORS, AND THEIR SKILLS

Speed All cursors and many saltators are swift runners. More than two dozen species of lizards are bipedal when running fast, and some can attain 25 km/h. Many thecodont dinosaurs were excellent cursors. Pheasants and the roadrunner likewise run about 25 km/h. The ostrich has been credited with 80 km/h (50 mph), which may be as fast as any biped has ever run (Figure 24.1).

The maximum speed of the kangaroo is about 65 km/h. The marsupial wolf and rabbitlike bandicoots are also quite fast. Humans are the fastest of primates, attaining about 37 km/h (23 mph) for 200 m. Jackrabbits are able to run 64 to 72 km/h, and some of the larger rodents (agouti, cavy) are good runners. Kangaroo rats and mice, jerboas, springhares, and some other rodents are ricochetal except when moving slowly; the behavior evolved four or five times within the order (Figures 24.2 and 24.3).

FIGURE 24.1 EXAMPLES OF A SALTATOR (above left), TWO BIPEDAL CURSORS, AND A QUADRUPEDAL CURSOR (below right), showing long hind legs and, for the bipedal cursors, balance of the body over the legs.

FIGURE 24.2 EXAMPLES OF RICOCHETAL MAMMALS shown at takeoff and landing and midway in a very high jump. Note body and leg proportions and use of tail for balance.

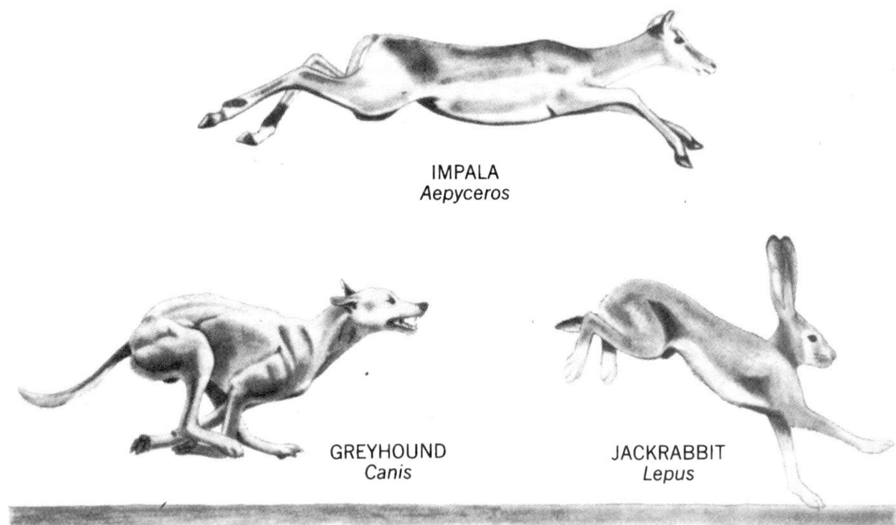

FIGURE 24.3 EXAMPLES OF MAMMALIAN CURSORS showing flexion and extension of spine and legs, passage of hind feet outside of forefeet, and varied positions of the shoulder.

The cats and dogs are cursorial, and many other carnivores moderately so: The whippet runs 55 km/h, the coyote 69 km/h, and the red fox 72 km/h. The cheetah seldom runs more than $\frac{1}{2}$ km, but for short distances is the fastest of animals, probably attaining 105 km/h (66 mph). The horse has sprinted to about 70 km/h, several antelopes run at 85 to 95 km/h, and the pronghorn has been paced with a car at 98 km/h.

Endurance A second skill of many cursors, endurance, is less well documented but also impressive. Humans run 30 km at 19.5 km/h. One fox, running before hounds, covered 240 km in $1\frac{1}{2}$ days. The horse has run 80 km at 18.2 km/h. One Mongolian wild ass is reported to have run 26 km at 48 km/h, and a pronghorn ran 11 km at 59 km/h. A camel traveled 186 km in 12 h.

Leaping Ability There is a report of a South African frog that averaged 3.28 m for three consecutive jumps. One kangaroo cleared a 2.7 m fence. The tarsier, which weighs only about 120 g, is a prodigious leaper, and the little galago can leap vertically more than 2 m. A jump of one rabbit measured more than 7 m, and the impala jumps 2.4 m high.

Acceleration and Maneuverability The ability of cursors and saltators to start fast, to speed over uneven terrain, and to dodge has not been quantified, yet is known to be outstanding for many small cursors and ricochetal mammals.

GENERAL REQUIREMENTS OF CURSORS

Saltators and bipedal cursors tend to be more specialized than quadrupedal cursors. Accordingly, we shall start with an analysis of adaptations for running on four legs. In order to run well an animal must (1) overcome the inertia of its body to attain speed; (2) overcome the movement and inertia of the legs, and any other oscillating parts, with every reversal in the direction of their motion; (3) support the body without benefit of wheels; (4) compensate for forces of deceleration, including wind resistance and the action of the ground against the feet as they come down; (5) control its course; and (6) maintain these functions as long as required.

A full cycle of motion of a running or walking animal is called a **stride.** Speed equals length of stride times rate of stride. The giraffe emphasizes length, and the warthog emphasizes rate. Great speed requires both. Endurance is speed sustained through economy of effort. Economy of effort is dependent on body shape, on the ways that parts of the body are moved, and on the magnitude and distribution of masses. For convenience we shall consider these related factors one at a time.

LENGTH OF STRIDE

A galloping racehorse covers about 7 m per stride. The faster, but smaller, cheetah covers at least as much distance per stride—about 10 times its shoulder height. How do runners manage such long strides?

Length and Proportions of Legs The longer the leg, the longer the stride. One might think, therefore, that speed could be increased merely by enlarging the body uniformly in all dimensions. Many cursorial animals are large; however, this must be for other reasons, because enlarging body size also reduces speed in several ways. To contribute to speed, it is necessary to make the legs *relatively* long in relation to the other parts of the body (Figure 24.4).

For reasons explained on p. 420 and noted later in this chapter, the distal segments of the legs usually lengthen more than the proximal segments. One cannot offer an exact formula, but cursorial ungulates usually have a radius that is as long, or a little longer, than the humerus, and a tibia that is as long, or sometimes markedly longer, than the femur. The foot skeleton, which in walkers and climbers is shorter than the corresponding middle limb segment, is, in cursorial ungulates, about equal to, or even longer than, the middle limb segment. It is the metacarpals or metatarsals that lengthen most (Figure 24.5). The carpals never lengthen, and the tarsals lengthen only in several jumpers (see figure on p. 483).

(Among birds, cursorial ability is not easily determined from the length or proportions of the legs. The legs are longest among wading birds, not runners. Furthermore, although cursorial lizards have relatively long hind legs, their limb proportions do not follow the general rule stated above. Seemingly, this is because of their small size and because their legs move in elliptical arcs lateral to the body.)

Foot Posture We have just seen that runners have relatively long leg bones. But it is the *effective* length of the leg—the part that contributes to stride length—that is important, and this can be further increased in other ways. The human foot contributes little to the length of the leg unless one rises on "tiptoes." The heel is on the ground as one stands, and strikes first in each stride. Bears, opossums, raccoons, and most other vertebrates that walk well but seldom run, have similar feet. Such feet are called **plantigrade** (= *sole* + *walking*).

Running dinosaurs, birds, carnivores, and extinct ancestors of hoofed mammals increase effective leg length by standing on what corresponds to the ball of the human foot. These animals are **digitigrade** (= *finger* + *walking*).

Perissodactyls, artiodactyls, and the cursorial representatives of several extinct orders of mammals have, like the ballet dancer, further increased effective leg length by standing

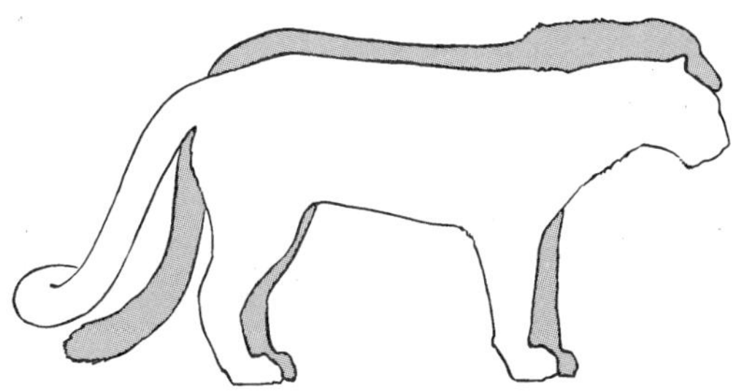

FIGURE 24.4 THE RELATIVELY LONGER LEGS OF THE CURSOR shown by the very fast cheetah, *Acinonyx*, standing behind the moderately fast mountain lion, *Felis*.

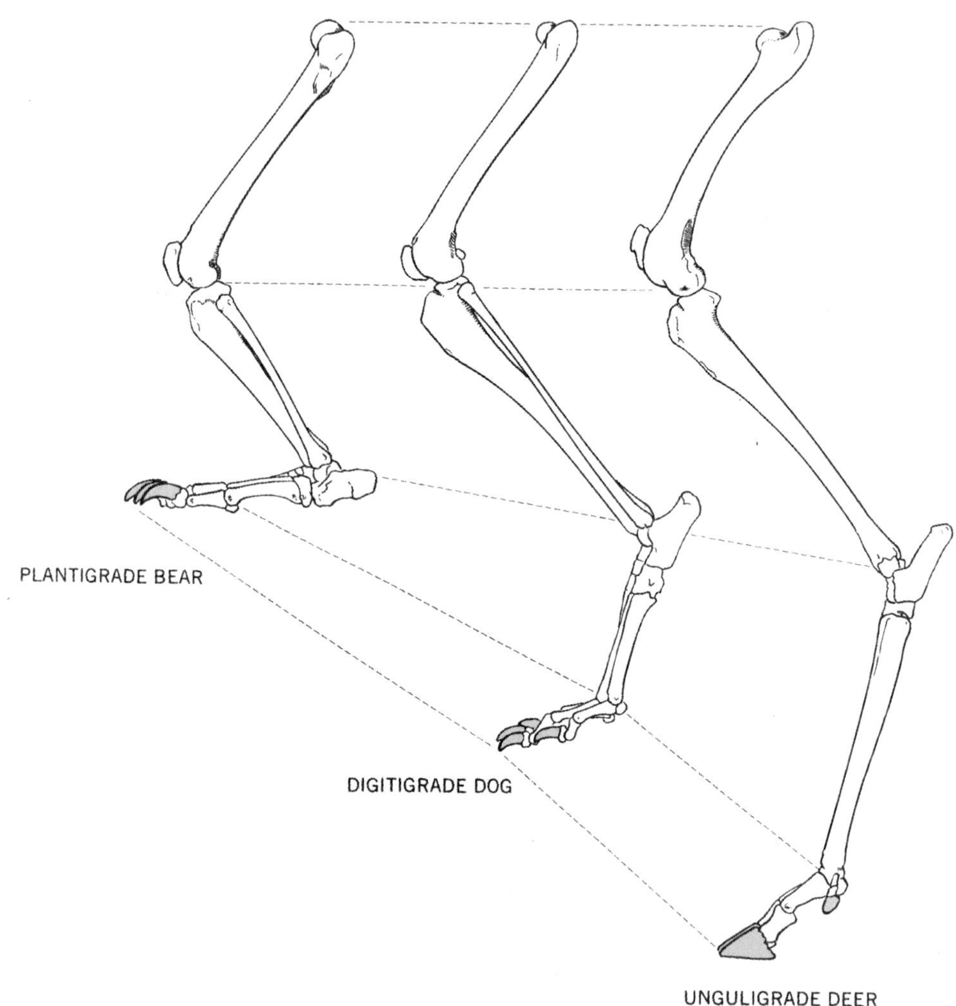

FIGURE 24.5
CONTRAST IN PROPORTIONS AND FOOT POSTURE in the left hind leg skeleton of a noncursor (left), moderate cursor (center), and highly specialized cursor (right).

on the tips of the digits. This foot posture is **unguligrade** (= *hoof* + *walking*), which is why these animals are called ungulates.

Where foot posture and limb proportions are each modified for the cursorial habit, the enhanced length and slenderness of the leg skeleton is striking (Figure 24.5).

Role of the Shoulder The effective length of the forelimb of many runners is increased further by altering the structure and function of the shoulder. In amphibians and birds, and in some reptiles and mammals, the position of the shoulder joint is virtually immobilized by the clavicle and coracoid (if present), which run like struts from the sternum to the scapula. For some lizards, however, stride is lengthened by sliding of the coracoid back and forth along a groove in the sternum. The scapula of most mammals is free to move somewhat, and cursors increase this freedom by (1) reducing the clavicle to a vestige (carnivores) or abandoning it altogether (ungulates), and (2) reorienting the scapula so that it lies not flat against the back of a broad chest (as in humans) but flat against the side of a deep, narrow chest, where it is free to rotate in the same plane in which the leg swings. The shoulder joint then moves in the sagittal plane, which is the equivalent of lengthening the leg by moving its pivot from the shoulder joint to a point partway up the scapula (Figures 24.6 and 24.11). Inability of the cursorial dinosaurs to use the shoulder in this way may have contributed to their "preference" for the bipedal habit.

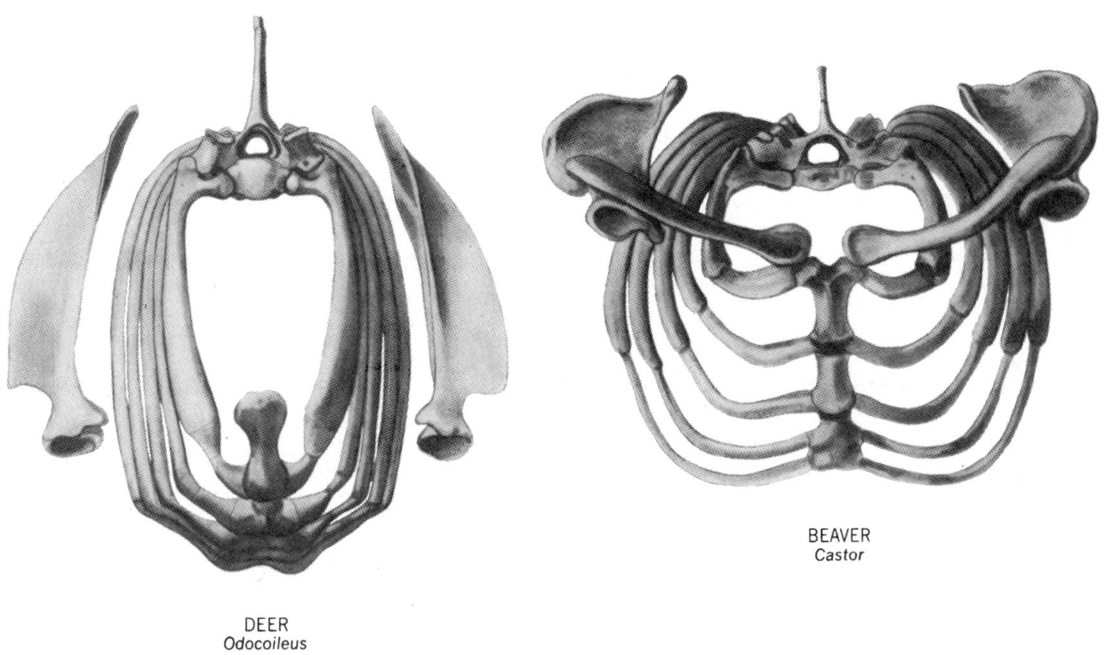

FIGURE 24.6 CONTRAST BETWEEN THE SHAPE OF THORAX, POSITION OF SCAPULAS, AND DEVELOPMENT OF CLAVICLES IN A CURSOR (left) AND A NONCURSOR (right). Only the first five thoracic segments are shown.

Role of the Spine Cursorial lizards undulate the spine in the horizontal plane as they walk or run. (Amphibians, and to a lesser extent some mammals, do the same when walking.) The pivoting of each girdle is timed to help advance the related unweighted leg and to help swing the weighted leg to the rear.

The smaller and swifter quadrupeds among mammals—particularly carnivores—have their legs positioned under the body instead of to the side, and consequently undulate the spine in the vertical plane; the back advances like the body of a measuring worm at the same time that the legs are swinging back and forth. The body of the animal is longer when the back is extended than when it is flexed. Were the animal to extend its back while its body was suspended in the air, the hindquarters would move backward as the forequarters moved forward, and the center of mass of the body would not be affected (an expression of Newton's Third Law). However, the galloping animal extends its back only when its hind feet are on the ground. Muscles of the legs pressing on the ground (and friction at the ground) prevent the hindquarters from moving backward (decelerating), so all of the increase in body length is added to stride length. Similarly, the forelimbs prevent, or at least reduce, deceleration of the forequarters as the back is flexed; therefore, the shortening of the body is also added to stride length. The cheetah is so adept at this maneuver that it could theoretically run nearly 10 km/h without any legs at all (Figure 24.7). (Flexion and extension of the spine, and its timing relative to the thrust of the legs, also ensures that the speed of the animal, that is, of its center of mass, is slightly greater than the speed of the associated girdle during the time that a pair of legs is propelling the body.)

The extra rotation of the hip and shoulder girdles that is added by flexion and extension of the spine increases the swing of the legs proper so that the limbs reach out farther, front and back, and strike and leave the ground at more acute angles than they would if the spine were held rigid. Again, this increases stride length.

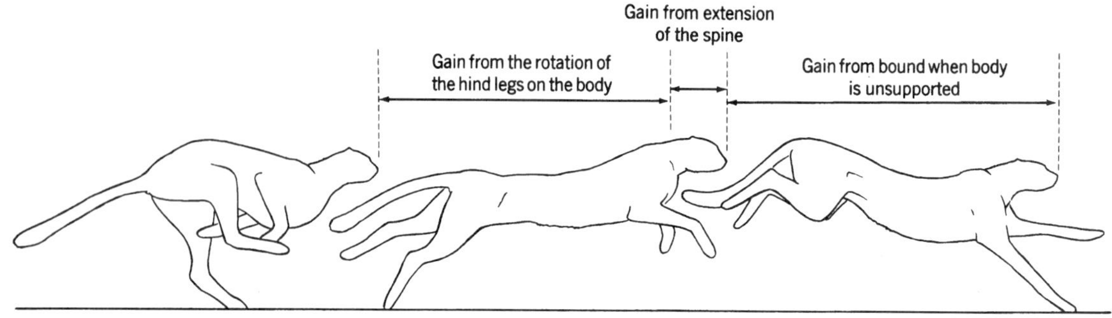

FIGURE 24.7 SOURCES OF THE LENGTH OF STRIDE OF A FAST-RUNNING CHEETAH shown for half of a cycle. Each factor is repeated in the other half cycle except that flexion of the body substitutes for extension.

Unsupported Intervals There is another important way to lengthen stride. Running gaits usually include periods of suspension when all feet are off the ground. The distance the body moves forward while it is unsupported is added to the length of the step to give the length of the stride. The bear (which runs fairly well, but is hardly cursorial) scarcely gets all its feet off the ground at the same time when it runs. Most ungulates have one unsupported period in each stride. The galloping canid has two unsupported periods, one when the body is flexed and another when it is extended. The proportion of the stride interval during which all feet are off the ground is nearly twice as great for the cheetah as for the horse. Bipedal runners are also unsupported for much of the duration of the stride. The hopping African springhare may have its feet off the ground 85 percent of the time!

Muscle Mechanics: The Most Motion for the Least Shortening Muscle can move the joints through wider angles when they insert close to the joints than when they insert farther away. Thus, in Figure 24.8, muscle A moves the foot only over distance a, whereas

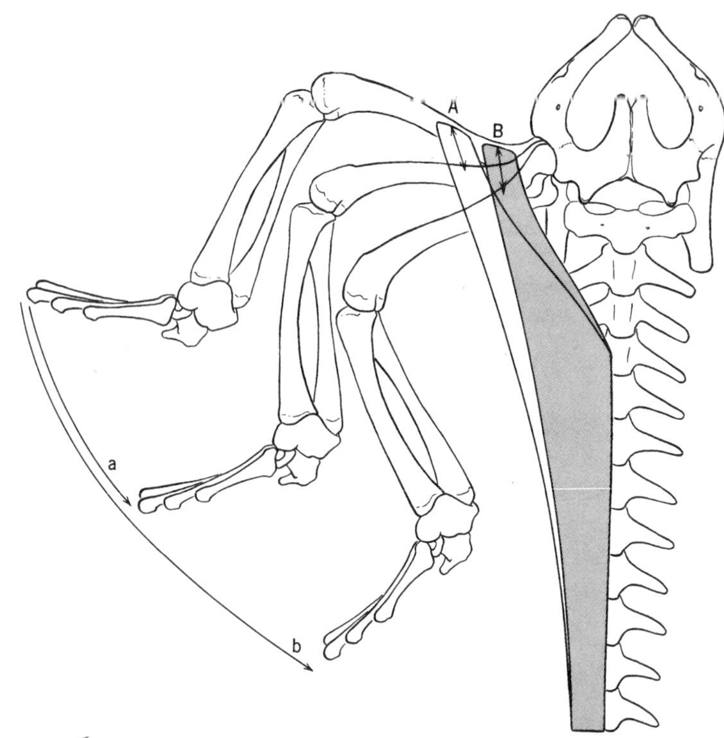

FIGURE 24.8 EFFECT OF DISTANCE OF MUSCLE INSERTION FROM THE JOINT TURNED ON THE RATE AND DEGREE OF ROTATION. The drawing shows in ventral view the caudofemoralis muscle, base of tail, pelvis, and right leg skeleton of the tegu lizard, *Tupinambis*. (Position B approximates the actual condition. A small insertion of the same muscle onto the knee is omitted.)

muscle B, contracting an equal distance, moves the foot over the greater distance b. If the load is not too great, this arrangement also benefits the rate of the stride (see below) and enables the muscle to function with less shortening, so it is characteristic of cursorial animals to have limb muscles that insert relatively close to joints.

RATE OF STRIDE

The short-legged mouse can travel faster than the long-legged frog. Speed is the product of length of stride times rate of stride, and the mouse takes many strides for each plop of the frog. Fast runners must take long strides but take them rapidly. The galloping racehorse completes about $2\frac{1}{4}$ strides/s, and the fast-running cheetah completes about $3\frac{1}{2}$ strides/s. How are cursorial animals adapted to take rapid strides?

Rate of Muscle Contraction As an animal gathers speed, it at first increases the rate of its stride by causing its muscles to contract faster. One might infer that cursorial animals as a group would have evolved muscles with relatively faster intrinsic contraction times or different-shaped force-velocity curves compared with noncursors of equal body size, but this does not seem to be the case. We must look elsewhere for ways to significantly increase rate of stride.

Muscle Mechanics: The Most Velocity for the Least Shortening It was noted above that muscles can move joints through wider angles when they insert close to joints than when they insert farther away. Following principles given on p. 418, $\mathbf{v}_o = \mathbf{v}_i\mathbf{l}_o/\mathbf{l}_i$, and therefore out-velocities are also increased by moving insertions close to joints (i.e., by reducing the value of $\mathbf{l}_i$). Thus, it is shown in Figure 24.8 that muscle B, contracting at the same rate as muscle A and having equal loading, moves the foot through the distance b in the same time interval required for muscle A to move the foot through the shorter distance a. Because it inserts closer to the joint, muscle B can move the foot a greater distance in the same time or an equal distance in less time. The muscle also gains an important physiological advantage by doing the job without shortening as much.

The ratio $\mathbf{l}_o/\mathbf{l}_i$ for limb muscles of cursors and saltators tends to be larger than for corresponding muscles of unspecialized tetrapods, and much larger than for those of diggers and swimmers. "Larger" and "much larger" are useful terms only when one is speaking in generalities; with specific bones at hand for analysis, more specific values are required. Figure 24.9 shows characteristic differences between cursors and noncursors in regard to representative bone-muscle systems.

(Because different kinds of animals emphasize different mechanisms to attain similar ends, it is preferable, when assessing cursorial ability, to consider the mechanical advantages of several muscles, or to compare the same leverages in related animals, as in Figure 24.9: For the teres major muscle, the swift horse has a lower $\mathbf{l}_o/\mathbf{l}_i$ value than the slow porcupine, but for the gastrocnemius muscle it has a higher value than even the kangaroo.)

As shown on p. 418, there is a reciprocal relationship between the velocity and the force that a given muscle can produce through a series of skeletal levers. Like the car, which in high gear can move fast but is correspondingly handicapped in climbing grades, the cursorial animal can add further speed only by sacrificing torque. However, runners, like race cars, retain some relatively low gears. For example, Figure 24.10 shows that the semimembranosus, with its relatively long effective lever arm, is a low-gear system relative to the middle gluteus muscle system, which has the same action but a much shorter effective lever arm. Finally, cursors have reduced requirements for force by reducing the loads on their muscles. We will come to that shortly, but first there is another way of increasing rate of stride.

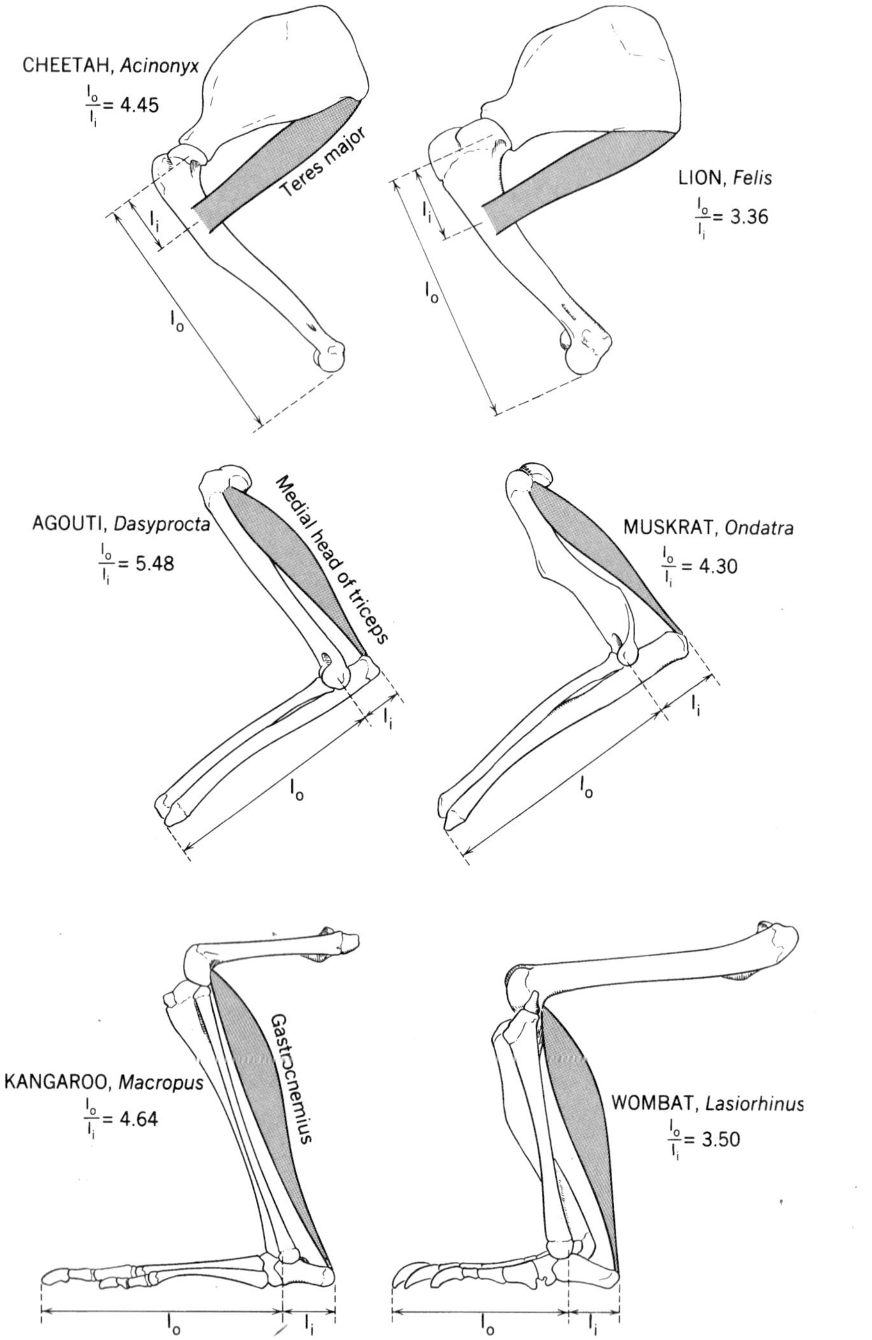

FIGURE 24.9 CONTRAST BETWEEN THREE BONE-MUSCLE SYSTEMS OF MAMMALIAN CURSORS (left) AND NONCURSORS OF THE SAME ORDERS (right). The ratio of the out-lever ($\mathbf{l}_o$) to the in-lever ($\mathbf{l}_i$) is greater for cursors.

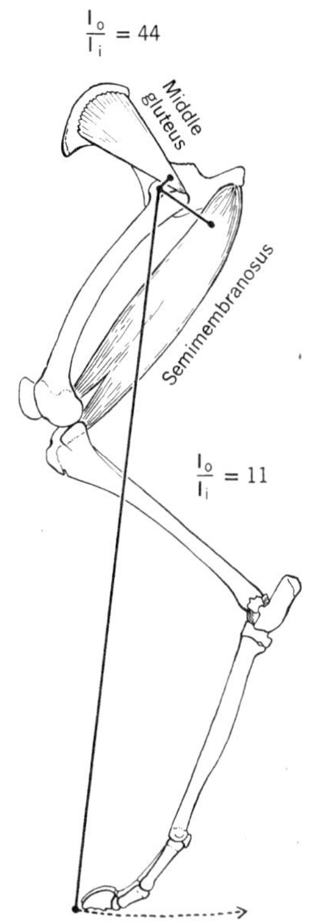

FIGURE 24.10 HIGH-GEAR AND LOW-GEAR RATIOS OF LEVER ARMS shown by a lateral view of the left hind leg of the vicuna, *Vicugna.* The muscles are diagrammatic.

Summation of Velocities Having achieved leverage that is optimum for speed, there is little further that the musculature, acting on a single joint, can do to increase the velocity of the particular action it controls. If these muscles become larger, or if additional muscles are added to help them, force is increased, but the velocity of the action remains about the same. (Several men can lift more weight together than one can lift alone, but

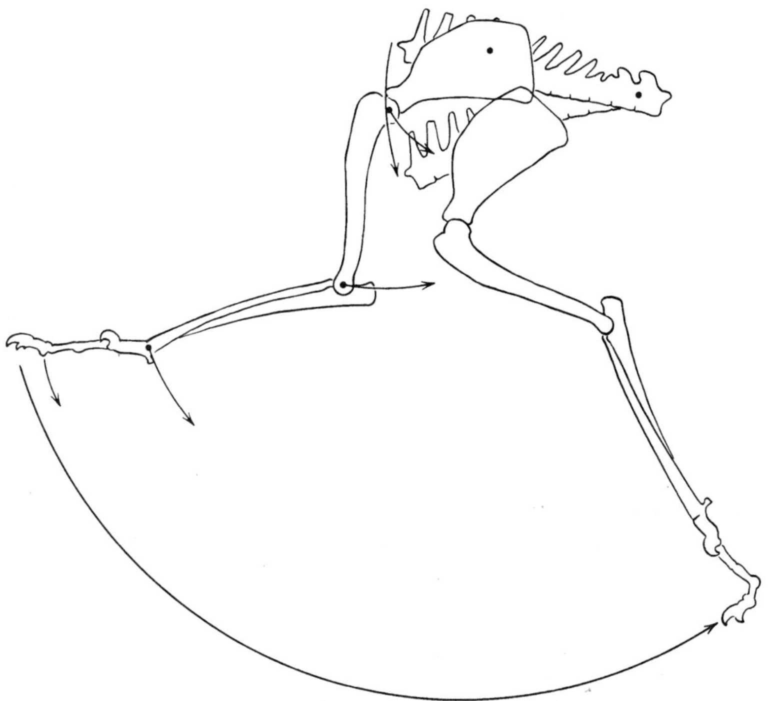

FIGURE 24.11 PRINCIPLE OF THE SUMMATION OF INDEPENDENT VELOCITIES shown by the travel of the leading forefoot of a running cheetah from the initiation of the backswing to the instant the foot leaves the ground. Arcs show approximate amount of rotation around the respective pivots (indicated by spots), not relative velocities.

several equally skilled sprinters cannot run faster together than one of them can alone.) But if different limb muscles move different joints in the same direction to achieve a greater total motion, the independent velocities they produce are added to derive the total velocity at the foot (Figure 24.11). (When a woman walks down an escalator, her motion in relation to the steps and their motion in relation to the building are added to give her rate of advance in relation to the building.) The trick is to move as many joints as possible in the same direction at the same time without interfering with the support role of the limbs. We have already seen that cursorial vertebrates add an extra pivot to the limbs by abandoning the flat-footed, plantigrade foot posture in favor of digitigrade or unguligrade posture. Furthermore, their scapulas can rotate through 20° to 25° on the chest. Finally, cursorial carnivores time the flexing of the spine in such a way that the chest and pelvis are always rotating in the direction of the swinging limbs. The net benefit to speed is considerable.

(Figure 24.11 shows *net* rotation. It may not be possible, or desirable, for all joints to rotate simultaneously; the wrist and elbow may need to *flex* during part of the support phase of the leg before extending again.)

MASS, ENDURANCE, AND DESIGN FOR ECONOMY OF EFFORT

The relationship between body size and the requirements for strength of body framework was given in Chapter 23, where it was shown that if body size were increased without altering body proportions, then the load on the locomotor system would increase faster than its capacity to provide support and power. This principle is significant to the analysis of the structure of large cursorial vertebrates. It explains why elephants (and the larger dinosaurs) cannot gallop or jump, why some small runners, such as foxes, can travel as fast as racehorses without having marked structural adaptations for speed, why no saltatorial animal is as large as typical ungulates, and why the larger cursors must have great adaptation for speed and endurance in order to run well. The adaptations of the larger runners must include not only many already discussed, but others that reduce the loads on

COMMENT 24.1

LOCOMOTOR GEARS? YES, PERHAPS, AND IT DEPENDS

High gear is used when a car is driven fast, and low gear is used for starting, pulling a load, or climbing a hill. Do vertebrates have gears? There are several ways to address this question. We have seen in Figure 24.9 that cursors have relatively high values of $\mathbf{l}_o/\mathbf{l}_i$ for many locomotor muscle systems. In the next chapter it is explained that diggers have low values. So yes, animals having different specialties tend to have different mechanical gear ratios. Note, however, that in the main text the comparison was qualified by citing muscles "having equal loading" and "contracting at the same rate." Cursors, being trim animals, may have muscles with relatively small cross-sectional areas (hence reduced capacity to accelerate loads) and shorter length (hence reduced shortening distance). Overall, animals tend to have gearing appropriate to their habits, but the range among specialists is not accurately revealed by contrasting representative ratios of lever arms.

As mentioned on p. 175, muscles have slow twitch fibers and fast twitch fibers (in varying proportions). A cruising fish uses the former, a startled fish the latter. It has been shown for a carp that during fast starts the slow fibers do not even assist the fast fibers—they cannot keep up. This is intramuscular gearing.

Within a species, might synergistic muscles provide lower and higher gears for accomplishing the same action? Contrasting the muscles shown in Figure 24.10 (an extreme example), it appears so, but one must be cautious. The putative higher gear muscle (middle gluteus), being smaller than the other, may lose some of its mechanical advantage by not shortening as fast. At peak performance, do the two muscles contract in sequence, thus shifting gears?

Another way to analyze gears is to measure the changes observed in gearing at specific joints as they extend during the support phase of a stride. For cursors, favorable change would occur if the gear ratio increases during the extension. (If the ratio decreases, muscles must compensate by contracting progressively faster.) D.R. Carrier and his associates tested this concept of dynamic gearing in trotting and galloping dogs. They found that it depends: Gear ratios decreased at four joints and increased at two. The researchers suggested that it may be mechanically impossible for gear ratio to increase at all joints simultaneously, and that the different demands of starts, level travel, and climbing may make such gearing undesirable. They note that interpretation is complicated by the storage of elastic spring energy (see Chapter 29) and the transfer of energy across two-joint muscle systems.

locomotor structures, providing economy of effort. The evolutionary process has been so effective at fulfilling these requirements that the energy cost of locomotion is, in fact, inversely related to body weight, and large cursors have superior endurance. What are the elements of design for economy of effort?

First, large cursors reduce or eliminate many oscillating motions. The legs must swing back and forth, but the feet are not lifted so high, the back is relatively stiff (Figure 24.12), and the center of mass has less vertical displacement. Unsupported periods

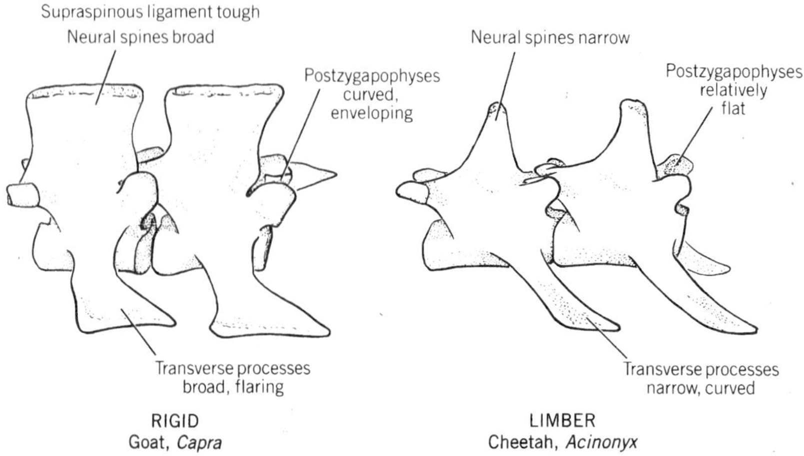

FIGURE 24.12
LUMBAR VERTEBRAE OF LIMBER AND PASSIVELY RIGID SPINES.

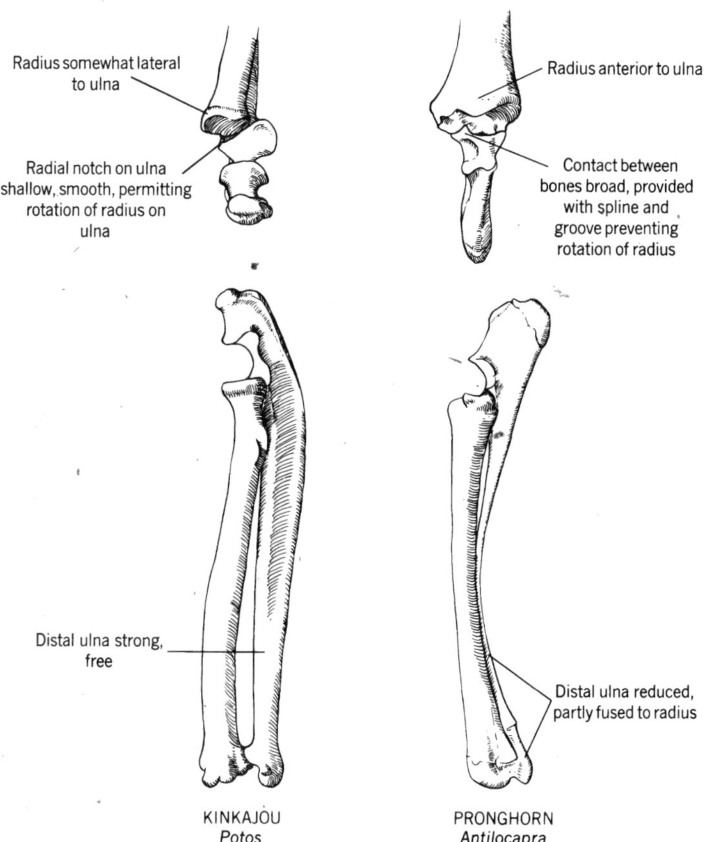

FIGURE 24.13 CONTRAST BETWEEN THE FOREARM SKELETONS OF A NONCURSOR (left) AND A CURSOR (right) shown by the left radius and ulna in lateral view (below) and the same bones at the elbow seen from above and behind (above).

can follow only vertical acceleration, which is costly, so suspensions are reduced or avoided.

Next, the mass of the limbs is reduced in several ways. Adductors and abductors of the legs are reduced or are converted to move the limbs in the direction of travel. Muscles that manipulate the digits, rotate the forearm, or twist the feet inward or outward are also reduced or even eliminated. The ulna and fibula, which in other animals take part in foot twisting and rotation mechanisms, are reduced as these functions are lost. The ulna must be retained where it completes the elbow joint, but distally may become a sliver that fuses to the radius (Figure 24.13). The fibula may become an attenuated splint (even in small mammalian cursors), and in artiodactyls is sometimes represented only by a nubbin of bone at the ankle (Figure 24.14).

Furthermore, since the forces that must be developed in oscillating systems with every reversal in the direction of their action are also in proportion to the square of their angular velocities (pp. 420 and 550), the loads on muscles causing such motions can be reduced not only by reducing the masses of the systems but also by reducing their velocities. This is a principal reason why it is the lower limb segments that lengthen in the evolution of the long legs of the runner: When devoid of the muscles and bones related to twisting, rotating, and digit manipulation, those segments are relatively light; fleshy parts of the limbs are kept close to the body, where they do not move so far, and hence not so fast, as the more distal segments. The advantage gained is particularly important during acceleration.

Other elements in design for economy of effort provide further saving of weight without loss of strength. The feet of unspecialized vertebrates tend to be broad and

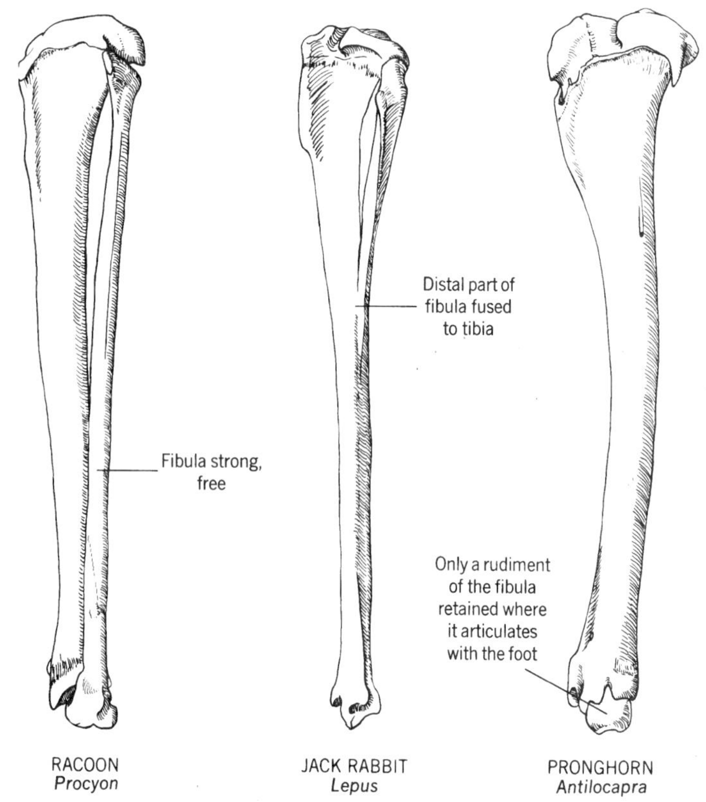

FIGURE 24.14 CONTRAST BETWEEN THE FIBULAS OF A NONCURSOR (left) AND TWO CURSORS (center and right) shown by the left leg in lateral view.

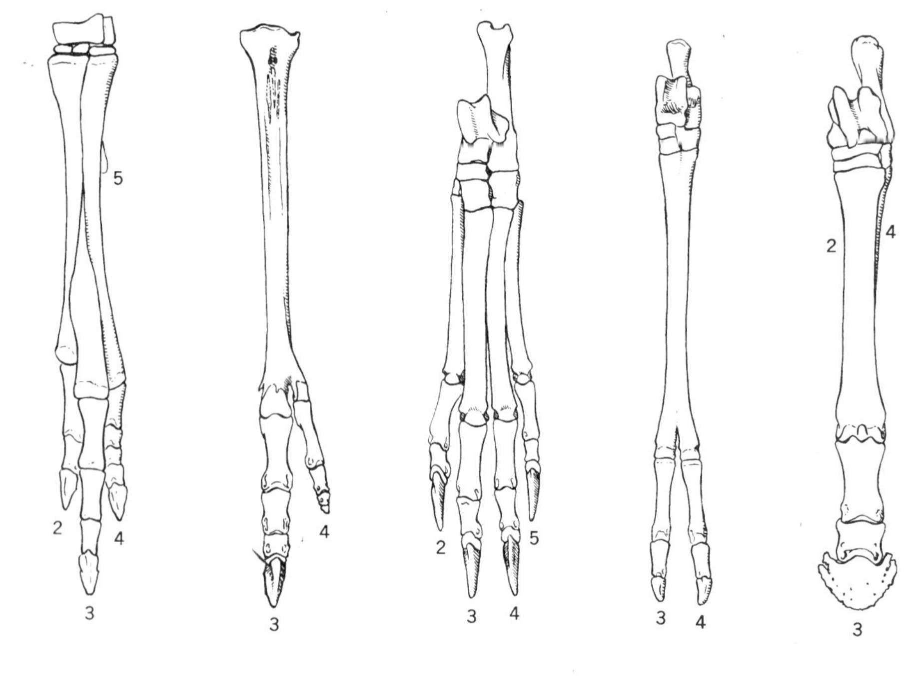

FIGURE 24.15 LENGTHENING, COMPACTION, AND FUSION OF METATARSALS, LOSS OF LATERAL DIGITS, AND SQUARING OR FUSION OF TARSALS in the left hind foot of selected cursors of three classes. Digits are numbered.

pliable. Their metapodials are rounded in cross section and are well separated. Some runners (some dinosaurs, carnivores) provide more strength by crowding these bones together into a compact unit; each bone becomes somewhat square in cross section. Still more strength can be provided with the same weight (or the same strength with less weight) if the skeletal material is distributed among fewer bones; hence, some of the best runners and jumpers (kangaroos, jerboas, perissodactyls, artiodactyls), and particularly the large ones, tend to lose the lateral toes and to fuse the basal elements of the remaining toes into a single bone, of compound origin. This process has produced the cannon bone of ungulates and, in response to a hopping habit, the tarsometatarsus bone in the ancestors of all birds. The result is a slender, light, strong foot (Figure 24.15 and figure on p. 17).

To compensate for the bracing lost as bones and muscles of the lower limbs are reduced or eliminated, and to guard against dislocations, cursorial vertebrates have evolved joints modified to function as hinges, allowing motion only in the line of travel. This has been done by introducing (or enlarging) interlocking splines and grooves in the joints (digits, ankle or hock, elbow) (Figure 24.16) or by substituting flat or cylindrical shapes for spherical shapes at the articulations (wrist and, to lesser extent, shoulder) (Figure 24.17).

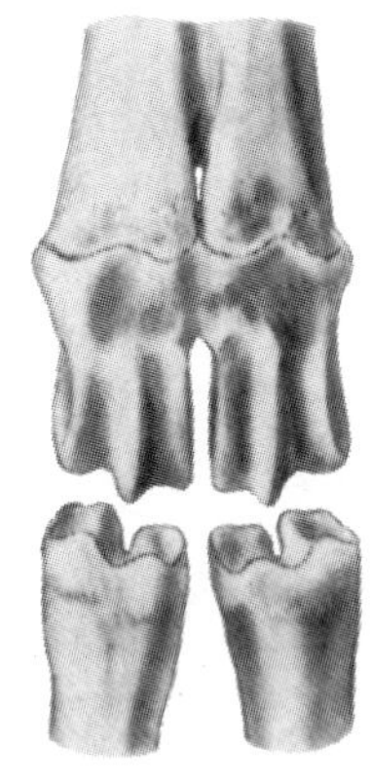

FIGURE 24.16 FETLOCK JOINT OF A PRONGHORN, *Antilocapra,* SHOWING STRENGTHENING OF A HINGE JOINT BY SPLINES AND GROOVES.

Moving tetrapods also save much energy by either recycling between gravitational potential energy and kinetic energy or by storing and releasing elastic spring energy. This important topic is discussed in Chapter 29. Furthermore, physiological adaptations are critical for endurance. The superior ventilation of birds was noted on p. 232. Among mammals, distance runners such as pronghorns and jackrabbits have relatively enormous lungs, large tracheas, large hearts, much hemoglobin in the blood, and many mitochondria in the muscles.

STABILITY AND MANEUVERABILITY

Large grazers and browsers that are relatively free from predation need not be particularly agile. Water buffalos, camels, giraffes, and rhinoceroses cannot start, wheel, dodge, and stop quickly. Small antelopes and certain carnivores, on the other hand, can maneuver their bodies with almost unbelievable skill. A fairly light and supple body, alertness, and rapid neuromuscular coordination are primary requisites.

A consequence of the mechanics of curvilinear motion is that when a moving object turns, there is an outward, or centrifugal force that must be opposed by an equal inward, or centripetal force if the object is to avoid skidding out of its curved course (Figure 24.18). Centrifugal force is directly proportional to mass and the square of velocity, and is inversely proportional to the radius of the turn. In order to turn sharply, a running animal must lean into the turn, adjusting the angle of its body so that the outward component of the thrust of its feet onto the ground equals the centrifugal force. This is opposed by the inward component of the force of the ground on the feet. There must be sufficient friction between feet and ground to prevent slipping—a requirement easily met by hoof and claw on rough ground, though a turning mouse may slip badly on a polished floor. Some animals can turn in several body lengths, even when running fast. Their bodies may make an angle of only 25° to 30° with the ground as they spin around.

Any force, $\mathbf{F}$, that acts against the side of an animal (e.g., centrifugal force, wind pressure, or the jostling of another animal) tends to upset the animal with a torque of $\mathbf{F}h$, where h is the height of its center of mass above the ground (Figure 24.19). In order for stability to be maintained, $\mathbf{F}h$ must be less than $\mathbf{M}w$, where $\mathbf{M}$ is the mass of the animal and w is half the width of the animal's stance. It follows that an animal that stands or moves straight ahead (keeping feet of both sides of the body on the ground at once) is most

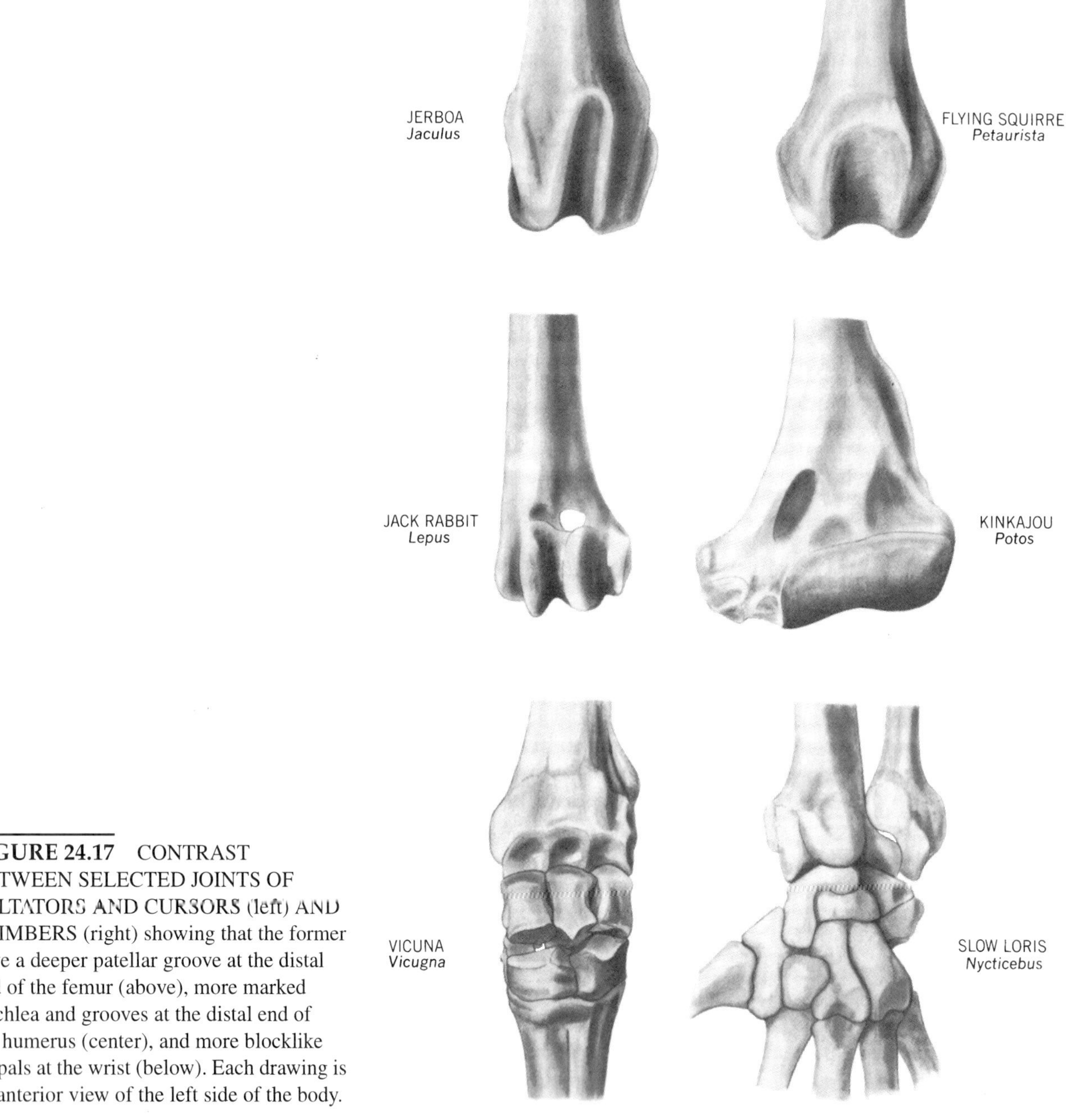

FIGURE 24.17 CONTRAST BETWEEN SELECTED JOINTS OF SALTATORS AND CURSORS (left) AND CLIMBERS (right) showing that the former have a deeper patellar groove at the distal end of the femur (above), more marked trochlea and grooves at the distal end of the humerus (center), and more blocklike carpals at the wrist (below). Each drawing is an anterior view of the left side of the body.

stable if it is broad and short-legged like the hippopotamus. Cursors must be long-legged, however, and often must lift both right or both left feet at once to move fast. Furthermore, as the previous paragraph explained, they must lean into turns when running. This is difficult for a highly stable animal. It follows that stability must be sacrificed for maneuverability. Agile cursors control lateral forces actively by adjusting posture rather than passively by virtue of body proportions.

Moreover, in order to turn sharply, a galloping mammal must lead with its inside front leg. That is, the foot toward the inside of the turn must strike the ground after its opposite in each couplet of footfalls. This provides that successive footfalls are in the direction of

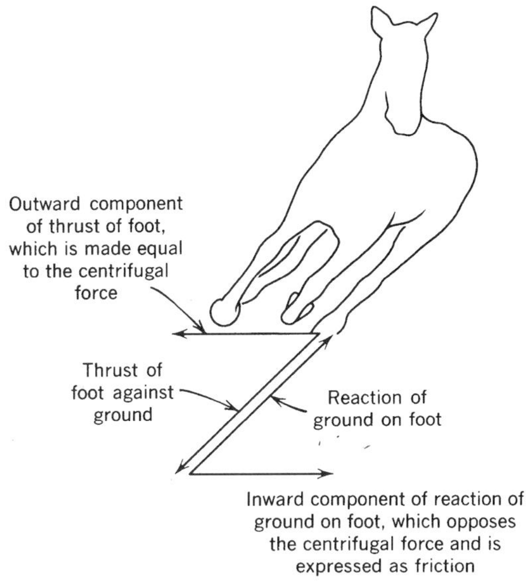

FIGURE 24.18 FORCES RELATED TO STABILITY WHEN TURNING.

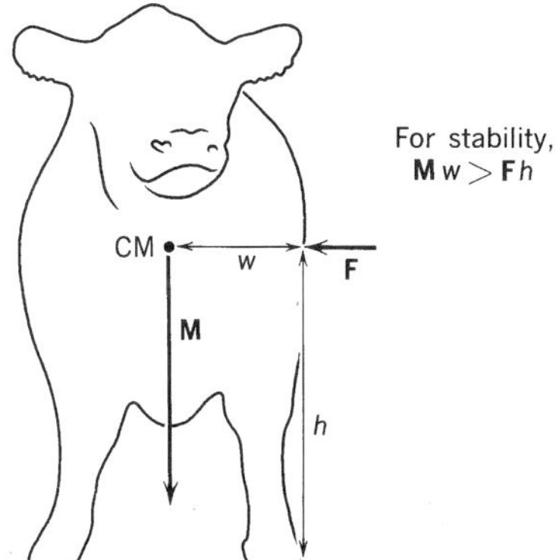

FIGURE 24.19 FACTORS INVOLVED IN STABILITY (CM = center of mass).

the turn and the animal can balance over its support. The hapless antelope cannot evade the cheetah's dash by dodging, for the cheetah follows almost in its tracks.

GAITS

A regularly repeating sequence and manner of moving the legs in walking or running is called a gait. Gait selection relates to energetics, rate of travel, maneuverability, stability, and the size and structure of the body. If the footfalls of a pair of legs, fore or hind, are evenly spaced in time, as in pacing, walking, and trotting, the gait is said to be symmetrical. In walking gaits, each foot is on the ground more than half the time; in running gaits, each foot is on the ground less than half the time.

In the pace, the two feet on the same side of the body swing more or less in unison, which avoids interference between fore and hind feet. The camel family and some large dogs pace naturally when moving moderately fast, and some harness horses are trained to race at this gait (Figure 24.20). All pacers are long-legged; the gait would be unstable for short-legged animals.

In the trot, which like the pace is usually performed at moderate speed by mammals, a fore and hind foot on opposite sides of the body swing approximately in unison. Since a line between the supporting feet passes close under the center of mass, the gait is favorable for animals with broad bodies or, as for lizards, with legs splayed to the side of the body.

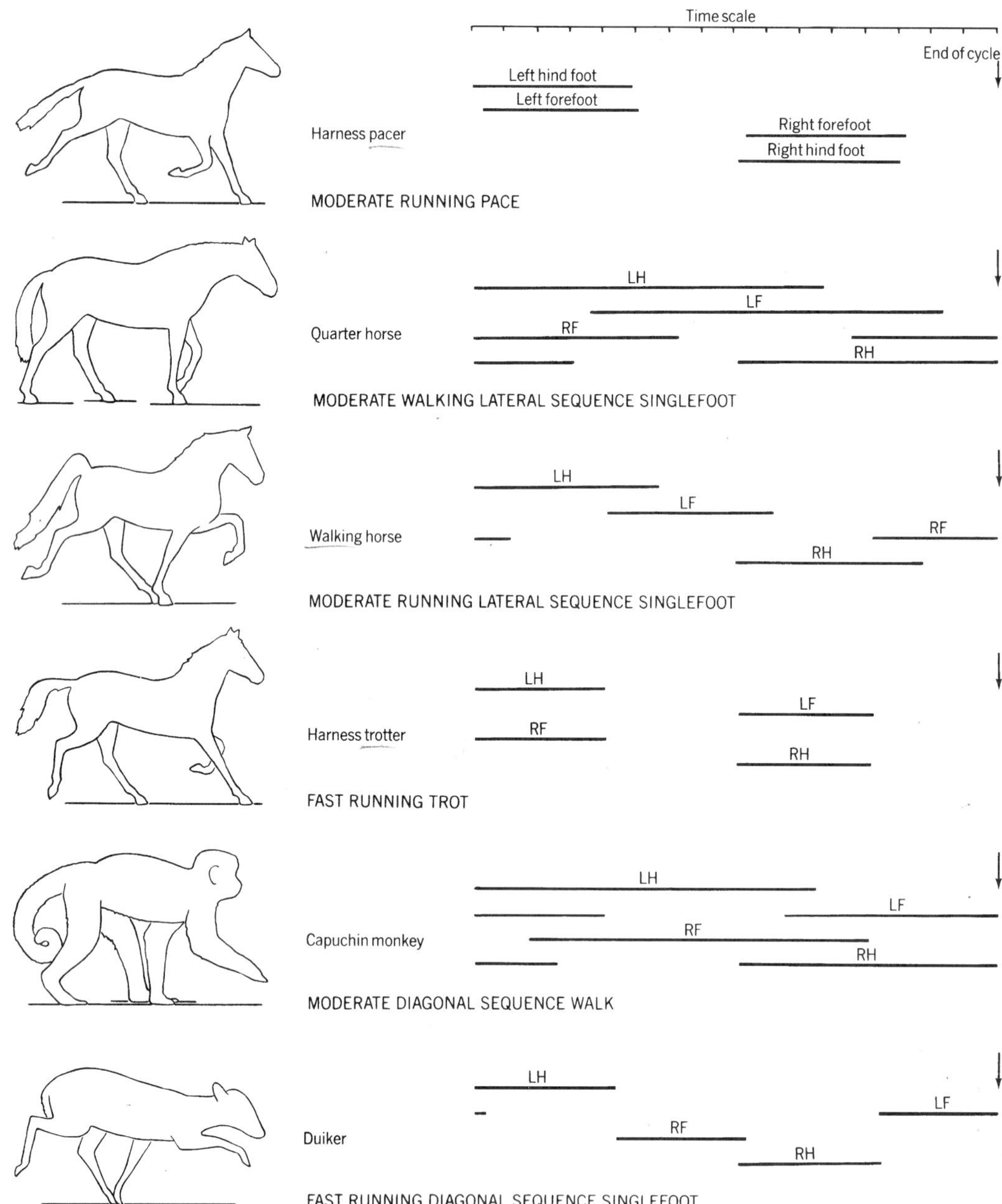

FIGURE 24.20 REPRESENTATIVE SYMMETRICAL GAITS OF TETRAPODS. All drawings show position of the body the instant the left hind foot strikes the ground. In moving from top to bottom, the two forefeet rotate counterclockwise relative to the hind feet. The gait diagrams show by the length of the lines the duration of contact of the respective feet with the ground. Each diagram shows one complete cycle starting with the instant the left hind foot strikes the ground.

At the usual walk, the four footfalls are independent; a forefoot strikes the ground next after the hind foot on the same side of the body (lateral sequence) in most tetrapods except primates, because interference between forefeet and hind feet is then avoided. The same gait can be done at the run but is unusual except for certain show horses. Gaits having independent footfalls provide the most stability.

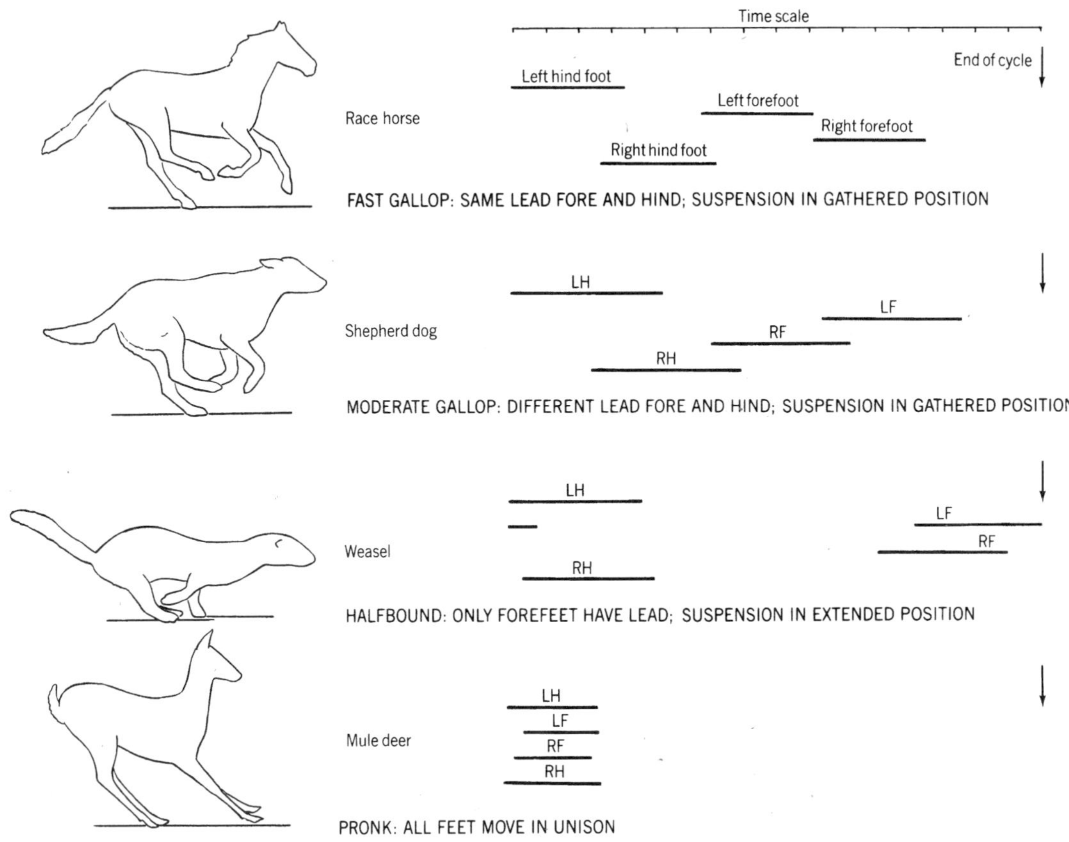

FIGURE 24.21 REPRESENTATIVE ASYMMETRICAL GAITS OF MAMMALS. All drawings show position of the body the instant the left (and trailing) hind foot strikes the ground. Notations for the gait diagrams are the same as for Figure 24.20.

Most primates and several other tetrapods do a diagonal-sequence walk; a forefoot strikes the ground next after the hind foot on the opposite side of the body. The hind foot passes to the side of the forefoot to avoid interference. Some lemurs and small artiodactyls do the same gait at the run.

Galloping and bounding gaits are said to be asymmetrical because the footfalls of the two feet of a pair are unevenly spaced in time (Figure 24.21). The foot of a pair that strikes the ground first in each couplet of footfalls is called the trailing foot; the other is the leading foot. If the lead is the same, fore and hind, the gait is a transverse gallop (horse); if it is different, fore and hind, the gait is the more maneuverable but less stable rotary gallop (cheetah). Unless performed slowly, asymmetrical gaits increase length of stride by introducing periods of suspension when all feet are off the ground. The suspension may come when the legs are gathered under the animal (horse), stretched out fore and hind (deer), or both (cheetah, pronghorn). Rabbits, weasels, and many other mammals of comparable size place the forefeet on the ground alternately, but place the hind feet on the ground more or less in unison. This gait is the half bound. Many squirrels, mice, and other small mammals place each pair of feet in unison, thus doing the bound. Bounding gaits are favorable for small mammals on rough terrain. In the half bound and bound, each hind foot must be swung forward lateral to its corresponding forefoot to avoid interference; that is, the hind feet have a wider track than the forefeet. Several artiodactyls may place all four feet in unison—a gait called a pronk.

Comments on gait selection in relation to energetics are found on pp. 550 and 551.

COMMENT 24.2

HOW MANY GAITS ARE THERE?

Riders know what gait their mount is using. Horses also know, for they shift gaits quickly when directed, and predictably when not being ridden. However, when slow-motion photography is used to plot footfalls against a timeline, as in the accompanying illustrations, it is found that successive strides are rarely identical, that different individuals of a kind often move differently, and that when all tetrapods are considered, the gaits merge into one another. There is no such thing as *the* walk or *the* gallop; any scheme of naming must deal with a continuum. How should one parcel the continuum?

The successive combinations of supporting feet in a locomotor cycle have been used as a basis for distinguishing certain gaits. In the gait diagram for a dog in Figure 24.21, it is seen that there are 8 combinations of support (LH, LH + RH, RH, RH + RF, RF, RF + LF, LF, and none). *All* quadrupedal gaits (symmetrical and asymmetrical, fore and hind contacts equal or unequal) have 8 combinations per cycle unless two or more feet chance to strike or leave the ground simultaneously (see the duiker and harness trotter in Figure 24.20). It has been calculated that, all told, the various quadrupeds might use more than 200 combinations of support per cycle; horses alone could use about 70. This is surely not a basis for naming gaits.

Trained observers can distinguish by eye (on the basis of speed and relative timing of footfalls) about 45 symmetrical gaits and somewhat fewer asymmetrical gaits (more variables, faster action). Specialists have provided descriptive names for this many gaits, but most of the designations are cumbersome and unfamiliar, and who needs them? This brings us back to horseback riders (also dog breeders, camel racers, elephant trainers, and naturalists), who recognize only as many gaits as are useful to them.

SALTATION AND BIPEDAL RUNNING

Ricochetal mammals can accelerate faster from rest and can alter both the speed and direction of their motion faster than their quadrupedal relatives. These are important escape mechanisms that may be purchased at the price of efficiency. Since the body must repeatedly be lifted against gravity, much energy is required. Saltators rarely maintain fast progression for long (although some compensate by recycling spring energy—see again pp. 550 and 551).

The height (h) to which an animal jumps equals $(\mathbf{v}^2 \sin^2\theta / 2g$, where $\mathbf{v}$ is the upward velocity at takeoff, θ is the takeoff angle, and g is the pull of gravity. $\sin^2\theta$ equals 1, which is maximum, when $\theta = 90°$. For a vertical jump, therefore, $h = \mathbf{v}^2 / 2g$. A 250 g galago has been accurately observed to jump vertically 2.26 m (7 ft $4\frac{3}{4}$ in) from a crouch. Since its center of gravity was lifted more than 2 m, this performance is remarkable and must be close to a record for any animal, although a rat kangaroo, in jumping 2.4 m high, did about as well. The human high jumper runs up to the takeoff in order to translate horizontal velocity to vertical velocity, yet lifts the center of gravity only a little more than 1 m. The takeoff velocity needed to lift the center of gravity 2 m is 625 cm/s; that needed to lift it 1 m is 442 cm/s.

The range (R) of an animal's jump depends on takeoff velocity and the angle of takeoff: $R = (\mathbf{v}^2 \sin^2\theta) / g$. Theoretically, maximum range is attained when θ is 45° (then $R = \mathbf{v}^2 / g$), which is about the angle adopted by frogs and galagos for long jumps. The vertical jump of the galago as described above is the equivalent of a standing long jump of about $4\frac{1}{4}$ m. The human long jumper runs up to the takeoff, and because at maximum speed he or she cannot change the direction of motion enough to take off at 45°, takes off instead at 25° to 35°. Kangaroo rats and springhares usually ricochet with a takeoff angle that is also below 35°.

The acceleration required to achieve takeoff velocity is $\mathbf{v}^2 / 2s$, where s is the distance through which the applied force acts. This, in turn, is the difference in the functional length of the springing hind leg between its initial flexed and final extended positions. As their proportions attest, long hind legs are a great asset to saltators.

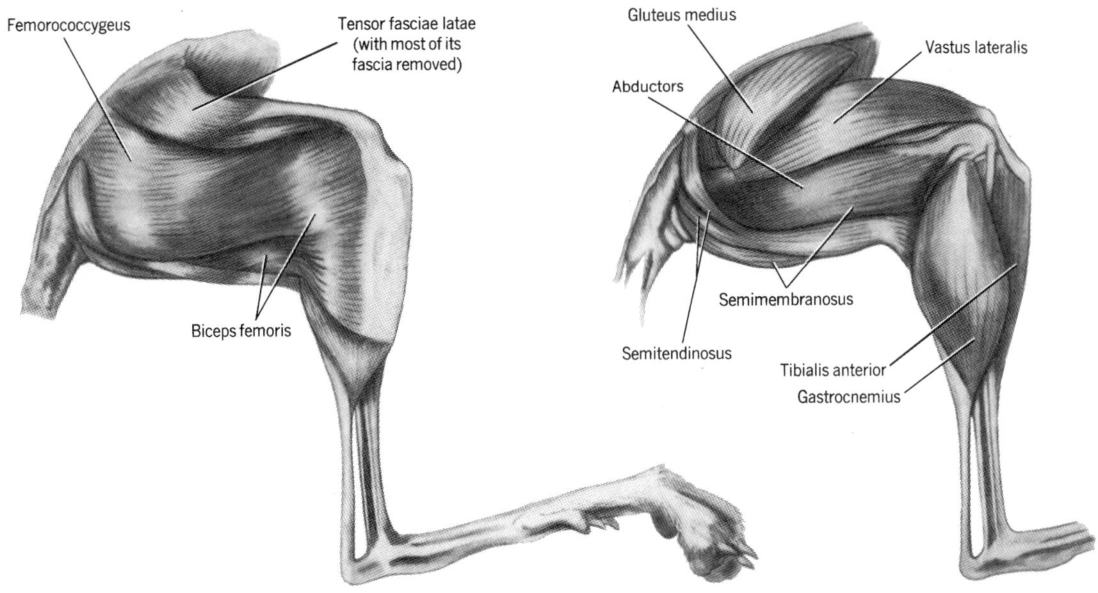

FIGURE 24.22 RIGHT HIND LEG OF A RICOCHETAL RODENT, the jerboa *Allactaga,* showing proximal position of musculature, lengthened distal limb segments, reduced lateral digits, and some principal superficial muscles (left) and deeper muscles (right).

The mass (m) of the body does not enter directly into the formulas for height and range of jump. However, the force that must be applied to the ground to accelerate the body to takeoff velocity equals $m\mathbf{v}^2 / 2s$. As mass increases, the power plant increases in rough proportion, so that within limits, animals of different size but equal proportions can perform about the same. Nevertheless, the body cannot endure excessive forces, and when animals widely disparate in size are compared, complex physiological factors must be considered. It is not possible to decide if the locust or kangaroo is the better jumper.

Relative lengthening of the hind limbs and of their distal segments is more extreme for saltators than for cursors (Figure 24.22). The tibia may be one and one-half or even two times as long as the femur (see kangaroo, Figure 24.9). Several tarsal bones may lengthen (frogs, tarsier, galago—see figure on p. 483). Lengthening of the hind limbs of ricochetal mammals is the more striking because the forelimbs are not modified for speed. They are usually used for slow progression and for handling food, so they do not become vestigial but often they are reduced in size. Some bipeds that swing the legs alternately, increase the length of the stride by oscillating the pelvis around the long axis of the vertebral column—humans are a notable example.

Bipedal reptiles share with ricochetal mammals the need for a long tail. It is elevated while running so that it will counterbalance the fore part of the body and thus bring the center of mass over the hind feet. Bipedal lizards are unable to run if part of the tail is amputated; rabbits could probably not become predominantly ricochetal with so short a tail; absence of a tail correlates with our upright posture. Most ricochetal mammals also use the tail as a prop, thus providing a third point of support when standing on the hind legs. Although once a jumper has left the ground its center of mass must move in a predetermined parabolic path until the animal touches the ground again, a saltator can change the orientation of its body in midjump by lashing its tail. The kangaroo rat can even reverse its field in the air, ready to bounce back along its own track on the next hop. The tail is often tufted with hair; its greater distal weight and air resistance then improve its effectiveness for controlling the jump.

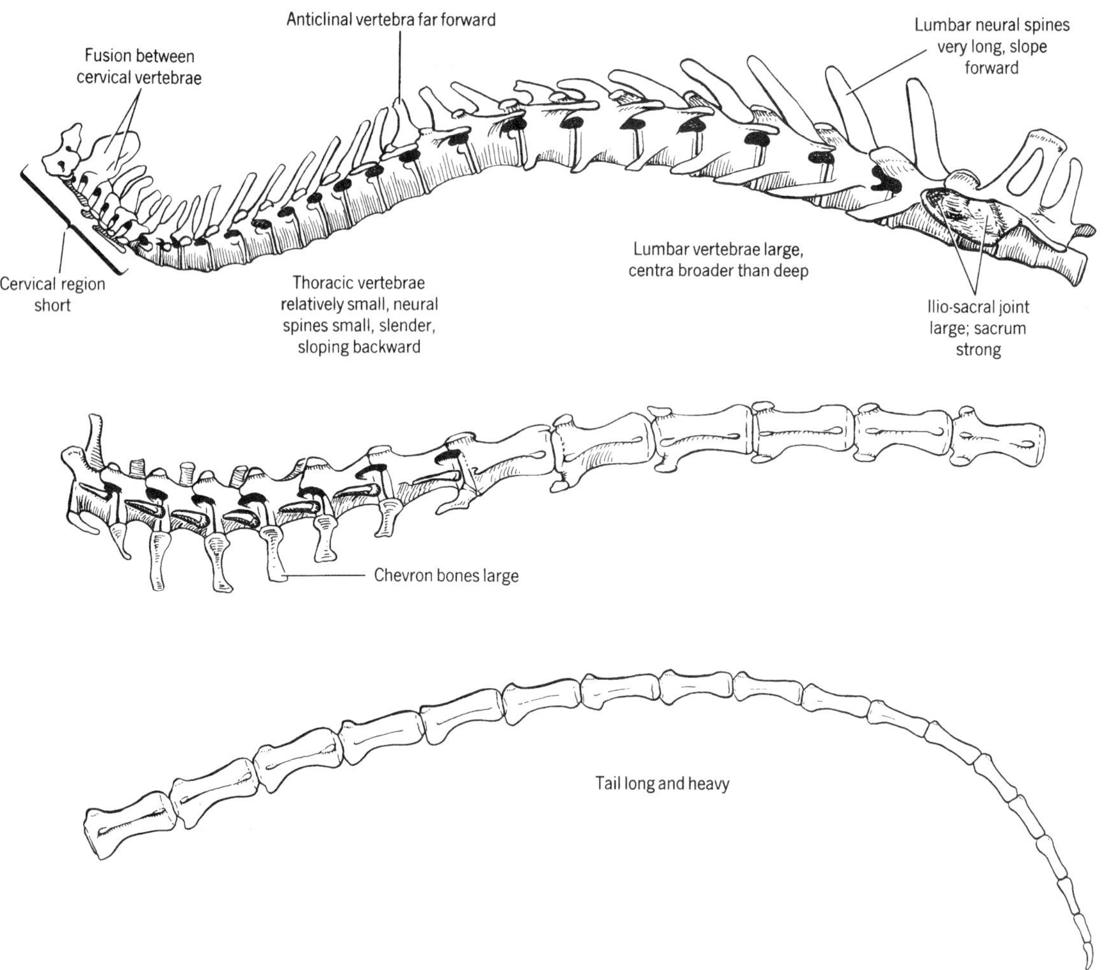

FIGURE 24.23 SOME CHARACTERISTICS OF THE VERTEBRAL COLUMN OF A RICOCHETAL MAMMAL shown by a springhare, *Pedetes.*

Concentration of weight in line with the thrusting hind legs is also achieved by shortening somewhat the presacral part of the spine, particularly in frogs and mammals. (When frogs jump, the vertebral column extends at the sacroiliac joint; see Jenkins and Shubin in References.) The lumbar region of the spine of ricochetal mammals is robust and has prominent neural spines. The thoracic region is slight and has small neural spines. Associated with these proportions is the relatively anterior position of the **anticlinal vertebra** (Figure 24.23), which is transitional between those with backward-sloping neural spines and those with forward-sloping neural spines. The cervical vertebrae of ricochetal rodents tend to fuse, apparently to reduce bobbing of the head.

Rodents that hop or bound tend to develop two ligamentous shock absorbers that limit whiplash of the spine. One is a fan of ligaments that runs down and forward from the summit of the enlarged neural spine of the second thoracic vertebra to support the cervical vertebrae. The other is an enlarged supraspinous ligament that runs from the forward-sloping neural spine of the last lumbar vertebra to the backward-sloping neural spine of the first (or second) sacral vertebra (Figure 24.23).

REFERENCES

Alexander, R. McN. 1991. How dinosaurs ran. *Sci. Am.* 264(4):130–136.

Bennet-Clark, H.C. 1977. Scale effects in jumping animals, pp. 185–201. *In* T.J. Pedley (ed.), *Scale effects in animal locomotion*. Academic Press, New York.

Biewener, A.A. 1990. Biomechanics of mammalian terrestrial locomotion. *Science* 250:1097–1103.

Biewener, A.A., J. Thomasan, and L.E. Lanyon. 1983. Mechanics of locomotion and jumping in the forelimb of the horse (*Equus*): *in vivo* stress developed in the radius and metacarpus. *J. Zool. Lond.* 201:67–82.

Camp, C.L., and N. Smith. 1942. Phylogeny and function of the digital ligaments of the horse. *Univ. Calif. Mem.* 13:69–124.

Carrier, D.R., C.S. Gregersen, and N.A. Silverton. 1998. Dynamic gearing in running dogs. *J. Exp. Biol.* 201:3185–3195.

Emerson, S.B. 1978. Jumping and leaping, pp. 58–72. *In* M. Hildebrand et al. (eds.), *Functional vertebrate morphology*. Harvard Univ. Press, Cambridge, MA. Provides a theoretical framework and valuable discussion of methodologies of how jumps are measured. Integrates morphology, biomechanics, and size of jumpers.

Gambaryan, P.P. 1974. *How mammals run: anatomical adaptations*. Wiley, New York. 367p. (Originally published in Russian in 1972.)

Hall-Craggs, E.C.B. 1965. An analysis of the jump of the lesser galago (*Galago senegalensis*). *Zool. Soc. Lond. Proc.* 147:20–29.

Hatt, R.T. 1932. The vertebral columns of ricochetal rodents. *Am. Mus. Nat. Hist. Bull.* 63, Article 6:599–738. Extensive survey with functional interpretation.

Hildebrand, M. 1980. The adaptive significance of tetrapod gait selection. *Am. Zool.* 20:255–267.

Hildebrand, M. 1985. Walking and running, pp. 38–57. *In* M. Hildebrand et al. (eds.), *Functional vertebrate morphology*. Harvard Univ. Press, Cambridge, MA.

Howell, A.B. 1965. *Speed in animals: their specializations for running and leaping*. Hafner, New York. 270p. (Originally published in 1944.)

Irschick, D.J., and B.J. Jayne. 1999. Comparative three-dimensional kinematics of the hindlimb for high-speed bipedal and quadrupedal locomotion of lizards. *J. Exp. Biol.* 202:1047–1065. Model study of limb kinematics. Demonstrates the correlation between limb length and the animal's stride length and speed.

Jenkins, F.A., and S.M. Camazine. 1977. Hip structure and locomotion in ambulatory and cursorial carnivores. *J. Zool. Lond.* 181:351–370.

Jenkins, F.A., Jr., and N.H. Shubin. 1998. *Prosalirus bitis* and the anuran caudopelvic mechanism. *J. Vert. Paleontol.* 18:495–510.

Marsh, R.L. 1994. Jumping ability of anuran amphibians, vol. 38B, pp. 51–111. *In* J.H. Jones (ed.), *Advances in veterinary science and comparative medicine*. Academic Press, New York. Comprehensive study of the kinematics, mechanics, and performance of jumping.

Seyfarth, A., R. Blickhan, and J.L. Van Leeuwen. 2000. Optimum take-off techniques and muscle design for long jump. *J. Exp. Biol.* 203:741–750. Illustrates how to tease out the musculoskeletal contributions to human jumping performance.

Snyder, R.C. 1962. Adaptations for bipedal locomotion of lizards. *Am. Zool.* 2:191–203.

Chapter 25

Digging, and Crawling without Appendages

Vertebrates that spend all or most of their lives underground are said to be **subterranean.** Most of these animals make tunnels and are also called **burrowers.** Many other vertebrates are active above ground, yet are likewise highly adapted to dig for food or shelter. All of these animals are described as **fossorial** in terms of digging ability. Some vertebrates live in burrows dug by other animals, and many tetrapods can dig somewhat, even without marked structural adaptation. Thus, the alligator lizard pushes its way into litter to hide, the thrush scratches leaves away to uncover food, the caribou paws snow from the lichens it eats, and the elephant scrapes holes with its forefeet to reach groundwater where surface sources have gone dry. This chapter will stress the more highly adapted diggers.

The fossorial habit may have evolved in every class of vertebrates, though little is known of digging by ostracoderms and placoderms. Fossorial adaptations evolved independently in many orders of fishes and mammals, and more than once in several orders. Legless progression on land is often related to burrowing and is described at the end of the chapter.

ADVANTAGES OF DIGGING

Fossorial vertebrates have various advantages:

1. Digging establishes microhabitats that are suitable for resting, estivating, or hibernating. The burrow is cooler and more humid than the desert air (most desert rodents are active on the surface of the ground only at night, and the remainder must retreat periodically to cool off), warmer than the winter storm (mountain chipmunks could not sleep above the snow without freezing), and relatively safe from lightning fires (many small forest tetrapods survive the flames).

2. Many diggers, and most that are small, secure from the ground foods such as insects, insect larvae, earthworms, roots, and tubers. Several predators dig to secure smaller diggers as food.

Above: Freeze-dried dissection of a badger arm.

3. Diggers can store food underground, where it is safe from other animals and the weather, and where it will be available during another season. Pikas and many kinds of rodents make large stores of dry grass or seeds; the arctic fox buries caches of birds and eggs.

4. Nearly all fossorial vertebrates escape from predators by retreating underground. Many do not venture far from their holes and scamper back on the slightest sign of danger.

5. Digging provides protected nests and dens in which to lay eggs or rear young. Many vertebrates, from the 10 g shrew mole to the 60 kg aardvark, raise their families underground.

FOSSORIAL VERTEBRATES

Agnathans and Fishes The flattened bodies and dorsal eyes of cephalaspids and antiarchs indicate that they were bottom feeders. Perhaps some dug to find food or get out of sight. Skates and rays are dorsoventrally flattened and may cover themselves up lightly, leaving eyes and spiracles free. Flounders are bilaterally flattened and lie on one side. The eye of the down side migrates to the up side during ontogeny. Light digging may supplement protective coloration in making these fishes hard to see. Among the more fossorial bony fishes are some gobies and catfishes, the jaw fish, and various eels. Cichlid and centrarchid fishes dig breeding holes or depressions in the substrate. Synbranchiform eels dig deep and extensive burrows. Dipnoans dig vertical burrows into the mud to wait out times of drought.

Amphibians The most constant and modified amphibian burrowers are the caecilians (see figure on p. 53). Their adaptations include loss of limbs, slender snakelike form, a blunt or flattened and wedge-shaped head, a relatively firm skull, and strong muscles for lifting and turning the head as it is forced against the substrate to create a burrow. Some short-legged urodeles wriggle through litter and loose soil. Others, including the ambystomids, or mole-salamanders, have stout bodies and legs with which they dig. Anurans commonly dig into mud, and various species dig holes in soil using their hind legs or (more rarely) their forelegs and heads.

Reptiles and Birds Many snakes are fossorial. Amphisbaenians and legless lizards (Figure 25.1) have adopted similar habits and convergent structure. Many skinks and other lizards also

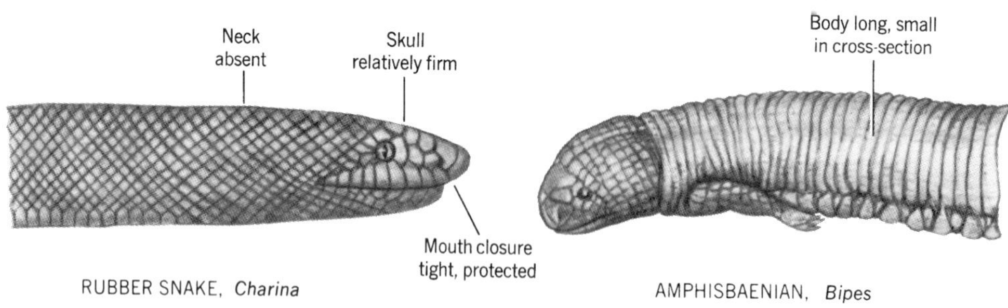

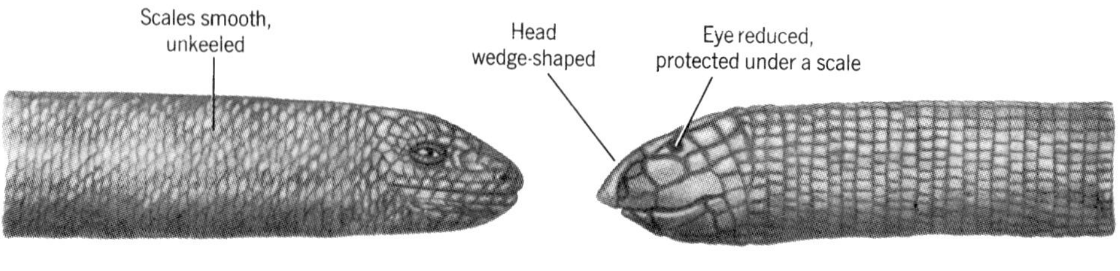

FIGURE 25.1 REPRESENTATIVE REPTILES THAT CRAWL THROUGH THE SOIL showing some adaptations for this mode of digging.

burrow or cover themselves up, and the tuatara retreats underground. All turtles bury their eggs using the hind feet as scoops. Tortoises have modified the forelimbs as effective digging tools.

No bird has evident structural adaptations for digging, but shearwaters, puffins, some penguins, an owl, and several other birds nest in holes or burrows that they either appropriate from mammals or laboriously construct themselves with beak and feet.

Monotremes and Marsupials The platypus and the echidnas are powerful diggers; the platypus exposes its claws by folding away the webs used when swimming. The marsupial mole is among the most highly adapted of subterranean mammals (Figure 25.2). One bandicoot digs, and the wombat, which weighs about a third as much as a person, may excavate a burrow 30 m long.

Insectivores About a dozen genera of true moles (family Talpidae) are extremely effective diggers. Moles can progress just under the surface in the damp soil they prefer at 2 body lengths/min and 200 body lengths/day. The unrelated golden moles (family Chrysochloridae) are among the most specialized of subterranean diggers that scratch the earth with strong claws. Hedgehogs, mole shrews, and tenrecs also make burrows.

Edentates, Pangolins, the Aardvark, and Carnivores Armadillos (nine genera), pangolins, and the aardvark are the most powerful of scratch-diggers (defined below). Anteaters do not burrow but rip into termite nests and the soil to secure insects. The six genera of badgers and the ratel are fossorial; some of them dig out burrowing rodents for food. Various canids excavate dens or dig to cache food, yet have scant structural adaptation for digging.

Rabbits and Rodents The plains pika and the less cursorial of the rabbits are moderately good diggers. The order Rodentia has more burrowing representatives than any other; only some of them are mentioned here. The mountain beaver digs in stream banks. Ground squirrels, marmots, and prairie dogs (all of the family Sciuridae) are avid diggers. Many of them make extensive burrow systems in hard soil. Gophers (eight genera in the family Geomyidae) are subterranean rodents of North America that make large burrow systems. Kangaroo rats, jerboas, and springhares are all ricochetal, yet manage to dig daytime retreats in sandy soil. African blesmols (family Bathyergidae) are expert at burrowing with their teeth. The Mediterranean mole rat (Spalacidae), bamboo and root rats (Rhizomyidae), and all but one of the mole voles (Muridae) also dig with their teeth. The Asian mole rat, or zokor (Cricetidae), has enormous foreclaws and is a scratch-digger. The related shrew mouse, burrowing mouse, and mole mouse, and the unrelated tuco-tuco (Ctenomyidae), all of South America, are also scratchers. Their neighbor, the coruro (Octodontidae), probably uses its teeth.

To better appreciate the prowess of fossorial animals, imagine a sports event that would determine which human contestant could first dislodge 30 times his or her body weight of firm soil using only fingernails and then, using hands and feet, transport all the dirt 10 m distant and pile it onto a platform that is as high as can be reached. The little tuco-tuco does this, not in competition but in daily activity. And who knows, in the animal olympics the tuco-tuco might not make it past the gophers, mole rats, and blesmols to the finals.

GENERAL REQUIREMENTS OF DIGGERS

Fossorial vertebrates must be effective in meeting certain requirements: (1) All diggers must exclude sand, dust, and earth from the mouth, eyes, ears, respiratory passages, and cloaca or anus. (2) Most diggers must maneuver within the soil or inside confined spaces. Many must find their way, detect and avoid predators, and for some, locate mates and protect eggs in complete darkness. (3) All that do not dig in sand, litter, or soft mud need tools

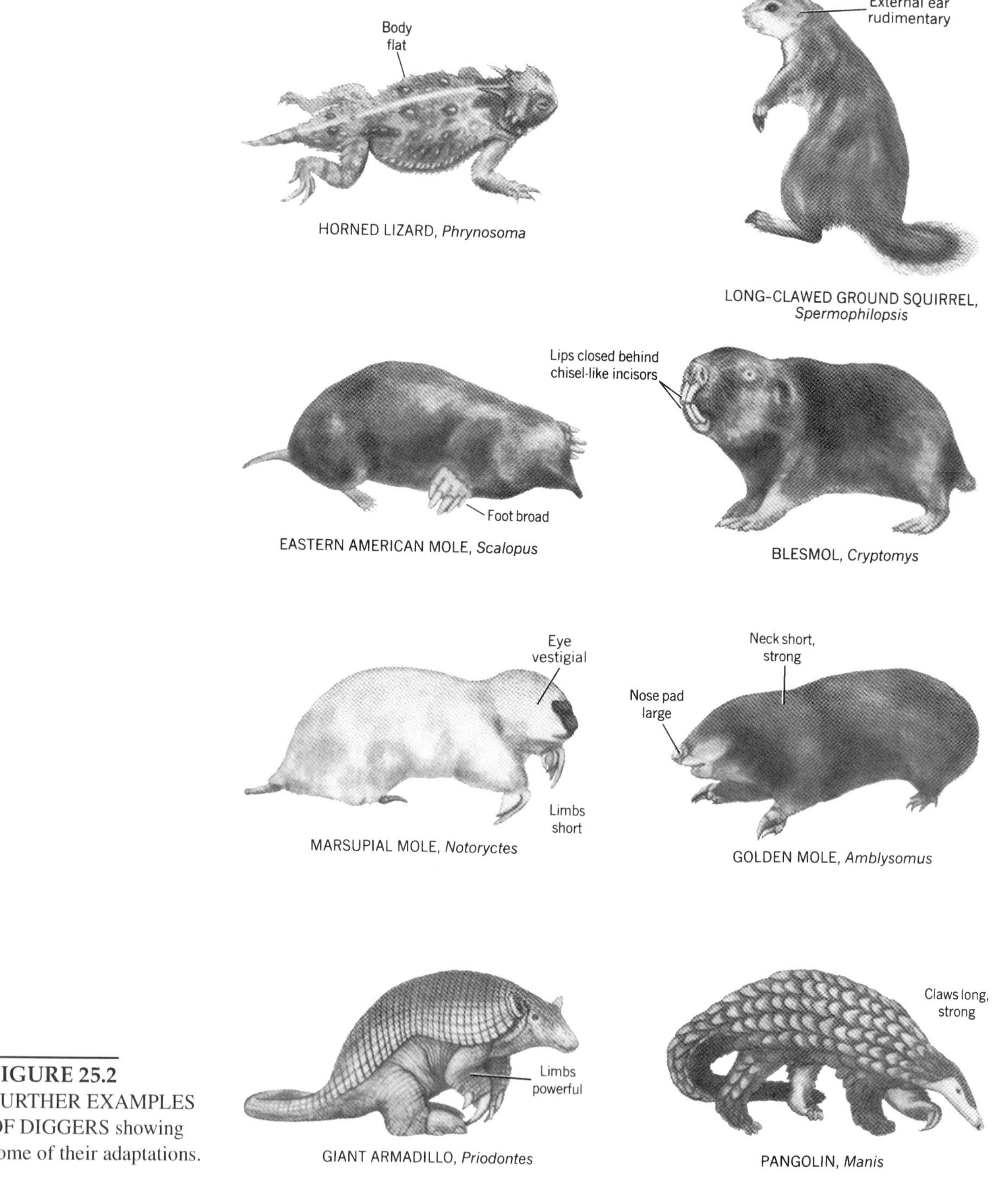

FIGURE 25.2 FURTHER EXAMPLES OF DIGGERS showing some of their adaptations.

to break and loosen the soil. (4) Most must exert great force against the earth according to manner of digging. (5) Most must strengthen their joints against hyperextension. (6) All that make open burrows must either compact soil or transport and dispose of it. (7) The energy cost of digging being very high, the better diggers must conserve energy and dissipate heat. Those exerting themselves in closed burrow systems must manage with scant oxygen. (These last subjects are postponed to Chapter 29.)

SEVERAL WAYS OF DIGGING

Vertebrates dig in various dissimilar ways. A first method may be called **cover-up digging.** The animal merely covers itself with sand or soft mud. It does not make an open hole or progress through the substrate, and rarely digs deep. The animal may cover itself to escape from the environment at the surface, but usually does so either to lie in wait for prey (with protruding eyes uncovered) or to escape or hide from predators. The digger may shuffle into sand or mud (some fishes and anurans), vibrate its body for the few seconds it takes to submerge in sand (some desert lizards), run or swim rapidly and then dive into sand (several lizards, various fishes), undulate its body to penetrate a soft substrate (amphioxus, ammoceotes, some bony fishes), or sway its body from side to side, thus creating a trench into which it sinks (several snakes). Most of these animals have flat bodies and somewhat modified sense organs. Nevertheless, their adaptations are largely behavioral, and cover-up diggers are usually not said to be fossorial.

Soil-crawling is the term we shall give to a second general technique that is used by synbranchiform eels, caecilians, some salamanders, amphisbaenians, legless and short-limbed lizards, and burrowing snakes (Figure 25.1). The animal moves through soil that is usually soft or sandy but may be quite firm. Often the substrate closes behind the animal so that a permanent burrow is not established, but the soil may instead be compacted, thus opening a burrow. A high degree of specialization is required. The body is long and slender, which reduces the amount of soil that must be displaced. Legs are reduced or absent. The head usually serves as the digging instrument. It is commonly wedge-shaped, and the skull is compacted and heavily ossified.

Another method of digging is **scratch-digging.** By alternately flexing and extending its limbs in the manner of a dog with a bone to bury, the animal cuts and loosens the soil with its claws and pushes or flings it to the rear. Some turtles, some birds, armadillos, pangolins, carnivores, ground squirrels, and a variety of other mammals dig in this way. Nests, dens, and burrows are thereby excavated, often in hard soil.

A fourth method, **chisel-tooth digging,** is followed by gophers, mole rats, and various other rodents. Huge gnawing incisors and powerful jaw and neck muscles are used to dislodge the soil, which is then moved with head or feet. Hard soil can be excavated, but somewhat damp or otherwise tractable soil is usually preferred.

A fifth method is **humeral-rotation digging.** The method is best exemplified by the true moles. A mole placed on the surface can dig out of sight in the damp soil it prefers in about 6 s. These subterranean digging machines have broad shovel-like forefeet and short powerful forelimbs. There is no pronation and supination of the forearm, and (in sharp contrast to scratch-diggers) movement at the elbow merely positions the hand without providing a forceful stroke. The elbow is positioned high above the shoulder, and the power for digging comes from the rotation of the uniquely short but broad humerus around its own long axis (Figure 25.3). This is accomplished in moles primarily by the relatively

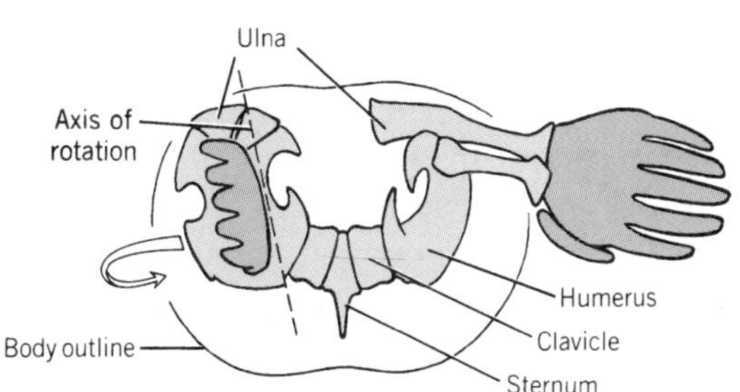

FIGURE 25.3 THE MECHANISM OF HUMERAL-ROTATION DIGGING shown by the mole, *Scapanus,* in anterior view.

enormous teres major muscle. The echidna holds its wide humerus horizontal to the ground and rotates the bone around its long axis when it walks. It is probable that these strong animals are also humeral-rotation diggers.

Golden moles and several rodents use **head-lift digging** to make shallow tunnels or to compact loose soil. Some of these creatures can lift 15 to 20 times their body weight with their heads. Neck muscles and extensors of the arm are powerful. **Hook-and-pull digging** is the method used by anteaters. They hook a huge claw into a crevice in a termite hill or ant nest and pull it apart. Flexors and supinators of the arm are enlarged.

Some vertebrates dig by more than one method. Many rodents are both scratchers and tooth-chiselers; several reptiles either scratch with short fore-limbs or fold them away and use soil-crawling. Also, various diggers do not fit into this classification (e.g., the dipnoan that digs into the mud using body and fins, the jaw fish that moves and arranges rocks with its powerful jaws, and the young crocodile that bites chunks of clay from a muddy bank). Some vertebrates make themselves uninvited guests in burrows (vacant or occupied) dug by their more fossorial cousins, doing only house cleaning and slight alterations on their own.

KEEPING DIRT OUT OF MOUTH, SENSE ORGANS, AND LUNGS

Large diggers (wombat, aardvark, badger) probably can keep their mouths out of the dirt, at least most of the time. Fossorial amphibians, reptiles, and insectivores have close registration of the closed jaws; the margin of one jaw often fits into a groove in the other to make a tight seal. The lower jaw of some burrowing reptiles is recessed behind the upper jaw. The furred lips of chisel-tooth diggers meet behind the protruding incisors, thus enabling the animal to exclude dirt from the mouth at the same time it is gnawing.

Diurnal vertebrates that burrow to nest, rear young, or hibernate, yet do their foraging above ground, have eyes of normal size (tortoises, hedgehogs, ground squirrels, canids). Nocturnal rodents that burrow to escape daytime heat have large eyes (kangaroo rats, jerboas). Presumably these animals close their eyes as needed when digging. Possibly some have evolved improved mechanisms for cleansing the eyes. Snakes and certain lizards have fused but transparent eyelids. Several genera of burrowing snakes have "horns" that protrude over the eyes and are thought to provide protection. The real specialists have small to minute eyes (monotremes, armadillos, pangolins, gophers, African blesmols, root rats, tuco-tuco) or vestigial eyes that may differentiate light from dark but form no image and often are hidden under the skin (caecilians, amphisbaenians, marsupial mole, true moles, golden moles, Mediterranean mole rat).

Many burrowing amphibians and reptiles have no external auditory canal. If there is a tympanum, it is thick. Sound reception is often mediated by a special mechanism involving the skin, jaw, or other structure and that is adapted for sensing low-frequency ground vibrations. The external auditory canal of burrowing mammals tends to be small. It is probable that some diggers can close the canal.

The external nares of diggers are also small. In digging reptiles the outer part of the nasal passage is narrow and slopes upward from the nares. Moisture around the nostrils may cause adhesion of sand and prevent the inhalation of single grains. The openings can be closed or at least constricted by muscular valves or erectile tissue and may be covered by a fold. The armadillo is able to suspend breathing for 3 or 4 min while digging vigorously in dust and sand. Some shrews have peculiar diverticula from the lungs that seem moderately effective at trapping and disposing of foreign material.

MANEUVERING UNDERGROUND

Maneuvering in the confines of a narrow burrow is facilitated in several ways. Some fossorial amphibians and reptiles and all fossorial mammals are short-legged. The spine is relatively stiff in burrowing snakes, but flexible in fossorial mammals. Some can turn very sharply by doubling into a ball and then unrolling in the new direction. Several fossorial vertebrates have such loose skin that the animal can to a degree turn within its skin and then let the skin follow. All burrowers are adept at backing up.

Some diggers have special reasons for having long tails: Ricochetal species use theirs for balance, the pangolin uses its as a grasping organ when climbing, and numerous species use theirs as a prop when digging. Otherwise, the evolutionary process seems often to have found a tail to be in the way underground, and it tends to be short, even in burrowing snakes and some other legless diggers (though some amphibians and lizards are exceptions). Some small subterranean vertebrates have little or no tail (caecilians, golden mole, Mediterranean mole rat, and various others).

How subterranean animals find their way in the absolute darkness of a deep burrow system is inadequately known. An inherent sense of direction and the memory of an accurate map of the home range have been demonstrated for various diggers. Touch is highly developed in some burrowers. Sensory whiskers are usual on the head and may occur at the edges of the feet, on the tail, and elsewhere. Many burrowers are particularly sensitive to the low-frequency sound waves that can reach them through the ground. The mole rat *Spalax* communicates underground by bursts of drumming with its head against the earth.

TOOLS FOR DIGGING

When a person shovels dry sand or forest litter, energy must be expended to transport the material but virtually none to break or loosen it. The same is true of diggers that confine themselves to loose sand or soil. Accordingly, they may have no tools for breaking compact soil.

When people shovel damp earth they must break it free before they move it, but this is not difficult to do. A large shovel blade is satisfactory. Similarly, moles confine their activities to moderately soft soil. They compact, break, and cut the soil by pushing on it or scraping it with their stout claws. Again, a large "shovel blade" is effective: Moles have broad claws and very wide forefeet (Figures 25.3 and 25.9). Similarly, amphisbaenians dig by ramming the wedge-shaped head into the soil and then (according to species) lifting the head to compact the soil into the wall of the burrow.

The person who digs in dry compacted soil must expend much energy in breaking the soil prior to moving it. The shovel cannot be forced into the undisturbed material, so a pick is first used to loosen it. The pick is effective because it delivers great force to a restricted area; that is, with each blow it applies high pressure to a limited area. Chisel-tooth and scratch-diggers among vertebrates tend to avoid rocky and otherwise intractable soils, yet many of them burrow in remarkably hard earth. They gnaw with their incisors or scratch with their sharp claws, thus applying great pressure to a small area before going on to another spot. The blade must be long and strong to do its job. The badger has five claws, the longest of which is half the length of its forearm between the elbow and wrist joints. The mountain beaver, tuco-tuco, gophers, and ground squirrels emphasize three or four claws, the longest of which may (in some gophers) be three-quarters of the length of the forearm. The anteaters, marsupial mole, and golden moles emphasize one or two claws, the longest of which may (in some golden moles) be longer than the forearm (Figure 25.7). The "shovels" of moles can exert against soft soil more force in relation to body weight than can the "blades" of ground squirrels, but the ground squirrel exerts against harder soils about twice as much pressure as does the mole.

The tools of diggers are subject to tremendous wear. Burrowing reptiles compensate by molting successive generations of the epidermis, thus exposing new surfaces to the substrate. The upper incisors of some gophers grow out at the rate of 248 mm/yr. The lower incisors, which are maneuvered more as the condyle of the mandible slips in its loose groove, may grow at the rate of 445 mm/yr. These rates are two and one-half to three times those recorded for some nonfossorial rodents of comparable size. Likewise, the center foreclaw of a gopher may grow out at 90 mm/yr, and that of a tuco-tuco at 72 mm/yr.

The incisors of rodents and of the wombat have enamel only on their forward surfaces. The softer dentine wears away behind, thus providing a self-sharpening mechanism.

The spadefoot toad digs with a different kind of tool. There is a horny epidermal tubercle at the edge of the hind foot that is used as the animal progresses backward into loose soil (Figure 25.4). The marsupial mole, golden moles, and several rodents have tough nose pads with which they dislodge and move soil, and several burrowers use the top of a broad head (Figure 25.10).

DESIGN FOR LARGE OUT-FORCES

Fossorial vertebrates that dig in firm soil must be capable of applying great force against the substrate. Therefore, unlike cursors and climbers, they are constructed so that their relevant bone-muscle systems (particularly of the forelimb) produce large out-forces ($\mathbf{F}_o$). In Chapter 22 it was shown that $\mathbf{F}_o = \mathbf{F}_i\mathbf{l}_i / \mathbf{l}_o$, where $\mathbf{F}_i$ is the in-force and $\mathbf{l}_i$ and $\mathbf{l}_o$ are, respectively, the in- and out-levers.

It is evident that one way to increase $\mathbf{F}_o$ is to reduce $\mathbf{l}_o$. Consequently, the more expert diggers all have short legs and necks. In sharp contrast to the limbs of cursors, the limbs of diggers have relatively short distal segments. The radius is nearly always shorter than the humerus, and the manus, exclusive of the terminal phalanx with its claw, is markedly shorter than the radius. Although strong, the metacarpals may be very short (tortoises, echidna, moles) (Figure 25.5) and the proximal phalanges may be even broader than long (echidna, pangolin, anteaters, moles).

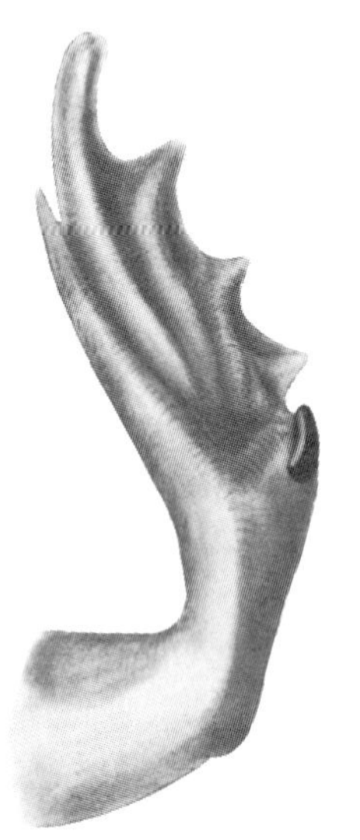

FIGURE 25.4 FOOT OF THE SPADEFOOT TOAD, *Scaphiopus*, showing tubercle for digging.

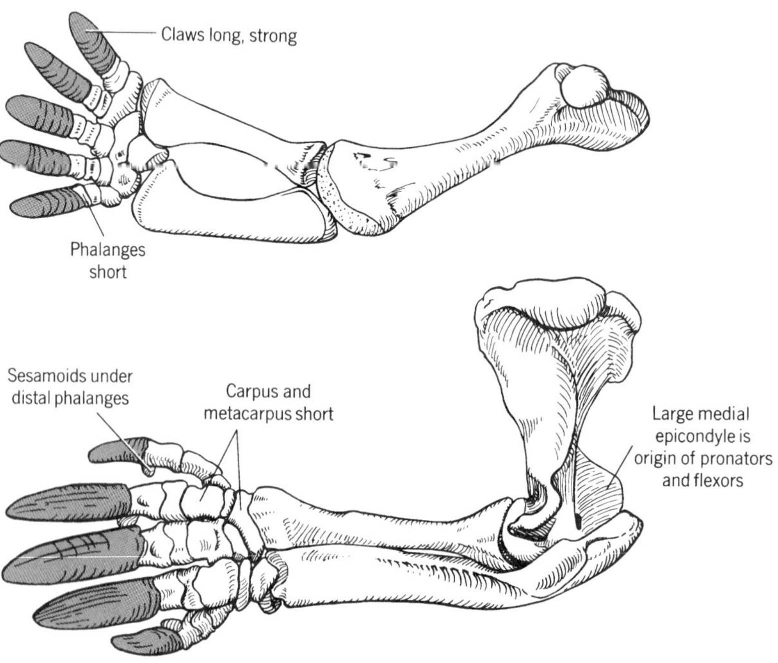

FIGURE 25.5 SOME ADAPTATIONS FOR DIGGING shown by dorsolateral views of the left forelimb skeletons of the tortoise, *Gopherus* (above), and echidna, *Tachyglossus* (below).

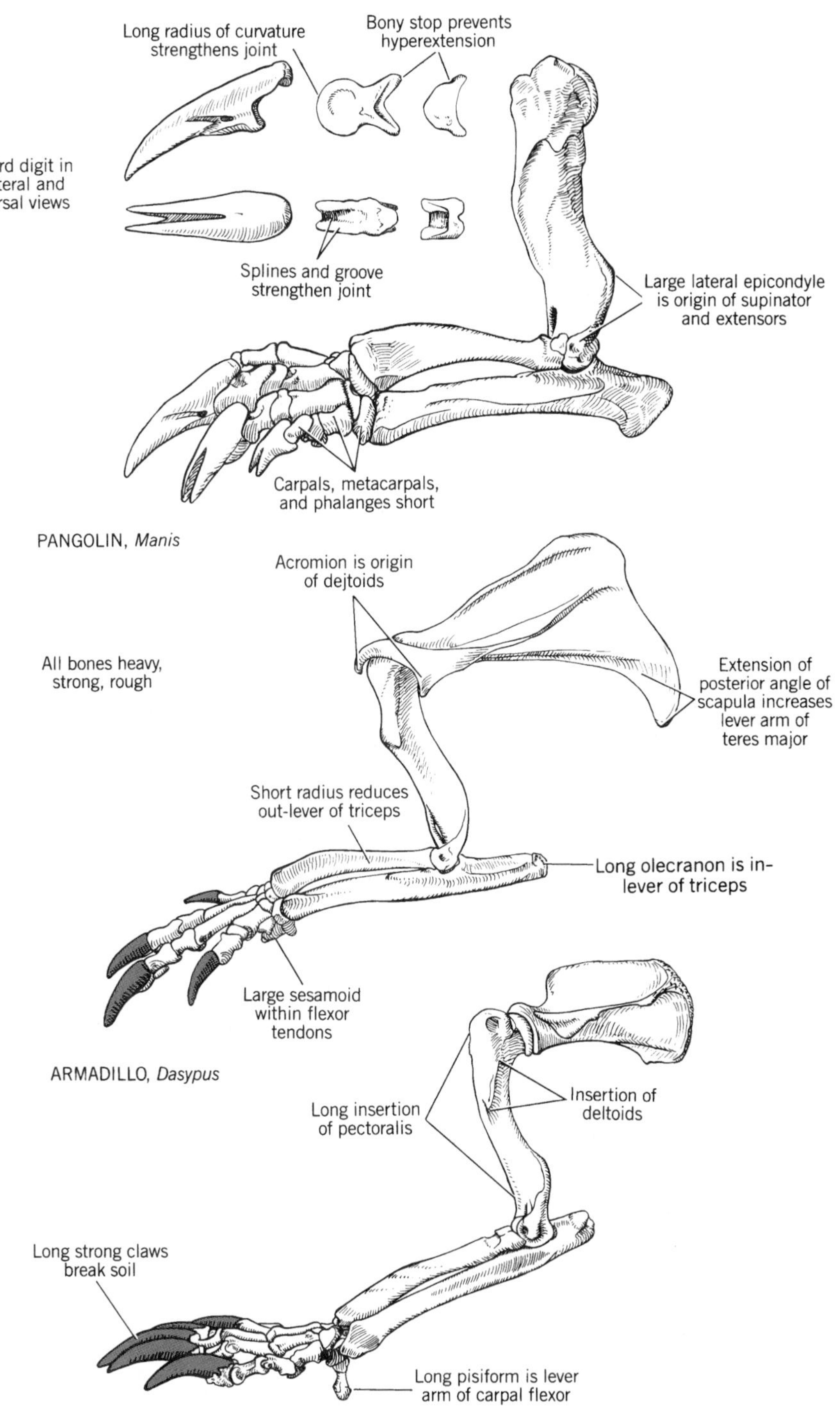

FIGURE 25.6 SOME ADAPTATIONS FOR SCRATCH-DIGGING shown by lateral views of left forelimb skeletons.

A second way to increase an out-force is to increase the related in-lever. Accordingly, muscles used in digging tend to insert relatively far from the joints they turn. The insertion of the deltoid muscles of mammalian diggers commonly extends more than halfway down the length of the humerus (away from the shoulder joint, which pivots) (Figure 25.6, lower drawing). Part of the latissimus dorsi of the golden mole increases its in-lever to the shoulder joint by shifting its insertion from the proximal part of the humerus (the usual position) nearly to the elbow joint (Figure 25.7). The wide medial

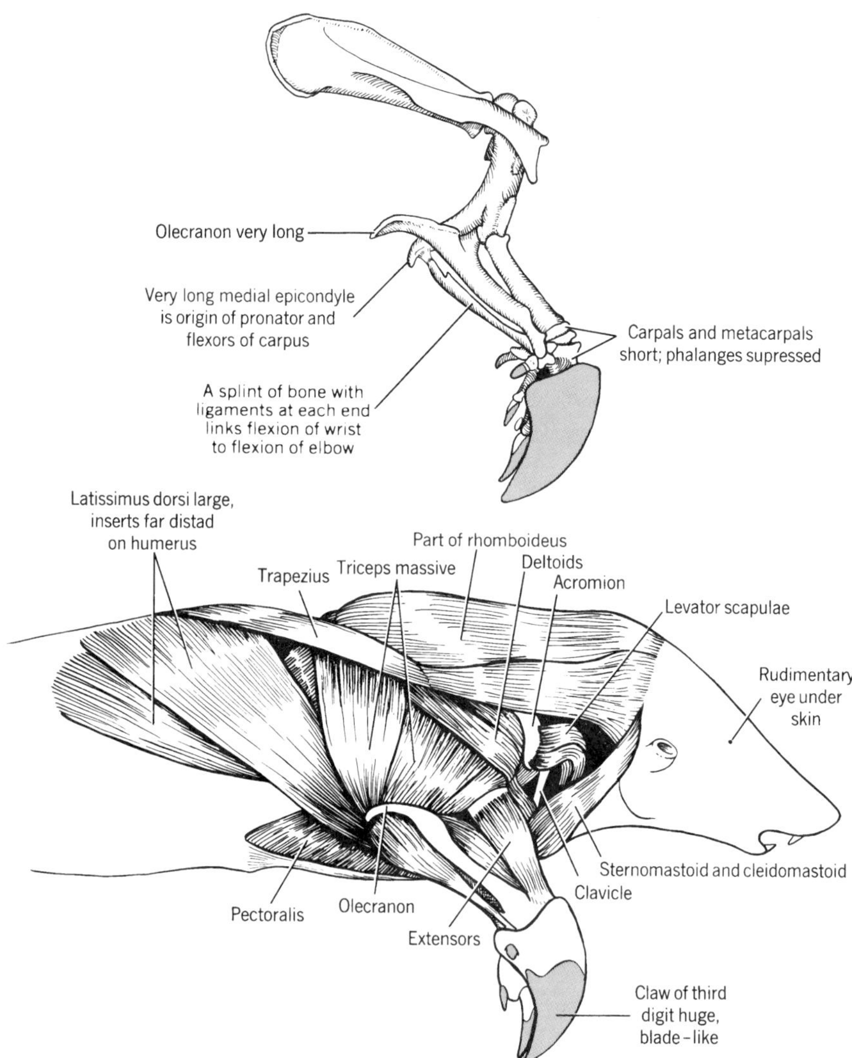

FIGURE 25.7
STRUCTURE ASSOCIATED WITH THE SCRATCH-DIGGING OF THE GOLDEN MOLE, *Amblysomus.*

epicondyle of the humerus, which is a feature of all scratch-diggers, increases the in-lever of the pronator of the forearm. A relatively proximal origin on the humerus of the long supinator muscle increases its in-lever and enables it to flex the manus as well as supinate the forearm (Figure 25.8). A relatively long pisiform bone at the carpus increases the in-lever of one of the flexors of the manus. Crucial to the special mechanism of the humeral rotation digging of moles is a very wide flaring tubercle for the insertion on the humerus of the enormous teres major muscle. This carries the insertion away from the central long axis of the bone and thus increases the in-lever, enabling powerful rotation of the humerus around its own long axis (Figure 25.9).

These kinds of adaptations of diggers are particularly striking when in-levers are expressed as fractions of their related out-levers. Thus, in measuring representative skeletons of 27 genera of expert diggers belonging to seven mammalian orders, the olecranon process (the in-lever of the triceps) is found to be about one-fifth (a ground squirrel), one-third (gopher, African mole rat), one-half (aardvark, pangolin, mole), two-thirds (Mediterranean mole rat, armadillos), or even three-fourths (marsupial mole, golden moles) the length of the ulna distal to the pivot at the elbow joint (Figure 25.7). A complete listing (if the data were available) would doubtless show that diggers differ from

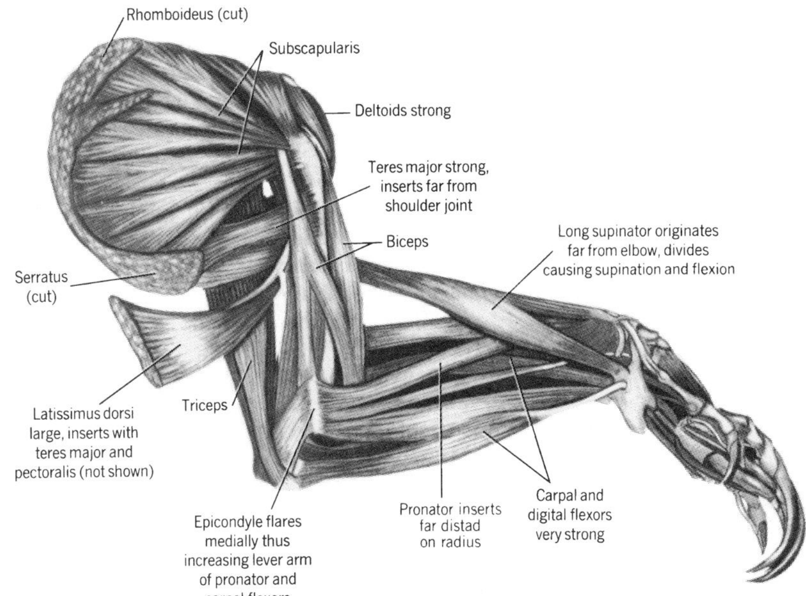

FIGURE 25.8
STRUCTURE ASSOCIATED WITH THE HOOK-AND-PULL DIGGING OF THE GIANT ANTEATER, *Myrmecophaga,* shown by a medial view of the left forelimb. (Drawn from an air-dried dissection and hence somewhat shrunken.)

their nonfossorial relatives in having larger values of $\mathbf{l}_i/\mathbf{l}_o$ for every bone-muscle system used in digging.

A third way to increase out-force is to increase in-force. The relevant muscles of diggers are enormous. To accommodate such muscles, origins and insertions are large. This, together with their proportions, makes the forelimb bones of diggers rugged and rough. The medial epicondyle of the humerus (origin of flexors of digits) and deltoid crest (insertion of deltoids) are particularly prominent. The posterior angle of the scapula may be enlarged to accommodate the origins of the teres major and long head of the triceps. The anterior segment of the sternum of true moles and golden moles is long and deep to receive their great pectoral muscles (Figure 25.9). Chisel-tooth diggers have large areas of origin and insertion for their powerful jaw muscles. Diggers that push dirt with their heads have a large flat occipital area for the insertion of strong neck muscles (Figure 25.10).

There is another factor in design for large out-forces. When starting a burrow, some soil-crawlers loop the body over the head, thus weighting it so it can be thrust into the soil. Large scratch-diggers hunch the back over the forelimbs and may prop the body with the tail, thereby applying the weight of the body to the digging tools. Subterranean diggers, however, are usually small and hence light. (They can then dissipate heat more easily, feed on the small food found in the earth, keep within one stratum of the soil, and avoid rocks and large roots.) In order to prevent motions of digging from merely pushing them away from the soil, they must force their bodies against their digging tools. Humeral-rotation diggers brace against one side of the burrow with one forepaw while digging with the other. To make this possible, the forelimbs are positioned laterally, opposite to one another. Human weightlifters lift overhead about twice their own body weight. It has been shown that by its thrust a mole can move as much as 32 times its own body weight.

If an amphisbaenian presses up with its head, it presses down with its "chest." As chisel-tooth diggers press up with their lower incisors, they press down with their forefeet. Small burrowers also brace themselves with their hind legs. Xenopid frogs can at the same time lengthen the back at the sacroiliac joint, thus forcing the head forward in

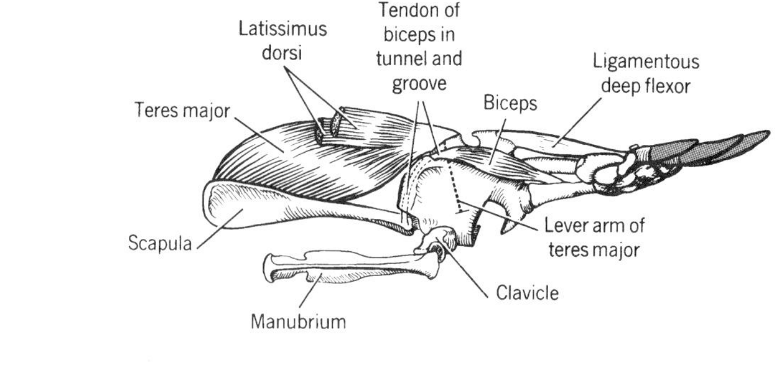

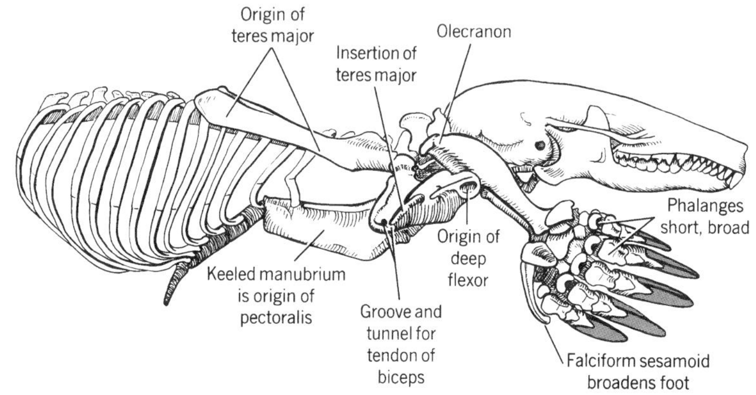

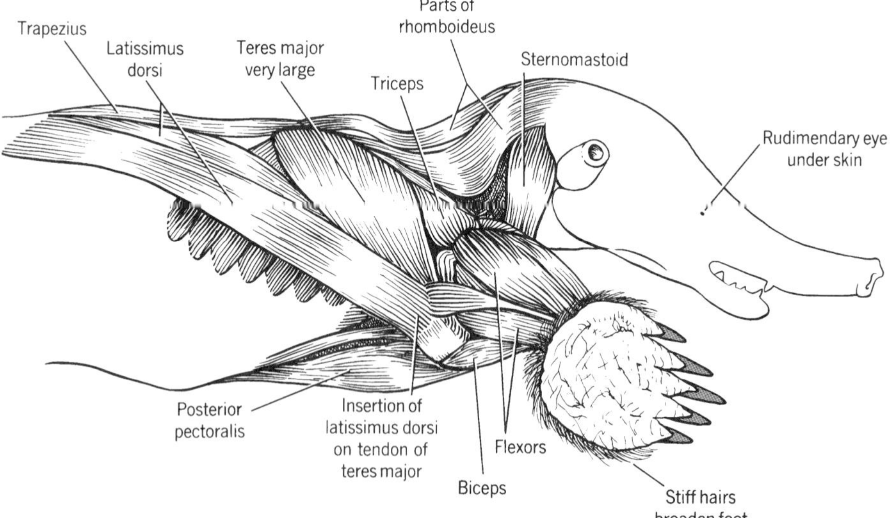

FIGURE 25.9
STRUCTURE ASSOCIATED WITH THE HUMERAL-ROTATION DIGGING OF MOLES. Above, ventral view; center and below, lateral views. Above and center, the Western American mole, *Scapanus;* below, the Old World mole, *Talpa.*

the mud. In response to the use of the hind legs for bracing, the innominate bones of the mammals tend to be nearly horizontal (in line with a forward thrust). The hip joint is relatively far dorsal to be on a level with the spine. This reduces compressive forces at the pelvic symphysis, which is nearly always weak and sometimes absent (moles, anteaters, pangolins, and some gophers) (Figure 25.11). The innominate bones are firmly sutured with, or fused to, a relatively large number of vertebrae, and the sacrum is long.

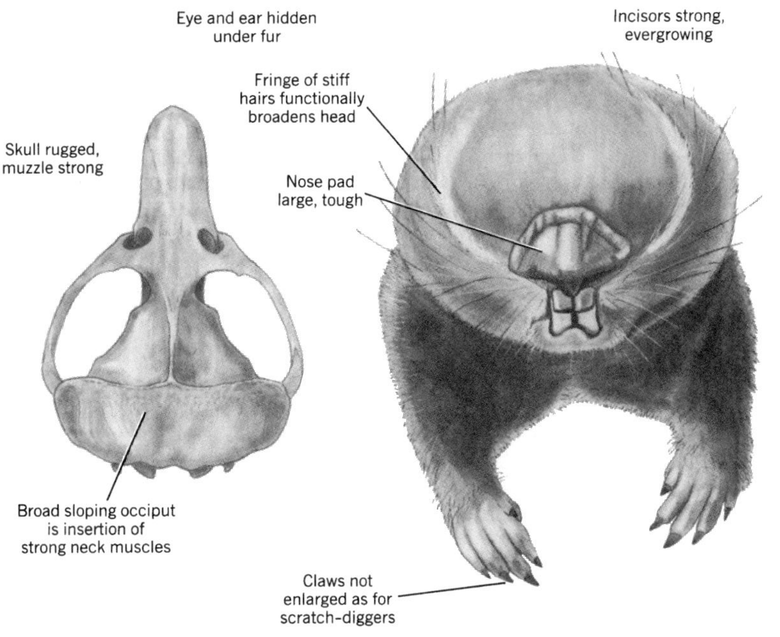

FIGURE 25.10
SOME ADAPTATIONS OF A CHISEL-TOOTHED DIGGER THAT MOVES DIRT WITH ITS HEAD, the Mediterranean mole rat, *Spalax*.

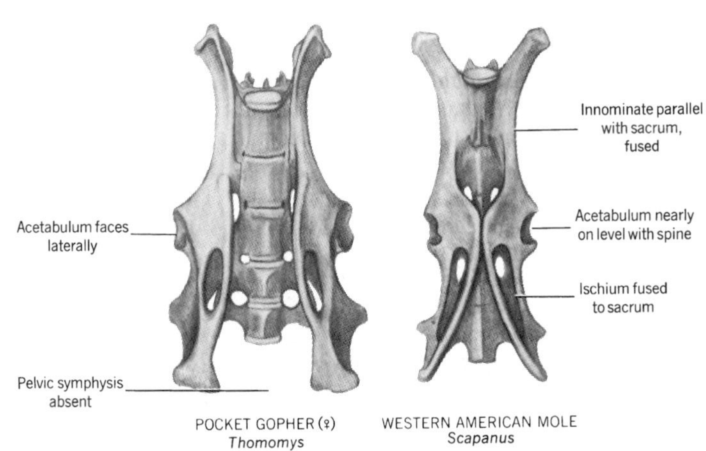

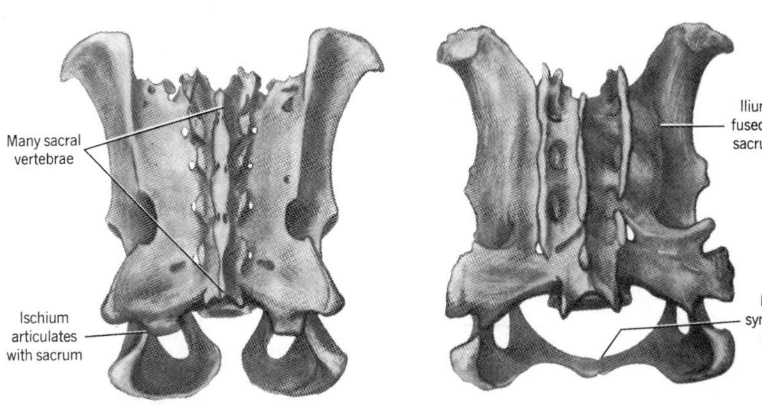

FIGURE 25.11 SOME CHARACTERISTICS OF THE PELVISES OF FOSSORIAL MAMMALS. Ventral views above; dorsal views below.

RESISTANCE TO THE FORCE OF THE SOIL ON THE BODY

Soil-crawlers reduce or resist the force of the soil on the body in various ways. The skins of burrowing fishes are commonly particularly rich in mucous glands. Scales are absent (various fishes, amphibians) or, if present, smooth and unkeeled (reptiles), to reduce the friction of soil against the body. Friction is further reduced in uropeltid snakes by microscopic ridges that reduce wetting in damp soil. The head is short and narrow. The skull is relatively solid; most sutures are obliterated and others are serrate. There is no neck or shoulders. The reduction and loss of legs is itself a major accommodation to streamlining. In order to displace as little soil as possible, the trunk becomes long and slender—in other words, snakelike. Unspecialized lizards have about 23 presacral vertebrae, whereas fossorial lizards may have 60 and amphisbaenians more than 100. Caecilians and snakes may have 250 or more such vertebrae. Cervical and lumbar ribs are usual.

Vertebrates that move along narrow burrows must avoid snagging and abrading themselves against burrow walls. Subterranean diggers have small external ears or none at all. Fur is lax, often short, and sometimes nearly upright so it can brush in any direction. At least some of the mammals clean their fur by shaking the body, dog fashion. Others groom away mud and dirt with the paws.

There must be provision against dislocation and hyperextension of joints of the forelimb and manus when a digit snags on a rock or root. Hyperextension of the phalanges of echidnas, pangolins, and anteaters (at least) is prevented by squared articulatory surfaces or bony stops that limit rotation of the joints (Figure 25.6). Dislocation of the digits is prevented in pangolins and anteaters by large areas of bone-to-bone contact at the joints and by deep interlocking splines and grooves, recalling those of the phalanges of ungulates. The scapula of gopher tortoises is braced against the plastron (ventral part of shell) more firmly than is that of turtles. Protection is also afforded by generally heavy and rugged construction, and in some instances by structural unity (i.e., common firmness) of the palm (tortoises, echidna, moles). Some ligamentous checks against dislocation have also been described.

TRANSPORTING AND DISPOSING OF SOIL

Diggers that excavate dens or burrows must transport and dispose of earth after they loosen it. Many snakes use coils of the body to sweep sand and other debris out of burrows. Lizards and turtles sweep with their feet. Some reptiles create an air space under the body so they can breathe in spite of an overburden. As mammals break the soil with forefeet or teeth, it is pushed underneath or beside the body. From time to time the animal kicks this loose soil back out of the way with the hind feet. When the burrow behind becomes choked with soil, it is time to take it away, and several methods are used. Some diggers back up in the burrow, vigorously kicking the dirt back as they go. The hind feet are used simultaneously (tuco-tuco, African blesmols, armadillo) or alternately (hedgehogs). Other burrowers turn around to push the dirt forward using the forefeet only (moles); forefeet, chest, and chin (gophers); or nose and top of the head (gopher tortoise, Mediterranean mole rat). Leaf-nosed and hog-nosed snakes probably also use their heads as shovels, as do synbranchiform eels and amphisbaenians. Most dirt is moved to the surface, but some is used to plug abandoned side tunnels of the burrow system.

Whichever foot, fore or hind, is used for moving the soil, it is made broad in one or more ways: The toes may be webbed (toads, sea turtles and others, moles, golden moles), the pad of the foot may be widened by cartilages or bones placed lateral to the first digit (mountain beaver, moles, gophers, tuco-tuco, etc.), or the pad of the foot may be fringed with stiff hairs (nearly all subterranean mammals). The Mediterranean mole rat similarly increases the effectiveness of its broad, flat head as a dirt pusher by adding lateral fringes of stiff hairs (Figure 25.10).

CRAWLING WITHOUT APPENDAGES

Terrestrial locomotion without limbs has evolved independently several times among amphibians and lizards, but is best exemplified by snakes and amphisbaenians. Several thousand species are involved, so the adaptation must be considered highly successful. It is probable that the ancestors of these animals first evolved long, slender bodies to enable them to enter crevices or move by undulation, and then secondarily lost their limbs.

There are four principal ways in which legless vertebrates move in natural habitats, though more than one may be used at a time. (An additional "slide-pushing" method has been observed on very smooth surfaces, and variants, thrashing, and even jumping may occur.) The most common principal method is **lateral undulation.** The body is thrown into serpentine loops, right and left. The animal locates with its coils several projections such as pebbles or plant stems. The body then presses sideways (not downward) against these objects in a direction that is obliquely backward in relation to the direction the snake is to move. The mechanical analysis for the forces at any one projecting object resembles that for the action of a fish tail: The thrust of the snake against the object ($\mathbf{F}_t$ in Figure 25.12) is opposed by an equal and opposite force ($\mathbf{F}_o$) exerted by the object against the body. This force has a forward component ($\mathbf{F}_f$) in the direction that the snake as a whole is moving, and a lateral component ($\mathbf{F}_l$). (The value of $\mathbf{F}_f$ is somewhat reduced by friction—see below.) Motion is continuous as waves travel along the body. Each coil stays in the same place, the body following in the trail established by the head.

A snake cannot move forward by sliding a coil past a single object against which it pushes as it goes. This is because that particular coil does not move in the direction in which the snake as a whole is moving, but instead moves along its own long axis as shown in Figure 25.12. That direction being at right angles to $\mathbf{F}_o$, $\mathbf{F}_o$ can have no component in the coil's direction of motion. For continuous motion, the snake requires three or more objects that cannot all be on the same side of the body. The action is more efficient if few objects are used (three to five unless the snake is very long and slender). The lateral components of the various thrusts add up to zero, and the sum of the forward components (less frictional resistance) is the propulsive force of the animal. It is seen that all the forces are interconnected.

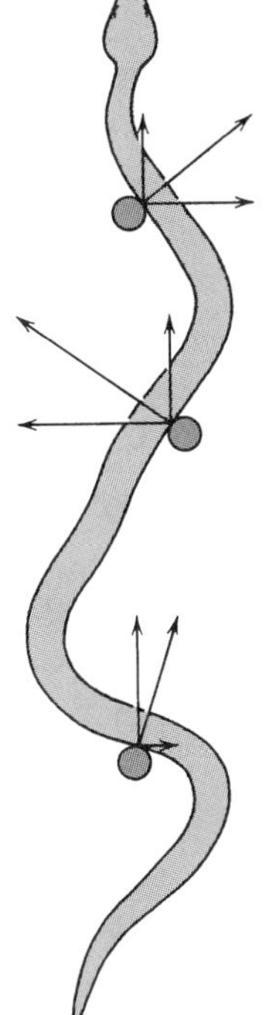

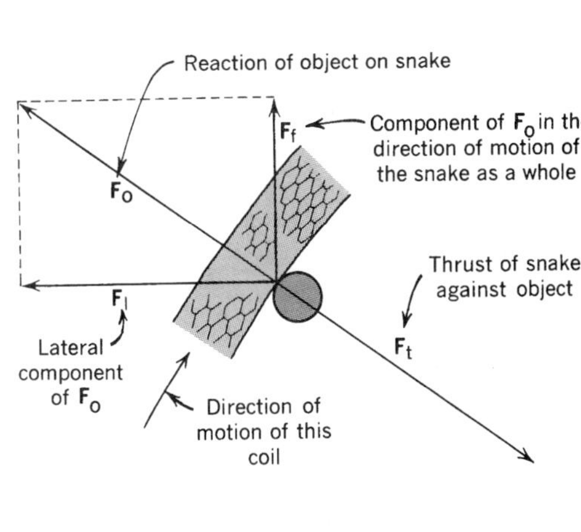

FIGURE 25.12 DIAGRAM OF LATERAL UNDULATION OF A SNAKE. (The value of $\mathbf{F}_f$ is somewhat reduced by sliding friction.)

As the "neck" contacts new objects and the tail slides away from others, and as any object slips, the animal must instantly adjust both the magnitude and direction of the thrusts of all active coils. One must be impressed by the complexity of the feedback mechanism and nervous control required.

Snakes are unable to progress along narrow burrows by lateral undulation because then they cannot thrust obliquely backward with their coils. Furthermore, they cannot progress by this method on smooth surfaces because they must thrust laterally, not vertically. There is, of course, a vertical force from the animal's weight. This causes friction, which in this instance is undesirable. The coefficient of friction (see p. 480) is minimized by the smooth nature of the snake's ventral scutes. There is also sliding friction against the projections where the snake thrusts, and this also counters the propulsive force of the animal. Snakes can move very freely among rotating pegs placed on lubricated glass.

The second method of legless progression is **rectilinear movement.** It is used by various snakes and all amphisbaenians—particularly if short-bodied. The skin fits loosely over the ventral part of the body and is very distensible. Muscles that slant back and down from the ribs to the scutes cause the ventral skin to bunch at several regions so that the scutes overlap (Figure 25.13). Between these regions the skin is stretched. Where scutes are bunched they rest on the ground; where stretched they are lifted clear of the ground. One by one, additional scutes are drawn into each bunched region from behind as others are stretched away in front. Thus, the scutes move along somewhat in the manner of the feet (or prolegs) of caterpillars—each starts and stops as it goes. The bunched scutes thrust obliquely backward against the substrate, and friction is required to prevent slipping. Muscles that slant backward and up from the scutes to the ribs haul the body along within its skin in continuous motion. The body is held in a straight line. Motion is symmetrical and directly ahead. It is slow. This kind of motion may be used for stalking prey or for moving in a narrow tunnel.

The third way of moving without limbs is **concertina movement,** which is commonly used by caecilians, snakes, and amphisbaenians. The animal draws itself into one or more S-shaped coils. The posterior coils then press downward and backward against the substrate, relying on friction to prevent slipping (Figure 25.14). The forward component of this thrust is used to advance the head and anterior part of the body, which is held clear of the ground to avoid resistance from sliding friction and to increase the static friction of the stationary posterior coils by increasing the loading on them. Before stability is threatened, the anterior part of the body touches the substrate, builds coils, and ceases motion, so that it, in turn, can draw up the posterior part of the body before the cycle is repeated. The firmness of the base provided by the stationary coils is increased if coils are forced outward to wedge the animal within a tunnel or between rocks or crevices in bark, or, if they are forced inward, to constrict a branch. Concertina movement is common, particularly among climbing and burrowing species, and often is combined with lateral undulation.

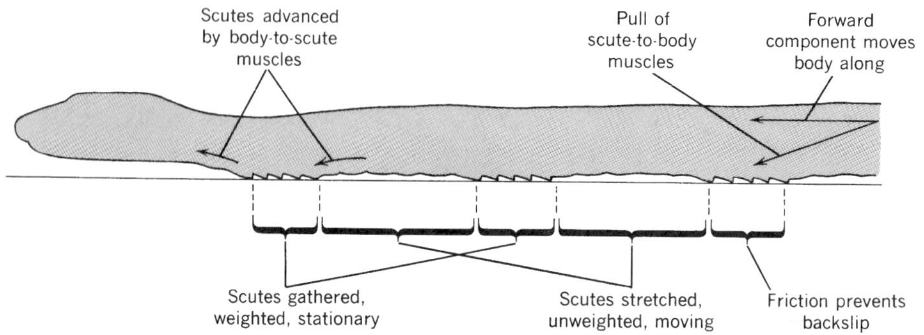

FIGURE 25.13
DIAGRAM OF RECTILINEAR MOVEMENT OF A SNAKE.

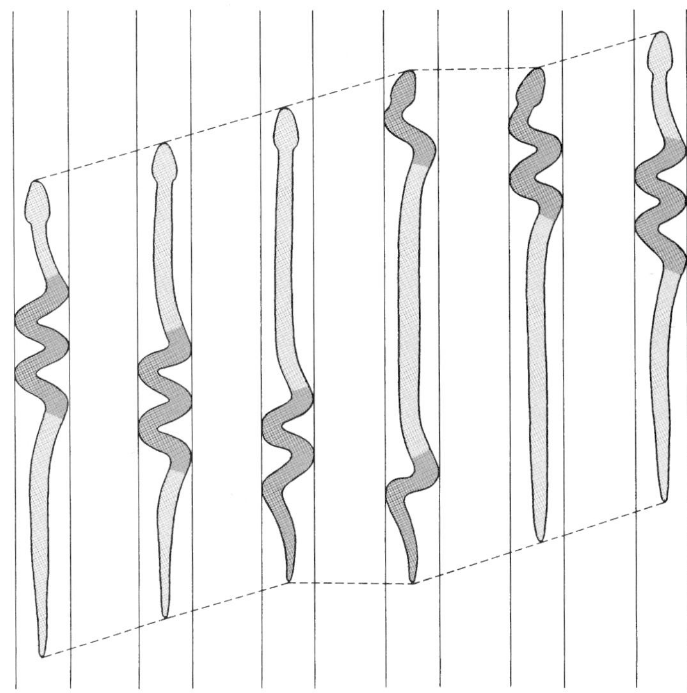

FIGURE 25.14
DIAGRAM SHOWING SUCCESSIVE POSITIONS IN THE CONCERTINA MOVEMENT OF A SNAKE PROGRESSING IN A TUNNEL.
Dark-shaded parts of the body are stationary; light-shaded parts are moving.

Gans and colleagues described a variant of concertina movement ("internal concertina") used by some snakes when burrowing. The anterior third of the spine coils or bunches within the "loose" integument, which remains in contact with the burrow wall, and the body contour widens but does not bend. The widened region braces the body within its burrow, providing a purchase for the head, which then thrusts forward into the soil. In their comparison of two caecilians, one terrestrial and one aquatic, Summers and O'Reilly reported that only the terrestrial form is able to use this modified concertina method of burrowing. Rather than coiling just the anterior third of the spine, as in snakes, caecilians throw the entire spine, from atlas to terminal vertebra, into waves of small amplitude. O'Reilly and colleagues proposed that by applying hydrostatic forces to the peculiar arrangement of the body wall musculature (oriented as a crossed-helical array of tendons that surrounds the body cavity) terrestrial caecilians can generate far higher forces than burrowing snakes. Subsequent X-ray videography coupled with strategically placed markers revealed that as the anterior coils form in the burrowing caecilian, the vertebral muscle mass moves forward up to five vertebral segments relative to the marker in the subdermal tissue. Thus, this study suggests that the presence of a loose connection between the vertebral column (axial musculature) and the skin (body wall musculature) is a morphological predictor of the ability to perform internal concertina movement. Summers and O'Reilly have proposed that internal concertina locomotion evolved in the ancestors of all extant caecilians and was secondarily lost in the aquatic forms.

The last method is **sidewinding,** a kind of progression that probably evolved from the concertina method and is adapted for fast and efficient travel over loose or sandy soil. Figure 25.15 shows that the snake makes a series of tracks that are more or less straight lines, parallel to one another, angled to the direction of travel, and each about as long as the animal. The snake contacts two or three tracks at a time. Parts of its body are within the tracks and parts arch between tracks. As successive segments of the body are laid down to extend one track, successive segments are released from the previous track. The parts of the body that are along tracks are stationary, whereas those that span between tracks are moving and

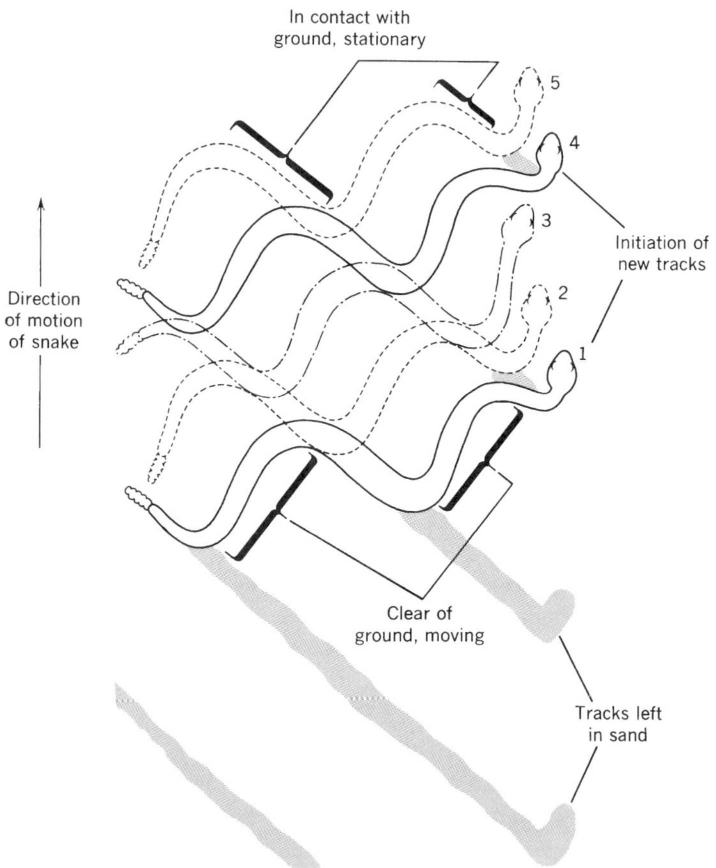

FIGURE 25.15 DIAGRAM SHOWING SUCCESSIVE POSITIONS IN THE SIDEWINDING OF A SNAKE.

are held clear of the ground (which may be hot). Periodically, the head and neck reach forward to initiate new tracks. The entire action is very rapid. Most snakes can move in this way, but desert species are the most adept.

REFERENCES

Agrawal, V.C. 1967. Skull adaptations in fossorial rodents. *Mammalia* 31:300–312.

Chapman, R.N. 1919. A study of the correlation of the pelvic structure and the habits of certain burrowing mammals. *Am. J. Anat.* 25:185–219.

Edwards, J.L. 1985. Terrestrial locomotion without appendages, pp. 159–172. *In* M. Hildebrand et al. (eds.), *Functional vertebrate morphology.* Harvard Univ. Press, Cambridge, MA.

Ellerman, J.R. 1959. The subterranean mammals of the world. *Roy. Soc. S. Africa Trans.* 35:11–20.

Emerson, S.B. 1976. Burrowing in frogs. *J. Morphol.* 149:437–458.

Gans, C. 1975. Tetrapod limblessness: evolution and functional corollaries. *Am. Zool.* 15:455–467.

Gans, C. 1984. Slide-pushing—a transitional locomotor method of elongate squamates. *Symp. Zool. Soc. Lond.* 52:13–26.

Gans, C., H.C. Dessaur, and D. Baic. 1978. Axial differences in musculature of uropeltid snakes: the freight-train approach to burrowing. *Science* 199:189–192.

Gasc, J.P., F.K. Jouffroy, and S. Renous. 1986. Morphological study of the digging system of the Namib Desert golden mole (*Eremitalpa granti namibensis*): cineflourographical and anatomical analysis. *J. Zool. Lond.* 208:9–35.

Gupta, B.B. 1966. Fusion of cervical vertebrae in rodents. *Mammalia* 30:25–29.

Hildebrand, M. 1985. Digging of quadrupeds, pp. 89–109. *In* M. Hildebrand et al. (eds.), *Functional vertebrate morphology.* Harvard Univ. Press, Cambridge, MA.

Lehmann, W.H. 1963. The forelimb architecture of some fossorial rodents. *J. Morphol.* 113:59–76.

Nevo, E. 1979. Adaptive convergence and divergence of subterranean mammals. *Ann. Rev. Ecol. Syst.* 10:269–308.

O'Reilly, J.C., D.A. Ritter, and D.R. Carrier. 1997. Hydrostatic locomotion in a limbless tetrapod. *Nature* 386:269–272.

Reed, C.A. 1951. Locomotion and appendicular anatomy in three soricid insectivores. *Am. Midland Nat.* 45:513–671.

Rose, K.D., and R.J. Emry. 1983. Extraordinary fossorial adaptations in the Oligocene palaeanodonts *Epoicotherium* and *Xenocranium. J. Morphol.* 75:33–56.

Summers, A.P., and J.C. O'Reilly. 1997. A comparative study of locomotion in the caecilians *Dermophis mexicanus* and *Typhlonectes natans* (Amphibia: Gymnophiona). *J. Linnean Soc.* 121:65–76.

Taylor, B.K. 1978. The anatomy of the forelimb in the anteater (*Tamandua*) and its functional implications. *J. Morphol.* 157:347–368.

Wake, M.H. 1993. The skull as a locomotor organ, vol. 3, pp. 197–240. *In* J. Hanken and B.K. Hall (eds.), *The skull.* Univ. Chicago Press, Chicago.

Yalden, D.W. 1966. The anatomy of mole locomotion. *J. Zool. Lond.* 149:55–64.

Chapter 26

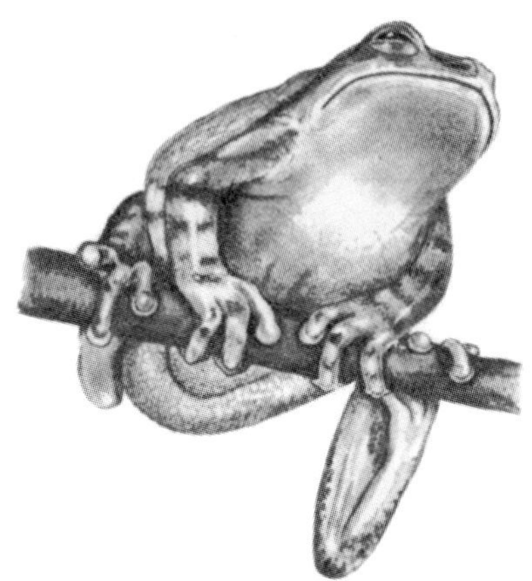

Climbing

Tetrapods that are adept at climbing may be called **scansorial.** Climbers are often **arboreal,** but this term means "living in trees" and does not directly indicate manner of locomotion; most birds are arboreal, yet few climb.

Many tetrapods are expert climbers and many more climb moderately well on occasion. The climbing habit evolved independently more times than can be traced, and several times in each of several orders. Adaptations for climbing by primates and by some of the other, more strikingly modified climbers have been analyzed. However, the adaptations of many small scansorial tetrapods have scarcely been studied.

ADVANTAGES OF CLIMBING

The selective advantages of climbing include the following:

1. Climbers can secure in shrubs and trees such foods as leaves, shoots, flowers, fruits, cambium, honey, spiders, insects, and birds' eggs.
2. Many climbers avoid predation by remaining off the ground or by returning to the safety of rocks or vegetation when danger threatens. Also, climbing affords vantage points from which to look out for danger.
3. Several predators follow their prey into the trees: The fisher captures tree squirrels, and the arboreal viper lies in wait for scansorial rodents. The leopard hauls his kill into a tree partly to keep it safe from jackals and hyenas when he is not in attendance.
4. By climbing, many animals find sheltered places to rest during the part of the day when they are inactive. Similarly, they find or make safe secluded nests in which to rear their young.
5. Where ground vegetation is dense, climbers may be able to travel more freely and rapidly in the open upper story of the trees than they could on the ground.
6. Animals that glide must climb to reach takeoff points.

SCANSORIAL VERTEBRATES

Fishes and Amphibians Several kinds of air-breathing fishes move about on land and may scramble into low vegetation using strong mobile fins and perhaps fin spines as well. None, however, is really scansorial. Among amphibians, the many species of tree frogs in at least seven families are expert climbers (Figure 26.1). There are many arboreal salamanders (family Plethodontidae), some of which are so skilled that they can walk along a string.

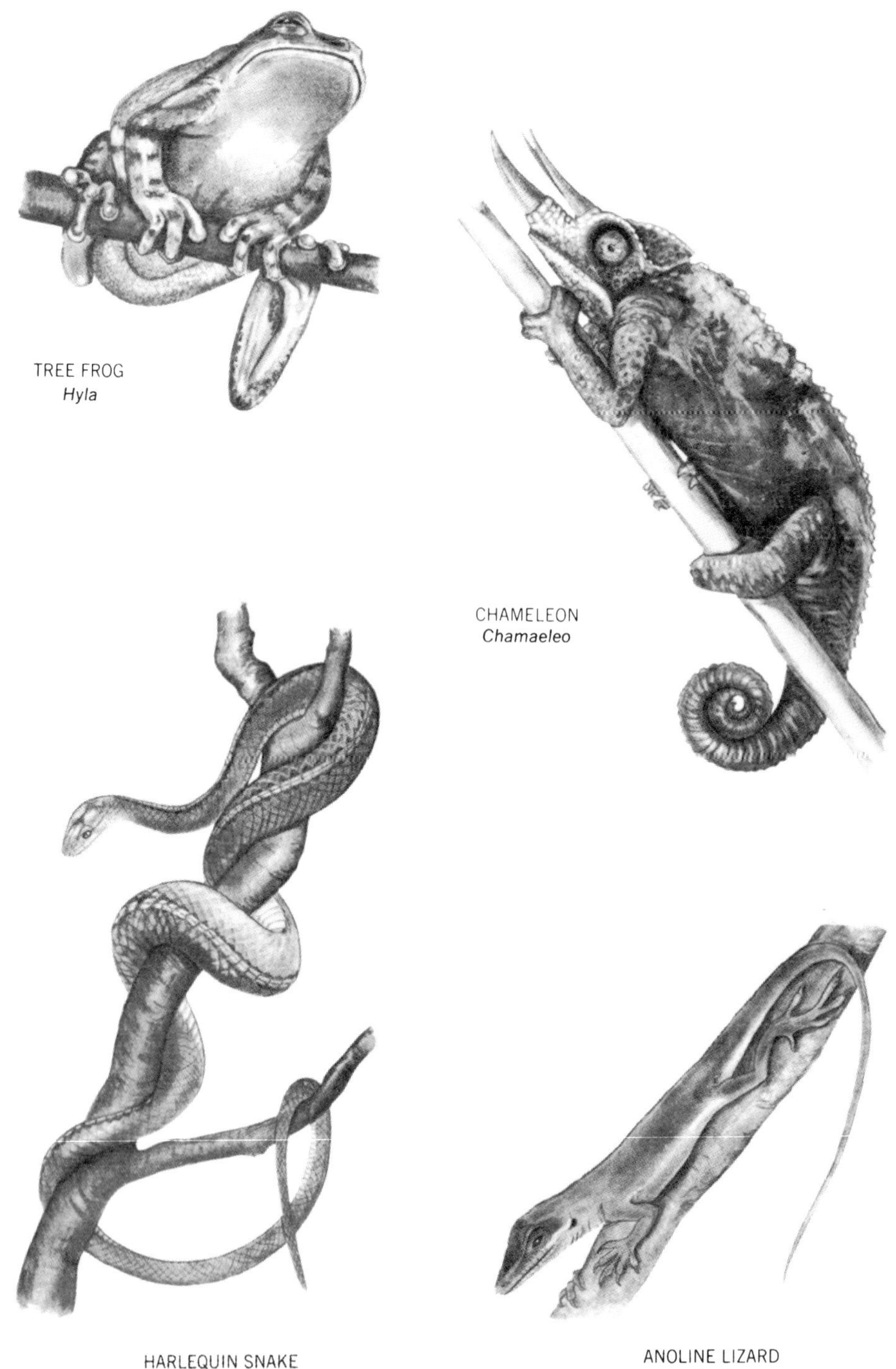

FIGURE 26.1
REPRESENTATIVE SCANSORIAL AMPHIBIANS AND REPTILES.

Reptiles and Birds Many lizards and snakes are climbers; some rarely descend from shrubs or trees. Noteworthy are the chameleons (family Chamaeleontidae), geckos (Gekkonidae), various iguanids (Iguanidae) including the anoline lizards, and the tropical tree snakes.

Excluding birds that merely perch in trees or forage by flitting from twig to twig, some remain that are truly scansorial (Figure 26.2). These are the woodpeckers, woodhewers, creepers (of two families), nuthatches (of three families), parrots, crossbills, some of the ovenbirds, and the hoatzin. In addition, rock nuthatches and wall creepers climb on rocks. The nuthatch climbs with feet only, the woodpecker with feet and tail, the parrot with feet and beak, and the hoatzin with feet and wings.

Marsupials and Insectivores Eleven of the 12 genera of opossums are fine climbers, and all 17 genera of phalangers climb with ease (Figure 26.3). Some marsupial mice climb, and, surprisingly, one genus of kangaroo has secondarily become a climber. Among insectivores, the 5 genera of tree shrews are highly scansorial.

FIGURE 26.2 REPRESENTATIVE AVIAN CLIMBERS.

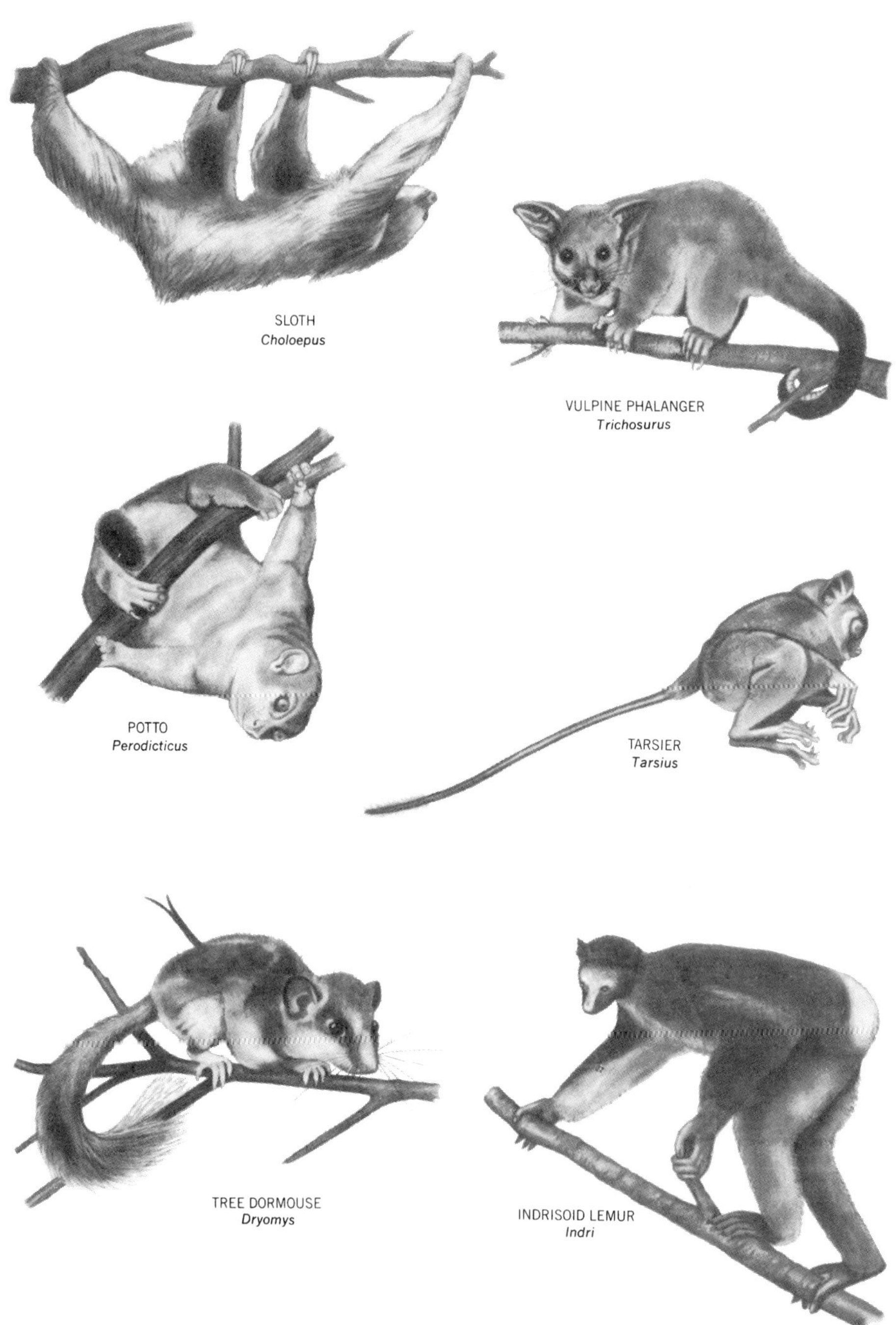

FIGURE 26.3
REPRESENTATIVE MAMMALIAN CLIMBERS.

Colugo, Bats, and Primates Like other gliders, the colugo is an arboreal climber (see figure on p. 521). Many bats roost in trees without doing much climbing, but several genera are nimble climbers.

Nearly all primates are skilled climbers, and the few that climb little or not at all (baboons, some lemurs, gorilla, humans) had arboreal ancestors. Particularly noteworthy for their structural adaptations or climbing behavior are certain lemurs (family Lemuridae); indris (Indridae); lorises, pottos, and galagos (Lorisidae); tarsiers (Tarsiidae); spider and woolly monkeys (Cebidae); langurs (Cercopithecidae); and gibbons and the orangutan (Pongidae).

Edentates, Pangolins, and Rodents The two smaller genera of anteaters are climbers, as are the two genera of sloths. Some pangolins are arboreal.

Scansorial rodents are in general less structurally modified than are the better climbers of other orders. Nevertheless, there are more climbers in this order than in any other. The common names of many are unfamiliar or poorly established. Climbing rodents include tree squirrels, "flying" squirrels, and chipmunks (Sciuridae); scaly-tailed squirrels (Anomaluridae); vesper rats, harvest mice, gerbil mice, red-nosed mice, tree mice, pine mice, and wood rats (Cricetidae); climbing rats and mice, forest rats and mice, and cloud rats (Muridae); dormice (Gliridae); New World porcupines (Erethizontidae); and echimyid rats (Echimyidae).

Carnivores, Hyraxes, and Ungulates Climbing carnivores are little modified structurally for the habit, yet they include all members of the raccoon family, some members of the weasel, bear, mongoose, and cat families, and two kinds of foxes. One genus of hyrax is remarkably skilled at climbing smooth tree trunks, and others scramble among rocks. Excepting the occasional acrobatics of some goats, no ungulate climbs trees, yet mountain goats and sheep, chamois, tahrs, and the klipspringer are master rock scramblers.

REQUIREMENTS AND BASIC MECHANISMS OF CLIMBERS

Climbers have two basic requirements: (1) They must propel themselves on a uniquely discontinuous and three-dimensional substrate, and (2) they must avoid falling, both when moving and at rest, under particularly difficult circumstances.

Runners and diggers must also propel themselves on a somewhat uneven substrate and must also avoid falling, so it is not surprising that the structural and behavioral adaptations of climbers differ in degree, but usually not in kind, from the adaptations of many other animals. Indeed, many climbers are not markedly modified for their locomotor habit. The tree mouse looks about like the terrestrial mouse, the scansorial and terrestrial species of murine opossums are similar, and the climbing fennec and nonclimbing red fox are constructed in nearly the same way. Furthermore, of all the locomotor specializations, ability to climb combines with the most other specializations: The climbing leopard also sprints, the tree kangaroo also hops, the anteater also digs, the tree frog also swims, the colugo also glides, and the parrot also flies. Nevertheless, some climbers (tree frogs, various salamanders, geckos, anoles, several bats) do utilize unique mechanisms, and scores of others have modified more familiar mechanisms to a striking degree.

The requirement that climbers propel themselves over a discontinuous substrate leads to adaptations for leaping, springing, swinging, and reaching and pulling. The requirement that climbers avoid falling under difficult circumstances leads to adaptations for grasping, balancing, bracing, cushioning, applying suction, clinging, hooking, and adhering. These adaptations will be analyzed, but since they are numerous and combine in many ways (springing with grasping, swinging with hooking, running with clinging, walking with adhering, running and leaping with grasping and cushioning, etc.), it will be useful to discuss first the principles a climber can utilize to remain in contact with sloping rocks, branches, or twigs. Basically, there are only two such principles—interlocking and bonding—but friction, which combines the two, is of importance.

The Role of Friction When a monkey stands on a horizontal log with each foot directly under its girdle, the downward force of the body is countered by an equal upward force of the support. There is compression at the interface between foot and log, but virtually no shear. Accordingly, there is no tendency for the foot to slip. The thrust of the foot against the log (**T**), and the force that is normal (i.e., perpendicular) to the surface of the log (**N**)

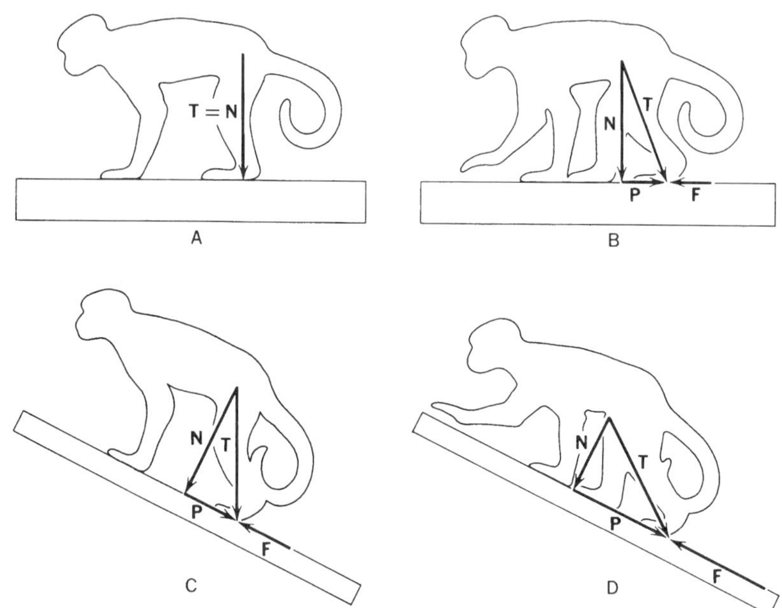

FIGURE 26.4 THE ROLE OF FRICTION IN PREVENTING SLIPPAGE.

are the same (Figure 26.4A). This would also be true of a standing horse, and the monkey is at the moment no more in the act of climbing than the horse.

If the monkey walks along the horizontal log, then the foot pushes against the surface at an angle to the vertical. This thrust can be divided into a component that is normal to the surface and one that is shearing, or parallel to the surface (**P** in Figure 26.4B). If the monkey is not to slip, **P** must be countered by an equal frictional force (**F**), which is the mechanical resistance to motion of the foot along its support. Likewise, the horse is able to walk only because friction opposes the propulsive forces along the ground.

If the monkey stands on a branch that is inclined to the horizontal, then **T** can be divided into **N** and **P** even though there is no motion (part C), and friction is required to keep the foot from sliding down the branch. If the monkey walks up the inclined branch, **N** is decreased and **P** is increased (part D). Consequently, friction must also increase if the monkey is not to slip. The climbing monkey and terrestrial horse have now parted company. It is evident that adaptations for climbing include mechanisms for maximizing friction.

How can that be done? Friction is a complicated phenomenon that does not lend itself to exact analysis. The maximum friction that can be developed before sliding starts depends on the kinds and textures of the materials in contact, and on the force acting between them. The approximate relationship for dry and rigid surfaces is $\mathbf{F} = \mu\mathbf{N}$, where μ is the coefficient of friction, which must be empirically determined for each combination of materials. However, if one of the surfaces is curved and viscoelastic (i.e., has both viscous and elastic properties), such as footpads and finger balls, then $\mathbf{F} = \mu\mathbf{N}^{\alpha}$, where $\alpha < 1$. Now as **N** increases, μ decreases, and slippage becomes more likely. In this circumstance the animal cannot avoid a fall by gripping harder. As drivers may learn, μ also decreases if sliding starts; it is easier to prevent a skid than to stop one.

It is obvious from the formula $\mathbf{F} = \mu\mathbf{N}$ that climbers could increase **F** by (1) selecting substrates that would give high values for μ, (2) evolving integumentary surfaces that would increase μ, and (3) developing mechanisms for increasing **N.** Methods 1 and 2 invite further study. Presumably, extra care is taken when the substrate is wet because lubrication greatly reduces friction (and alters the formula for its calculation). As we shall see, climbers are efficient at increasing **N** in various ways. These do not include, however, marked increase in body weight. Climbers are of medium to small size so they will not break the branches that must support them and so they can be agile.

As the formula shows, maximum friction tends to be independent of apparent area of contact. This is why the klipspringer (an antelope) can perform feats of rock climbing on the tips of very tiny hooves. Nevertheless, some climbers have large footpads. This reduces abrasion per unit area of the integument and, because large pads that are also flexible tend to touch the substrate in several planes, increases stability by preparing the foot to resist, with friction, disrupting forces coming from various directions in space. Large pads may also increase interlocking (see below).

The Role of Interlocking If the flat surface between an object and its support is inclined to the horizontal, then, as noted above, there is a force, **P,** parallel to the surface that must be opposed by an equal frictional force, **F,** if slipping is to be avoided. Force **P,** however, has a horizontal component (**H** in Figure 26.5A), and slipping could also be prevented by a counterforce (**H′**). If the support has a sidewall, this counterforce is provided (part B). Interlocking has then substituted for friction in providing stability. When the claw of an iguana, parrot, or chipmunk lodges in a crevice of a rock or branch, this method of support is operative.

If the object instead contacts its support on a dozen or so small sloping surfaces and on as many small sidewalls, then the same interlocking principle applies (part C). The numerous stiff tail feathers of a woodpecker and the horny plates under the tail of a scaly-tailed squirrel exemplify this kind of support when they press into rough bark (Figure 26.2).

Intermeshing of fine points of contact is also a basis for frictional force, so as interlocking surfaces become quite small (e.g., scales on the foot of a small lizard, or

COMMENT 26.1

FORTUNATELY, CLIMBERS NEED NOT UNDERSTAND FRICTION TO USE IT

The variables affecting friction at the macroscopic level of footpad and hoof are presented in the main text and have been known for several hundred years. Nevertheless, friction was previously not understood in terms of surface roughness and molecular adhesion. Research at the atomic level awaited the development of instrumentation and was in its infancy when this text was first published. It remains a challenging field. Even if sliding surfaces can be perfectly characterized, it is not yet possible to predict exactly the friction that will occur. Some experimental observations are unexpected or counterintuitive.

Experience shows that friction is proportional to load but not to apparent area of contact. However, even flat, smooth surfaces actually touch only at microscopic high points that constitute a small fraction of the visible area of contact. When the visible area is decreased (load remaining constant), pressure on the remaining area is increased and more microscopic points are forced together, thus maintaining about the same area of actual interaction between the surfaces. Friction *is* proportional to the *true* area of contact.

Friction of sliding surfaces was long stated to be independent of velocity. At nanoscale this should be true when the actual contact points do not heat up, but in the macroscopic world of automobile brakes, friction increases at very slow speeds. Likewise, it is common knowledge that lubricants decrease sliding friction, yet at nanoscale a sufficiently thin layer of lubricant increases friction. Also, we all know that dry sliding friction always results in wear, but in theory, at least, appropriate materials of adequately small size could slide without wear because of synchrony of the vibrations of atomic lattices. Climbers avoid these complications by being macroscopic and by avoiding sliding.

Ultimately, friction must be explained in terms of molecular behavior at surfaces of contact. How is it that friction between two surfaces is sometimes less if one of the surfaces if rougher than the other? Or that friction can increase when the surfaces are made smoother? It appears that friction correlates less with the strength of the adhesive bonds than on how easily the surfaces become stuck relative to becoming unstuck. It is a complicated issue.

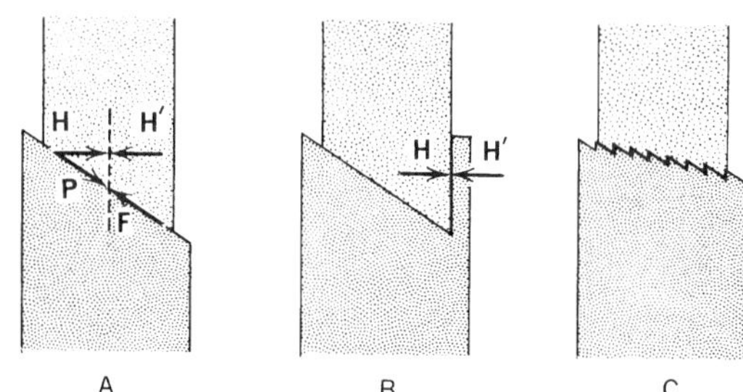

FIGURE 26.5 THE ROLE OF INTERLOCKING IN PREVENTING SLIPPAGE.

"fingerprint" ridges on the finger balls of a galago), the boundary between friction and interlocking as methods of support becomes fuzzy. A cushionlike footpad that contacts both microscopic and macroscopic projections on a branch utilizes both friction and interlocking to prevent slipping. The hooklike claw of a sloth might be prevented by friction from slipping along a smooth pole, but if its sharp edge were to cut slightly into the pole, interlocking would also contribute support.

The Role of Bonding Bonding is the consequence of molecular attraction. Two kinds are used by climbers. **Capillary adhesion** (or wet adhesion) occurs if the area of contact between two surfaces is dampened by a suitable adhesive. Thus, a glass coverslip adheres to a vertical windowpane when bonded by a thin film of water. Common experience tells us that sticky materials are better adhesives than water, and the fingerpads of some amphibians secrete a sticky material that bonds them even to vertical leaves that are smooth and glossy.

The other kind of bonding is **dry adhesion.** When one smooth metal slides over another, great pressure and high temperature at the microscopic points of contact may cause molecular bonds to form. This is another basis for frictional force. Ordinarily, however, dry materials cannot be brought close enough together at enough points, even if polished and clean, for intermolecular forces to establish a significant amount of bonding between them. The ability of the dry-footed geckos and anoline lizards to walk upside down on glass long defied explanation. There is now experimental evidence that the highly specialized structure of their toes (see below) enables them to establish intermolecular attractions (or van der Waals forces) with the substrate, that is, adhesion without glue or sticky adhesive.

ADAPTATIONS FOR PROPULSION

Walking, Running, Leaping, and Springing Animals that commonly walk or run along more or less horizontal branches (iguanas, tree mice, anteaters) have no problems of propulsion that are not shared by their terrestrial relatives. The feet, and sometimes the tail, may be modified to grip the substrate, but the remainder of the body is not distinctive.

Some climbers commonly, though not exclusively, propel themselves by leaping from one support to another. The jump may be somewhat upward, but usually is outward or partly downward. The animal may be moving when it leaps, and the body is more or less horizontal at takeoff. Examples are certain arboreal snakes; numerous lizards; tree squirrels; capuchin, howler, vervet, and proboscis monkeys; langurs; and arboreal mangabeys. These climbers (except the snakes) tend to have long limbs, slender bones, and muscle mechanics similar to that of cursors, though without relative shortening of proximal limb segments. The back is relatively long, strong, and flexible.

Several primates propel themselves primarily by springing. When at rest they tend to hold the body in a vertical position, and they are often stationary before takeoff. The jump

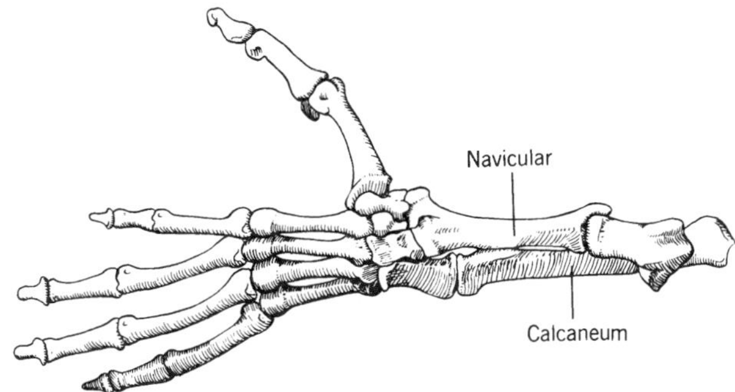

FIGURE 26.6 LENGTHENED TARSUS OF A SPRINGER shown by the left foot of the primate *Galago*.

may be in any direction including steeply upward. These highly specialized animals are the tarsiers, galagos, indris, and two genera of lemurs. The smaller of these prodigious springers can jump 2 m straight upward, and the larger ones may jump 10 m out and down from one tree to another. All have the long hind legs, flexed knee posture, and general limb mechanics of saltators (see pp. 450–452), although the femur is not shorter than the tibia. The foot (unlike that of jerboas and kangaroo rats) must be adapted for gripping the substrate, but in tarsiers and galagos it is also much elongated for springing by lengthening of the navicular and calcaneum bones (Figure 26.6).

Reaching, Pulling, and Bridging Many of the more adept climbers propel themselves entirely or in part by reaching from one support to another and pulling themselves along. Large gaps may be smoothly bridged without loss of secure support. The orangutan, pottos, lorises, and (in their upside-down way) sloths are noteworthy examples, though some frogs, chameleons, opossums, phalangers, various monkeys, the colugo, echimyid rats, and many others also use this method of propulsion. There is some evidence that orangutan-like climbing was in the ancestry of humans.

These animals must meet three principal requirements. The first is a long reach. Hence (with birds excepted) the limbs are longer than for any other locomotor adaptation except gliding, springing (hind limbs), and flying (forelimbs). Proximal and middle limb segments are nearly equal in length. The feet are large, yet must respond more to the needs of gripping than propulsion, so they do not lengthen as they do in cursors and terrestrial saltators. The thorax also tends to be long.

Various tree snakes provide for reaching in a very different way. The morphology of the anterior vertebrae, together with a complex musculature, prevent the spine from sagging as it is extended from one support toward another.

The second requirement is flexibility and agility. To gain strength with a wide range of movement, the heads of humerus and femur not only are spherical in curvature but also represent larger portions of complete spheres than is usual. (This contrasts with the conditions in cursors, diggers, and flyers.) The girdles, even of some scansorial reptiles, are modified to allow freedom of movement. Toward this end, the scapula and clavicle of the mammals tend to become modified in ways that, being even more extreme for arm swingers, are described in the next section. To ensure maximum pronation and supination of the forearm, the ulna and radius are free and about equally developed, the proximal head of the radius is round, the radial notch on which the radius rotates is evenly curved and is lateral in position (not anterior as for cursors) (see figure on p. 443), and a styloid process at the distal end of the ulna usually forms a pivot around which the carpus turns. Similarly, the fibula is free and relatively large. The wrist joint is ellipsoid, not hingelike. Considerable rotation, adduction, and abduction may be possible within the tarsus. Splines and grooves are relatively little developed at limb and foot joints (see figure on p. 446).

Third, these climbers require appropriate bone-muscle mechanics. Marked strength is not needed, so muscles and muscle attachments are not prominent, and the bones are light and slender. Many of these animals (and particularly the sloths) commonly assume postures in which extensor muscles do not oppose gravity in the usual manner. Hence, these muscles are less developed, and their in-levers are shorter than is characteristic in terrestrial mammals. Thus, the in-lever of the triceps (the olecranon process) ranges from about one-eighth (an opossum) to one-twelfth or less (lorises, sloths) of the out-lever (the length of the remainder of the ulna). Flexors, pronators, supinators, and abductors are better developed.

Arm Swinging Some primates propel themselves by arm swinging (or **brachiating**) under the branches. Unlike other methods of propulsion, only the forelimbs are used for support. Arm swinging is best exemplified by gibbons and the related siamang, though it is used on occasion by spider, woolly, howler, langur, colobus, and proboscis monkeys and by the chimpanzee, orangutan, and (more rarely) gorilla.

With one hand around a support that is above the body and ahead in the line of travel, the climber swings pendulum-fashion down under the support, rotates the body nearly 180° on the supporting arm (i.e., advances the unweighted shoulder), and, at the end of the upswing, stretches the other arm to another overhead support. The free arm and (in the siamang) also the legs are extended downward on the downswing. This moves the animal's center of gravity away from the pivot at the supporting hand, thus maximizing the velocity and kinetic energy gained. On the upswing the free arm and legs are flexed to shorten the pendulum, decrease its moment of inertia, and increase its angular velocity. As with a child pumping on a swing, there is a net gain in momentum. In rapid locomotion each support may be reached only after a period of free-floating travel. Other secrets of the complex physics of brachiation remain to be learned; the performances of these remarkable acrobats still exceed full understanding.

Swingers, like springers, are highly adapted for their specialty. They have the same adaptations for reaching, agility, and use of arms under tension as the reach-and-pull climbers, only the modifications are more extreme (Figure 26.7). The hind limbs are long relative to the trunk, particularly in gibbons and spider monkeys, but the forelimbs are disproportionately long, becoming even two or more times as long as the trunk in the orangutan, gibbons, and siamang. Because the arms are primarily in tension when loaded, there is little tendency for arm bones to buckle; they can be more slender than bones loaded in compression.

The fossa on the humerus that accommodates the very short olecranon process is deep so that the elbow can be completely straightened.

Supination of the supporting forearm coupled with twisting of the trunk advances the leading (nonsupporting) shoulder during a swing, adding reach and force to the action. To help accomplish this the supinator is strong, its in-lever is increased by bowing of the radius, the sternum is broad, the chest is broad rather than deep (Figure 26.8), and there is a rotary midcarpal joint.

The clavicle is long, reaching over the broad chest to the large acromion process. The scapula lies relatively flat on the back, rather than against the side of the thorax as for animals that do not arm-swing. The glenoid cavity is oriented forward and sideward, not downward (Figure 26.9). The latissimus dorsi, pectoralis, biceps, and long head of triceps brace the shoulder against tension. Also strong are the trapezius, which pulls the acromion process in toward the neck, and the anterior serratus, which pulls the posterior angle of the scapula out toward the side of the chest. The insertion and orientation of these muscles are also modified so that together they rotate the scapula on the trunk to a unique degree. This raises the arm on the body and, during the swing, probably rotates the body on the outstretched arm.

In contrast to cursors and quadrupedal leapers, arm swingers have short, compact backs so that the trunk can swing as a unit. The lumbar area contributes the least to loco-

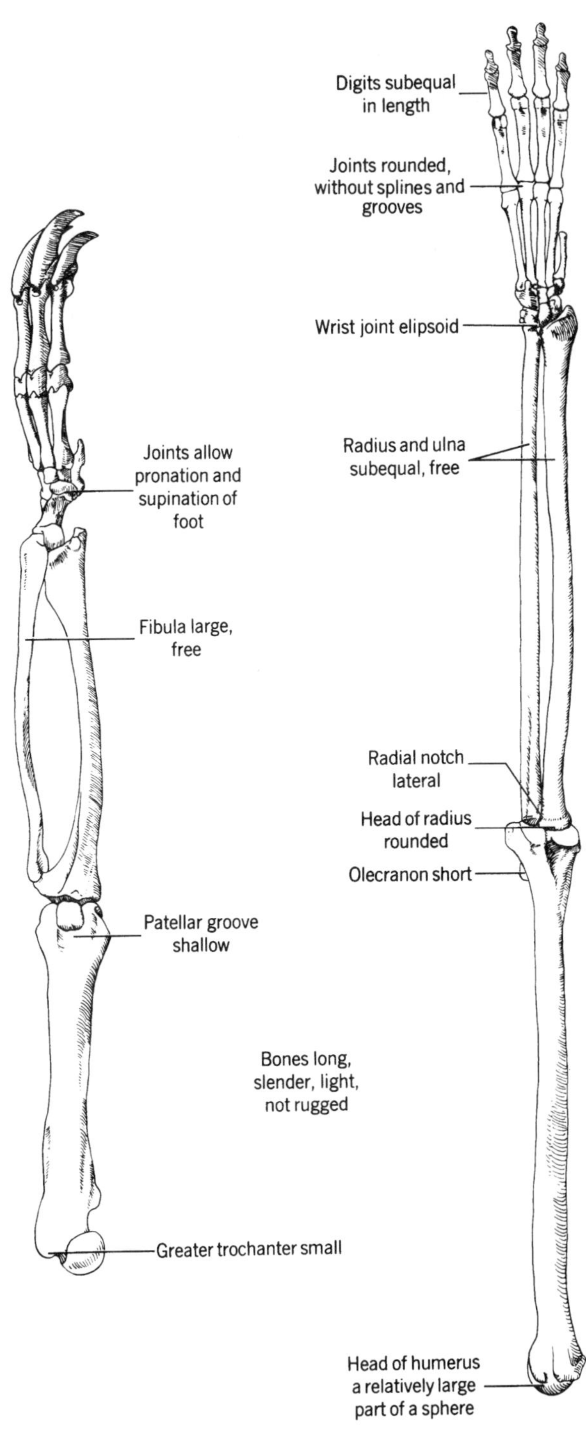

FIGURE 26.7 FEATURES OF THE APPENDICULAR SKELETON OF CERTAIN CLIMBERS shown by the left leg of a sloth, *Choloepus* (left), and left arm of a spider monkey, *Ateles* (right).

motion, so it tends to be inflexible, has relatively few vertebrae, and these vertebrae have short centra (Figure 26.8). The zygapophyses of mammalian lumbar vertebrae are constructed to allow flexion and extension but to limit rotation of the spine around its long axis. The zygapophyses of anterior thoracic vertebrae do allow rotation. The transition occurs within one vertebra. This vertebra, the **diaphragmatic vertebra,** is farther posterior in swingers (which rotate the spine with each swing) than in cursors and leapers (which flex and extend but do not rotate the spine).

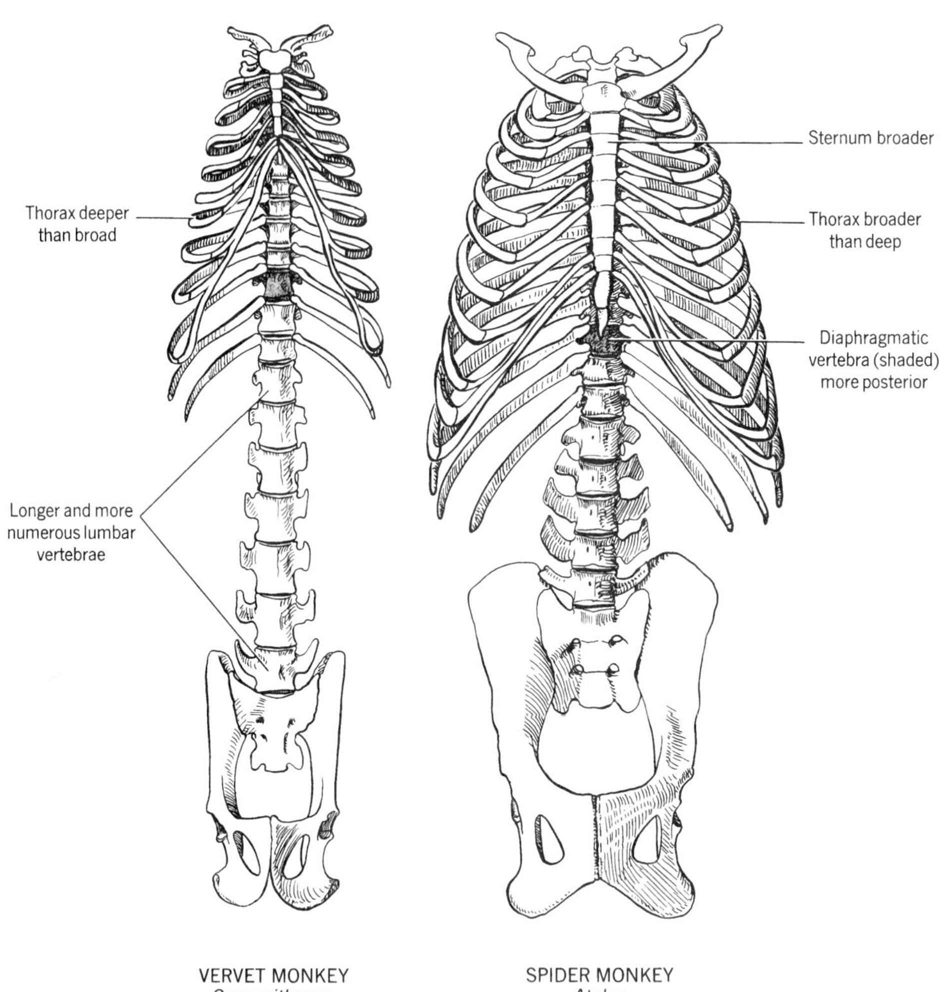

FIGURE 26.8
CONTRAST BETWEEN THE TRUNK SKELETONS OF A MONKEY THAT LEAPS (left) AND ONE THAT ARM-SWINGS (right).

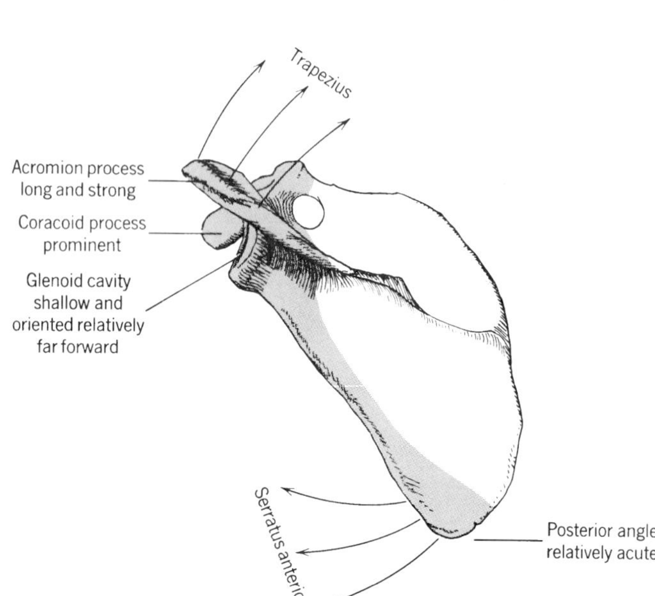

FIGURE 26.9
ADAPTATIONS OF THE SCAPULA FOR REACHING AND ARM SWINGING shown by the spider monkey, *Ateles.* Shading indicates areas of relatively heavy bone, which includes the lever of the mechanism that pivots the scapula.

Some climbers are slow-moving (chameleons, lorises) whereas many runners, leapers, and arm swingers are very quick. The latter require remarkably rapid and precise neuromuscular responses. The morphological bases for such control may be assumed to include the relative prominence of cerebellum, olivary nuclei, ruber nucleus, primary motor and sensory cortex, and optic tracts and centers. The eyes are large and face forward to provide overlapping fields of vision, and hence depth perception.

ADAPTATIONS FOR MAINTAINING CONTACT WITH THE SUBSTRATE

Grasping When the fingers and palm of a chameleon, potto, or a person encircle a twig or pole and grip firmly, muscular effort creates forces that are normal to the surface of the support. These forces increase frictional resistance to slipping: the tighter the grip, the more resistance. An animal with strong digital flexors can thus supplement the normal forces that it can develop using only its body weight. Furthermore, by grasping it can resist slipping in any direction.

Grasping is a particularly versatile and effective way of maintaining contact with the substrate, and the different climbers have independently evolved many grasping mechanisms. Various snakes grasp branches with coils of the body to provide support. The first digit is opposed to the others in one or both pairs of feet of some tree frogs, salamanders, birds, opossums, and many primates (Figure 26.10). The second digit of the hand tends

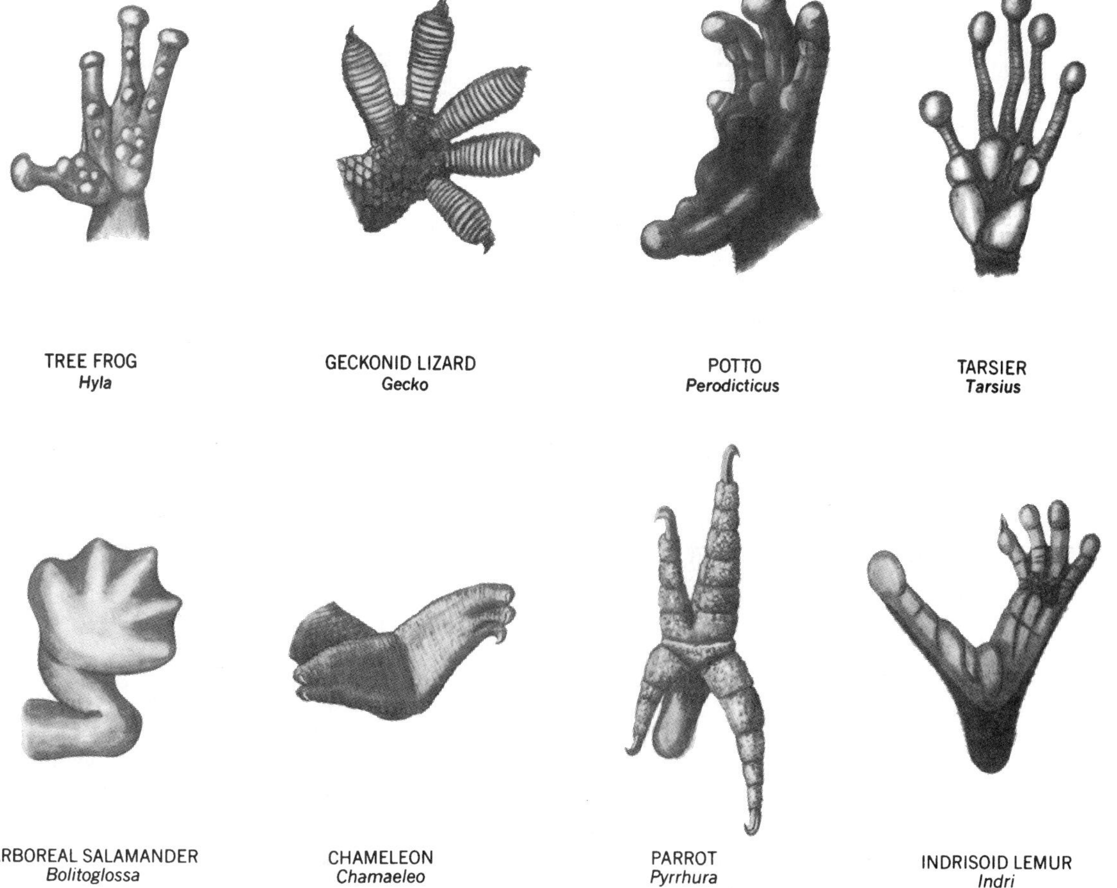

FIGURE 26.10 LEFT HANDS (above) AND FEET (below) OF SOME CLIMBERS.

not to be very effective in a power grip (even in humans), and in the potto and lorises has become short and weak. The koala and some other phalangers grasp between the second and third digits, as do the chameleons with the hind foot. The forefoot of these lizards grasps between the third and fourth digits, and echimyid rats do the same. Parrots and some other avian climbers oppose digits 2 and 3 with digits 1 and 4. The palms, soles, and digits of graspers are naked and sensitive.

Terrestrial mammals often irretrievably lose one or more lateral digits. If their descendants become secondarily arboreal, those digits cannot be used for grasping, and nature has produced interesting compensations: The two-toed anteater can depress its heel into strong opposition to its remaining digits, and one Central American porcupine can fold the pad of its foot lengthwise and forcefully grasp between the lateral edges of the footpad.

The tail often evolves into a grasping organ. Such a tail is said to be **prehensile** and is characteristically long, strong, sensitive, and curled at the end. Animals with prehensile tails include some salamanders; chameleons, and several other lizards; some arboreal snakes; opossums; some phalangers; capuchin, spider, and woolly monkeys; anteaters; pangolins; the kinkajou; various rats and mice; and one porcupine. Most of these animals curl the tail ventrally, but the porcupine curls its tail dorsally. Prehensile tails tend to be flexible at the base and to have short, broad vertebrae near the end. Often they have naked, sensitive pads where they grasp.

Balancing, Bracing, Cushioning, and Sucking Climbers that move quickly must be effective balancers. It follows from principles given on p. 445 that scansorial mammals that walk or run on top of branches can increase stability by lowering their center of gravity. Arboreal salamanders, snakes, and lizards keep their center of gravity very low. Tree squirrels, rats and mice, marmosets, and the kinkajou either have short legs or flex the legs to hold the body low. Birds that forage on tree trunks have short legs (the tarsometatarsus shortening the most) to keep their center of gravity close to their support. The girdles of the chameleon permit the legs to come vertically under the body (unusual in reptiles) so the animal can balance over narrow stems. Climbers that swing or hang under their support using hooklike appendages (see below) are in stable equilibrium—like rocking chairs they tend to maintain position. Most climbers have long tails that contribute to the maintenance of balance; climbers without tails swing, hang, or move slowly.

Numerous climbers use their tails as braces, struts, or props. Woodpeckers, woodhewers, and creepers prop themselves with their stiff tails; the terminal segment of the spine (pygostyle) and its musculature enlarge for the purpose. Some species of tarsier have a naked and roughened area at the base of the tail to make it a better prop. The scaly-tailed squirrel has horny ventral scales on its tail.

The appendages of most climbers have broad, soft, cushionlike pads. These are often roughened by small grooves or fingerprints that increase friction and interlocking. Footpads are particularly well developed in primates, anteaters, and porcupines. Fingerpads are conspicuous in tree frogs, some arboreal salamanders, opossums, phalangers, indris, pottos, galagos, and tarsiers. Climbers with prehensile tails have tail pads. Feet having such cushions tend to be broad and loose. Metapodial bones are well spaced, round in cross section, rounded on their distal ends, and devoid of splines at the joints. Phalanges are similarly rounded except that the terminal phalanx may be somewhat spatulate. Claws or nails are positioned so as not to interfere with the action of the pads.

The suction cup of human technology is usually a shallow cup of rubber, with the rim pressed against a smooth surface. The elasticity of the rubber keeps the rim tightly pressed against the surface and thus reduces pressure inside the cup so that atmospheric pressure presses the cup against the surface. Shearing force is resisted by friction. Disk-winged bats of two genera have suction cups on knuckles and ankles that function in the same way (Figure 26.11). Elastic tissue, not muscular tension, maintains suction within the disk once it is seated.

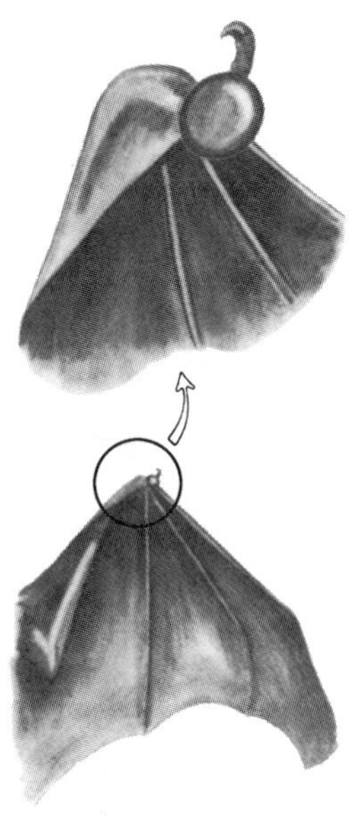

FIGURE 26.11 SUCTION DISK ON THE WING OF THE BAT, *Thyroptera.*

Clinging and Hooking Most climbers that do not grasp use strong, much-curved claws to cling to the substrate. The tips of the claws interlock with small cracks and crevices. On large stems, claws are more secure than grasping digits. When the weight of the animal is insufficient to maintain firm interlocking, one set of claws must be pulled against another. The two feet of a pair may sprawl wide on opposite sides of the body (rock lizard), or the hind feet may be turned toes-backward so that hind claws can pull against foreclaws or (in the head-down position) against gravity. Hind foot reversal is seen in squirrels, various marsupials, lower primates, tree shrews, and several carnivores. This important ability results from adaptations of the hip and ankle joints, and of joints within the tarsus. In addition, scansorial birds pull with one or two claws of the foot against the other claws of the same foot. The digital flexors are strong, and the terminal phalanx is designed to provide a good in-lever (Figure 26.12).

Some climbers modify the appendages as hooks and swing or hang under their support. Sloths and pangolins use one to three very long, strong, curved claws as their hooks. Bats and the colugo use five nearly equal claws. Primates that swing use the four fingers of the hand together as a hook. The hand is very long, and the phalanges are curved to conform to the round cross section of the branches. Prehensile tails can be used as hooks as well as grasping organs. In all these instances, the flexor tendons of the hook are short enough to passively prevent the hook from opening up: A dead gibbon can be hung by an upstretched arm; a dead spider monkey can be hung by the end of its tail; and some bats have a ratchetlike locking mechanism between flexor tendons of the claws and their sheaths to prevent the claws from opening as the bats sleep.

Adhering The expanded finger balls of tree frogs enclose glands having a sticky secretion that these little frogs use to glue themselves to rocks, leaves, and stems. They usually press their moist bellies against the substrate so that these will also adhere.

Many tropical salamanders have webs between their well-spread digits. The result is a large common adhesive pad. To break contact, the pad is curled up from its margin or lifted from the back. Frogs may have an "extra" segment of cartilage or bone just proximal to the terminal phalanx. This seems to assist the animal in feeling about with the tips of its toes for the best spot to make contact with the substrate.

The gecko has sharp claws with which to cling to rough substrates. If a smooth support is moderately inclined, the animal can maintain its position by friction. On steep, smooth surfaces and overhangs, however, this creature brings into play one of nature's most remarkable adaptive mechanisms. Under each toe are 16 to 21 broad imbricated lamellae (Figure 26.13).

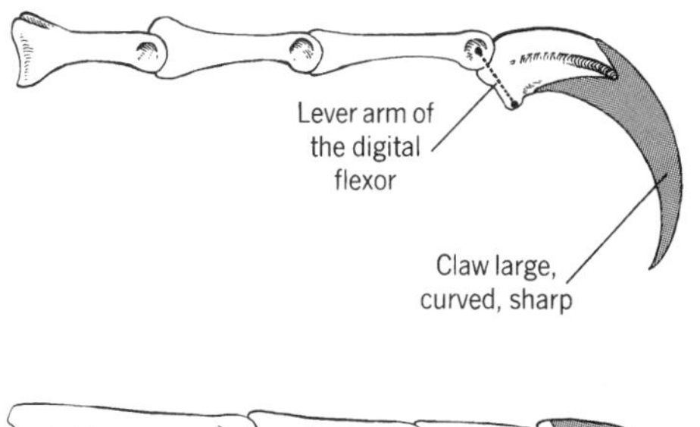

FIGURE 26.12 ADAPTATIONS OF THE AVIAN FOOT FOR CLIMBING shown by third toes of a flicker, *Colaptes* (above), and for contrast, of the nonclimbing, nonperching merganser, *Mergus* (below).

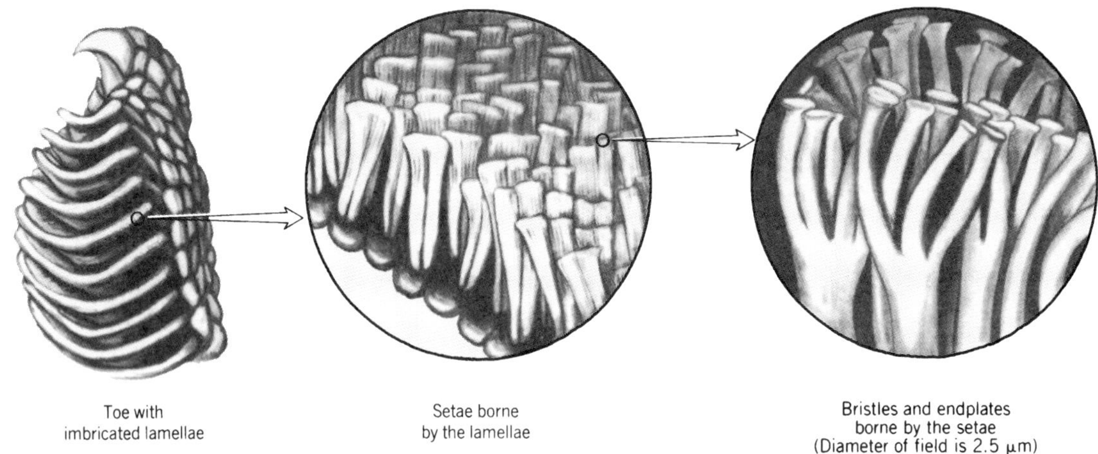

FIGURE 26.13 ADAPTATIONS FOR CLIMBING BY DRY ADHESION SHOWN BY A TOE OF THE LIZARD, *Gecko*.

On the exposed surfaces of the lamellae of each toe are up to 150,000 hairlike setae ranging from 30 to 130 μm long. Each seta branches into about 2000 bristles, and each of these has a saucerlike endplate measuring about 0.2 μm in diameter. There are in all some 100 million of these endplates that touch the substrate at points on their rims. (All these numbers vary by species.) The points of contact are so close that intermolecular attraction (van der Waals forces) comes into play, creating a very firm bond between toes and substrate. A 2000 study by Autumn and associates quantified the adhesive forces of single setae and demonstrated that in order to establish the bond, the setae must first be pushed against the supporting surface and then drawn parallel to it. Blood sinuses under the lamellae cushion the toes and adjust pressure so that a maximum number of endplates (yet never all at the same time) can reach into any irregularities of the surface.

The contact is firm: When one investigator tried to pull a large gecko from a vertical pane of glass, the glass broke. Adhesion to vertical glass continues even when the animal is dead and the body is in a vacuum. However, the lizards adhere with difficulty to materials having low surface tension (e.g., Teflon), and if the setae become dirty or mussed, climbing ability is impaired until after the skin is shed and new setae replace old. To break the contacts, the lizard peels its extra-flexible toes off the substrate by rolling them up "backward," starting from the tips.

The unrelated anoline lizards have shorter, unbranched setae, but they function the same way. This is an amazing example of convergent evolution.

REFERENCES

Autumn, K., et al. 2000. Adhesive force of a single gecko foot-hair. *Nature* 405 (Jun 8):681–685.

Bertram, J.E.A., et al. 1999. A point-mass model of gibbon locomotion. *J. Exp. Biol.* 202:2609–2617.

Bock, W.J., and W.D. Miller. 1959. The scansorial foot of the woodpeckers, with comments on the evolution of perching and climbing feet in birds. *Am. Mus. Novit.* 1931:1–45.

Bock, W.J., and H. Winkler. 1978. Mechanical analysis of the external forces on climbing mammals. *Zoomorphologie* 91:49–61. Free-body analysis of climbing animals in equilibrium.

Cartmill, M. 1974. Pads and claws in arboreal locomotion, pp. 45–83. *In* F.A. Jenkins, Jr. (ed.), *Primate locomotion.* Academic Press, New York.

Cartmill, M. 1979. The volar skin of primates: its frictional characteristics and their functional significance. *Am. J. Phys. Anthropol.* 50:497–510.

Cartmill, M. 1985. Climbing, pp. 73–88. *In* M. Hildebrand et al. (eds.), *Functional vertebrate morphology.* Harvard Univ. Press, Cambridge, MA.

Emerson, S.B., and D. Diehl. 1980. Toe pad morphology and mechanisms of sticking in frogs. *Biol. J. Linnean Soc.* 13:199–216.

Fleagle, J.G. 1974. Dynamics of a brachiating siamang [*Hylobates (Symphalangus) syndactylus*]. *Nature* 248:259–260.

Green, D.M. 1981. Adhesion and the toe-pads of treefrogs. *Copeia* 1981:790–796.

Hiller, U. 1968. Untersuchungen zum Feinbau und zur Funktion der Haftborsten von Reptilien. *Z. Morphol. Tiere* 62:307–362. Some publications in English have equally fine illustrations, but this study put dry adhesion climbing of reptiles on a sound physical basis.

Jenkins, F.A., Jr., P.J. Dombrowksi, and E.P. Gordon. 1978. Analysis of the shoulder in brachiating spider monkeys. *Am. J. Phys. Anthropol.* 48:65–76.

Jenkins, F.A., Jr., and D. McClearn. 1984. Mechanisms of hind foot reversal in climbing mammals. *J. Morphol.* 182:197–219.

Jones, F.W. 1953. Some readaptations of the mammalian pes in response to arboreal habits. *Zool. Soc. Lond. Proc.* 123:33–41. Adaptations in porcupines and anteaters.

Krim, J. 1996. Friction at the atomic scale. *Sci. Am.* 275(4):74–80.

Napier, J.R., and A.C. Walker. 1967. Vertical clinging and leaping—a newly recognized category of locomotor behavior of primates. *Folia Primatol.* 6:204–219.

Richardson, R. 1942. Adaptive modifications for tree-trunk foraging in birds. *Univ. Calif. Publ. Zool.* 46:317–368.

Russell, A.P. 1981. Descriptive and functional anatomy of the digital vascular system of the Tokay, *Gekko gecko. J. Morphol.* 169:293–323.

Spring, L.W. 1965. Climbing and pecking adaptations in some North American woodpeckers. *Condor* 67:457–488.

Walker, J. 1989. How to get the playground swing going: a first lesson in the mechanics of rotation. *Sci. Am.* 260(3): 106–109.

Williams, E.E., and J.A. Peterson. 1982. Convergent and alternative designs in the digital adhesive pads of scincid lizards. *Science* 215:1509–1511.

Wimsatt, W.A., and B. Villa. 1970. Locomotor adaptations in the disc-winged bat, *Thryoptera tricolor.* I. Functional organization of the adhesive discs. *Am. J. Anat.* 129:89–119.

Chapter 27

Swimming and Diving

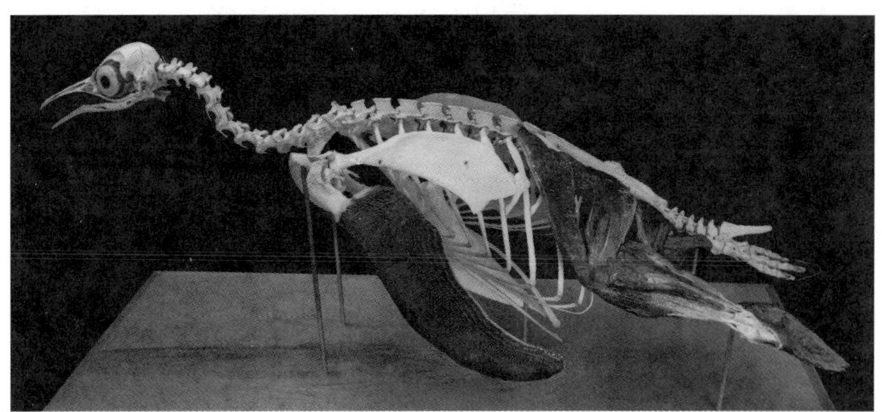

Vertebrates that live in water are said to be **aquatic.** Unfortunately, there is no term for the experts among them that swim with particular skill, speed, or endurance. All fishes are **primary swimmers**—their ancestors also swam. Other swimming vertebrates are **secondary swimmers**—their ancestors passed through a terrestrial state, and consequently they have structural and physiological handicaps that have prevented most of them from becoming entirely aquatic again.

Nearly all vertebrates can swim somewhat, and there are many expert swimmers in every class; there can be no sharp division between swimmers and nonswimmers. However, most that will receive attention here seek food and refuge in the water and can swim below the surface.

Relatively much has been published about swimming, yet detailed analysis is difficult, and many opportunities for future study remain. We report not only results from quantitative studies of structure and performance, but also general features of swimming and diving vertebrates *hypothesized* to correlate with function but not yet tested. Taken alone, structure can be misleading (see Comment 27.1), which makes the detective work that much more interesting.

ADVANTAGES OF SWIMMING AND DIVING

It is academic to ask about the survival value of an aquatic life to primary swimmers: It is a successful way of life, and nature has provided no alternative to most of them. As we saw in Part II of this book, the change to terrestrial life was so profound that it took about 100 million years to complete. The reverse trend back to water has been "easier" and has occurred many times. Secondary swimmers and divers may have any of the following advantages over nonswimmers:

1. They gain access to a wide variety of aquatic foods including fishes, plankton, and larger invertebrates and plants.

Above: Skeleton of an emperor penguin, with freeze-dried leg and wings.

2. They escape terrestrial predators. They may also subject themselves to aquatic predators, of course, but the process of evolution has in specific instances moved in the direction of greater safety.

3. The oceans and major inland waterways are favorable avenues for dispersal and migration.

4. Water often affords a relatively constant environment.

THE SKILLS OF SWIMMERS AND DIVERS

Many aspects of performance contribute to the survival of swimmers and divers. Speed, endurance, and maneuverability may be necessary, even at substantial cost.

Speed A first skill shared by many aquatic vertebrates is speed. The mako shark is probably the fastest cartilaginous fish, and the families Scombridae (tunas, mackerels), Istiophoridae (marlins), and Xiphiidae (swordfish) hold the honors for bony fishes (Figure 27.1). Accurate records are difficult to obtain for maximum performances, but marlin,

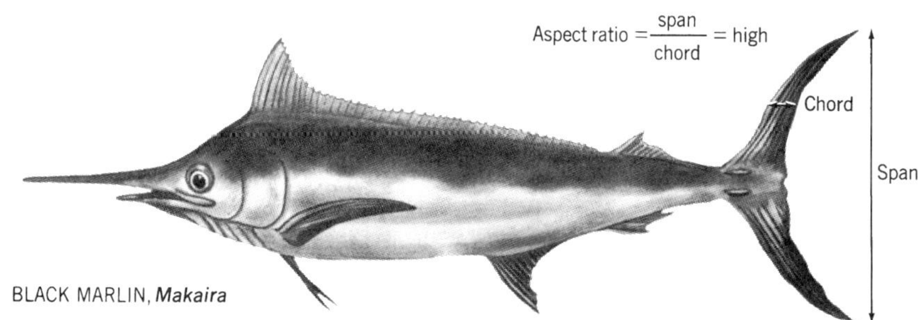

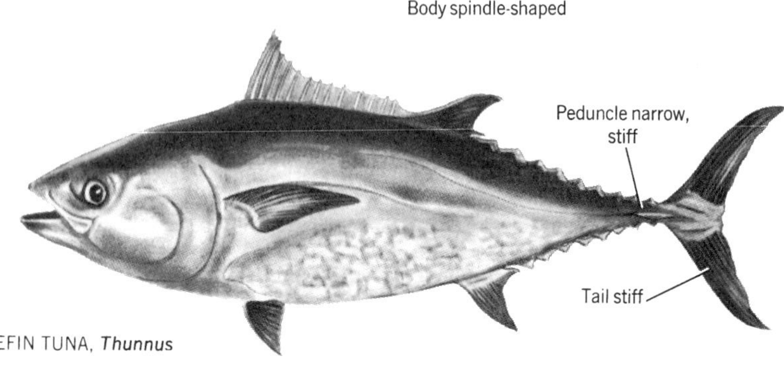

FIGURE 27.1 EXAMPLES OF FISHES SHOWING ADAPTATIONS FOR FAST SWIMMING.

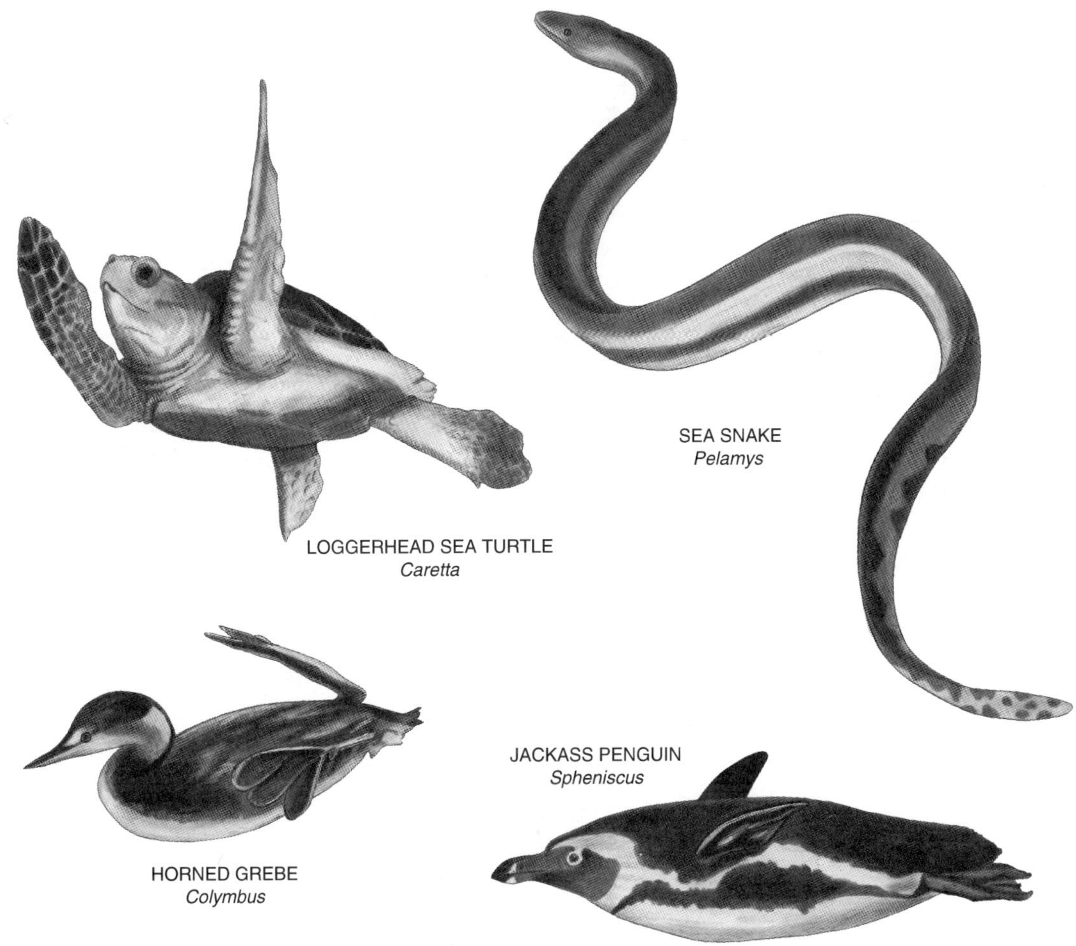

FIGURE 27.2 EXAMPLES OF UNDERWATER SWIMMERS AMONG REPTILES AND BIRDS.

wahoo, and tuna take baits trolled at about 28 km/h (17.5 mph) and all have been estimated to be capable of bursts at more than twice that speed. Their cruising speeds are commonly 3 to 5 km/h.

Many living amphibians and reptiles are aquatic, but none is remarkably speedy. The extinct sea lizards called mosasaurs and the dolphinlike ichthyosaurs (see figure on p. 58) may have been much faster. Some penguins attain 36 km/h (Figure 27.2).

Among aquatic mammals also, even the best swimmers of various orders—the platypus; water opossum; water shrews, tenrec, and desmans; beaver, capybara, and nutria (Figure 27.3); manatee; and hippopotamus are not speedsters. Sea lions and the related fur seal can sprint at about 22 km/h. The spotted dolphin can approach 40 km/h (25 mph), and the finback whale and rorquals are thought to be able to sprint faster. Again, cruising speeds are much slower.

Diving Various primary swimmers dive to the depths of the ocean. In spite of their dependence on breathing air at the surface, many secondary swimmers dive deeply and remain submerged for long periods. The green sea turtle dives to 290 m, and sea snakes remain underwater for up to two hours. Leatherback turtles dive much deeper, to 1200 m, and can remain submerged for more than half an hour.

FIGURE 27.3 EXAMPLES OF AQUATIC MAMMALS of three orders: Sirenia (upper left), Carnivora (upper right), and Rodentia (bottom).

Pelicans, diving petrels, tropic birds, boobies, and terns dive from the wing. Penguins use their paddlelike flightless wings underwater; the emperor penguin routinely dives to 400 m, and a dive of 534 m has been recorded. The king penguin allows its abdominal temperature to drop as much as 11°C on long dives, which conserves oxygen and energy. Auks, murres, puffins, and diving petrels both swim and fly with their narrow wings. Cormorants, loons, grebes, and some ducks swim underwater using their feet. Both feet and wings are used in swimming by some of these birds. Loons and puffins dive to at least 55 m and murres to 180 m.

Various rodents can remain submerged for 6 to 10 min, and the manatee for 16 min (Figure 27.3). The bush dog, otter-civet, water mongoose, river otter, and sea otter are among the diving carnivores. The sea lion can dive to 300 m, the harp seal to 273 m, and the Weddell seal to 600 m. The latter can range at least 12 km from its blow hole under the ice, commonly remains submerged for 30 min, and can stay under for at least 70 min. The remarkable elephant seal dives some 64 times a day to about 300 m and can reach 1200 m. It remains at the surface only about three minutes between dives. Dolphins can remain submerged for 10 to 20 min. Other records are 50 min for the blue whale, 90 min for the sperm whale, and 120 min for the bottlenose whale. Fin whales dive to at least 355 m. Sperm whales have gone as deep as 1100 m (where the pressure is 110 atm), and a bottlenose whale, carrying a depth recorder, dove to 1453 m in a submarine trench near Nova Scotia.

Endurance Salmon swam 1000 km up one river with an expenditure of energy equivalent to an average speed in quiet water of 4.2 km/h. Most scombrid fishes and the mako shark swim continuously. Fur seals commonly migrate 12,000 km per year. Gray whales cruise at about 5.5 km/h on their annual round trip migration of about 19,000 km. The blue whale is reported to be able to swim at 27 km/h for 2 h (with a harpoon wound).

Acceleration and Maneuverability Other skills of aquatic vertebrates are less well documented. Trout can accelerate at 40 m/s^2, achieving maximum speed of 9 body lengths/s in 0.10 s. The spotted dolphin can reach 40 km/h in only 2 s. The agility of schooling fishes, reef fishes, sea-snakes, penguins, otters, and many other swimmers is both beautiful and astounding.

GENERAL REQUIREMENTS OF SWIMMERS AND DIVERS

All proficient swimmers and divers must (1) reduce the resistance that water offers to motions of the moving body, (2) propel themselves in a relatively dense medium, (3) control vertical position in the water, and (4) maintain orientation and steer the body. Secondary swimmers must also (5) exclude water from their respiratory passages and ears, (6) avoid harm from crushing of gas-filled spaces, (7) alter ears and eyes to function (again) underwater, and (8) modify their respiratory and circulatory physiology to permit suspension of breathing and avoidance of the bubbling of gas in the blood (bends) on returning to the surface after a dive. Some aquatic birds and mammals must also (9) control body temperature in a medium with high thermal conductivity, and (10) adapt their reproductive biology to life in the water.

How have the various swimmers and divers met these many requirements?

DRAG

Origins and Nature of Drag The resistance force that a medium (here water) exerts to impede the motion of an object is called drag. There are several sources, or kinds, of drag; energy is required to overcome each, and each acts parallel to and opposite the direction of movement. The different kinds of drag are interdependent, but can best be presented one at a time. First, there is **pressure drag** (also termed *inertial drag*). Imagine a smooth, spindle-shaped, rigid object moving through water (Figure 27.4). The water

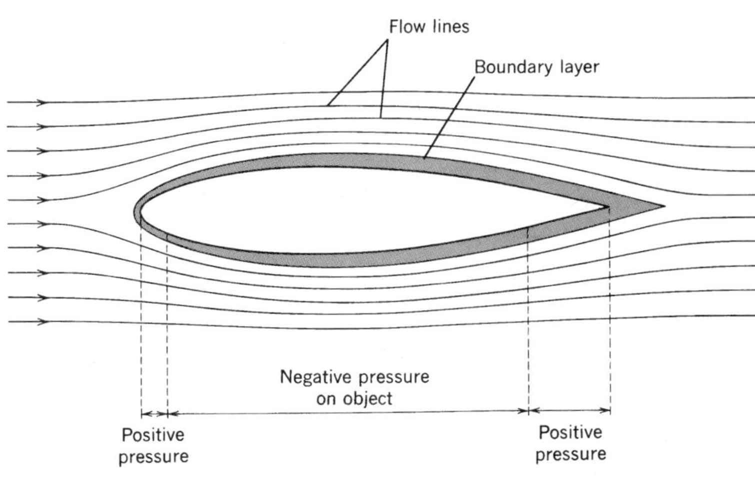

FIGURE 27.4 FLOW LINES, BOUNDARY LAYER, AND PRESSURE DISTRIBUTION WHEN FLOW AROUND A STREAMLINED OBJECT IS LAMINAR.

molecules themselves possess position inertia relative to one another. The movement of the object through them exerts a pressure that affects their inertia and causes a deflection of the molecules above and below the object's surfaces. More precisely, since the object deflects fluid, it adds kinetic energy to the fluid system. This energy is drawn from the object, which therefore slows down. Pressure is highest at the most anterior point of contact between the object and the moving fluid, and is lowest behind the object. Pressure drag varies with fluid density (more or fewer molecules to deflect), velocity of the object (more or fewer deflections per unit time) and the object's shape (streamlined vs. clunky).

The second kind of drag is **frictional drag** (also termed *viscous drag*). The surface area of the object and the viscosity or "internal stickiness" of the fluid are of key importance. As the object moves through the water, a film of water wets (adheres to) its surface and moves with the object. Yet, a short distance away, the water appears still or undisturbed, and does not move with the object at all. Between the object and the still water is the thin **boundary layer,** where successive layers of water slide past one another; those nearest the object move nearly as fast as it does, and those more and more distant move slower and slower relative to the object, which sets up a velocity gradient (Figure 27.4). The boundary layer gets thicker toward the posterior end of the moving object. Viscous forces (intermolecular forces) ensure resistance to flow between adjacent layers of any fluid (more for syrup, less for water), and are the source of the frictional drag. Frictional or viscous drag varies not so much with density (as pressure drag does), but primarily with the total wetted surface area of the object, fluid viscosity, and relative fluid velocity.

Finally, if the object moves on the surface of the water, like a ship or a duck, or close enough to the surface to cause surface waves, then energy is extracted from the object to create the waves and **wave drag** occurs. Resistance is greatest when the object moves just below the surface. The variables associated with wave drag are not well known.

Together, these kinds of drag act on an object moving through a medium to resist that movement and independently sum to yield the total fluid drag. For organisms "designed" to move through a fluid, the relative magnitude of these forces has substantial impact on their morphology.

Consider the boundary layer further. If a smooth spindle-shaped object moves slowly through the water, successive layers, or lamina, of the boundary layer slip past one another without any eddies or vortices. Flow is said to be **laminar.** However, if the object moves fast, then where the boundary layer reaches a certain thickness, or where there is even slight roughness on the surface of the object, the water curls into complex eddies (Figure 27.5). The energy that moves the water in these eddies comes from the moving object; eddies greatly increase drag. Such flow is said to be **turbulent.** Turbulent flow produces a thicker boundary layer and much more drag than does laminar flow. Many fishes have nearly laminar flow when moving slowly, but when moving fast flow becomes turbulent

FIGURE 27.5 FLOW LINES SHOWING THE TURBULENCE CREATED BY A SWIMMING FISH as seen from above. Various other patterns might also be created, depending on the variables.

over the posterior part of the body. Turbulence is usually considered undesirable, but we shall see that it can be advantageous.

To be realistic, let us examine the variables in some detail before seeking a simplification. Pressure and viscous drag (expressed in dynes) are both dependent on relative fluid velocity (or the speed of the object) and on the object's dimensions (cross-sectional area or length, for example), but differ in their respective dependence on fluid density and viscosity. Thus, for the comparison of swimmers with different morphologies and locomotor habits, it is useful to have an index that is proportional to the ratio of inertial forces to viscous forces acting upon it. The simplest index incorporating these parameters is the **Reynolds number,** a ratio of inertial and viscous forces that varies with the density and viscosity of the fluid, and incorporates the object's dimensions and its velocity relative to the surrounding fluid. It is a dimensionless number of high magnitude (about $10^{5.5}$–$10^{7.5}$) for large fast-swimming vertebrates in water.

There are specific formulas for the calculation of the Reynolds number and the various kinds of drag (Vogel in References); some of these will be considered in Chapter 28. To calculate all the parameters in every instance, however, can be very involved, and, in fact, theory does not always correlate with our empirical observations. Although theory is often appropriate for constructing models and making predictions, keep in mind that engineers continue to explore new designs for the America's Cup sailing vessels and biomimetic underwater vessels. This is not because they do not understand the theory, but because even man-made objects made to tight specifications do not always behave as predicted.

The effects of drag can be measured empirically for objects of different shapes and sizes using force transducers mounted in flow tanks. This is not always possible when the object is an animal, but we can make reasonable hypotheses to explain the morphologies we encounter. Relative velocity (speed) of the surrounding fluid is important in calculations of fluid drag as well as the Reynolds number. It follows (and is intuitive) that slow-swimming vertebrates have negligible drag no matter how the other variables change: witness the unstreamlined bodies of the sluggish sea horse and trunkfish. Conversely, drag on rapid swimmers increases very fast with each increment of speed, and in order to compensate for the increase in drag, metabolic rate must be approximately doubled every time speed is increased by 1 body length/s. It appears that the fastest swimmers closely approach the biological limit imposed by their metabolic output. Furthermore, in order to swim fast, the experts must reduce as much as possible all factors other than speed that increase drag.

With few exceptions, swimmers can do little or nothing to alter the density or viscosity of the water, so these elements of the formula usually can be disregarded. Drag increases with body size, but so does the output of the animal's power plant, and these factors nearly cancel one another. The consequence of some rather complicated physiological considerations seems to be that moderately large swimmers have some advantage. The fastest swimmers are large fishes and small whales. There remain the important variables of body shape and the nature of the surface of the body. Furthermore, since swimmers are not rigid, drag can also be reduced by certain behavioral adaptations.

Reduction of Drag by Adaptations of Body Form Pressure drag is low when the body is long and slender, like that of a snake or eel, because there is then little displacement and backfill. Frictional drag is minimal, however, when the body is short and plump, because surface area is then minimal. The best compromise is a spindle shape that is circular in cross section and thickest near the center of its length, where its diameter is one-fourth to one-fifth of its length. The bodies of tunas, swordfishes, and dolphins closely approach this shape. Absence of a functional neck (primary swimmers, cetaceans, sirenians), symmetry of the head, molding of thorax and body musculature, and the distribution of fat and blubber all may contribute to streamlining.

Projections from the basic spindle shape usually cause turbulence and eddies, and increase drag. Accordingly, expert swimmers reduce or eliminate all major projections not needed for propulsion and steering: Swimmers other than mammals have no external ears or external genitalia in their ancestry. Aquatic mammals secondarily lose their external ears and move the testes back into the abdomen. Nipples or teats and the penis may be withdrawn within the body contour when not functioning. Fast primary swimmers have no limb segments between their fins and bodies. Fast secondary swimmers have very short proximal limb segments to again bring the feet or flippers close to the body. The humerus of cetaceans may be only about as long as it is wide; the femur of pinnepeds may be less than twice as long as it is wide; the femur of diving birds is short and most of the leg musculature is contained within the contour of the body. Cetaceans and sirenians have reduced the pelvic appendages to internal vestiges, and some other swimmers position the hind limbs in such a way that they do not protrude but instead extend the contour of the spindle-shaped body. The knee joints of pinnipeds, beavers, and many diving birds are constructed to allow the necessary reorientation of the limb.

Salamanders, crocodilians, and aquatic lizards hold their limbs against the body as they swim with tail and trunk. Lateral fins and flippers that propel the body, on the other hand, must protrude and present a flat surface to the water. Most swimmers with lateral paddles have a power stroke and a recovery stroke. Drag is reduced in various ways during the latter. Rotation of the entire appendage at its base may cause the appendage to cut the water edge-on (flipper of sea lion). The median lobes on the toes of grebes are similarly rotated on the recovery stroke, and the lateral lobes passively fold. Such flippers and lobes, and also median fins and paired appendages that are used primarily for steering, are streamlined in cross section so that the flow of water over them is nearly laminar when they are presented to the water edge-on. Pectoral flippers of pinnipeds, wings of most diving birds, and paired fins of fishes are pressed against the body when the animal glides. Bony fishes can also reduce the area of their fins by folding. Dorsal and anal fins (including the "sail" of the sailfish) may be retracted into grooves on the body surface during fast swimming. Web-footed tetrapods flex their limbs and adduct and curl their toes on the recovery stroke.

Reduction of Drag by Adaptations of Body Surface and Behavior It is advantageous for most swimmers to achieve laminar flow over as much of the body as possible. To maximize laminar flow, fishes evolve small smooth scales, or none at all, and become covered with slime. The small scalelike feathers of penguins and the hair of seals and otters form remarkably smooth coverings. Cetaceans and sirenians (and probably ichythyosaurs) are (or were) secondarily naked and slick-skinned.

Larger and faster swimmers are unable to prevent turbulent flow over most of the body and, in fact, benefit from it because moderate turbulence greatly reduces separation of the boundary layer and backfill in the wake of the body. The "strategy" of the swimmer, therefore, is to cause, but control, turbulence. In sharp contrast to the large eddies associated with pressure drag, eddies of the desired turbulence are very small and close to the body. As with the dimples on a golf ball, the increase in frictional drag that is caused by such turbulence is more than offset by the reduction in pressure drag. Associated adaptations appear to be numerous, but experimental verification of theory is as yet scanty.

Many fishes have projections on their scales that are calculated to be large enough to cause turbulence. These are found even on the sword of the swordfish. Other scales commonly have microrelief that forms longitudinal "runoff grooves" that are thought to control flow in the boundary layer. The skin of cetaceans and the basking shark has a spongy layer that is capable of elastic deformation and probably dampens pulsations of turbulence. The finlets on the caudal peduncle of many fast-swimming fishes (Figure 27.1), and the lateral keels on their tails, are described as damping screens and deflectors that direct the flow of water past the caudal fin.

It is not only the smoothness of fish slime that reduces drag. Slime is soluble in water but only when stirred, and hence is not easily washed away. In concentrations as low as 1 percent it reduces flow friction by as much as 60 percent. That is, it reduces the viscosity of water in the boundary layer. Fishes that accelerate fast, such as trout, have the most slime.

Importantly, fast swimmers outperform submarines and torpedos in ways that involve behavior. The factors are complex and are understood only in general terms. It was noted that separation of the boundary layer creates suction that causes water to follow after a swimmer. However, in propelling the body, the tail fin pushes water back, thus tending to cancel this source of drag. The opercula of the fastest bony fishes open alternately in synchrony with the undulations of the body. The result is that water may be ejected from the gills fast enough to reduce drag, or slow enough to initiate (desirable?) turbulence. Ducklings reduce drag by swimming in formation.

When animals swim just below the surface, wave drag increases total resistance by as much as five times the minimum. It is unlikely that many vertebrates attempt to swim rapidly in that position. The "playful" leaps of dolphins may be made in part to avoid swimming at the surface while breathing.

FORM, FUNCTION, AND MANNER OF PROPULSION

Vertebrate swimmers propel themselves only with undulatory or oscillatory mechanisms; they have no analogs of sails, screw propellers, or jet engines (except for small forces acting at the gills). Propulsive locomotion in swimmers is a fertile and challenging area of research requiring integration of anatomy, physiology, and biomechanics.

Source of Propulsive Force All undulatory and most oscillatory swimmers move themselves forward by thrusting a propulsor (i.e., fin, paddle, or body segment) diagonally against the water. This movement arises as the consequence of numerous forces acting on the swimmer. Figure 27.6 illustrates some of the forces acting on a fish tail as forward motion is initiated. The fin is broad and flat, so water cannot easily flow around it but instead resists its lateral motion. In other words, it has high drag when presented broadside to the water. Side-to-side pivoting of the fin at its base, and of the peduncle supporting the tail (parts A and B of the figure) are timed so that, except at the limits of its travel, the fin is constantly thrusting obliquely against the water (part C) with a force $\mathbf{F}_t$ (part D). The inertia of the water causes it to push with equal force in the opposite direction ($\mathbf{F}_w$ in the figure). This force can be broken down into a forward component ($\mathbf{F}_f$) and a lateral component ($\mathbf{F}_l$). Because of the streamlined shape of the fish, the water offers little resistance to $\mathbf{F}_f$, so the body glides forward. Force $\mathbf{F}_l$ tends to cause the fish to pivot around its center of mass; the posterior part of the fish moves right (in the illustration) and the anterior part moves left (the in-torque equaling the out-torque). However, the water offers considerable resistance to sideways motion of the body. The entire body does move from side to side, but lateral motion of the heavier, stiffer, flatter, anterior part of the body is much less than the lateral motion of the peduncle of the tail.

In fact, the function of the tail is complicated by forward motion of the fish, change in the velocity of the fin, motion of the water column moved by the tail, "lift" (as explained

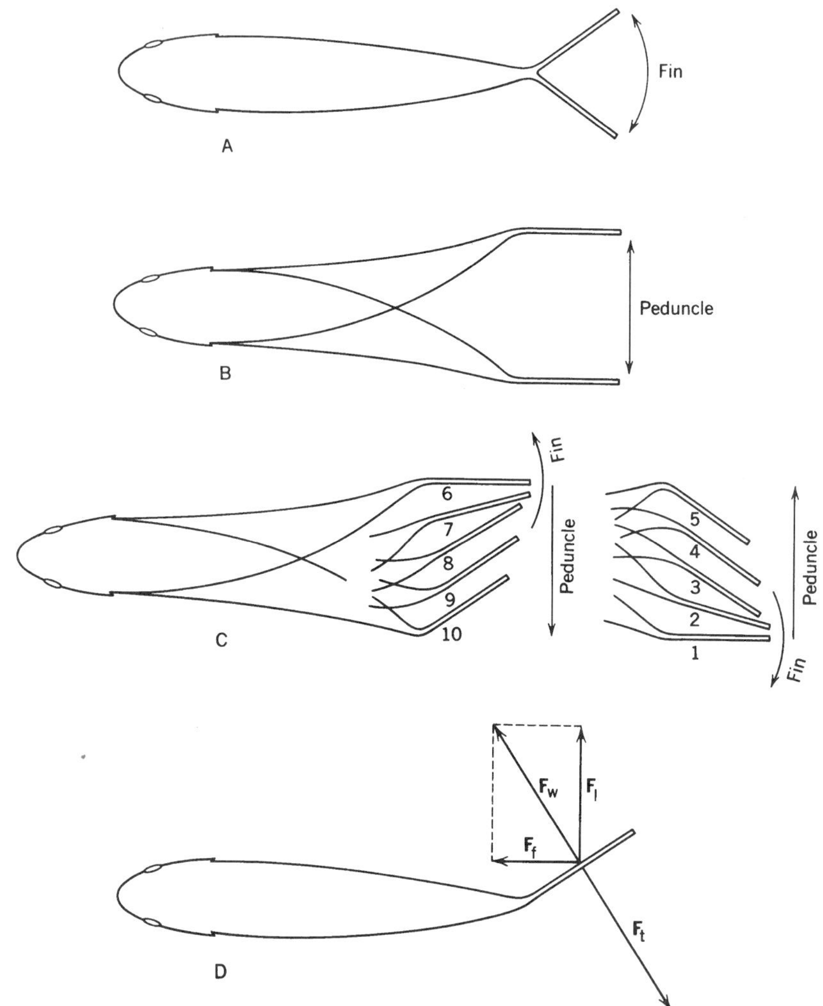

FIGURE 27.6 MOTION OF THE CAUDAL FIN OF A SWIMMING FISH AND SOME PRINCIPAL FORCES ACTING AT THE FIN. (Other forces may also be present.)

below), frictional forces, and pressure fields. These are difficult to quantify and can be of considerable magnitude.

Undulatory Propulsion In undulatory swimming, traveling waves pass along the propulsor, anterior to posterior, a little faster than the animal moves forward. The propulsor is the axial part of the body and tail, or long dorsal and ventral fins, or both, the fins then acting as extensions of the body. Each part of the propulsor thrusts, in turn, against the water. The entire body may undulate conspicuously, as for lampreys, eels (Figure 27.7), and sea snakes (Figure 27.2); the anterior body may undulate only moderately, as for carp and trout (Figure 27.8); or the undulation may be virtually confined to the caudal fin and its peduncle, as for tuna and marlin (which reduces drag on the body when the fish swims fast).

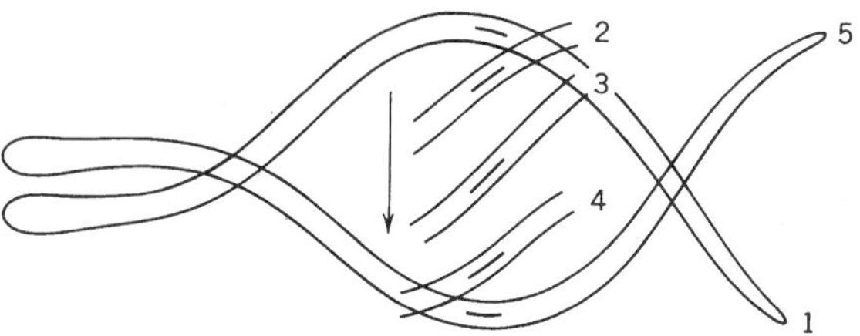

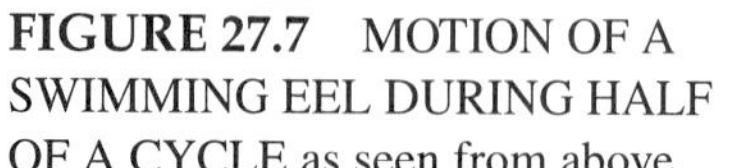

FIGURE 27.7 MOTION OF A SWIMMING EEL DURING HALF OF A CYCLE as seen from above.

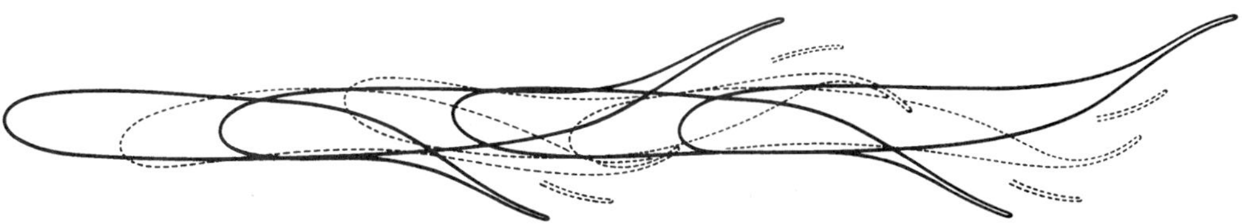

FIGURE 27.8 PATH OF A SWIMMING TROUT as seen from above.

Undulatory swimming is described as either periodic or transient. In **periodic swimming** the animal sustains propulsion for at least several seconds, and often for minutes or hours. Some scombrid fishes and sharks never stop swimming. The anteriormost part of the propulsor thrusts against still water, but as it thrusts it gives the water motion in the opposite direction. Thus, a next posterior segment of the propulsor must push against receding water, and it must have greater velocity if its thrust is to be adequate. Accordingly, the amplitude of undulation must increase progressively from anterior to posterior; hence the wider sweep of the posterior part of the undulating body or fin. Analysis uses **blade-element theory:** Forces acting on successive elements (or segments) of the propulsor, being different, must be calculated independently, and then integrated to determine the overall forces.

To begin with the less complex of two categories, some periodic undulatory swimmers use the median dorsal and ventral fins. Examples are some ribbon fishes, the sea horse, and knife fishes. The adaptation is for precise maneuvering in the confines of water plants or coral. By reversing the direction of the traveling waves, these animals can swim backward. Motion is very slow, so drag is negligible and there is no need for streamlining (Figure 27.9).

Most periodic undulatory swimmers instead propel themselves using the body and tail. The (very different) adaptation is for sustained cruising and sometimes for bursts of high speed. The prowess of some scombrid fishes, active sharks, and cetaceans was noted at the beginning of this chapter. Other swimmers of this category are salamanders, crocodilians, sea snakes, and some aquatic rodents. Seals trail the hind feet behind the body and use them much as though they were a single vertical fin. The tail usually moves symmetrically from side to side, but in cetaceans and sirenians instead sweeps up and down. Perhaps this better suits their need to breathe at the surface and dive.

Drag is high for cruising and sprinting swimmers. Accordingly, for the best of them the body is streamlined, circular in cross section, and sometimes roughened posteriorly to create the controlled turbulence that reduces pressure drag (see above). The shape of the trailing edge of an undulatory propulsor, here the caudal fin, is of great importance. The span of the fin from tip to tip divided by its chord, or average width in the direction of forward motion, is called the **aspect ratio** (Figure 27.1). (If it is easier to calculate, the equivalent equation span^2/area can be used.) The fastest swimmers have an aspect ratio of 4 to 6, whereas slow swimmers have values of 1 to 2 (see bowfin, figure on p. 46). By having a large span, much of the tail fin gains effectiveness by extending above and below the turbulence that follows the swimmer. The caudal fin of fast swimmers is streamlined in cross section, has low mass relative to the body, and is stiff—particularly on its leading edges. Such tails oscillate with low amplitude. Frequency, but not amplitude, increases with speed; during a burst of speed the tail virtually vibrates (12 Hz being common for a 2 m tuna). The peduncle is narrow, and finlets and keels may guide water over the fin in an advantageous manner.

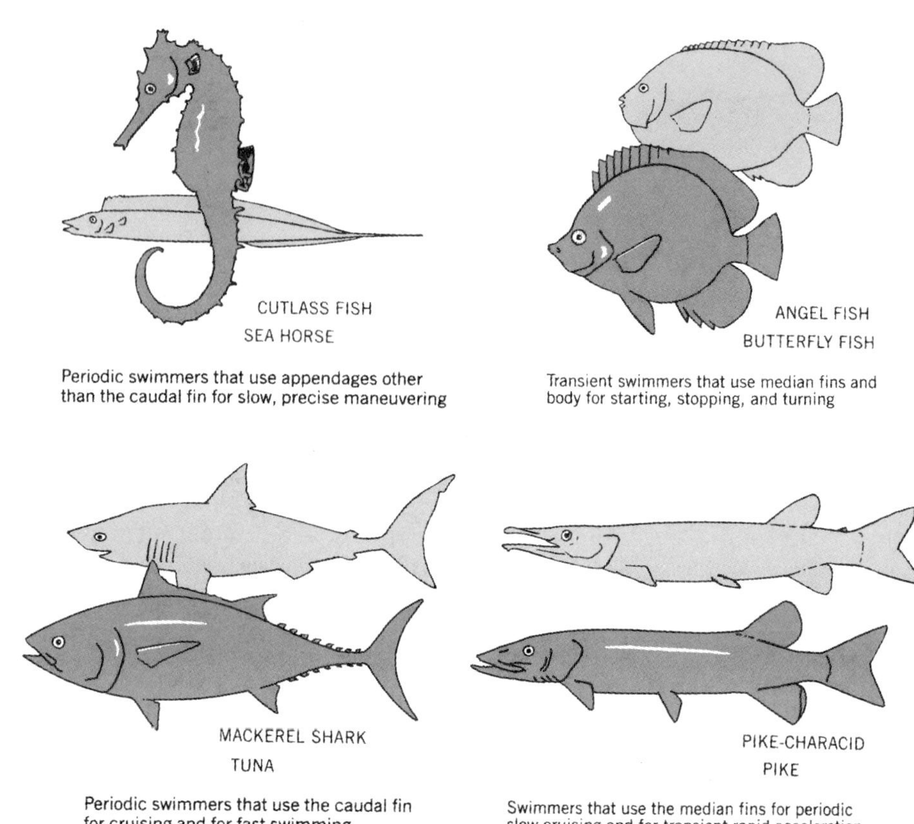

FIGURE 27.9 BODY FORM OF UNDULATORY SWIMMERS in relation to manner of propulsion and swimming adaptation.

Rapid swimmers that propel themselves with the tail tend to have stiff spinal columns: Centra may be long to reduce the number of intervertebral joints (sailfish), the centra may be large and platyan to reduce flexibility at the joints (some cetaceans), among fishes the zygapophyses, or spines (or both), may be unusually broad and strong to brace the joints (marlin, tuna, Figure 27.10). Stiffness coupled with resilience of the trunk is important for undulatory swimmers but both are a challenge to measure. Stiffness may be added by the myosepta in fishes, and by collagenous fibers in the skin (sharks, cetaceans) that are wound into helices around the body. Pabst illustrates that not only do dolphins and scombrid fishes (tunas and relatives) share swimming style and body shape, but they also share a pattern of force transmission through a complex, three-dimensional system of collagenous fibers, which are stiffened by muscular hydrostatic pressure. This force-transmission system increases both the displacement advantage and moment arm of contracting axial muscle. Long and Nipper concluded that large-mouth bass minimize the mechanical cost of bending by increasing their body stiffness to tune the body's natural frequency to match tailbeat frequency.

There must be adequate flexibility, however, at the base of the tail and peduncle. Bony fishes usually have diarthroses where the tail fin joins the spine, and the spine of cetaceans is dorsoventrally compressed in the anal area to allow vertical flexibility at that place.

The myomeres of fishes fold to form zigzags as seen on the surface of the body, but a series of interrelated cones as seen in three dimensions (Figure 27.11). The structure and mechanics of these myomeres is complex and has only been investigated for a few species, and function is still debated. Their shape may vary both longitudinally within a single individual and across different species (see Jayne and Lauder, and Westneat et al. in References.) The cones may extend the force of contraction of one muscle segment over several segments

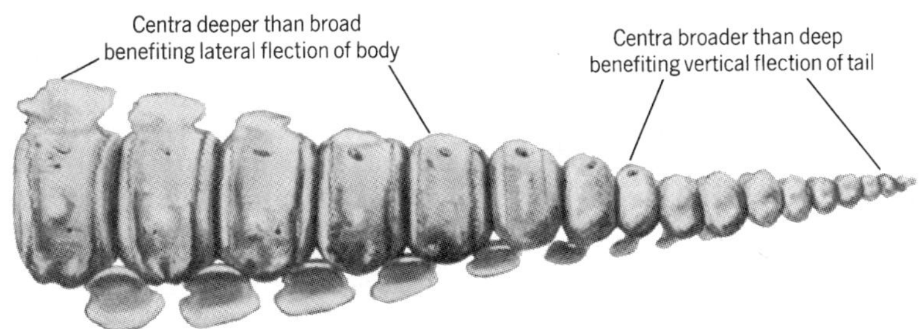

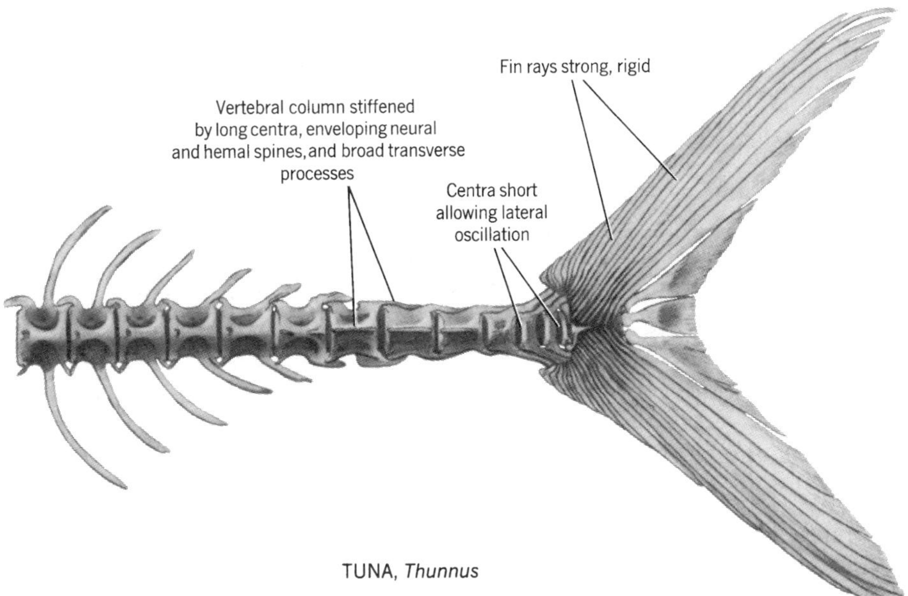

FIGURE 27.10 CAUDAL SKELETONS OF TWO VERY DIFFERENT FAST SWIMMERS.

of the skeleton. They allow muscle fibers at different distances from the body axis to shorten equally in flexing the body. They also ensure that muscle fibers will insert on the myosepta at oblique angles, thus providing some of the advantages of pinnate muscles. In many fishes, the red "aerobic" muscle fiber bundles are arranged superficially and in parallel just beneath the lateral line, and the more extensive, but deeper, white "anaerobic" fibers make up the bulk of the cones and run in a helical orientation closer to the vertebral axis. Alexander theorized that this helical architecture enables the white fibers on one side of the fish to maintain a near-constant sarcomere length irrespective of their distance from the backbone. In addition, the helical arrangements results in a high gear ratio for increased power (work/time). Rome and his colleagues verified both predictions for swimming carp. They showed that during low-speed cruising, the relatively slow-contracting red muscle fibers undergo small length changes and contract to lengths corresponding to myofilament overlap coincident with generating 96 percent or more of their maximum isometric tension. During an escape response, however, when rapid, high-amplitude movements of the backbone are intitiated, the white fibers are recruited and, on average, undergo only about one-quarter the sarcomere length changes experienced by the red fibers. Thus the white fibers operate at myofilament overlaps coincident with greater than 90 percent maximum force production. In addition,

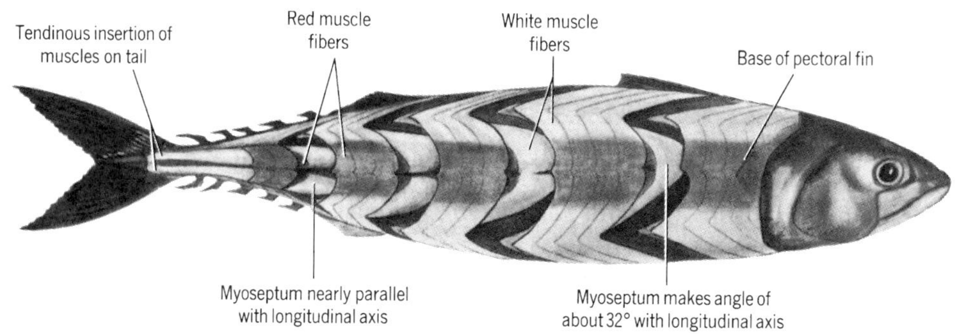

FIGURE 27.11 AXIAL MUSCULATURE OF THE FAST- SWIMMING MACKEREL, *Scomber*, with myomeres removed at successive levels.

both fibers operate at or near an optimal V/V_{max}. Small bones, which are a great nuisance at the dinner table, may run from the apex of one cone to the apex of another in the myosepta. The apexes of cones of the more posterior myomeres of the fastest fishes are extended into the tail as longitudinal tendons. The peduncle is therefore slender and tendinous, rather than broad and fleshy as in other fishes. (Contrast Figure 27.11 with the figure on p. 181.)

The other kind of undulatory swimming is **transient swimming.** Here the adaptation is for acceleration. The fish spurts ahead, usually turning sharply as it does so, glides, stops, and starts again. Such swimmers include the bluegill and kelp and reef fishes. As for an automobile in city traffic, it is inertia, not drag, that consumes the most energy. Consequently, body shape is responsive to the requirements for maneuverability, not streamlining. It is compressed laterally but deep, with fins often extending the body vertically (Figure 27.9). The body is short and flexible, giving it a short turning radius. Because the fish accelerates from rest, and only for a fraction of a second at a time, all parts of the propulsor thrust against water that is at rest. Therefore, blade-element theory does not apply.

Most undulatory swimmers compromise, according to habit, between form that is optimum for periodic cruising and form optimum for transient bursts. Thus, the pike is capable of very rapid acceleration coupled with periodic swimming at only moderate speed (Figure 27.9). The body is longer and more streamlined than that of the transient specialist, yet more flexible and with broader peduncle, lower caudal aspect ratio, and more posterior dorsal and anal fins than expert periodic cruisers. Trout are similarly adapted and are capable of exceedingly fast starts in whatever direction is needed to capture prey or escape from danger. A sharp flexure of the body into a "C" configuration initiates the fast start and is followed by rapid extension to send the fish off in various directions (Figure 27.12). Because of the importance of the fast start maneuver to survival, in aquatic amphibians as well as fish, its execution has received attention from the perspective of both neural control (see Fetcho in References) and evolution. Hale studied the

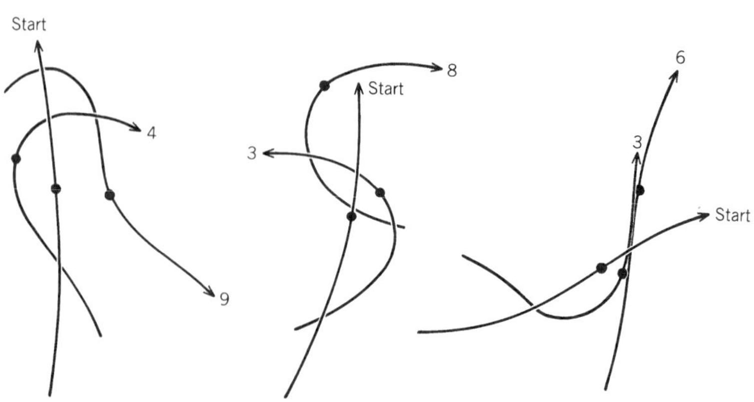

FIGURE 27.12 FAST STARTS OF TROUT in various directions relative to starting position. Arrows show centerline of fish as seen from above. Spot shows center of mass. Numbers indicate number of 0.015 s time intervals elapsed from start. (Based on Webb.)

effects of size on fast start kinematics during growth in three species of salmonid fish and found that velocity and acceleration are dependent on each species' ontogenetic state. These variables are maximum, however, at the end of yolk-sac absorption, a time corresponding to the initiation of search behavior for food, and the life history stage most hazardous for survival.

Oscillatory Propulsion Most oscillatory swimming is done by secondary swimmers using their paired appendages as propulsors. These pivot as a unit without traveling waves. Oscillatory propulsion may be drag-based or lift-based.

In **drag-based oscillation** the appendages function like oars or paddles. Unlike undulatory propulsors, they have a power stroke and a recovery stroke. On the power stroke there is a large angle of incidence (i.e., the paddle is oriented with its broadside nearly crosswise to the direction of travel). (The paddle cannot thrust in quite the same direction throughout the stroke because it pivots at its base and hence describes an arc.) Drag-based oscillatory swimmers include frogs, most turtles, ducks and other birds that swim on the surface, the sea lion, beaver, capybara, polar bear, and various other mammals. Most of these animals maneuver well but do not swim fast. Some fishes use their pectoral fins as oscillatory propulsors when moving slowly. Such fins tend to be of moderate length and constricted at the base.

On the power stroke the paddle must be large, broad, and stiff enough to stand against water pressure without muscular effort. Pinnipeds flatten and greatly lengthen the usual complement of metapodials and phalanges, particularly at the leading edge of the flipper (Figure 27.13). Ichthyosaurs evolved extra digits to broaden the paddle, one species having nine digits in all. Ichthyosaurs, plesiosaurs, and cetaceans add phalanges to one or more digits, bringing each series to from 4 to as many as 26 units. Pinnipeds extend some of the bony digits with cartilages. The integumentary membrane extends beyond the skeleton in the flippers of some pinnipeds (Figure 27.14). Diving birds and cetaceans incorporate the forearm into the paddle; the radius and ulna become short, flat, and positioned in the same plane. Small aquatic mammals usually have fringes of long, stiff hairs that functionally broaden the foot. These paddles are made rigid in various ways, though some resilience remains. In flippers of cetaceans, amphiarthroses replace diarthroses, and bones are flat-ended. Spaces between the bony digits are sufficiently filled with firm tissue to brace the digits and make the surface contour of the paddle smooth.

On the recovery stroke the paddle may be canted edge-on to the water stream. It is then streamlined in cross section, and rotates around its long axis from its base. Alternatively, the paddle may fold on the recovery stroke. This is facilitated by the webs between three toes of ducks, flamingos, gulls, auks, loons, and penguins; between four toes of cormorants, boobies, and pelicans; and between all five toes of frogs, pond turtles, platypus, beaver, and sea otter. Grebes, mudhens, and finfoots have folding lobes on the toes instead (Figure 27.15).

The other kind of oscillatory propulsion is **lift-based oscillation.** The function of the appendage is similar to that of a wing in air, but here support against gravity is not required, and usually propulsion occurs on both the upstroke and downstroke. The angle of incidence between paddle and oncoming water is relatively small. Lift-based swimmers include some skates and rays, teleosts in several families, sea turtles, plesiosaurs, some mosasaurs, penguins, auklets, murres, puffins, diving petrels, and (on occasion) sea lions.

On the downstroke of the paddle, water streams against it as shown in Figure 27.16. This produces a lift force perpendicular to the water stream in a manner that will be described in the next chapter. The lift has a forward component that propels the swimmer. The mechanism is repeated on the upstroke (the force diagram being turned bottom to

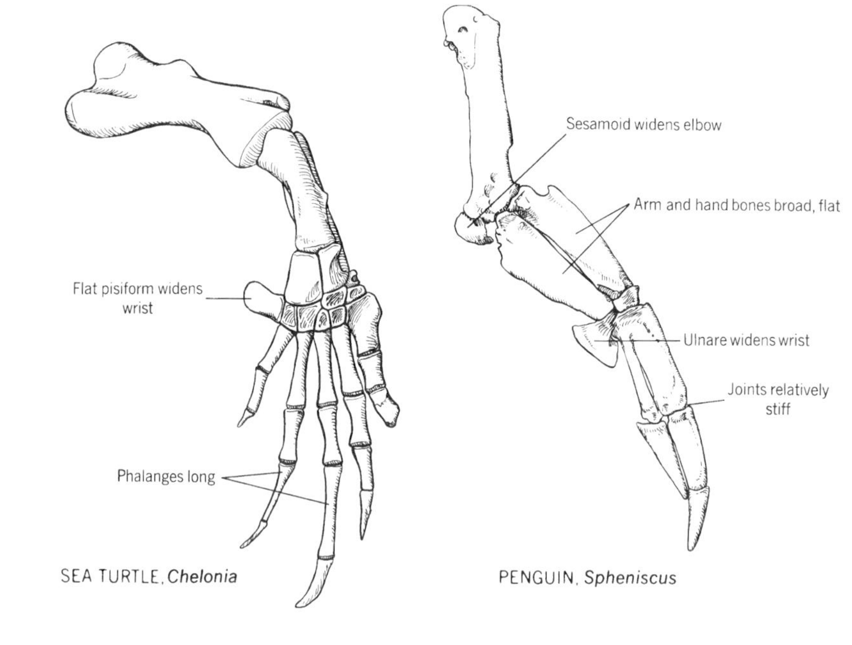

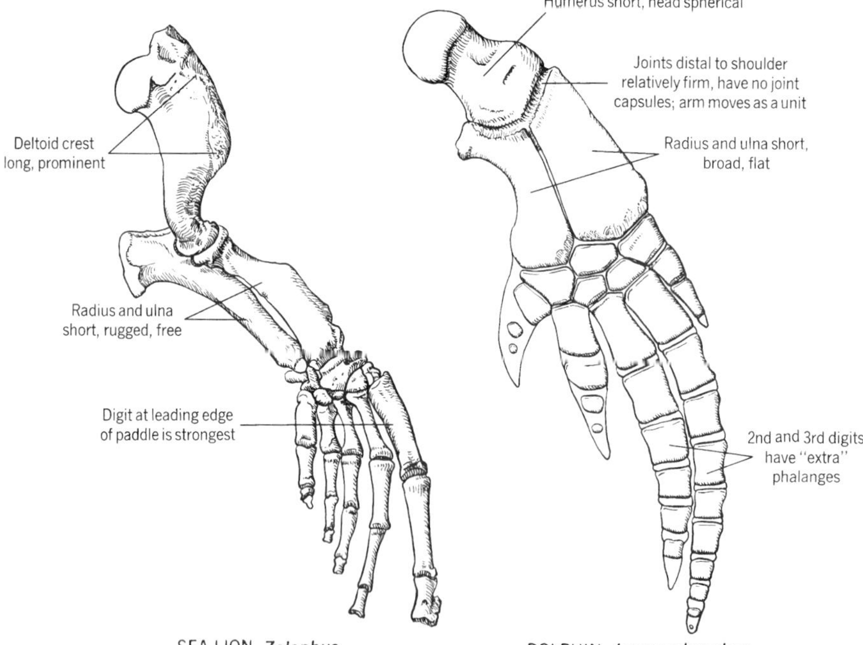

FIGURE 27.13 ARM SKELETONS OF SOME AQUATIC VERTEBRATES that use the pectoral appendage in oscillatory propulsion. Dorsal (lateral) views of right appendage.

top). As for drag-based oscillators, the paddle is stiff and broad, though it tends to be longer, narrower, less constricted at the base, and more tapered near the end. The wing paddle of penguins is broadened by sesamoid bones at the elbow and by a lateral extension of the ulnare of the wrist. The pisiform of sea turtles is similarly extended (Figure 27.13).

[It can be difficult to determine if oscillatory propulsion is drag-based or lift-based, and approaches to the problem vary. For example, Reiss and Frey relied heavily on the

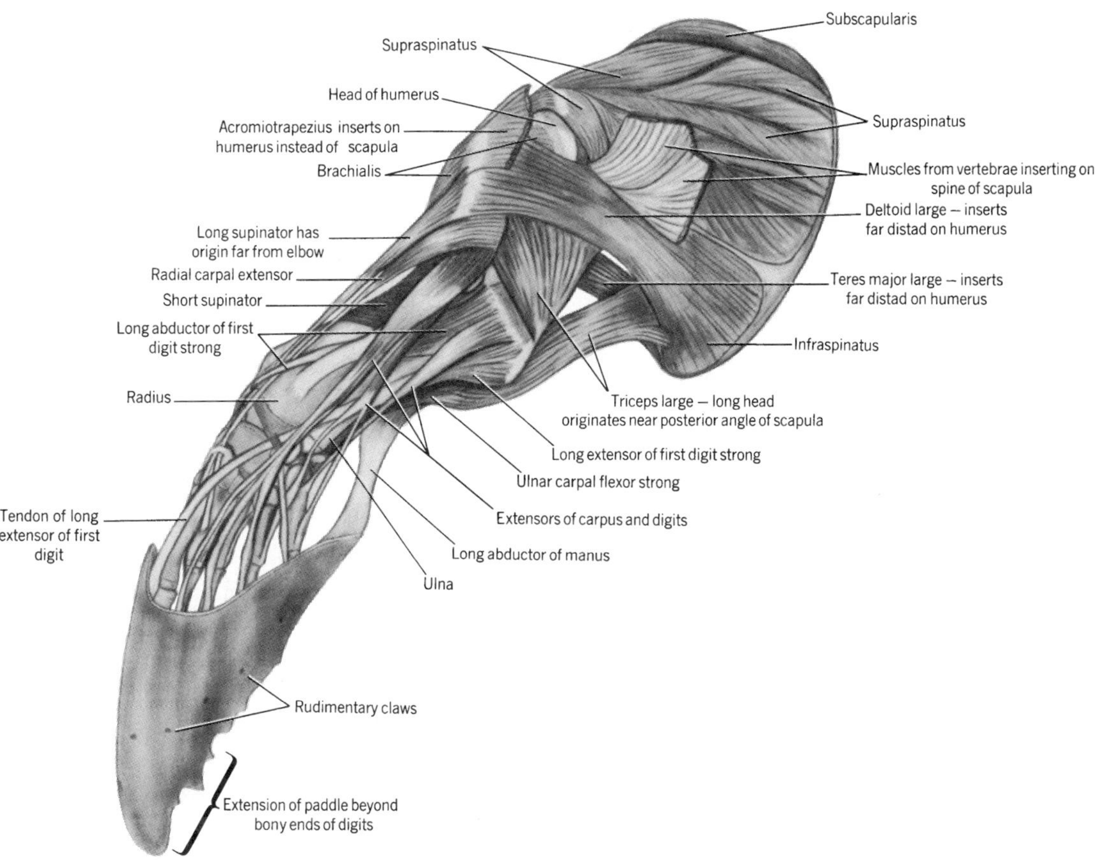

FIGURE 27.14 LEFT FORELIMB OF THE SEA LION, *Zalophus,* seen in lateral view. (Drawn from an air-dried dissection and hence somewhat shrunken.)

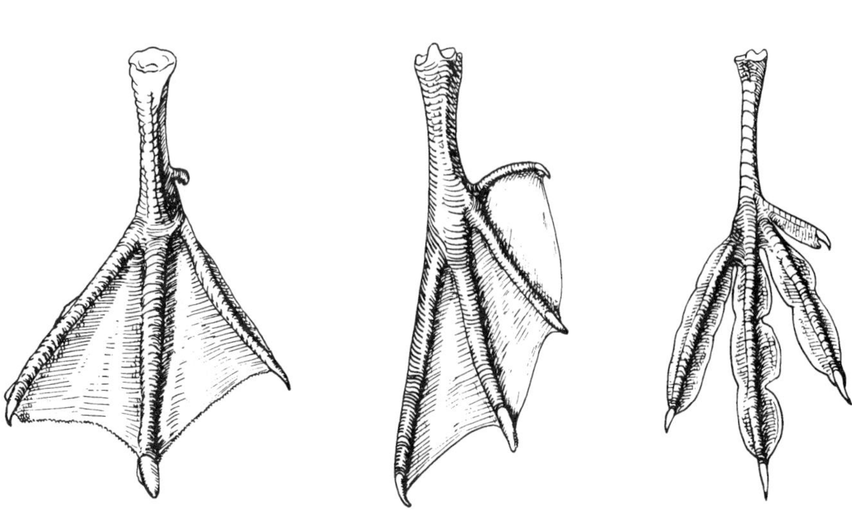

FIGURE 27.15 FEET OF SOME AQUATIC BIRDS.

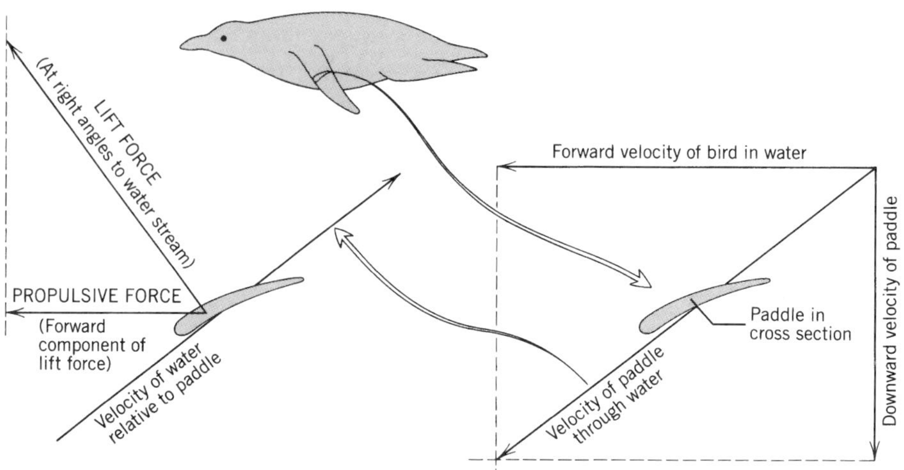

FIGURE 27.16 GENERATION OF PROPULSIVE FORCE IN LIFT-BASED OSCILLATION shown for the downstroke of the paddle of a penguin.

swimming movements of modern turtles and comparative morphology to propose the transition stages from drag-based rowing to underwater "flight" in extinct Mesozoic plesiosaurs. In an effort to reconstruct the secondary transition of mammals from terrestrial to aquatic habitats, Fish used modern semiaquatic (drag-based) forms as analogues and compared their morphologies to fully aquatic (lift-based) mammals. Considered within a context of performance (speed and efficiency), Fish was able to hypothesize probable transition stages. It has been noted that among some bony fishes, particularly the large assemblage of reef-dwelling perciform fishes (wrasses, damselfishes, surfperches), the pectoral fins are used for locomotion. In an integrative study of a high-performance fish, the bird wrasse, assumed to use drag-based propulsion, Westneat examined fin morphology, fin kinematics, and electromyographic patterns of key muscles at several speeds. He then employed mechanical models of fin structure to determine how the fins were used. Surprisingly, he discovered that the bird wrasse does not use the expected drag-based propulsion, particularly at high speeds, but relies instead on a lift-based system.]

Freeloading Another kind of propulsion is used on occasion by some swimmers. If the animal can find water that has either motion (a velocity field) or a suitable pressure gradient (pressure field), it can then freeload, at least in part. One fish or whale may station itself beside and a little behind another, often larger, fish or whale, and thus benefit from the pressure drag created by the lead animal. It is probable that waves caused by the wind are sometimes briefly used in similar fashion.

The most spectacular example of aquatic freeloading is the wave-riding of dolphins. Groups of dolphins may move along for many kilometers in the bow wave of a ship, seemingly without exertion. The pressure field in the front slope of the wave is parallel to the surface of the water, not to the horizontal, and hence has a forward component. Upwelling water thrusting against the dolphins' obliquely oriented tail flukes also provides a forward impetus. The tendency to pitch forward that is created by this pressure on the tail may be compensated by the pectoral flippers. It is clear that the animal is remarkably sensitive to the pressure and velocity fields of its immediate environment and instantly compensates for every change.

CONTROL OF VERTICAL POSITION

Vertebrates that swim only on the surface are light so that they will float high in the water like swans and gulls. Nondiving ducks have a specific gravity of only about 0.6. A light skeleton, fat deposits, gas in air sacs or lungs, and air trapped in feathers or fur contribute to buoyancy. Vertebrates that rest on the bottom, by contrast, need to be more dense than water to maintain their position. Flatfishes and skates, which have no gas bladders, have a specific gravity of about 1.09. The bones of diving birds are less pneumatic. Their air sacs are reduced (loons, penguins). They press their feathers against the body to exclude air: Auks bubble constantly when underwater, and the feathers of some diving birds become wet. Penguins achieve a density of 0.98.

Mammals that dive deep may hyperventilate before submerging, but they do not fill their lungs. Indeed, they may exhale before diving. Deep-diving whales have relatively small lungs. Sirenia, which may feed while resting on the bottom or standing on their tails, have unusually heavy skeletons; their ribs are swollen and solid. Likewise, the skeleton of the hippopotamus is unusually heavy.

Swimmers that vary their vertical position in the water maintain place in any of several ways. The hydrostatic function of the gas bladder of bony fishes was noted on p. 228. Some sharks may exercise some control by selectively producing in the liver either of two metabolites that have different densities. A similar selective production of lipids for storage in muscle, skin, or skull occurs in a number of teleosts and in the surviving coelacanth, which has a fat-filled gas bladder. Various sharks and bony fishes that have no gas bladders and are slightly more dense than water (e.g., leopard shark, mackerel) maintain position by swimming slowly all the time, just as tetrapods breathe all the time. The source of their lift is discussed below.

STABILITY, BRAKING, AND STEERING

Rotation of a swimmer (or ship) around its long axis is called **roll,** rotation around its transverse axis is **pitch,** and rotation around its vertical axis is **yaw.** The body is stable if it passively tends to correct for displacements from a given position; it is unstable if a small displacement tends to increase to become a larger displacement. As for cursors (see p. 445), increased stability of swimmers reduces muscular effort but also reduces maneuverability.

Of several factors influencing stability, one is independent of forward motion. Two forces act on any submerged object: Gravity tends to make it sink, and buoyancy tends to make it rise. If the object has the same density as water, then the two forces are equal and the object neither sinks nor rises. Gravity acts on an object as though all its mass were at its **center of gravity** (CG). Buoyancy acts as though all lift were applied at the object's **center of buoyancy** (CB). The CB is located where the CG would lie if the object were uniformly dense throughout (like the displaced water). Vertebrates, however, are not uniformly dense: Bone, cartilage, and muscle are more dense than water, whereas fat, oil, and gas in lungs or gas bladders are less dense. The CG and CB are, therefore, usually in different places, and gravity and buoyancy act to turn the object in the water (Figure 27.17).

The diaphragm of cetaceans is oriented diagonally under long lungs placed high in the body, rather than behind short lungs placed forward in the body. This raises the CB relative to the CG and places it near, and usually a little above, the CG, thus giving the animal

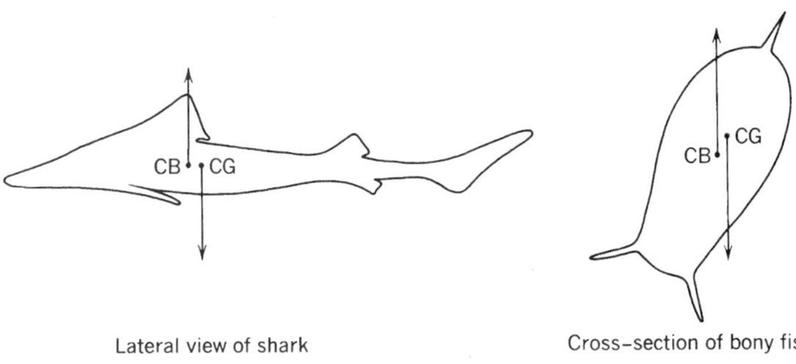

FIGURE 27.17 INSTABILITY RESULTING FROM DIFFERENT LOCATIONS OF CENTER OF GRAVITY (CG) AND CENTER OF BUOYANCY (CB).

slight positive stability. The CG and CB of sharks are at about the same vertical level, so there is scant tendency to roll, but the CB may be a little more anterior, thus tending to lift the head. The gas bladders of bony fishes are located high in the coelomic cavity, but the heavy spine and epaxial muscles are still more dorsal, so the CG tends to lie above the CB, and the fish is unstable in regard to roll. Dead fishes float belly-up.

Any fins or flippers may be extended to function as brakes, but the pectoral fins are most commonly used for this purpose. If they are lower than the CG, then the head tends to pitch downward. This tendency may be countered by canting the pectoral fins to give lift (the entire body then rising as it slows), by extending the dorsal fin, or by actions of other fins. The caudal fin may be curled into a hook as an effective supplemental brake.

Caudal fins of various shapes were named according to evolutionary relationships in the figure on p. 153. Following much experimentation, tails of different shapes have also been classified according to function (see Comment 27.1). Assuming the notochordal or spinal axis to be adequately rigid and the fin membranes to be passive and uniformly flexible, symmetrical (and certain asymmetrical) caudal fins do not cause pitch (parts A–D of Figure 27.18). Caudal fins with a down-tilted spinal axis, or with more membrane below than above the longitudinal axis of the body, or with the dorsal lobe stiffer than the ventral lobe, cause the tail to rise and the head to pitch downward (parts E–G). Tails with the reverse structure have the reverse function (parts H–J).

The traditional view has been that tails of sharks resemble part G; constant lift by the pectoral fins is then needed to counter a tendency for the head to pitch downward. If an edge of the fin membrane does not necessarily trail passively in the water but instead can

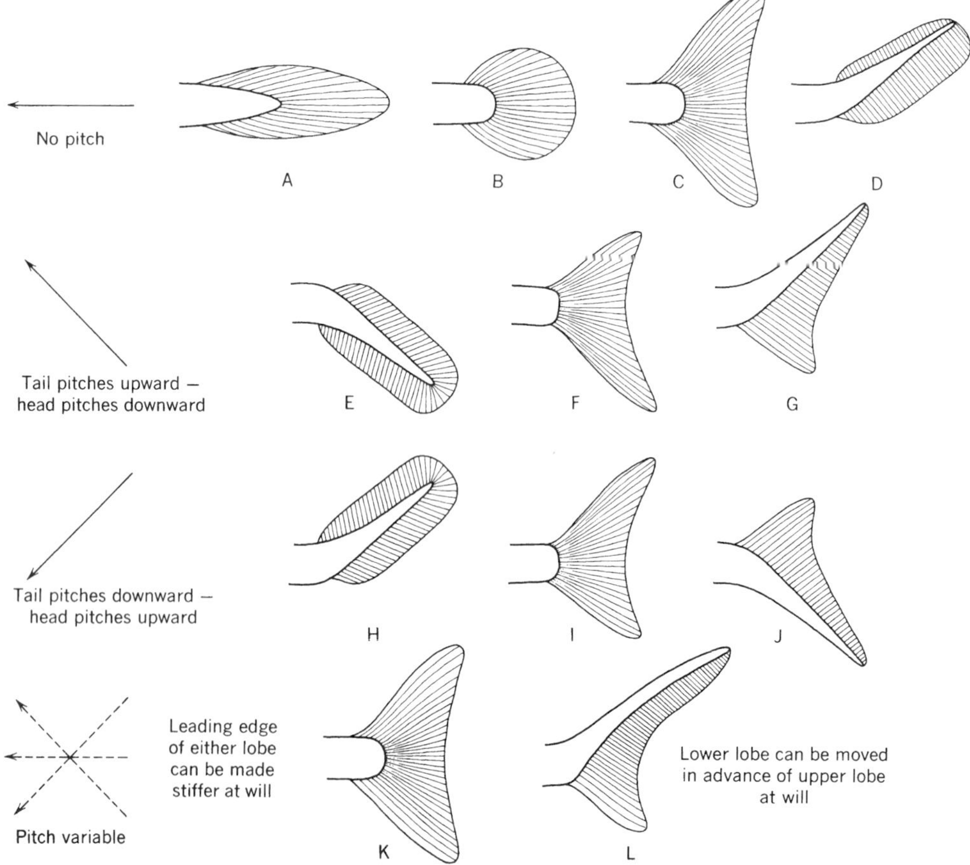

FIGURE 27.18
FUNCTION OF THE CAUDAL FIN IN RELATION TO SHAPE AND OTHER VARIABLES. The membrane is assumed to be passive in A through J and active in K and L.

be made temporarily more rigid, or can be actively advanced with, or ahead of, the axis of the tail (parts K and L and Figure 27.19), then the tail might cause no pitch, or pitch either up or down. Shapes D, G, and L get the tail up and out of the way as the fish swims over the bottom. Most sharks are a little heavier than water, so some lift is required much of the time. See Comment 27.1.

The density of bony fishes that have hydrostatic gas bladders closely approximates that of water. Hence, the tail need not be constructed to provide constant lift. Likewise, the pectoral fins need not provide constant lift, so they can fold against the body to reduce drag. Even though the density of these fishes is about the same as that of water, their equilibrium tends to be unstable, so slight corrections are constantly needed. These fishes can alter the area, extension, curvature, shape, period, and amplitude of all their fins in subtle ways; their sense organs and nervous systems instantly make computations that humans would find difficult to approximate.

Flexibility of the entire body is advantageous for maneuvering. Pinnipeds have supple necks and backs; penguins have secondarily regained much of the flexibility of the trunk that was lost in the ancestry of birds. As explained above, the most maneuverable fishes are the transient swimmers that live among plants and in coral reefs. Their bodies are short, deep, bilaterally compressed, and flexible.

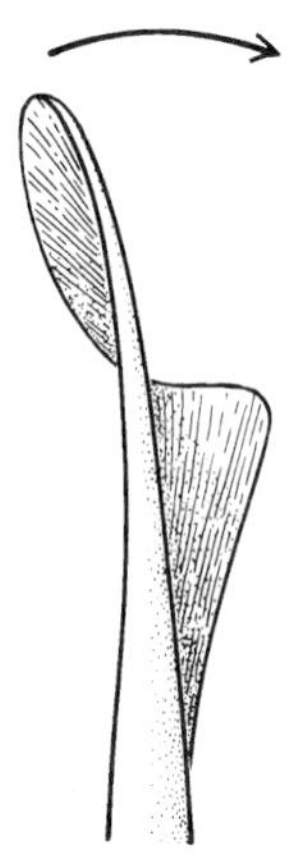

FIGURE 27.19 DORSAL VIEW OF HETEROCERCAL TAIL WITH AN ACTIVE MEMBRANE.

OTHER ADAPTATIONS OF SECONDARY SWIMMERS

Protection of Skin, Ears, and Respiratory System Many aquatic mammals and birds trap enough air in the fur or plumage to shield the skin from wetting. The sebaceous glands of pinnipeds secrete quantities of waterproof sebum, and aquatic birds have large oil glands. The skin of cetaceans is resistant to water but not to drying. It is said that even the pouch of the water opossum is waterproof.

Cormorants and pelicans have lost the external nares. The external nares of other aquatic tetrapods, from frogs and alligators to beavers, hippopotamuses, and dolphins, are always dorsal in position, and the owner seems always to know when they are barely out of the water. A ridge deflects water from the blowhole of many whales. When underwater, the nares are automatically tightly closed. This is usually accomplished by sphincter muscles, but baleen whales use a large valvular plug, and toothed whales add an intricate system of pneumatic sacs, so that great pressure can be resisted in each direction. Respiratory exchange may be surprisingly rapid: Whales of moderate size require only 1 to 2 s to exhale and inhale. Tidal air is relatively great and residual air is relatively small.

Various aquatic tetrapods, from crocodilians to rodents, have modified the palate and glottis to permit chewing and swallowing underwater without interference to the airway: The epiglottis is up within the nasal chamber in beavers as well as dolphins. The pelican, which plunges into water from the wing, has modified the larynx to exclude water under pressure.

The external auditory canal can be plugged or furled by pinnipeds. Cetaceans admit water to the outer part of the canal; the inner part grows shut in baleen whales. Protection of the middle ear from collapse is mentioned in the next section.

Sirenia, and Cetacea and Pinnipedia that dive to moderate depths, have a succession of 8 to 40 valves of smooth muscle in each bronchiole to hold air in the alveoli against pressure. Lung tissue of whales and manatees is relatively rigid because of the presence of cartilage and muscle. Their tracheas and bronchi are thick and strong. The tracheas of penguins, some petrels, sea lions, and the dugong are braced by a longitudinal partition.

Deep-diving mammals do not carry much air down. The lungs are not an oxygen store. If oxygen were exchanged, nitrogen would dissolve in the blood, only to bubble and cause great distress on the return to the surface. Alveolar collapse is probably complete at

COMMENT 27.1

NEW VIEWS OF CAUDAL FIN EVOLUTION AND FUNCTION

Traditional views of evolution and action of the fish tail are being challenged and modified through experimental studies. Lauder studied the action of tails of various shapes of freely swimming fishes using three-dimensional kinematic analysis to record tail movements, and electromyography to monitor muscle activity. He also studied quantitative flow measurements of the wake created by the tail using digital particle image velocimetry. Vortices in the wake shed by a tail cause fluid movements that are proportional to the forces generated by the animal. Thus if one can measure the direction and velocity of individual molecules (or particles suspended in the water) as they are shed from the tail, one can determine their vectors and calculate the forces required to generate them. Small neutrally buoyant reflective glass beads are added to the water, a laser images an "orthogonal slice" that is captured on video, and rapid computational processing of the time-variant patterns of the movements of the individual reflective beads enables quantification of the caudal fin during steady swimming. Lauder's studies of a chondrichthyian fish (leopard shark) and two ray-finned fishes (sturgeon and bluegill sunfish) swimming in a flume (Figure 27.20) found that the caudal fin of leopard sharks functions in a manner consistent with the classical model of heterocercal tail function; that is, in addition to generating a positive forward thrust it generates vertical lift forces and torques that must be counteracted anteriorly by the body and pectoral fins. Although also heterocercal, the sturgeon tail, in contrast, is extremely flexible, and the upper lobe trails the lower during the beat cycle. Although the sturgeon tail also produces a positive thrust throughout the beat, it does not appear to generate net vertical forces. The functional analysis of the homocercal tail of the bluegill was also surprising; the dorsal and ventral lobes do not function symmetrically as expected. Rather, the dorsal lobe undergoes greater lateral excursions and moves at higher velocities than the ventral lobe, suggesting that these asymmetries were actively produced. Recordings of the electrical activity in the muscles traversing from the caudal skeleton to insert on the first four fin rays of the dorsal lobe confirmed this prediction. These data prompted Lauder to propose that the homocercal tail generates lift and torque in the manner expected of a heterocercal tail. The findings from this study further emphasize that morphological characterizations of caudal fins do not accurately reflect function, and that it is likely that future studies will reveal considerable functional diversity among morphologically similar tails.

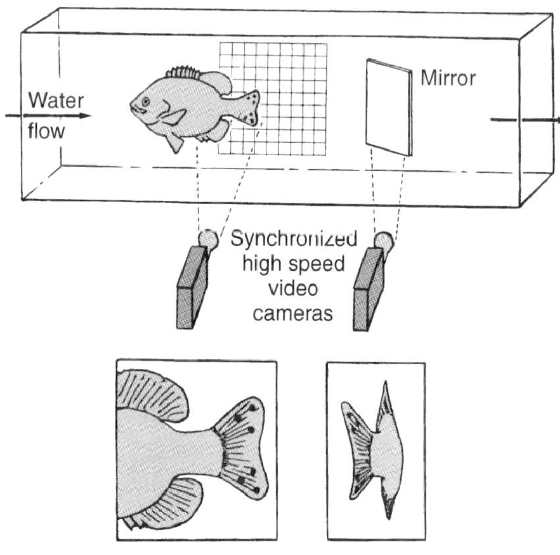

FIGURE 27.20 EXPERIMENTAL SETUP OF G.V. LAUDER FOR RESEARCH ON THE DYNAMICS OF TAIL FUNCTION IN SWIMMING FISHES. Markers on the tail are registration points. Apparatus for electromyography not included here. (Redrawn from G.V. Lauder, 2000, *Am. Zool.* 40:101–122.)

30 m in the Weddell seal and at 70 m in the bottlenose dolphin, forcing the air into larger, stronger, and nonabsorbing airways. Whales have a short sternum and few fixed ribs. The remaining ribs have a single head. Thus, the thorax can also "collapse" without damage.

Adaptations of Sense Organs The sense of TASTE appears to be "normal" in most aquatic vertebrates, but is rudimentary in Cetacea. OLFACTION is considered to be poor in Pinnipedia and very poor in Cetacea. The structure of the nervous system clearly indicates the regression of these senses. Deep-diving mammals have small cranial sinuses, or none.

HEARING is acute in the more highly adapted aquatic tetrapods. Many whales have an amazing repertoire of sounds ranging from low resonant honks to high flutelike tones. In favorable circumstances some whales probably can communicate over distances of 160 km or more, though little is known of their "language." At least most toothed cetaceans and pinnipeds are capable of remarkably discriminating echo-ranging. They make sounds up to 200,000 Hz emitted in pulses of from 16 to 400/s. There is indirect evidence that penguins use the cavitation clicks produced by the turbulence of their own swimming as the sound source for echo-ranging (they can quickly locate fish in absolute darkness).

Underwater hearing is not dependent on a superficial tympanum or external auditory canal, provided that a gas-filled space functions as a resonator (see Chapter 19). The tough, fibrous auditory canal of cetaceans may be partly occluded. The oil-filled cavities of the heads of certain cetaceans may beam sounds and presumably return sound to the ear by specific pathways.

The tympanum and ossicles of cetaceans function in the usual way, yet are of distinctive structure: The ligamentous tympanum is mounted by a horny cone, and the ossicles are large, heavy, and firm. It is essential that the tympanum vibrate against an air space in the middle ear and, of course, that this space not collapse during dives. The middle ear of pinnipeds (and possibly parts of the external canal), and air sinuses communicating with the middle ear of cetaceans, are lined by highly vascular tissue that engorges during dives. The volume lost by compression of air is thus replaced by blood. Furthermore, in cetaceans the sinuses and parts of the middle ear not adjacent to the drum are filled with a foam consisting of small air bubbles in an oil-mucus emulsion. Experiments show that these bubbles do not collapse, even under a pressure of 100 atm. Finally, the tympanic bulla of whales is strengthened by some of the thickest, most dense bone known. The bullas are loosely attached to the remainder of the skull and are cushioned in foam and blood sinuses. This permits them to function independently of each other and of the body. Directional hearing of aquatic mammals is excellent and probably is based largely on intensity discrimination.

Sirenians have poor VISION, as would be expected from their sluggish habits, stationary (plant) food, and often murky environment. Baleen whales have moderately good eyesight, but have a limited field of vision. Their food is passive, and some of them dive below the level of light penetration. The food of toothed whales and pinnipeds is active, and they have excellent vision, both below and above water.

Eyes of aquatic tetrapods with good vision have secondarily acquired characteristics of the eyes of their remote ancestors among primary swimmers: The eye is large, the eyeball short along its optical axis, the lens large and spherical, and the cornea flattish or elliptical for streamlining. Lacrimal glands are reduced (pinnipeds) or absent (sirenians, cetaceans). To protect the eye from saltwater, the cornea is cornified and is bathed by the secretion of large glands in the lids. In order to withstand wave pressure, the sclera of cetaceans is very thick and tough. Pinnipeds can change the shape of the lens more than is usual to permit vision both in and out of water.

Thermoregulation and Response of the Circulatory System Since the thermal conductivity of water is about 20 times greater than that of air, endothermic aquatic vertebrates must protect themselves from heat loss, particularly when inactive and when in cold seas. Air trapped in plumage or dry underfur (as of the sea otter, beaver, or fur seal) is an effective insulator. Large mammals have relatively little heat loss because of their low surface-to-volume ratio. Blubber is an effective insulator for them, coming, in extreme instances, to one-quarter of the body weight. Flippers of cetaceans have slow circulation and countercurrent exchange, so warm outgoing blood gives its heat to the cold incoming blood (see p. 265). It is probable that some whales require moderate activity to maintain body temperature in arctic waters.

Conversely, swimmers must be able to dissipate heat during periods of activity when heat production may rise tenfold. The countercurrent exchange mechanism can be bypassed, and the large flat flippers (which have little blubber) then serve as radiators. Also, vascular papillary ridges in the epidermis of whales dissipate heat when needed.

The circulatory physiology of air-breathing vertebrates during dives adjusts to supply oxygen to the brain and heart, and otherwise to avoid stress from lack of oxygen or buildup of carbon dioxide and lactic acid. Bradycardia, or slowing of the heart, is universal and occurs on submergence. The rate commonly is reduced to one-tenth or one-fifteenth of normal, and the slowing is in part preventive; its onset is faster when a deep dive is anticipated. The aorta dilates near the heart to help maintain blood pressure during bradycardia, but all arterioles constrict except those of the brain and heart. Excretion stops. The veins of pinnipeds and cetaceans (like those of fishes) have no valves. Blood volume of these swimmers, and of some diving turtles and birds, reaches two times (even $2\frac{1}{2}$ times in elephant seals) that of comparable terrestrial vertebrates. The hepatic portal system is large, and venous sinuses may be present in the thorax and abdomen. The result is that quantities of blood stagnate in the body cavities.

Deep divers are usually not very active. Their hearts tend to be relatively small (though that of a blue whale may still weigh 600 kg!). The metabolic rate falls off a little (pinnipeds, cetaceans, alligators, ducks). The blood of fast-swimming and deep-diving dolphins is able to carry up to three times as much oxygen as that of their less-active relatives. The myoglobin content of divers' muscles is high. They tolerate twice as much carbon dioxide in the blood as do humans. Lactic acid is stored in the muscles until breathing resumes. These various adaptations are so effective for elephant seals that their long dives are aerobic; lactic acid does not build up and they need only 3 min between dives.

The circulatory systems of swimmers show convergence in other ways that remain enigmatic. Why is the postcava doubled (a turtle, pinnipeds, cetaceans, and sirenians, but also the nonaquatic edentates and slow loris)? Why do pinnipeds have a sphincter in the postcava at the level of the diaphragm? Why are the intervertebral vessels enlarged and the jugular veins reduced (pinnipeds, cetaceans)? Why is there a venous plexus in the drainage of the kidney, or a rete to damp the flow of blood to the brain (cetaceans)?

Reproductive Biology Most sea snakes are viviparous and give birth at sea, as did the ichthyosaurs. Other reptiles and all birds lay their eggs or give birth on land. Among aquatic mammals, the walrus and hippopotamus sometimes give birth in the water, and Cetacea and Sirenia always do so. A single, large, precocious young is born at a time. Cetacea deliver rapidly; the calf emerges tailfirst. The mother whirls in the water, thus snapping the relatively short umbilical cord at a predetermined point of weakness. The newborn swims to the surface to breathe, sometimes with maternal assistance. The tail flukes are soft and curled at birth, but harden in about two days, by which time the calf can keep up with the herd.

Whale milk is thick and rich in fat. It collects in sinuses and is forced out in mouthfuls during underwater nursing. Growth of young pinnipeds and cetaceans is rapid. The calf of the blue whale, which is 7 m long at birth, gains about 90 kg/day on its mother's milk!

REFERENCES

Alexander, R. McN. 1967. *Functional design in fishes.* Hutchinson Press, London. 160p.

Bannasch, R. 1994. Functional anatomy of the "flight" apparatus in penguins, pp. 163–192. *In* L. Maddock, Q. Bone, and J.M.V. Rayner (eds.), *Mechanics and physiology of animal swimming.* Cambridge Univ. Press, New York. Comprehensive review of the wing anatomy of penguins and how it relates to wing kinematics during swimming.

Black, B.A. 1992. Direct measurement of swimming speeds and depth of blue marlin. *J. Exp. Biol.* 166:267–284.

Daniel, T.L., C. Jordan, and D. Grunbaum. 1992. Hydromechanics of swimming, pp. 17–49. *In* R. McN. Alexander (ed.), *Advances in comparative and environmental physiology,* Vol. 11. Springer, Berlin.

Denison, D.M., and G.L. Kooyman. 1973. The structure and function of the small airways in pinniped and sea otter lungs. *Resp. Physiol.* 17:1–10.

Drucker, E.G., and G.V. Lauder. 1999. Locomotor forces on a swimming fish: three-dimensional vortex wake dynamics quantified using digital particle image velocimetry. *J. Exp. Biol.* 202:2393–2412. A comprehensive review and analysis of forces produced by hydrofoils and a detailed explanation of the principles of digital particle image velocimetry.

Feldkamp, S.D. 1987. Swimming in the California sea lion: morphometrics, drag and energetics. *J. Exp. Biol.* 131:117–135.

Fetcho, J.R. 1991. Spinal network of the Mauthner cell. *Brain Behav. Ecol.* 37:298–316. The role of the Mauthner cell in the initiation of the fast start escape behavior of swimming fishes and amphibians.

Fish, F.E. 1996. Transitions from drag-based to lift-based propulsion in mammalian swimming. *Am. Zool.* 36:628–641.

Fish, F.E. 1998. Comparative kinematics and hydrodynamics of odontocete cetaceans: morphological and ecological correlates with swimming performance. *J. Exp. Biol.* 201:2867–2877. Analyzes body and fluke morphology of three species of cetaceans and examines their effects on performance.

Hale, M.E. 1996. The development of fast-start performance in fishes: escape kinematics of the chinook salmon (*Oncorhynchus tshawytscha*). *Am. Zool.* 36:695–709.

Jayne, B.C., and G.V. Lauder. 1994. Comparative morphology of the myomeres and axial skeleton in four genera of centrarchid fishes. *J. Morphol.* 220:185–205. Excellent morphology.

Kooyman, G.L., et al. 1992. Diving behavior and energetics during foraging cycles in king penguins. *Ecological Monographs* 62:143–163. Review of diving biology of penguins and the use of technology to obtain measurements in the field.

Lauder, G.V. 2000. Function of the caudal fin during locomotion in fishes: kinematics, flow visualization, and evolutionary patterns. *Am. Zool.* 40:101–122.

Lauder, G.V., and B.C. Jayne. 1996. Pectoral fin locomotion in fishes: testing drag-based models using three-dimensional kinematics. *Am. Zool.* 36:567–581.

Le Boeuf, B.J. 1989. Incredible diving machines. *Nat. Hist.* 1989(2):35–40. About the elephant seal.

Long. J.H., Jr., and K.S. Nipper. 1996. The importance of body stiffness in undulatory propulsion. *Am. Zool.* 36:678–694.

Pabst, D.A. 2000. To bend a dolphin: convergence of force transmission designs in cetaceans and scombrid fishes. *Am. Zool.* 40:146–155.

Reiss, J., and E. Frey. 1991. The evolution of underwater flight and the locomotion of plesiosaurs, pp. 131–144. *In* J.M.V. Rayner and R.J. Wooten (eds.), *Biomechanics in evolution.* Cambridge Univ. Press, New York.

Rome, L. 1994. The mechanical design of the fish muscular system, pp. 75–98. *In* L. Maddock, Q. Bone, and J.M.V. Rayner (eds.), *Mechanics and physiology of animal swimming.* Cambridge Univ. Press, New York.

Rosenberger, L.J., and M.W. Westneat. 1999. Functional morphology of undulatory pectoral fin locomotion in the stingray *Taeniura lymma* (Chondrichthyes: Dasyatidae). *J. Exp. Biol.* 202:3523–3539. Informative study of the kinematic patterns and muscle activity patterns of the pectoral fin of the blue-spot ray. Includes a comprehensive review of this fascinating form of locomotion.

Vogel, S. 1981. *Life in moving fluids: the physical biology of flow.* 2nd ed. Princeton Univ. Press, Boston. 467p.

Wainwright, S.A., F. Vosburgh, and J.H. Hebrank. 1978. Shark skin: function in locomotion. *Science* 202:747–749.

Webb, P.W. 1984. Form and function in fish swimming. *Sci. Am.* 251(1):72–82.

Webb, P.W. 1988. Simple physical principles and vertebrate aquatic locomotion. *Am. Zool.* 28:709–725.

Webb, P.W. 1994. The biology of fish swimming, pp. 45–62. *In* L. Maddock, Q. Bone, and J.M.V. Rayner (eds.)., *Mechanics and physiology of animal swimming.* Cambridge Univ. Press, New York. An excellent general review of the various types of locomotion observed in fishes: directions for future studies offered.

Webb, P.W., and R.W. Blake. 1995. Swimming, pp. 110–128. *In* M. Hildebrand et al. (eds.), *Functional vertebrate morphology.* Harvard Univ. Press, Cambridge, MA.

Westneat, M.W. 1996. Functional morphology of aquatic flight in fishes: kinematics, eletromyography, and mechanical modeling of labriform locomotion. *Am. Zool.* 36:582–598.

Westneat, M.W., et al. 1993. The horizontal septum: mechanisms of force transfer in locomotion of scombrid fishes (Scombridae, Perciformes). *J. Morphol.* 217:183–204. A combination of dissection and modeling to determine how forces are transmitted to the backbone and tail.

Zapol, W.M. 1987. Diving adaptations of the Weddell seal. *Sci. Am.* 256(6):100–105.

Chapter 28

Flying and Gliding

If an animal is capable of sustaining itself in the air, we say that it can **fly** and is **volant.** The principles of vertebrate flight are understood in general, but analysis is complex and simplifications are necessary at almost every level. Much of what we know about the aerodynamics of flight comes from our understanding of fixed-wing aircraft, and when these principles are applied to animals with flapping wings, errors are invariably introduced. Even when vertebrates fly under relatively uniform conditions (in the field or in a wind tunnel), they must make frequent slight adjustments to compensate for alterations of external variables. When a gull maneuvers in a changeble wind, major adjustments, many of kinds that are not possible for man-made aircraft, must be made constantly and nearly instantaneously. We are challenged to determine the morphological and behavioral bases of such performance. Thanks to ingenious methods, we have made much progress in the last several decades. It can be expected that careful observation and the imaginative application of sophisticated technologies will continue to reward us with discoveries about this complicated activity.

Some lightweight climbers can retard a fall and travel horizontally as they move downward. If control of the path followed is minimal and the line from takeoff to landing is steeper than about 45° to the horizontal, then the animal is said to **parachute.** If some maneuvering in the air is possible and the line from takeoff to landing is less than 45° to the horizontal, the animal is said to **glide.**

ORIGIN AND ADVANTAGES OF FLYING AND GLIDING

Although much more primitive than modern birds, the first known birds (suborder Archaeornithes) were apparently already moderately good flyers (see figure on p. 59). There are several theories on the origin of avian flight. One is that birds evolved from small, bipedal, arboreal archosaurs that hopped from branch to branch, steadying themselves with outstretched forelimbs. As feathers enlarged on the margins of the forelimbs, the hops were

Above: Grappling eagles. (Redrawn from V. Gargett, 1990, *The black eagle,* Academic Press.)

likely extended. The animals may have utilized gravity to increase air speed as they planed to a lower perch. A second theory is that avian ancestors were small, bipedal, cursorial dinosaurs that steadied themselves with outstretched arms. An alternative is that the nonflying, ground-living ancestor gained stability and control from its protowings as it hopped into the air to catch flying insects. The origin of flight was related to the origin of birds in Comment 4.1 on p. 61.

The first known bat lived 50 million years ago, but was modern in appearance and flying ability. Presumably bats evolved from small, agile mammals that scrambled about in trees seeking insects. The flying reptiles, or pterosaurs (an order of the infraclass Archosauria), survived from about 180 until 65 million years ago. The earliest known representatives were already competent flyers, so nothing is certainly known of the origin of their ability to fly.

As we shall see, birds, bats, and pterosaurs differ considerably in structure. Nevertheless, in order to fly at all, certain conditions must be met, and these three groups display much convergent evolution. Flyers are among the most specialized of vertebrates.

Parachuters may drop from tree to ground to escape from less versatile predators or to move quickly to another location. Also, an inadvertent fall is not damaging. Gliders have the same advantages to a greater degree, and by gliding can forage more quickly and over a wider area than would otherwise be possible. They have the advantage over flyers that the forelimbs, being little specialized, remain useful for climbing and the manipulation of food.

The various flyers benefit from flight in different ways. Potential advantages include the following:

1. Flyers gain access to food that is in the air (flying insects), that must be reached from the air (terminal flowers), or that can be located from the air (rodents, fishes).
2. Great mobility and maneuverability enable flyers to search rapidly and efficiently for food and shelter.
3. Escape is provided from nonvolant predators.
4. By migrating, flyers can travel, according to season, to regions where climate, food supply, and nesting sites are favorable.
5. Dispersal is possible over distances and geographic barriers that would otherwise be insurmountable.

VERTEBRATES THAT PARACHUTE AND GLIDE

Parachuters and Gliders All parachuters are arboreal. Various tree frogs are included. These launch themselves with a jump, hold the limbs out to the side, and control their orientation so that their flat ventral surface is presented to the airstream. One species (family Hylidae) has no other adaptations for slowing its fall, yet achieves an angle of descent of about 60°. Several other tree frogs (family Rhacophoridae) are more expert, approaching an angle of descent of 45°. Their huge feet are fully webbed, and small membranes fringe the arms and span the angle between thighs and body wall (Figure 28.1).

A genus of tree snakes is capable of controlled parachuting at fairly flat angles. The body is held horizontal, the ribs are spread to the sides, and the belly is drawn in to present a concave surface to the airstream.

Some lizards can descend at about 70°, relying mainly on behavioral adaptations. Others, having fringed tails, do a little better. Several genera of geckos (family Gekkonidae) have broadly webbed toes and fringes on the head and body. They can descend at nearly 45° to the horizontal with considerable maneuvering. Various small tree squirrels

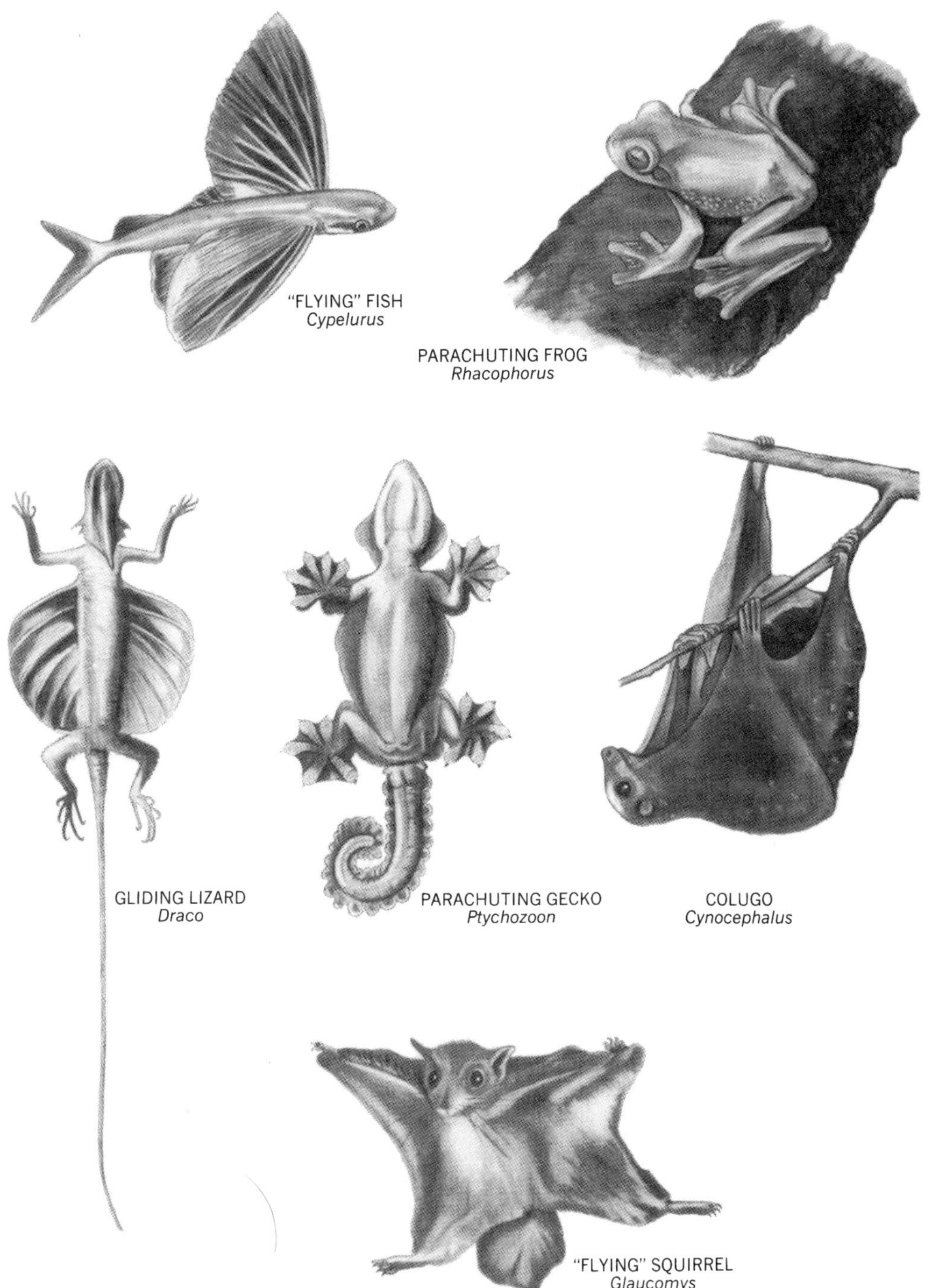

FIGURE 28.1
EXAMPLES OF PARACHUTERS AND GLIDERS.

also parachute to break a fall by spreading legs and tail and coming down flat to the airstream.

Several fishes, at least three genera of lizards, and representatives of three orders of mammals are gliders. All have evolved broad membranes to catch an airstream, but these can function only when adequate air speed is attained. The lizards and mammals climb a tree to gain height, and then they jump. They fall steeply until sufficient speed is attained by gravity, then they sail off following a parabolic path. Doubling the height of the takeoff more than doubles the horizontal travel.

The fishes attain adequate air speed in a very different way. A "flying" fish (family Exocoetidae) first swims rapidly just under the surface and then emerges until only the large lower lobe of its hypocercal tail is in the water. The pectoral fins, which in one genus are as long as the body, are spread to the sides as the tail thrusts rapidly back and forth in the water. The fish skims along in this manner for from 1 to 6 m until its air speed increases to an estimated 40 to 70 km/h. It then spreads the somewhat smaller pelvic fins and glides free of the water, usually for 2 to 4 s, but occasionally for 10 s or more, sometimes traveling 100 m. On landing, the fish either submerges or immediately skims again with only the tail in the water preparatory to another glide. From 2 to 12 glides may be made in succession. The fishes are unable to glide if there is no wind, but they can sail several meters high over the water if there is a good breeze. The fins are fixed during most of the glide, but they may be seen and heard to vibrate at takeoff and landing.

Some fishes of another family (Characinidae) leap high in the air and then spread their large pectoral fins stiffly as they fall back to the water. They sometimes jump into boats. Other members of the same family have larger pectoral fins that they flutter, using enormous ventral muscles as they make long arcing excursions from the water. The aerodynamics of this near flight is unknown.

Gliding lizards of the genus *Draco* have extensive membranes reaching on each side from the thorax to the base of the hind leg. The membranes are supported by six pairs of much-lengthened ribs. When they are not in use for gliding or display, they are folded against the body. These lizards commonly glide at 20° to 30° to the horizontal, but can even gain elevation if they move into an updraft. Glides of 24 m have been observed. (There were gliding lepidosaurs in the Permian period that also supported their membrane with ribs.)

The colugo (mammalian order Dermoptera) is a cat-sized Asiatic glider with the largest "flight" membrane of all: It extends from the throat to the wrists to the ankles to the tip of the tail. Even the toes are webbed. One animal sailed 136 m at an angle of only 5° to the horizontal.

Five species of phalangers (order Marsupialia) in three genera are gliders. They range in weight from 14 to about 1360 g. All are vivacious forest dwellers with soft fur, long bushy tails, and membranes extending from elbow to knee. They launch themselves with a leap, and commonly glide up to 100 m.

Fifteen genera of rodents are expert gliders. They range from chipmunk size to cat size. The very large membranes of scaly-tailed squirrels (family Anomaluridae) are supported in part by long cartilaginous struts (or calcars) from the elbows. Some "flying" squirrels (family Sciuridae) have shorter struts from the wrists. Usual glides are 6 to 10 m in length at angles of 30° to 50° to the ground. However, an American species was seen to glide 50 m with a drop of only 18 m. Such figures really mean little; in usual circumstances all gliders seem able to glide as far as they "want" to.

Most birds frequently intersperse gliding with flying, sometimes with wings open and sometimes with wings folded. The saving of energy may be considerable.

FLYERS AND THEIR SKILLS

PTEROSAURS had short bodies, long necks, large birdlike heads, and long narrow wings supported by the arms and elongated fourth fingers (Figures 28.2 and 28.3). Eyes and brains were also birdlike.

There were no scales. It was claimed that some had hair, but a relatively good fossil described in 1994 was interpreted as hairless. The wing membrane was stabilized by

FIGURE 28.2
RESTORATION OF THE GIANT PTEROSAUR, *Pteranodon.*

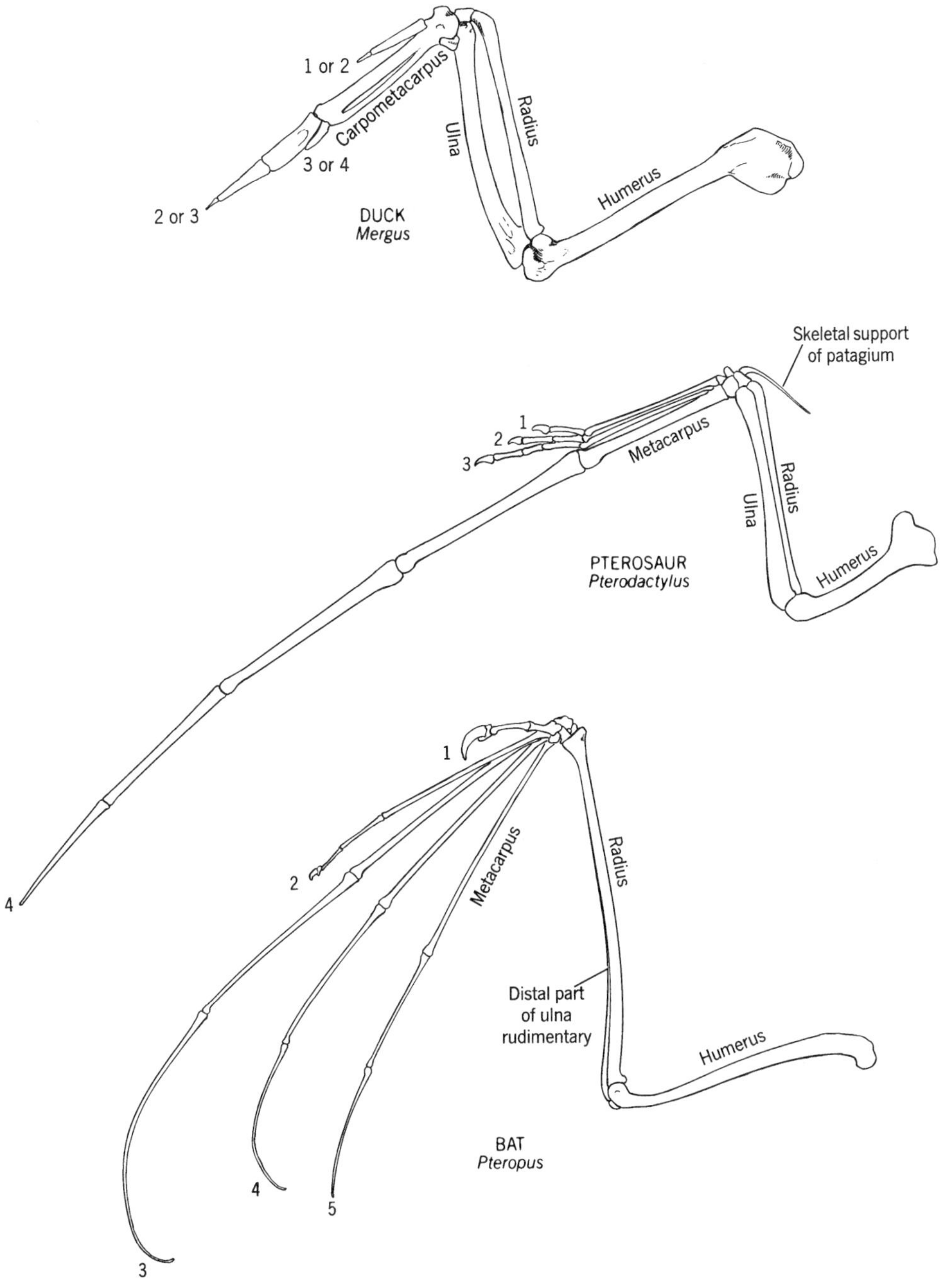

FIGURE 28.3
COMPARISON OF THE RIGHT WING SKELETONS OF A BIRD, A PTEROSAUR, AND A BAT. The digits are numbered.

thin fibers that might be mistaken for hair. It is probable that pterosaurs were able to elevate the body temperature, at least when flying. One group had long tails and many homodont teeth; the other group had short tails and few or no teeth. The more than 25 known genera ranged from the size of a starling to the largest of all flyers, which had a wingspan estimated to be 11 to 12 m. Most pterosaurs lived along seacoasts and ate fish. It is probable that some could alight and take off into the wind from water. Perhaps some could even swim a little with their wings. Although many fossils are in existence, the biology of pterosaurs is contentious. There is a tendency to interpret their morphology in a birdlike or batlike framework. It has been claimed, for example, that pterosaurs were bipedal, birdlike creatures with legs not joined by the flight membrane below the knee (see Padian and Rayner in References). Others (see review by Unwin) make a strong case for a quadrupedal, more batlike animal that was capable of flying long distances.

BATS are very successful flyers. There are about 175 living genera and more species than in any other mammalian order except Rodentia. The smallest bats weigh only 4 g. The largest weigh 900 g and have a wingspan of 1.7 m. Bats are variously adapted for eating insects, fruits, flowers, nectar and pollen, blood, and fish. Flight membranes of skin are supported by the arms, greatly elongated second through fifth fingers, hind limbs, and usually all or part of the tail. Some bats have fast and direct flight and others have slow and erratic flight.

BIRDS are relatively uniform in structure compared with other vertebrate classes, yet are diverse in habits and habitats. The 2 g bee hummer is the smallest. Excluding flightless birds, the extinct teratorn vultures were the largest, perhaps reaching 80 kg and a wingspan of about 7 m.

Flight feathers are supported by the long arms and one robust digit; those borne by the hand are **primary feathers** and those on the forearm are **secondary feathers.** A feathered membrane called the **patagium** spans the angle in front of the elbow.

Speed Many flyers are capable of high speed, yet their true capabilities are difficult to ascertain. The fastest reliably measured speed is 58 m s^{-1} (210 km/h) for a diving gyrfalcon. Songbirds fly 16 to 40 km/h, ducks cruise at 50 to 65 km/h, and several species probably can fly 100 km/h when pressed. In terms of body lengths per second, a songbird might move 5 times faster than a cheetah and 20 times faster than a human.

Endurance In contrast to cursors and swimmers, it is small to medium-sized flyers that have the greatest endurance. The mastiff bat remains on the wing continuously for 6 h or more. The sooty shearwater and sanderling migrate 11,000 to 13,000 km one way between Arctic America and Patagonia. The golden plover flies 3800 km nonstop from Labrador to South America. A wandering albatross (tracked by satellite) flew 15,000 km on one foraging trip. Various species may remain in the air for 90 h when migrating. It was calculated that if the fat stores of the blackpoll warbler are equated to gasoline on a weight basis, the little bird gets 720,000 mpg.

High Flying and Lift Certain bats fly at least 3000 m high. Most birds fly below 1500 m, but migrants occasionally travel as high as 6400 m (21,000 ft). Birds have been observed above 9000 m in the Himalayas. Even when at rest, mammals become unconscious at lesser altitudes: To compensate for low oxygen, they hyperventilate. This flushes carbon dioxide from the body, which makes the blood alkaline, and that in turn causes blood vessels, to the brain and elsewhere, to constrict. An unknown mechanism enables birds to hyperventilate without vasoconstriction.

Numerous bats can fly carrying young up to 50 percent of their own weight. One species can lift 73 percent of its weight. Various birds can carry loads of prey or fat stores that about equal their unladen weight.

Acceleration and Maneuverability With new technology, we are just beginning to document these remarkable skills of flyers. Certain bats and birds start, stop, and turn at impressive rates. Flocking birds do these things in unison. Many bats and birds can hover in place, and hummingbirds can fly backward.

GENERAL REQUIREMENTS OF FLYERS

Flyers, like swimmers, move within a fluid medium. Aerodynamics and hydrodynamics are closely related fields, so the requirements of flyers parallel those of swimmers. However, air being much less dense than water, the relative importance of the variables is altered and flyers are deprived of support by flotation. All flyers must (1) derive sufficient upward force from their muscles or from the environment to counter the pull of gravity; (2) reduce drag, particularly if flights are long or fast; (3) propel themselves at various speeds and sometimes in restricted spaces; and (4) retain stability, maneuver, brake, and land according to habit. These primary requirements establish some rigid secondary requirements that focus on needs for (5) strength with light weight, (6) firmness of the trunk, and (7) the efficient production and utilization of power.

MOSTLY ABOUT LIFT

Since flyers are more dense than air, an upward force must act on them in order for flight to be sustained. During level flight, this force must just counter the pull of gravity, which is to say it must equal the flyer's weight. During ascending flight, flight in a downdraft, and flight while carrying young or prey, the upward force must exceed the weight of the flyer. Where does the upward force come from?

We shall first consider only level flight in still air with no flapping of the wings. These conditions are set so we can clearly distinguish upward force from backward force (drag) and forward force (propulsion), which are considered in following sections.

As a first approximation, imagine a crude model of a flying bird with wings cut from thin slats of wood. If moved in an airstream in such a way that the wings meet the wind exactly edge-on, then air flows equally over and under the wings and there is no upward force (Figure 28.4A).

Now suppose that the model is improved by tilting the leading edge of the wing upward. The angle the wing makes with the airstream is called the **angle of attack,** or α. When α is small, air flows over the wing as shown in part B. (This is the result of combining the oncoming airflow with the tendency of air to flow forward under a wing.) Air passing over the wing travels farther and faster than air below. Consequently (following Bernoulli's theorem, which relates velocity to pressure in "ideal" fluids), the pressure is lower above the wing than below. The disturbed part of the airstream is thinnest near the leading edge of each wing. Air moves fastest there and creates the lowest pressure. Forces on all other parts of the wing are also in proportion to the adjacent air pressures (part C). All the forces acting on the wing that are derived from its motion can be divided into a component called **drag (D),** which by definition is in line with the airstream and opposite to the direction of flight, and a component called **lift (L),** which is at right angles to **D** (part D). These forces act at the center of pressure, **X,** which is usually one-quarter to one-half of the way back from the leading edge of the wing. In other words, all the actual forces together have the same effect on the wing as **D** and **L** acting at **X.**

In level flight in still air, **L** is directly upward and is the force needed to counter the pull of gravity. In ascending or descending flight on fixed wings, "lift," in spite of the

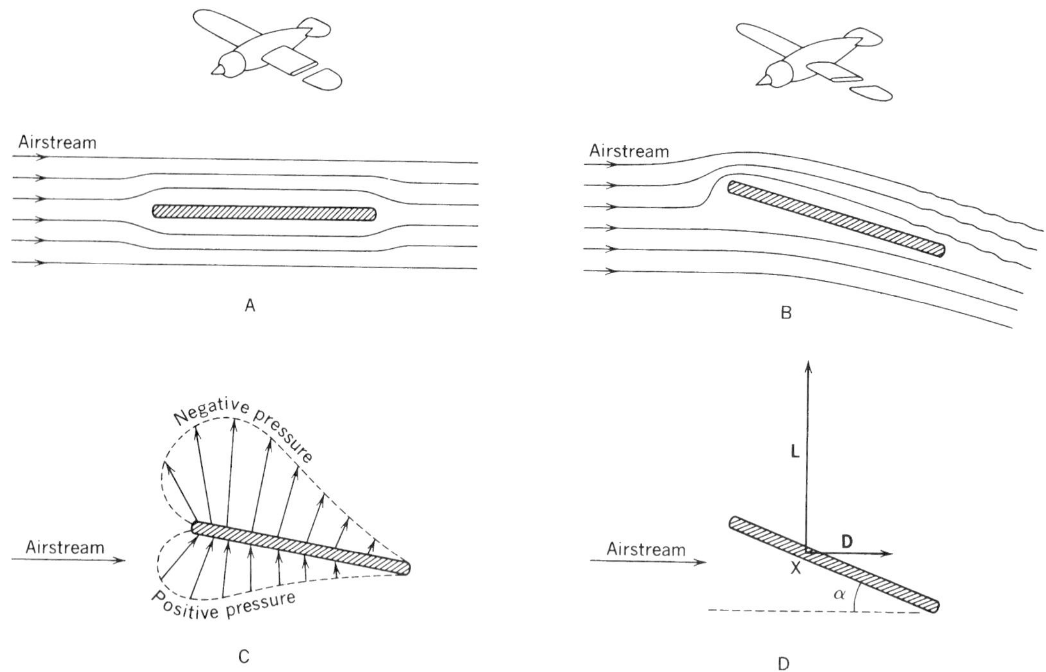

FIGURE 28.4 SOME FACTORS RELATED TO LIFT.

term, is not vertical but at right angles to the airstream. It does always have a vertical component (unless the flyer flies upside-down or depresses the leading edge of the wing, making α negative), but it also has a horizontal component that may be forward (descending flight) or backward (ascending flight).

As α increases from 0°, **L** also increases (Figure 28.5, left), and the center of pressure moves forward to about one-quarter of the way back from the leading edge of the airfoil. However, as (α) exceeds about 15°, the airstream above the wing suddenly lifts away from the leading edge of the wing, ceases to flow smoothly, and instead forms strong eddies. Lift is then lost, and the flyer is said to **stall.**

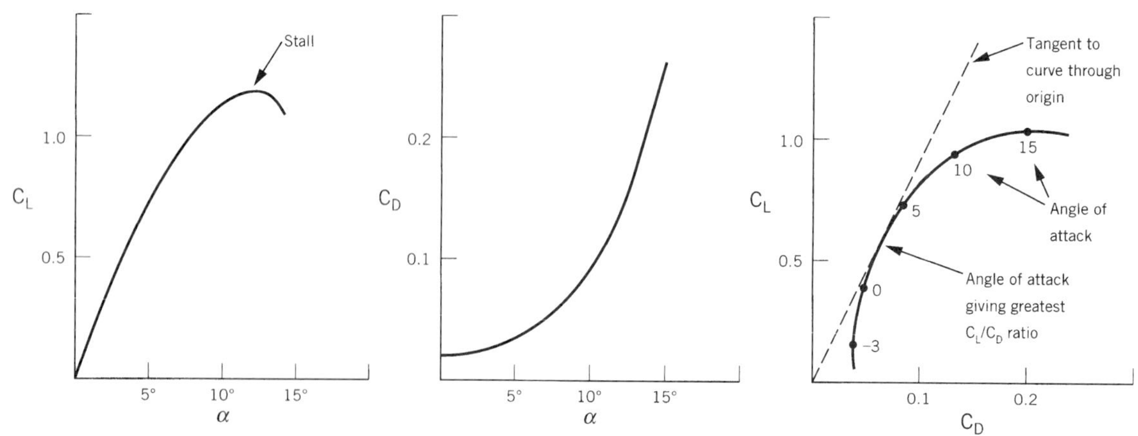

FIGURE 28.5 RELATIONS BETWEEN THE COEFFICIENT OF LIFT, C_L, COEFFICIENT OF DRAG, C_D, AND ANGLE OF ATTACK, α. Curves vary according to size and shape of airfoil.

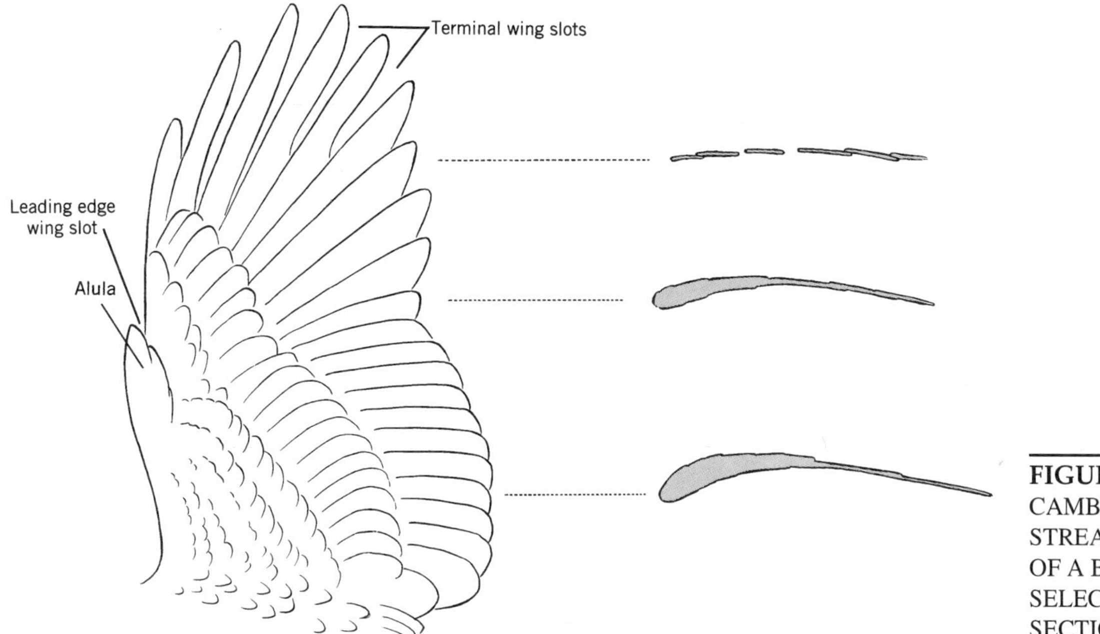

FIGURE 28.6
CAMBER AND STREAMLINING OF A BIRD WING AT SELECTED CROSS SECTIONS.

So far we have considered only an airfoil (or wing) that is flat, like a thin slat of wood. The wings of aircraft and vertebrate flyers are instead convex on their upper surfaces as seen in cross section (Figure 28.6); they are said to have **camber.** Camber increases the velocity of the air flowing above the airfoil (relative to that below), thus significantly increasing lift. The straight line joining front and back edges of a cambered airfoil is called the **chord** (Figure 28.7). Moreover, the placement of bone and flesh in a vertebrate wing is such that it is thicker along its leading edge than elsewhere, making the wing "streamlined" in cross section. This is particularly true of bird wings (Figure 28.6). Finally, if a small second airfoil is positioned just above the leading edge of the first, a **wing slot** is created. This slot deflects air down onto the upper surface of the main wing, thus energizing the region near the boundary and preventing separation. As a result, **L** is increased and α can be greater before the wing stalls. The first digit in the wing of a bird, together with its feathers, is called the **alula.** It ranges from one- to three-tenths as long as the wing, and when lifted creates a wing slot over the "palm," which is the first region to stall as α is increased. (Some birds also have slots between the tips of the primary feathers at the end of the wing. Their function is mentioned later on.) Bats do not have slots, but by depressing the thumb some create a leading-edge flap. Similarly, by spreading a broad tail some birds form a trailing-edge flap.

We can now relate these variables more precisely. Lift $= \frac{1}{2}\rho V^2 S C_L$, where ρ = density of air, V = air speed, S = projected area of the wing (i.e., the area of its shadow), and C_L = coefficient of lift. The two wings and the part of the body that joins them function as a unit in creating lift, so they are best taken together in calculating S. When the wings are not flapped, lift is greater toward the center of the unit than toward the extremities, or wing tips. The coefficient C_L is a dimensionless number that usually equals about 1.5 for birds but can range above 2 if the wings are slotted. Its value depends on the angle of attack, the camber and streamlining of the wing, the presence and nature of wing slots or flaps, texture of the wing surface, and the Reynolds number. Recall that the Reynolds number is a dimensionless number, $\mathrm{Re} = \rho l V/\mu$, where l = a characteristic length (here standardized as the average chord of the wing), μ = viscosity of air, and the other symbols are as above.

For vertebrate flyers, usual values are 25,000 to 140,000 (which is lower than corresponding numbers for swimmers).

For our purposes, this complex of variables can be simplified. Since the density and viscosity of the air are small in value and scarcely subject to control by the flyer, we can discount them when considering the generation of lift. The camber and shape of the wing influence lift largely through their relation to the angle of attack. This leaves angle of attack, wing area, and air speed as of particular importance to lift. Air speed is squared in the basic formula for lift, and occurs also in the calculation of the Reynolds number. It follows that if flyers are fast, they have enough lift even when their wings are narrow, small in area, have little camber, no slots, and function with a small angle of attack. Conversely, slow flyers need large wings with high camber and maximum angle of attack in order to ascend quickly.

(There is, of course, an upward component to ascending flapping flight. This will be considered below. Also, a flyer may derive upward force from updrafts. This is discussed under soaring, gliding, and formation flying.)

DRAG

Drag (the resistance the air offers to the motion of the flyer) acts horizontally backward when flight is level and in still air. It is convenient (for animals with wings) to divide the total drag (**D**) into two categories, profile drag and induced drag.

Profile drag ($\mathbf{D}_p$) (or parasite drag) is all the drag acting on a hypothetical airfoil of infinite length. It is also all the drag acting on the wings and intermediate body (again taken as a unit) of a real flyer, exclusive of the drag induced by motions of air around the tips of the wings. Profile drag is produced by energy lost to the environment through the friction of the air against the body, the displacement of air, the formation of pressure gradients in the air, and the creation of eddies or vortices. (Profile drag is composed of the pressure drag and the frictional drag discussed for swimmers in the previous chapter.) The value of $\mathbf{D}_p$ is equal to $\frac{1}{2}\rho V^2 S C_{dp}$, where C_{dp} is the coefficient of profile drag and the other variables are the same as those in the formula for **L.** The value of C_{dp} varies with the camber, streamlining, slotting, and outline of the wing-body unit, and with the Reynolds number. (These are the same variables that influence the value of C_L, but the relationships are different here.) On the upstroke of the wings, C_{dp} may be much reduced by birds by flexing the wrist and elbow to reduce surface area, and allowing air to pass down through the feathers. Flyers, like swimmers, can sense and control the flow of the medium over the body. In wind tunnels the energy loss by models of birds is about twice as great as that by live birds. Analysis has shown that a vulture can achieve laminar flow over much of its body.

Since air pressure below a wing that is lifting is positive relative to atmospheric pressure, and the air pressure above the wing is negative, there is a flow of air outward under the wing, around the wing tip, and inward over the wing. This flow induces a large eddy, or vortex, which causes loss of lift from the wing tip surface (Figure 28.7). (These vortices also affect airflow elsewhere, equivalent to a slight increase in α.) This takes energy from the flyer, and causes an additional drag, called **induced drag ($\mathbf{D}_i$)**. $\mathbf{D}_i = \frac{1}{2}\rho V^2 S C_{di}$, where C_{di} is the coefficient of induced drag and the other variables are as before.

Induced drag can be minimized in either of two ways. High-speed flyers (mastiff bat, falcon, albatross) have narrow wings with pointed wing tips. Airflow around small wing tips is limited, which reduces drag, but also lift; these flyers cannot fly slowly. Birds that fly relatively slowly (eagle, vulture, crow) have broader wings with wide wing tips that have fingerlike projections (the outer halves of the primary feathers) between which are spaces, or slots (Figures 28.6 and 28.8). The result is that only the spaced feather tips are a source of induced drag. The portion of the vane of each feather that is posterior to its shaft

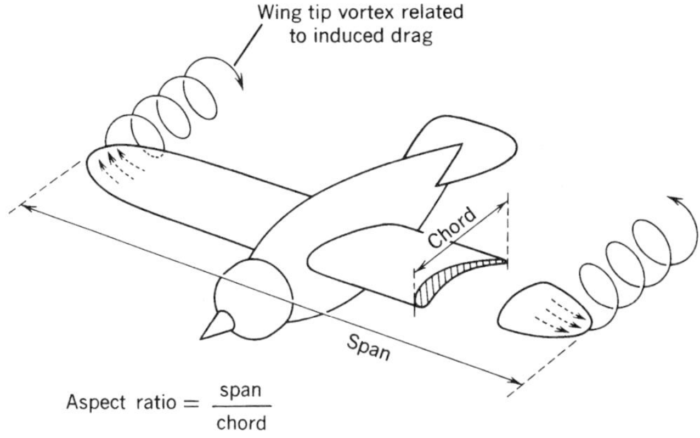

FIGURE 28.7 THE NATURE OF THE ASPECT RATIO AND WING TIP VORTEX for a gliding flyer.

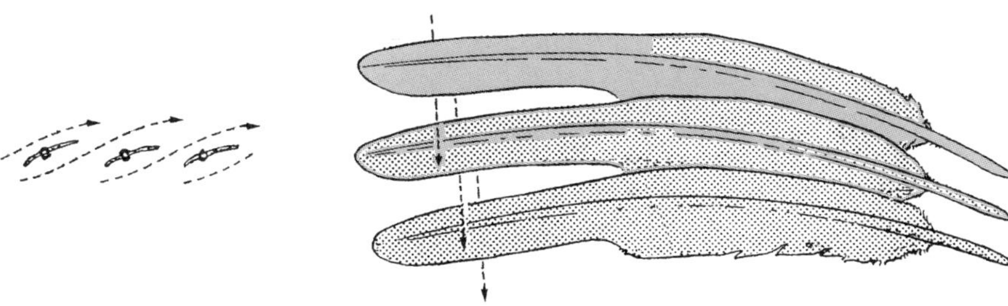

FIGURE 28.8 THREE PRIMARY FEATHERS OF AN EAGLE, seen in dorsal view and cross section, showing the path of air in relation to the slots.

is broader than the anterior portion (Figure 28.8). Consequently, the greater air pressure below the wing tip causes each primary feather to rotate around its shaft (as longitudinal axis), thus opening adjacent feathers like the slats of a window blind. Onrushing air flows up through the slats on the downstroke, and each feather acts as a small airfoil. The configuration seemingly reduces turbulence and increases lift and thrust; the complexities remain imperfectly known.

Total drag, like lift, increases with angle of attack, but the relationship differs (Figure 28.5, center). A useful curve (called a *polar diagram*) plots C_L against C_D and includes representative angles of attack. The tangent to this curve that passes through the origin now identifies the angle giving the highest ratio of lift to drag (Figure 28.5, right).

Narrowness of wing is expressed by the **aspect ratio,** A, which is the span of the wings (tip of one wing to tip of the other) divided by the average width or chord. (Since the average chord cannot be directly measured, the equivalent formula, $A = \text{span}^2/\text{area}$, may be used instead.) The value of C_{di} is equal to kC_L^2/A, where k is an empirical constant equal to about 2.

PROPULSION

Formulas in the preceding paragraphs make it clear that there can be no lift or drag without wind rushing over the wing or the wing having forward velocity. So far, forward velocity has been assumed without establishing its source. Flyers usually propel themselves by flapping the wings, but we will start with gliding and soaring because they are somewhat less complex.

Gliding and Soaring One way for an animal to derive or maintain forward velocity is to glide downward using gravity to convert potential energy to kinetic energy. Assume that the animal (whether bird, reptile, or mammal) glides with constant speed and direction in still air. The angle that its path makes with the horizontal is called the **gliding angle,** and

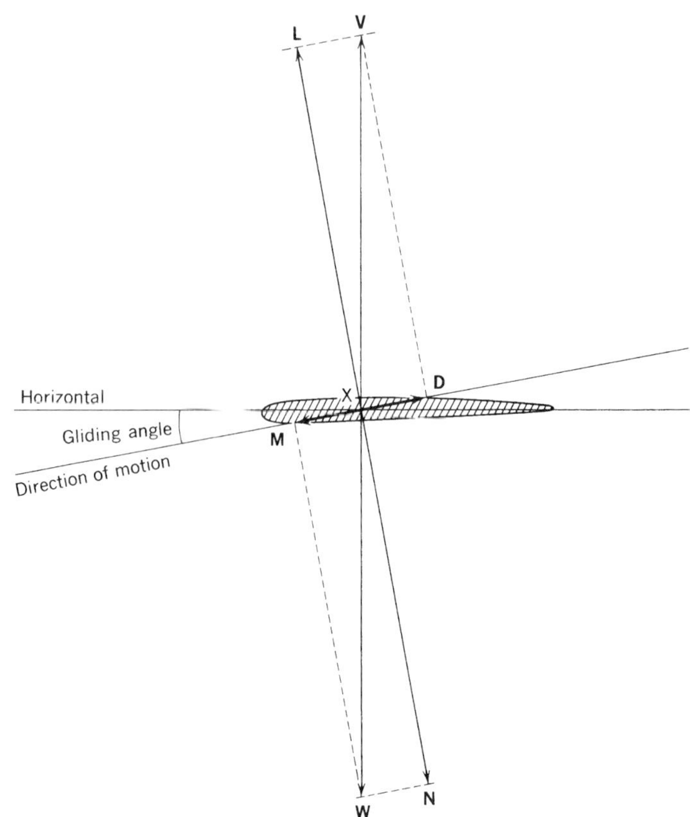

FIGURE 28.9
FORCES RELATED TO THE AIRFOIL OF A GLIDING ANIMAL moving with constant speed and direction when **L/D** = 5; X is the center of pressure.

the rate of vertical descent is called the **sinking speed.** It is as though the glider were sliding down an incline. As for a wagon that coasts down a hill, the glider's weight, **W** (Figure 28.9), has a component **M** that is in the direction of motion, and a component **N** that is normal to the slope of the incline.

The glider's airfoils create lift and drag. Lift, drag, and gravity are the external forces that act on the glider. In order for speed and direction to remain constant, these forces must be in equilibrium. The glider therefore adjusts the angle of attack and the area of its airfoils so that **L** = **N** and **D** = **M.** Similarly, **V,** the resultant of **L** and **D,** is vertical and equal to **W.** It is evident from Figure 28.9 that the greater the value of **L/D,** the flatter the angle of glide will be. A far-ranging, fast-flying albatross may sink 1 unit of distance vertically while moving 12 to 18 units horizontally, and a vulture may do as well. It is usually desirable for small birds to achieve minimum sinking speed. This is accomplished with a steeper angle of glide (about 1:8) but with flight speed only a little faster than the stalling speed.

When a flyer glides low (usually less than a wing length) over a flat surface, air under the body cannot be displaced downward and, to a degree, acts as a cushion. Drag is reduced, and adequate lift is achieved at a reduction of about 15 percent in the energetic cost of transport. This is called *ground effect.* Pelicans, skimmers, and some bats commonly glide over water in this way.

Sustained flight without flapping of the wings is soaring. There are several kinds of soaring, depending on circumstance. The first, **static soaring,** is dependent on updrafts. A flyer can stay aloft if its sinking speed relative to the surrounding air is equal to, or less than, the rate of ascent of the surrounding air relative to the ground. Like a person who walks continuously down a rising escalator, the flyer glides continuously down through

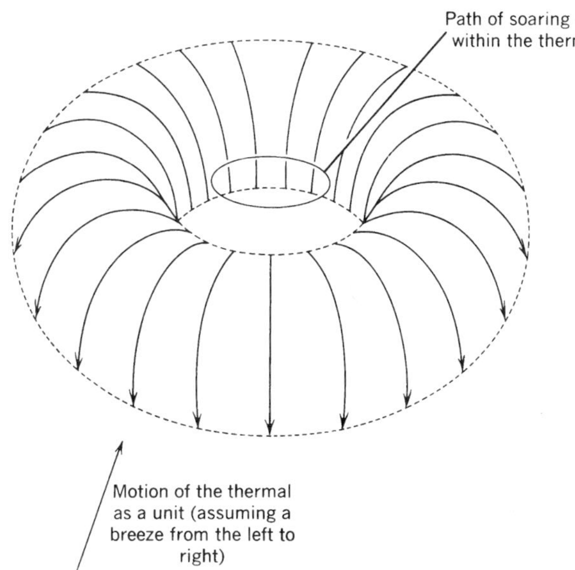

FIGURE 28.10
SIMPLIFIED DIAGRAM OF THE CIRCULATION OF AIR IN A THERMAL as seen from above and to one side. Static soaring is done in the air rising throughout the center of the "doughnut."

rising air. Air rises when wind is deflected upward by a hill, coastline, wave, or ship. Gulls, pelicans, and some other sea birds often soar in such updrafts. (This has been called slope soaring.) Similarly, columns of air sometimes rise by convection over warm water or land. If a breeze is blowing, the columns tilt over, becoming inclined to the ground.

Vultures, hawks, and most other soaring land birds (and even large diurnal bats) instead do their static soaring in **thermals.** The morning sun warms the ground, which warms a layer of air next to the ground. This air flows into bubbles that arch up and break away from the ground layer to rise like balloons through the higher, cooler air, expanding as they go and drifting horizontally if there is any breeze. Unlike the gas in a balloon, however, the air in a thermal is not stagnant but circulates in a vortex shaped like a doughnut (Figure 28.10). Air constantly rises in the center of the thermal (the entire center, not just on the surface of the doughnut), radiates outward at the top of the floating bubble, descends in the margins, and turns in to rise again. The vulture soars only in the center of a thermal, wheeling constantly to remain within the hole of the doughnut. After rising with the thermal, sometimes for several thousand meters, the bird glides slowly away to the ground or into another thermal. In this way it remains on the wing for long periods and covers much distance with little expenditure of energy.

Thermals rarely rise over water. Albatrosses and some related oceanic birds are masters of **dynamic soaring,** which is dependent on the wind. The wind (which is brisk and steady over vast areas of the oceans) has ever-decreasing velocity from about 15 m over the water down to water level. This is because of the friction between air and water. The difference in wind velocity between the top and bottom of this shear layer commonly amounts to about 60 km/h, and the bird soars only in this zone (Figure 28.11).

Let us follow a basic flight cycle of an albatross starting with the bird flying into the wind near the top of the shear layer. Its air speed might be 70 km/h, but its absolute speed over the water might be only 10 km/h because of a wind speed of 60 km/h. Now the bird turns downwind. At the completion of the turn the velocity of the wind is added to, rather than subtracted from, the bird's air speed to derive its absolute speed downwind, which might be 130 km/h. The albatross next partially flexes its wings to increase wing loading and further increase speed as it glides steeply down to the water, converting potential energy to kinetic energy as it drops. The bird now banks sharply to execute a windward

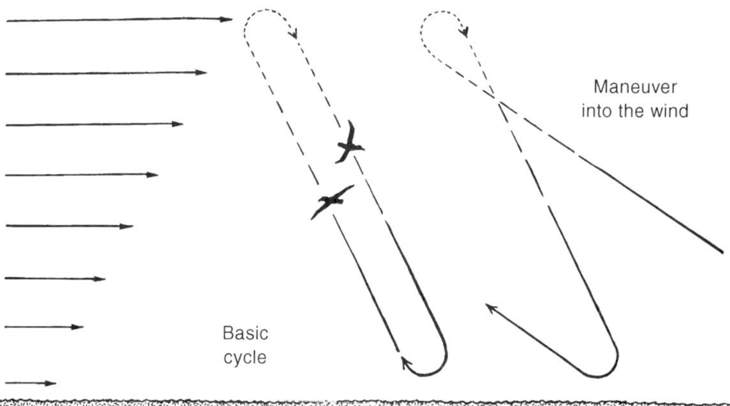

FIGURE 28.11 THE DYNAMIC SOARING OF THE ALBATROSS. Horizontal arrows represent vectors of wind speed in the shear layer. Dotted lines in the path of the bird indicate relatively slower flight speed over the water, though air speed is maintained.

turn. It still has high absolute speed when it completes this turn, because the wind has less velocity at water level. Fully extending its wings and increasing their angle of attack, the bird now completes the basic flight cycle by rising to the top of the shear layer, converting kinetic energy to potential energy again as it climbs. Absolute speed falls off, but ever-increasing wind speed maintains adequate air speed to prevent stall.

The energy seemingly extracted from the wind is sufficiently greater than the energy lost during the basic flight cycle to provide an excess that can be used for gradual maneuvering—for instance, by gliding crosswind when halfway through a turn, or by lengthening each climb to progress directly into the wind. These birds are also expert at static soaring in the updrafts over wave fronts. In these ways albatrosses travel thousands of kilometers with only infrequent flapping of the wings.

Vultures sometimes interrupt long glides downwind to loop into the wind, thus using the principle of dynamic soaring to gain elevation for further travel downwind.

Introduction to Flapping Flight We noted on p. 507 that penguins use lift-based oscillatory propulsion in water. The downbeat of the wing ensures that the oncoming water streams upward and backward (see figure on p. 510). Lift, which is at right angles to the water, is therefore inclined upward and forward, and has a forward, propulsive component. Substitute air for water, and a crow for the penguin, and we have the elements of powered, flapping flight. Keep in mind that in flapping flight, it is the movement of the wing through the air that generates air speed over the wing sufficient to produce lift and propulsion (and drag).

However, the full story is more complicated. Although the various kinds of flapping flight are understood in general terms, its complexity continues to provide numerous challenges. The relatively recently (but sparsely) employed techniques of cineradiography, electromyography, in vivo measurements of muscle length and bone strain, and analysis of vortices behind free-flying birds and bats have answered some questions but raised many others. The empirical data from observation and experimentation are generally considered within the context of blade element theory, as for airplane wings, and momentum theory, as for propellors. The biology of flapping flight is fascinating, and there are several excellent sources for review (see Norberg, Pennycuick, Rüppell in References). The descriptions given below are representative but not exhaustive.

Muscles of Depression and Elevation It was noted on pp. 176 and 177 that the pectoralis and supracoracoid muscles of birds are the principal depressor and elevator of the wing, respectively. Kinematic and electromyographic studies of starlings (and a few other species) in free flight reveals the onset of electrical activity of the pectoralis in late upstroke, and continued activity through the upstroke–downstroke transition into mid-downstroke (see Figure 10.6 on p 177). The supracorcacoid begins its contraction in late downstroke, sustains through the downstroke–upstroke transition and ceases in early downstroke. These muscles are not merely depressor and elevator muscles, however. The pectoralis (through certain of its fascicles) also contributes a substantial component of pronation or retraction to the humerus, depending on phase of the wingbeat cycle and mode of flight. Pronation in late downstroke provides additional thrust; retraction is important for keeping the wing oriented properly during takeoff and landing, when the body axis assumes a vertical orientation relative to the ground. The supracoracoid, with its tendon coursing over the bony pulley of the coracoid to insert on the anterodorsal aspect of the proximal humerus, imparts a high-velocity, longitudinal rotation to the humerus in addition to elevation. This rotation is important for reorienting the wing at the end of upstroke in preparation for the subsequent downstroke.

The shoulder of bats retains a generally mammalian organization, but the subtleties of its morphology are specialized. The humerus and scapula form an interlocking or "stop" mechanism when the humerus is elevated. Thus, extreme humeral elevation requires rotation of the scapula about its longitudinal axis. The pectoralis and anterior serratus are the principal downstroke (adductor) muscles, although other muscles (subscapularis, clavodeltoid, latissimus dorsi) provide control. The upstroke is produced mainly by the deltoid muscles, but others (trapezius group, triceps) are also active. The larger number of muscles involved in the wingbeat cycle of bats, as compared with birds, reflects the use of the forelimbs also in terrestrial locomotion and climbing. As in birds, the electrical activity of these muscles occurs at the transition phases between downstroke and upstroke.

Pterosaurs show remarkable convergence with birds in skeletal morphology. Although the clavicles are absent in pterosaurs, the coracoid is stout and the sternum keeled. Because pterosaurs of the Triassic tended to be small, it is thought that they actively flapped their wings. The later Cretaceous forms were relatively large and *Pteranodon,* at least, possessed a humeral-scapular locking device. This may have been an adaptation for gliding, but it may also have aided in flapping. Padian and Wellnhofer provide thoughtful anatomical comparisons between pterosaurs, and birds and bats.

Attempts to find correlations between muscle fiber structure and the diversity of wing morphologies and flight styles have been both rewarding and disappointing (see Hermanson and Norberg, 1990, for reviews). The pectoralis of birds and bats is primarily composed of fast-twitch fibers. (Review fiber types on p. 175.) Some species with high wing loading beat their wings at a high frquency. As predicted, those that fly only short distances (pheasant, quail, vampire bat) tend to have primarily fast glycolytic (aerobic) fibers, whereas those that fly long distances (ducks, free-tailed bat) tend to have fast oxidative (aerobic) fibers. Other species, however, have fiber types that do not seem to reflect mode of flight. Rosser and George provide a useful survey and thoughtful discussion of the histochemical profile of pectoralis muscle fibers for a large number of bird species, and Hermanson does the same for bats.

Hovering Hovering, or stationary flight in still air, represents a specialization in the evolutionary sense. The small flyers that can hover make the body motionless in midair, usually to feed on the nectar of flowers that are not accessible in another way. Hummingbirds are prominent among such flyers, though some other birds and many kinds of bats can also hover.

The body is usually held nearly vertical. The wings, therefore, beat backward and forward instead of up and down, propelling a vortex of air downward with each stroke. The

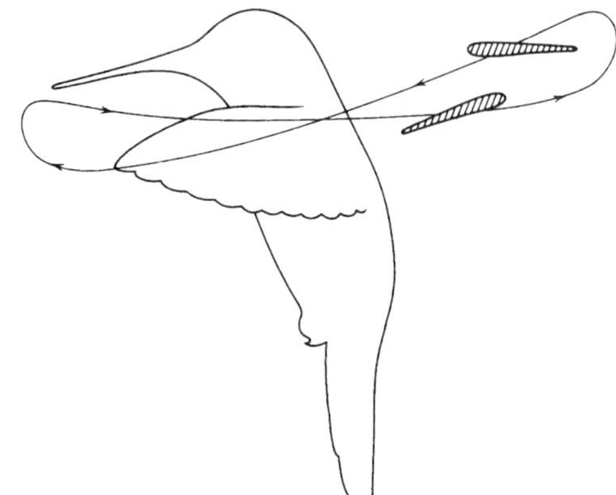

FIGURE 28.12 HOVERING FLIGHT OF A HUMMINGBIRD showing approximate path of the wing tip and angle of attack of the wing on forward and backward strokes.

wing tips describe a distorted figure eight (Figure 28.12). On the forward stroke, the wing, having a positive angle of attack, creates much lift, which is at right angles to the airstream but has a large vertical component. At the forward limit of the long wing stroke the wing rotates at the shoulder (the wrist and elbow are quite stiff in these birds), thus turning over and making its anatomical underside uppermost. With a positive angle of attack in the new direction, the wing then sweeps backward, with the ventral aspect of the wing now providing lift with a large vertical component. The horizontal components of the two strokes tend to cancel, yet by slightly altering the power of one stroke in relation to the other, the flyer can move horizontally forward or backward as needed when feeding. When the flyer is not hovering, it flies with the body horizontal.

Since the body of a hovering hummer is stationary relative to the air, the bird must move its wings very fast to establish enough air speed at the flight feathers to create lift. The frequency of the beat is 35 to 50 Hz. The wings (and therefore the bird) are small so that their oscillations will not generate excessive inertia. The hand part of the wing is relatively long and the slower-moving arm part is short (Figure 28.19). Since the backstroke of the wings is active, the muscles that elevate the wings are relatively larger than for other birds. The entire power plant is large: The breast muscles account for 25 percent of the weight of the entire bird, and the little flyer may eat about twice its body weight in food per day.

Bats that hover do not follow the pattern of hummers in turning the wing over at the shoulder. They flick the hand wing in the manner described below for pigeons and gulls ascending steeply.

Slow Ascending and Descending Flight Most small birds of bush and forest make frequent short flights that often include steep ascents and descents. They fly slowly but flap their wings rapidly (commonly 15 to 25 Hz) to attain the wing air speed necessary for the flight feathers to produce lift. The body axis is held between 45° and vertical while ascending and descending. At such times the wings, therefore, move nearly backward and forward, with their tips describing flat ovals or loops. Small bats with erratic flight presumably fly much as small birds do. The following account, however, is based on birds.

On the forward stroke the wing is fully extended and has a positive angle of attack. Marked lift (to raise the flyer or retard its descent) and propulsion result. The backstroke, by contrast, is merely a recovery stroke having little reaction with the airstream. In order to reduce drag on the backstroke, the wing is partially flexed, reducing its area. The muscles that elevate the wing tend to be much smaller than those controlling the downstroke.

Wing loading is the weight of the flyer divided by the area of its wings. As explained on p. 423, small animals have more surface area in relation to volume than do larger animals of identical proportions. Small flyers, therefore, have relatively low wing loading without having particularly large wings. Representative values for small bats are 0.07 to 0.17 g/cm^2 and for small birds are 0.11 to 0.23 g/cm^2. (There are various methods of measuring wing area, and the effect of age, sex, and season on body weight should be considered in making comparisons; see Pennycuick in References.)

Pigeons, gulls, ducks, hawks, owls, pheasants, and various other strong flyers of medium size can also ascend and descend at steep angles while flying slowly. Their larger size means that their wings must beat more slowly. Frequencies of 3 to 10 Hz are usual. Furthermore, because of the relationship between surface-volume ratio and body size, and because of the nature of the fast flight of these birds, they have moderate to high wing loadings (i.e., 0.40 to 1.30 g/cm^2).

When the outstretched wing sweeps downward and forward, it produces much lift and a little propulsion. Since the wing loading is large and the frequency of beat is only moderate, these birds cannot afford a passive backstroke (as small birds can). Their backstroke is complicated, but three motions are particularly characteristic: (1) The wing is flexed so that the tip travels close to the body, (2) the outer half of the wing (only) is turned over with what is called a flick, and (3) the primary feathers rotate individually like the slats of a venetian blind so that each feather acts as a strong airfoil with positive angle of attack, yet air can pass between the feathers (Figure 28.13). The opening of this venetian blind is the automatic consequence of the facts that (1) the shafts of the feathers are not central in their respective vanes so air pressure tends to make the vanes rotate, and (2) the flick

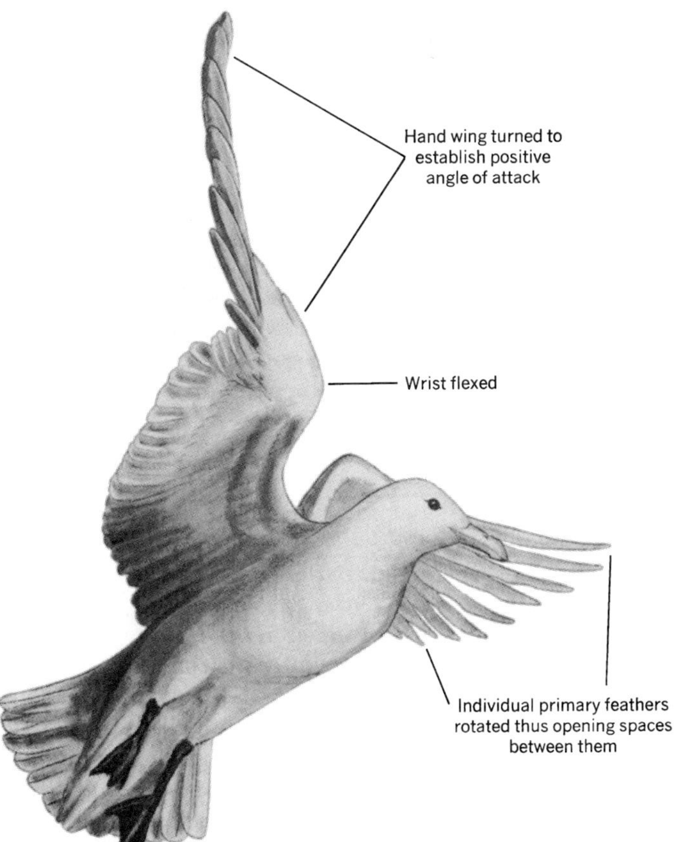

FIGURE 28.13 BACKSTROKE OF WINGS IN ASCENDING FLIGHT OF A MEDIUM-SIZE BIRD shown by a gull, *Larus*.

reverses the direction of air pressure on the feathers, enabling them to rotate in the direction not prevented by their overlap. This backstroke produces even more lift than the forward stroke and also some propulsion. The slower-moving inner wing is nearly useless. Pigeons can take off and land without the inner, or secondary, feathers. The wing tip describes a figure eight in each full cycle.

Some birds of comparable size, and many larger birds, having even slower wingbeats and higher wing loadings (swan, 1.7 g/cm^2; heron, 2.5 g/cm^2), are unable to ascend and descend steeply while flying slowly. Such birds often run to gain speed for a fast, flat take-off and land fast on water, which cushions the impact.

The twisting of the primary feathers merits further explanation. For flapping flight, the tip of the wing moves up and down farther, and hence faster, than the base. Accordingly, the direction of the airstream varies along the length of the wing, and in order for all parts of the wing to have the optimum angle of attack, the wing must twist like an airplane propeller (Figure 28.14). But unlike the propeller, which turns only in one direction, the tip of a wing should twist leading-edge-down on the downstroke and up on any powered upstroke. Nature's solution for numerous birds is the twisting of the primary feathers according to whether the airstream strikes them from above or below.

Fast Level Flight Fast level flight is characteristic of the large mastiff bat and many birds of moderate or large size such as ducks, geese, shorebirds, falcons, toucans, and gulls. The body is carried horizontally, and the wings beat up and down. High speed of the body provides high air speed over the wings. Consequently, the slower-moving inner wing produces much lift on both the upstroke and downstroke. The wingbeat is relatively slow, and the amplitude is small. Wing loading is usually high. Since a propulsive force is needed only to oppose drag and not to power acceleration or ascent, or to retard descent, the demand for power is relatively low. Virtually all propulsion is provided by the outer half of the wing on the downstroke. The upstroke is passive; negative air pressure over the wing may lift it without muscular effort. The outer wing does not flick over on the upstroke as for slow flight of some of the same kinds of birds. The position of the inner wing changes little on the upstroke; the outer wing is flexed somewhat at the wrist. The wing tip never moves backward in relation to the ground.

As noted in a previous section, there is an updraft of air at the wing tips. By flying in formation, ducks, geese, and some other birds utilize the upwash created by their neighbors to reduce the lift they must generate with their muscles. Induced drag is diminished by 30–40 percent. Nearly maximum advantage is attained when there are 10 or more birds and they

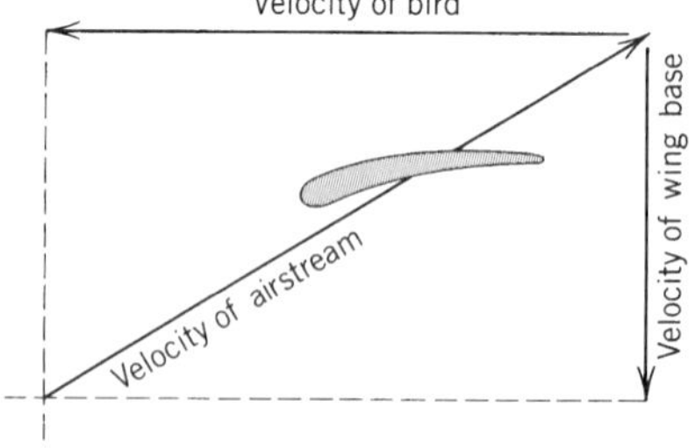

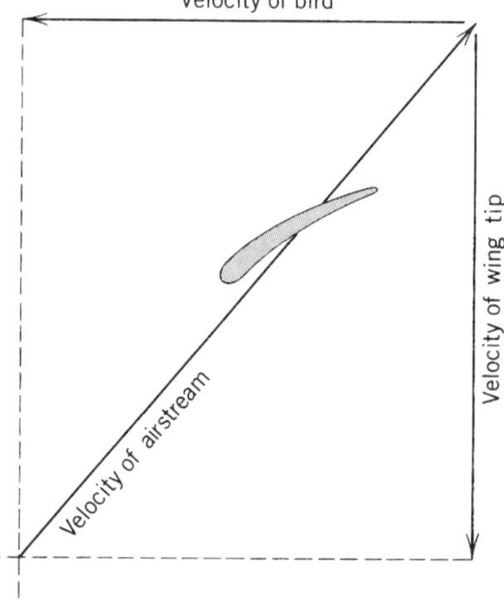

FIGURE 28.14 RELATION OF THE VELOCITY OF THE PARTS OF A FLAPPING WING TO THE VELOCITY OF THE AIRSTREAM AND TO THE OPTIMUM ORIENTATION OF THE WING as seen in cross section. Angle of attack is constant. Compare with figure on p. 510.

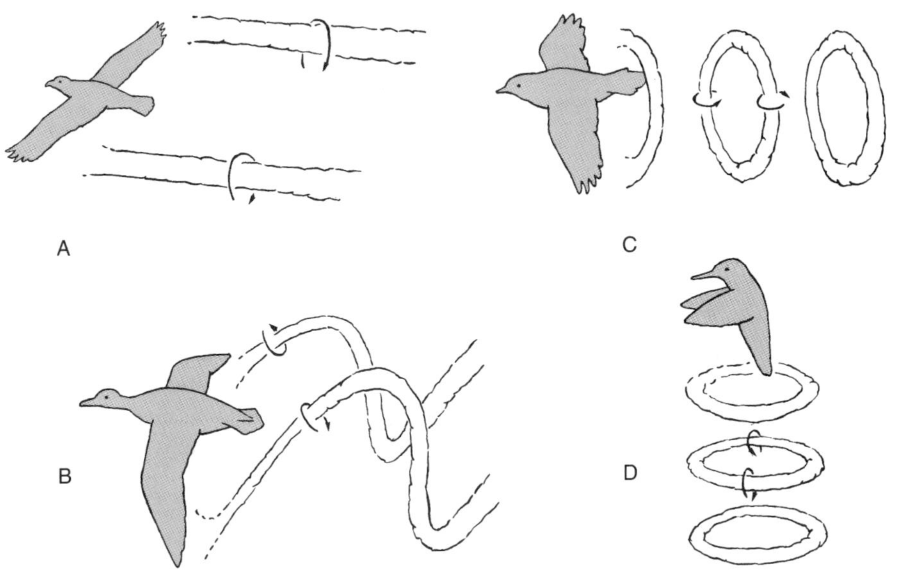

FIGURE 28.15 VORTICES IN THE WAKE OF FLYERS. A, a gliding hawk; B, a fast-flying, flapping duck; C, a slow-flying, flapping songbird; D, a hovering hummingbird.

fly with less than half a wingspan separating them horizontally. The energy saved is considerable. Echelon flight also allows each bird to see all other birds. Migrating geese compensate for winds that are not severe, keeping direction of flight and ground speed constant.

Vortex Theory Since aerodynamic lift results from differential pressure gradients above and below a moving wing, there is a net circulation of air. Flapping bats and birds, in level flight, generate compelex, unsteady air movements around their wings that impart momentum to the air below and behind the animal in reaction to the lift forces generated. The "strength" of these vortices is proportional to the mass of air moved and to its velocity. Visualization and analysis of the vortex wake patterns, preferably in three dimensions, enables a quantitative estimate of the aerodynamic function of the airfoil. This is not a trivial problem, however, as mentioned earlier (Comment 27.1). Spedding, Rayner, and Pennycuick all trained birds and bats to fly through a cloud of neutrally buoyant helium-filled soap bubbles and recorded airflows in the wake by photographing the movements of the bubbles in stereo. Although the technique reveals little of the airflow immediately around the wings, or of the formation of the trailing vortex cores close behind the wing, the vortex patterns left behind are instructive for clarifying forces generated in flight (see Figure 28.15 and Comment 28.1).

FLIGHT CONTROL

Stability and Maneuverability As noted in previous chapters, animals must choose between stability and maneuverability. The elephant opts for stability; many flyers (more than designers of aircraft) opt for maneuverabilty. If the body is moved slightly out of position, aerodynamic forces usually do not restore the original orientation but instead tend to quickly force the flyer farther out of position. This makes it mandatory that the sensory–nervous–motor control system correct constantly and almost instantaneously for minor displacements. The same circumstances, however, permit rapid maneuvering. Some birds occasionally do loops or rolls, and such displays are surpassed by the aerobatics of many bats and birds as they pursue flying insects or engage in courtship flight. (See the chapter-opening figure).

A flyer can correct for roll by increasing the angle of attack of the lower wing, by increasing its surface area, or by flapping it harder than its opposite, any of which will

COMMENT 28.1

VORTEX WAKE AND THE ANALYSIS OF GAITS OF FLYERS

Early observations (by Spedding, Rayner, and Pennycuick) of pigeons and small passerine birds flying slowly through suspended helium bubbles revealed a series of circular, closed vortex rings, each formed by a single downstroke (Figure 28.15). The independence of successive rings indicates that aerodynamic force is generated only during the downstroke and that the upstroke does not contribute to lift. Subsequent observations of the wake of a small falcon revealed a very different pattern: The wake behind each wing was continuous, which suggests the generation of lift during the upstroke. Further observations of the wake vortex structure of a number of birds and bats, coupled with studies of wing kinematics, led these investigators to conclude that the two patterns represent two distinct gaits. The vortex ring gait seems ubiquitous among birds and bats with short, round wings and is widely used during ascending and descending flight. (See below for further descriptions of wing shapes.) During the upstroke, when the surface area of the wing is reduced and there is no lift, the wing is rotated and elevated by active contraction of muscles. The continuous vortex gait is characteristic of longer-winged birds and bats at all speeds and for others during fast, level flight. (Gulls, for example, use the vortex ring gait at low speeds and the continuous vortex gait at high speeds.) The wrist is partially flexed during upstroke and extended during downstroke; the continuous vortex wake is shed from the distal wing. Lift (but not positive thrust) *is* generated during the upstroke to elevate the wing aerodynamically without the necessity of muscle contraction. That birds and bats are restricted to two gaits led the investigators to hypothesize that the aerodynamics of flapping flight is similarly constrained for each group. Accordingly, one can propose theoretical models of flapping flight, make predictions about the energetics associated with various morphologies, and attempt to reconstruct the evolution of flapping flight. For example, Rayner proposes that since fast flight with the continuous vortex gait is less demanding mechanically than slow flight, *Archaeopteryx,* possessing a small sternal keel and no evidence of a derived pulleylike arrangement of the supracoracoid muscle for elevating the wing, may have been capable of steady cruising flight, but could not fly slowly. Finally, pigeons use a vortex ring gait in slow flight and a continuous vortex gait in fast flight, and their pectoralis consists of two populations of fibers distinct in histochemical profile and size. It is proposed that the large, anaerobic fibers are used for takeoff and landing and the small, aerobic fibers for cruising flight (as reviewed by Sokoloff et al.). The experiment to verify this remains to be done.

---

increase its lift. (Since the same actions influence drag, compensations are needed to prevent yaw.) Also, the tail, if large, might be formed into a screw shape to correct for roll. The same kinds of behavior can, of course, be used to cause roll

Unlike the swimmer and man-made aircraft, the flyer has no vertical fins or rudders to control yaw. Instead, it increases the drag on the wing that is tending to advance faster than its opposite. This can be done by increasing its angle of attack. The wing's greater angle of attack, however, and greater air speed (since it is moving ahead of the other wing) increase its lift and hence tend to cause roll. To compensate, the advancing wing is flexed to reduce its area. Dropping one foot out of the streamlined contour of the body also increases drag on that side. Furthermore, if a bird with a large, stiff tail is banking to turn, it can open and twist the tail to form a vertical rudder.

To correct for downward pitch of the body (or to initiate upward pitch), a flyer moves its wings forward, thus placing its support farther forward in relation to its center of gravity. Also, if the tail is suitably constructed, it is bent upward. The reverse behaviors correct for upward pitch.

Braking and Turning Gliders, and flyers approaching an elevated perch, may brake by swooping upward into a stall, thus using gravity for deceleration. As noted, flyers can also flap their wings to create an upward force that slows descent preparatory to landing. In sustained level flight it is desirable for the ratio of **L**:**D** to be maximum consistent with moderate to high speed. For braking, by contrast, **D** should be high and speed should be low. This is achieved by making the angle of attack, the wing area, and the camber maximum (Figure 28.5, right). Birds also depress their fanned tails and bats depress the tail membrane. This is the equivalent of lowering the wing flaps of man-made aircraft. The feet are thrust forward (toes spread if they are webbed), not only to be in position for landing but also to provide drag (Figure 28.16). Moreover, when birds brake, they lift the alula

FIGURE 28.16
EXAMPLES OF FLIGHT CONTROL.

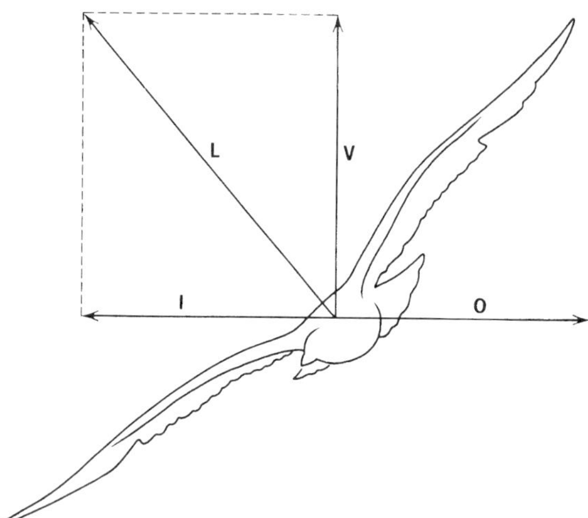

FIGURE 28.17 FORCES RELATED TO A FLYER DURING TURNS.

to create a wing slot. This increases the angle of attack and decreases the velocity that can be attained before stalling occurs.

Adaptations of the legs of flyers for absorbing the shock of landing have been little studied. The impact does not shear the ball-and-socket joint at the hip, but instead is transmitted to the ilium as compression by the distinctive neck and trochanter of the femoral head. Many fast-flying water birds land into the wind and plane briefly on their webbed feet as they strike the water.

Since moving bodies tend to continue to move in straight lines, when animals turn they tend to slip sideways toward the outside of the turn. The force out of the turn is **centrifugal force,** which varies directly with the square of the velocity and the mass of the body and inversely with the radius of the turn. This force must be opposed by an equal and opposite **centripetal force** into the turn. Sharply turning cursors resist centrifugal force by friction at the ground (see p. 447), and swimmers rely on drag produced by vertical fins and a deep body. Flyers must use another method. They bank into the turn, causing **L** to tilt inward (Figure 28.17). The horizontal component of **L,** or **I,** must equal the outward centrifugal force, **O.** If the flyer is not to lose altitude, the vertical component of **L,** or **V,** must equal the full value of the lift that existed before the turn started. This means that lift, and also wing loading, must increase during turns: If a 60° bank is executed, lift must be doubled. It is clear from the relation of the magnitude of **O** to mass and velocity that flyers that dodge quickly must be small, and that large flyers cannot turn sharply at high speed.

WING STRUCTURE

The various animal flyers have strikingly different flight habits. Since form must correspond to function, wing structure is also varied. Saville has described four general types of wings, it being understood that they intergrade and that within types there is great variation of detail. The aerodynamics of flight is not yet well enough understood to fully interpret the morphology of wings.

First is the **elliptical wing** (Figure 28.18). It is characteristic of most bats and of most small and medium-sized birds of shrub and forest (e.g., sparrows, robins, crows, quails, pigeons). The specialization is for high maneuverability and precise control, often in confined spaces, and for minimum induced drag. The flyer is able to fly slowly and to ascend and descend rapidly. The wing has an elliptical outline (except where it meets the body). It is short and broad, making the aspect ratio low (commonly 3–6). Camber is moderate to marked (particularly in bats). Flapping flight is usual, though there may also be some glid-

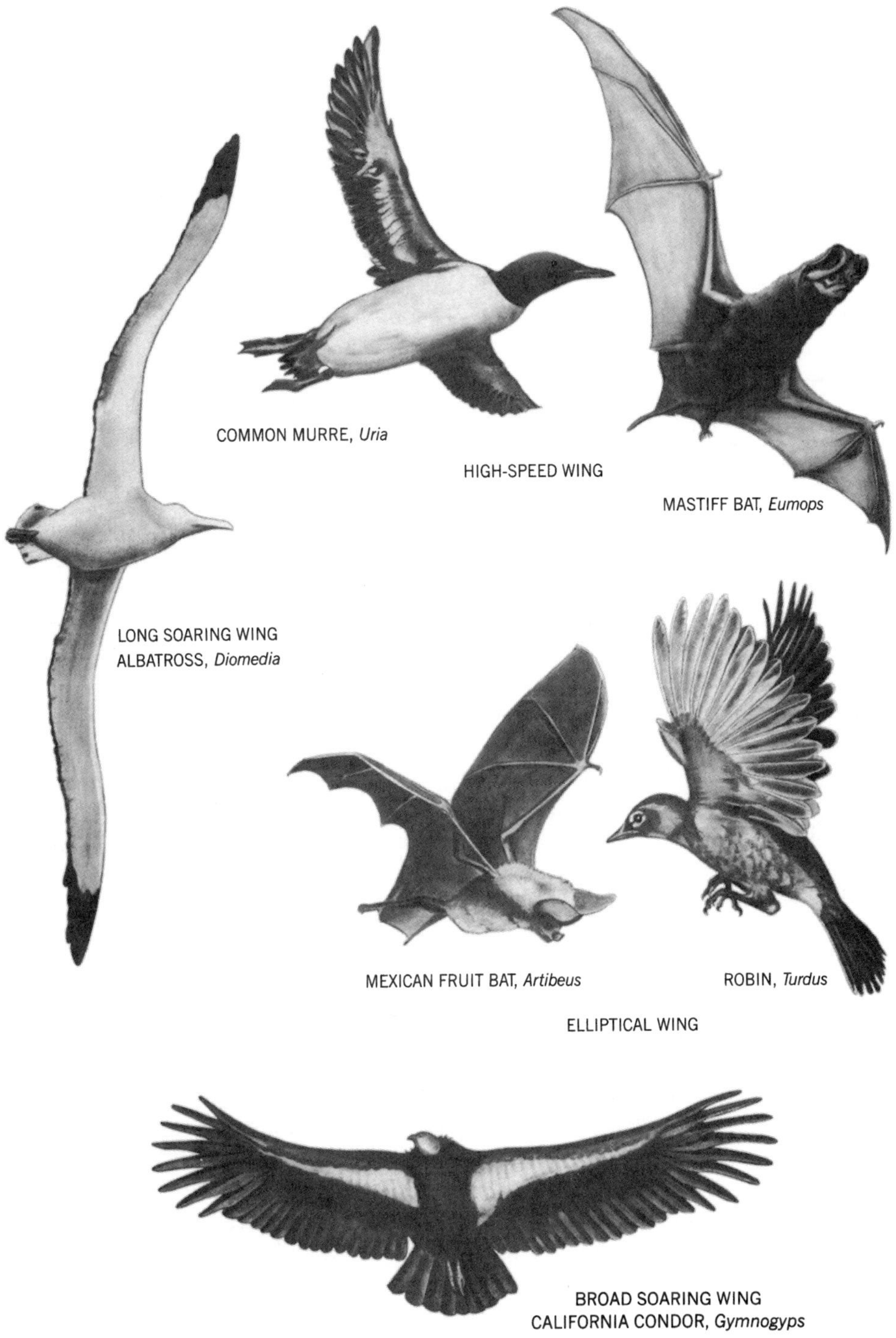

COMMON MURRE, *Uria*
HIGH-SPEED WING
MASTIFF BAT, *Eumops*
LONG SOARING WING
ALBATROSS, *Diomedia*
MEXICAN FRUIT BAT, *Artibeus*
ROBIN, *Turdus*
ELLIPTICAL WING
BROAD SOARING WING
CALIFORNIA CONDOR, *Gymnogyps*

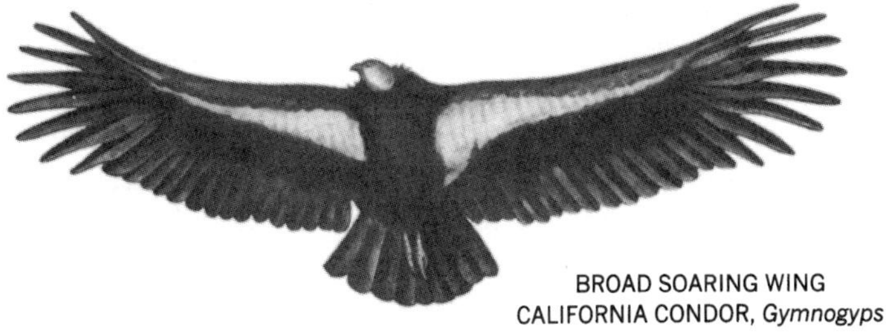

FIGURE 28.18
THE FOUR PRINCIPAL TYPES OF WINGS.

ing or swooping. The wingbeat is moderately fast, and the amplitude is relatively great. The birds have a large alula, and in many birds the primary feathers separate to form additional wing slots. In all, the slots may extend along the leading edge of 15 to 30 percent of the length of the wing. Each slot speeds up the air passing over the next posterior feather, thus preventing air from breaking away in eddies. The slots serve as an antistalling device at low speeds. They can be closed in faster, level flight. The primary feathers may rotate individually when the slots open. Slots may increase lift by as much as 60 percent.

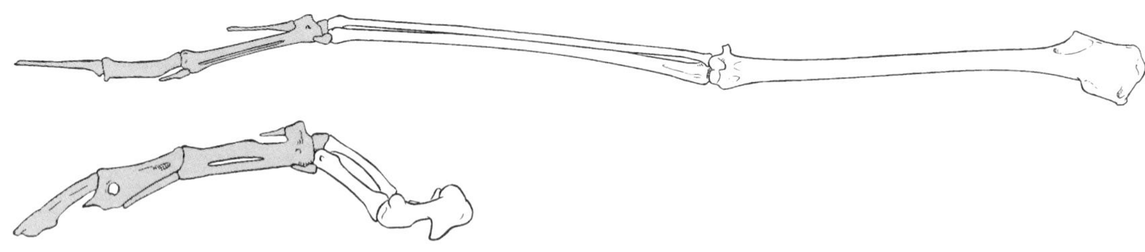

FIGURE 28.19 CONTRAST BETWEEN THE RIGHT WING SKELETONS OF A DYNAMIC SOARING BIRD AND HOVERING BIRD shown by an albatross, *Diomedea* (above), and a hummer, *Lampornis* (below). Hand skeletons drawn to the same length.

The **high-speed wing** is characteristic of mastiff and free-tailed bats and of swifts, swallows, falcons, shorebirds, hummers, and ducks. The specialization is for high flight speed with low drag and low expenditure of energy. The wing is relatively small, so wing loading is large. The wing tapers to a slender tip, may be somewhat swept back along the leading edge, and is faired into the body behind without a sharp angle. The aspect ratio is moderately high (commonly 5–9). The hand skeleton is relatively long, and strikingly so in hummers (Figure 28.19). In cross section the wing is thin and has little camber. Flapping is constant except perhaps for short glides when descending. The beat is rapid in relation to the size of the flyer, and the amplitude is small. There are no wing slots.

The **long soaring wing** is characteristic of albatrosses, frigate birds, gannets, terns, and gulls. Some large pterosaurs probably had similar wings. Most such birds fly over water, where long wings are not a handicap. The specialization is for a high ratio of lift to drag permitting soaring at high speed with low expenditure of energy and a low gliding angle. The wing is long, slender, and pointed; the aspect ratio ranges from 9 to 18; and the span of the wings reaches more than five body lengths. The hand skeleton is relatively short in these birds (Figure 28.19), though it is not in pterosaurs. Wing loading is high. Camber is low. There are no wing slots. Landing and takeoff speed is high. Most of the birds land on water and take off into the wind—often after running to gain speed.

The **broad soaring wing** is seen in vultures, eagles, and buteo hawks. The wings of the raven and pelican are similar. The specialization is for soaring at low speed, takeoffs and landings in confined areas, high lift, and low sinking speed. The wing is moderately long and broad, giving it large area; has only moderate wing loading for such large birds; and has moderate aspect ratio (e.g., 6 or 7). Camber is marked. The alula is prominent, and terminal wing slots are conspicuous. The slots seem to correlate with broad wing tips and correspondingly large tendency to induced drag. The slots usually have U- or ⊔-shaped bases rather than the less highly evolved V-shape of some elliptical wings. Each spaced primary feather functions independently as a winglet, and each is streamlined in cross section. Muscles that elevate the wing are less developed than in birds that emphasize flapping flight: Contrast the supracoracoid muscle in the figures on pp. 187 and 188.

FURTHER STRUCTURAL ADAPTATIONS

We have now considered the primary requirements for flight. These, however, are not enough: Even if they were provided with wings, lizards and rats (and angels) could not fly. One secondary requirement is for light weight. Flyers have light skeletons. (The skeleton of an eagle accounts for less than 7 percent of the total body weight—about half as much as for a human.) The bones of birds and pterosaurs are hollow and air-filled. This permits the bones to have a maximum diameter and hence maximum resistance to bending forces with minimum weight. Many bones have thin inner and outer lamellae joined by spicules that tend to follow stress lines. Furthermore, sutures tend to become obliterated. These constructions provide maximum strength with minimum hard tissue. The bony axis of the avian tail is short. Reduction or loss of digits also saves weight for birds and pterosaurs. Modern birds and some pterosaurs lack teeth, which are heavy. (The feeding habits of most bats require that they have large ears and teeth. The fast-flying mastiff bat has enormous horizontally oriented and cambered ears. It is probable that these ears generate enough lift to help support the heavy head!)

The gonads of flyers regress when they are not active. Good flyers eat foods that, being nutritious, are lightweight in relation to the energy provided. Also, digestion is very rapid (often within an hour). Some flyers (hummers, swifts, many bats) use the legs only for perching or roosting. The legs are then very small and light. Feathers, being stiff yet elastic, are remarkable for the loads they can bear in spite of their light weight. They provide a smooth, streamlined contour and are resistant to damage (see p. 94 and 95).

A second requirement is for compactness and firmness of the body so that the thrust of the wings will be transmitted to the body near its center of mass and will propel the body as a unit without deforming it or causing it to flap. Flyers have a short trunk that is stiff (Figure 28.20). Fusions between adjacent trunk vertebrae are common in bats and usual in other flyers. Pterosaurs had many sacral vertebrae and usually had fusions in the thoracic region. Birds have 12 to 20 vertebrae fused together in the synsacrum (see figure on p. 148) and may

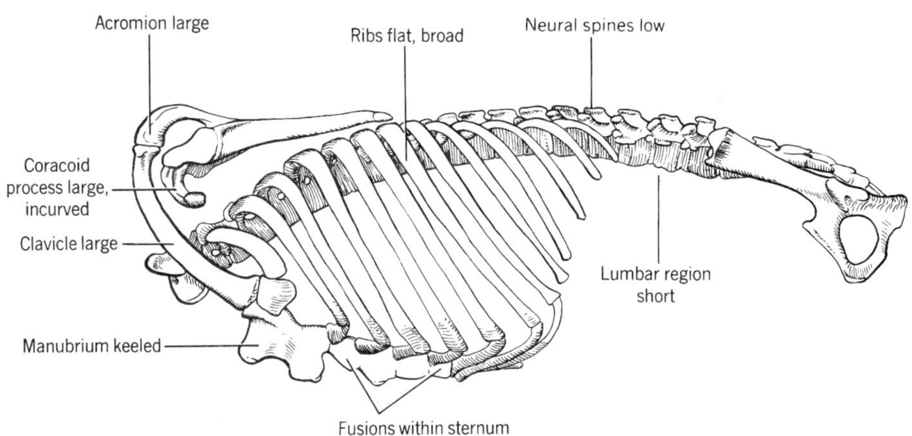

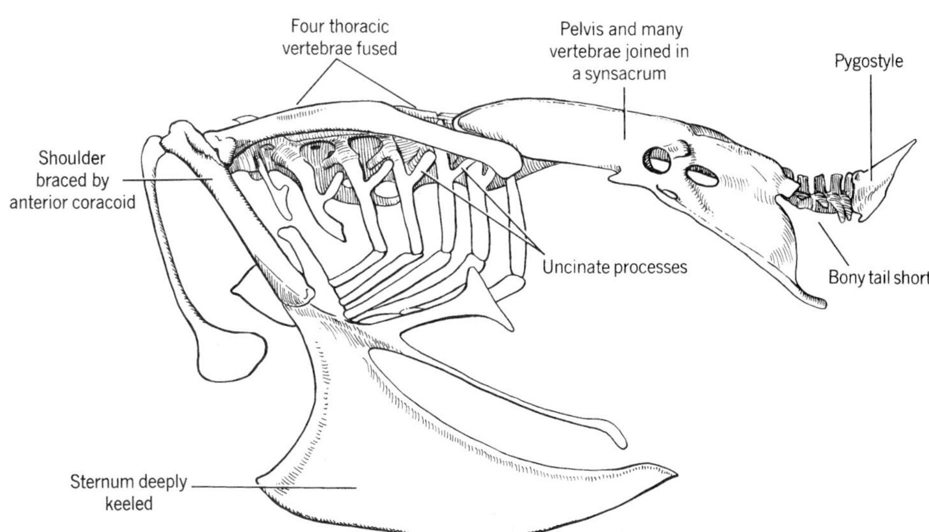

FIGURE 28.20 SOME CHARACTERISTICS OF THE BODY SKELETON OF FLYERS shown by a fruit bat, *Pteropus* (above), and a pheasant, *Phasianus* (below).

have additional fusions in the thoracic region (pheasants, falcons). Even in the absence of fusions, the shape of the joints between trunk vertebrae is such as to limit or prevent motion. Ribs tend to be broad and flat so they nearly touch edge to edge. They may fuse together proximally (some bats, many pterosaurs) or to the spine (some pterosaurs). The uncinate processes of avian ribs probably provide bracing. The sternal ribs tend to ossify. The number of units in the sternum is reduced to three (bats) or one (birds, pterosaurs).

The body is made compact by the shortness of the bony tail (birds, some pterosaurs, and some bats) and by sharp flexion of the neck during flight. The pectoral girdle is large and strong, and has unusually firm attachment to the body by the anterior coracoid (birds, pterosaurs), clavicle (bats), or even the scapula (some pterosaurs). The entire girdle is virtually immovable in birds and pterosaurs, though not in bats. The concentration of flight muscles (including elevators of the wings) on the breast of birds positions their great weight near, but a little below, the point of support of the body by the wings. The same is true to a lesser extent of bats.

A third requirement is for automatic mechanisms that save muscular effort and, therefore, both weight and energy. The joints of the wings of both birds and bats are constructed so as to limit motion that is not in one plane, although certain avian carpals do automatically ensure some pronation on the downstroke and supination late in the upstroke. The ulna (which would be required for rotation of the forearm) adheres to the radius in pterosaurs, and its distal portion is nearly lost in bats. It is retained in birds because it has a new function: The radius and ulna join the humerus and carpus in such a way as to form a parallelogram. When the elbow opens out, the wrist opens automatically (Figure 28.21).

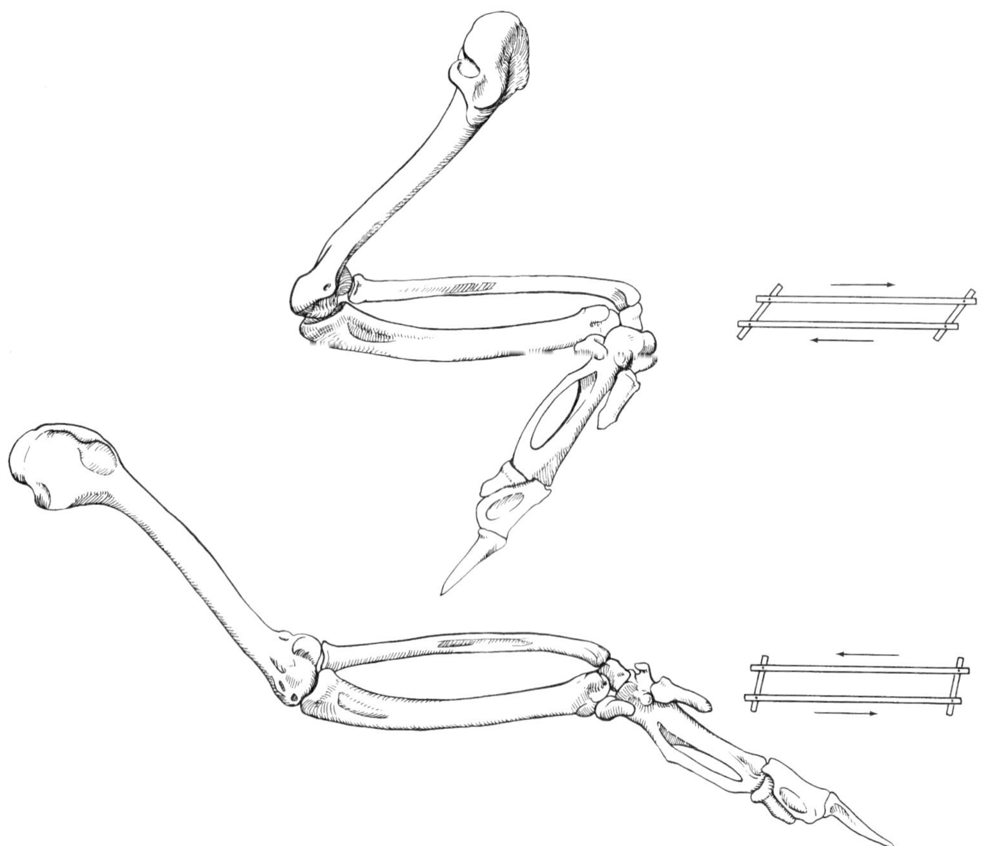

FIGURE 28.21 USE OF A PARALLELOGRAM TO EXTEND AUTOMATICALLY THE WRIST WITH THE ELBOW, shown by a pheasant, *Phasianus*.

Various bats have locking devices to assist in holding the shoulder, elbow, and wrist joints at the correct angles. It is said that a bone at the elbow in the albatross can be placed in the joint to hold it open. Only a small part of the wing membrane of bats is attached to the small second digit, yet many bats use this digit and its ligamentous extension as a bow to keep the leading edge of the wing stiff.

Bats and birds of many orders (and the colugo) have a mechanism (remarkably convergent in the different groups) for locking the digits in the flexed position: Tendons of the toes have ventral tubercles that can be meshed with plications on the tendon sheaths to hold the tendons firmly. The mechanism functions not only in hanging (bats) and perching but also in climbing, swimming, and grasping prey.

Another requirement of flyers is for an efficient power plant. Birds have a relatively high body temperature, and it is likely that even pterosaurs could elevate the body temperature, at least when flying. Birds and bats have a high metabolic rate and efficient respiratory systems. Ventilation of the lungs can be synchronized with the wingbeats in some species. The heart is large, blood pressure high, and circulation rapid. Even making allowance for great efficiency, it seems certain that the energy release in avian muscles is higher than in nonflying vertebrates. Fat is the main source of energy.

For reasons explained earlier, flyers have a large cerebellum and (bats excepted) large eyes. Bats roost upside down and hence have specializations of the feet to form hooks and of the pelvis and knee to allow suitable orientation of the legs.

REFERENCES

Bennett, S.C. 1997. Terrestrial locomotion of pterosaurs: a reconstruction based on *Pteraichnus* trackways. *J. Vert. Paleontol.* 17(1):104–113.

Brower, J.C. 1983. The aerodynamics of *Pteranodon* and *Nyctosaurus,* two large pterosaurs from the upper Cretaceous of Kansas. *J. Vert. Paleontol.* 3:84–124.

Dial, K.P., G.E. Goslow, Jr., and F.A. Jenkins, Jr. 1991. The functional anatomy of the shoulder in the European starling (*Sturnus vulgaris*). *J. Morphol.* 207:327–344.

Goslow, G.E., Jr., K.P. Dial, and F.A. Jenkins, Jr. 1989. The avian shoulder: an experimental approach. *Am. Zool.* 29:287–301.

Hainsworth, F.R. 1987. Precision and dynamics of positioning by Canada geese flying in formation. *J. Exp. Biol.* 128:445–462.

Hainsworth, F.R. 1988. Induced drag savings from ground effect and formation flight in brown pelicans. *J. Exp. Biol.* 135:431–444.

Hermanson, J.W. 1998. Chiropteran muscle biology: a perspective from molecules to function, pp. 127–139. *In* T.H. Kunz and P.A. Racey (eds.), *Bat biology and conservation.* Smithsonian Inst. Press, Washington, DC. Excellent review of how muscle fiber types relate to modes of flight in ecologically distinct bats.

Jenkins, F.A., Jr. 1993. The evolution of the avian shoulder joint. *Am. J. Sci.* 293A:253–267. An outstanding analysis of the functional morphology of the shoulder joint from Paleozoic amphibians to modern birds. Well illustrated.

Norberg, U.M. 1985. Flying, gliding, and soaring, pp. 129–158. *In* M. Hildebrand et al. (eds.), *Functional vertebrate morphology.* Harvard University Press, Cambridge, MA.

Norberg, U.M. 1990. *Vertebrate flight: mechanics, physiology, morphology, ecology, and evolution.* Springer, New York. 291p. Comprehensive and scholarly treatment. In-depth presentation of the mathematical basis of the principles of flight and performance; numerous examples provided. Generously illustrated.

Norberg, U.M., and J.M.V. Rayner. 1987. Ecological morphology and flight in bats (Mammalia; Chiroptera): wing adaptations, flight performance, foraging strategy and echolocation. *Philos. Trans. R. Soc. Lond. Biol.* 316:335–427.

Padian, K. 1983. A functional analysis of flying and walking in pterosaurs. *Paleobiology* 9:218–239.

Padian, K. 1991. Pterosaurs: were they functional birds or functional bats?, pp. 146–160. *In* J.M.V. Rayner and R.J. Wooton (eds.), *Biomechanics in evolution.* Cambridge Univ. Press, New York.

Padian, K., and J.M.V. Rayner. 1993. The wings of pterosaurs. *Am. J. Sci.* 293:91–166. A detailed analysis of the skeletal and soft tissue components of the wing membrane of pterosaurs as determined from the fossil record. Arguments for a birdlike posture of pterosaurs are reviewed.

Pennycuick, C.J. 1988. On the reconstruction of pterosaurs and their manner of flight, with notes on vortex wakes. *Biol. Rev.* 63:299–331.

Pennycuick, C.J. 1989. *Bird flight performance: a practical calculation manual.* Oxford Univ. Press, New York. 153p. A practical book that reviews the factors important in the design of wings in birds, explains how to make meaningful measurements important to flying forms, and provides a computer program for calculating power during flight.

Quinn, T.H., and J.J. Baumel. 1990. The digital tendon locking mechanism of the avian foot (Aves). *Zoomorphology* 109:281–293.

Rayner, J.M.V. 1988. The evolution of vertebrate flight. *Biol. J. Linnean Soc.* 34:269–287.

Rayner, J.M.V. 1991. Avian flight evolution and the problem of *Archaeopteryx,* pp. 183–212. *In* J.M.V. Rayner and R.J. Wooton (eds.), *Biomechanics in evolution.* Cambridge Univ. Press, New York.

Rayner, J.M.V. 1995. Dynamics of the vortex wakes of flying and swimming vertebrates. *J. Exp. Biol.* 49:131–155. Good discussion of wake vortices.

Rayner, J.M.V. 1996. Biomechanical constraints on size in flying vertebrates. *Symp. Zool. Soc. Lond.* 69:83–109.

Rosser, B.W.C., and J. George. 1986. The avian pectoralis: histochemical characterization and distribution of muscle fiber types. *Can. J. Zool.* 64:1174–1185.

Rüppell, G.R. 1977. *Bird flight.* Van Nostrand Reinhold, New York. 191p. (Translation by M.A. Biederman-Thorson. Originally published in 1975 under the title *Vogelflug,* Kindler Verlag GmbH, München.) Outstanding contribution to the analysis of bird flight. Rüppell was one of the first to use high-speed photography to freeze birds in flight. Illustrations are magnificent.

Saville, D.B.O. 1957. Adaptive evolution of the avian wing. *Evolution* 11:212–224.

Sokoloff, A.J., et al. 1998. Neuromuscular organization of avian flight muscle: morphology and contractile properties of motor units in the pectoralis (pars thoracicus) of pigeon (*Columba livia*). *J. Morphol.* 236:179–208.

Spedding, G.R. 1984. Momentum and energy in the wake of a pigeon (*Columba livia*) in slow flight. *J. Exp. Biol.* 111:81–102. Technically superb study.

Unwin, D.M. 1999. Pterosaurs: back to the traditional model? *TREE* 14:263–268. Critical review of the arguments put forth over the past 200 years concerning the wing morphology, flight capabilities, and terrestrial posture of pterosaurs. This paper illustrates the process of how our interpretation of previously described morphology can be influenced by new data from phylogenetics and new finds of fossil trackways.

Vaughan, T.A. 1959. Functional morphology of three bats: Eumops, Myotis, Macrotus. *Univ. Kans. Publ. Nat. Hist.* 12:1–153.

Vazquez, R.J. 1992. Functional osteology of the avian wrist and the evolution of flapping flight. *J. Morphol.* 211:259–268. Excellent anatomical study of a complex structure.

Vazquez, R.J. 1994. The automating skeletal and muscular mechanisms of the avian wing (Aves). *Zoomorphology* 114:59–71.

Vogel, S. 1981. Life in moving fluids: the physical biology of flow. 2nd ed. Princeton Univ. Press, Princeton, NJ. 467p.

Wellnhofer, P. 1991. *The illustrated encyclopedia of pterosaurs.* Crescent Books, New York. 192p. A scholarly work that considers the evolutionary changes in pterosaurs from the Jurassic through the Cretaceous; beautifully illustrated.

Wilson, J.A. 1975. Sweeping flight and soaring by albatrosses. *Nature* 257:307–308

Chapter 29

Energetics and Locomotion

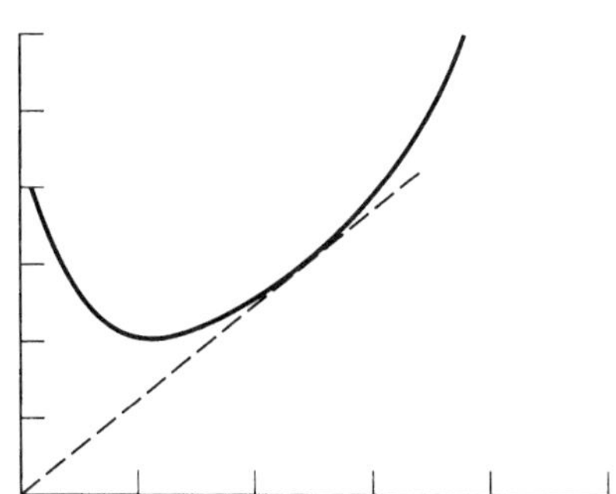

Energetics is the field of study, spanning biomechanics and physiology, that deals with energy and its transformation. Locomotion is costly in terms of energy, and the cost increases rapidly with speed. Accordingly, one might expect that body structure and manner of locomotion are closely linked to energetics. Indeed they are. But the precise relationship for different forms of locomotion among closely related, and distantly related, species is the focus of many current studies. This short chapter will summarize some fundamentals and present or cross-reference various ways that vertebrates adapt to their energy requirements, leaving the physiological considerations for another course of study.

SOME BASICS

Even when at rest, vertebrates use energy to maintain circulation, excretion, digestion, and (for endotherms) body temperature. The maintenance level of energy utilization, measured as volume of oxygen burned per unit of body weight in a unit of time, is the **standard metabolic rate** or, for endotherms, the **basal metabolic rate.** The rates of resting mammals are six or more times greater than those of reptiles. (Metabolic rates double or triple when body temperature increases 10°C). It is assumed that even when not moving, tetrapods also use energy to maintain posture (Figure 29.1A).

The maximum expenditure of energy that a vertebrate can sustain is roughly 10 times the resting level, but there is considerable variation among taxa, and endotherms have a greater range than ectotherms. **Aerobic metabolism** is the complete combustion of carbohydrate, delivered by the circulatory system, to form carbon dioxide and water. Speed (or effort needed to climb or dig) can be increased severalfold beyond sustainable levels by **anaerobic metabolism,** in which there is incomplete combustion of energy stores within the tissues, with the production of lactic acid. Anaerobic metabolism is less efficient and quickly leads to fatigue, but usually not before the burst of effort has served its purpose for escape, defense, or pursuit. Following the effort there must be a recovery period while the animal pays its oxygen debt and removes the lactic acid.

Some kinds of energy (electromagnetic, chemical) do not directly concern us here even if they are involved in muscle physiology. The heat energy produced during locomotion usually represents wasted energy, although it may be a desirable consequence of activity for small endotherms in cold weather, or for ectotherms that need to move faster. Of direct relevance to locomotion are potential energy and kinetic energy, which are described below. Finally, studies of animal energetics during locomotion are approached in two distinct ways. Some studies measure the amount of oxygen consumed per unit time and use this as an indicator of metabolic cost and mechanical power output. Other studies, in contrast, estimate mechanical power output directly from measurements of the forces acting in muscles or on the limb, and the trajectory of the animal during its locomotor activity.

Mechanical efficiency is the ratio of the useful energy taken out of a system (usually as force or work that produces potential or kinetic energy) to the energy put in (usually as fuel). The efficiency of an inorganic machine can be exactly measured: For a hydraulic turbine it might be 90 percent, for a steam engine 20 percent, and for many engines less. With practice, rhythmic human muscular effort, as measured by an ergometer such as a bicycle-pedaling apparatus, has an efficiency of 25 percent or a little more. However, when considering normal behavior of animals, efficiency is a less exact though equally important concept. Thus, the cheetah's way of running is a more extravagant use of energy than that of the blackbuck, yet the cheetah overtakes the blackbuck often enough to survive, and in that (different) sense is the better runner. The weight of the hydraulic turbine is unimportant, but for flying machines, animal or man-made, the ratio of useful energy output to weight may be more important than its ratio to fuel input. The ratio of thrust to weight is greater for the flight muscles of birds than for aircraft engines, which in turn rate far ahead of steam engines. It might be expected that an animal's morphology enables it to perform the needed activity, however demanding or exacting, with sufficient economy of effort to make the activity advantageous (see Comment 29.1).

COMMENT 29.1

SYMMORPHOSIS: BALANCING DESIGN WITH COST

Before focusing on the energetics of locomotion, we touch on a broader concept. Body structure and manner of locomotion are closely linked to energetics. Organisms tend to be designed for economy of effort: Anatomical structures typically approximate, but do not exceed, their functional requirements. This concept, called **symmorphosis** (= *together* + *shaping*), is sometimes used to form hypotheses about morphology (from microscopic to macroscopic levels) that relate to biomechanics and energy. For example, when a rattlesnake vibrates its tail at a frequency of 90/s, for the duration of a threat from a predator, are the morphological and physiological aspects of its tail muscles precisely matched? The escape response of a frog is a single, but maximal, jump, whereas that of a turkey consists of multiple, maximal strides for several seconds. Is the morphology of their respective limbs optimal for these functions? If not, at what level (anatomical or physiological) is the system either overbuilt or underbuilt? Scientists grappling with these questions have reported their progress in a symposium volume (see Weibel, Taylor, and Bolis in References) and discuss symmorphosis as a viable guiding principle. For example, how does one falsify a symmorphosis hypothesis? Optimality models in any area of evolutionary biology are controversial. Through natural selection, organisms are not "designed" to satisfy a goal, nor is natural selection mere engineering. Engineers tend to design structures for a single purpose; structures of organisms often serve several purposes. Nevertheless, symmorphosis effectively frames some comparative studies, particularly ones directed toward patterns of evolutionary convergence, helping to determine the relative importance of different functions to structural design.

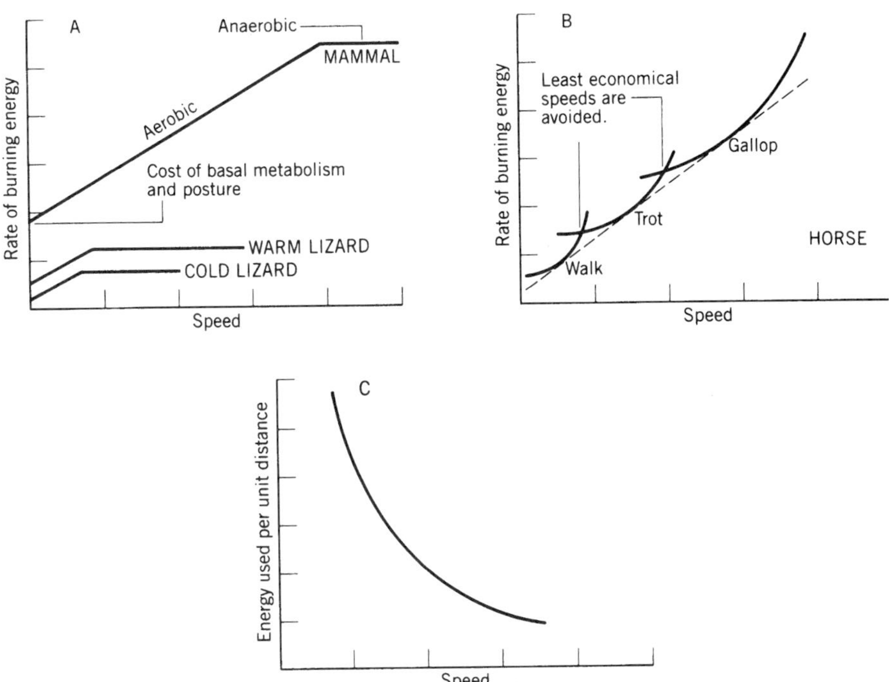

FIGURE 29.1
LOCOMOTORY ENERGETICS OF TETRAPODS showing the general nature of net cost of transport (A and B) and total cost of transport (C).

TERRESTRIAL LOCOMOTION

Utilization of Energy The aerobic rate of burning energy (above the resting metabolic level) divided by speed, or the (equivalent) total oxygen burned divided by the distance traveled, is the **net cost of transport.** It is commonly stated that the relationship is linear; the fuel needed to traverse a given distance is independent of time and speed (Figure 29.1A). However, for the horse, and probably for other cursors, there is a separate curving regression line for each gait (part B of the figure). The animal tends to move at the gait and speed that are relatively economical. (However, gait may also be changed as speed increases to maintain musculoskeletal forces within safe limits.) An overall straight regression line (dashed on the figure) is a satisfactory approximation if drawn through the preferred slow and moderate speeds, but it would be expected to curve somewhat if maximum speed were included.

Total cost of transport is energy burned at a given speed divided by that speed (part C). This relationship is clearly speed dependent: within aerobic limits, the cost of moving a unit of mass a unit of distance decreases with rate of travel. When moving faster, the animal pays the constant cost of maintenance and posture for a shorter period. (Higher speed also benefits the conservation of energy—see below.)

The cost of bipedal locomotion appears to be about the same as that of quadrupedal locomotion (although, as noted below, the comparison is complicated by the recycling of energy). Hopping has probably not evolved to save energy. Snakes that crawl by lateral undulation use about as much energy as lizards, and when crawling by concertina motion they use nearly 10 times more. A load carried (antlers, young, prey) seems to increase the cost of transport in proportion to the mass of the load as a percentage of the animal's unloaded mass (though there are apparent exceptions).

Digging can cost several hundred to several thousand times as much as moving the same distance overground. Burrowers must tolerate high workloads in atmospheres having low oxygen tension, high carbon dioxide tension, high humidity, and limited temperature variation. All but the smallest mammalian burrowers have a low basal metabolic rate and a wide range of thermoneutrality (temperature at which oxygen consumption is minimal). Compared with white rats, the highly fossorial mole rat (see figure on p. 467) has about 30 percent greater capillary density in its muscles, and a nearly 50 percent higher density of mitochondria (which process oxygen in muscle cells). There are physiological adaptations of the respiratory, circulatory, and thermoregulatory systems.

Conservation and Recycling of Energy Previous chapters have told of ways in which energy is conserved: Cursors and saltators may avoid unnecessary oscillations, modify the skeleton to reduce weight without loss of adequate strength, and concentrate the mass of moving parts near the body to reduce inertia (pp. 441–445). Fossorial vertebrates passively resist various forces by the structure of their joints and ligaments, and by bony stops that restrict motion (p. 463).

Cursors and saltators also save significantly on fuel by recycling energy. First, they convert back and forth between potential and kinetic energy. **Gravitational potential energy** (E_p) is energy of position. In climbing a tree, a flying squirrel does work against gravity, thus gaining potential energy for its glide. The limbs of running animals gain potential energy when lifted. Quantitatively, $E_p = mgh$, where m is mass, g is the pull of gravity, and h is the height. **Kinetic energy** is energy of motion. In rectilinear motion (the charging rhinoceros and striking falcon), $E_k = \frac{1}{2}mv^2$. In rotational motion (the beating wing, snapping jaw, and swinging limb), $E_k = \frac{1}{2}I\omega^2$, where I is the moment of inertia and ω is angular velocity (see pp. 419 and 420).

Energy is never destroyed: The E_p of the flying squirrel at the top of a tree becomes E_k as it glides down; the E_k of a jumping salmon becomes E_p as the fish comes to rest in a higher pool. When a *pendulum* swings, it changes E_p to E_k on the downstroke and E_k to E_p on the upstroke. Disregarding friction and air resistance, no new energy need be introduced to keep it going. Oscillating appendages of moving tetrapods do the same, to a degree, as they swing clear of the ground. The rate at which a pendulum naturally swings relates to its length: the greater its length, the longer its period. This is a reason that each person can walk most comfortably at a certain rate, and why long-legged animals tend to walk with long, slow strides whereas short-legged animals tend to walk with short, fast strides. If an animal does not oscillate its legs at just their natural periods, it may still recycle much energy at slow and moderate rates of travel.

When an extended leg of a walking animal is planted on the ground, the body tends to vault up over it to the vertical position and then fall again in the manner of an *inverted pendulum.* This is why the hips of walking humans tend to go up and down. Over a series of steps, the gravitational E_p and E_k of the center of mass oscillate between maximum and minimum values, decelerating as it rises and accelerating as it falls. In important and influential research, Cavagna, Heglund, and Taylor used a force plate and synchronized cinematography to measure these oscillations in two species of birds and three diverse species of mammals. They showed that E_p is highest at the midpoint of the stance phase of each foot, and lowest when one foot of a pair is just striking the ground and the other just leaving it. Values for E_k are out of phase with those for E_p. Because the total energy changes only a little throughout a walking step, the result is a recovey of 65–70 percent of the energy changes in each stride, leaving only 30–35 percent to be supplied by muscles.

The story for speeds faster than a walk, however, is different. Cavagna and colleagues demonstrated that running bipeds, trotting quadrupeds, and hopping animals use their limbs twice in a stride to push upward and forward, once to break their fall when landing, and again later during propulsion. Thus the oscillations of E_p and E_k are substantially in phase, which results in no energy transfer and no energy saved. Energy *is* conserved, however, in many moderate- to large-sized vertebrates by another mechanism.

Elastic potential energy is exemplified by a loaded spring, a drawn archer's bow, a stretched tendon or ligament, or an inflated lung. It equals $\frac{1}{2}ks^2$, where k is the spring constant, or restoring force per unit displacement of the particular material stretched, and s is the displacement. The first evidence for the storage of elastic energy was shown for kangaroos in a metabolic study by Dawson and Taylor: Oxygen consumption initially increased with increased hopping speed (as expected), but then leveled off and even decreased at high speed. This startling observation has since been shown to be due to the recovery of elastic energy. As kangaroos land from each hop, the tendons of their actively contracting (hence rigid) gastrocnemius and plantaris muscles stretch, thus storing energy for release during the subsequent propulsion phase. The energy saving reaches an impressive 40 percent. When the foot of a running horse (or any ungulate) strikes the ground, the impact bends the joint between the phalanges and the metapodial bones (the fetlock joint). This stretches the ligaments that are called suspensory, or springing, ligaments (see figure on p. 416). The energy of the deformation is recovered as the system is unloaded, thereby relieving muscles in straightening the joint and giving an initial upward impetus to the entire body (Figure 29.2). Nearly 700 kg is required to break the major springing ligament in the front leg of the horse. Similar ligaments supporting the head and spine also contribute to the extremely high efficiency of galloping horses (see Minetti et al. in References).

The recovery of spring energy is now known to be widespread. Another example: Elastic tissue in the floor of the mouth, pharynx, and lungs recycles energy as vertebrates ventilate or pant.

Morphological Correlates to Energy Conservation Investigators are sorting out the morphological and behavioral correlates of the costs of locomotion. For example, energy constraints impact on muscle architecture. Roberts and colleagues made direct measurements of the force and length of the gastrocnemius of running turkeys and determined that when the foot strikes the ground, the muscle contracts isometrically (the most economical kind of contraction) and its tendon is stretched. When the foot comes off the ground, the tendon recoils and stored energy is recovered to propel the animal forward. The hypothesis

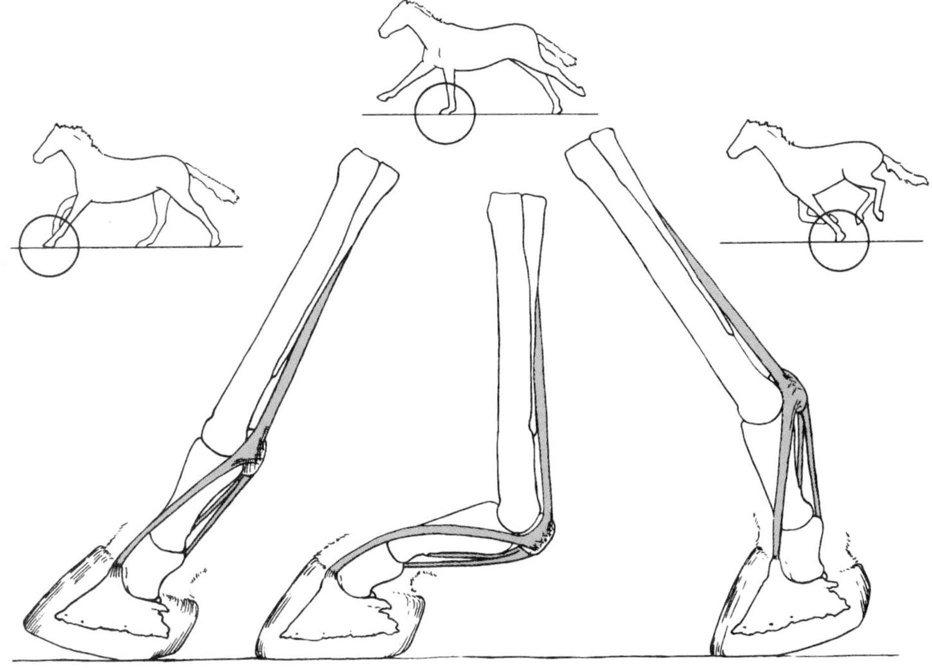

FIGURE 29.2
SPRINGING ACTION OF THE PRINCIPAL SUSPENSORY LIGAMENTS IN THE FOOT OF THE HORSE.

that emerged from this experiment, and similar ones on hopping wallabies (see Biewener in References), is that pinnate limb muscles with long tendons provide an economical, elastic strut that contributes to high recovery of energy.

Even with increasingly sophisticated methods of measurement, the energy costs of locomotion are difficult to tease out, and competing hypotheses abound. For example, contrary to the hypothesis that animals switch gaits to save energy, Farley and Taylor present evidence that change of gait has to do with minimizing the internal forces of the musculoskeletal system. The cost of the metabolic work of the muscles driving the limb was initially considered to explain many of the size-related differences in the energetic cost of terrestrial locomotion among different groups (Figures 29.1A and 29.4B). But recent studies of quadrupeds (see Kram and Taylor) and bipeds (see Roberts et al.) suggest that cost relates to the duration of the time of foot contact (and the application of muscle force) during the propulsion phase.

SWIMMING

The net cost of transport for swimmers (Figure 29.3) differs from that of tetrapods in being clearly speed dependent. The cost increases about as the square of swimming speed. A line drawn through the origin of the graph and tangent to the curve gives the **maximum range speed,** or speed of minimal cost of transport. This most economical speed might be selected for migration. If temperature is lowered for fishes, the curve shifts down on the graph and the speed of minimum cost is reduced. Warm-bodied cetaceans, on the other hand, have a high metabolic rate, even for mammals. Their blubber sustains them as a source of fuel when food is not available.

For fishes, anaerobic bursts of speed may be three or four times greater than sustainable speed. The body form of the fish (e.g., streamlining) is much more critical at burst speeds than at cruising speeds. Swimming costs turtles and lizards up to three times more than it costs fishes, and it costs warm-bodied humans and ducks up to 30 times as much. It is probable that schooling saves some energy for certain fishes, and it is possible that repeated coasting between short bursts of swimming may also save energy. Deep-diving seals and whales conserve significant amounts of energy by prolonged gliding, particularly in the descent.

Some swimmers also recycle energy. The tough, fibrous skin of sharks is stiffened by the hydrostatic pressure of underlying muscles when the animal swims fast. Thus it becomes a whole-body exotendon and stretches on the convex side of the undulating body to store energy for release as that side flexes. After studying the vertebral and intervertebral morpohology of the saddleback dolphin, Long and colleagues concluded that the vertebral column stores significant elastic spring energy, particularly when the flukes move with high amplitude. Beneath the blubber of cetaceans there is a wrapper of collagen fibers wound in helices around the body and peduncle. This layer may also store and release spring energy. Similarly for bony fishes, elastic elements associated with the spine,

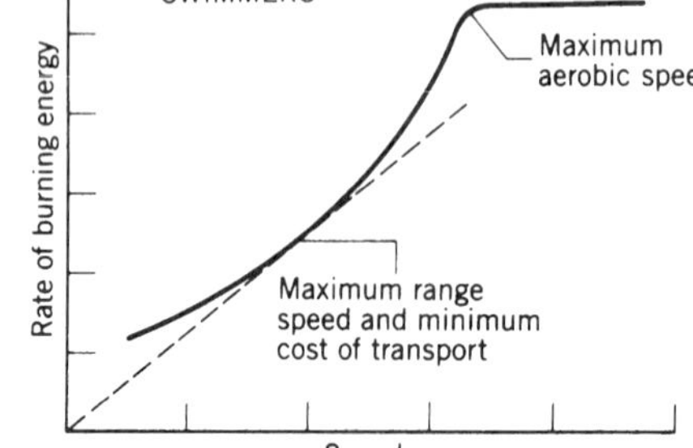

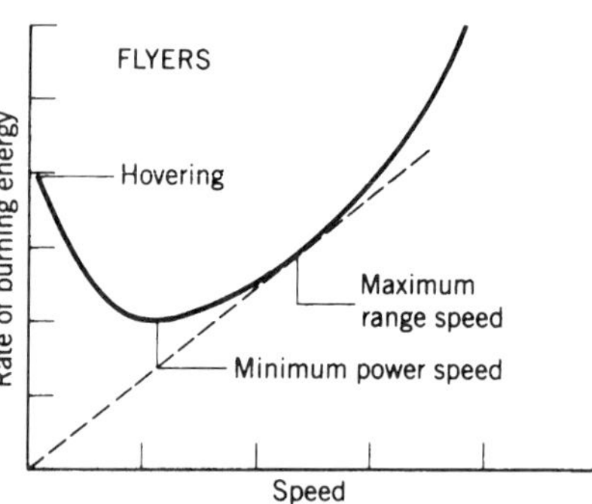

FIGURE 29.3 LOCOMOTORY ENERGETICS OF SWIMMERS AND FLYERS. The latter is theoretical pending more research.

and remnants of the notochord (between the centra) are stretched as the body flexes to one side and then snaps back, thus recycling elastic spring energy.

The axial musculature, which is the power plant of the tail swimmer, may weigh half as much as the entire animal. Fishes that swim constantly (tunas) or migrate long distances (salmon) have on their flanks red muscle that contracts relatively slowly, contains myoglobin, is aerobic, and has an unusually rich blood supply. The temperature of such muscle is commonly 10°C above water temperature and is not dissipated because of a countercurrent exchange mechanism (see pp. 265 and 266). The white muscle, which accounts for most of the bulk of the muscular system, is capable of faster, anaerobic contraction (see Altringham and Block in References).

FLYING

Cursors and swimmers use the least energy when stationary, but it is costly for flyers to hover. Several specialists have calculated that for aerobic horizontal flight the power curve (which is proportional to the rate of energy use) should look about as shown in Figure 29.3. Unlike the corresponding curves for terrestrial vertebrates and swimmers, there is a **minimum power speed** at which fuel is burned the slowest. The faster maximum range speed (determined as for swimmers) allows the longest travel on a given amount of fuel. However, the curve depicted is theoretical; it is supported by some experiments and not others. In a recent study where mechanical power was calculated from direct *in vivo* measurements of force and wing movements of magpies flying at several speeds, support for the shape of the curve for hovering and slow flight was observed (see Dial et al. in References). At speeds ranging from slow flight to the fastest flight, however, values of mechanical power output were statistically indistinguishable. Thus, at these speeds, the arm of the curve does not ascend at the higher speeds (Figure 29.3), but remains flat.

Soaring costs a third or less as much as flapping flight. Undulating flight, which alternates flapping with gliding with wings folded (as for woodpeckers), may save energy. Overall, the cost of flying is two or three times that of swimming, but only one-third to one-fourth that of running (Figure 29.4A).

SCALING OF LOCOMOTORY COST

The cost of transport does not increase in direct proportion to body mass, but instead as about $M^{0.7}$. It is cheaper for large animals than for small ones to move a unit of mass a given distance. In overground locomotion there is an inverse relationship between the rate of using energy and the time during which each foot applies force to the ground. The slower strides of large mammals give them more time to generate the force required to support the body. This gives them more endurance than small mammals. Large fishes can swim farther and can sustain maximum aerobic speed longer (Figure 29.4B). The maximum running speeds of mammals are about two times their maximum respective aerobic

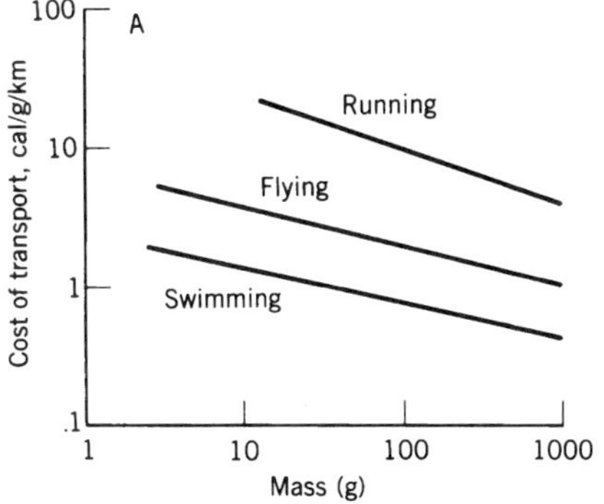

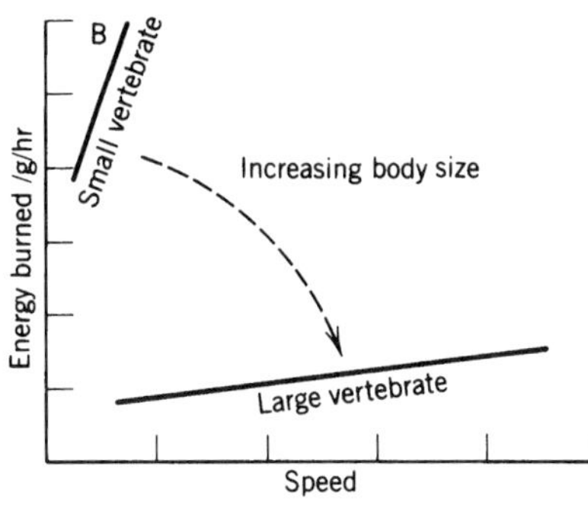

FIGURE 29.4 COST OF TRANSPORT IN RELATION TO TYPE OF LOCOMOTION AND BODY SIZE

speeds. The maxima vary about as $M^{0.17}$ but peak at approximately 120 kg. The pronghorn antelope, second only to the cheetah in maximum speed, is a distance runner rather than a sprinter. It can run 11 km in 10 min at an average speed of 65 km h^{-1}. Performance studies in the laboratory (at lesser speeds) suggest that pronghorns have the ability to consume and process oxygen at a much higher rate than predicted for mammals of their size (see Lindstedt et al. in References). They accomplish this not by novelty, but by amplification of structures present in all mammals (increased number of mitochondria/muscle volume, for example).

There is one circumstance in which the small vertebrate has the advantage. It is energetically cheaper for small climbers than large to move vertically. Imagine yourself running up a long, steep staircase as fast and easily as the small squirrel scampers up the trunk of a tall tree.

REFERENCES

Alexander, R. McN. 1988. *Elastic mechanisms in animal movement.* Cambridge Univ. Press, New York. 141p.

Altringham, J.D., and B.A. Block. 1997. Why do tuna maintain elevated slow muscle temperatures? Power output of muscle isolated from endothermic and ectothermic fish. *J. Exp. Biol.* 200:2617–2627.

Bennett, A.F. 1985. Energetics and locomotion, pp. 173–184. *In* M. Hildebrand et al. (eds.), *Functional vertebrate morphology.* Harvard Univ. Press, Cambridge, MA.

Bennett, A.F. 1988. Structural and functional determinates of metabolic rate. *Am. Zool.* 28:699–708.

Biewener, A.A. 1998. Muscle function *in vivo:* a comparison of muscles used for elastic energy savings versus muscles used to generate mechanical power. *Am. Zool.* 38:703–717.

Bone, Q. 1975. Muscular and energetic aspects of fish swimming, vol. 2, pp. 493–528. *In* T.Y.-T. Wu, C.J. Brokaw, and C. Brennen (eds.), *Swimming and flying in nature.* Plenum, New York.

Cavagna, G.A., N.C. Heglund, and C.R. Taylor. 1977. Mechanical work in terrestrial locomotion: two basic mechanisms for minimizing energy expenditure. *Am. J. Physiol.* 233:R243–R261.

Dawson, T.J., and C.R. Taylor. 1973. Energetic cost of locomotion in kangaroos. *Nature* 246:313–314.

Dial, K.P., et al. 1997. Mechanical power output of bird flight. *Nature* 390(6 Nov.):67–70.

Farley, C.T., and C.R. Taylor. 1991. A mechanical trigger for the trot-gallop transition in horses. *Science* 253:306–308.

Jones, J.H., and S.L. Lindstedt. 1993. Limits to maximal performance. *Annu. Rev. Physiol.* 55:547–569. A practical guide to how variables of locomotor performance are measured and compared across mammals of different size.

Kram, R., and C.R. Taylor. 1990. Energetics of running: a new perspective. *Nature* 346:265–267.

Lindstedt, S.L., and R.G. Thomas 1994. Exercise performance of mammals: an allometric perspective. *Adv. Vet. Sci. Comp. Med.* 38B:191–217. Excellent introduction to the literature on comparative energetics of mammals.

Lindstedt, S.L., et al. 1991. Running energetics in the pronghorn antelope. *Nature* 353:748–750.

Long, J.H., et al. 1997. Locomotor design of dolphin vertebral columns: bending mechanics and morphology of *Delphinus delphis. J. Exp. Biol.* 200:65–81.

McNab, B.K. 1979. The influence of body size on the energetics and distribution of fossorial and burrowing mammals. *Ecology* 60:1010–1021.

Minetti, A.E., et al. 2000. The relationship between mechanical work and energy expenditure of locomotion in horses. *J. Exp. Biol.* 202:2329–2338.

Roberts, J.J., et al. 1977. Muscular force in running turkeys: the economy of minimizing work. *Science* 275:1113–1115.

Schmidt-Nielsen, K. 1984. Scaling: why is animal size so important? Cambridge Univ. Press, London. 241p.

Thompson, S.D., et al. 1980. Energetic cost of bipedal hopping in small mammals. *Nature* 287:223–224.

Weibel, E.R., C.R. Taylor, and L. Bolis. 1998. *Principles of animal design: the optimization and symmorphosis debate.* Cambridge Univ. Press, New York. 314p. Researchers report ongoing research projects and theoretical arguments.

White, T.D., and R.A. Anderson. 1994. Locomotor patterns and costs as related to body size and form in teiid lizards. *J. Zool. Lond.* 233:107–128. Carefully clarifies the often confusing terms used regarding cost of locomotion and transport.

Woledge, R.C., and N.A. Curtin. 1990. The price of being a snake. *Nature* 347:619, 620.

Chapter 30

Feeding

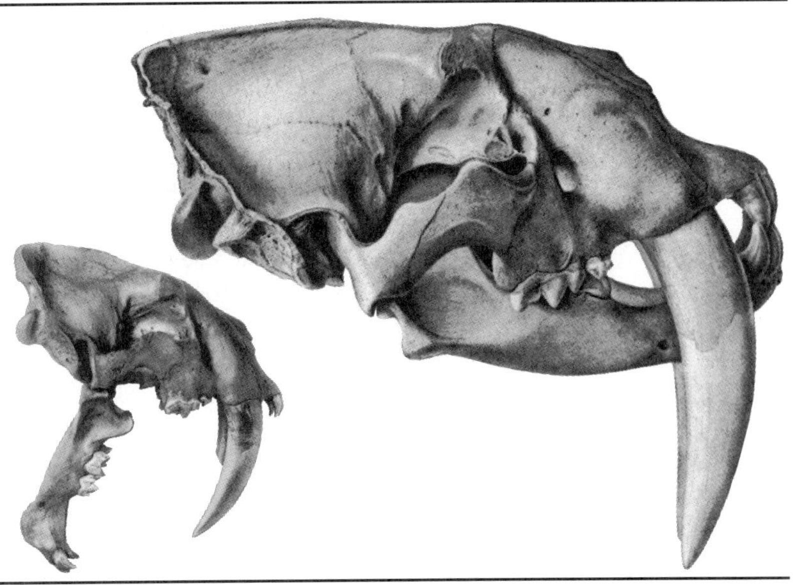

The food of vertebrates ranges from microscopic (diatoms, algae) to very large, from passive (plants) or nearly so (most mollusks) to very active, from nutritious (insects, worms) to low in food value (stems), from defenseless to protected, and from extensively available to highly restricted (e.g., the nectar of only certain flowers). Food may be seasonal or constantly accessible, abundant or rare, and available with or without competition. Small wonder that the ways that vertebrates locate and process their food are exceedingly varied.

This is, therefore, a good place to emphasize that each animal adapts not to a random combination of independent activities, but instead to a coordinated lifestyle. Manner of locomotion is nearly always related to feeding habits, and reproductive, defensive, and other behaviors are usually correlated with the manner of feeding and locomotion. Parts of preceding chapters on the teeth, digestive system, sense organs, and locomotor adaptations relate directly to feeding.

Analysis of feeding has become a complex, challenging field. Most studies completed in the first half of the twentieth century treated the jaw as a simple lever pivoting at the mandibular condyle. Dissection, manipulation, and linear measurements were used to make correlations between form and function. Current research supplements those still-important techniques with behavioral observation, stress analysis using strain gauges affixed to functioning jaws, motion analysis using accelerometers and tracers, physical analysis using pressure transducers, monitoring of muscle action by electromyography, and the study of trajectories, velocities, and accelerations using fiberoptics and high-speed cine- or videophotography (standard and X-ray, and often simultaneously from different angles).

AQUATIC FEEDING

The biomechanics of aquatic feeding is sharply set off from that of feeding on land because the medium, water, is 830 times more dense, and 80 times more viscous, than air. This has important consequences: Motion of a predator as it approaches its prey may

Above: Skull of the sabertoothed cat, *Smilodon.* (From J.C. Merriam and C. Stock, 1932, *The Felidae of Rancho La Brea*, Publ. 422, Carnegie Inst., Washington.)

deflect the prey out of reach. Much energy is needed to move the (often considerable) amounts of water taken in with food, and there is high resistance to the passage of water through any filter of small mesh used to capture food particles.

On the other hand, because high quantities of prey may be suspended in water, and other prey can be sucked into the mouth with water, aquatic feeding can be more versatile and opportunistic than terrestrial feeding. Since terrestrial feeding of tetrapods is a departure from feeding in water by their fish ancestors, studies of these two groups frequently differ (see Schwenk in References).

Among aquatic vertebrates, a relationship is seen between food preference and feeding morphology. The classic studies by Liem of cichlid fishes living in Lake Tanganyika (Africa) illustrated fascinating examples. But Liem and his students recognized that any relationships noted must be influenced by many biological as well as physical factors, and that generalizations must be viewed with caution. Aquatic feeding provides a rich area for inquiry regarding the details of individual performance, the relationship between environment and morphology, and the mechanisms by which feeding systems evolve.

Suspension Feeding Most aquatic vertebrates use suspension feeding, or suction feeding, or both. In **suspension feeding** (also called **filter feeding**) minute food particles are removed from water as it passes through a filter or as the particles impact in mucus. This was probably the feeding mode of ancestral vertebrates and evolved multiple times among fishes. Feeding bouts are adjusted to the concentration of food in the water. Although the animal senses the presence of food, selection of food is made only at the filter in one or more ways. The pores of the filter may form a simple sieve; small particles pass, while larger ones are held back. For most suspension feeders the filter is the gill apparatus, which is coated with mucus. Particles may stick to the mucus even if smaller than the pores. They impact into the mucus because of inertia gained from the waterstream, from their swimming motions (if they are alive), or from gravitational deposition. Furthermore, particles having an electrostatic charge (positive or negative) may be drawn into the mucus. For one minnow (and probably various other fishes) the gill bars do not trap food particles, but instead direct water to the roof of the mouth, where particles stick to mucus.

Most suspension feeders are also sustained **ram feeders:** to take water into the mouth and throat they merely open the mouth wide as they swim slowly forward, allowing the water to escape through the gills. Examples are the paddlefish (Figure 30.1), shad, herrings, anchovies, sardines, manta ray, the huge basking and whale sharks, and bowhead and right whales. The heads of such feeders are enormous, ranging to one-quarter or even one-third the total length of the body. The gill rakers of the fishes are modified to form filters. They are long, slender, closely set along the gill bars, and have secondary or even tertiary branches. The environmental water is a thin soup: The animal must spend much time feeding (feeding and respiration are linked), and must have a high filtration rate. The menhaden (which is about 30 cm long) pumps 20 L of water per minute; the basking shark takes in at least 1850 m^3 of water per hour. The efficiency ranges from 25 to 80 percent removal of food from water. Selection at the filter is more or less fixed, yet may be influenced by filtration rate, water pressure, slight alteration of pore size (e.g., by degree of crowding of gill rakers), and by rate of cleaning. How fishes clean their filters and move food into the esophagus is poorly understood. Coughing (sudden, forceful reversal of the waterstream) may contribute.

The baleen whales feed mostly on krill, which consists of swarming crustaceans measuring 75 mm and less in length. The whale has up to 400 plates of baleen along each side of the upper jaw (Figure 30.1). Each plate is ribbonlike, being flat and from 0.25 to 3 m long, according to species and position in the mouth. A plate is made up like a sandwich

Trolling for small food particles
PADDLEFISH, *Polyodon*

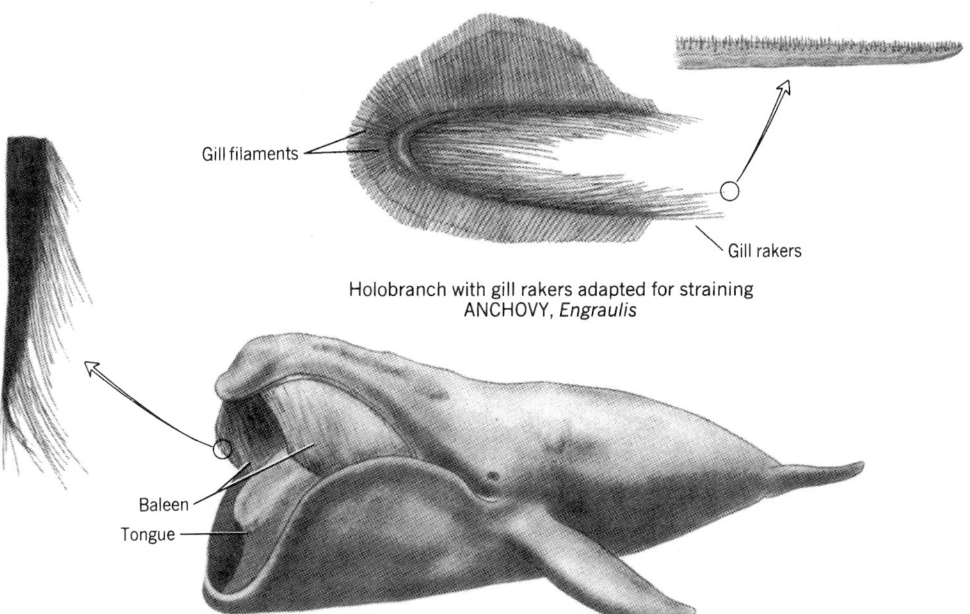

Holobranch with gill rakers adapted for straining
ANCHOVY, *Engraulis*

Rows of baleen plates adapted for straining
RIGHT WHALE, *Eubalaena*

FIGURE 30.1
SOME ADAPTATIONS FOR SUSPENSION FEEDING.

consisting of a horny lamella on each surface and a core of horny hollow tubes running lengthwise within. This makes it light and flexible but very strong. The lamellae gradually wear off of the tip and inside margin of each plate, releasing the hundreds of tubes that tangle like stiff hairs to make the filter. The plates are ever-growing at their bases to compensate for wear. Some whales (right and bowhead) swim slowly along with the mouth open and water spilling out at the corners through the baleen. The ram feeding of the humpback and rorqual whales is instead intermittent, the animal taking one great gulp at a time (see Comment 30.1).

Other suspension feeders use suction instead of ramming to take water into the mouth. This type of feeding, primitive for vertebrates, is retained by ammocoetes, the larva of the lamprey, which circulates water with a buccopharyngeal pump. Algae and detritus are trapped by impaction into mucus on the gill apparatus. Cilia slowly move strands of the mucus toward the intestine. The tadpoles of various frogs also feed this way and are remarkably effective. Like fishes, they use their mucus-covered gills. The tadpole of *Xenopus* can filter to 0.13 μm, which challenges human technology.

Among ducks, the shoveller and other dabblers strain plant or animal food from water. The bill is swung back and forth sideways, while the mandible pumps very fast. One hundred to 300 lamellae at the edge of the bill form the strainer. Flamingos are much more highly specialized and can remove diatoms and algae measuring only 0.1 to 0.2 mm in

COMMENT 30.1

THE INTERMITTENT RAM MODE OF SUSPENSION FEEDING

The rorqual whales (blue, fin, sei, Bryde's, minke) and the related humpback whale are unique among vertebrates in being intermittent ram feeders. When they sense a concentration of krill in the water, they swim rapidly, open their huge mouths very wide and take in a prodigious gulp of water (up to one-third the volume of the body), which expands the throat and a saclike pouch under the body (Figure 30.2). Numerous external grooves, extending from chin to umbilicus, open out, accordian fashion, to accommodate the expansion. The mouth is then quickly closed as water gushes out through the baleen apparatus.

It is relevant that these streamlined whales range from mid-size to the largest mammals that have ever lived (roughly 10,000 to 110,000 kg), and that they include the fastest of whales (which cruise at 20 km/h, bursts to perhaps 40 km/h). Accordingly, they have enormous kinetic energy ($= \frac{1}{2}$ mass $\times$ velocity^2).

The dynamics of their feeding has been studied by (among others) Lambertsen, Ulrich, and Straley (see References). They used photography of free-ranging animals, study of skulls, dissection, and manipulation of the jaws of carcasses by winches. The long, bowed jawbones are joined at the mandibular symphysis by a flexible pad of fibrocartilage. The temporomandibular joint has no synovial space; it is a fibrous cushion of dense connective tissue. The principal elevator of the jaws is the conical temporalis muscle, which is surrounded by a tough, funnel-shaped *stay apparatus*. Distally this structure is tendinous, inserting on the coronoid process of the jaw; its mid-region is a thick meshwork of oblique connective tissue.

As the feeding maneuver starts the jaw begins to open—perhaps passively, if fast-flowing water under the head creates negative pressure (like lift over a wing). Immediately, the forward motion of the whale establishes an inertial rush of water into the mouth and pouch, and rotates their mandibles around three axes: the jaws open about 85°, their condyles separate somewhat, and each mandible rotates around its long axis by at least 40° to bow outward. Each of these actions is completely passive, and all increase the volume of food-laden water taken in.

During most of jaw opening the jaw supports are slack and the mandibles loose, but at the end (jaws passing 70° of depression) the fibers of the stay apparatus come under strain and the mandibles stiffen. The stay apparatus then instantly recoils, releasing spring energy that initiates jaw closure. The pouch likewise recoils and starts to empty. At the moment that distension is maximum the water mass bounces forward, accelerated by the forward motion of the animal. This forces water through the baleen apparatus and may contribute further to passive jaw closure.

Overall, an effective, efficient, and spectacular mechanism. For a fascinating review of feeding in all major groups of marine mammals, see Werth in References.

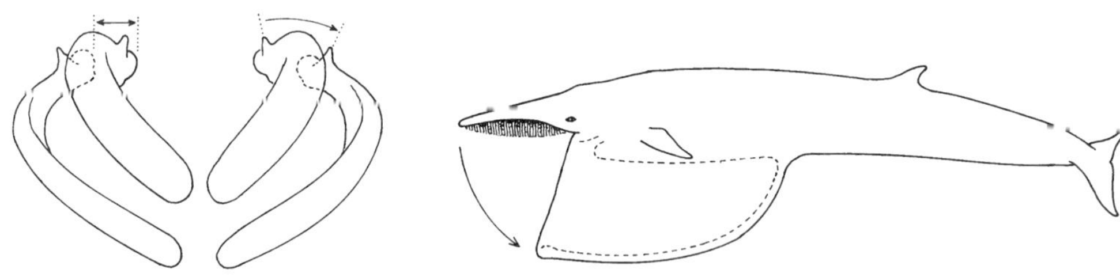

FIGURE 30.2 JAW MOTIONS OF A RAM-FEEDING RORQUAL WHALE. Right: jaw depression and pouch inflation. Left: lateral motion of condyles and outward rotation of mandibles around their long axes, both viewed from a constant position in relation to symphysis of jaw. (Adapted from Lambertsen, Ulrich, and Straley, 1995. See References.)

their longest dimension. During feeding, the large curved bill is positioned under the water "upside down," pointing toward the bird's feet. Upper and lower bills have up to 20 parallel rows of horny platelets that intermesh to form a fine screen. The platelets are less than 1 mm long, and frayed into dozens of hairlike projections on one edge (but with variations according to species). The muscular tongue is the pump. Horny hooks on the tongue comb the food from the platelets.

Suction Feeding In **suction feeding,** or **gape-and-suck feeding,** water is drawn into the buccopharynx with such force that prey is drawn in with it. Suction feeding differs from suspension feeding in that a particular macroscopic prey is taken. Ventilation of the gills is interrupted by the feeding act. The feeding event is so fast that the principles of steady-state physics do not apply: Pressure can be different in different parts of the same chamber at the same instant. The inertia of the prey determines the flow velocity needed for its capture. The feeding sequence lasts about 0.025 s for the gar, 0.015 s for the anglerfish, and only 0.006 s for certain frogfishes.

The buccopharynx is the suction pump. The mouth is usually round and without "corners" from which food could escape. The gape is small or medium. The hyoid is large and rotates down and to the rear. The explosive expansion may enlarge the head 40 percent by volume. Small teeth may be present (to shred food that rushes past them) but in the fishes often the only teeth are in the pharynx.

Suction feeding is observed in some sharks and many bony fishes. Examples are sturgeons, suckers, carp, and perch. Sea horses and pipe fishes feed on active small crustaceans by whisking them into a tiny mouth at the end of a pipettelike rostrum. Some salamanders and turtles (snappers and matamata) are suction feeders (Figure 30.3). The walrus purses its lips and sucks mollusks into a mouth having a domed hard palate, and there is evidence that the pilot whale feeds by suction.

[Study of events inside the mouth and pharynx of suspension- and suction-feeding fish has required ingenious research protocols and advanced technologies. Several examples: Sanderson and her coworkers inserted polyethylene cannulas (close to 1 mm in diameter) into holes drilled into the pharynx through the hyomandibular and temporal regions of the skulls of anesthetized blackfish. A microthermister probe was threaded through the cannulas to measure flow velocity, and a fiber-optic endoscope was inserted for remote videotaping of the paths followed by styrene microspheres attached to food items. They learned that the mucus-covered roof of the mouth is the site of particle retention, not the gill rakers as previously thought. A Dutch team of researchers made simultaneous lateral and dorsoventral X-ray cine recordings of bream fish feeding on food particles to which 1 mm iron spheres had been attached. Subsequent stereomicroscopic examination revealed entrapment in an unexpected part of the pharyngeal apparatus. Nemeth inserted a pressure transducer through the snout of a suction-feeding fish (a greenling), and found that reduction in buccal pressure is faster and greater when sucking in sedentary (i.e., grasping) prey than elusive prey, and is least for inactive prey.]

Jaw Protrusion The upper jaw of fishes is said to be protrusible if it not only can be opened and closed, but also moved forward and backward in relation to the remainder of

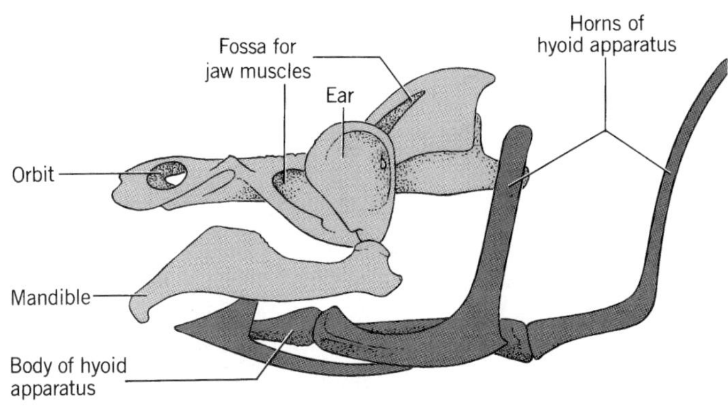

FIGURE 30.3 SKELETON OF THE SUCTION PUMP OF A TURTLE shown by the huge hyoid apparatus of a matamata, *Chelus*. Lateral view.

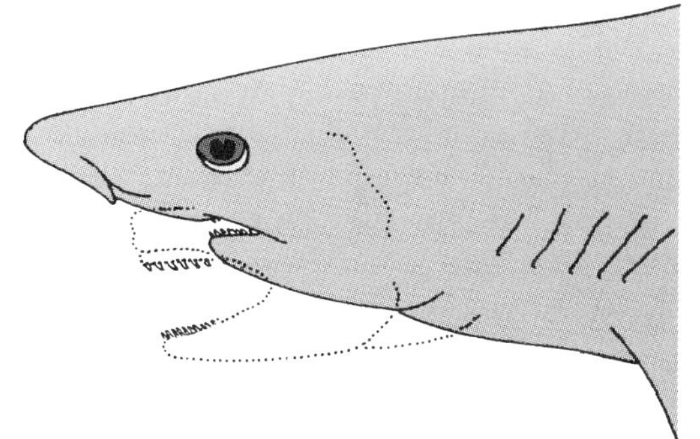

FIGURE 30.4 JAW PROTRUSION IN A SHARK, *Carcharhinus.* (Redrawn from T.H. Frazzetta and C.D. Prange, *Copiea* 1987:990. ©American Society of Ichthiology and Herpetology.)

the head. Protrusible jaws have evolved independently several times. Except in predaceous sharks, they are usually associated with suction feeding.

As noted in Chapter 8, many cartilaginous fishes are *hyostylic;* a movable hyomandibular cartilage joins the jaws to the chondrocranium (see figures on pp. 119 and 122). When the lower end of the hyomandibula swings forward (and often outward), the jaws protrude (Figure 30.4). Sharks do this during prey capture, bringing the jaws closer to their prey, exposing the upper teeth, and making them more nearly vertical. Perhaps this reduces the time required to close the mouth by lowering the upper jaw as the lower jaw is raised. By analysis of kinematics and electromyography, Wilga and Motta determined that the control mechanism of jaw protrusion and retraction differs in the spiny dogfish and the lemon shark, members of different families. The reason for this unexpected finding awaits further research.

Acanthodians and the most primitive of actinopterygian fishes had large mouths. The hinge of the jaws was under the back of the braincase. Premaxilla and maxilla were not movable, and the upper jaw was not protrusible (Figure 30.5A). These fishes apparently opened the mouth wide and snapped at prey; the rapidly moving lower jaw completed closure partly by inertia.

A subsequent evolutionary stage is exemplified by the bowfin and salmon. The hinge of the jaw is nearly as far back as before, so the mouth remains large. The premax-

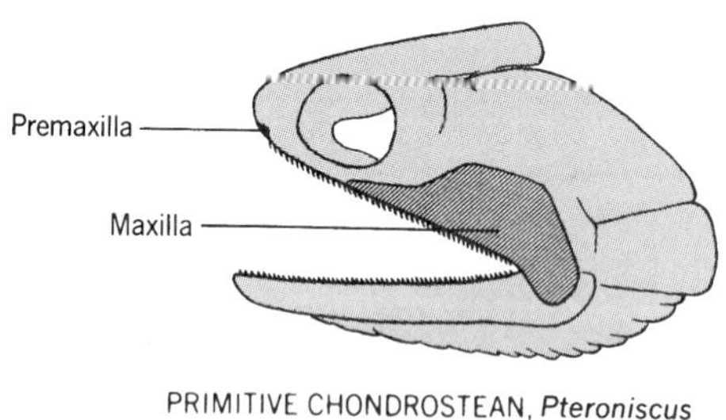

PRIMITIVE CHONDROSTEAN, *Pteroniscus*
A

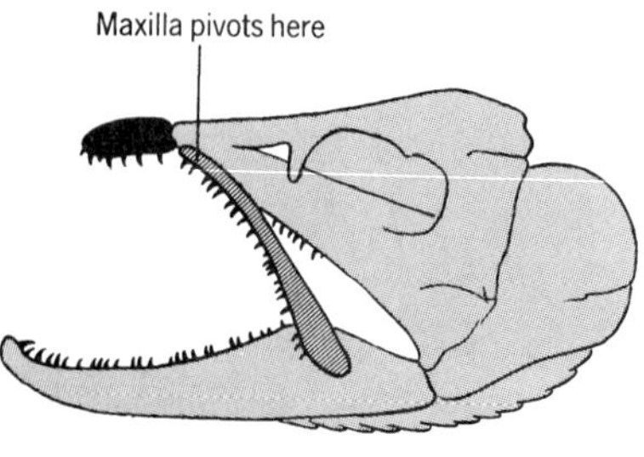

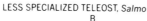

LESS SPECIALIZED TELEOST, *Salmo*
B

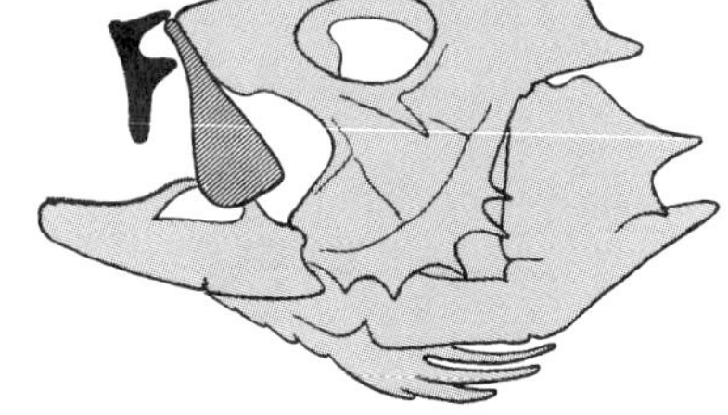

MORE SPECIALIZED TELEOST, *Sebastes*
C

FIGURE 30.5 STAGES IN THE EVOLUTION OF THE JAWS OF RAY-FINNED FISHES.

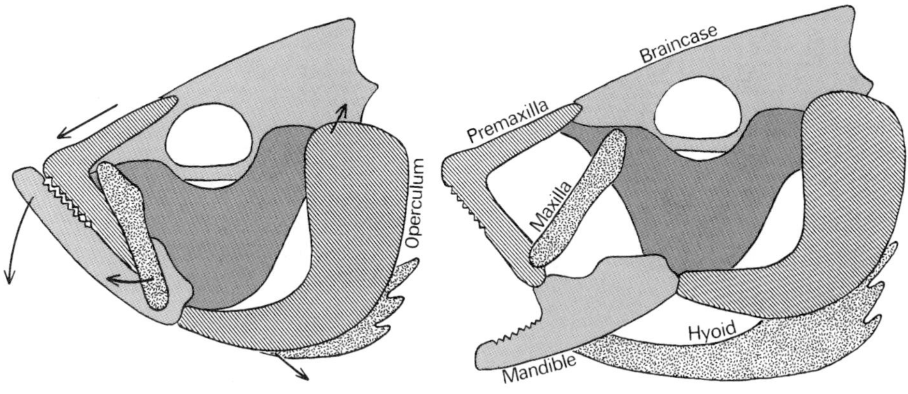

FIGURE 30.6
DIAGRAM SHOWING JAW PROTRUSION OF A TELEOST FISH.

illa is still fixed, but the maxilla is now free. Its posterior end is joined to the mandible by a ligament, and its anterior end pivots on the rostrum (part B). Consequently, when the mandible opens, the lower end of the maxilla rocks downward and forward, thus preventing food from escaping at the corners of the mouth. The upper jaw still protrudes very little or not at all.

At the next stage in the specialization of the jaws of bony fishes, the hinge of the mandible moves forward under the orbit, so the mouth is smaller. The premaxilla, like the maxilla, pivots forward as the mandible is opened (part C). Furthermore, the premaxilla is L-shaped; one arm forms part of the margin of the mouth, and the other, shorter arm forms an ascending process running backward toward the ethmoid region. The premaxilla can slide forward along the axis of this ascending process, thus making the upper jaw truly protrusible. The lengthening of the head amounts to 10 to 20 percent (Figure 30.6). A complicated system of ligaments controls the motions of the bones, but the mechanism is not the same in all fishes. The position of the maxilla is commonly altered by motion of the mandible. Motion of the maxilla, in turn, moves the premaxilla down and forward. The upper jaw usually can be closed on the lower jaw both when it is protruded and when it is retracted.

Protrusible jaws have various advantages according to species: They convert the mouth into a tube, thus making suction feeding possible. Importantly, they permit the mouth to be closed during or immediately after sucking food into the mouth, without blowing the food back out again (which might occur if a long lower jaw were raised). Sometimes the protruded mouth points more downward than the unprotruded mouth—an advantage when taking food, including algae, from the substrate. The jaws probably can be closed a little faster when protruded because the mandible does not have as far to travel. In some fishes protrusion helps to winnow ingested food from inedible material. Finally, protrusion places the mouth a little closer to food. Perhaps half of all teleosts have protrusible jaws. They can be recognized by a transverse fold of skin just behind the margin of the upper jaw. This fold is pulled tight as the premaxillas move forward.

Various aquatic vertebrates also spear, bite, tear, grind, and crush food, and swallow large prey whole. Because these activities are little influenced by feeding in water, the adaptations are shared with terrestrial counterparts, and are noted below.

CRANIAL KINESIS

Cranial kinesis is the relative motion of some parts of the cranium on others. The upper jaw moves up and down, but differs from jaw protrusion in that it does not move forward and backward. There are four or more units in the system, which always includes pivoting

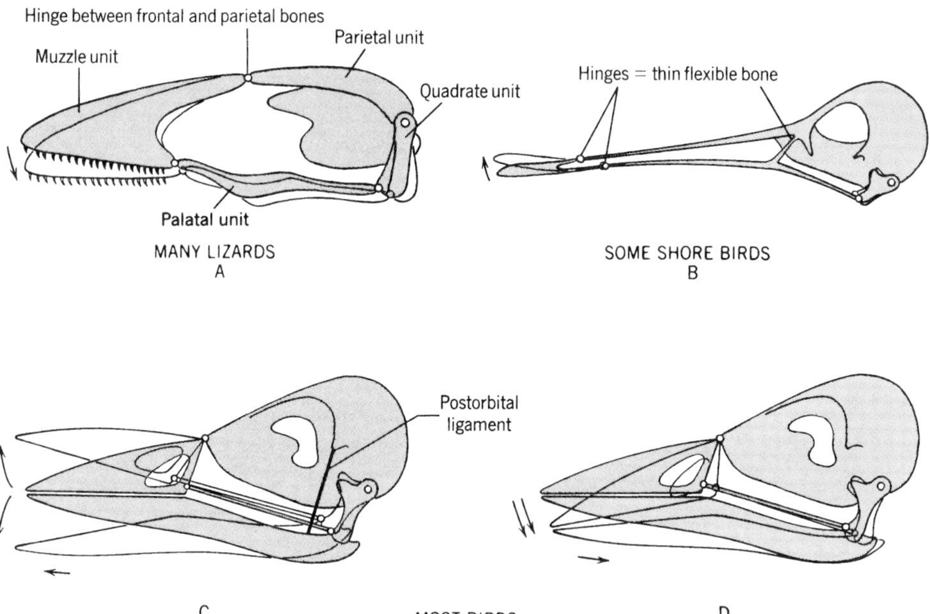

FIGURE 30.7
CRANIAL KINESIS.

of the quadrate on the braincase and sliding of a palatal unit on the basicranium. Crossopterygians, several amphibians, some lizards, and all snakes and birds have kinetic skulls.

The hinged braincase of crossopterygians (see figure on p. 124) is unique to that group of fishes. There was some related lateral motion of membrane bones of the skull; the hyomandibula served to integrate various of the complex movements. The mechanism increased the gape of the mouth and possibly also the power of the bite.

Most lizards have kinetic skulls. Four principal units (some paired) are joined on each side of the head by four principal joints (Figure 30.7A). (Some struts, not shown in the figure, articulate with the mechanism.) Motion of any one unit requires motion of two others and of all four joints. Movement on each side of the head is nearly restricted to one plane, and the two sides usually must function together. Kinesis may allow the upper jaw to move relative to the lower jaw, but seems not to be used to increase gape. The detailed structure and flexibility of the mechanism varies among lizard species and is not clearly adaptive in every instance. Some crevice-dwelling lizards employ cranial kinesis, reducing the vertical dimension of the head.

The kinetic mechanism of snakes includes many units (eight in the boa) that are loosely joined together, and are independent on the two sides of the head. In consequence, an infinite variety of relative motions is possible—a characteristic of the feeding of snakes. They depend on cranial kinesis to draw prey into and through the mouth.

Cranial kinesis is limited in paleognathous birds (ostrich, etc.). The skull of birds has four principal units and numerous joints, the mechanism being more complex than that of lizards. Some birds can move the quadrate units on the two sides of the head somewhat independently, and some can slide as well as pivot the mandible on the quadrate. Motion of upper and lower bills is not necessarily confined to one plane. Furthermore, the mandible is joined to the braincase by the stout postorbital ligament. This links the motions of upper and lower bills, although in some birds the mechanism can be uncoupled, permitting the jaws to move independently. The single dorsal hinge of the mechanism is always farther forward than any joint of lizards, but may be at the base of the upper bill or near the tip (compare parts B and C of Figure 30.7). Kinesis in birds allows great diversity of function: It provides for skilled manipulation of food by sliding one part of the bill lengthwise in relation to the other and by creating a wedge between the parts that opens backward into the mouth, thus preventing food from escaping (part D). Kinesis permits the bill to open wide, and in so doing to maintain its longitudinal axis (part C). Kinesis probably also cushions the shock of pecking for some birds, and may function not only in feeding but also in preening and nest building.

PROJECTILE FEEDING

Some predators run, swim, or fly to overtake prey, whereas others merely spring forward after stalking or waiting in ambush. Still others intercept the prey by striking with the head or tongue. Being much lighter than the entire body, these can be accelerated much faster. Hence the term projectile feeding.

If the head is the projectile, then it is propelled by a long neck, which is cocked in an S-curve before the strike. The head itself is relatively light. Various slender-necked turtles use the head as a catapult in this way. Many snakes, substituting the anterior body for a neck, achieve a long strike. Some are so skilled that the direction of the strike can be corrected while it is in progress, should the prey move. The long-necked, aquatic plesiosaurs (see figure on p. 58) may have lashed out with their small heads to strike fish. Herons and egrets, which wade, and snapping turtles, cormorants, and loons, which swim under water, strike at fish by suddenly straightening the long curved neck (Figure 30.8). Most of these birds use their long straight beaks as forceps to grasp the prey, but the tropical anhinga spears fish; fine barbs at the edge of the bill prevent loss of the food.

The food of pigeons and chickens is not likely to flee if not stabbed. However, each item is small, so the bird must forage quickly. Many seed-eating birds peck in the manner of true projectile feeders, though not quite as fast.

Most frogs and many salamanders, as well as the chameleon, project their tongues to capture worms or insects. The speed and range of some is astonishing. Various birds and mammals use their tongues with corresponding effectiveness to secure insects that are otherwise out of reach. Tongue projection has evolved independently numerous times; the

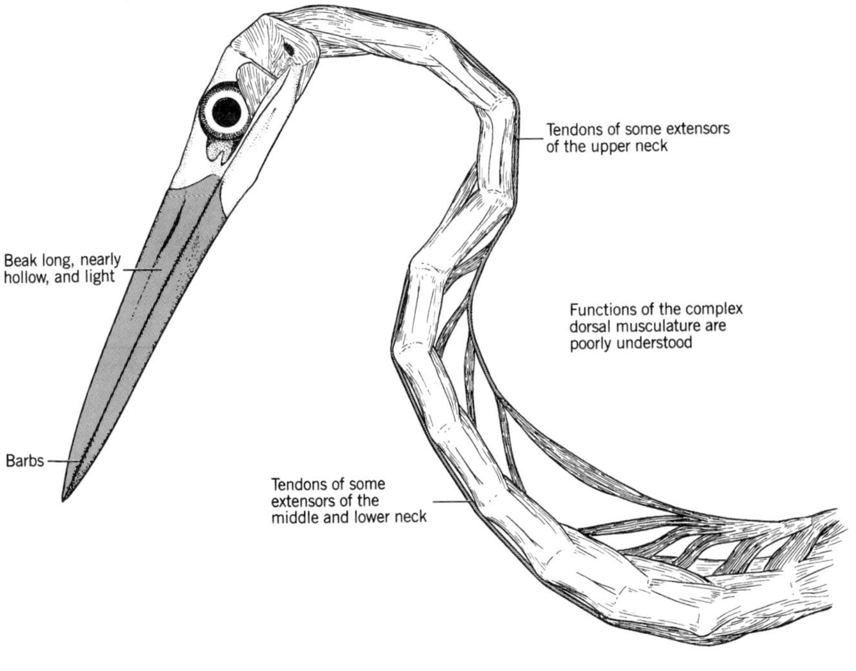

FIGURE 30.8 HEAD AND NECK OF A PROJECTILE FEEDER, the blue heron, *Ardea,* shown with some dorsal muscles separated.

COMMENT 30.2

PLUNGE-DIVING BIRDS HAVE REMARKABLE COMPUTER PROGRAMS

Plunge-diving birds that locate their fishy prey from a perch or on the wing include pelicans, gannets, boobies, terns, and kingfishers. Herons and egrets locate prey while wading. All must cope with the refraction of light at the air-water boundary, which magnifies the fish, makes it less bright, and causes it to appear to be above its actual position. (The discrepancy is minimal when the fish is viewed nearly vertically, but can exceed 10 cm for waders; see Figure 30.9.) There are also surface reflections, which create lenses that distort and displace images of underwater objects. Furthermore, the strike must be adjusted for the depth and movements of prey, distance to prey, and angle of sighting.

Quite an order, yet in the field herons are successful in 40 to 90 percent of their attempts, apparently depending on such factors as size and behavior of prey, turbidity of water, and surface ripples. In the laboratory, under controlled conditions, egrets maintained high success rates over a wide range of strike angles and prey depths. They favored shallow prey and intermediate sighting angles (from the horizontal), but capture rate was best for acute angles, even though refraction then displaces the image the most. (At acute angles the bird may be more hampered by reflections, but the fish has the poorest view of the bird, thus reducing escape maneuvers.)

Capture starts with a pre-strike phase during which the head moves slowly straight forward. At this time the parameters of the strike are determined. (During the strike the bird is blinded by bubbles and closure of nictitating membranes.) The strike is very fast and is directed at the prey, not at its displaced image. The bill is opened only $\frac{1}{60}$ to $\frac{1}{30}$ s before contact.

Researchers have developed an equation for determining the required strike angle when apparent depth of prey, horizontal distance from eye to prey, and angle of sighting are known. How does the heron measure these variables? And does it, like the researchers, compute trigonometric functions, multiply by one constant, and subtract another?

For information about this and other attributes of avian vision, see the book edited by Zeigler and Bischof in References for Chapter 19.

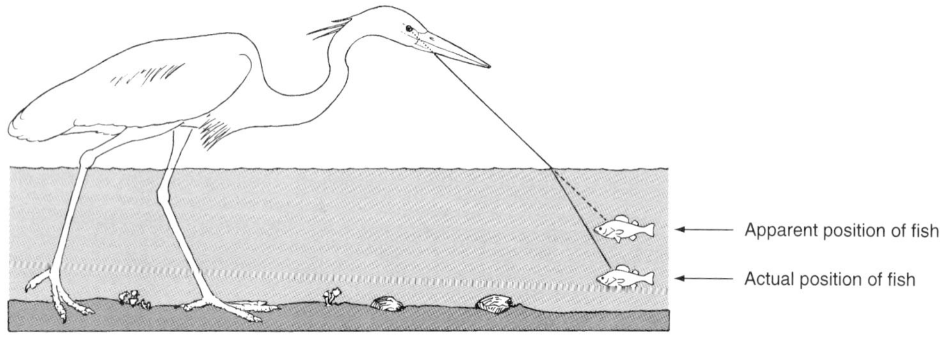

FIGURE 30.9 DISPLACEMENT OF IMAGE OF PREY BY REFRACTION.

mechanisms are diverse. Muscles can pull but not push, so how can a soft object be projected far out of the mouth?

Among the most adept tongue projectors are certain salamanders. There is a wide diversity in the morphology of the tongue, its attachment to the hyoid, and its projection distance. In a representative example, the hyoid apparatus has a pair of horns extending far back into the neck (Figure 30.10). Each protractor muscle has its origin on a relatively fixed, lateral piece of the hyoid, wraps tightly around its horn (to make it longer so it can shorten more), and inserts on the posterior end of the horn. When the protractors contract, the unfixed, jointed part of the hyoid folds inward and shoots forward, sliding on lubricated guide rails that were aimed during a preparatory phase. The soft part of the tongue stretches by its inertia to impact the prey. It may project as far as 30 percent of the animal's snout-vent length.

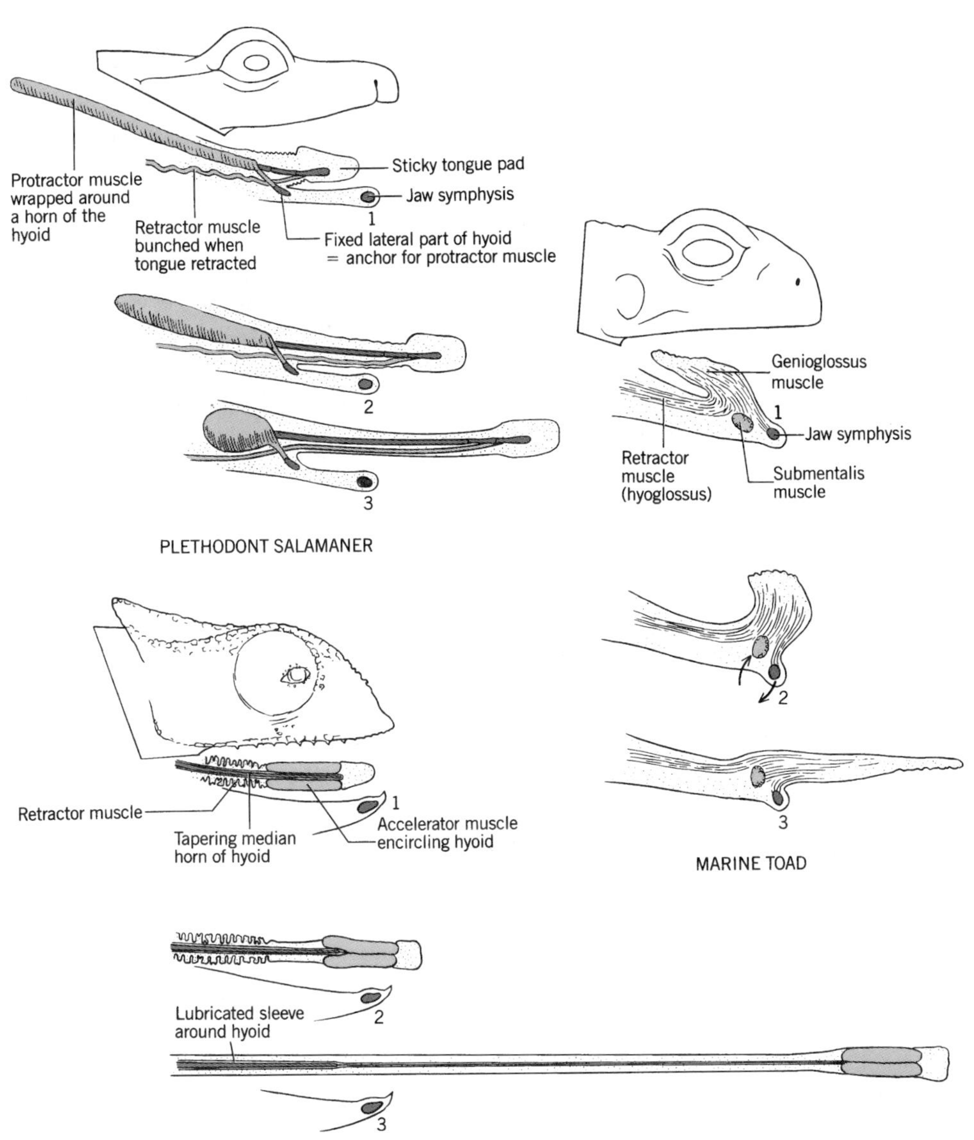

FIGURE 30.10
THREE MECHANISMS FOR PROJECTING THE TONGUE. Jaws and tongues are shown in section, and somewhat simplified.

The anuran tongue, in contrast, is relatively simple. Nevertheless, Nishikawa and her students identify three basic mechanisms that differ in biomechanics and neural control. The first, *inertial elongation,* is exemplified by the marine toad (Figure 30.10, and figure on p. 5). The fleshy tongue is anchored at the front of the mouth. Its intrinsic musculature has relatively long fibers and is in part attached to the symphysis of the jaw. When the muscle contracts, it draws the tongue forward, and shortens and stiffens it. Under the base of the tongue is a mass of several muscles (including the submentalis, which lies transversely in the mouth). When they contract, they depress the symphysis (separate, small bones there make this possible), thus pulling sharply down on the front of the tongue root. At the same time, this mass stiffens and rises, thus wedging up on the back of the tongue root. The tongue is flipped forward in about 0.05 s. It extends and elongates by its inertia, turning over in the process, and contacts the prey with its sticky pad. The hypoglossal nerve of many such feeders has afferent fibers (which is unusual) from mechanoreceptors on the tongue. Inertial elongation is the fastest but least accurate of the anuran feeding modes.

The second mechanism is *mechanical pulling*. The tongue is spherical in shape, relatively high in mass, and the morphology of the genioglossus and hyoglossus muscles is the most like that of other vertebrates. As the genioglossus contracts, the tongue shortens and is pulled upward and forward toward the jaw symphysis. Because the tongue shortens during protraction, the frog must lunge forward with its entire body to make contact with the prey. Mechanical pullers are considered intermediate in accuracy. The third mechanism, *hydrostatic elongation,* is the rarest of the feeding systems. The genioglossus possesses vertical fibers in addition to the more typical fibers that are parallel to the tongue's long axis. During protraction, the volume and width of the tongue remain constant, but when the vertical fibers contract, they change the tongue's thickness, which results in hydrostatic elongation. Although the tongue elongates (up to 200 percent of its resting length), tongue protraction is relatively slow. This system is thought to be the most accurate. Nishikawa and her students concluded that mechanical pulling was ancestral to the other mechanisms; that hydrostatic elongation evolved once, or perhaps twice, and inertial elongation evolved up to seven times; and (importantly) that small changes in anatomy can lead to large changes in biomechanics.

The chameleon (see figure on p. 476) is a bizarre lizard that catches insects on the end of a sticky tongue, which it can protrude a distance equal to its own body length in about 0.04 s. How the animal accomplishes this feat has long been of interest to morphologists, and there are various hypotheses. The precise anatomical and modeling studies of van Leeuwen, and kinematic and electromyographic studies of Wainwright and Bennett, provide some answers. The hyoid apparatus has a long horn that tapers near its tip. The horn fits within a lubricated, tendinous sleeve that is surrounded by an architecturally complex accelerator muscle of the tongue (Figure 30.10). The lizard opens its mouth slowly, advances the hyoid into position, aims by orienting its head just right, and by differentially contracting portions of the accelerator muscle, squeezes its tongue off of the tapering hyoid horn like a tiny man pinching himself away from a giant watermelon seed.

Having exposed an insect with its drill-like bill, a woodpecker captures it with a tongue that is long, cylindrical, sharp, barbed, and sticky. The mechanism resembles that of the salamander, but the flexible hyoid horns wrap around the back of the skull, when retracted, and reach forward into a nostril. The tongue can be extruded as much as five times the length of the long bill.

The larger mammals that feed on swarming insects have a tongue that is cylindrical, sticky, mobile, and exceedingly long (reaching three times the length of the head). In several anteaters of two orders it is anchored at the posterior end of the breastbone rather than in the throat (Figure 30.11). Vascular spaces in the tongue of echidnas and pangolins

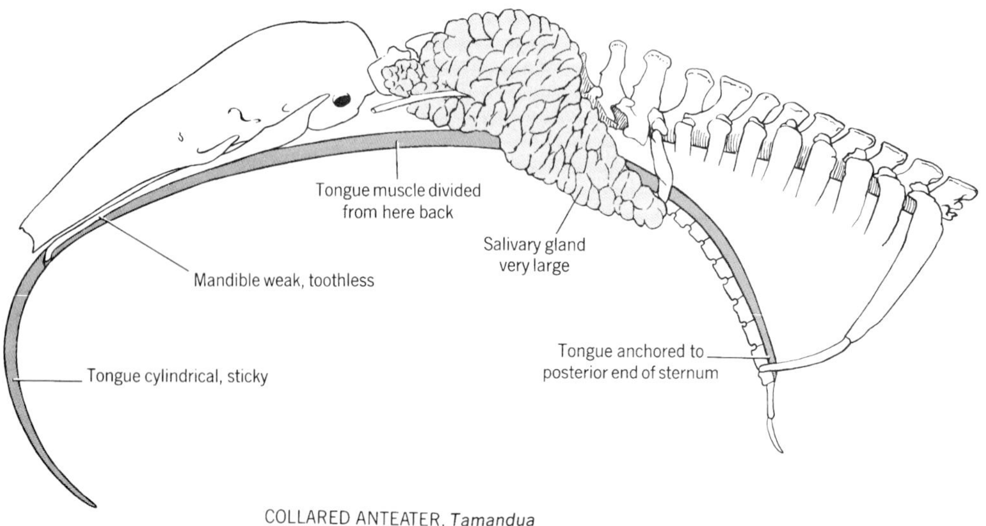

FIGURE 30.11
SOME ADAPTATIONS OF A MAMMALIAN ANTEATER.

function as erectile tissue. Although not quite a projectile, the tongue flicks in and out very fast.

Finally, a bizarre kind of projectile: The archerfish looks up from under the water, locates insects as much as a meter away on overhanging vegetation, and then, with unerring aim, strikes them down with a forceful jet of water squirted from the mouth. The gill chamber is the force pump, and a groove on the roof of the mouth is the barrel of this water pistol.

OTHER MEANS OF SECURING FOOD

We have seen that suspension, suction, and projectile feeding are means of securing food. Other means are so numerous and diverse that a sampling must serve to illustrate.

Many animals move some object to uncover food: The turnstone (a bird) rolls stones with its beak in search of invertebrate food, the walrus digs clams from mud with its tusks, the towhee kicks away litter to reveal insects, the bear rips logs to find animal food, the warthog uproots tubers with its tusks, and the aye-aye (a primate) cuts bark with its strong curved incisors to get at insect larvae. Some woodpeckers drill into solid wood to secure insects: The horny beak is extra long, strong, and rapid-growing, and the bony beak is reinforced. Force is transmitted to the thick and well-ossified interorbital septum of the skull. Certain bats fly low over water with the large curved claws of their feet trailing in the water ready to gaff any small fish encountered. Similarly, the bird called the skimmer flies over water with its huge mandible cutting the surface ready to scoop up any small fish in its path.

Numerous vertebrates make food accessible by using a long reach (see Smith and Kier in References). The arms of apes and monkeys and the trunk of the elephant serve this purpose. The browsing giraffe reaches with a neck so long that a plexus of arteries has evolved at the base of the brain to even the blood pressure as the head swings up and down. The strikingly long neck and legs of the gerenuk (an antelope) similarly increase its reach (Figure 30.12). Several vertebrates have structures for reaching into restricted spaces. Examples are the bills of hummingbirds and creepers, and the tongues of anteaters. The third finger of the aye-aye (Figure 30.13) and the fourth finger of two kinds of phalanger show remarkable convergence in that each is slender and longer than the other fingers to serve as a probe for removing insects from crevices.

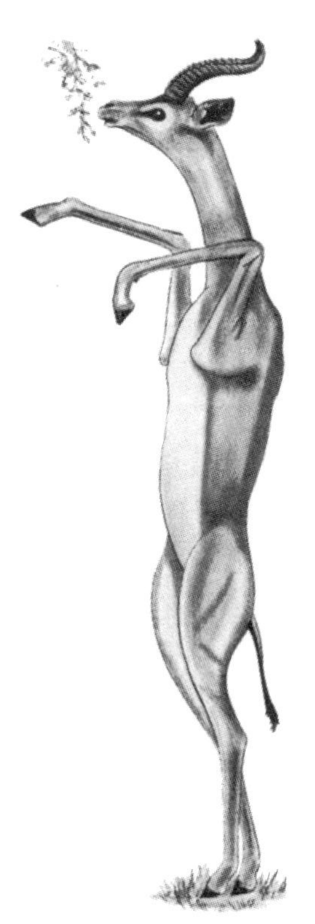

FIGURE 30.12
A BROWSING GERENUK, *Litocranius.*

Tools other than parts of the body are used by several vertebrates: The chimpanzee may use a stick as a probe and leaves as a sponge, a Galapagos finch uses cactus spines to poke into holes, an Egyptian vulture and the banded mongoose drop or throw rocks to break large eggs, and as it swims on its back, the sea otter pounds shellfish on a rock anvil placed on its chest for that purpose.

Jays hold acorns against chopping blocks as they hammer at the hulls, gulls drop crabs and mollusks onto rocks from the air to crack them open, and the beaver cuts down trees to get at the bark. Alligator snapping turtles have a lure at the tip of the tongue that attracts fishes. Anglerfishes attract prey by dangling in front of their jaws a lure derived from a dorsal fin spine. Those living deep in the ocean even have luminescent lures.

Most hawks and owls dive on rodents from the air, securing the booty with their talons. Similarly, the osprey and fish owl plunge feet first into water to capture fish. The undersurface of the toes of these birds is rough, to prevent fish from slipping, and the scutes of the legs overlap so as to enter the water the "smooth way." Several falcons dive, or "stoop" at flying birds, delivering a glancing blow with the feet and raking the victim with their strong hind talons. Various birds (pelicans, boobies, gannets, tropic birds, kingfishers, etc.) dive head first into the water to catch fish. It is thought that the air sacs of the pelican cushion the impact. The eye is probably covered by the lower lid, and the nostrils

FIGURE 30.13 THIRD FINGER OF THE AYE-AYE, *Daubentonia,* MODIFIED AS AN INSECT PROBE.

and throat are modified to exclude water. The mandibles bow outward on impact so that the pouch temporarily holds about 10 liters of water and fish.

Birds that take insects on the wing have large broad mouths fringed by stiff bristles. The night-feeding frogmouths, poorwills, and nighthawks have small horny bills but a prodigious gape. Another group of night feeders, the insectivorous bats, locate individual insects by echo-ranging so they do not need such large mouths. Experimental evidence from the laboratory of James Simmons has shown that one species can distinguish between objects only 0.3 mm apart (they process overlapping echoes arriving 2 millionths of a second apart). Several species of bats use the flight membranes in catching insects. One small bat captured 175 mosquitos in 15 min.

Some large prey must be immobilized before it can be swallowed: The secretary bird stamps reptiles to death with its long heavy legs; some fishes paralyze prey with electric shocks (see p. 191); some snakes kill with a poison bite and others by constriction; and birds that eat flesh or fish may kill with their claws or beaks. Some flesh eaters bleed or strangle their hapless victims, whereas others (African hunting dogs, several sharks, piranhas) cripple or surround the prey and then eat it alive—death comes from shock and bleeding. Underwater video recordings by Domenici and colleagues have shown that killer whales use a tail-slap to stun herring for subsequent ingestion.

FOOD MANIPULATION AND TRANSPORT

Once taken into the mouth, food must be transported to the esophagus for swallowing, and often must first be manipulated to facilitate sorting, severing, or chewing.

In water, food is supported in the buccopharynx by buoyancy, and can be moved by the drag of a waterstream. Fishes accomplish most manipulation and transport by moving the water that supports the food, although teeth on the jaws, roof of mouth, tongue, and fifth gill arch often assist. The upper and lower pharyngeal teeth may be moved synchronously or alternately. Some fishes (moray, toadfish) have teeth that are hinged so that they tilt back toward the throat as prey enters the mouth, but stand erect when pressure is in the other direction.

Out of water, gravity acts on food held in the mouth, and the drag of any airstream is insufficient to move it. Manipulation and transport are accomplished primarily by traction of the tongue, or by inertia. Because tetrapod feeding is such a marked departure from aquatic feeding, studies have been done concerning the maintenance of motor control during the transitions from aquatic larva to terrestrial adult as well as from fish to tetrapod.

Lingual Feeding and the Bite Cycle Most terrestrial vertebrates use the tongue and jaws in a rhythmic biting, or chewing, cycle to move food within the mouth for processing, and then through the mouth to be swallowed. The phases of the cycle, and its muscular control, are remarkably consistent among amphibians, reptiles, and mammals, and there is minimal change during amphibian metamorphosis. Common ancestry and conservatism of the basic mechanism are indicated.

In contrast to aquatic feeders, which tend to have a small, stiff tongue, smooth flat palate, and large hyoid apparatus, the lingual feeders tend to have a large, mobile tongue, a domed and often rough palate, and a small hyoid. In each cycle the tongue, functioning as a ratchet, moves forward under the food, engages it, draws it backward, and releases it. Researchers divide the cycle into four phases (and several subphases omitted here).

During the **slow open** phase, the jaws open slowly and moderately. The tongue moves upward and forward, advancing under the food that is held in place by the teeth and by bonding and interlocking with the palate. At the end of the phase the tongue, accommodating its shape to that of the food, engages the food by its sticky saliva and its roughness. For an instant the food is held both above and below.

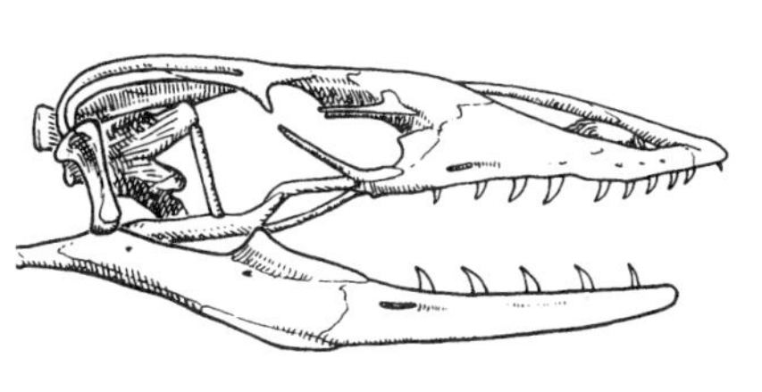

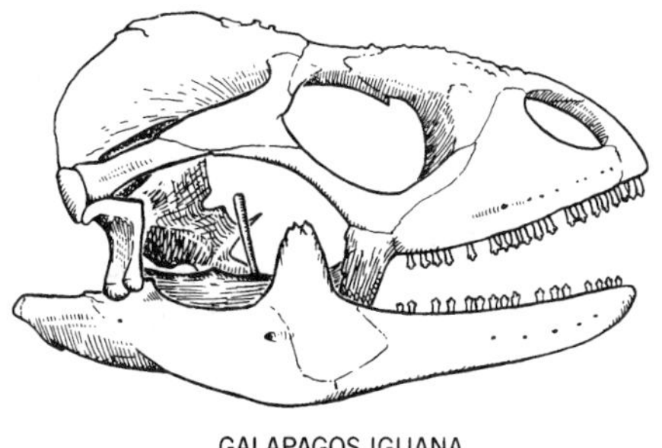

FIGURE 30.14 SKULLS OF A FLESH-EATING, INERTIAL-FEEDING LIZARD (left) AND A PLANT-EATING, LINGUAL-FEEDING LIZARD (right).

During the **fast open** phase, the jaws open faster and wider, helping to disengage the food from the palate, and the tongue transports the food backward and downward. If the food items are large, then at the same time the head may tilt up and the neck extend. The jaws close rapidly during the **fast close** phase as the tongue continues to move back. If the head has tilted up on the neck, it now drops again. Finally, during the **slow close** phase, the jaws crush down on the food, and the tongue moves forward to the intermediate position of the start of the cycle.

Inertial Feeding Inertial feeding employs a modified bite cycle in which motions of the head and neck are exaggerated to transport food by its inertia, rather than by the tongue. The animal opens the jaws to release the food held there and, in the same instant, lunges the head forward, thus darting the throat around the nearly stationary food. A backward motion of the head, before the jaws are opened, but ending the instant the food is released, may instead throw the food farther back in a nearly stationary mouth. These two actions are commonly combined, the head jerking first backward and then forward. We have seen dogs gulping food this way. It is used also by some grain-eating birds and by many lizards, birds, and mammals as they eat fish or flesh. The head tends to be light, making it easier to accelerate (Figure 30.14). (The feeding of snakes, noted below, is a special kind of inertial feeding; the inertia and friction of the food hold it in place as the snake slowly surrounds it.)

ADAPTATIONS FOR EATING SOFT, TOUGH FOODS

Adaptations of the gut were summarized on pp. 210 and 213. Adaptations for the transport of food having been noted, specializations of the teeth and jaws remain for review. This is best done according to kind of food eaten, although the variety of diets and versatility of some feeders make distinctions difficult.

Vertebrates are said to be **carnivorous** if they eat other animals. We will at first pass over those that usually swallow prey whole (because their adaptations are different—see below) and consider carnivores that tear or shear flesh into hunks, which are then swallowed without being finely divided in the mouth. If the animal eats mostly carrion, it is also a **scavenger.** Carnivorous vertebrates include many sharks; some bony fishes; most labyrinthodonts and caecilians; various stem reptiles, mammal-like reptiles, dinosaurs, lizards, and crocodilians; raptorial birds (hawks and owls) and such avian scavengers as albatrosses, fulmars, vultures, and caracaras; a family of marsupials; and most members of the order Carnivora.

These animals often have talons or conical teeth adapted for killing and temporarily holding their prey. The canines and other anterior teeth of the dog are examples (see figure on p. 109). The remaining teeth tend to be blades (Figure 30.15; p. 130, lower figure). The margins of the blades may even be serrate (p. 108, upper right figure). In some instances upper and lower teeth shear past one another. Carnivora have one pair of teeth on each side of the

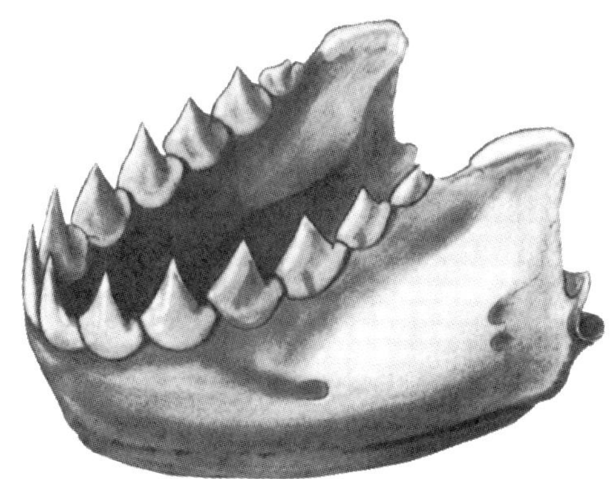

FIGURE 30.15 MANDIBLE AND SHEARING TEETH OF THE PIRANHA FISH.

FIGURE 30.16 TONGUE OF A MOUNTAIN LION showing rasping surface.

mouth (upper fourth premolar and lower first molar), called **carnassials,** that are enlarged and positioned so as to form a powerful shearing mechanism (Figures 30.17, 30.18, and figure on p. 135). The carnassials of hyenas are virtually as effective as tin snips. These shears are so efficient that other cheek teeth are little needed and may be reduced in number—particularly in cats (see cheetah in figure on p. 398). The big cats supplement their teeth with horny projections on the tongue, which enable them to rasp the flesh from bones (Figure 30.16).

Raptorial and scavenging birds, having no teeth in their recent ancestry, must divide their food another way. Hawks and owls have the horny upper bill curved down at the tip to make a formidable meat hook. Pushing down with its talons and pulling up with its beak, the bird tears its prey apart. Some extinct reptiles had similar beaks.

Carnivora bite with a chopping motion. Upper and lower tooth rows come together without sliding past one another either front to back or side to side. The joint between mandible and skull consists of a transversely oriented, cylindrical condyle on the mandible that rotates, hingelike, in a fossa on the temporal bone. However, when the carnassial shearing mechanism on one side of the mouth is in use, the carnassials on the other side are not quite aligned and cannot function. The mandibular condyle is about in line with (not higher than) the tooth row. The significance of this is explained in a following section.

The jaw muscles of carnivorous mammals differ from those of herbivores. There are three adductors of the mandible. The **masseter** takes its origin from the zygomatic arch and often also from the orbit or maxillary region. It inserts on the outside of the ramus and angle of the mandible (Figure 30.17). The **temporalis** muscle originates on the braincase and sagittal crest, if present, and inserts on both the outside and inside of the coronoid process of the mandible. The **pterygoid** muscle originates at the base of the skull beside the palate and

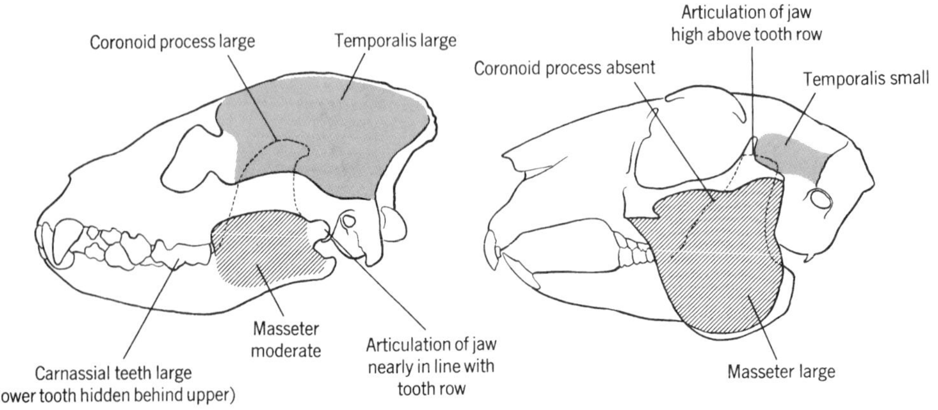

FIGURE 30.17 CONTRAST BETWEEN THE JAW MECHANICS OF A CARNIVORE (left) AND A HERBIVORE (right).

COMMENT 30.3

WHAT DID SABERTOOTHED CATS *DO* WITH THOSE BLADES?

Morphologists enjoy deciphering the relationships between form and function, and the game is particularly challenging for structural complexes of extinct animals having no living analogs. Sabertoothed cats fit the bill. How did they—how *could* they—use those protruding canines in killing prey and feeding (see chapter-opening figure)? For nearly 100 years paleontologists have been trying to figure this out. We mention here selected studies that illustrate progress of the analysis over time.

Early on it was not known if sabertooths could open the mouth wide enough to clear the canines. If not, were the serrated blades used (mouth closed) to kill by stabbing and slashing? If the mouth could be opened wide enough to bite, how strong was the bite? These questions were addressed in 1980 by Emerson and Radinsky (see References). They manipulated the jaws of anesthetized big cats in zoos, measured 20 indexes on extensive skeletal material, made mechanical models and identified derived character-states shared by many genera of sabertooths. They demonstrated that sabertooth cats *could* open the mouth wide enough to clear the canines by about as much as modern cats. Analysis of bite force at the carnassial teeth is complex, and the force varied among genera, but these authors concluded that the bite was comparable to that of lions and tigers. The kill, they speculated, was made by a deep throat slash (this conclusion is shared by some recent investigators).

Analysis of killing and feeding behavior was carried further by Akersten (1985), who studied the genus *Smilodon*, known by extensive (though disarticulated) remains from the La Brea tar pits in southern California. He believed that these cats were cooperative hunters (as with lions, injured individuals often survived on the kills of others) that stalked their quarry, rushed, and pulled the prey down with their powerful forequarters and huge claws. He believed that they then attacked the abdomen of the subdued animal, thus avoiding biting onto bone with their fragile sabers. Compared to lions, the symphysis of the jaw is stronger, the lower incisors larger, the roots of the lower canines bigger, and the head-depressor mechanism more powerful. The bladelike sabers are thinnest, and have the most serrations, on their posterior edges. From these clues Akersten surmised that the cats opened their jaws wide, seated lower front teeth firmly against the hide of the prey, and then lowered the head, forcing the sabers to penetrate deeply. The tips of the sabers being farther from the jaw joint than their bases, each tooth would slice a wound several centimeters long as the jaws closed.

Various paleontologists have extended the analysis of feeding in sabertooths to clues from the incisors and carnassials (see References). The upper incisors of *Smilodon* are more robust and more worn than in lions. Instead of forming a straight line across the mouth as in modern cats, the six upper incisors are placed in an arc, approaching the canid condition. These features indicate that the incisors had relatively heavy use, were better adapted for food manipulation, and contributed to stabilizing food, thus protecting the sabers from lateral bending forces. The carnassials were large, effective shears. Patterns of microwear indicate heavy use, but the near absence of pits indicates little contact with bone. It appears that sabercats ate the viscera and large muscles of their kills and then left the carcasses for scavengers to clean.

The last word may never be written, but the analysis game has been played well.

---

behind the orbit, and inserts on the inside of the ramus and angle of the mandible. Each muscle is commonly divided into several parts. These may be distinct, but they often merge with each other and (particularly for the masseter) with a complex of internal tendinous sheets.

The temporalis muscle comprises more than half the total adductor mass of carnivores. The coronoid process provides the lever arm of this muscle, and in these animals is large. The masseter is somewhat smaller and serves in part to stabilize the articulation of the jaw. The pterygoid muscle positions the carnassials, but adds little to the force of the bite and is small.

ADAPTATIONS FOR EATING TURGID, BRITTLE, AND VARIED FOODS

Before considering adaptations for processing other kinds of foods, two cautions should be noted. First, we do not really know much about the physical strengths of the different kinds of foods, which surely affects chewing strategy. Second, Herring points out that the traditional assumption that occlusal force is maximized is not always justified; the extent of the mammalian skull's dynamic nature during food manipulation has not been appreciated.

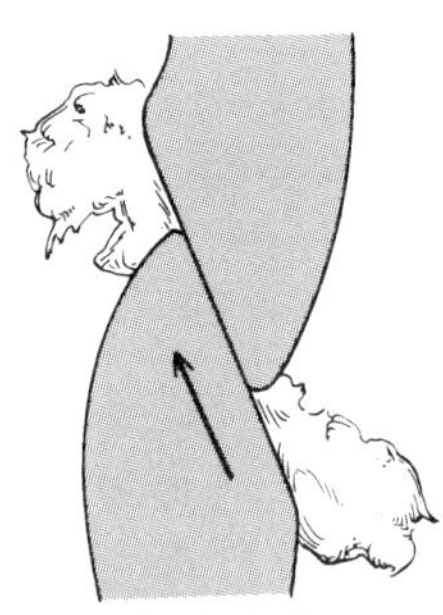

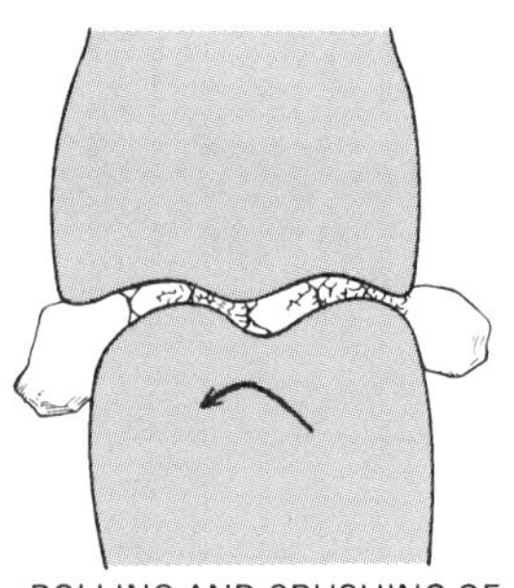

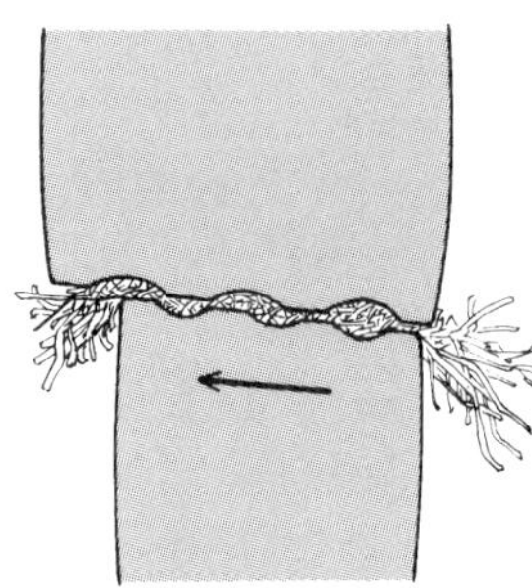

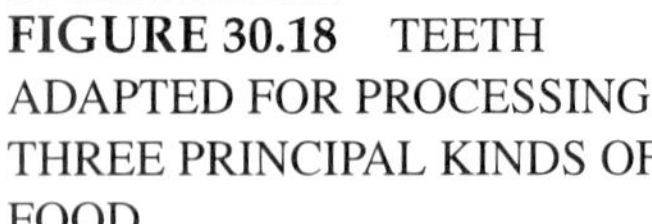

FIGURE 30.18 TEETH ADAPTED FOR PROCESSING THREE PRINCIPAL KINDS OF FOOD.

Foods that are turgid, like the cells of fruits, or hard but brittle, like large seeds and nuts, must be burst open or fractured to prepare them for digestion. Blades would serve (as for flesh), but would necessitate many bites. More effective is the crushing and rolling action of a tooth combination resembling a mortar and pestle (Figure 30.18). A low cone on one tooth fits into a basin on an opposing tooth. Usually the mortar and pestle do not conform tightly, the pestle having the shorter radius of curvature. This permits some transverse grinding of the pestle, and facilitates the cleaning of crushed food from the mortar. The cheek teeth of such fruit eaters as the kinkajou and fruit bats, and of nut eaters like squirrels, conform to this design. (Nut-eating birds crush their food in the gizzard.)

Animals that eat insects, other small arthropods, and worms are said to be **insectivorous.** Many amphibians, reptiles, and birds swallow insects whole, but often the prey must be severed to ready it for swallowing (e.g., the shrew or mouse that captures a large beetle). Furthermore, tearing or puncturing the prey hastens its digestion. Teeth with sharp cones and blades are best for penetrating and shearing tough exoskeletons, whereas mortars and pestles are superior for bursting larvae and worms. Consequently, bats, shrews, moles, and hedgehogs tend to have numerous teeth including some of each type. The gape is usually wide, and the jaws snap quickly shut. (As explained in a following section, large mammals that feed on swarming insects have different adaptations.)

Tetrapods that have a varied diet, perhaps including flesh, insects, eggs, seeds, berries, and tender vegetation, are said to be **omnivorous** (= *all* + *to eat*). Familiar examples are the opossum, house rat, bear, pig, and human. Their teeth tend to be numerous and varied, but not highly specialized for shearing or grinding. Cheek teeth are moderately broad, with low cusps and basins suitable for crushing. Such teeth are termed **bunodont** (= *mound* + *tooth*).

ADAPTATIONS FOR EATING TOUGH, FIBROUS FOODS

The leaves, stems, and roots of plants are tough and fibrous. Animals that eat them are said to be **herbivorous.** Relatively few adult fishes and amphibians are primarily herbivorous, but among reptiles, the giant dinosaurs, duck-billed dinosaurs, horned dinosaurs, some lizards, and many turtles are, or were, herbivorous. The ostrich, hoatzin, geese, and some parrots, grouse, and finches are plant eaters. Kangaroos, wallabies, wombats, langurs, sloths, rabbits, many rodents, elephants, hyraxes, manatees, and ungulates are herbivorous mammals.

Structures that Sever and Crop Birds, sloths, langurs, and most reptiles cut, shred, or crush their food, but do not grind it in the mouth. The horny margins of the mouth function as shears for tortoises and parrots (Figure 30.19). Serrate lateral teeth serve as cutting edges for the iguana (Figure 30.14). Geese have numerous lamellae at the margins of the bill that act as cutters. Most mammalian herbivores crush and grind food with their cheek-

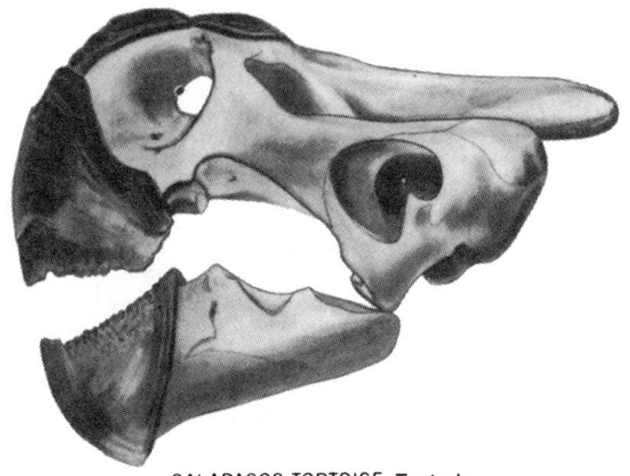

GALAPAGOS TORTOISE. *Testudo*

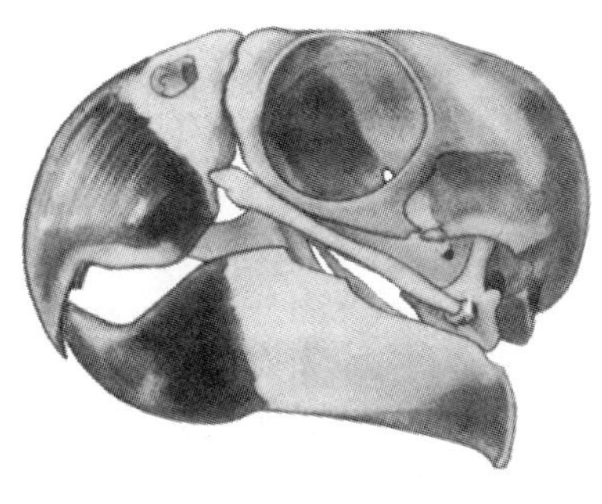

COCKATOO. *Cacatua*

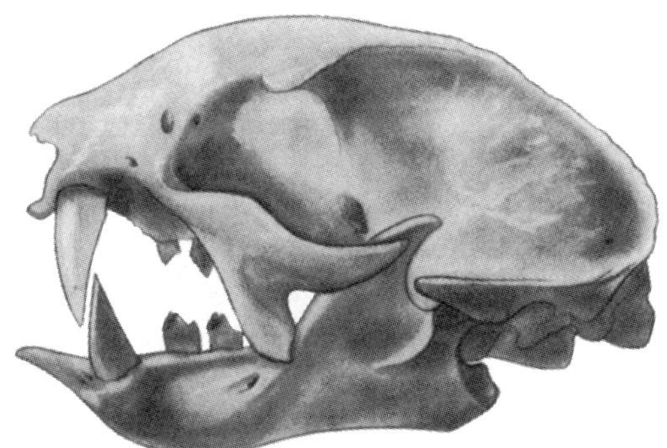

SLOTH. *Choloepus*

HORSE. *Equus*

FIGURE 30.19 EXAMPLES OF SHEARING AND CROPPING MECHANISMS OF PLANT EATERS.

teeth, but their anterior teeth are instead specialized for shearing, gnawing, or cropping. This apparatus is separated from the cheek teeth by a space called the **diastema.** The two mechanisms do not function simultaneously: when a rabbit chews, its incisors do not touch; when a beaver gnaws, the mandible is moved forward and the cheek teeth do not occlude. A specialized pair of upper and lower front teeth form a strong shearing apparatus in wombats, some sloths, hyraxes, rabbits, and rodents. These teeth have open roots and are ever-growing (Figure 30.20). Horses pinch grass between the upper and lower rows of contiguous incisors and then break it off with a sideways jerk of the head. Most artiodactyls instead break off the grass after pinching it between their lower incisors and a horny plate on the upper jaw. The analogous mechanism of kangaroos consists of a row of upper incisors that crop down against a pair of broad, forward-slanting lower incisors.

Teeth That Grind Once it is free in the mouth, plant food is swallowed with little or no chewing by some tortoises and birds, but is very thoroughly chewed by most herbivores.

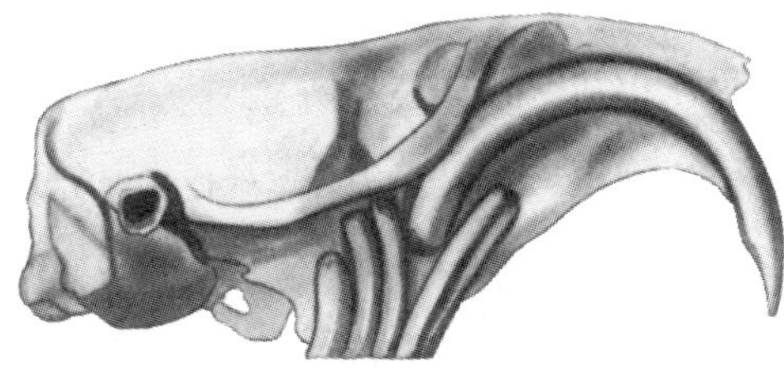

FIGURE 30.20 DISSECTION OF THE SKULL OF A GOPHER, *Thomomys,* SHOWING OPEN-ROOTED, EVER-GROWING TEETH, AND LARGE DIASTEMA.

This means that feeding must be slow. Most artiodactyls swallow their food a first time with little chewing, but regurgitate it, one bolus at a time, to be chewed as cud at the animal's leisure. As much as 7 to 10 h/day may be spent chewing. (Some marsupials regurgitate and chew an occasional bolus, but spend little time at it.) Large quantities of saliva must be secreted, particularly if the food is dry. See comment on p. 214.

In order to triturate coarse vegetation, a grinding mill is needed. As explained on p. 424, the requirements are most stringent for large animals, so the adaptations of ungulates and elephants will be mentioned first. Because there is one kind of job to do, the cheek teeth are similar to one another; the functional premolars become molariform in nature. The occlusal surfaces of the teeth are flat and large in area: A single mammoth tooth may present a grinding surface of more than 250 cm^2 (Figure 30.21). Such large and heavy teeth must be supported by multiple roots.

Furthermore, provision is made for long wear because the teeth are used so much of the time and because the material ground is itself coarse and often has an admixture of grit. To provide for long wear, the roots are deep in the jaw of the young animal and the crown (the part of the tooth above the roots, not just above the gums) is high. Such teeth are called **hypsodont** (= *high* + *tooth*). As the exposed part of the tooth wears down, the roots slowly rise higher in the jaw, and bone fills in under them. This exposes more of the crown. When

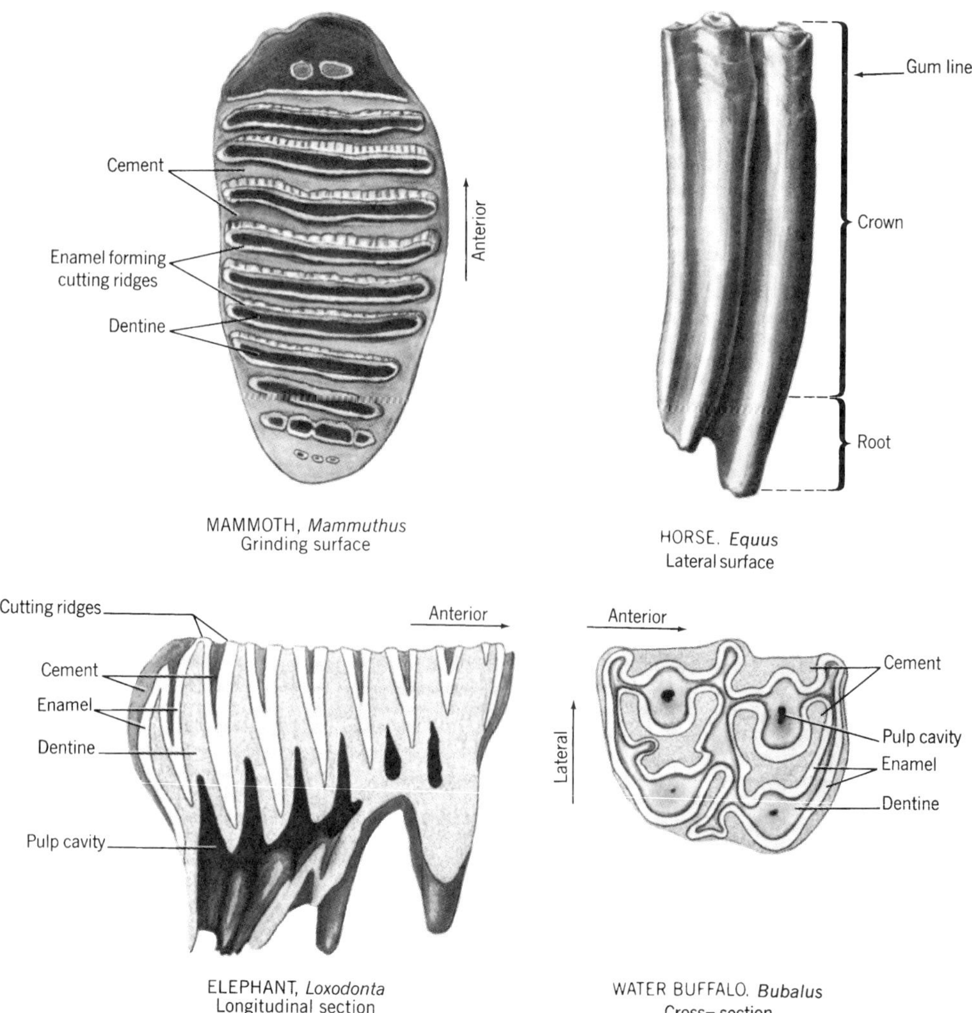

FIGURE 30.21
CHEEK TEETH ADAPTED FOR GRINDING.

most of the crown is gone and the roots are just under the gums, the animal, probably now very old, may suffer from malnutrition.

Elephants have evolved a different mechanism that assures complete wear of all of the crown. The teeth lie in a groove in the jaw rather than in sockets. Each groove is deep at the back of the jaw and slopes upward toward the front. Six teeth form in succession at the back of each groove and migrate slowly forward. Each erupts first at its forward corner and starts to wear there long before the posterior part of the tooth is in service. As forward migration continues, a tooth rises in the sloping groove to compensate for wear. Finally, the crown of each part of each tooth wears off completely as it reaches the anterior end of the tooth row. The short roots are pushed up out of the gums and fall away. Only two of these great teeth function at one time in each jaw. As the last tooth moves along, the groove is filled in from behind by spongy bone. The series of six teeth lasts the elephant the 70 or more years that it lives.

Hypsodont teeth have another adaptation. Their occlusal surfaces are broad and flat, like millstones. But in order to grind effectively, millstones must not be smooth. The enamel of these teeth folds up and down, in and out of the substance of the tooth, thus forming a succession of hills and valleys. The pattern of folding is precise for each species, but varies widely among species. The valleys may remain open (many artiodactyls) or may be filled with cement shortly before the tooth erupts (horses, elephants). In the latter case, as the tooth wears there will be a horizontal succession of enamel–dentine–enamel–cement (Figure 30.21). Since enamel is harder than dentine or cement, it is a bit more resistant to wear and projects a little above the other materials to form cutting ridges, or blades of low profile. This provides and maintains the needed roughness.

The cheek teeth of small herbivores are less specialized in that they are smaller and usually have less complicated patterns of infolded enamel, yet they may be more specialized in being open-rooted, evenly curved in their sockets, and ever-growing (rabbits, some rodents) (Figure 30.20). Individual teeth of herbivorous reptiles never achieve the complexity of those of the large mammalian grazers and browsers, but nature provided the duckbilled dinosaurs with an effective analog. Several hundred relatively simple teeth became pressed together to form a large rough grinding plate. Some vegetarian fishes have crowded thousands of small teeth together in a comparable way. Others have no teeth in the mouth, but shred plant food with pharyngeal teeth that are closely packed blades.

Jaw Mechanics Most mammalian herbivores chew by sliding the lower tooth rows over the upper tooth rows. Motion may be back to front or side to side according to species. Grinding occurs during one phase of the cycle—usually either as the mandible moves forward or from one side toward the center line. During the other, slower phase of the cycle, the jaw opens a little as it is returned to the starting position; the tongue meanwhile places more food between the tooth rows. If motion is side to side, as is usual, the lower tooth rows are commonly closer together than the upper rows, though the reverse is sometimes true. Either way, the animal must chew on one side of the mouth at a time. Whatever the relative motion of upper and lower tooth rows, it is nearly always crosswise or oblique (not parallel) to the orientation of the cutting ridges on the teeth.

Seen in side view, a tooth row may be straight or arched; seen in cross section, a row may be horizontal or may slant inward or outward (Figure 30.22). The rows of one jaw (upper or lower) may be either parallel or diverging toward the back of the mouth (though usually less so than for carnivores). These factors may guide and restrict the relative motions of upper and lower tooth rows. It is clear that the chewing motions of herbivores are exceedingly diverse.

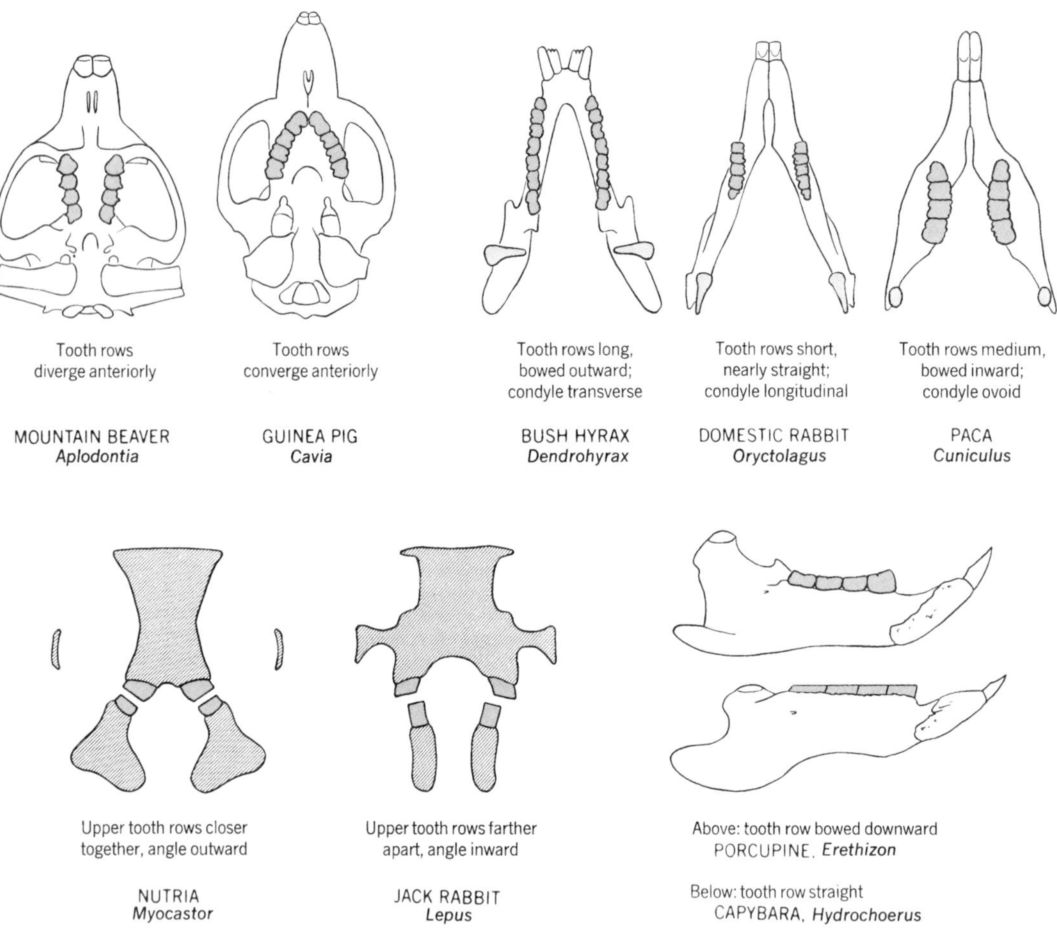

FIGURE 30.22 SOME VARIATIONS IN TOOTH ROWS AND MANDIBULAR CONDYLES AMONG HERBIVORES IN THREE ORDERS OF MAMMALS. Upper left, skulls in ventral view; upper right, mandibles in dorsal view; lower left, cross sections of skulls and mandibles at the level of orbits; lower right, medial views of left mandibles.

The structure of the articulation between mandible and skull is correspondingly varied. The condyle may be transverse as for carnivores. When it moves instead in a less restricting fossa (wombat, hyrax, horse), it may be longitudinally oriented; when it rocks and slips in a longitudinal groove (rabbit), it may be bluntly rounded (beaver, paca), or relatively flat (many artiodactyls). The reader might find it useful to infer from the structure and functions of the tooth rows (preceding two paragraphs) how many different kinds of motions may be required at this joint.

The articulation of the jaw is about in line with the tooth rows in carnivores. Its position is various in rodents, but usually is somewhat above the tooth rows. It is much above the tooth rows in most other herbivores that grind their food (Figure 30.17). The advantage of the high position has been debated. The masseter muscle is relatively large in herbivores, commonly comprising about two-thirds of the adductor mass. The pterygoid is next in size and joins the masseter in producing lateral grinding motions of the mandible. The temporalis is relatively small, and the coronoid process, though variable, tends to be small and may be absent (rabbit, capybara). A high mandibular condyle often increases the lever arm of the masseter (Figure 30.23A). A high position *shortens* the lever of the weaker

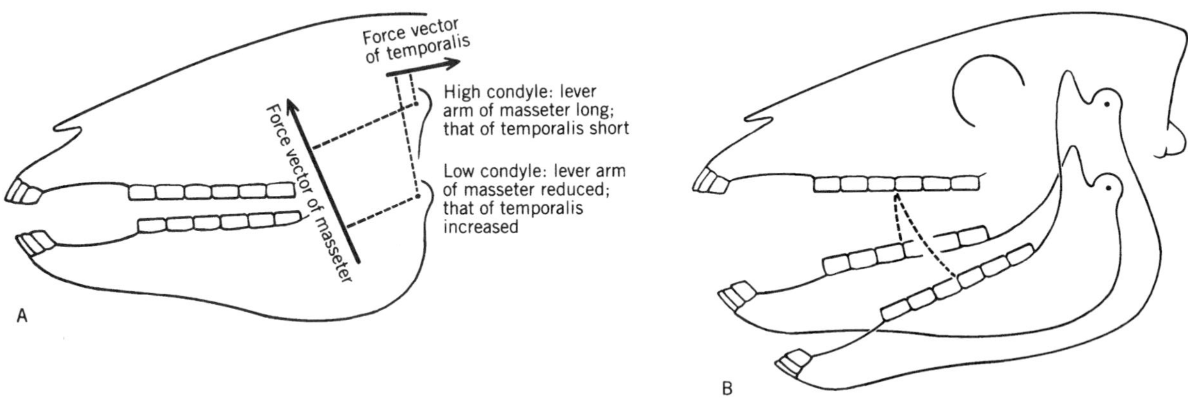

FIGURE 30.23 JAW MECHANICS IN RELATION TO THE POSITION OF THE MANDIBULAR CONDYLE.

temporalis, but this can be advantageous. Being a small muscle, it could not shorten enough to cause adequate motion were its lever arm not also shortened.

When the mandibular condyle is in line with the tooth rows, lower teeth approach upper teeth on an arc that brings them opposite as they near contact (Figure 30.23B). This is desirable if they are to function as blades or as mortars and pestles. When the condyle is high, the teeth approach more obliquely, causing food to be wedged and rolled as it is crushed. This is desirable for milling fibrous food.

Finally, the distances from the condyle to the anterior and posterior ends of a tooth row are the lever arms for grinding done in those respective positions. If the condyle is in line with the tooth row, then the difference in the levers equals the length of the row. If the condyle is high, then the difference is less than the length of the row, and force on the food varies less with position. Other factors in jaw mechanics were noted on p. 128.

OTHER FEEDING ADAPTATIONS

Gulping Fish and Other Large Prey It is the hapless fate of many fishes and frogs, and of some terrestrial vertebrates, to be swallowed whole. Predators that eat mostly fish are said to be **piscivorous.** About 30 percent of all fishes eat other fishes. Penguins, loons, pelicans, cormorants, auks, herons, and kingfishers are examples of birds that eat fishes. Roadrunners, secretary birds, and storks gulp lizards and mice. Seals, sea lions, and dolphins are piscivorous mammals.

Usually the prey must be oriented end-on to the throat, with any large scales pointing the "smooth" way. Pelicans (and, to a lesser extent, some other birds) have a distensible gular, or throat, pouch to hold large fishes until they can be positioned for swallowing. Birds may toss a fish in the air and catch it with the desired orientation.

If the food eaten is active or slippery, then teeth tend to be small and of simple structure (not blades, as for carnivores), yet sharp and numerous to prevent the escape of prey before it can be swallowed. Examples are the teeth of many salamanders, lizards, crocodilians, ichthyosaurs, plesiosaurs, seals, sea lions, and toothed whales (Figure 30.24). Mergansers (aquatic birds) have horny tooth analogs at the edge of the bill to prevent the escape of fishes. Sea turtles have numerous horny, backward-pointing spines in the gullet.

Slow Swallowing of Large Prey Some vertebrates swallow prey that is so large, relative to the head and throat, that it cannot be gulped, but must be engulfed slowly (Figure 30.25). One fish with a distensible belly forces down prey as large as itself. Some fishes can protract and retract the upper jaw mechanism, with the small, backward-slanting

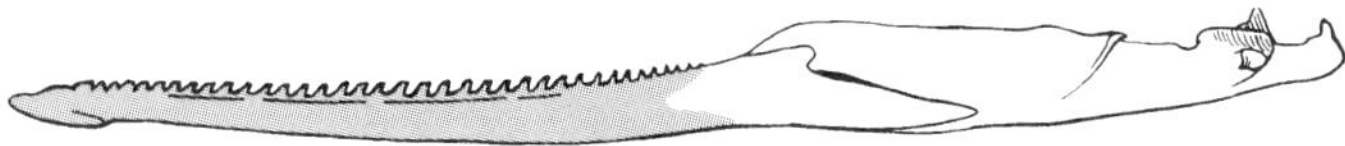

MERGANSER, *Mergus*

CROCODILE, *Crocodylus*

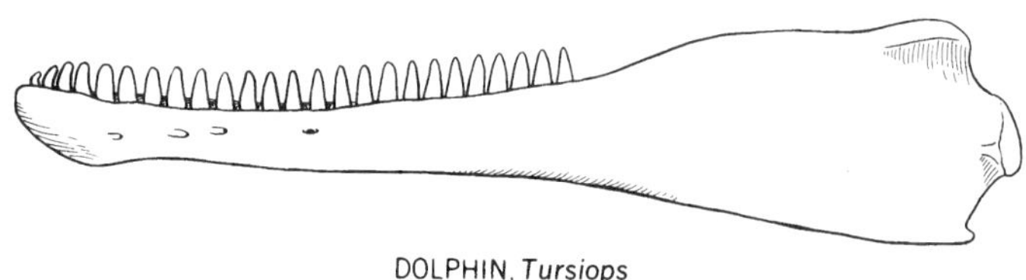

DOLPHIN, *Tursiops*

FIGURE 30.24 LEFT MANDIBLES OF VERTEBRATES THAT SWALLOW ANIMAL FOOD OF MODERATE SIZE REQUIRING HOLDING BUT LITTLE SHEARING.

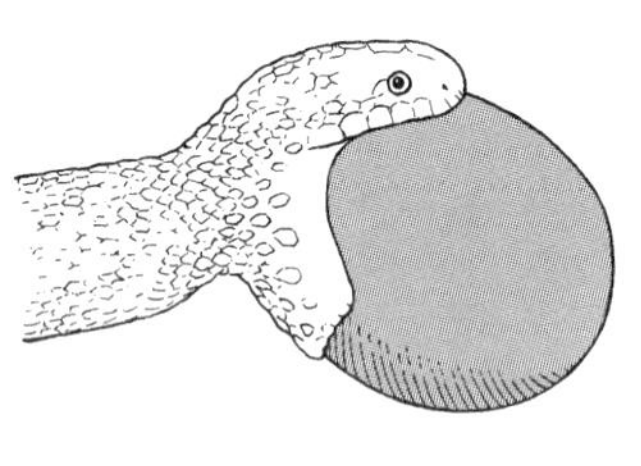

EGG-EATING SNAKE, *Dasypeltis*

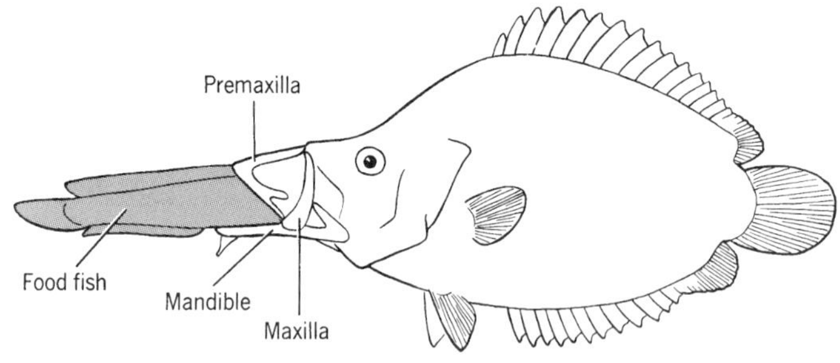

NANDID FISH, *Monocirrhus*

FIGURE 30.25 EXAMPLES OF VERTEBRATES THAT ENGULF LARGE FOOD OBJECTS.

marginal teeth of right and left sides alternately pulling on the prey, releasing and reaching forward, and pulling again. When the prey reaches the pharynx, pharyngeal teeth function in a similar way.

Snakes swallow large food, and several kinds that are adapted for eating eggs can manage eggs with a diameter three or more times the diameter of the head. The head of a snake is remarkably specialized. The teeth not only curve toward the throat, but the bones bearing them can be rotated so the teeth angle inward or outward at will. (Egg-eating snakes, however, tend to reduce or lose the teeth.) The lower jaw has no symphysis, so the mandibles can separate widely. The angles of the jaw pivot laterally from the skull, and the upper and lower jaws of many species can work forward and backward one side at a time, thus pulling the head around the food.

Swallowers of large prey have extra large and distensible throats and huge stomachs. Roadrunners, secretary birds, and some large frogs may swallow part of a large snake or lizard and wait for it to digest so there will be room to swallow the remainder. Snakes have no sternum so that large food items can distend the body by spreading the ribs. The airway to the lungs is kept open during the slow swallowing process by protruding the epiglottis out of the mouth, like a snorkel, below the prey being ingested. Several egg-eating snakes have evolved a remarkable method of breaking an egg after it is swallowed: Sharp, midventral projections from about a dozen vertebrae actually penetrate the esophagus. When the egg is in position, powerful muscles squeeze the egg against the projections, thus breaking the shell. Some species pass the crumpled shell back through the gut, whereas others hold the egg contents in place with special valves and regurgitate the shell. Various snakes pass large food along the gut using peristalsis of the axial musculature.

Crushing and Cracking Some animals must crush or crack shells, hulls, woody seeds, or other hard materials to make digestible food available. These animals are termed **durophagus** (= *hard* + *to eat*). Examples of fishes that crush shells to secure the soft meat within are the stingray, chimaeras, dipnoans (see figure on p. 108), some cichlid fishes, and the porcupine fish. These fishes have powerful jaws that are usually autostylic. Most have few teeth, and they are pavementlike and strong. Stingrays have numerous teeth that fit together like bathroom tiles to form crushing plates (Figure 30.26). Parrotfishes crush quantities of sand and rock as they eat algae; a powerful pharyngeal mill divides the material to a fine grit.

As their name implies, the extinct sauropterygian reptiles called placodonts had large flat teeth for crushing mollusks. Several lizards and several turtles crush snails. The molars of the sea otter and walrus are flattened for the same purpose. The platypus has horny plates for crushing snails and other food.

Grain- and nut-eating birds have particularly muscular gizzards for grinding their food. Grit is eaten to make the mill more effective. Similarly, sturgeons, the gizzard shad, and the mullet, which are detritus eaters, have muscular stomachs with tough linings for crushing miscellaneous invertebrates.

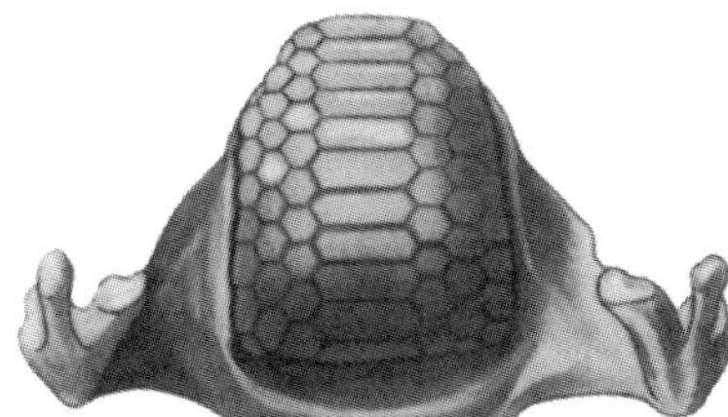

STINGRAY, *Myliobatus*

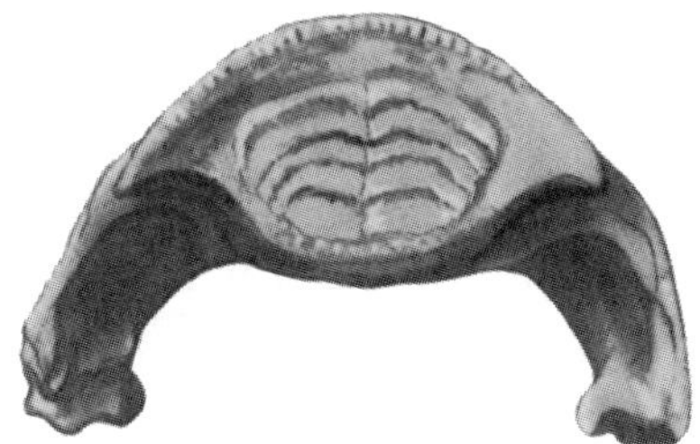

PORCUPINE FISH, *Diodon*

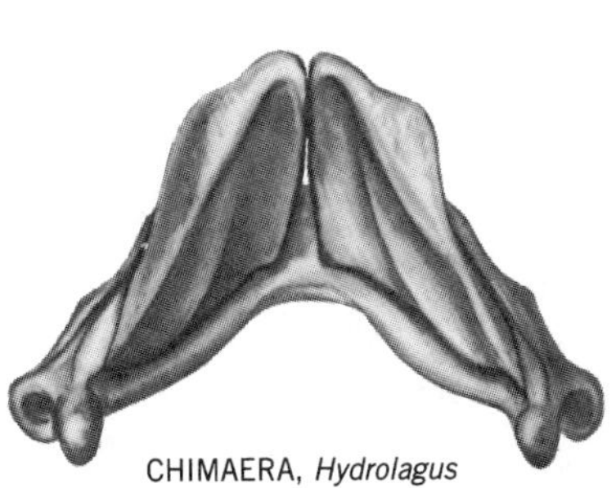

CHIMAERA, *Hydrolagus*

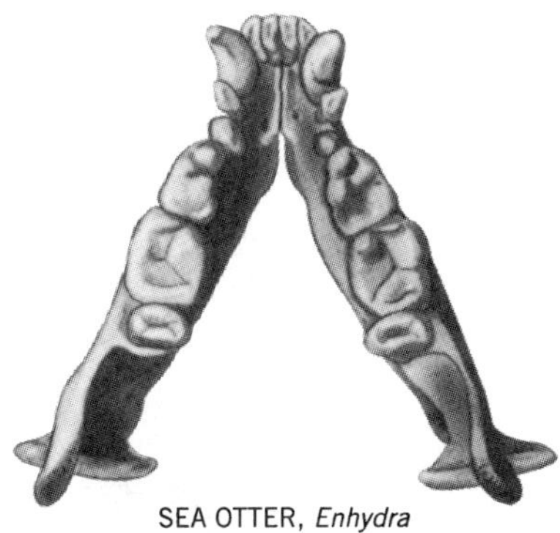

SEA OTTER, *Enhydra*

FIGURE 30.26 MANDIBLES AND LOWER TEETH OF DUROPHAGUS VERTEBRATES.

Some other vertebrates crack rather than crush their hard foods. The grackle (a bird) cuts acorns open against a keel on the inside of the upper bill. The Asian hawfinch cracks cherry and olive pits. The beak is very stout and deep at the base, and the jaw adductors are remarkably extensive. It has been claimed that this bluebird-sized finch can bite a pit with a force of more than 45 kg. The dusky shark has a biting pressure of 3000 kg/cm^2. Hyenas crack open the bones of large ungulates to secure the marrow. Their adaptations include extra heavy jaws and teeth, enormous jaw muscles and sagittal crest, and early closure of cranial sutures. Since the food is touched only by the points of the carnassials and several other teeth, great pressure is generated.

Additional Diets Nectar and pollen form all or most of the diets of hummingbirds, honeyeaters, sunbirds, honeycreepers, and some other birds, and also of six genera of bats, and the little honey possum. Most of the birds have long slender bills, variously curved according to the structure of the preferred flowers. The mammals have much longer rostrums than related species of other habits. Teeth are small and weak (bats) or reduced in number (honey possum), and jaw musculature has regressed. The tongue is invariably long, slender, and protrusible. It usually terminates in a brush consisting of rows or tufts of hairlike projections that slope toward the throat. This device effectively transports nectar to the mouth (Figure 30.27). The tongues of several nectar feeders have in addition, or instead, one or two narrow tubes through which nectar can be sucked into the mouth; muscles of the throat act as a suction pump. Pollen-feeding bats have unique hair adapted to collect pollen, which is then groomed from the pelt.

Vampire bats feed exclusively on blood. They make a shallow wound on the prey using a pair of sharp, curved, upper incisors. Other teeth are few and small. The blood meal is prevented from clotting by action of the saliva. When blood oozes from the wound, it is lapped up; when it flows freely, the edges of the tongue are curled up to make a tube, and the blood is sucked into the mouth. The tongue is pumped in and out to establish the suction, but the details of the mechanism have not yet been learned.

The larger mammals that feed on swarming insects are highly specialized. These include the spiny echidna, rat-sized marsupial anteater, aardvark, pangolins, and the giant and smaller "true" anteaters. The rostrum is usually long, the mouth small, and the mandibles weak. Because the ants or termites are not chewed, the teeth become small (marsupial anteater); then peglike, wanting in enamel, and single-rooted or open-rooted (aardwolf,

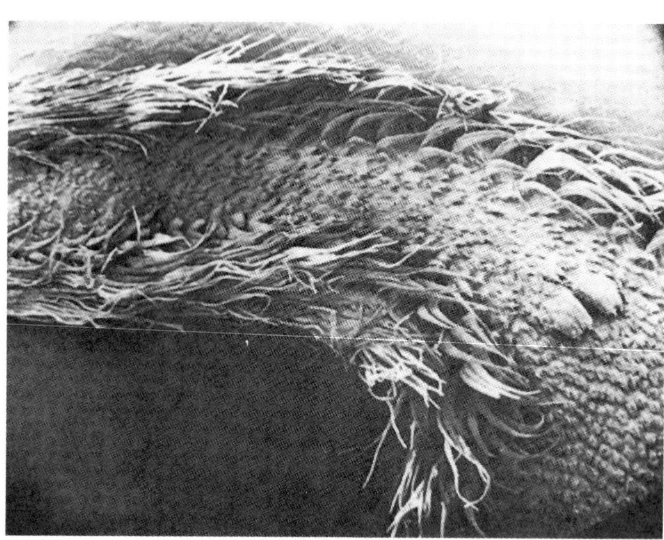

FIGURE 30.27
TONGUE OF A NECTAR-FEEDING AND POLLEN-FEEDING BAT, *Leptonycteris,* under 19× magnification. Dorsal surface, with tip of tongue out of picture to left. (Photograph taken with a scanning electron microscope by John Mais.)

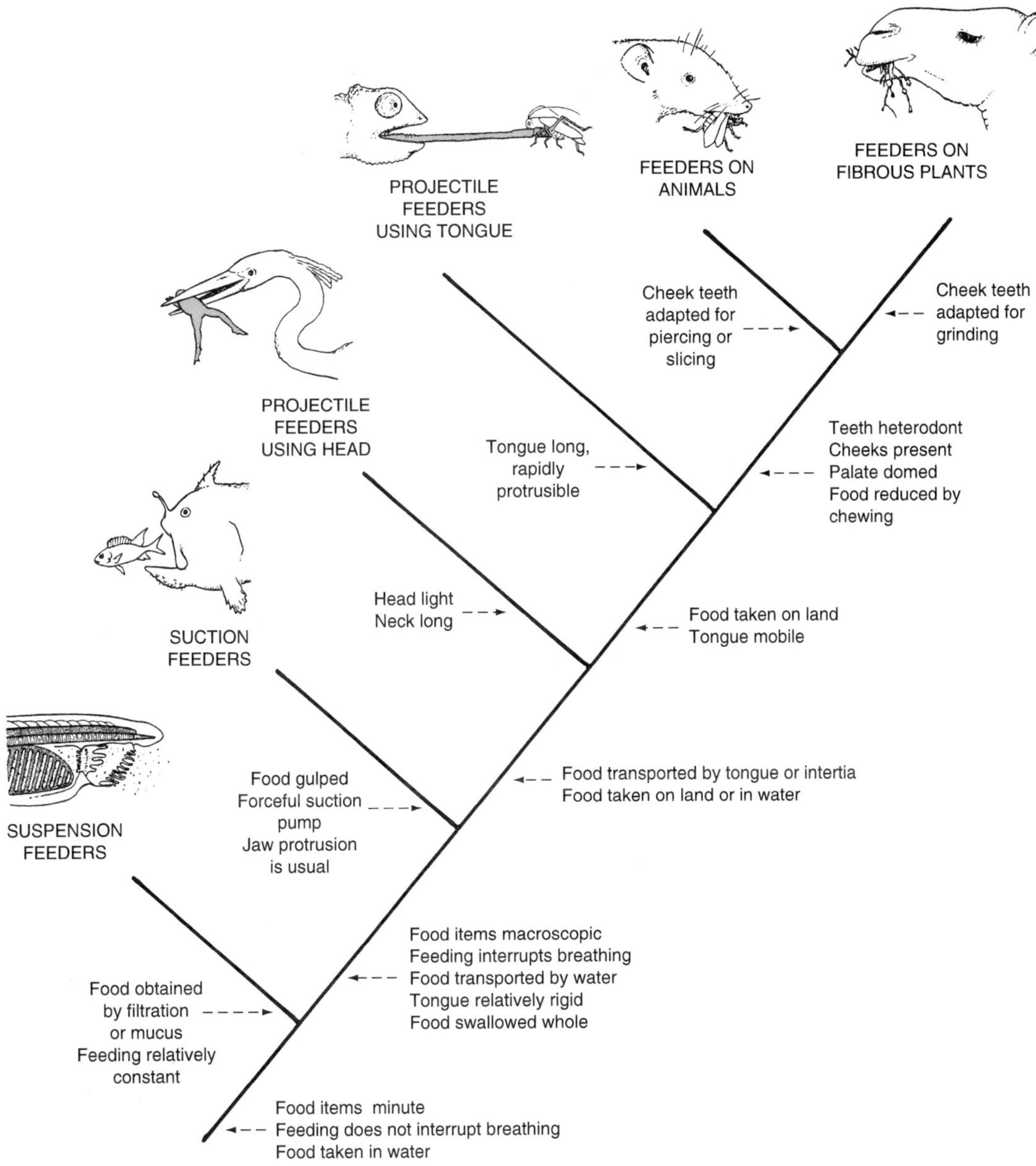

FIGURE 30.28 CLADOGRAM SHOWING CHARACTERISTICS OF SELECTED MODES OF FEEDING. (Many vertebrates, including sharks, piranhas, vampires, most birds, and toothed whales, do not fit this particular scheme.)

aardvark, armadillo); and finally lost (echidna, pangolin, anteaters). The nature and use of the tongue were mentioned above under projectile feeding. These animals are protected in several ways from ant bites: Pangolins (and others?) can close the nostrils. Most anteaters have a very thick, tough skin, and the eyelids and even the cornea may be thickened.

We have not described the filelike teeth of the fishes and tadpoles that feed on algae, or many other specializations. Although we have learned much, one chapter cannot include it all. Figure 30.28 summarizes part of the material covered.

REFERENCES

Akersten, W.A. 1985. Canine function in *Smilodon* (Mammalia; Felidae; Machairodontinae). *Contributions in Science* 356:1-22. Natural History Museum of Los Angeles County, CA.

Arnold, E.N. 1998. Cranial kinesis in lizards: variation, uses, and origins. *Evol. Biol.* 30:323–357.

Bels, V.L., M. Chardon, and P. Vandewalle (eds.). 1994. Biomechanics of feeding vertebrates. *Advances in comparative environmental physiology*, vol. 18. Springer, New York, 362p.

Biknevicus, A.R., B. Van Valkenburg, and J. Walker. 1996. Incisor size and shape: implications for feeding behaviors in saber-toothed "cats." *J. Vert. Paleontol.* 16(3):510–521.

Bramble, D.M., and D.B. Wake. 1985. Feeding mechanisms of lower tetrapods, pp. 230–261. *In* M. Hildebrand et al. (eds.), *Functional vertebrate morphology*. Harvard Univ. Press, Cambridge, MA.

Crompton, A.W., and K. Hiiemae. 1969. How mammalian molar teeth work. *Discovery* 5:23–34.

Domenici, P., et al. 2000. Killer wales (*Orcinus orca*) feeding on schooling herring (*Clupea harengus*) using underwater tail-slaps: kinematic analyses of field observations. *J. Exp. Biol.* 203:283–294.

Emerson, S.B., and L. Radinsky. 1980. Functional analysis of sabertooth cranial morphology. *Paleobiology* 6(3):295–312.

Frazzetta, T.H. 1966. Studies on the morphology and function of the skull in the Boidae (Serpentes). Part II. Morphology and function of the jaw apparatus in *Python sebae* and *Python molurus*. *J. Morphol.* 118:217–296.

Frazzetta, T.H. 1988. The mechanics of cutting and the form of shark teeth (Chondrichthyes, Elasmobranchii). *Zoomorphology* 108:93–107.

Frazzeta, T.H., and C.D. Prange. 1987. Movements of cephalic components during feeding in some requiem sharks (Carcharhiniformes: Carcharhinidae). *Copeia* 1987(4):979–993.

Gans, C. 1952. The functional morphology of the egg-eating adaptations in the snake genus *Dasypeltis*. *Zoologica* 37:209–244.

Gans, C., and G.C. Gorniak. 1982. Functional morphology of lingual protrusion in marine toads (*Bufo marinus*). *Am. J. Anat.* 163:195–222.

Greene, H.W. 1997. *Snakes: The evolution of mystery in nature*. Univ. of California Press, Berkeley. 351p. A good discussion of feeding in Chapter 3 (pp. 51–73); instructive photographs.

Herring, S.W. 1993. Functional morphology of mammalian mastication. *Am. Zool.* 33:289–299. Noteworthy for outlining problems and for a long bibliography.

Hiiemae, K.M., and A.W. Crompton. 1985. Mastication, food transport, and swallowing, pp. 262–290. *In* M. Hildebrand et al. (eds.), *Functional vertebrate morphology*. Harvard Univ. Press, Cambridge, MA.

Hoogenboezem, W., et al. 1991. A new model of particle retention and branchial sieve adjustment in filter feeding bream (*Abramus brama*, Cyprinidae). *Can. J. Fish. Aquat. Sci.* 48:7–18.

Janis, C.M., and M. Fortelius. 1988. On the means whereby mammals achieve increased functional durability of their dentitions, with special reference to limiting factors. *Biol. Rev.* 63:197–230.

Kooloos, J.G.M., et al. 1989. Comparative mechanics of filter feeding in *Anas platyrhynchos, Anas clypeata* and *Aythya fuligula* (Aves, Anseriformes). *Zoomorphology* 108:269–290.

Lambertsen, R., N. Ulrich, and J. Straley. 1995. Frontomandibular stay of Balaenopteridae: a mechanism for momentum recapture during feeding. *J. Mammal.* 76(3):877–899.

Larsen, J.H., Jr., J.T. Beneski, Jr., and D.B. Wake. 1989. Hylolingual feeding systems of the Plethodontidae: comparative kinematics of prey capture by salamanders with free and attached tongues. *J. Exp. Zool.* 252:25–33.

Lauder, G.V. 1985. Aquatic feeding in lower vertebrates, pp. 210–229. *In* M. Hildebrand et al. (eds.), *Functional vertebrate morphology*. Harvard Univ. Press, Cambridge, MA.

Liem, K. 1980. Adaptive significance of intra- and interspective differences in the feeding repertoires of cichlid fishes. *Am. Zool.* 20:295–314.

Liem, K.F. 1990. Aquatic *versus* terrestrial feeding modes: possible impacts on the trophic ecology of vertebrates. *Am. Zool.* 30:209–221.

Nemeth, D.H. 1997. Modulation of buccal pressure during prey capture in *Hexagrammos decagrammus*. (Teleostei: Hexagrammidae). *J. Exp. Biol.* 200(15):2145–2154.

Nishikawa, K. 2000. Feeding in frogs, pp. 117–147. *In* K. Schwenk (ed.), *Feeding: form, function, and evolution in tetrapod vertebrates*. Academic Press, New York.

Pietsch, T.W., and D.B. Grobecker. 1990. Frogfishes. *Sci. Am.* 262(6):96–103.

Sanderson, S.L., J.J. Cech, Jr., and M.R. Patterson. 1991. Fluid dynamics in suspension-feeding blackfish. *Science* 251:1346–1348.

Sanderson, S.L., and R. Wassersug. 1990. Suspension-feeding vertebrates. *Sci. Am.* 262(3):96–101.

Sanderson, S.L., and R. Wassersug. 1993. Convergent and alternative designs for vertebrate suspension feeding, pp. 37–112. *In* J. Hanken and B.K. Hall (eds.), *The skull*, vol. 3. Univ. of Chicago Press, Chicago. Excellent review and summary.

Schwenk, K. (ed.). 2000. *Feeding: form, function and evolution in tetrapod vertebrates*. Academic Press, New York.

537p. Comprehensive treatment; chapter 1 effectively places feeding studies into the context of integrative morphology.

Smith, K., and W.M. Kier. 1989. Trunks, tongues, and tentacles: animal movement with muscular skeletons. *Am. Sci.* 77:28–35. Discusses use of the tongue and trunks as muscular-hydrostat.

Taylor, M.A. 1987. How tetrapods feed in water: a functional analysis by paradigm. *Zool. J. Linnean Soc.* 91:171–195.

van Leeuwen, J.L. (1997). Why the chameleon has spiral shaped muscle fibers in its tongue. *Phil. Trans. R. Soc. Lond. B* 352:573–589.

Van Valkenburgh, B., M.F. Teaford, and A. Walker. 1990. Molar microwear and diet in large carnivores: inferences concerning diet in the sabertooth cat, *Smilodon fatalis. J. Zool. Lond.* 222:319–340.

Vincent, J.F.V., and P.J. Lillford (eds.). 1991. *Feeding and the texture of food.* Society for Experimental Biology Seminar series 44. Cambridge Univ. Press, New York. 247p.

Wainwright, P.C., and A.F. Bennett. 1992. The mechanism of tongue projection in chameleons. I. Role of shape change in a muscular hydrostat. *J. Exp. Biol.* 168:23–40.

Wainwright, P.C., and S.S. Shaw. 1999. Morphologial basis of kinematic diversity in feeding sunfishes. *J. Exp. Biol.* 202:3101–3110.

Werth, A. 2000. Feeding in marine mammals, pp. 487–526. *In* K. Schwenk (ed.), *Feeding: form, function and evolution in tetrapod vertebrates.* Academic Press, New York.

Westneat, M.W. 1994. Transmission force and velocity in the feeding mechanisms of labrid fishes (Teleostei, Perciformes). *Zoomorphology* 114:103–118.

Wilga, C.D., and P.J. Motta. 1998. Conservation and variation in the feeding mechanism of the spiny dogfish, *Squalus acanthias. J. Exp. Biol.* 201:1345–1358.

Zusi, R. 1967. The role of the depressor mandibulae muscle in kinesis of the avian skull. *U.S. Natl. Museum Proc.* 123:1–28.

Zweers, G., et al. 1995. Filter feeding in flamingos (*Phoenicopterus rubber*). *Condor* 97:297–324. An integrative analysis of a specialized mode of feeding.

Appendix

Anatomical Preparations

Milton Hildebrand

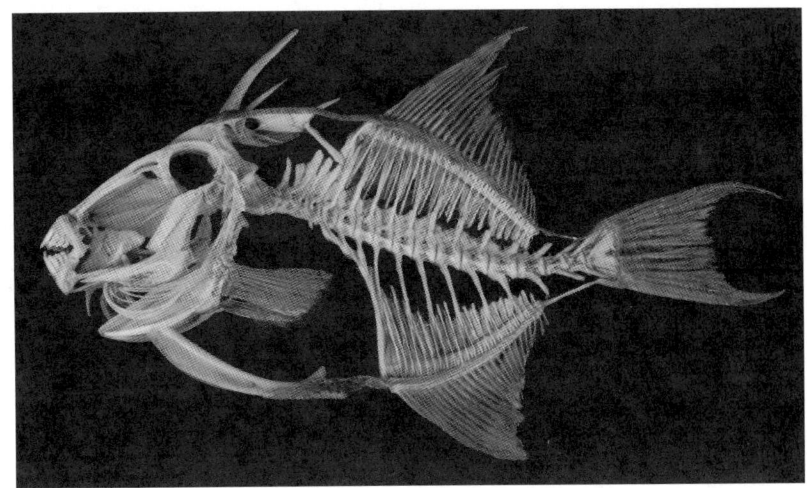

I started collecting anatomical specimens when I was 6 years old: a pelican skull picked up on a beach, opossum bones found beside a trail, coyote remains donated by an older boy. By the time I was 14 I had taught myself to clean and mount skeletons. If you are a student, you might also discover that acquiring specimens can be an interesting and informative hobby. From the time I started teaching vertebrate morphology, my collecting focused on materials to use as hands-on demonstrations in student labs. If you are a teacher, you might also find that even a modest collection can go far to interest and instruct your students. The making of an anatomical preparation can be a special assignment that is beneficial and well received—particularly for students with experience making models. For student and teacher alike, the focus and manipulation required to reassemble a spine or foot skeleton surely benefits understanding and interpretation. A well-prepared specimen also provides a gratifying aesthetic reward; those who look closely find that the body has great beauty.

My book *Anatomical Preparations*, published in 1968, is now long out of print. It emphasized the preparation of specimens for use in teaching. Unfortunately, no subsequent publication fits the level and coverage sought. Therefore, to encourage initial efforts, we offer here concise instructions for the more useful methods. The equipment needed is minimal, inexpensive, and usually available from local stores or the biology stockroom. (My only expensive equipment was a freeze-drying apparatus.)

It takes time to develop a teaching collection, but the first specimens are useful right away, and securing material may be easier than you think. Friends and colleagues make donations. Students return from far-flung vacations with offerings purchased at a curio shop, bargained for in a marketplace, or found near a campsite. You might obtain casualties from an animal shelter, pet shop, or zoo (but in case of disease, consult a veterinarian). Beachcombing usually turns up skeletons. Some hunters and trappers are cooperative.

Above: Triggerfish skeleton, beach-cleaned and hand-finished.

Your state game commission might contribute confiscated carcasses. You can trap and hunt your own trophies (but obtain all required permits, and respect private property). Species that are locally abundant but endemic to your part of the world can be traded to collectors elsewhere.

The preparation of anatomical specimens is somewhat of an art; example, dedication, ingenuity, and experience are great assets. Begin with a dog skull, not a mouse skeleton, and these instructions will help you get started.

(Most of the photographic chapter-opening illustrations in this book are of specimens from my teaching collection, and many of the halftone illustrations were drawn by my wife from my materials. The collection is now at the Museum of Vertebrate Zoology at the Berkeley campus of the University of California.)

CLEANING SKULLS AND SKELETONS

Cleaning by simmering gives excellent results for skulls of adult mammals and birds the size of a rat or larger. Results are good for skeletons of dog size or larger, but all bones become disarticulated and cartilage (as between ribs and sternum) is destroyed. The method is not suitable for small animals, juveniles, or fishes.

Skin the specimen, remove viscera, cut away larger muscles and fat deposits (taking care not to scar the bones), and dismember carcass into pieces of convenient size. Pulp the brain using a wire or probe inserted through the foramen magnum, then flush the brain out with running water, shaking the skull as needed. Alternatively, a jet of water from the needle of a large hypodermic syringe, inserted into the foramen magnum, is very effective for skulls up to cat size. Drill a hole 1 to 3 mm in diameter (depending on size of material) into each end of the shaft (not epiphyses) of each long bone, and flush out the fatty marrow using water from a large hypodermic syringe, or (better) an air jet (join tubing to an air line and use an eye dropper as a nozzle). Remove blood by soaking fresh bones for 24 hours in 1 part household ammonia to 3 to 4 parts cold water. It is usually best to cook the bones fresh, but if they must be temporarily stored, at this stage the skull or skeleton can be air-dried; rinse, blot, and use a fan but no artificial heat.

Select a large enough pot so that bones will be completely submerged. Add a small amount of detergent or household ammonia to the water. Simmer (not rapidly boil) for $\frac{1}{2}$ to 2 hours. Small, young, fresh specimens need the least time; large, old specimens that have been dried take longer. Let cool slowly to avoid cracking of teeth and long bones. Pour off fat and broth. Disarticulate vertebrae, ribs, and long bones (taking care to save all small pieces), and if you intend to reassemble the skeleton, take any needed precautions (sketches, labels) to ensure that you will know how to fit the pieces together. Do not dismember the feet at this time. Hold bones individually under the tap and use your fingers to remove as much flesh as comes away easily. This cleaning is preliminary. Tie each foot separately in a piece of cloth (to prevent intermixing) and return everything to the pot.

Simmer again until all flesh and ligaments are soft. Again cool slowly. Clean each bone under the tap using brushes and scrapers, and an air jet if available. A nail brush and toothbrush are good, and the small blade of a pocketknife is adequate. Dental scrapers of varying size and shape are useful (ask your dentist for a source). Take loose teeth out of skulls before they are lost; they will be glued in place later. Foot bones can be laid out in order on paper towels as they are cleaned. To prevent small bones from disappearing down the drain, fit the opening with a screen of small mesh. Allow the clean bones to dry slowly and thoroughly. Degreasing is usually needed (see below).

Cleaning by maceration can be complete, producing a disarticulated skeleton, or controlled, producing a skeleton that remains partly articulated by its ligaments. *Complete maceration by chemical action* is not suitable for small skeletons but is the method of choice for large specimens by most preparators of teaching collections. Many preparators of archival or research collections fear that delicate bones (turbinates) and projections (styloid process) may be weakened or eroded by this process and that bones so cleaned may deteriorate after long-term storage. Hand labor is minimal, quality is excellent, and subsequent degreasing and bleaching are not needed. Prepare specimens as for simmering (except that it is less important to soak out blood). Use a glass, ceramic, or stainless steel container (not aluminum), and affix any tags with string, not wire. Cover specimens with a 0.5 percent solution of potassium hydroxide and maintain at about 45°C (115 °F) for 5 to 8 days. (Alternatively, use a 1 percent solution at room temperature for a longer period.) Inspect the specimens from time to time to see when they are ready for removal; leave in the solution no longer than necessary. Wear rubber gloves and an apron. Rinse thoroughly in dilute chlorine bleach to remove scum and lingering tissue fragments. Dry slowly. In most jurisdictions the used alkali solution may be washed down the drain. In case of doubt, a stronger solution can be neutralized before disposal.

Complete maceration by bacterial action avoids the use of a strong chemical and can be used for smaller specimens, but is slow (usually several weeks) and produces a putrid odor. Prepare specimens as for chemical maceration. Immerse in water; large glass jars with lids are favorable. Place in a warm spot and let stand until soft tissue has decayed away. Change the solution several times, but retain a little liquid to maintain the bacterial action. When finished, rinse, soak in dilute household ammonia or chlorine bleach to remove odor, again rinse thoroughly, and dry.

Controlled maceration is used for the commerical preparation of cat, dog, chicken, and perch skeletons that remain held together by their ligaments and joint capsules. Joints remain partly obscured. More hand labor is required for cleaning, less for mounting. The posture of the mounted skeleton may be less natural than for disarticulated skeletons reassembled by an expert. Skin and eviscerate. Remove head and legs (saving the clavicles of mammals), but leave the body in one piece. Remove brain as for simmering, and use a hooked wire to remove as much as possible of the spinal cord. Flesh thoroughly by cutting and scraping, but preserve all joint ligaments. Remove eyes and tongue (saving the delicate hyoid). Flesh between the ribs. Skin toes out to the claws. Soak in water as for complete bacterial maceration, but examine and change the water more often. Macerate just long enough to soften flesh without destroying the more resistant ligaments. Rinse and brush away remaining scraps of flesh. Soak briefly in dilute household ammonia or chlorine bleach to control odor. Position the units of the skeleton during drying so that as the joints stiffen they will have the form desired for the mounted skeleton. Some gluing, and often wiring, will be needed.

Cleaning by "bugging" is the term given to skeletal preparation using the carnivorous larvae of the beetle *Dermestes vulpinus*. This is by far the best way to clean all skeletons of squirrel size or smaller that are to be mounted or boxed as fully or partly articulated specimens. Bones are meticulously cleaned of flesh, yet if removed from the larvae at the appropriate time, they remain articulated by their ligaments, which are virtually invisible on the finished skeleton. There is no warping or displacement of parts: teeth are not loosened, sternal cartilages are preserved, and ribs are properly spaced. Even fish heads and very small skeletons can be cleaned perfectly. The downside is that maintaining a colony of the insects takes considerable effort. You will need at least 25 beetles to get started. Obtain them from a friendly museum preparator, a carcass found in the open, or the floor of a bat cave. The beetles are oval, 8 to 9 mm long (the female being larger than the male), black above, and silvery below. Larvae are brown above, whitish below, fuzzy, active, and range from 2 to 14 mm long. The colony must be housed, far from your collection, in a container of metal, plexiglass, or glass (not wood). An aquarium is satisfactory. The lid must fit tightly, but ventilation is required to prevent mold and discourage parasitic mites, so fit the lid with a large hole covered with fine metal screen. In order to provide places for the larvae to tunnel and hide, cover the bottom of the container with several layers of cotton batting. Keep the colony in the dark, and maintain a temperature of 70 to 85°F. If an electric light is used for warmth, wrap it in aluminum foil. The level of activity increases with temperature. At 85° generation time is 45 days. The ravenous larvae must be fed, even when you run out of specimens. The Museum of Vertebrate Zoology recommends sheep heads—skinned, halved, with brain removed, and dried enough to avoid spoilage. The "bugs" like the crevices for hiding and laying eggs. The colony needs fat to thrive. If the food you provide is lean, dry, and hard, dip it in cod-liver oil (better than other fats) and then blot so the insects will not get wet. From time to time clean the detritus from the bottom of the container.

To prepare specimens for the colony, skin (including the legs and toes of birds and footpads of mammals), eviscerate, remove eyes of animals larger then a rat, and remove the brains (as decribed above). The foramen magnum can be exposed without severing the head from the body: Depress the head and make a tranverse cut just behind the skull. Keep away from fixatives and flies. Dry until the flesh is firm but not hard, and then place in the colony. An active colony moves specimens about; to prevent mixing and loss, place specimens (or parts of specimens, such as feet) in separate cardboard box lids or the pockets of egg crates. Cover with sheet cotton. Inspect daily and remove each specimen (or part specimen) promptly as soon as it is ready to avoid disarticulation. Delicate specimens (newborns, shrews, hummingbirds) should be confined in open jars with a few small larvae, which can be replaced as they grow; cleaning will be slow but perfect.

When specimens are removed from the colony, the adhering insects must be immediately killed by fumigation (chloroform is satisfactory), or by freezing, or heating to 160°F. Heat may crack the teeth of the larger specimens, but has the advantage that larvae come out of crevices before they die. Account for every insect, for they are *very* destructive in a museum (or clothes closet). The further cleaning that is needed is described below. Some people are very sensitive to "bug dust"; it causes sneezing and makes the eyes burn. Wash your hands carefully after each contact and wear a mask if necessary.

DEGREASING

Skulls and skeletons cleaned by complete chemical maceration are degreased and bleached by the cleaning process. Nearly all material cleaned by other methods should be degreased. Even bones that seem to be free of grease immediately after cleaning may become stained in time as fat slowly spreads from within. Oily bones are unsightly, collect dirt, may attract pests, and stain their containers. If holes have not already been drilled in the long bones (see above), this should be done now to facilitate circulation of the solvent. (If holes are objectionable, they can be plugged later with wood dough.)

Many reagents have been used for degreasing bones. I pass over some (benzene, lacquer thinner, sodium perborate, and others) because I have not tried them, and also white gasoline, which I have used with success for large bones, because it is too hazardous to recommend.

Household ammonia in several parts of water was mentioned above for removing blood prior to simmering and for controlling odor after maceration. It also removes some grease. It is safe, cheap, easy to obtain and use, and is the method of choice for the small skulls and skeletons that come out of a "bug" colony. Use at full strength or somewhat diluted. Soak specimens 24 hours, changing the solution once or twice, and rinse for 12 to 24 hours. Some cleaning by hand may be needed to remove tissue fragments or to dislodge dead larvae from crevices. Ammonia removes grease and also lightens color, and (importantly) removes adhering "bug dust" (feces, larval fuzz, cotton fragments, etc.). Should it be desirable to avoid wetting (thus warping) the sternal cartilages (more of a problem with larger skeletons), place the specimen on its back in a pan and soak in ammonia with sternum protruding. The cartilages can be cleaned by brushing with a soft, infant's toothbrush, and drying quickly before the cartilages become wet through.

For larger bones ammonia may be tried and provides initial degreasing, but other methods are usually also required. Three solvents give good results. They are used in about the same way. Two of them (trichloroethylene and methylene chloride) may be difficult to obtain, so I focus on the third, acetone, which is available where paint is sold. **Acetone** is clear, volatile, fragrant, and highly flammable. Avoid skin contact, but in case of minor contact wash and don't worry. Avoid prolonged inhalation of vapor, but a few whiffs are of no concern. A fumehood is useful. Fire hazard *is* serious; take great care to avoid sparks, flames, and high heat. Immerse bones at room temperature for 2 to 3 weeks according to hardness of bone and nature of fats. Glass containers are recommended because specimens are then visible. Replace acetone when it becomes too discolored. It is economical to start the degreasing in used solvent and finish in clean. Remove finished bones from solvent and dry for 24 hours or more. Dispose of acetone by turning it over to your institutional or municipal hazardous waste disposal unit.

BLEACHING

Soaking bones in cold water and household ammonia swells and softens tissues, removes some fat, and bleaches by leaching out blood and pigment. Degreasing removes stains and whitens further. Additional bleaching is usually not needed, or even advisable: Excessive bleaching obscures sutures and muscle scars, and gives bones a flat, chalky appearance. If more bleaching *is* desired, immerse bones in a 1 to 3 percent solution of hydrogen peroxide (available at drugstores). Use a glass container. Small specimens may take 10 minutes; others may need several hours. Watch and remove when ready, rinse, and dry. To bleach a stain or limited region of a specimen, paint full-strength peroxide on the area with a cheap watercolor brush.

SECTIONING BONES AND TEETH

Sectioned materials can be very instructive: Longitudinal sections of long bones show cylindrical shaft, cancellous epiphyses, and trabeculae; a sagittal section of a mammal skull shows nasal septum, sphenoid sinus, pocket for pituitary, and secondary palate; cutaway jaws show roots of teeth and teeth not yet erupted (see photo on p. 103); sectioned teeth of ungulates show complex infolding of enamel and dentine (see figure on p. 574). Hardware and hobby shops sell powered hand tools fitted with tiny circular saws, burrs, and drills. It is easy to section bones, but it takes patience and practice to do it well. Hold the bone firmly or steady by pressing it against a sandbag. Work slowly. Do not let a tool bind or "climb" out of its groove. Make a preliminary cut adjacent to the desired final cut. Strive for smooth, even surfaces. If delicate bones (e.g., turbinates) are to be cut, imbed in paraffin, section, and melt or dissolve away the paraffin. To have teeth sectioned, take them to a rock shop, or geology or paleontology department, and ask the technician to make the cuts with a diamond saw.

GLUING AND MOUNTING

Gluing is necessary to return loose teeth to their sockets and to assemble disarticulated bones. The adhesive should be clear and quick drying. I use Duco cement, which is available at hardware and stationery stores. Apply sparingly and neatly with a toothpick. Porous bones can be hardened, and fragile bones strengthened, by painting with glue thinned half-and-half with acetone. Completely disarticulated skeletons can be daunting puzzles. Most skeletons are much more interesting and instructive if assembled into units: Glue together the bones of each foot, anterior spine (preferably with ribs), posterior spine with pelvis, and tail, leaving skull, jaw, and long bones loose. You may wish to clean and assemble the feet of one side of the body at a time, leaving those on the other side as models. Modeling clay (from an art or toy store) is very useful for positioning bones as glue dries. Buy a kind that is not oily to the touch.

The complete mounting of small skeletons is an extension of the partial assembly suggested above. Commerical skeletons between chicken and dog size usually have the vertebrae strung on a rod and the limbs wired to pelvis and ribs. I find gluing to be adequate. Draw on paper the desired curvature for the spine (best done from the articulated spine before cleaning). Arrange the clean vertebrae along the curve and glue them together, first in pairs, then fours, and so on. A cork can be cut into spacers for use between scapula and ribs. Large skeletons require rods, wires, and resourcefulness. Some of the references have suggestions.

BONE-LIGAMENT PREPARATIONS

The feet and leg joints of large animals can be dissected completely free of all tissues except bone, ligaments, and tendinous insertions and then air-dried. Shrinkage separates the hardening ligaments and tendons, thus opening up the dissection and facilitating the study of mechanical relationships. Specimens are clean, tough, and easy to handle and store. They are novel and instructive, and some are marvels of biomechanics (see photo on p. 383, and drawing on p. 391).

Dog, pig, sheep, cow, and horse are favorable. A cooperative butcher, veterinarian, or slaughterhouse may provide specimens. Use knife and saw to isolate the selected part and make the desired dissection. This will challenge your technique, patience, and knowledge of anatomy. Use sharp cutters, toothed forceps, and medium-sized scissors. Remove skin, fat, muscle, tendon sheaths, and joint capsules. Reveal the mechanism. There is no one right way; use judgment and imagination. Drill one small hole into each bone of moderate size (e.g., phalanx of ungulate) and a hole into each end of any large bone (e.g., cannon bone) that has not been sawed across. If the specimen was not fixed before dissection, fix it now for a week using 7 percent formalin. (Wear rubber gloves, and handle outdoors or under a fumehood.) Remove, wash thoroughly, and tidy up any rough spots.

Drying comes next. Ligaments pull straight and taut with tremendous force, and the tendons, having free ends, tend to kink. Some specimens benefit from being securely tied within an improvised wooden frame. Weights may be devised to keep tendons straight. Make adjustments as needed. When dry, degrease thoroughly. Finally, smooth any rough surfaces with clean, fine steel wool or by the gentle use of a burr on a handheld mechanical tool. Paint hooves and any large and particularly hard ligaments with thinned neat's-foot oil (from a hardware store) to reduce further shrinkage; wipe off the excess after several days. Storage under fumigation is preferred.

Should a freeze-dryer be available it can substitute for air drying; there will be less shrinkage and darkening of ligaments and tendons (see drawing on p. 416).

BONE-MUSCLE PREPARATIONS

The preparation of dry bone-ligament displays, described above, can be extended to include muscles. Specimens of rabbit to sheep size are best. Select functional units: A forelimb with scapula is excellent; a hind limb with half of the pelvis, sacrum, and several vertebrae is good but more difficult. Other units can be devised. Drying causes shrinkage, which opens the dissection, thus reavealing all origins, insertions, and ligamentous ties, and showing the locomotor mechanism to excellent advantage. See the photos on pp. 169

and 405, and the drawings on pp. 188, 465, and 509. (The same dissections stored wet in museum jars and tanks would be rarely seen and unpleasant to handle.)

Skin the animal and isolate the part selected. Specimens can be dissected fresh, but muscles are firmer if first preserved. Soak in 7 percent formalin for a day or so (penetration is about 6 mm in 12 hours). Wash several days with several changes of water to reduce irritation from the toxic formalin. Provide good ventilation and wear rubber gloves. The heart of the technique is a careful, clean, complete dissection. Separate muscles all the way to their origins and free tendons all the way to their insertions, taking great care with the small, complex tendons in the feet. Remove joint capsules but preserve all ligaments, including the loops that guide tendons where they angle. Include deep and out-of-the-way places because they will open up during drying. Scrape the periosteum from exposed bone surfaces. Muscles cut in isolating the unit selected (e.g., for a forelimb, the rhomboideus, pectoralis, etc.) can be dissected completely away, or the insertions can be saved (contrast photos on pp. 455 and 381). In the latter case, do not trim them short at this stage. When the dissection is complete, drill small holes in the bones. Tie the specimen firmly in the desired position in an improvised frame. Next, fix the specimen, if this has not already been done.

Drying the specimen again gives an opportunity for skill and ingenuity. The objective is to have all muscles dry, with smooth contours and appropriate spacing from one another. Separate the more superficial muscles and tendons using spacers cut from foam rubber, clamps, rubber bands, strings, popsicle sticks, or other devices. Suspend the specimen in warm, circulating air, but avoid sun and oven; slow, even drying is best. Tend frequently: As superficial muscles dry, move the spacers to deeper positions. Additional dissection may be desirable in newly opened spaces. As muscles become firm, but not yet hard, they can be shaped by pinching with the fingers. At that stage trim the edges of any extrinsic muscles. Dry thoroughly.

The preparation is now hard, greasy, and discolored. Degrease thoroughly, then dry thoroughly. Exposed bones can be bleached by painting on full-strength hydrogen peroxide. If too much connective tissue remains in deep places, remove with a burr on a hand tool. If large muscles are fuzzy, singe them carefully over an open flame. If edges are weak or frayed, harden them with Duco cement thinned with acetone. If there are breaks, repair with epoxy cement. If there are offending irregularities, fill them with wood dough.

Finish by painting the muscles with artist's acrylic paints. Use a background reddish color (preferably selected when the dissection was fresh), and appropriate touching up with a lighter color simulating the connective tissue with which muscles are streaked. Tendons can be coated with iridescent nail polish. Finally, spray lightly with clear plastic.

Much to do, big job; much satisfaction, big reward.

AIR-DRYING HOLLOW VISCERA

Lungs are quickly and easily preserved by air drying, and the results are excellent (see photo on p. 219 and drawing on p. 231). Start with fresh material. Animals from the size of a bullfrog or iguana to that of a dog or sheep are all satisfactory (except birds; their lung structure is best revealed by corrosion casts). Remove lungs and trachea from the body. Dissect free of heart, esophagus, fat, and glandular tissue, taking care not to nick the respiratory system. To remove blood, either perfuse the pulmonary vessels with cold water, or immerse in cold water for several hours (use the lid of a full jar to hold the lungs under). Remove and gently blot with a towel. Attach a tube to an air source and insert an eye dropper into the free end as a nozzle. Pass a thread through the end of the trachea. Insert the nozzle into the trachea, and tape the ends of the thread to the glass to prevent the lungs from being blown from the nozzle. Suspend the lungs from a ring stand so they hang free.

Turn the air on slowly. Air will sizzle from the surface of mammalian lungs; it may be necessary to make pinpricks in an amphibian or reptilian lung to allow air to escape. Do not overinflate. Gently massage any spot that is slow to inflate. The lungs will drip at first. Tend them now and then, adjusting airstream as needed. Continue for some time after the lungs look dry. Dry lungs can be dissected to reveal chambers and airways (see drawing on p. 231).

Stomachs, caeca, and spiral intestines can also be air-dried. Isolate the organ and flush out contents with water. Dissect free of fat and other extraneous material, but do not trim closely; it is too easy to make holes. If the material is fresh, inflate moderately with 7 percent formalin and immerse for a day. Remove, rinse thoroughly, and blot with paper towels. Attach to an air source and dry as for lungs. Again, do not overinflate; if stretched too thin, the dry viscus is crinkly. After drying, windows can be cut in a spiral intestine or ruminant stomach to show internal structure. Air-dried viscera should be stored under fumigation. The finished product can be sprayed with clear plastic to give it body and make it resistant to moisture.

FUMIGATION

All the preparations mentioned so far should be protected from damage by insects. Keep them in glass display cabinets or in storage cabinets with gaskets and draw-tight latches. Selection of a fumigant is a compromise between effectiveness at killing insects and acceptable risk to human health and the environment. Of the many alternatives, none is completely satisfactory. *Napthalene,* also called mothballs or moth flakes, is a satisfactory repellant, but not an adequate fumigant if an infestation has occurred. A tablespoon of the flakes, in a saucer or box lid, placed in every cabinet every few months (or when the previous application has disappeared) is usually satisfactory for a small collection if coupled with routine inspection for signs of insects. Naphthalene is registered for nonrestricted use. The "signal word" is CAUTION. Avoid prolonged inhalation, which could cause eye irritation and dermatitis.

Paradichlorobenzene smells like mothballs, is a solid, and is a repellant. It is used like naphthalene. Prolonged inhalation gives some persons nausea, headache, and loss of coordination. It is registered for general use, but not for use in museums. The signal word is WARNING.

Carbon disulfide is a very effective insecticide. It is a faintly yellow liquid with a rotten-egg odor. Every 6 months put about two tablespoons in a saucer in the top drawer or shelf of each cabinet. Sensitivity varies, but prolonged contact (inhalation or skin) may cause narcosis, and can affect behavior and internal organs. Carbon disulfide is combustible and ignites at only 212° F. A fumigated cabinet would not explode, but a nearly empty bottle might. It is registered for general use, but not in public institutions. The signal word is WARNING.

STAINING SKELETONS OF WHOLE MOUNTS

The skeletons of small vertebrates can be demonstrated by staining and clearing specimens, which are then preserved wet in vials or jars. Favorable materials include late fetuses, hatchlings, tadpoles, and juveniles, and also adults of small fishes (see photo on p. 597) and other vertebrates up to mouse size. Bone becomes wine red, cartilage bright blue, and surrounding tissue transparent, thus revealing skeletal structure in minute detail. The general recipe can be varied: There can be some modification in fixation; only the bone can be stained, only the cartilage, or both; an enzyme can be used to assist clearning, but need not be; if a step in the process is later deemed to have been insufficient, it can often be repeated. Accordingly, and because the general technique is old, many investigators have published many articles reporting their preferences. The formulary that follows is, with some emendations, that of Ronald E. Cole (curator, Museum of Wildlife and Fisheries Biology, University of California, Davis); it is itself a blend of other recipes, and the one that has given him and his students the most success.

Procedure

1. Fix specimen in 10 percent formalin (solution 1, below) for several days, but not more than two weeks.

2. Soak in distilled water for several hours. (Prolonged or vigorous washing removes mucopolysaccharides, onto which the cartilage stain is deposited.)

3. If need be, the specimen can be stored in 70 percent alcohol; otherwise, proceed to step 4.

4. The excellence of the finished specimen depends not only on the completeness with which the tissues are cleared, but also on the amount of tissue present. Hence this process works best with smaller specimens. Small fishes, reptiles, and amphibians can usually be cleared and stained without removing their skin, although the finished product might be improved by skinning. Birds and mammals, however, should be skinned. Skinning can be done before fixation or after, when the tissues are tougher. Skin should be left on feet and tails to reduce maceration at these delicate areas. Tails of salamanders and lizards can be slit lengthwise to reduce the chance that the tail will swell, split, and fall off. Evisceration is optional and can be done before fixation to speed the fixation process.

5. Specimens of fish, reptiles, and most amphibians should be bleached to remove melanin. Birds and mammals need not be bleached, as they have been skinned and are free of pigment. Bleach in KOH and H_2O_2 (solution 4) until this dark pigment has become creamy yellow (10 minutes to an hour). Do not exceed the prescribed concentration of H_2O_2, as rapid oxidation will result in oxygen bubbles forming in the tissue. If this happens, these bubbles can be removed only by teasing them away with a needle probe or by using a vacuum chamber, or (hopefully) they will dissipate in time.

6. If yellowish-brown fat is present, it should be removed. It may be possible to remove larger masses by careful dissection under the KOH solution (the specimen is becoming very soft and delicate) using very fine instruments. Otherwise, soak in 40, 60, 80, and 99 percent isopropyl alcohol, for 24 hours each; then in xylene for 1 to 7 days, as needed; then again in 99, 80, 60, and 40 percent alcohol for 24 hours each; ending with 24 hours in distilled water.

7. If cartilage is to be stained, proceed to step 8; if only bone is to be stained, proceed to step 9.

8. Stain in freshly prepared Alcian Blue (solution 7) for 24–48 hours.

9. The enzyme trypsin can be used to further the "clearing," although its action is not to destain tissue but to digest it away. Place the specimen in the buffered enzyme (solution 6). Change the solution every 3 days for as long as it continues to take on a bluish color. Continue in the solution until only about one-fourth of the muscle remains and the skeleton is clearly visible. Use caution, however, because over-long treatment will expose and damage the skeleton.

10. Transfer to Alizarin Red S (solution 9) for 24 hours.

11. Place in 2 percent KOH (solution 2) until excess stain has leached out of the tissues. Several changes may be required. Small specimens destain in 1 to 2 days; larger ones take longer.

12. Reptiles and fishes having scales that have not yet been removed can now be cleaned by very gently wiping scales away with a soft cotton swab, stroking from head to tail. Do this while floating the specimen in 2 percent KOH or water.

13. Place the specimen successively in 30, 60, and 90 percent glycerin for 24 hours each. Further clearing will occur. Finally, store in pure glycerin containing several crystals of thymol to prevent mold.

Stock Solutions

1. Ten percent buffered formalin: Mix 1 part 37 percent (technical grade) formaldehyde with 9 parts water. Unbuffered formalin is highly acidic and quite damaging to vertebrate material. Buffer with borax (10 g/L), ammonium hydroxide (150 mL/L), or marble chips (3–4 tbsp/L).

2. Two percent potassium hydroxide (KOH): Mix 20 g of KOH pellets with 980 mL of distilled water.

3. Three percent hydrogen peroxide (H_2O_2): Dilute one part technical-grade hydrogen peroxide (27 percent) with 9 parts distilled water, or use full strength if purchased at the reduced 3 percent concentration.

4. Bleaching solution: Mix 7–8 parts 2 percent KOH (solution 2) with 2–3 parts 3 percent H_2O_2 (solution 3).

5. Buffer solution: Mix sodium tetraborate or laundry-grade borax with distilled water to make a saturated solution. Allow to settle until clear.

6. Buffered enzyme solution: Mix 3 parts saturated buffer solution (solution 5) with 7 parts distilled water. Mix $\frac{1}{4}$ tsp. trypsin with 400 mL of the buffer. Trypsin powder should be kept refrigerated during storage and only added to the buffer solution when needed.

7. Alcian Blue stain: Mix the following: ethanol, 95 percent—80 mL; glacial acetic acid—20 mL; Alcian Blue 8GN—15 mg.

8. Stock Alizarin Red S stain: Mix the following: 50 percent glacial acetic acid—5 mL; glycerine—10 mL; Alizarin Red S. Saturate to dark red-purple color.

9. Alizarin stain: Mix 1 mL of stock solution (solution 8) with 100 mL of 2 percent KOH (solution 2), or add Alizarin Red S dye to 2 percent KOH until a red purple color is obtained.

OTHER TECHNIQUES

Other techniques, less used, are beyond the scope of this appendix. Instructions for some of them are found in the references that follow. Three are illustrated by photos in the main text: on p. 141, the dry preservation of cartilage by impregnation with paraffin; on p. 319, 326, and 334, the staining of thick brain slices to contrast white and gray matter; on p. 239, the casting of vessels and ducts with vinyl resin and the subsequent corrosion of surrounding tissues.

REFERENCES

Anderson, R.M. 1965. Methods of collecting and preserving vertebrate animals. 4th ed. *Nat Mus. Can. Biol. Ser.* 18, Bull. 69. 199p.

Borell, A.E. 1938. Cleaning small collections of skulls and skeletons with dermestid beetles. *J. Mammal.* 19:102–103. Preparation, soaking, "bugging," and cleaning with ammonia.

Burdi, A.R. 1965. Toluidine Blue–Alizarin Red S staining of cartilage and bone in whole-mount skeletons *in vitro. Stain Technol.* 40:45–48.

Dingkerkus, G., and L.D. Uhler. 1977. Enzyme cleaning of Alcian Blue stained whole small vertebrates for demonstration of cartilage. *Stain Technol.* 52:229–232. Obtained better results than with Toluidine Blue.

Filipski, G.T., and M.V.H. Wilson. 1985. Staining nerves in whole cleared amphibians and reptiles using Sudan Black B. *Coepia* 1985(2):500–502.

Hall, E.R. 1962. Collecting and preparing study specimens of vertebrates. *Univ. Kans. Mus. Nat. Hist. Misc. Publ.* 30. 46p. Field techniques for collecting and preparing skins for research (not display) collections.

Hall, E.R., and W.C. Russell. 1933. Dermestid beetles as an aid in cleaning bones. *J. Mammal.* 14:428–431. A pioneering paper and still a valuable reference.

Hangay, G., and M. Dingley. 1985. *Biological museum methods, vol. 1: vertebrates.* Academic Press, New York. 379p. Directed toward the preparation of museum displays, but includes directions for collecting specimens, embalming, injecting, and cleaning skeletons.

Hildebrand, M. 1968. *Anatomical preparations.* Univ. California Press, Berkeley. 100p. Covers many topics; the only reference on making bone-ligament-muscle preparations. Emphasis on use in teaching.

Hower, R.O. 1979. *Freeze-drying biological specimens: a laboratory manual.* Smithsonian Inst. Press, Washington, DC. 196p. Theory, instrumentation, and procedures for freeze-drying vertebrates for preservation and display.

Kampmeier, O.F., and E.W. Hospodar. 1951. Mounting of stained serial slices of the brain as wet specimens in transparent plastic. *Anat. Rec.* 110:1–15. Bibliography.

Mahoney, R. 1966. *Laboratory techniques in zoology.* Butterworths, Washington, DC. 404p. Wide ranging, but includes 21p. on fixation, embalming, injection, and corrosion-casting, and 10p. on cleaning skeletons (including enzyme digestion), degreasing, and mounting.

Ojeda, J.L., E. Barbosa, and P.G. Bosque. 1970. Selective skeletal staining in whole chicken embryos: a rapid Alcian Blue technique. *Stain Technol.* 45:137–138.

Russell, W.C. 1947. Biology of the dermestid beetle with reference to skull cleaning. *J. Mammal.* 28:284–287. A classic reference.

Stohler, R. 1945. Preparation of shark chondrocronia for class use. *Science* 102:403–404.

Tompsett, D.H. 1970. *Anatomical techniques.* 2nd ed. E. & S. Livingston, Ltd., Edinburgh and London. 420p. Many techniques including the staining of brain slices with ferric ferrocyanide.

Tucker, J.L., Jr., and E.T. Krementz. 1957. Anatomical corrosion specimens. *Anat. Rec.* 127(4):655–676. A vinyl resin technique.

Wagstaffe, R., and J.H. Fidler. 1968. *The preservation of natural history specimens, vol. II, pt. 2. Zoology: vertebrates.* H.F. and G. Witherby, Ltd., Philosophical Library of London. 404p. 115p. on vertebrates; very good on field techniques and preparation of study skins with notes on animals of all sizes and shapes; skeletal preparation dated and brief.

Wassersug, R.J. 1976. A procedure for differential staining of cartilage and bone in whole formalin-fixed vertebrates. *Stain Technol.* 51(2):131–134.

Watson, A.G. 1977. *In toto:* Alcian Blue staining of the cartilaginous skeleton in mammalian embryos. *Anat. Rec.* 187:743.

Williams, T.W., Jr. 1941. Alizarin Red S and Toluidine Blue for differentiating adult or embryonic bone and cartilage. *Stain Technol.* 16:23–25.

Williams, S.L., R. Laubach, and H.H. Genoways. 1977. *A guide to the management of recent mammal collections.* Carnegie Museum of Natural History, Special Publication no. 4. Primarily devoted to the acquisition, cataloging, storage, maintenance, and utilization of specimens for large research collections, but includes material (with references) on "bugging," fumigation, and fat solvents.

Zycherman, L.A., and J.R. Schrock. 1988. *A guide to museum pest control.* Assn. of Systematics Collections, Washington, DC. 205p.

Glossary

A-, Ab- Prefix meaning *without, from, away.*

Abducens (*away* + *to lead*). Denoting the sixth cranial nerve.

Abductor (*away* + *to lead*). A muscle that moves a part away from the sagittal plane, or separates two parts.

Acceleration. Rate of increase of velocity; force divided by mass.

Acetabulum (*vinegar cup*). The cup-shaped depression in the innominate bone that holds the head of the femur.

Acoelous (*without* + *hollow*). Said of a centrum that is more or less flat at each end; platyan.

Acousticolateralis (*listen* + *side*). Denoting the lateral line system plus the inner ear.

Acrodont (*extremity* + *tooth*). Having rootless teeth fused at their bases to the jawbone.

Acrosome (*tip* + *body*). The structure that caps the head of a sperm cell and functions in fertilization.

Actinopterygium (*ray* + *fin*). A fin having bony radials and no fleshy stalk or skeletal axis.

Ad-, af-, ag-, etc. Prefix meaning *toward.*

Adaptation (*to fit*). A structural or behavioral feature that contributes to the adjustment of an organism to its environment, usually favoring survival of the species through natural selection.

Adductor (*toward* + *to lead*). A muscle that moves a part toward the sagittal plane, or draws two parts together.

Adenohypophysis (*gland* + *under* + *growth*). The part of the hypophysis derived from the hypophyseal pouch.

Adhesion. The sticking together of dissimilar materials. The molecular attraction exerted between surfaces in contact.

Adrenal (*near* + *kidney*). An endocrine gland adjacent to the kidneys.

Aerobic metabolism. The derivation of energy by the complete combustion of foodstuffs, in the presence of oxygen, to form carbon dioxide and water.

Afferent (*toward* + *to bear*). Conducting toward or into.

Agnathan (*without* + *jaw*). Any jawless vertebrate.

Alar (*wing*). Winglike.

Allantois (*sausage* + *form*). The fetal membrane of amniotes that is derived from the hindgut and is functional in respiration and excretion.

Allometry (*different* + *to measure*). Analysis of the correlation between form and size.

Alula (*small wing*). The first digit of the bird wing together with its feathers.

Alveolus (*small cavity*). A small lobular cavity.

Ameloblast (*enamel* + *germ*). A cell that forms enamel.

Amnion (*fetal membrane*). The innermost fetal membrane of amniotes.

Above: 65 mm sunfish. Cleared, stained with Alizarin, and stored in glycerin.

Amniote. Those vertebrates whose embryos are surrounded by an amnion; reptiles, birds, and mammals.

Amphiarthrosis (*both* + *joint*). A joint allowing limited motion.

Amphicoelous (*both sides* + *hollow*). Said of a centrum that is concave at both ends.

Amphistyli (*both* + *pillar*). Suspension of the jaws from the chondrocranium in part directly and in part through the hyomandibula.

Ampulla (*flask*). A dilation of a canal, such as a semicircular canal of the inner ear.

Amygdala (almond). A cluster of nuclei derived from the archistriatum.

An-. Prefix meaning *without, not.*

Anaerobic metabolism. The derivation of energy by the incomplete combustion of foodstuffs in the absence of oxygen.

Analogy. Structural correspondence based on common function.

Anamniote (*without* + *small lamb*). Those vertebrates whose embryos have no amniotic membrane; fishes and amphibians.

Anapsid (*not* + *arch*). Having no opening in the temporal region of the skull.

Anastomosis (*coming together*). A communication between two blood vessels.

Androgen (*male* + *produce*). Any male sex hormone.

Angle of attack. Angle between the chord of an airfoil (or waterfoil) and the oncoming airstream (or waterstream).

Animal pole. The region of the egg where the nucleus is located and metabolic activity is highest.

Ant-, anti-. Prefix meaning *against, opposite.*

Antagonist (*against* + *fight*). A muscle the action of which opposes that of another.

Anticlinal (*against* + *to lean*). Said of a thoracic vertebra having its neural spine transitional between backward-leaning and forward-leaning.

Antler (*before* + *eye*). The bony, deciduous outgrowth from the head of deer.

Aorta (*to lift*). The main arterial trunk.

Apomorphic character (*away from* + *form*). A character that has changed from its former condition; a derived character.

Aponeurosis (*away from* + *tendon*). A tough flat sheet of connective tissue serving to distribute the tension of a muscle.

Apophysis (*away from* + *growth*). A process of a vertebra. Specific processes are indicated by a prefix.

Arachnoid (*spiderlike*). The middle meninx of the brain of mammals.

Arch-. Prefix meaning *first, primitive, ancestral, chief, ruler.*

Archenteron (*first* + *gut*). The embryonic digestive tube.

Archicortex (*primitive* + *bark*). The more medial part of the cerebral cortex.

Archinephros (*ancestral* + *kidney*). A hypothetical ancestral kidney that develops from all of the nephrotome; holonephros.

Archipterygium (*ancestral* + *fin*). A fin having a fleshy stalk and a central skeletal axis flanked by radials on both sides.

Archistriatum (*primitive* + *striped*). The more ventral of the basal ganglia.

Artery. A large vessel that carries blood away from the heart.

Aspect ratio. The ratio of span to average chord of a waterfoil or airfoil.

Aspidin (*shield*). Acellular bone.

Atrium (*entrance hall*). A cavity; the division of the heart between the sinus venosus and ventricle.

Auditory (*to hear*). Relating to hearing.

Autonomic (*self* + *law*). Relating to the part of the peripheral nervous system that supplies visceral motor nerves to vascular, nutritive, reproductive, and some other involuntary organs.

Autostyli (*self* + *pillar*). Suspension of the jaws directly from the chondrocranium without participation of the hyomandibula.

Axon (*axis*). The process of a neuron that transmits impulses away from the nerve cell body.

Baleen (*whale*). Horny plates in the mouths of toothless whales that filter food from the water.

Basal membrane. The thin membrane that separates epidermis from dermis.

Basal metabolic rate. The minimum rate at which homeotherms burn energy.

Basal nuclei. The complex of ganglia in the brain of mammals that corresponds to the corpus striatum.

Bi-. Prefix meaning *two*, *twice*, *double*; *life*.

Blade element theory. An interpretation of powered swimming or flying based on the integration of the different force vectors acting on the different parts of the propulsor.

Blastocoel (*germ* + *hollow*). The cavity of the blastula.

Blastocyst (*germ* + *bladder*). The blastula of mammals, which is characterized by a large blastocoel.

Blastoderm (*germ* + *skin*). The blastula derived from macrolecithal eggs, which consists of a disc of cells spread on the yolk.

Blastomere (*germ* + *part*). Any one of the cells into which the egg divides during cleavage.

Blastopore (*germ* + *opening*). The opening into the gastrocoel.

Blastula (*small germ*). The early embryo consisting of one tissue layer of several hundred cells.

Boundary layer. The region surrounding a swimming or flying object within which shearing forces occur; the region responsible for frictional drag.

Boundary lubrication. Lubrication having the number and pressure of the contacts between moving parts reduced by the intervention of a lubricant.

Brachial (*arm*). Relating to the arm.

Brachiation. Climbing by arm swinging.

Branchial (*gill*). Relating to gills.

Brev-. Prefix meaning *short*.

Bronchus (*windpipe*). An airway within the lung that is supported by cartilage.

Bunodont (*mound* + *tooth*). Having low cusps on the molar teeth, as for most omnivorous mammals.

Caecum (*blind*). A pouchlike extension from the digestive tract.

Calcar (*spur*). Cartilaginous rod supporting a flight membrane.

Camber. The front-to-back curvature of an airfoil; the transverse bowing of a wing.

Cantilever. A projecting beam or similar member that is supported at only one end.

Capillary (*relating to a hair*). A microscopic blood vessel through which diffusion takes place.

Capillary adhesion. The bonding of two surfaces as a consequence of the surface tension of a film of liquid that covers the contact surface.

Carapace (*hard covering*). A dorsal bony covering, as of a turtle or armadillo.

Cardiac (*heart*). Relating to the heart.

Cardinal (*red*). Relating to the primitive system of veins that drains the head, dorsal body wall, and kidney.

Carnassial. Denoting the teeth of carnivorous mammals that shear past one another: upper fourth premolor and lower first molar for living forms.

Carnivorous (*flesh* + *to eat*). Eating meat.

Carotid (*heavy sleep*). One of the large arteries of the neck.

Caudal (*tail*). Relating to the tail.

Cavernosus (*cavern*). Having internal cavities, as the erectile tissue of the penis.

Cement. The acellular, fibrous bone that joins teeth to their sockets.

Center of buoyancy. The point in an immersed body that represents the center of gravity of the displaced water; the point through which the resultant force of buoyancy acts.

Center of gravity. The point in a body through which the resultant force of gravity acts; that point from which a body can be suspended in equilibrium in any position.

Centrifugal (*center* + *to flee*). Said of a force that tends to impel a moving object outward away from a center of rotation.

Centripetal (*center* + *to approach*). Said of a force that tends to impel a moving object inward toward a center of rotation.

Centrum (*center*). The body of a vertebra.

Cephalic (*head*). Relating to the head.

Cephalization (*head* + *state of*). The tendency to concentrate neurosensory and feeding mechanisms in a head.

Ceratotrich (*horn* + *hair*). Slender, horny, unsegmented fin ray.

Cerebellum (*small brain*). The derivative of the dorsal part of the metencephalon.

Cerebrum (*brain*). The hemispheric derivatives of the telencephalon.

Cervical (*neck*). Pertaining to the neck.

Chiasma (*figure X*). A crossing of fibers, as of the optic nerves.

Chondrocranium (*cartilage* + *skull*). The part of the head skeleton, other than the splanchnocranium, that consists of cartilage or replacement bone.

Chondrocyte (*cartilage* + *cell*). A cartilage cell.

Chord. The straight-line distance between the leading and trailing edges of a waterfoil or airfoil in the line of travel.

Chordamesoderm. The roof of the archenteron, which induces the neural plate, and itself forms notochord and mesoderm.

Chorion (*skin*). The outermost fetal membrane of amniotes.

Choroid (*skinlike*). A vascular layer of the brain or eye.

Chromaffin (*color* + *affinity*). Denoting endocrine tissue that is functionally related to the adrenal medulla but is diffuse.

Chromatophore (*color* + *bear*). A pigment cell.

Ciliary (*eyelash*). Relating to a hairlike structure.

Clade (*branch*). A group of animals sharing uniquely evolved features and therefore common ancestry; a monophyletic group.

Cladistic (*branching*) **classification.** Classification that reconstructs phylogenetic sequences by deductive processes that analyze primitive and derivative character-states of related organisms to generate dichotomously branched sister groups.

Cladistics (*branching* + *pertaining to*). The field of taxonomy that ranks organisms into a succession of nesting, monophyletic sister groups in going to ever more inclusive levels of the evolutionary hierarchy.

Cladogram. A diagram of evolutionary relationships according to the principles of cladistic classification.

Cleavage. The cell divisions that convert the zygote to a blastula.

Cleidoic (*locked up*). Descriptive of the eggs of amniotes that can survive in air.

Clitoris. The sensitive female homolog of the male penis.

Cloaca (*sewer*). A common passageway for products of the digestive and urogenital systems.

Cochlea (*snail shell*). The spiraled auditory part of the inner ear of mammals.

Coelom (*hollow*). Any body cavity that is derived (in vertebrates) by splitting of the hypomere, and hence is lined by mesoderm.

Colic (*colon*). Relating to the colon.

Collagen (*glue* + *producer of*). The substance of collagenous fibers, which are present in all connective tissue.

Commissure (*connection*). A tract joining equivalent structures on the two sides of the central nervous system.

Compression. Stress in an elastic solid resulting from a load directed toward the object and perpendicular to its surface.

Conjunctiva (*to connect*). The membrane covering the front of the eyeball.

Conodonts (*cone* + *tooth*). Small toothlike fossils. Also, the animals that had these structures.

Conus. The most anterior of the primitive heart chambers.

Convergence. Evolutionary change in two or more lineages such that corresponding features that were formerly dissimilar became similar.

Coprodeum (*dung* + *way*). The dorsal part of a partially divided cloaca.

Coprophagy (*dung* + *to eat*). Feeding on dung.

Copulation (*to join*). The act that accomplishes internal fertilization.

Cornea (*horny*). The transparent superficial part of the eyeball.

Corona radiata (*crown* + *radiating*). The cells derived from the ovarian follicle that surround the egg at ovulation. The branching of the pyramidal tract in the cerebral hemispheres.

Corpus (*body*). Any mass or solid part of an organ.

Cortex (*bark*). The outer part of an organ.

Cosmine (*ornament*). Dentine characterized by internal tufts of radiating canals.

Crista (*crest*). Sensory cells of the ampullae of semicircular canals.

Crossopterygium (*fringe* + *fin*). A fin having a fleshy stalk and a skeletal axis flanked by radials on one side.

Ctenoid (*comblike*). Denoting a fish scale having a serrated margin.

Cupula (*small cask*). The gelatinous structure in which the sensors of a neuromast are embedded.

Cursorial. Adapted for running.

Cutaneous (*skin*). Relating to the skin.

Cuticle. A thin, noncellular, external covering of the skin of some animals.

Cycloid (*circle*). Denoting a fish scale having a smooth margin.

Cystic (*bladder*). Relating to a bladder or pouch.

De-. Prefix meaning *away, down, from.*

Decussation (*cross*). A tract joining unlike structures on the two sides of the central nervous system.

Deferent (*away* + *carrying*). A duct that carries away, as the sperm duct.

Delamination (*from* + *layering*). The formation of a tissue layer by the separation and subsequent aggregation of cells from a preexisting tissue layer.

Dendrite (*tree*). The processes of a neuron that receive and propagate impulses toward the nerve cell body.

Denticles (*small teeth*). Small toothlike structures that may either project from dermal armor and scales or occur independently as small scales.

Dentine. A type of dentinuous tissue, characteristic of tetrapods, that has cells external to the matrix.

Dentinous tissue. The material of teeth, scales, and armor that is harder than bone and softer than enamel. The type found in tetrapod teeth is termed dentine.

Derived character. A character that was relatively late to evolve in a monophyletic group.

Dermal bone. Membrane bone that ossifies in the integument.

Dermatocranium (*skin* + *skull*). The part of the head skeleton that consists of membrane bone.

Dermatome (*skin* + *to cut*). The outer division of the epimere.

Dermis (*skin*). The inner part of the skin, derived from mesoderm.

Diaphragm (*partition*). The muscular partition of mammals separating the pleural and peritoneal cavities.

Diaphragmatic (*partition*). Said of a thoracic vertebra having prezygapophyses that tend to face dorsally, but postzygapophyses that tend to face laterally.

Diaphysis (*through* + *growth*). The shaft of a long bone.

Diapsid (*two* + *arch*). Having two openings in the temporal region of the skull.

Diarthrosis (*two* + *joint*). A freely movable joint having a joint cavity.

Diencephalon (*through* or *between* + *brain*). The posterior derivative of the embryonic prosencephalon; becomes the anterior part of the brainstem.

Digitigrade (*finger* + *walking*). Having only the digits and distal ends of the metapodials in contact with the ground when standing or moving, as for cats and dogs.

Dioecious (*two* + *house*). Having the male gonads in one individual and the female gonads in another.

Diphycercal (*twofold* + *tail*). Said of a fish tail that is about symmetrical and has the spinal axis extending to its tip.

Diphyodont (*two* + *to grow* + *tooth*). Having two developmental sets of teeth.

Diplospondyly (*double* + *vertebra*). Having two vertebrae per primary body segment, in the caudal regions of certain fishes.

Distal. Located away from the central axis of the body.

Dorsal (*back*). Said of the back, or vertebral side of the body.

Drag. Resistance to the motion of a body through water or air.

Duodenum (*twelve*). The first segment of the small intestine.

Durophagous (*hard + to eat*). Eating that requires the breaking up of hard materials such as shells or nuts.

Dynamic soaring. Flight that is sustained by extracting energy from wind in a shear layer—usually over the ocean.

Dynamic strain similarity. Scaling of posture and behavior so that safety factors remain unchanged as body size changes.

Ec-. Prefix meaning *out, outside.*

Ectoderm (*outside + skin*). The outermost of the three embryonic germ layers.

Ectomesenchyme. Mesenchyme derived from neural crests.

Ectotherm (*outside + heat*). An animal that derives its body heat primarily from the external environment.

Efferent (*away + to bear*). Conducting away from or out of.

Elasmobranch (*plate + gill*). A fish having septal gills.

Elasticity. Capacity of a strained (or deformed) elastic solid to recover its original size and shape after a load is removed.

Elastic similarity. Scaling that adjusts the diameters of supporting members to masses so that sag and bending remain unchanged as body size changes.

Enamel. The exceedingly hard, acellular tissue of ectodermal origin that caps teeth, denticles, and some fish scales.

Enameloid. An enamel-like tissue derived from mesectoderm.

Endocrine (*within + separate*). Secreting into the blood stream.

Endoderm (*within + skin*). The innermost of the three embryonic germ layers.

Endolymph (*within + clear fluid*). The fluid within the labyrinth of the ear.

Endometrium (*within + uterus*). The soft glandular tissue that lines the uterus.

Endostyle (*within + pillar*). A mucous gland lying below the pharynx of lower chordates.

Endotherm (*within + heat*). An animal that maintains a high body temperature using metabolic heat and control over heat loss.

Enterocoely (*gut + hollow*). The formation of coelom by pouching from the archenteron.

Epaxial (*upon + center line*). Said of muscles of the trunk lying dorsal to the lateral septum.

Ependymal (*outer garment*). Relating to the cells that line the cavities of the central nervous system.

Epicercal (*upon + tail*). Said of a fish tail having the spinal axis extending into the dorsal, larger lobe.

Epidermis (*upon + skin*). The outer part of the skin, derived from ectoderm.

Epididymis (*upon + testes*). The coiled part of the sperm duct that is adjacent to the testis.

Epiglottis (*upon + laryngeal opening*). The valvelike closure of the glottis.

Epimere (*upon + part*). The somite, or segmented dorsal division of the embryonic lateral mesoderm.

Epimysium (*upon + muscle*). The membrane surrounding a muscle.

Epiphysis (*upon + growth*). A separate ossification forming the end of a long bone.

Epithelium (*upon + nipple*). A layer of cells covering a surface or lining a cavity.

Erythrocyte (*red + cell*). A red blood cell.

Estrogen (*mad desire + produce*). A female hormone responsible for secondary sexual characteristics.

Euryapsid (*broad + arch*). Having one opening in the temporal region of the skull that is bordered below by the postorbital and squamosal.

Evolutionary (*unrolling*) **classification.** Classification that reconstructs phylogenetic sequences by judgments based on all available data, including the fossil record, and that incorporates both linear and branching evolutionary processes.

Evolutionary trend. A gradual, adaptive change in the evolution of a feature within a phyletic line.

Ex-. Prefix meaning *out, beyond.*

Exocrine (*out + separate*). Denoting a gland that discharges into a duct.

Facial (*face*). Relating to the face.

Falciform (*sickle + shape*). Having the shape of a sickle.

Fascia (*band*). Fibrous connective tissue.

Fascicle (*small bundle*). A bundle of nerve or muscle fibers.

Faunivore (*animal life + to eat*). An animal that eats primarily animal food.

Fenestra (*window*). Any large opening, as in the innominate bone.

Fermentation (*yeast + process*). Enzymatic breakdown of an organic compound. Digestion of cellulose by enzymes provided by symbiotic bacteria and protozoa.

Fissure (*a split*). A cleft or groove, as on the ventral surface of the spinal cord.

Fluid film lubrication. Lubrication having the moving parts forced out of contact by a film of the lubricant.

Follicle (*small bag*). A structure having a cavity.

Force. The product of mass times acceleration; a push or pull that causes, or tends to cause, motion.

Fossorial. Highly adapted for digging.

Fovea (*pit*). Depression in the retina where the sharpest image is formed.

Free-body diagram. A drawing of an isolated mechanical system showing as vectors all external translational and rotational forces acting on the system.

Friction. Mechanical resistance to relative motion. Dry static friction is the product of the normal force and the coefficient of friction.

Frictional drag. Drag on a moving object caused by friction within the boundary layer.

Frontal. Said of planes that divide the body into dorsal and ventral parts.

Funiculus (*string*). A bundle or region of nerve cell fibers of the spinal cord.

Gait. A regularly repeating manner and sequence of moving the feet when walking or running.

Gamete (*spouse*). An egg or sperm.

Ganglion (*swelling*). An aggregate of nerve cell bodies—particularly when located outside of the central nervous system.

Ganoine (*brightness + made of*). Thick, lamellar enamel.

Gastralia (*belly*). Bony supports of the abdomen of some tetrapods.

Gastrocoel (*stomach + hollow*). The cavity of the gastrula.

Gastrula (*small stomach*). The early embryo consisting of two, and potentially three, tissue layers.

Gastrulation. The events that convert a blastula to a gastrula.

Geometric similarity. Scaling that retains unchanged body proportions as size changes.

Glide. A controlled descent in air at an angle of less than 45° to the horizontal.

Glomerulus (*small ball*). The tuft of capillaries within a renal capsule; any of several aggregates of nerve fibers.

Glossal (*tongue*). Relating to the tongue.

Glossopharyngeal (*tongue + throat*). Denoting the ninth cranial nerve.

Glottis. The opening from the pharynx into the larynx.

Gnathostome (*jaw + mouth*). Any vertebrate having jaws.

Gonad (*seed*). A sex gland; the ovary or testis.

Gonopodium (*seed + foot*). The copulatory organ on the anal fin of some male teleosts.

Graviportal (*heavy + to carry*). Adapted for supporting great body weight.

Ground effect. Energetic advantage gained by gliding or flying low over a flat surface.

Gyrus (*circle*). An elevated convolution of the cerebrum or cerebellum.

Habenula (*strap*). A small cluster of nuclei in the epithalamus.

Hemal arch. The part of certain caudal vertebrae that arches under the caudal vessels.

Hemibranch (*half + gill*). A gill bar bearing filaments on one surface only.

Hemipenes (*half + penis*). The paired male copulatory organs of lepidosaurs.

Hemopoiesis (*blood + create*). The production of blood cells.

Hepatic (*liver*). Relating to the liver.

Herbivorous (*plant + to eat*). Eating the leaves or stems of plants.

Hermaphrodite (*Hermes + Aphrodite*). An individual having both male and female sex organs.

Heterocercal (*different + tail*). Said of a fish tail having the spinal axis extending into the dorsal, larger lobe.

Heterochrony (*different + time*). Change in the timing of developmental events relative to an ancestor.

Heterocoelous (*different + hollow*). Said of a centrum having saddle-shaped ends, as in birds.

Heterodont (*different + tooth*). Having several kinds of teeth.

Hippocampus (*sea horse*). A derivative of the archicortex in the shape of an arching band impinging on the lateral ventricle.

Holoblastic (*whole + germ*). Said of cleavage that is total, that is, that divides the entire egg.

Holobranch (*whole + gill*). A gill bar bearing filaments on both anterior and posterior surfaces.

Holonephros (*whole + kidney*). A hypothetical ancestral kidney that develops from all of the nephrotome.

Homeobox. A cluster of *Hox* genes.

Homeotherm (*same + heat*). An animal that maintains a nearly constant body temperature.

Homocercal (*same + tail*). Said of a fish tail having dorsal and ventral lobes of about the same size, and both extending beyond the spinal axis.

Homodont (*same + tooth*). Having one functional kind of teeth.

Homology (*same + ratio*). Structural correspondence based on common ancestry.

Homoplasy (*equal + molding*). Similarity of structure between parts of different organisms that is not due to homology.

Hormone (*excite*). A chemical released in one part of the body, transported by the circulatory system, and causing a response in another part of the body.

Hox gene. A homeotic gene. A regulatory gene that controls such fundamental developmental processes as the establishment of polarity, segmentation, and appendages.

Hyoid (*U-shaped*) **arch.** The second visceral arch.

Hyomandibula (*U-shaped + jaw*). The dorsal, and principal, segment of the hyoid arch.

Hyostyli (*U-shaped + pillar*). Suspension of the jaws from the chondrocranium primarily through the hyomandibula.

Hypaxial (*under + center line*). Said of muscles of the trunk lying ventral to the lateral septum.

Hyper-. Prefix meaning *above, beyond, over.*

Hypobranchial (*under + gill*). Muscles of the throat derived phylogenetically from hypaxial musculature.

Hypocercal (*under* + *tail*). Said of a fish tail having the spinal axis extending into the ventral, larger lobe.

Hypoglossal (*below* + *tongue*). Denoting the twelfth cranial nerve.

Hypomere (*under* + *part*). The unsegmented ventral division of the embryonic lateral mesoderm.

Hypophysis (*under* + *grow*). An endocrine gland lying below the hypothalamus; the pituitary.

Hypsodont (*high* + *tooth*). Having cheek teeth with high crowns, as for ungulates and some rodents.

Impedence. The ratio of pressure to volume displacement at a given surface in a sound-transmitting medium.

Induced drag. The drag at the ends of an airfoil that is a consequence of the difference in pressure above and below the airfoil.

Induction. The effect of one embryonic part on another through a chemical stimulus.

Inertia. The tendency of an object to remain at rest or in uniform motion in a straight line unless acted upon by external forces.

Infra-. Prefix meaning *below, under.*

Infundibulum (*funnel*). The ventral outgrowth of the diencephalon that forms the neurohypophysis.

Inner cell mass. The inner part of the mammalian blastocyst from which the embryo is derived.

Insectivorous (*insect* + *to eat*). Eating insects and similar food.

Insulin (*island*). A hormone of the pancreas.

Intercentrum (*between* + *center*). The more anterior unit of the centrum of certain labyrinthodonts.

Invagination (*in* + *sheath*). The folding of tissue from the vegetal pole of the blastula inward to establish the archenteron of the gastrula.

Involution (*to roll up*). The migration of cells into the gastrula at the blastopore.

Isometric (*equal* + *measure*). Muscle contraction without shortening.

Isotonic (*equal* + *strain*). Muscle contraction without change of tension.

Jugular (*throat*). Relating to the throat or neck.

Keratin (*horn*). A hard, nearly insoluble protein or albuminoid, present in the epidermis and some of its derivatives.

Kinematic similarity. Scaling of the lengths and periods of oscillating parts so that joint angles remain unchanged as body size changes.

Kinetic. Relating to motion. Said of tetrapod skulls having several somewhat movable units.

Kinetic energy. Energy derived from motion.

Kinocilium (*move* + *eyelid*). The longest and most complex hairlike sensor of a neuromast cell.

Labyrinth (*tortuous passage*). The membranous structure of the inner ear.

Labyrinthodont (*tortuous passage* + *tooth*). Having teeth with complicated patterns of infolded enamel on their sidewalls.

Lagena (*flask*). An extension from the saccule of the inner ear.

Lamella (*thin plate*). A thin membrane or layer.

Laminar flow. Passage of water or air over a moving body without the formation of eddies or turbulence.

Larynx. The cartilaginous box at the anterior end of the trachea.

Lepidotrich (*scale* + *hair*). Bony, segmented fin ray.

Leucocyte (*white* + *cell*). A white blood cell.

Lever. A rigid structure that transmits forces by turning, or tending to turn, at a pivot.

Lever arm. The perpendicular distance between the line of action of an applied force (*or component of such force*) and the associated pivot.

Lift. The force generated by an airfoil (or waterfoil) that is at right angles to the oncoming airstream (or waterstream).

Ligament (*band*). A cord or sheet of connective tissue serving to join two or more skeletal parts.

Lingual (*tongue*). Relating to the tongue.

Load. Any burden or force applied to a solid object.

Lumbar (*of the loins*). Pertaining to the region of the back between the ribs and the pelvis.

Lymph (*clear fluid*). The fluid of the lymphatic system and tissue spaces.

Machine. A mechanism that transmits force from one place to another, usually changing its magnitude and for a useful purpose.

Macrolecithal (*large* + *yolk*). Said of large eggs having much yolk.

Macula (*spot*). Sensory epithelium of the saccule or utricule.

Mammary (*breast*). Relating to the milk glands.

Mandibular (*jaw*) **arch.** The first visceral arch.

Mass. The numerical measure of an object's inertia, or resistance to being accelerated; the quantity of matter an object contains.

Mediastinum (*medial*). The septum in mammals that separates right and left pleural cavities.

Medulla (*pith*). The inner part of an organ; the posterior part of the brainstem.

Melanophore (*black* + *to bear*). A pigment cell that contains the black pigment melanin.

Membrane bone. Bone that ossifies directly without replacing cartilage.

Meninx (*pl.* **meninges**) (*membrane*). Membranous envelope surrounding the central nervous system.

Meniscus (*small moon*). A pad of fibrous cartilage located within a joint capsule, as at the knee.

Meroblastic (*part + germ*). Said of partial cleavage, that is, that does not penetrate the yolk mass.

Mesectoderm. Mesenchyme derived from neural crests.

Mesencephalon (*middle + brain*). The middle primary brain vesicle; the midbrain.

Mesenchyme (*middle + infusion*). Embryonic connective tissue composed of branched, loosely organized cells, often with the capacity to migrate.

Mesentery (*middle + gut*). A membrane that supports an internal organ from the body wall.

Mesoderm (*middle + skin*). The middle one of the three embryonic germ layers.

Mesolecithal (*middle + yolk*). Said of eggs having a moderate amount of yolk.

Mesomere (*middle + part*). The small middle division of the embryonic lateral mesoderm.

Mesonephros (*middle + kidney*). The functional kidney of fetal amniotes, which develops from a middle part of the nephrotome.

Mesorchium (*middle + testis*). A mesentery that supports the testis.

Mesovarium (*middle + ovary*). A mesentery that supports the ovary.

Metamere (*after + part*). One of serially repeated structural units along the body axis.

Metamorphosis (*to transform*). Acclerated development that is preceded and followed by relative developmental quiescence and that transforms a larva into an adult.

Metanephridium (*later in time + kidney*). An excretory organ with tubules having at one end cells (podocytes) specialized for filtration from the blood and discharging at the other end into coelom derivatives.

Metanephros (*after + kidney*). The adult kidney of amniotes, which develops from a short posterior part of the nephrotome.

Metencephalon (*after + brain*). The anterior derivative of the embryonic rhombencephalon; becomes the cerebellum and pons.

Microlecithal (*small + yolk*). Said of small eggs having little yolk.

Modulus of elasticity. Stress divided by strain; a measure of the deformation caused per unit load.

Momentum. The capacity of an object moving in a straight line to overcome resistance; the product of mass times velocity.

Monophyletic group. One that includes a common ancestor and all of its descendants.

Morphocline. An evolutionary trend.

Morphology (*form + science*). The science of relating and interpreting observed structures.

Mucosa (*juice*). A tissue that contains or secretes mucus.

Mucus (*juice*). A clear, slippery secretion.

Myelencephalon (*spinal cord + brain*). The posterior derivative of the embryonic rhombencephalon; becomes the medulla.

Myelin (*marrow*). The fatty sheath of a nerve fiber.

Myocardium (*muscle + heart*). The muscle layer of the heart.

Myocoel (*muscle + hollow*). The transitory cavity of the myotome.

Myomere (*muscle + part*). The axial muscle of one body segment.

Myometrium (*muscle + uterus*). The muscular part of the uterus.

Myoseptum (*muscle + barrier*). The partition of connective tissue separating myomeres.

Myotome (*muscle + to eat*). The middle division of the epimere.

Neocortex (*new + bark*). The medial, and in mammals the largest part of the cerebral cortex.

Neostriatum (*new + striped*). The more dorsal of the basal ganglia.

Neoteny (*recent + to stretch*). The retardation of the development of a somatic feature so that it remains juvenile in later developmental stages.

Nephric (*kidney*). Relating to the kidney.

Nephridium (*kidney*). Any tubular excretory organ.

Nephrocoel (*kidney + hollow*). The cavity of the mesomere.

Nephron (*kidney*). The functional unit of a kidney.

Nephrostome (*kidney + mouth*). A ciliated opening leading from the coelom into an excretory tubule.

Nephrotome (*kidney + to cut*). The nephrogenic part of the mesomere.

Neural arch. The part of the vertebra that arches over the spinal cord.

Neural crests. Aggregates of cells, derived from ectoderm, that flank the embryonic neural tube before migrating to form skeletal and other structures.

Neural folds. The longitudinal folds that flank the neural plate and arch together during neurulalation.

Neural plate. The thickened part of the ectoderm, lying over the chordamesoderm, which will form the central nervous system.

Neural tube. The embryonic central nervous system.

Neurilemma (*nerve + sheath*). The thin sheath that surrounds the myelin of a nerve fiber or, if the fiber is unmyelinated, the axon cylinder.

Neurocoel (*nerve + hollow*). The cavity of the neural tube.

Neurocranium (*nerve + skull*). The part of the skull derived from the chondrocranium.

Neuroglia (*nerve + glue*). The supportive tissue of the central nervous system.

Neurohypophysis (*nerve + under + to grow*). The part of the hypophysis derived from the infundibulum.

Neuromast (*nerve* + *hill*). One of the hair-cell organs of the inner ear or lateral line system.

Neuron (*nerve*). A nerve cell.

Neurotransmitter. A chemical, released by nerve endings, that modulates the firing of other neurons.

Neurulation. The process that converts the neural plate into a neural tube.

Notochord (*back* + *chord*). The fibrocellular rod that forms the skeletal axis of embryonic, and some adult, vertebrates.

Nuchal (*neck*). Relating to the neck.

Nucleus (*kernel*). The inner part, as of a cell; an aggregation of nerve cell bodies within the central nervous system.

Octavolateralis (*eight* + *side*). Denoting the lateral line system plus the inner ear.

Oculomotor (*eye* + *motion*). Denoting the third cranial nerve.

Odontoblast (*tooth* + *germ*). A cell that produces dentine.

Odontode (*tooth* + *like*). An isolated superficial tooth or scalelike structure, derived from a single dermal papilla, consisting of dentinous tissue and sometimes capped by enamel or enameloid.

Olfactory (*to smell*). Relating to the sense of smell.

Omasum (*paunch, tripe*). A muscular part of the ruminant stomach.

Omentum (*membrane*). A membrane that joins one internal organ to another.

Omnivorous (*all* + *to eat*). Eating a variety of plant and animal food.

Ontogeny (*a being* + *become*). The development of the individual from fertilized egg to adult.

Oogonia (*egg* + *generation*). The proliferating cells of the ovary that will become ova.

Operculum (*cover*). The flap covering the gills of actinopterygians and holocephalians.

Ophthalmic (*eye*). Relating to the eye.

Opisthocoelous (*behind* + *hollow*). Said of a centrum that is concave posteriorly and convex anteriorly.

Opisthonephros (*behind* + *kidney*). The adult kidney of anamniotes, which develops from all or most of the nephrotome posterior to the pronephros.

Optic (*vision*). Relating to the eye.

Oral (*mouth*). Relating to the mouth.

Oscillatory propulsion. Propulsion of a swimmer resulting from back-and-forth paddling or flapping motions of paired appendages.

Os cornu (*bone* + *horn*). A permanent, bony horn core.

Ossicone (*bone* + *cone*). Giraffe horn; a permanent, skin-covered, bony projection.

Osteoblast (*bone* + *bud*). A cell that deposits bone.

Osteoclast (*bone* + *to break*). A cell that resorbs bone.

Osteocyte (*bone* + *cell*). A bone cell.

Osteoderm (*bone* + *skin*). A bone in the dermis of some reptiles.

Osteon. A cylinder-within-cylinder structural unit of dentine or bone.

Ostium (*mouth*). A small opening into a duct or space.

Ostracoderm (*shell* + *skin*). An agnathous vertebrate having bony armor or scales.

Otolith (*ear* + *stone*). A calcified body within the inner ear.

Outgroup. In cladistics, organisms related to, but not part of, a group under study with which comparisons are made to identify shared derived characters.

Ovary (*egg*). The female gonad.

Oviduct (*egg* + *duct*). The duct that conveys eggs to the cloaca or uterus.

Oviparous (*egg* + *to bear*). Animals that lay eggs.

Ovisac (*egg* + *sac*). A part of the reproductive tract where eggs are retained, but not nourished, prior to laying.

Ovoviviparous (*egg* + *alive* + *to bear*). Animals that retain their eggs in the body until the time of hatching, yet do not nourish their embryos while in the ovisac or uterus.

Ovum (*egg*). The egg cell

Paedomorphosis (*child* + *form*). The retention of ancestral juvenile characters by later developmental stages or descendants.

Palatoquadrate (*palate* + *squared*). The dorsal segment of the mandibular arch.

Paleocortex (*ancient* + *bark*). The more lateral part of the cerebral cortex.

Paleostriatum (*ancient* + *striped*). The more central of the basal ganglia.

Parachute. A partially controlled fall at an angle of more than 45° to the horizontal.

Paraganglia (*beside* + *knot*). Chromaffin tissue that is adjacent to sympathetic nerve ganglia.

Parallelism. Evolutionary change in two or more lineages such that corresponding features undergo equivalent alterations without becoming markedly more or less similar.

Paramesonephric (*beside* + *kidney*). Denoting the primordium of the female reproductive tract.

Paraphyletic group. One that includes a common ancestor and some, but not all, of its descendants.

Parasympathetic (*beside* + *with* + *suffering*). Denoting the part of the autonomic nervous system having craniosacral outflow.

Parathyroid (*near* + *shield* + *form*). An endocrine gland near the thyroid gland that controls calcium and phosphate metabolism.

Parietal (*wall*). Relating to an outer wall, as of the chest or skull.

Parthenogenesis (*virgin* + *origin*). Development of the embryo without fertilization.

Patagium (*a border*). A membrane of skin used as an airfoil in flying or gliding.

Pecten (*comb*). The nutritive organ within the eyeball of birds.

Peduncle (*small foot*). One of the tracts supporting the cerebellum. The constriction in front of a fish tail.

Penis. The male organ of copulation of amniotes.

Pericardial (*around* + *heart*). Surrounding the heart.

Perichondrium (*around* + *cartilage*). The fibrous membrane covering cartilage.

Perilymph (*around* + *clear fluid*) The fluid surrounding the labyrinth of the ear.

Periosteum (*around* + *bone*). The fibrous membrane covering bone.

Peritoneal (*around* + *stretch*). Relating to the cavity containing the digestive viscera.

Phagocyte (*eat* + *cell*). A cell capable of ingesting foreign matter.

Pharynx (*throat*). The part of the gut between the mouth and esophagus.

Pheromone (*to carry* + *excite*). A chemical, produced by an animal, that elicits a behavioral response in a conspecific.

Photophore (*light* + *to bear*). A light-producing cell.

Phyletic line. A lineage that is relatively continuous and complete in the fossil record.

Phylogeny (*a race* + *become*). Evolutionary history of a lineage.

Physoclistous (*bladder* + *closed*). Having a gas bladder not joined to the gut by a duct.

Physostomous (*bladder* + *mouth*). Having a lung or gas bladder joined to the gut by a duct.

Piezoelectric effect (*pressure* + *electricity*). Electric polarity resulting from the deformation of a crystalline material.

Pineal (*pine cone*). A glandular outgrowth of the epithalamus.

Pinnate (*feather*). Said of muscles having their fibers sloping in toward one or more central tendons.

Piscivorous (*fish* + *to eat*). Eating fish.

Pitch. Rotation of a swimmer or flyer around its transverse axis.

Pituitary (*slime*). An endocrine gland lying below the hypothalamus; the hypophysis.

Placenta (*small flat cake*). An organ of fetal and maternal tissues that are associated for physiological exchange between the respective blood streams.

Placode (*flat round plate*). A local thickening of embryonic ectoderm.

Plantigrade (*sole* + *walking*). Having the sole of the foot in contact with the ground when standing or moving, as for man and bears.

Plasma (*something formed*). The acellular component of blood or lymph.

Plastron (*breastplate*). The ventral part of a turtle shell.

Plate tectonics. The concept that large areas of the earth's crust float independently on the underlying molten mantle.

Platyan (*flat*). Said of a centrum that is more or less flat at each end; acoelous.

Plesiomorphic character. A primitive or ancestral character.

Pleural (*side*). Relating to the wall of the thoracic cavity.

Pleurocentrum (*side* + *center*). The paired, more posterior units of the centrum of certain labyrinthodonts, and any centrum derived therefrom.

Pleurodont (*side* + *tooth*). Having rootless teeth that are connected to the jawbone on their outer surfaces.

Plexus (*a braid*). An interweaving of nerves or vessels.

Podocyte (*foot* + *cell*). Excretory cell, having fingerlike projections, that is adapted for ultrafiltration.

Poikilotherm (*changeful* + *heat*). An animal that does not maintain a constant body temperature.

Polyphyodont (*many* + *to grow* + *tooth*). Having many successive developmental sets of teeth.

Pons (*bridge*). The ventral derivative of the metencephalon.

Pontine (*bridge*). Relating to the pons of the brain.

Portal (*gate*). Said of a venous circuit that joins two capillary beds.

Potential energy. Energy derived from position, either in relation to gravity or elastic loading.

Preadaptation. Evolutionary change of function for a given structure with minimal change of form.

Prehensile. Adapted for grasping by wrapping around.

Pressure. Force per unit area.

Pressure drag. Drag on a moving object caused by displacement of the fluid medium, backflow, and the formation of pressure gradients.

Primitive character. A character that was present early in the evolution of a monophyletic group.

Primitive streak. The structure of amniotes that forms most of the embryonic dorsolateral mesoderm by the migration and involution of cells.

Procoelous (*in front* + *hollow*). Said of a centrum that is concave anteriorly and convex posteriorly.

Proctodeum (*anus* + *way*). The posterior part of a partially divided cloaca, derived from ectoderm.

Profile drag. All the drag on a flying or gliding object other than that associated with the ends of the wings.

Progenesis (*before* + *be born*). Paedomorphosis produced by precocious sexual maturation of an organism that is still in a morphologically juvenile stage.

Progesterone (*before* + *to bear*). A female hormone that maintains pregnancy.

Pronator (*to bend forward*). A muscle that rotates the palms or soles downward.

Pronephros (*in front* + *kidney*). The most anterior of vertebrate kidneys and the first to develop in ontogeny.

Proprioceptor (*one's own* + *take*). An organ in muscle, tendon, or joint that senses tension.

Prosencephalon (*in front + brain*). The most anterior of the three primary brain vesicles; the forebrain.

Protonepridium (*first + kidney*). An excretory organ having tubules that are blind at one end and discharge outside the body at the other end.

Proximal. Located toward the central axis of the body.

Pterygiophore (*fin + bearer of*). The skeletal supports of dorsal and anal fins.

Pulmonary (*lung*). Relating to the lungs.

Putamen (*shell*). A lateral part of the mammalian neostriatum.

Pygostyle (*rump + pillar*). The bladelike bone forming the posterior end of the spine of birds.

Pyloric (*gatekeeper*). Relating to the part of the stomach adjacent to the intestine.

Ram feeding. The taking of small suspended food particles into a large mouth held open as a swimmer moves forward.

Recapitulation. The repetition of ancestral adult stages in embryonic or juvenile stages of descendants.

Rectum (*straight*). The terminal segment of the large intestine.

Renal (*kidney*). Relating to the kidney.

Replacement bone. Bone that replaces cartilage as it ossifies.

Resultant. The single vector that is equivalent to a given set of vectors.

Rete (*network*). A network of small vessels or fibers.

Reticular (*net*). Netlike in structure.

Reynolds number. The ratio of the inertia of a fluid medium to its viscosity. For fast-swimming vertebrates this dimensionless number ranges from $10^{5.5}$ to $10^{8.5}$, depending on length and velocity.

Rhinal (*nose*). Relating to the nose.

Rhombencephalon (*rhomboid + brain*). The most posterior of the three primary brain vesicles; the hindbrain.

Ricochet. A bipedal hopping gait using the hind legs in unison, as for the kangaroo.

Roll. Rotation of a swimmer or flyer around its horizontal (anteroposterior) axis.

Rumen (*throat*). The largest chamber of the ruminant stomach.

Ruminate. To chew the cud.

Saccule (*small sack*). The more ventral chamber of the inner ear.

Sacral (*sacred*). Pertaining to the region where the spine articulates with the pelvic girdle.

Sagittal. Said of planes that divide the body into right and left parts.

Saltatorial. Adapted for jumping or hopping.

Sarcolemma (*flesh + husk*). The membrane enclosing a muscle fiber.

Sarcomere (*flesh + part of*). The unit of contraction of striated muscle.

Sarcopterygium (*flesh + fin*). A fin having a fleshy stalk.

Scaling (*staircase*). The graduated relationship between body size and form.

Scansorial. Adapted for climbing.

Schizocoely (*split + hollow*). The formation of coelom by splitting of the hypomere.

Sclera (*hard*). The tough outer envelope of the eye.

Sclerotome (*hard + to cut*). The inner division of the epimere.

Scute (*shield*). Any large, flat, horny plate on the surface of a reptile.

Sebaceous (*tallow*). Relating to oil or fat.

Secondary palate. A palate that separates the respired airstream from the mouth cavity.

Serial homology. The correspondence of structures that occupy different spatial positions in a series of like structures.

Sesamoid bone (*resembling a sesame seed*). A bone embedded in, or interrupting, a tendon.

Sexual homology. The correspondence of male and female structures that develop from identical embryonic primordia.

Shear. Stress in an elastic solid resulting from loads directed in opposite directions along parallel, closely adjacent lines.

Sinus (*cavity or hollow*). A cavity in an organ or tissue.

Sinusoid (*cavity or hollow*). An expanded capillary, as found in the liver and certain glands.

Sinus venosus. The most posterior of the primitive heart chambers.

Sister group. In cladistics, either of the two lineages established by one dichotomous evolutionary branching.

Solenocyte (*channel + cell*). Excretory cell having a flagellum surrounded by a sheath formed by slender filaments.

Somatic (*body*). Relating to parts of the body other than the viscera.

Somatopleure (*body + side*). A membrane derived from ectoderm and the outer sheet of the hypomere.

Somite (*body + part*). The epimere, or segmented dorsal division of the embryonic lateral mesoderm.

Somitomere (*body + part*). A thickening in the embryonic, continuous, dorsolateral mesoderm. The precursor, in the trunk but not the head, of a discrete somite.

Specialized character. A feature that has become modified to perform a restricted function, often with great effectiveness.

Species (*kind or sort*). A group of actually or potentially interbreeding natural populations that is reproductively isolated from other such groups.

Sperm (*seed*). The male gamete.

Spermatogonia (*seed + generation*). The proliferating cells of the testis that will become sperm.

Sphincter (*to bind*). A flat, washer-shaped muscle that restricts an orifice.

Spiracle (*breathing hole*). An opening into the pharynx derived from the first gill cleft.

Splanchnocranium (*viscera* + *skull*). The skeleton of the visceral arches.

Splanchnopleure (*viscera* + *side*). A membrane derived from endoderm and the inner sheet of the hypomere.

Stall. Loss of lift by an airfoil.

Standard metabolic rate. The minimum rate at which resting vertebrates burn energy.

Static soaring. Flight that is sustained by updrafts.

Statoacoustic (*standing* + *hearing*). Denoting the eighth cranial nerve.

Stereocilia (*solid* + *eyelid*). The numerous hairlike sensors of a neuromast cell.

Sternum (*chest*). The breastbone.

Stomodeum (*mouth* + *road*). Anterior invagination of the ectoderm that comes to line the mouth.

Strain. Deformation of an elastic solid that is caused by a load.

Stratum. A tissue layer.

Streaming potential. The electric potential that occurs when an ionized fluid flows through channels in a solid that has charged surfaces.

Strength. Capacity of supportive material to resist force without breakage or permanent deformation; capacity of muscle to produce much force.

Stress. Pressure within an elastic solid that results from strain.

Stroma (*bed*). The connective tissue framework of an organ, as of the ovary.

Subclavian (*under* + *collar bone*). Located in the shoulder.

Sulcus (*furrow*). A groove, as on the dorsal surface of the spinal cord or cerebrum.

Super-, supra-, sur-. Prefix meaning *over, above*.

Supinator (*to bend backward*). A muscle that rotates the palms or soles upward.

Suture (*to sew*). The union, or seam, between the bones at an immovable joint.

Symmorphosis (*together* + *shaping*). The concept that the structure of an organism matches, but does not exceed, its functional requirement.

Sympathetic (*with* + *suffering*). Denoting the part of the autonomic nervous system having thoracolumbar outflow.

Symphysis (*together* + *growth*). An amphiarthrosis having a pad of collagenous fibers or fibrous cartilage separating the bones.

Symplesiomorphy (*together* + *near* + *form*). The sharing of primitive characters by descendant groups.

Synapomorphy (*together* + *separate* + *form*). The sharing of derived characters by descendant groups.

Synapse (*junction*). The junction of nerve with nerve or nerve with muscle for the transmission of an impulse.

Synapsid (*together* + *arch*). Having one opening in the temporal region of the skull that is bordered above by the postorbital and squamosal.

Synarthrosis (*together* + *joint*). An immovable joint.

Syndesmosis (*together* + *tie*). An amphiarthrosis having a band of collagenous fibers between the bones that allows moderate motion.

Synergist (*together* + *work*). A muscle having about the same action as another.

Synovial. Relating to the viscid lubricating fluid occurring within joint capsules and tendon sheaths.

Synsacrum (*together* + *sacred*). The unit of the avian skeleton formed by the fusion of numerous vertebrae.

Syrinx (*tube*). The vocal organ of birds, located near the bifurcation of the primary bronchi.

Systemic. Relating to the parts of the circulatory system that are not associated with lungs or gills.

Taxon (*pl.* taxa) (*to arrange*). A group of organisms recognized as a unit in classification.

Tectum (*roof*). The roof of the midbrain.

Telenecephalon (*end* + *brain*). The anterior derivative of the embryonic prosencephalon; becomes the cerebrum.

Telolecithal (*end* + *yolk*). Said of eggs having the yolk massed toward the vegetal hemisphere.

Tendon (*cord*). A tough cord or band of connective tissue serving to join a muscle to a bone.

Tension. Stress in an elastic solid resulting from a load directed away from the object and perpendicular to its surface.

Testis (*testicle*). The male gonad.

Thalamus (*bed*). A lateral wall of the diencephalon.

Thecodont (*sheath* + *tooth*). Having teeth rooted in sockets.

Thermal. A doughnut-shaped bubble of circulating warm air.

Thoracic (*chest*). Pertaining to the chest, or region of the ribs.

Thrombocyte (*clot* + *cell*). A kind of blood cell associated with the clotting process.

Thymus (*sweetbread*). A gland in or near the neck that relates to the immune system.

Thyroid (*shield* + *form*). An endocrine gland in the neck; a cartilage of the larynx.

Torque. Moment or turning force; the product of a force times the perpendicular distance between its line of action and the pivot around which an object turns, or tends to turn, as a consequence of its application.

Trabecula (*small beam*). A small rod, bar, or support; a certain part of the chondro-cranium; a spicule of bone.

Trachea (*windpipe*). The airway from the larynx to the bronchi.

Transverse. Said of planes that divide the body into anterior (head end) and posterior (tail end) parts.

Trigeminal (*three* + *originating* + *together*). Denoting the fifth cranial nerve.

Trochlear (*pulley*). Relating to the fourth cranial nerve, or to the bony grooves within certain joints.

Trophoblast (*nourish* + *germ*). The outer wall of the mammalian blastocyst.

Truss. An assemblage of solid parts arranged to form a rigid framework.

Tunica (*coat*). A surrounding layer, as of a testis or blood vessel.

Tympanum (*drum*). The eardrum.

Undulatory propulsion. Propulsion of a swimmer resulting from the passage of traveling waves along the body or median fins.

Unguis (*hoof*). The tough, outer or lateral material of a hoof, claw, or nail.

Unguligrade (*hoof* + *walking*). Having only hoofs in contact with the ground when standing or moving, as for horses and deer.

Ureter (*urinary canal*). The duct of a metanephric kidney.

Urethra (*canal*). The duct that discharges urine from the bladder.

Urodeum (*urine* + *way*). The ventral part of a partially divided cloaca.

Urophysis (*tail* + *grow*). Ventral swelling on the caudal part of the spinal cord of many fishes.

Urostyle (*tail* + *pillar*). The rodlike bone forming the posterior end of the spine in Anura.

Uterus (*womb*). The organ in which the egg or fetus develops.

Utricule (*small bag*). The more dorsal chamber of the inner ear.

Vagina (*sheath*). The canal that receives the penis during copulation.

Vagus (*wandering*). Denoting the tenth cranial nerve.

Vector. A quantity that has both magnitude and direction, such as force or velocity; a line that represents such a quantity by its length and orientation.

Vegetal pole. The region of the egg where yolk is concentrated and metabolic activity is lowest.

Vein. A large vessel that carries blood toward the heart.

Velocity. Rate of change of position in a given direction.

Ventral. Said of the belly, or underside, of the body.

Ventricle (*small cavity*). A cavity of the heart or brain.

Vesicle (*small bladder*). A small sac or space.

Vestigial (*a trace*). The state of a structure that was functional in an ancestor but is no longer useful and is reduced in size or complexity.

Visceral (*entrail*) **arch.** One of the bars separating adjacent gill slits or pharyngeal pouches; pertaining to viscera.

Viscus (*pl.* **viscera**). An internal organ, particularly of the abdominal cavity.

Vitelline (*yolk*). Relating to the yolk of an egg.

Viviparous (*alive* + *to bear*). Animals that give birth to "living" young.

Volant. Capable of sustained flight.

Vortex theory. An interpretation of powered flight based on the interaction between trailing-edge vortices and the movement of the wings.

Wave drag. Drag on a boat or swimmer caused by the formation of surface waves.

Weeping lubrication. Lubrication haing the lubricant squeezed from the spongelike surfaces of the moving parts into the area of contact between them.

Wing loading. The weight of a flyer divided by the area of its airfoils.

Work. The product of the tension of a muscle times the distance through which it acts.

Yaw. Rotation of a swimmer or flyer around its vertical (*dorsoventral*) axis.

Yolk sac. The fetal membrane that surrounds and absorbs yolk.

Zona pellucida (*girdle* + *clear*). A prominent egg membrane of mammals.

Zygapophysis (*yolk* + *away from* + *growth*). A process that joins one vertebra to the next.

Zygote (*yolked*). The fertilized egg.

Index

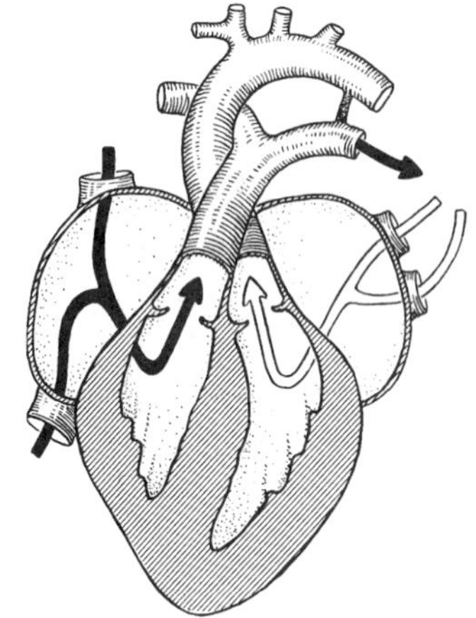

Numbers in boldface indicate pages where the entry is illustrated.

A CLASSIFICATION OF VERTEBRATES

This classification has been adapted to the needs of the book; several extinct taxa are omitted. Technical names are listed first, followed by common names and familiar examples, if such are available. Taxa preceded by an asterisk are extinct.

Class **Agnatha:** jawless vertebrates
- Subclass **Myxinoidea:** hagfishes
- Subclass **Petromyzontia:** lampreys
- *Subclass **Conodonta:** conodonts
- *Subclass **Pteraspidomorpha:** including pteraspids (Heterostraci)
- *Subclass **Cephalaspidomorpha:** including cephalaspids (Osteostraci), and anaspids

*Class **Placodermi:** including arthrodires, antiarchs, and several smaller orders

Class **Chondrichthyes:** cartilaginous fishes
- *Subclass **Cladoselachii:** cladoselachians
- Subclass **Elasmobranchii:** including *pleuracanths, sharks, and rays
- Subclass **Holocephali:** chimaeras

*Class **Acanthodii:** acanthodians

Class **Osteichthyes:** bony fishes
- Subclass **Actinopterygii:** ray-finned fishes
  - Infraclass **Chondrostei:** including *palaeoniscoids, bichirs, sturgeons, and paddlefishes
  - Infraclass **Neopterygii:** including gars, bowfins, and teleosts
- Subclass **Sarcopterygii:** lobe-finned fishes
  - Infraclass **Actinistia:** coelacanths
  - Infraclass **Dipnoi:** lungfishes
  - *Infraclass **unnamed:** osteolepids and panderichthyids